"The broad pathway to feeding everyone a diet that is healthy and sustainable is clear, but to make this happen will take enormous local and regional creativity and perseverance. This can be greatly facilitated by learning from the experiences of each other, and this handbook is a giant step in this direction."

Walter Willett, *Professor of Epidemiology and Nutrition, Harvard T.H. Chan School of Public Health, USA*

"Our recognition that we are now in the Anthropocene era has highlighted that many activities of humans are not sustainable; therefore, we should recognize and learn from our past mistakes and successes. This involves re-evaluating our past behaviour, considering new pathways to sustainability, educating ourselves, and finding opportunities from crises. This book, co-edited by Kathleen Kevany and Paolo Prosperi, helps each of us to do that re-evaluation, through meeting the four interwoven critical challenges of our time: global demand for food; weather chaos due to climate change; decreasing quality and quantity of freshwater; and human livelihood disruption. A positive future is possible, and the authors of this comprehensive and wide-ranging text help point the way."

CD Caldwell, *Prof. Emeritus, Faculty of Agriculture, Dalhousie University, Canada*

"This comprehensive, all encompassing handbook on sustainable diets, curated by Kathleen Kevany and Paolo Prosperi, provides us with the theory, the measuring, the meaning and the cultivating of sustainable diets around the world. If you were unclear on *what* exactly are sustainable diets in all their facets, this book will provide the nuanced answers from different experienced perspectives."

Jessica Fanzo, *Johns Hopkins University, USA*

ROUTLEDGE HANDBOOK
OF SUSTAINABLE DIETS

This handbook presents a must-read, comprehensive, and state of the art overview of sustainable diets, an issue critical to the environment and the health and well-being of society.

Sustainable diets seek to minimise and mitigate the significant negative impact food production has on the environment. Simultaneously they aim to address worrying health trends in food consumption through the promotion of equitable access to healthy diets that reduce premature disability, disease, and death. Within the Routledge Handbook of Sustainable Diets, creative, compassionate, critical, and collaborative solutions are called for across nations, disciplines, and sectors. To address these wide-ranging issues, the volume is arranged into sections dealing with environmental strategies, health and well-being, education and public engagement, social policies and food environments, transformations and food movements, economics and trade, design and measurement mechanisms, and food sovereignty. Substantial contributions from up-and-coming and established academics, this handbook provides a global, multi-disciplinary assessment of sustainable diets, drawing on case studies from regions across the world. The handbook concludes with calls to action that provide readers with a comprehensive map of strategies that could dramatically increase sustainability and help to reverse global warming, diet-related non-communicable diseases, and oppression and racism.

This decisive collection is essential reading for students, researchers, practitioners, and policymakers concerned with promoting sustainable diets, and for establishing a sustainable food system to ensure access to healthy and nutritious food for all.

Kathleen Kevany is an associate professor and director of Rural Research Collaborative with Dalhousie University, Canada. She has decades of experience in community building along with research expertise in increasing well-being for individuals and communities, as well as strategies to foster greater food security, sovereignty, and sustainability through sustainable diets. She is also the editor of *Plant-Based Diets for Succulence and Sustainability* (Routledge, 2019).

Paolo Prosperi is a senior lecturer and researcher of agricultural and food economics and scienctific co-coordinator of the MSc Sustainable Agri-Food Value Chains at the Mediterranean Agronomic Institute of Montpellier. He researches on agricultural and food economics, with a particular focus on sustainable food systems, sustainable agri-food value chains, sustainability and resilience assessment frameworks, and sustainable diets.

ROUTLEDGE HANDBOOK OF SUSTAINABLE DIETS

Edited by Kathleen Kevany and Paolo Prosperi

Routledge
Taylor & Francis Group
LONDON AND NEW YORK

earthscan
from Routledge

Cover image: Getty Images

First published 2023
by Routledge
4 Park Square, Milton Park, Abingdon, Oxon OX14 4RN

and by Routledge
605 Third Avenue, New York, NY 10158

Routledge is an imprint of the Taylor & Francis Group, an informa business

British Library Cataloguing-in-Publication Data
A catalogue record for this book is available from the British Library

Library of Congress Cataloging-in-Publication Data
Names: Kevany, Kathleen May, editor. | Prosperi, Paolo, 1979- editor.
Title: Routledge handbook of sustainable diets/edited by Kathleen Kevany and Paolo Prosperi.
Description: New York, NY: Routledge, 2023. | Includes bibliographical references and index.
Identifiers: LCCN 2022022903 (print) | LCCN 2022022904 (ebook) | ISBN 9781032004860 (hardback) | ISBN 9781032004976 (paperback) | ISBN 9781003174417 (ebook)
Subjects: LCSH: Food–Moral and ethical aspects. | Food habits–Environmental aspects. | Sustainable living.
Classification: LCC GE196 .R68 2023 (print) | LCC GE196 (ebook) | DDC 304.2--dc23/eng20221004
LC record available at https://lccn.loc.gov/2022022903
LC ebook record available at https://lccn.loc.gov/2022022904

ISBN: 978-1-032-00486-0 (hbk)
ISBN: 978-1-032-00497-6 (pbk)
ISBN: 978-1-003-17441-7 (ebk)

DOI: 10.4324/9781003174417

Typeset in Bembo
by Deanta Global Publishing Services, Chennai, India

The Open Access version of chapter 40 was funded by The European Union.

This publication has received funding from the European Union's Horizon Europe Research and Innovation Program under Grant Agreement of Excel4med project, No 101087147.

We dedicate this handbook to all people who are struggling every day for food and to all people across social and economic sectors, who are putting forward efforts, to make diets and food system more equitable and sustainable, with the aim of respecting nature and all living beings, improving everyone's well-being and health, and working for peace.

In memoriam of Galyna Medyna, colleague and co-author within this collection. She will be sorely missed.

CONTENTS

Contents

FIGURES

TABLES

BOXES

ABOUT THE EDITORS

Kathleen Kevany is an associate professor of systems thinking at Dalhousie University Faculty of Agriculture, Canada. She is a leading authority on sustainable diets and plant-rich living. She is trained in social psychology and specialises in individual and collective well-being and social change. As a certified psychotherapist, she ran her own counselling and consulting firm. Kathleen has worked in many countries on facilitating vibrant communities and collaborative relationships. She undertook a post-doctoral fellowship at United Nations University and an OECD Research Fellowship at CIHEAM-IAMM. She is the co-editor of the *Routledge Handbook of Sustainable Diets* and editor of *Plant-Based Diets for Succulence and Sustainability*, also published with Routledge. https://orcid.org/0000-0002-3233-2873

Paolo Prosperi is a senior lecturer and researcher in agricultural and food economics, and scientific administrator at the Mediterranean Agronomic Institute of Montpellier (CIHEAM-IAMM). He holds a PhD in economics from the University of Montpellier (France) and a PhD in agri-food economics from the University of Catania (Italy). He is a permanent member of the Joint Research Unit "Montpellier Interdisciplinary Center on Sustainable Agri-Food Systems" (UMR Moisa, University of Montpellier, CIHEAM-IAMM, CIRAD, INRAE, IRD, L'Institut Agro, Montpellier, France). Prior to joining CIHEAM-IAMM, Paolo was researcher and assistant professor of agricultural economics at the University of Pisa. He has served in several European Union (Horizon 2020 and Horizon Europe) and international projects as researcher and principal investigator (e.g. SALSA, SUFISA, SHERPA, QESAMed, HealthyFoodAfrica, PRIMA Med-Links, etc.) Paolo's main research focuses on understanding and assessing the complex socio-economic dynamics of food system sustainability, including the analysis of stakeholders' strategies and innovation for creating sustainable agri-food value chains. He is the co-editor of the *Routledge Handbook of Sustainable Diets*. https://orcid.org/0000-0002-8494-0344

CONTRIBUTORS

Alexandra Adams is the director of the NIH funded Center for American Indian and Rural Health Equity (CAIRHE) at Montana State University. The foundation of her work is community-based participatory research, working in partnership with Indigenous communities to understand and solve health challenges using scientific rigour and crucial community knowledge. She has directed multiple clinical trials and received over $40 million of funding. Her work focuses on promotion of family and community wellness and healing trauma through community building.

Francis Adams, PhD in Government from Cornell University, is a professor of political science and international studies at Old Dominion University. Since joining the ODU faculty in 1995, Dr. Adams has served as director of the undergraduate International Studies program, associate and acting director of the graduate program in International Studies, and chair of the Department of Political Science and Geography. His teaching and research areas include: International political economy, international development, globalisation, and Latin American politics. He is the author of six books, including *The Right to Food: The Global Campaign to End Hunger and Malnutrition* (Palgrave Macmillan, 2021) and numerous scholarly articles. https://orcid.org/0000-0001-5685-1136

Elom K. Aglago is a postdoctoral scientist at the International Agency for Research on Cancer (IARC) in the Nutrition and Metabolism Branch. He received his PhD in Nutrition and Food Science from the Ibn Tofail University in Morocco, a Postgraduate Certificate in Public Health from Imperial College London, and a certificate in epidemiology and data analysis from Harvard University. Dr Aglago's research focuses on cancer risk associated with diet, lifestyle, and environment exposures, and he is a passionate speaker on dietary and lifestyle risk factors for cancers and behaviour changes for the public.

Sara Ahlberg, PhD, is a food technologist, holds a Masters in Law & Business and is the CEO and founder of Böna Factory, a Nairobi-based food company specialising in innovative plant-based foods, with a strong focus on food safety, food production systems, and industry development. Sara has worked in ILRI and CGIAR in Nairobi (KE) in food safety research and has performed extensive aflatoxin research in the FoodAfrica programme. She is passionate about changing the state of the global food industry and increasing diversified plant-based, healthy food options for all consumer groups. She is determined to build a technology-based company to change the food markets by providing jobs and serving plant-based, delicious, innovative, and high-quality foods for large consumer bases.

Selena Ahmed, an ethnobotanist, is the Global Director of the Periodic Table of Food Initiative (PTFI) at the American Heart Association, and an Associate Professor of Sustainable Food Systems at Montana State University. Her goal is to transform food systems through evidence-based innovations that support human and planetary health. Since 2003, Selena has engaged with diverse communities and stakeholders to examine food environment dynamics. Her long-term study in China's Yunnan Province examines effects of agroforestry practices and global change on tea biochemical composition, sensory attributes, and farmer livelihoods.

Grégori Akermann is a sociologist and researcher at the French National Institute of Research for Agriculture, Food, and Environment (INRAE). Specialising in economic sociology and social network analysis, his research focuses on emerging food consumption in alternative food networks. Recruited to INRAE, he researches social relations in changing food practices towards sustainability. He is interested in the mechanisms of change in social innovations that promote citizen participation, the impact of citizen solidarity initiatives in the resilience of food systems during the COVID-19 crisis, and analyses the dynamics of development of social innovations and the transformation of alternative food networks via the participatory implementation of a national observatory of local food systems (https://obsat.org). https://orcid.org/0000-0002-7195-2434

Simone Amadori is a research fellow at the Department of Agricultural and Food Science of the University of Bologna. His main research interests include business models development and food system sustainability.

Manish Anand is a policy researcher and Senior Fellow at The Energy and Resources Institute (TERI), New Delhi. In his professional career of over 16 years grounded in interdisciplinary research, project coordination and management, outreach and capacity building, he has worked in diverse areas of agriculture, energy and environment, and emerging technology policy, focusing on the governance of science, technology, and innovation to address sustainable development challenges. His research and policy analysis work focuses on generating evidence, strategies and methodologies based on systems thinking towards sustainable food and land use systems including sustainable agricultural practices, natural resource use, rural livelihoods, human health and nutrition, and climate change. Besides publishing in peer-reviewed international journals, he also contributes to policy briefs, newspaper articles, and blog-posts. https://orcid.org/0000-0001-5044-8020

Martha Anker is a researcher and expert on living wage and has devised, with Richard Anker, the Anker Methodology and Anker Research Institute to help governments support a living wage.

Richard Anker is a researcher and expert on living wage and has devised, with Martha Anker, the Anker Methodology and Anker Research Institute to help governments support a living wage. He has more than 90 publications and extensive international experience.

Marta Antonelli is an economist and a geographer. She has a PhD in Environment, Politics and Development and 12 years of experience as a researcher and project leader with a focus on food sustainability. Currently, she is a Senior Research Associate at the Euro-Mediterranean Centre on Climate Change and serves as Research Director of the Barilla Foundation. She is one of the 40 members of the European Commission's joint research centre expert group for the legislative framework of the Farm to Fork Strategy.

Gershim Asiki is a medical epidemiologist with over 13 years' experience in research on chronic diseases in Africa. As a Research Scientist at the African Population and Health Research Center, he leads the non-communicable diseases (NCD) research programme with a focus on epidemiology informing health systems strengthening for the management and prevention of NCDs. He leads implementation research on cardiovascular diseases management and undertakes regular policy analyses including food systems policies to create healthier food environments for NCD prevention.

Manlio Bacco received a PhD degree in information engineering and science from the University of Siena, Italy, in 2016. He has worked as a Researcher in the Institute of Science and Information Technologies (ISTI), National Research Council (CNR), Pisa, Italy, since 2012. He has been actively participating in European-funded, national-funded, and ESA-funded research and development projects. His research interests include wireless communications for aerospace and IoT scenarios, cyber-physical systems, and the use of digital technologies in agriculture, forestry, and rural areas.

Dinesh Balam is an Associate Director with Watershed Support Services and Activities Network (WASSAN). He has five years of experience in the financial sector and nine years of experience in the rural development sector. As a member of the WASSAN team in Odisha, he contributed significantly to the design and implementation of the Odisha Millet Mission. He works with organisations on agrobiodiversity, agroecology, and anthropological projects, and has authored several articles. He specialises in systems for in-situ agrobiodiversity and transforming food systems that meet the FAO's agro-ecological principles.

Mondira Bhattacharya is a Research Officer and Associate Lecturer at the School of Agriculture, Policy and Development at the University of Reading, UK. Her research focuses on food and nutritional security, rural development, and policy implications on farmers, especially in low- and middle-income countries. Before joining the University of Reading in 2020, she worked for non-profit organisations in India for several years. Mondira has a PhD in Economic Geography from Jawaharlal Nehru University, New Delhi, India and postdoc from the University of Reading.

Jennifer L. Black is an Associate Professor of Food, Nutrition, and Health in the Faculty of Land and Food Systems at the University of British Columbia where she leads the Public Health and Urban Nutrition Research Group and teaches courses related to applied nutrition, food systems, and research methods. Her research aims to understand how food is connected to the complex social and contextual factors that shape the health of individuals, communities, and the environment.

Jessica Bogard is an Advanced Accredited Practising Dietitian and Nutrition Systems Scientist in the Sustainability Program at CSIRO in Australia. Her research aims to leverage agriculture and food systems to improve nutrition and health outcomes, particularly for vulnerable populations. Her field experience has been largely focused in Asia and the Pacific region. She uses qualitative and quantitative research methods to understand the drivers of diet quality and food system interventions to support healthy and sustainable diets. She has a particular research interest in the role of aquatic foods for improving nutrition and health and has published widely on the links between fisheries and aquaculture and diet quality.

Bruno Bosari is Professor Emeritus in the Biology Department at Winona State University in Winona, Minnesota. An agroecologist by trade, he spent 34 years teaching, researching, and consulting in sustainable development, restoration ecology, soil/water conservation, and food security. He received his Laurea in agricultural sciences from the University of Bologna, Italy and a PhD in Curriculum & Instruction from the University of New Orleans, LA, US. He taught biology, agroecology, and restoration ecology in west and central Africa, in Bolivia, Italy, Panamá, Louisiana, and Pennsylvania. Between 2000 and 2004 Bruno taught in and coordinated the first graduate program in sustainability in the US.

William K. Bosu is a public health physician working at the West African Health Organisation. He worked for several years as the National Programme Manager for the Prevention and Control of Noncommunicable Diseases (NCDs) in Ghana and as the regional technical lead on nutrition and NCDs in the ECOWAS Region. His research interests include the evaluation of NCD-related policies and programmes, chronic NCD risk factor surveillance, nutrition transition, and adolescent nutrition.

Gianluca Brunori is Full Professor at the University of Pisa, where he teaches Food Policy and Bioeconomy. He has 30 years research experience in the fields of Agricultural Economics, Rural Sociology, and Food Policy, through several

European projects, in three of them (TRUC, GLAMUR, and DESIRA) as scientific coordinator. The research group he leads—Pisa Agricultural Economics (PAGE)—works on strategies with small farmers' and the impact of food supply chains, on the link between collective initiatives and rural development. He has been President of the Research Committee "Sociology of Agriculture and Food" (RC-40) of the International Sociology Association, and vice-president of the European Society of Rural Sociology. He has been Chief Editor of the journal *Rivista di Economia Agraria* and is editor-in-chief of the journal *Agriculture and Food Economics*. He has served and been chair of the expert panel of EU SCAR (Standing Committee for Agricultural Research) 2nd, 3rd, 4th and 5th foresight exercise. He also is chair of the Scientific Advisory Board of the Joint Programming Initiative "Agriculture, Food Security and Climate Change" (FACCE). https://orcid.org/0000-0003-2905-9738

Nicole Burton is currently living and learning in Peterborough/Nogojiwanong, Ontario, Canada. A recent graduate of Trent University's Sustainable Agriculture and Food Systems and Environmental Studies programs, her undergraduate work emphasised food and environmental justice, land rematriation, and the integration of cultural theory with food studies. Since graduating, Nicole has been working with Colette Murphy at Urban Harvest cultivating organic, endangered, and heirloom seeds. She values time spent on Trent's experimental farm and ways in which it has influenced her research on campus food systems.

Liesel Carlsson is an Associate Professor in the School of Nutrition and Dietetics at Acadia University, Canada, and a registered dietitian. She has focused her professional and scholarly work at the intersection of sustainability, food, and nutrition and has published 11 articles, four reports used by organisations to guide strategic work, and three book chapters. Her current research examines how RDs can best contribute to sustainability, barriers, knowledge needs, and standards for training. She co-chaired Dietitians of Canada's Sustainable Food Systems Leadership Team and was recognised in 2018 for leadership in advancement of the profession. https://orcid.org/0000-0001-8857-9608

Dario Caro is an Assistant Professor at Aarhus University in the Department of Environmental Science. His PhD in Chemistry, with an environmental specialisation, is from the University of Siena. He has a decade of research experience on environmental sustainability and a year as visiting researcher at Stanford University and one year as postdoctoral fellow at the University of California (Davis). He is co-leader of the strategic growth area: Resource flows in a circular economy. He has more than 60 publications including peer-review articles, encyclopaedias, and book chapters, including recent works on the correlation between pollution and COVID-19.

Yuna Chiffoleau is an engineer in agronomy, PhD in sociology, and research director at the French National Institute of Research for Agriculture, Food and Environment (INRAE). Specialising in economic and network sociology, she is interested in collective and social innovations that contribute to the transition of food systems towards greater sustainability. She has been conducting participatory research on short food supply chains and informing public policies at national and European levels and co-leading the national research-development and training network on short food chains and local food systems, funded by the French Ministry of Agriculture and Food. She also co-created a participatory and free trademark (Ici.C.Local) to identify local and sustainable food in open-air markets and retail, which makes it possible to experiment and assess a new type of participatory guarantee system. https://orcid.org/0000-0002-3120-9566

Long Chunlin is an ethnobotanist and Professor at Minzu University of China. He was a professor at the Kunming Institute of Botany, Chinese Academy of Sciences, of biodiversity, botany, ethnobotany, ethnomedicine, ethnopharmacology, plant genetic resources, phytochemistry, and plant taxonomy. https://orcid.org/0000-0002-6573-6049

Francesco Cirone is a PhD candidate (M) in Agricultural, Environmental and Food Science and Technology at the University of Bologna. His current research focuses on the sustainability assessment of food systems, with specific attention on life cycle thinking approaches, including Life Cycle Costing (LCC) and Territorial Life Cycle Assessment (T-LCA). He also is working on business models development. He contributed to several European projects including the Horizon 2020 FoodE and the Horizon Europe FishEUTrust. https://orcid.org/0000-0003-3909-7703

Jennifer Clapp is a Canada Research Chair and Professor in the School of Environment, Resources and Sustainability at the University of Waterloo. Among her recent books are *Food, 3rd Edition* (Polity, 2020) and *Speculative Harvests: Financialization, Food, and Agriculture* (with S. Ryan Isakson, Fernwood Press, 2018). She is Vice-Chair of the Steering Committee of the High-Level Panel of Experts on Food Security and Nutrition of the UN Committee on World Food Security and a member of the International Panel of Experts on Sustainable Food Systems.

Michael Classens is an Assistant Professor and the Coordinator, Certificate in Sustainability in the School of the Environment at University of Toronto. His teaching, research, and advocacy work in food systems are broadly motivated by commitments to social and environmental justice. His work on campus food systems has been inspired, informed, and supported by many incredible students, including Nicole Burton, Akua Agyemang, Sophia Srebot, Will Sistrom, Kaitlyn Adam, Matt Dutry, Rachel Lee, and Natasha Sheward. Michael lives in Toronto with his partner, three young kids, and his dog named Sue.

David A. Cleveland is a Professor with University of California, Santa Barbara, and The Centre for People, Food & Environment. He is a human ecologist who has done research and development project work on sustainable agrifood systems with small-scale farmers and gardeners around the world. His current focus is the role of local food systems and diets in mitigating climate change, improving nutrition and health, and promoting food justice.

Donald Cole is a Fellow of the Royal College of Physicians and Surgeons of Canada, with a specialty certification in Public Health & Preventive Medicine (1992). He has led/collaborated on multiple research, policy, and training grants and contracts from municipal to international funding organisations, with an emphasis on multi-stakeholder action research processes. He has co-authored over 40 book chapters and 275 peer-reviewed publications, and his 40 years of practice include: Primary care at multiple community health centres; environmental-occupational consultation services; global health with colleagues in Latin America and Africa; and public health with colleagues from all levels of government, in Canada and internationally. He currently practises rurally with a particular interest in ecological agriculture and alternative food networks. He remains a professor emeritus of the University of Toronto.

Alyson Dagang is the Chair of the Sustainable Development program in the SIT Graduate Institute. Based in Panama, she has worked for 25 years with rural and Indigenous peoples. As an agroforester, Alyson has spent much of her time learning from agriculturalists. Her most recent research interests include land use transformation as climate change adaptation, particularly the adoption of cattle ranching and pasture creation in Indigenous territories. As a founding member of the Alianza para la Conservación y Desarrollo, a Panamanian NGO, Alyson is also committed to conservation and development practice. She is buoyed by the flourishing of Ari Kuidte Waire Few for Change, a Panamanian NGO led by Ngabe families and SIT Panama alumni.

Ning Dai is a PhD candidate in the Department of Geography and Environmental Management at the University of Waterloo. His research focuses on government planning of sustainable food system transformation in China. He studies government roles in farmers' adoption of agroecological practices and the digitalisation of food retailing.

Marian Davidove, RD/N., MA, gained her BSc in Nutritional Science from the University of Connecticut at Storrs, then completed an internship at the University of Saint Joseph, West Hartford Connecticut. She has been a practising registered dietitian for over 27 years and a licensed dietitian/nutritionist in the state of Florida. She holds a MA from Patel College of Global Sustainability (PCGS) at the University of South Florida, Tampa in Food Sustainability and Food Security. She focuses on sustainable diets and breastfeeding, and through private practise, she offers nutrition education and ideas for living sustainably.

John de la Parra is an ethnobotanist and plant chemist with expertise in food crops and medicinal plants. He is the President of the Society for Economic Botany and Manager of the Global Food Portfolio at The Rockefeller Foundation. He also holds an appointment as an Associate of the Harvard University Herbarium. He has held additional appointments as a Research Scientist at MIT, a Lecturer of Environmental Studies at Tufts University, and a Lecturer of Biotechnology at Northeastern University.

Adama Diouf is a senior lecturer/researcher in Physiology, Food and Human Nutrition at the Faculty of Science and Technology of the Cheikh Anta Diop University of Dakar (UCAD). Her research focuses on maternal and child nutrition and health through the life cycle, using stable isotopes in human nutrition and diet-related non-communicable diseases. She researches public policies and the promotion of healthy and sustainable food environments in West Africa. She is a member of several scientific and technical committees and research networks, and has more than 15 years of experience as an expert and consultant in nutrition, research, and development.

Joseph Dorsey is Associate Dean for Academic Affairs, an Associate Professor of Instruction, and directs both the Food Sustainability and Security concentration and the Academic Capstone Experience Program in the Patel College of Global Sustainability at the University of South Florida, Tampa. He holds a PhD in Natural Resources and Environment from the University of Michigan. Dr. Dorsey is an experienced educator in the interdisciplinary fields of environmental justice and policy, international development, and human nutrition and food. His research interests include sustainability modelling, urban agriculture in developed and developing nations, food waste reduction strategies, and community economic empowerment initiatives.

Shauna Downs is an Assistant Professor at Rutgers School of Public Health. The overarching goal of her work is to provide evidence of the ways in which food systems can be reoriented to ensure that all populations, particularly the most marginalised, can consume diets that promote health and sustainability. Prior to coming to Rutgers, Shauna was a Hecht-Levi Fellow with the Global Food Ethics and Policy Program at the Johns Hopkins Berman Institute of Bioethics and an Earth Institute Fellow at Columbia University. Shauna received her PhD in Public Health from the Menzies Centre for Health Policy at the University of Sydney and a MSc in Nutrition from the University of Alberta, Canada.

Colin Dring is a PhD Candidate at the University of British Columbia in the Faculty of Land and Food Systems. His doctoral research explores the relationship between municipal governance, agricultural planning, and social justice outcomes. Colin teaches courses on food systems and social justice and conducts scholarship of teaching and learning on the subject of food system planning education. He has co-developed with undergraduate students an open-source Food Justice Educational Resource—https://justfood.landfood.ubc.ca/. He continues to conduct applied food system planning research for communities aiming for sustainable and just food futures.

Virgil Dupuis is the Extension Director at Salish Kootenai College, a Tribally controlled community college in Pablo, Montana. He implements community education and research projects on healthy and sustainable diets, invasive species, and land restoration techniques. Collaborations with the Confederated Salish and Kootenai Tribes Food Sovereignty Team and Tribal Health Department aim to improve food access for low-income community members, access local markets for tribal agricultural producers, support community and family gardening, and provide human and environmental health education opportunities.

Roland Ebel has a PhD in Organic Farming from the University of Natural Resources and Life Sciences, Vienna. From 2008 to 2014, Dr. Ebel was dedicated to the construction of a new university on the Yucatan Peninsula, the Intercultural Maya University of Quintana Roo with emphasis on agroecology, especially traditional Mesoamerican farming, and participatory research. Using a mixed-methods approach, he explored the impact of climate change on the agrobiodiversity, management, and food security of traditional farming systems. In 2020, he was appointed Assistant Research Professor at MSU where he is dedicated to the anaerobic digestion of food residues, the impact of COVID-19 on food systems, and several projects around agricultural and dietary diversification. Dr. Ebel is an active member of the American Society of Horticultural Sciences (ASHS) where he has served in several committees.

Katherine Eckert, MSc, PDt is a Doctoral Student and Arrell Food Institute Scholar in the department of Family Relations and Applied Nutrition at the University of Guelph. Katherine is developing strategies to promote healthy and sustainable diets among families with young children through mixed methods research. They have completed a MSc in Community Health and Epidemiology at Dalhousie University and a BSc in Nutrition at Acadia University. Katherine is a Registered Dietitian.

Sarah Elton is a critical food systems researcher who works at the intersection of health, ecosystems, and food in the city. She is an Assistant Professor in the Department of Sociology and a member of the Centre for Studies in Food Security at Ryerson University. She has been appointed to the Dalla Lana School of Public Health and is a fellow at the Centre for Critical Qualitative Health Research at the University of Toronto. Her scholarly research has been published in the journals *Gastronomica*, *Critical Public Health*, *Canadian Food Studies* and *Environmental Humanities*. Her two books, *Locavore: From Farmers' Fields to Rooftop Gardens, How Canadians are Changing the Way We Eat* and *Consumed: Food for a Finite Planet* were Canadian bestsellers.

Anna Farmery is a Senior Research Fellow at the Australian National Centre for Ocean Resources and Security (ANCORS) at the University of Wollongong in Australia. Her research focus is the intersection between human and planetary health, with a specific focus on aquatic foods. Anna has worked on aspects of sustainable food systems in Australia, Cambodia, East Timor, and countries in East Africa and the Pacific. She uses life cycle and value chain approaches to examine connections between sustainability, food and nutrition security and ways to improve diets through sustainable, equitable, and rational resource use. She has published widely on issues of food sustainability, fisheries management, and on the important role of aquatic foods in sustainable diets. https://orcid.org/0000-0002-8938-0040

David Fazzino is an Associate Professor and Chair of Anthropology at Bloomsburg University of Pennsylvania. He serves as the Chair of Anthropology, Criminal Justice, and Sociology at the Commonwealth University of Pennsylvania (Bloomsburg, Lock Haven, and Mansfield). His research interests include human potential, place, food sovereignty, complementary and alternative medicine, law and policy, and property. https://orcid.org/0000-0003-3548-7421

Georges Félix is a French-Puerto Rican activist and researcher on agroecology and agroforestry. His work focuses on the design of resilient food systems and the restoration of peasant farming systems in the aftermath of natural disasters in places of Europe, Africa, Latin America, and the Caribbean. He is a researcher at the Centre for Agroecology, Water and Resilience of Coventry University, in England. Additionally, he collaborates with several food collectives around the world, including Cultivate! in the Netherlands, Organización Boricuá/CLOC-Via Campesina in Puerto Rico, and SOCLA and CELIA in Latin America.

Christopher Fink is an associate professor in the Department of Health & Human Kinetics at Ohio Wesleyan University, where he focuses his teaching on interdisciplinary food studies, dietary health, community health, and issues related to dietary decisions, and the food system, including food insecurity. His research is primarily qualitative, exploring food and migration; community food insecurity and chronic disease; food education, culture, and tradition; and the role of community in dietary health and quality of life.

Evan Fraser is Director of Arrell Food Institute and a Professor of Geography at the University of Guelph. Evan helps lead the *Food from Thought* initiative, which is a $76.6 million research program based at the University of Guelph that explores how to use big data to reduce agriculture's environmental footprint. Evan is a co-author on over 100 academic papers and book chapters, and played a leadership role on teams that have raised over $100 M in research funding, and mentored around 50 graduate students. He is one of Canada's most cited social scientists working on food and sustainability and between 2010 and 2020 was a Canada Research Chair. He is Co-Chair of The Canadian Food Policy Advisory Council which advises the Minister of Agriculture on the implementation of Canada's Food Policy.

Alessandro Galli is a macro ecologist and sustainability scientist, with a passion for anthropology and human behaviour. He supports evidence-based decision-making via sustainability indicators and environmental accounting tools to address the challenges of living well within planetary limits. His work focuses on the sustainability of food systems, tourism, and education. Alessandro is co-author of several publications, including 50 articles in peer-reviewed journals, including in *Science*, various editions of WWF's Living Planet Reports, and Montenegro's National Strategy for Sustainable Development 2016–2030. Alessandro works as Mediterranean-MENA Program Director at the international think tank Global Footprint Network.

Saurabh Garg is high-ranking as the Chief Executive Officer of Unique Identification Authority, Government of India. He was Principal Secretary, Agriculture and Farmers Empowerment, Government of Odisha. He has been a member of Expert Committees set up by the Ministry of Finance, NITI Aayog, the Reserve Bank of India and Securities and Exchange Board of India. Dr. Garg was an Advisor to the World Bank in the post of the Executive Director for India and was the Chairman of Public Sector Companies. His PhD is in International Economics and Development from the Johns Hopkins University, US, and MBA from the Indian Institute of Management, Ahmedabad. He was a Gurukul Chevening Fellow at the London School of Economics and Political Science, London.

Terry Gibbs is Professor of International Politics at Cape Breton University with a focus on social justice and environmental issues. She is co-author with Tracey Harris of *The Vegan Challenge is a Democracy Issue: Citizenship and the Living World*, co-author of *Homeschool as a Living School* (2020), and author of *Why the Dalai Lama is a Socialist: Buddhism and the Compassionate Society* (Zed Books, 2017). Terry is also a founding member of the Animal Ethics Project, the Cape Breton University Community Garden and the Cape Breton/Unama'ki Climate Change Task Force. She is active in her local food movement and co-coordinator of the educational series Inspired by Nature. She is an animal lover, and enjoys gardening and vegan cooking.

L. Sasha Gora is a cultural historian and writer with a focus on food studies, the environmental humanities, and contemporary art. She earned a PhD from Ludwig Maximilian University of Munich and the Rachel Carson Centre on the subject of Indigenous restaurants in Canada, for which she received the 2021 Bavarian American Academy Dissertation Award. Based on her doctoral research she is writing her first book: *Culinary Claims*. In 2020 she joined Ca' Foscari University of Venice as an environmental humanities research fellow, where she is researching culinary reactions to climate change. https://orcid.org/0000-0002-0192-3574

Nithya Vishwanath Gowdru is an Assistant Professor at the National Institute of Rural Development and Panchayati Raj (NIRDPR), Under the Ministry of Rural Development, Government of India. Based at the Centre for Agrarian Studies of NIRDPR since 2016, she has led and coordinated a portfolio of developmental projects related to agriculture, nutrition, value chain, and rural development. In October 2021, Dr. Gowdru completed a post-doctoral fellowship in agriculture, nutrition, and health with IMMANA fellowship at the School of Agriculture Policy and Development at the University of Reading (UK). In 2016, Dr Gowdru received her PhD in Agricultural Economics from the Humboldt University of Berlin (Germany).

Allison Gray is a SSHRC postdoctoral fellow in the Department Geography and Environment department at Western University. She is the co-editor of *A Handbook of Food Crime: Immoral and Illegal Practices in the Food Industry and What to*

Do About Them (Policy Press, 2018). Allison's research examines food production and consumption as a key aspect of the nature-culture nexus, bringing together perspectives from Green Criminology, Environmental Sociology, Critical Animal Studies, and Food Studies. She was awarded a Tiny Beam Fund grant in 2021 in support of her research on the role of plant-based alternatives in mitigating dietary-based environmental harm and animal suffering.

Hélène Guétat-Bernard is professor of sociology at the ENSFEA, UMR CNRS LISST-Dynamiques Rurales, Toulouse University, France and associated with the French Institute of Pondicherry (IFP) Savoirs et Monde indiens USR CNRS 3330, IFP/UMIFRE 21 MEAE. She was the head of the social sciences department of the IFP from 2017 to 2021. She is working on gender issues in agriculture, food, and rural development. In 2019, she launched the Food Platform of Pondicherry and various projects on food justice and democracy based on a research-action and participatory approach. She coordinated three international symposiums and supervised six PhDs on gender issues. https://orcid.org/0000-0003-0842-1566

Vidhu Gupta is an Associate Fellow in Environment and Health Area at TERI, Delhi. She has been involved in multiple projects designing infographics, prelim data analysis, report and manuscript writing, samples analysis, and literature search related to environment, health, and nutrition. She and her co-authors are part of the FOLU India coalition. Her expertise is in nutrition and public health. She has also standardised the protocol for fatty acid analysis in foods and from dried blood spots. She has a Master in Food and Nutrition from Delhi University and has been involved in various nutritional epidemiological studies. She has published ten peer-reviewed articles, won various accolades for presentations, and reviewed peer-reviewed journals. https://orcid.org/0000-0001-9982-6264

Jess Haines, PhD, MHSc, RD is an Associate Professor of Applied Nutrition at the University of Guelph in Canada. Dr. Haines's research aims to bridge epidemiologic research on the determinants of health behaviours with the design, implementation, and evaluation of interventions to support children's healthy eating and growth. Dr. Haines is the Co-Director of the Guelph Family Health Study, a longitudinal family-based cohort, and the Director of the Parent-Child Feeding Laboratory, which focuses on identifying how parent-child feeding interactions influence children's eating behaviours.

Jason Hannan is Associate Professor in the Department of Rhetoric, Writing, & Communications at University of Winnipeg. He is the author of *Ethics Under Capital: MacIntyre, Communication, and the Culture Wars* (2020) and editor of *Meatsplaining: The Animal Agriculture Industry and the Rhetoric of Denial* (2020). His current book project is *Trolling Ourselves to Death: Democracy in the Age of Social Media* (under contract with Oxford University Press).

Tracey Harris is Associate Professor of Sociology at Cape Breton University. She is the author of several recent publications, including *The Vegan Challenge* *is a Democracy Issue: Citizenship and the Living World* (co-authored with T. Gibbs, Routledge, 2020), *The Tiny House Movement: Challenging Our Consumer Culture* (Lexington Books, 2018), and '*The Problem is Not the People, It's the System': The Canadian Animal-Industrial Complex* in (Praeger, 2017). She is a founding member of the Animal Ethics Project and co-coordinator of the ongoing educational series *Inspired by Nature.* Tracey enjoys gardening, baking plant-based sweets, and spending time in nature.

Anna Herforth works to improve global food and diet measurement to enable action toward healthy diets and true abundance in food systems. Her main projects are the *Global Diet Quality Project* and *Food Prices for Nutrition*, which serve to build global monitoring systems for diet quality and access to healthy diets, respectively. Dr. Herforth co-founded and co-leads the Agriculture-Nutrition Community of Practice (Ag2Nut), a professional community of over 7,500 members from 130 countries. She is a Senior Research Associate in the Department of Global Health and Population at the Harvard T.H. Chan School of Public Health, and a Visiting Senior Researcher at Wageningen University & Research.

Lucy Hinton is a PhD candidate in Global Governance at the University of Waterloo and the Balsillie School of International Affairs, in Waterloo, Canada. Lucy's research focuses on the intersections of food, governance, and the nutrition transition, as it affects the health of people and the planet. Her dissertation work analysed power in a process of standard setting in the Caribbean Community (CARICOM), which tried to produce a uniform, regional front-of-pack label to differentiate buying habits amongst consumers.

Nicholas M. Holden is Professor of Biosystems Engineering in the UCD School of Biosystems and Food Engineering, University College Dublin, Deputy Director of BiOrbic Bioeconomy, SFI Research Centre and the Food Chain Sustainability theme leader in the UCD Institute of Food and Health. His research is at the nexus of agriculture, food, soil, and bioeconomy using life cycle assessment, system modelling, and qualitative methods to address the impact and sustainability of these systems. https://orcid.org/0000-0002-0452-4632

Michelle Holdsworth is a Senior Researcher at the Montpellier Interdisciplinary centre on Sustainable Agri-food systems at the French National Institute of Research for Sustainable Development (IRD) in Montpellier, France; and an Honorary Professor of Public Health Nutrition at the University of Sheffield, UK. She has over 20 years post-doctoral research experience, which has involved working in different cultural contexts (especially in collaboration with African researchers). Her research investigates food environments and their interaction with the wider food system, particularly the determinants of dietary and food behaviours followed by studies to identify relevant policies and interventions that directly feed into informing preventive strategies.

Ryan Isakson is an associate professor of Global Development Studies and Geography at the University of Toronto. Broadly, his research focuses upon the political economy of food systems and agrarian change, particularly in Latin American contexts. His current research explores the links between the development of Guatemala's non-traditional agricultural export sector, peasant farmers' vulnerability to environmental change, and financial inclusion, especially in markets for index-based agricultural insurance. Along with Jennifer Clapp, he co-authored *Speculative Harvests: Financialization, Food, and Agriculture* (2018, Fernwood).

Johanna Jacobi is an Assistant Professor at ETH Zürich. Before that, she worked as a senior scientist at the University of Bern in different research projects on food system sustainability in Latin America and Africa. Her research focuses on agro-ecology as a transformative science, practice, and social movement, and on power relations in food systems with approaches and methods.

Ali Jafri is a lecturer at Mohammed VI University of Health Sciences' Faculty of Medicine, Casablanca, Morocco. His research in public health nutrition focuses on nutrition-related non-communicable diseases, food insecurity, nutrition education, and food composition. His experience in promoting healthier food systems comes from working with NGOs advocating for sustainable eco-friendly agriculture in Morocco.

Chhavi Jatwani is the Design and Innovation lead at Future Food with international experience in food design and systems thinking. Chhavi has consistently been the industry first mover, joining the first cohorts of all her educational experiences. Growing up in a developing country, she is guided by only one question: What is the impact I am creating? Through R&D projects and innovation workshops, Chhavi has helped major FMCG companies like Dole, Barilla, and Ab-Inbev in their human- and planet-centred agenda. She is currently developing a new human- and planet-centred methodology to drive systemic innovation.

Hugh Joseph has spent his career developing community-based food, agriculture, nutrition, and food security initiatives at the local, regional, and national levels. He is Director of the Institute for Social and Economic Development in Boston, where he currently runs several projects with refugee farming incubators nationwide to develop, disseminate, and provide innovative training and technical assistance resources for limited English-language audiences. Joseph holds an MS and PhD in Nutrition from Tufts University, where he is Assistant Professor (Adjunct) in the Agriculture, Food and Environment Program, focusing on food systems-related teaching and research.

Kathleen Kevany is an Associate Professor of systems thinking at Dalhousie University, Faculty of Agriculture. She is a leading authority on sustainable diets and plant-rich living. She is trained in social psychology and specialises in individual and collective well-being and social change. As a certified psychotherapist, she ran her own counselling and consulting firm, the Decentralised Intelligence Agency. Kathleen has worked in many countries on facilitating vibrant communities and collaborative relationships. She undertook a post-doctoral fellowship at United Nations University and an OECD Research Fellowship at CIHEAM-IAMM. She is the co-editor of the Routledge Handbook of Sustainable Diets. https://orcid.org/0000-0002-3233-2873

Azfar Khan is a Senior Economist at the Anker Research Institute and has worked with the International Labour Organisation (ILO) for over two decades. Previously, he taught at the Institute of Social Studies in The Hague, the Netherlands, where he was also the Director of the UNFPA funded Global Programme of Training in Population and Development. His major research and policy work has been on socio-economic security issues, employment and poverty, labour migration, and on the interplay of demographic and economic issues. He holds a DPhil in Development Studies from the University of Sussex, UK, and a BA (Honours) and MA in Economics from McGill University, Montreal, Canada.

Sara F. L. Kirk is a Professor of Health Promotion and Scientific Director of the Healthy Populations Institute, Dalhousie University, Canada. Her research uses a socio-ecological approach that considers how individual behaviours are influenced by other broader factors, such as income, education, and societal norms, including the role of food policies in shaping health. She is a member of the INFORMAS network (International Network for Food and Obesity/Non-communicable Diseases (NCDs) Research, Monitoring and Action Support), and INFORMAS Canada, which is benchmarking the Canadian food environment.

Amos Laar is a tenure-track academic at the University of Ghana School of Public Health. His research and professional practice focus on three areas of public health: Bioethics (including food ethics, ethics and public health; health equity, health and human rights); Public Health Nutrition (including nutritional epidemiology, food systems, food environment, and nutrition-related non-communicable diseases); and Social Public Health (including structural violence, social, and commercial determinants of health). Professor Laar's research contributes to a deeper understanding of how physical environments, social environments, and structural factors affect health.

Matilda Essandoh Laar is a Lecturer at the Department of Family and Consumer Sciences, University of Ghana. With a foundation in Human Nutrition and Public Health, she works on the development of monitoring and evaluation tools in the agriculture, nutrition and health Nexus, and the influence of food environments on maternal and child nutrition in low-resource settings. As an early career Queen Elizabeth Scholar, her research seeks to promote entrepreneurship in rural adolescents as well as strategies to address the multiple burden of malnutrition in women and children of disadvantaged populations.

David Laborde-Debucquet joined IFPRI, Washington DC, in 2007. He is a Senior Research Fellow in the Markets, Trade and Institutions Division and the Theme Leader on Macroeconomics and Trade for IFPRI. His research interests include globalisation, international trade, measurement and modelling of protectionism, multilateral and regional trade liberalisation, as well as environmental issues like climate change and biofuels. Recently, the focus is on costing the roadmap to achieving SDG2 in a globalised context while considering the role of goods, capital, and migration flows. https://orcid.org/0000-0003-3644-3498

Eleonora Lano after completing her studies as a dietitian, continued onto a master degree in Food and Wine Culture. She has experience in the design and implementation of food education and environmental sustainability programs in associations based in Italy and in developing countries. She is trainer in school programs related to environmental impact and sustainable and healthy food systems. She is the Focal Point for Slow Food international education projects and Coordinator of Slow Food and Health activities. She coordinates the international board of experts for Sustainable Food & Healthy diets, edits the contents of articles, papers, and communication materials related to sustainable food and health, and manages activities and projects related to this topic.

Aymara Llanque holds a Professorship for International Sustainable Development and Planning at Leuphana University. She works in the areas of social ecological transdisciplinary studies and sustainability education. She has collaborated on and produced many publications.

Gabriella Luongo MPH, is a PhD in Health candidate within the Faculty of Health at Dalhousie University. Her research examines the socio-ecological factors influencing dietary intake to inform the development of healthy public policies and reduce non-communicable disease risk. Gabriella's doctoral thesis aims to investigate the cost of a healthy diet in Canada. She holds a Canadian Institutes of Health Research Doctoral Research Award, a Nova Scotia Graduate Scholarship, and a Scotia Scholars Award for her doctoral research. Gabriella previously completed a Masters of Public Health in Social and Behavioural Health Sciences from the University of Toronto's Dalla Lana School of Public Health.

Catherine L. Mah MD FRCPC PhD, is Canada Research Chair in Promoting Healthy Populations and Associate Professor in the School of Health Administration at Dalhousie University. She directs the Food Policy Lab, a multidisciplinary program of research on the environmental and policy determinants of diet and consumption, with particular interests in consumer food environments and diet, household food insecurity, municipal and regional public policy, and health-promoting food systems. Dr. Mah is an expert appointee to Canada's federal Nutrition Science Advisory Committee reporting to Health Canada, and member of the inaugural Canadian Food Policy Advisory Council reporting to the federal Minister of Agriculture and Agri-food.

Julien Soliba Manga holds a doctorate in medicine, a master's degree in public health, and nutrition from the University of Montreal (Canada). He has ten years of experience in the Senegalese health system and has a background in public health focused on maternal and child health and the management of health facilities. Currently a PhD candidate in nutrition, his current interests are in public nutrition, particularly on food systems, public policies aimed at healthy eating and sustainability, by improving food systems and environments in West Africa to positively impact maternal and child health and fight against chronic diseases.

Paul Manning is an Assistant Professor in the Faculty of Agriculture at Dalhousie University. Paul's research is focused on understanding the importance of insect biodiversity to the health of agricultural ecosystems. His work draws elements from community ecology, ecotoxicology, agroecology, and citizen science.

Jennifer Marshman is contract faculty and a curriculum developer within the department of Geography and Environmental Studies and the Laurier Centre for Sustainable Food Systems at Wilfrid Laurier University. Jennifer's background as a Registered Nurse allows her to bring an interdisciplinary lens to her teaching and research in just and sustainable food systems.

Alicia Martin MA, is a Doctoral Student in the Department of Geography, Environment and Geomatics at the University of Guelph. Alicia is interested in how food literacy, if broadly conceptualised, can be a tool to empower citizens and policymakers to build more resilient, sustainable, and equitable food systems. She is also a policy analyst for Agriculture and Agri-Food Canada in the Food Policy Division and at the Public Health Agency of Canada in Chronic Disease Prevention. Alicia has an MA in Political Science with a specialisation in Environmental Sustainability focusing on food literacy.

Sonia Massari has decades of experience as a researcher, lecturer, consultant, and designer in sustainability education, food design, and innovative agri-food systems. She holds a PhD in Food Experience Design from the Engineering Department at the University of Florence, Italy. For 12 years, she was the Academic Director of the University of Illinois Urbana-Champaign Food Studies programs in Rome, and she designed and coordinated more than 50 academic programs and 150 educational activities on food and sustainability for prestigious international institutes. She is a board member of the Association for the Study of Food and Society and serves on the editorial board of the International Journal Food Design.

Federico Mattei is an Italian economist specialising in sustainable food systems, ecosystem services, and value chains. He has worked for many organisations including the CGIAR and Slow Food and has extensive experience in developing and implementing food system projects throughout Asia, Africa, and Latin America. Federico also owns and manages a 50-hectare organic farm in central Italy.

Rachel Mazac is a PhD student at the University of Helsinki in the Future Sustainable Food Systems research group. For her Master's in Integrated Studies in Land and Food Systems, Rachel worked on sustainability considerations in international food-based dietary guidelines at the University of British Columbia. Rachel's current research focuses on integrating interdisciplinary understanding into holistic models to assess the potential of protein alternatives for future, sustainable diets.

Stefanie McNerney manages the global Plant-Based Solutions team within the Farm Animal Welfare and Protection campaign of Humane Society International, which works with governments and institutions, both public and private, to reduce reliance on food made with animal products and increase acceptance of plant-based meals. HSI works to change consumption and spending habits to build food systems that are sustainable, protects the environment, respects animal welfare, and provides reliable access to healthy, nutritious foods. Stefanie received her BA from Cornell University and her MLitt from the University of Glasgow.

Breige McNulty is an Assistant Professor in Human Nutrition within the UCD School Agriculture and Food Science and UCD Institute of Food and Health. She completed her degree and PhD in Human Nutrition at Ulster University and has since been based in UCD managing the national food consumption surveys in Ireland. Breige's main research interests are in food consumption and understand the impact of nutrients, food ingredients, and chemicals on health to underpin food safety and policy. https://orcid.org/0000-0003-0841-063X

Galyna Medyna, PhD, is a researcher and Sustainability Science and Indicators group manager at the Natural Resources Institute Finland/Luonnonvarakeskus. She specialises in sustainability assessment of products and services linked to the bioeconomy, with a special focus on Life Cycle Assessment (LCA) and Social Life Cycle Assessment (SLCA). She is currently involved in several projects (H2020 SIMBA, H2020 HealthyFoodAfrica, CBC 2014–2020 EFSOA), assessing novel food products and agricultural practices. Previously, Galyna worked on early design phase environmental assessment at Aalto University (Finland) and environmental policy at the Joint Research Centre (European Commission).

Ashley Meredith is the National Cultural Anthropologist for the Federated States of Micronesia and housed in the National Archives of Culture and Historic Preservation. She carries out capacity building in carrying out ethnographic surveys with the country's five historic preservation offices, conducts research on traditional routes, wayfinding, inter-island connections, and food systems. Her interests include sustainable heritage conservation and management, community-based anthropology, and geography.

Bindu Mohanty is a writer and sustainability consultant based in Auroville, an alternative community and international town in South India. She has worked in the fields of rural development, waste management, sustainable agriculture, policy advocacy for climate justice, and environmental and social assessment of development projects. She also served as faculty for experiential education on sustainability for an international college program. Her writings span the wide range of her interests from sustainability to philosophy and poetry. She currently serves as the Research Coordinator for Revitalising Rainfed Agriculture Network.

J. Pablo Morales-Payan received a PhD in Horticulture with emphasis in weed science and plant pathology at the University of Florida-Gainesville (1999). He is currently a professor with the Department of Agro-Environmental Sciences at the Mayagüez Campus of the University of Puerto Rico. He teaches fruit crop production, fruit physiology, bioregulators, organic systems, recent advances in horticulture, and international topics in horticulture. His research interests include evaluation of varieties, applied physiology, and production, protection and post-harvest management of fruits and other horticultural crops. Dr. Morales-Payan has been Director of the State Agriculture Research Department of the Dominican Republic, and Research Director of the Dominican Team with the Network for Vegetable Research and Development of Central America, Panama, and the Dominican Republic (REDCAHOR).

Jean-Claude Moubarac is interested in food systems, the role of the kitchen, and the impact of food processing on nutrition, health, and society, public policies aimed at healthy eating, as well as the behaviour of actors in the food system, including the agri-food industry. He collaborated in the development of the 2014 Brazilian Food Guide and wrote a report on ultra-processed food purchasing and obesity in the US and Canada. Jean-Claude worked as a programme officer in the Food, Health and Environment programme at the Canadian government's International Development Research Centre in 2015 and collaborates with international organisations, including the WHO and FAO.

M. Muthukumar (IAS) currently serves as the Director of Agriculture, Government of Odisha. He has also served the Government of Odisha as the Collector and District Magistrate in the districts of Malkangiri and Balangir. He is well known in Odisha and India for his work in the design and implementation of innovative agricultural initiatives such as KALIA (Direct Benefit Scheme for farmers), Odisha Millets Mission, Integrated Farming Program, and Crop Diversification Mission. He has a PhD in Entomology from the Indian Agricultural Research Institute.

Antonina Mutoro, PhD, is a researcher specialising in maternal and child nutrition at the African Population & Health Research Centre Kenya (APHRC). She is a registered nutritionist and an honorary research fellow at the University of Glasgow.

Silver Nanema is a graduate student in Public Health with academic background in nutrition and currently serving as a research assistant for MEALS4NCDs project, an NCD prevention project in Ghana. She also served in undernutrition (SAM and MAM) management programme for under-five children and adolescents in Burkina Faso. She has a profound interest in transformation for healthier, equitable, more eco-friendly and sustainable Africa Food Systems.

Hung Nguyen-Viet is co-leader of the Animal and Human Health Program and was the regional representative for Southeast Asia of the International Livestock Research Institute (ILRI). His current research focuses on the link between health and agriculture, food safety, and infectious and zoonotic diseases with an emphasis on the use of risk assessment for food safety management with a One Health approach. He co-founded and led the Centre for Public Health and Ecosystem Research at Hanoi University of Public Health, Vietnam. He has experience working in developing countries, mainly in Southeast Asia and Africa.

Elizabeth Nix is an Assistant Professor in the department of Health & Human Kinetics at Ohio Wesleyan University, where she teaches courses on basic and advanced nutrition, food literacy, and food systems. Her research has focused on social influences of health. She has been part of a student-led program promoting food and health literacy for those suffering from food insecurity. At the university, she has used her background as a registered dietitian and enthusiasm for systems-thinking to build a program focused on sustainable nutrition.

María Teresa Nogales is the founder and Executive Director of Fundación Alternativas, a Bolivian institution dedicated to the design of multidisciplinary strategies for guaranteeing the Right to Food and the renewal of local and metropolitan food systems to ensure resilience in a context of urbanisation and climate change. This work revolves around participatory public policy development, the promotion of urban agriculture and education with a focus on food security. She studied at Baylor University (US) and holds a Master's degree in International Relations with a focus on Human Rights and Ethnic Conflict.

Phyllis Ohene-Agyei received her medical training from the Kwame Nkrumah University of Science and Technology in Ghana. She has further training in Public and Environmental Health from The University of Sheffield, UK, and French School of Public Health (EHESP). Her practice and research have focused on public health nutrition, having worked as a nutrition and health advisor for the international NGO Action Contre la Faim (ACF) in France and Sierra Leone.

Anna Okello is the Research Program Manager for Livestock Systems at the Australian Centre for International Agricultural Research. Anna is also an academic fellow at the University of Edinburgh Medical School, and is a Lancet Commissioner for One Health. Anna has worked in international livestock development and public health programmes in Africa and Southeast Asia since 2005, undertaking project management and technical advisory roles for international organisations. A registered veterinarian, Anna obtained her PhD in political science at the University of Edinburgh's School of Social and Political Science in 2012, and maintains a keen interest in how to ensure food safety and foodborne disease research outcomes are embedded into international development policy.

Hibbah Araba Osei-Kwasi is a Public Health Nutrition researcher with about 14 years research experience in Africa and Europe. She is currently an AXA Postdoctoral Research fellow at the University of Sheffield, exploring cooking practices of Ghanaians in the UK and also in Ghana, to leverage evidence to improve the traditional diets of Africans. Hibbah's research falls within the following themes: Dietary practices and its relationship with malnutrition; nutrition transitions in low- and middle-income countries; food environments; nutrition inequalities linked to migration and ethnicity; and food systems and food insecurity.

Aifric O'Sullivan is an Assistant Professor and Associate Dean for International Programmes in the UCD School of Agriculture and Food Science, and PI for UCD's Institute of Food and Health. Aifric's research aims to understand interactions between diet and metabolism, and sustainable nutrition strategies. Some recent and ongoing projects include: AgriDiet, Understanding agriculture-nutrition linkages in Tanzania. She is the Vice Chair of the Executive Committee of the UCD Childhood and Human Development Research Centre, a member of the Royal Irish Academy. https://orcid.org/0000-0002-7441-1983

Imana Pal received her PhD from the Department of Home Science (Food and Nutrition), University of Calcutta, Kolkata, India. She has ten years of teaching and research expertise in the field of Food and Nutrition, is actively involved in various research projects, and has published and presented nationally and internationally Her research interests include sustainable development, food security, reducing food waste, and reduction of malnutrition. https://orcid.org/0000-0001-5636-889X

Fiorella Picchioni is a Research Fellow in Gender and Social Differences at the Natural Resources Institute (NRI), University of Greenwich. Her research intersects food, nutrition, and development in the Global South. She uses the analysis of food systems as a gateway to understand gender, class, and ethnic inequalities. Her work mobilises social reproduction approaches to study how the world food economy promotes or hinders access to healthy, sustainable, and desirable diets. Before she completed her PhD at SOAS, University of London, she worked at the Food and Agricultural Organisation of the UN. https://orcid.org/0000-0002-3456-386X

Valeria Piñeiro is a Senior Research Coordinator in the Markets, Trade, and Institutions Division at the International Food Policy Research Institute (IFPRI). Her recent work includes modelling the impacts of agricultural support policies on emissions from agriculture and reviewing the evidence on incentives for the adoption of sustainable agricultural practices and their outcomes. She has significant experience working in economic development and growth using General Equilibrium Models (CGE) as an analyst. She is a faculty member at the Applied Economics Master's Program at Johns Hopkins University, and received her PhD in Agricultural Economics from the University of Maryland. https://orcid.org/0000-0002-4372-7141

Sarah Pittoello is a registered counselling therapist (RCT), organic farmer, writer, and mother living in Mi'kma'ki (Nova Scotia, Canada). She has a MSc in Holistic Science from Schumacher College and an MEd in Counselling from Acadia University. She is currently counselling at Acadia University, with a particular interest in mindfulness practices and Somatic Experiencing. She has been engaged in building sustainable food systems in a variety of ways, including as a grower, advocate, educator, researcher, and writer. https://orcid.org/0000-0003-3314-058X

Bernd Pölling is a researcher and teacher at Fachhochschule Südwestfalen University of Applied Sciences, Department of Agriculture in Soest, Germany. He holds a PhD in Agricultural Sciences (Humboldt University Berlin, 2018) on urban agriculture business models. His main research interests are in urban and peri-urban agriculture, new entrants into farming, rural-urban linkages, Green Infrastructure, business models, and rural development. He currently is involved in several national and EU projects, including FoodE (Food System in European Cities), EFUA (European Forum on Urban Agriculture), proGIreg (productive Green Infrastructure on post-industrial sites), and Newbie (New entrants network). ORCID: 0000-0001-7847-4255 Link ORCID: https://orcid.org/0000-0001-7847-4255

Paolo Prosperi is a senior lecturer and researcher in agricultural and food economics and scientific administrator at the Mediterranean Agronomic Institute of Montpellier (CIHEAM-IAMM). He holds a PhD in Economics from the University of Montpellier (France) and a PhD in Agri-food Economics from the University of Catania (Italy). He is a permanent member of the Joint Research Unit "Montpellier Interdisciplinary Center on Sustainable Agri-Food Systems" (UMR Moisa, University of Montpellier, CIHEAM-IAMM, CIRAD, INRAE, IRD, L'Institut Agro, Montpellier, France). Prior to joining CIHEAM-IAMM, Paolo was researcher and assistant professor of agricultural economics at the University of Pisa. He has served in several European Union (Horizon 2020 and Horizon Europe) and international projects as researcher and principal investigator (e.g. SALSA, SUFISA, SHERPA, QESAMed, HealthyFoodAfrica, PRIMA Med-Links, etc.) Paolo's main research focuses on understanding and assessing the complex socio-economic dynamics of food system sustainability, including the analysis of stakeholders' strategies and innovation for creating sustainable agri-food value chains. He is the co-editor of the Routledge Handbook of Sustainable Diets. https://orcid.org/0000-0002-8494-0344

Delia Grace Randolph is a Professor of Food Safety Systems at the University of Greenwich Natural Resources Institute. She was formerly Program Leader (joint) for the Animal and Human Health Program at the International Livestock Research Institute, based in Nairobi, Kenya. Originally trained as a veterinarian, Delia undertook a MSc at the University of Edinburgh, and a PhD at the Free University Berlin. Her post-doctoral position, with Cornell University and ILRI, focused on food safety. From 2006 to 2020 Delia undertook research at the intersection of the health of the environment, animals and people under One Health. She has co-authored the first book on food safety in informal markets. She has worked with the World Animal Health Organisation, the World Health Organisation, the Food and Agriculture Organisation, the World Bank and other international, regional and national organisations.

Jessica Evelyn Raneri is a Senior Nutrition Sensitive Agriculture Advisor to both The Australian Centre for International Agriculture Research (ACIAR) and to the Australian Department of Foreign Affairs and Trade (DFAT). Jessica's work focuses on improving nutrition and health outcomes of agriculture and food system initiatives, both by supporting research design and implementation, and providing evidence-based policy and programming recommendations. Jessica is passionate about food and nutrition security data, including novel indicator development and testing and supporting common indicator use to support global monitoring. She is currently completing her PhD in Bio-Science Engineering (Food and Nutrition) with the University of Ghent, Belgium.

Kerry Renwick is an Associate Professor in the Department of Curriculum and Pedagogy, Faculty of Education at the University of British Columbia. She teaches graduate students and into the Bachelor of Education preparing secondary educators specialising in home economics and K–12 teachers in health education. Her research interests include teacher education, critical pedagogies, health education and promotion, and food literacy. Dr. Renwick is a lead investigator in the Food Literacy International Partnership.

Silvia Rolandi received a PhD degree in Agricultural Law and International Law within the Doctoral Program in Politics, Human Rights and Sustainability at the Scuola Superiore Sant'Anna in Pisa. She is a lawyer, works in her law firm and is a post-doc at the University of Pisa within the Pisa Agricultural Economics (PAGE) Group, and a Lecturer in Food Law and IPR with the Law Group at the University of Wageningen, Netherlands. She is participating in European and national funded research and development projects. Her research interests include International Trade and Agri-Food Law, with a focus on Digitalisation and Internet Law using a multidisciplinary research approach.

Sara Roversi, President at the Future Food Institute, is an experienced entrepreneur and thought leader in the food ecosystem. As a seasoned growth expert with a flair for exploration and prosperity thinking, she works with globally recognised, high-profile think tanks on setting the agenda for the sustainable food industry. Over the past ten years, she has focused on a mission to take food leaders and youth to the next phase of action through education, research projects, and innovation challenges. In 2013, she founded Future Food: a purpose-driven ecosystem facilitating positive transformation in life on earth, promoting food innovation as a strategic element to achieve sustainable and impactful growth. She is the Italian Focal Point for the network of the Emblematic Communities of the Mediterranean Diet UNESCO.

Caitlin Scott is a Professor of Food Studies at George Brown College where she co-developed and teaches in the Honours Bachelor of Food Studies. Prior to joining GBC, Caitlin completed her PhD at the University of Waterloo in Social and Ecological Sustainability, where she explored the challenges of governing for sustainable diets, focusing on ideational debates and power. She is interested in food issues and politics at the intersection of health, environment, and justice.

Steffanie Scott is a Professor in the Department of Geography and Environmental Management at the University of Waterloo. She is the co-author of *Organic Food and Farming in China: Top-down and Bottom-up Ecological Initiatives* (2018). Her research unearths land-based sustainability practices and food system and organic sector developments in China and Canada. Steffanie is past president of the Canadian Association for Food Studies, past co-chair of the Food System Roundtable of Waterloo Region, and associate editor of *Frontiers in Sustainable Food Systems— Social Movements, Institutions and Governance*.

Brigitte Sébastia is a medical anthropologist affiliated with the French Institute of Pondicherry (IFP) and the CEIAS (Centre d'Etudes de l'Inde et de l'Asie du Sud) of the Ecole des Hautes Etudes en Sciences Sociales, Paris. She developed the programme "Food and nutrition in Indian contexts" at IFP from 2008 to 2017. She edited the volume *Eating Traditional Food: Politics, Identity and Practices* (Routledge, 2017) and is the author of three other books and 30 articles on her areas of expertise: religion, cultural psychiatry, siddha medicine, and health disorders resulting from dietary and agricultural change. https://orcid.org/0000-0002-9244-6944

Barbara Seed is a registered dietitian and consultant, focusing on integrating nutrition, food systems, and the environment. She led the development of the first dietary guidelines in Qatar in 2013—one of the first globally to include sustainability principles. In addition to speaking and publishing on sustainable diets, she was a lead writer for the Dietitians of Canada Sustainable Food Systems role paper. Barbara has completed projects on sustainable food systems and food security for Health Canada, the BC Provincial Health Services Authority, and the International Congress of Dietetics.

Meena Sehgal is a Senior Fellow and Area Convenor in Environment and Health Area of The Energy and Resources Institute (TERI), in New Delhi, India. She, and her co-authors, are part of the FOLU India coalition. She is an epidemiologist working on environmental exposures and nutrition security issues. Her interests are primarily in designing public health studies for evidence building and policy recommendations. Her projects have yielded insights for GOI policies for protecting public health and several scientific papers in peer-reviewed journals. Her projects have been funded by a variety of leading national and international organisations.

Mila Sell, PhD, is a senior researcher and research manager (Carbon neutral society) at the Natural Resources Institute Finland/Luonnonvarakeskus. She has worked for over 15 years with and in Africa in the NGO, UN, and research sector, focusing on food security, gender, agriculture, and sustainability. Her work currently includes management, coordination, and research activities within large-scale research for development programmes, especially in Africa, and includes strategic and technical programme design, implementation and monitoring, among other roles.

Amanda Shankland is a PhD Candidate in the Department of Political Science at Carleton University. Her dissertation looks at water governance discourse in agricultural communities in rural New South Wales, Australia. Her areas of research expertise include social ecology, agroecology, food sovereignty, water management, and rural development. Amanda was raised by a single mother committed to growing food, which gave her a drive toward cultivating sustainable and affordable food systems. Amanda lives on the unceded territory of the Algonquin-Anishinaabe with her three sons and her dog.

Carmen Byker Shanks is Principal Research Scientist at the Gretchen Swanson Center for Nutrition and Faculty at Montana State University. Carmen holds a PhD in Nutrition from Virginia Tech and is a Registered Dietitian Nutritionist. Her research aims to co-create and advance strategies that provide nourishing food for all people. She lives with her family in Bozeman, Montana, United States.

Zhenzhong Si is a Postdoctoral Fellow at Wilfrid Laurier University. His research examines food security, sustainable food systems, and rural development in China. He is working on a SSHRC-funded project about COVID-19's impacts on food security and a SSHRC Partnership Development Grant on using agroecology to advance the Sustainable Development Goals. Building on his doctoral research, he co-authored the book *Organic Food and Farming in China: Top-down and Bottom-up Ecological Initiatives*. He teaches courses about food systems at the University of Toronto and Wilfrid Laurier University.

Steven E. Smith (University of Arizona) is a plant breeder, botanist, and statistician whose research, training of students, and teaching cover those areas of expertise. His research interests reflect both his training in application-oriented plant breeding and his fascination with plant survival in natural plant communities in arid environments.

Daniela Soleri (University of California, Santa Barbara, and The Centre for People, Food & Environment) is an ethnoecologist whose research focus is local and scientific knowledge systems in small-scale agriculture and gardens. This research includes quantifying farmer practices, documenting risk assessment and cultural identity related to seeds, investigating new semi-formal seed networks, and community science.

Mark Spires is a Research Fellow at the Centre for Food Policy, responsible for establishing and leading new, interdisciplinary projects exploring public policy solutions for healthy diets in the UK and internationally. Mark's primary research interests centre on seeking to better understand peoples' lived experiences of local food environments, and how these findings can contribute to more effective and inclusive food policy.

Chittur S. Srinivasan holds a PhD in Agricultural Economics from the University of Reading, UK (2001), an MSc from the London School of Economics and Political Science UK, and an MBA from the Indian Institute of Management, Ahmedabad, India. He is currently a Professor of Agricultural and Development Economics in the Applied Economics and Marketing Department, School of Agriculture, Policy and Development, University of Reading, UK and leads the Agri-Food Economics and Social Science Research Division. His research interests include dietary and nutrition transitions in developed and developing countries, and is interested in the use of wearable digital technologies in empirical research on the analysis of agriculture-nutrition linkages in rural settings in developing countries.

Mary Stein is the Program Leader and Instructor within the Sustainable Food and Bioenergy Systems (SFBS) program at Montana State University (MSU). She has over 25 years of experience working and teaching in the fields of human nutrition and sustainable food systems. Between 2011 and 2015, Mary served as Deputy Director of the National Farm to School Network, successfully growing the organisation and expanding the integration of local food in schools, school gardens, and sustainable food system education in preK–12 schools across the US. Currently, Mary teaches, advises students, contributes to campus and community sustainability initiatives, and engages in sustainable food system pedagogical research.

Phoebe Stephens has been a SSHRC Postdoctoral Fellow at the University of Toronto and has recently taken up the post of Assistant Professor at Dalhousie University, Faculty of Agriculture. Her research focuses on the role of intermediaries in supporting transitions to more sustainable food systems. She has published articles in academic journals including *Agriculture and Human Values*, *Canadian Food Studies*, and *Globalizations* as well as several book chapters on finance and food systems.

Johan Swinnen is Director General of the International Food Policy Research Institute (IFPRI), and Global Director of Systems Transformation Science Group at CGIAR. Before, he was director of the LICOS Centre for Institutions and Economic Performance at KU Leuven, (visiting) professor at several universities, and was economic advisor at the World Bank and the European Commission. He has a PhD from Cornell University and received several awards for his work on global agricultural policies, food security, and international development. https://orcid.org/0000-0002-8650-1978

Margherita Tiriduzzi is a researcher at Future Food Institute. With a Masters in Social Sciences Applied to the Food Sector from the University of Toulouse (France), Margherita focuses her research on consumer's behaviour and food systems understanding. She worked in local development food plans for the city of Montpellier with research centres CIRAD and INRAE, as well as in Italian rural areas, such as the Umbria region with CESAR, to develop rural development plans in the area of consumer perception. Her goal is to bring together social sciences, design, and innovation to provide a planet-friendly food system.

Riana Topan is a campaign manager with Humane Society International/Canada, which is one of the largest animal protection groups in the world. She works to protect farm animals by advocating for higher-welfare policies and regulations, and manages HSI/Canada's innovative Forward Food program, which supports institutions and businesses across Canada to increase their offerings of delicious and nutritious plant-based options that are better for the animals, the environment, and human health. She is based in Ottawa, Ontario, Canada.

Theresa Tribaldos is Head of the Just Economies and Human Well-being Impact Area at the Centre for Development and Environment at the University of Bern in Switzerland. She is a geographer and political scientist by training and holds a PhD from ETH Zurich. Her research focuses on sustainability and justice in food systems, food system transitions and transformations, and interventions to make these systems more sustainable. Transdisciplinary research approaches for sustainable development, and the contributions science can make to support sustainability transformations, are at the core of her research. https://orcid.org/0000-0002-9050-8451

Joseph Tuminello is an Assistant Professor of Philosophy at McNeese State University in Lake Charles, Louisiana, US, and a Program Coordinator for the non-profit organisations Farm Forward and Better Food Foundation. Joseph's research interests include the philosophies of food, medicine, animals, and environment through the lenses of hermeneutics, pragmatism, and Jainism. His work has appeared in *Sofia Philosophical Review, Humana. Mente: Journal of Philosophical Studies, The Encyclopedia of Food and Agricultural Ethics,* and several edited volumes on food and animal ethics. https://orcid.org/0000-0002-7178-870X

Gabriel R. Valle is an Associate Professor of Environmental Studies at California State University, San Marcos, USA. He received his PhD in sociocultural anthropology from the University of Washington in 2016. His research explores environmental and food justice at the intersection of ethnicity, place, and power. His recent scholarship on the ethnoecology and agroecology of urban home kitchen gardens explores questions of labour, value, and conviviality in constructing alternative food systems. He is co-editor of the award-winning book -*Mexican-Origin Foods, Foodways, and Social Movements: A Decolonial Reader,* and has published in many journals.

Stephanie Van is a planner by education with a strong interest in Indigenous relations, climate change, and public policy. She graduated from the University of Waterloo's Planning program and is currently a planner for the County of Perth, Canada. From her previous experience working with the provincial and federal government, she is interested in creating policy to encourage a sustainable and accessible way of life.

Stefanie Vandevijvere is a senior public health nutrition scientist and her research focuses on nutrition policies for obesity prevention globally. Her work supports the development of policies to effectively and equitably improve the quality of population diets. In particular, she has an interest in the role of food environments in determining people's dietary habits and how to hold governments and food businesses accountable for their actions to create healthy food environments. She is a member of the leadership team of the International Network for Food and Obesity/NCDs Research, Monitoring and Action Support (INFORMAS). She has about 170 peer-reviewed publications.

Matteo Vittuari is an Associate Professor at the Department of Agricultural and Food Science of the University of Bologna. Since 2021 he has served as a delegate to International Cooperation and Development of the University of Bologna. He served as work package leader in a number of EU and international projects and as an international consultant for the evaluation of EU and FAO funded projects in Europe and Central Asia and an evaluator for several international institutions. He has extensive experience in teaching university courses on agricultural and food policies and rural development, with an international perspective. He is interested in food system sustainability and his main fields of expertise include: agricultural, food and rural economics and policy; agricultural and rural policy evaluation;

behavioural economics; life cycle thinking. He has published over 100 scientific articles and reports. https://orcid.org/0000-0003-4327-1575

Tony Weis is a Professor in the Department of Geography and Environment at Western University. He is the author of *The Ecological Hoofprint: The Global Burden of Industrial Livestock* (Zed Books, 2013) and co-editor of *A Line in the Tar Sands: Struggles for Environmental Justice* (BTL/PM Press 2014) and *Critical Perspectives on Food Sovereignty* (Routledge 2014). Tony's research is broadly located in the field of political ecology, with a focus on the social inequalities, biophysical instabilities, and interspecies relations associated with agriculture and food systems.

Giacomo Zanello is Associate Professor in Food Economics and Health, Head of the Agri-Food Economics and Marketing Department, and Deputy Director of the Graduate Institute for International Development, Agriculture and Economics at the University of Reading (UK). He is also a Fellow at the London Centre for Integrative Research on Agriculture and Health (London School of Hygiene and Tropical Medicine) and serves as a core member of the Independent Expert Group for the Global Nutrition Report. Giacomo's specific area of expertise is food and nutrition economics. His main research has been concentrated on efforts to understand the intersections of agriculture, food, nutrition, and health in low- and middle-income countries. https://orcid.org/0000-0002-0477-1385

ACKNOWLEDGEMENTS

We thank all those work in and on food systems: contributors, researchers, dreamers, and doers, who are out there advancing diets and food systems to more rapidly become sustainable.

We acknowledge the partnership with Sarah Pittoello, without whom this handbook would not be what it is. Sarah established our systems and enabled the high functioning flow of materials at every stage. We are deeply indebted to Sarah's elegant professionalism and tremendous humanity.

We thank Stephanie Van for the invaluable contributions that kept us progressing and positive.

We thank OECD for the funding for the fellowship in the Co-operative Research Programme.

Our colleagues at Dalhousie University and the Mediterranean Agronomic Institute of Montpellier also are appreciated for their support and engagement.

PART 1

Framing and vision

1

INSPIRING SUSTAINABLE DIETS

Kathleen Kevany and Paolo Prosperi

What are sustainable diets?

Over decades, there has been growing attention to, but insufficient action around, sustainable diets. Human activities, including our diets, are challenging with their recurrent complexities, contradictions, and conflicting interests; "why is it that agriculture world-wide not only produces more food than ever before, but also more ill-health, environmental destruction, and social disintegration?" (Waltner-Toews & Lang, 2000, p. 119). Agriculture, food production, and food consumption influence the trajectory of society and impact the well-being of citizens and planetary health. Governmental policies and practices, the powers of industry, and actions of citizens interacting through economic systems drive environmental outcomes. Our food systems influence us through environmental feedback loops, health outcomes, and economic and social hierarchies. How might humanity, with varying powers and privileges, contend with large trade-offs, overcome inadequate action, and foster sustainability? Instead of eating food for health, our foods are making many people ill, ecosystems are being destroyed, and future yields are in jeopardy with increasing destabilisation of weather patterns, exacerbated by increasing global warming. In diverse ways, in the Global South and Global North, many forms of malnutrition are afflicting millions of people. Roughly, 690 million people, or 8.9% of the global population, suffer malnutrition and inadequate food access and availability (FAO, IFAD, UNICEF, WFP, & WHO, 2020). Many others eat diets that are sub-optimal in supplying needed nutrients; 1.9 billion adults are overweight or obese, and similar patterns could increase. The WHO 2020 report reveals that now more people are suffering from being overweight than from hunger. It is alarming that humanity contends with such disparity in access to adequate food.

Advances in agricultural sciences, techniques, and practices have succeeded at producing more food to feed more people in the world. Over decades, technical

DOI: 10.4324/9781003174417-2

innovations have increased yields, enhanced reserves, and distributed more food. Advances in processing and making foods more "shelf-stable" have added to the availability of foods at lower prices. Today we clearly know that these gains, however, have come with great social, environmental, and economic costs and have engendered contradictions and inequalities. Compelling evidence reveals devastating outcomes with epidemic levels of non-communicable diseases, biodiversity loss, and environmental contamination, along with social inequity. The decade of 2010–2020 has seen the highest temperatures on record, disruptions in weather patterns, and tremendous volatility in food production. Over this decade, scientists from many fields have been urging action on these culminating crises. Our food systems are failing to produce food in ways that retain the integrity of life-supporting systems. Report after report are calling for responsible, urgent action and unprecedented change, including, for example, the Intergovernmental Panel on Climate Change (IPCC, 2018, 2019, 2022), the InterAcademy Partnership (2018), and the works of Dury et al. (2019), Fears et al. (2019), Searchinger et al. (2019), and Brunori et al. (2020). Also, Harwatt and her colleagues (2020), the often quoted EAT-Lancet report (Willett et al., 2019), and Clark and team (2020) have issued stark warnings that we have no time to waste. Actors across food systems—locally, regionally, and internationally—must transform agricultural production along with consumer choices and consumption patterns, as these are major levers for reducing climate change and reversing global warming. Limiting global warming to 1.5 °C has not proven possible with modest changes; we must undertake *"rapid, far-reaching and unprecedented changes in all aspects of society"* (IPCC, 2018, emphasis added).

This handbook is a "wake-up call" to world leaders, policy makers, media, practitioners, researchers, farmers, citizen consumers—all of us—to genuinely commit to averting catastrophic climate crisis. On what grounds would we justify not taking every action possible? For those reading this and inspired to act, we must guard against comparing our production practices or consumption habits to lower standards to feel better about insufficient progress or marginal improvements. We must hold bolder goals as our signposts for significant progress towards responsible food production and consumption. These wicked problems of unsustainable food systems are severe; our solutions need to rise to the challenge and set humanity on sustainable, equitable, nourishing, and prosperous pathways. In our quest for sustainability transformation for all, we should welcome discussions and constructive debates about the external and internal drivers, like values and worldviews and ways to operationalise sustainable diets as effectively as possible without waiting on perfect arrangements and outcomes. Actors in all sectors—legislators and decision makers; educators, media, and researchers; undergraduates, graduates, students at all levels; private sector and civil society; as well as policy makers and practitioners in public health, nutrition and dietetics, environment, music and the arts, trade, law, governance, justice—are essential in this collective journey to innovate, adapt, co-create, and implement sustainable diets.

Governments, most particularly, must act with urgency and conviction, by harnessing the comprehensive evidence that leaves no doubt that changes are essential.

Leadership is needed "*to create the vision, provide policy coherence, and focus the political will*" (Mason & Lang, 2017, p. 4, emphasis added). Local and national leaders should ensure the voice and the needs across all sectors are factored into our planning and transformative strategies. Around the world, we must work diligently to advance the Sustainable Development Goals (SDG), work towards sustainable dietary patterns, and replace dysfunctional outcomes. Through our food systems, we seek to improve society and individual and collective lives by sharing the bounty and benefits, while diminishing and distributing the hardship. All aspects of life interrelate with our food. Agriculture policy is *de facto* food policy, health policy, environmental policy, economic, social, and justice policy. But, is it *good policy*? Our food systems, food environments, and food choices interact with and influence health, culture, spirituality, economics, the environment, social development of humans, and well-being of animals, and law, justice, and politics (High Level Panel of Experts on Food Security and Nutrition [HLPE], 2017, 2020). Therefore, all actions along the pathway of making food, from regulating, subsidising, planting, producing, harvesting, storing, processing, packaging, marketing, distributing, retailing, purchasing, preparing, consuming, and wasting or composting determine whether, or not, it leads to greater sustainability, in the fullest sense of the word.

The increasing awareness of and debate on the sustainability of food systems raise crucial questions and challenge us to acknowledge the crises and commit to fair and steady change. We examine trade-offs, tensions, and contradictions as we strive for sustainable diets. For instance, plant-based diets are important when it comes to sustainability transformation, but these approaches also need further attention (e.g., regarding monocultures). What strategies have proven beneficial to prevent and avert detrimental practices like activating monocultures in Peru or Bolivia for quinoa, avocados in Mexico, almonds in California, USA, or coconut oil in Malaysia? Pulses may be promoted as cost effective and nourishing foods; however we face challenges in increasing their acceptability and addressing their production inputs. Problems are calling out to be solved. How might serious problems be averted for the environment, like pesticide use and water stress, as well as social tensions, like land grabbing and violence over land? How will we navigate larger data sets and make prudent use of technology and artificial intelligence? What are the implications for health, equity, and justice, such as addressing hunger and environmental justice, while considering ethical concerns around inefficiencies in animal-based agriculture and projections to double food production to feed growing populations? The complexity of the issues is evident, as is the need for context-sensitive strategies to accelerate the adoption of sustainable diets.

Facts alone won't fix things: curiosity, creativity, criticality, cooperation, comprehensive thinking, and commitment can. Committing to sustainable diets requires a reorientation. The complexity is not reduced through shining light on it, but it does afford valuable insights for making decisions and assessing trade-offs. Maintaining the status quo, we must agree, is untenable and unsustainable. When effectively and thoughtfully designed, food systems can be harnessed as pathways to reduce and, if sufficiently committed to act quickly, reverse global warming, air, water, and soil pollution, environmental destruction, and devastating biodiversity loss. Shifting to healthier diets necessitates profound shifts in ingrained beliefs about what foods humans need to be well, suitable actions to displace hunger, and optimal ways to feel nourished and energised.

Sustainable diets defined

The components that constitute sustainable diets have been well argued by many research teams, leading thinkers, and change activists (Burlingame & Dernini, 2012; Food and Agriculture of the United Nations (FAO), 2014; Gussow & Clancy, 1986; Mason & Lang, 2017, among others). The co-authors within this collection are building upon this good work and highlighting the essential elements needed to foster food patterns that enable much more of the beneficial outcomes and far fewer detrimental ones. The Brundtland Report (United Nations, 1987) on sustainable development and the FAO (2010) offer standards to which nations might aspire. Sustainable diets are those diets with low environmental impacts that contribute to food and nutrition security and to healthy living for present and future generations. Compelling definitions from FAO state that sustainable diets are patterns that enable all people to achieve minimum nutritional requirements, delivered by delightful food that is accessible, acceptable, affordable, always safe, and readily available. Such diets are delivered all while strengthening and protecting natural resources, thus guarding against the depletion of soils, wasting of water, polluting of ecosystems, or amplifying climate change (FAO, 2010). Sustainable diets are protective and respectful of biodiversity and ecosystems; culturally acceptable and accessible; economically fair and affordable; nutritionally adequate, safe, and healthy; while optimising natural and human resources (FAO, 2010).

We can call diets sustainable when they are health enriching, environment protecting, financially stimulating, socially appealing, and politically empowering. Sustainable diets a) produce and deliver nutrients consumers need and want for optimal well-being *and* b) preserve and improve life enhancing ecosystems, *while being* c) culturally appropriate and economically accessible to diverse tastes and respectful of the sacredness of food and religious requirements, *and* d) enable trade in food products that ensure producers and processors are fairly compensated, and strengthen local, regional, and national economies, *through* e) recognising and supporting the value of workers all along the food value chain, and treating humans, animals, ecosystems, and food itself with respect.

Imagine if populations were suitably nourished and the well-being of citizens and societies and planetary health were prioritised. When governments prioritise and accelerate sustainable diets, environmental and socioeconomic burdens could be replaced with the freeing of resources for human development. When we imagine a world with optimal health and well-being, what do we envision arising from our food and our food systems? Addressing the complexity of food systems' dynamics, Figure 1.1 provides a systemic model of sustainable diets to illustrate how these food-related human activities interact within local and global contexts and the essential role of sustainable practices to achieve the SDGs. Our depiction of sustainable diets (see Figure 1.1), utilises multidimensional concepts to

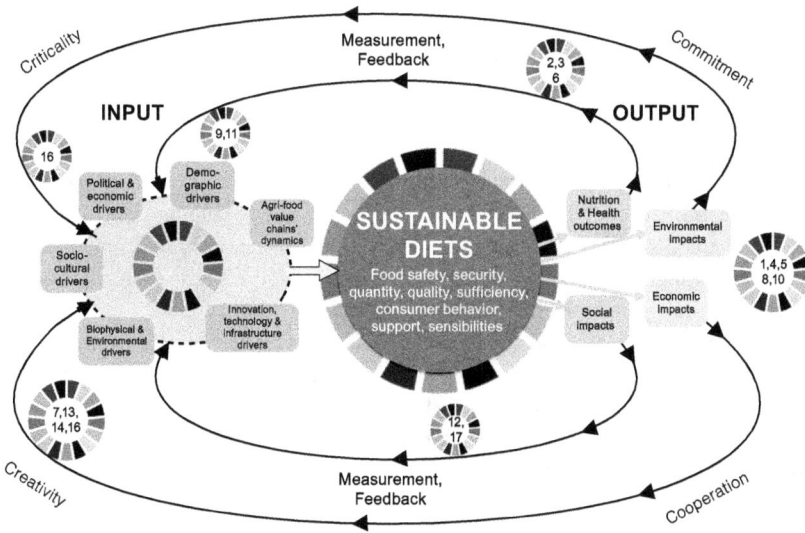

Figure 1.1 System and consumer dynamics of sustainable diets (authors' elaboration).

capture interactions with food consumers, their knowledge, competencies, values, and motivations, and the complexity of relationships involved in the production of their foods. Principles, policies, practices, technologies, and relationships all along the value chain influence whether or not sustainable food systems and sustainable diets become accessible or achievable. Interconnections can be identified between natural ecosystems, food policies, food production, food environments, food and nutrition programs, and support for food safety, security, sufficiency, and sovereignty. At a micro level each food stuff, through interdependent and dynamic interactions, causes and is affected by natural and human-driven cycles, pressures on energy and matter, and influenced by historical, economic, cultural, political, and psycho-social forces for stability and change.

Handbook organisation

This handbook aims to gather the best in scientific evidence, interdisciplinary debate, and creative collaboration to confront the wicked problems and fashion solutions for sustainable diets. Probing questions around how to foster sustainable diets have been catalysing the interest and energy of many researchers, planners, and policy makers. In this handbook, we explore multidisciplinary and transdisciplinary approaches, macro-, meso-, and micro-scales, quantitative and qualitative methods, and studies from diverse geographic areas across the Global South and Global North. We also illuminate emerging contradictions towards more sustainable diets that demonstrate the need for an increased diversity of food sources, agricultural and processing practices, business models, forms of coordination between actors, as well as focused policies and research. Through sharing research on, strategies for, and stories of sustainable diets, we envision food production and consumption that simultaneously eradicate systemic racism and colonialism, accentuate health outcomes, elevate social equity and animal welfare, while optimising trade relations and environmental rejuvenation.

The handbook is organised into ten sections. The first section provides our *framing and vision* with broad and significant principles guiding the work. As a collective, we advocate for the right to food for all and advancing the sustainable development goals. We call for commitments to anti-racism and decolonisation and a reframing of the discourse for greater accountability. Broader and bolder insights around global governance provide instructive perspectives.

The next section offers readers nine chapters on *environmental strategies*. We examine strategies around increasing food security through respectful measures and optimising sustainable production. Chapters on agrobiodiversity and agroecology offer proven approaches to food production with environmental conservation and rejuvenation at the core. Specialists offer valuable chapters on protecting biodiversity, insect populations, and ways to integrate low carbon foods, like insects, into culinary options. Readers also can explore chapters on the prospects of aquatic food for human consumption as well intriguing stories of molluscs and sustainable aquaculture. Two critical chapters elucidate the concerning trends of increasing industrial agriculture and the implications for environmental and social stability.

In *health and well-being*, the first two chapters provide evidence to inform food-based recommendations to reduce the burden of disease and to address threats of global pandemics. With illustrations from an organic family farm, authors of the next chapter illuminate the Ecological Determinants of Health. A chapter on the power of breastfeeding shows links to sustainable diets as does the subsequent chapter on nutrition transition faced by youth in rural communities in the Global South.

The fourth section, *education and public engagement*, integrates seven chapters, each telling stories of food literacy and pedagogy from diverse perspectives including food systems literacy, undergraduate studies, community collaboration and campus food production, community food gardens, and ways that policy and practices are shifting to include nonhuman animals.

Social policies and food environments provides examinations of city food systems, the role of local governments, and the power of local food initiatives. Food environments and international food-based dietary guidelines are complemented by transformative food systems with studies of Odisha Millet and sustainable diets as food consumption.

Under *transformations and food movements*, five chapters illustrate ways to consider the inner work of sustainable diets, transform the food service industry and food movements towards more planet-friendly food systems, as well as discussing alternative food networks and the influence of the slow food movement.

On *economics and trade*, the substantial contributions cover topics of the need for living wages, financialisation and sustainable diets, and pitfalls in seeking alternatives to large scale animal agriculture. Additional chapters inform readers on the connections between trade, food security, and nutrition, circular bioeconomy of agri-food value chains, and preventing food loss and waste.

Design and measurement offer leading edge approaches to evaluation, data collection, the fair use of technology and tools, and the importance of design in food environments and food systems.

Food sovereignty and case studies provide eight examples of national experiments, with some success stories and some lessons learned. The prominence of women in agriculture is revealed in these stories as well as the necessity to engage actors across all sectors for enduring change.

The final section is a *call to action* and an invitation to pay heed to the compelling evidence that the future of food and humanity rests on to urgently adjust food production and consumption.

Questions this handbook seeks to answer

Everything humans produce requires resources and generates a footprint. However, all food production and consumption are not equal in their impacts. What regional or local dietary patterns could optimise human, animal, and environmental health? What practices will enable us to protect biodiversity and natural resources, reverse global warming and mitigate impacts of climate change, update international, national, regional, and local trade policies and practices to ensure greater equity and well-being for consumers, food producers, and sector workers? How well have we reoriented our thinking and practices towards environmentally-friendly strategies that contribute positive impacts on climate change and the environment? What are the models and methods we can deploy at local, regional, and global scales in production, processing, and distribution to achieve greater sustainability? How might we, from among a range of diets, choose ones that could be most suitable for the region, sustainable for the planet, and succulent for the consumer?

The EAT-Lancet report advises that for the average diet, we need to increase our consumption of fruits, vegetables, pulses, and nuts by 200% and reduce our consumption of red meat, dairy, and added sugars by more than half. Investing in strategies to reduce food waste and prioritise plant-rich diets are proving to be two of the most promising approaches to not only reduce, but to reverse, global warming (Hawken, 2017). Internationally, scientists call for demonstrations of leadership in agriculture, health, and environmental policy. Citizens are urged to read the reports and to work together to transform global food systems into sustainable food systems.

Many inspiring examples abound of governments and citizens taking action. The Government of Finland has helped many dairy farmers become berry farmers, known as *the berry project*. It helped achieve significant reductions in cardiac death by 80% and all-cause mortality by 45% throughout the country (Puska, 2002)—investments in agriculture paid off many fold in reduced illness costs and enhanced quality of life. In Scotland, the Rural Economy Secretary Ewing promoted low carbon agricultural practices, including organic practices, among other strategies, to take advantage of the green economy. In the Netherlands, the Council for the Environment and Infrastructure (CEI), an advisory group to the government, made similar recommendations: "Giving farmers and parties in the value chain clarity about future emissions reductions will encourage them to bring forward innovations, develop new business models and/or shift their activities towards more plant-based food products" (CEI, 2018, p. 8). In Canada, past industry influence on public health dietary guidelines was replaced with the best available evidence (Dai et al., 2020) when Health Canada produced a guide to effectively advise citizens to make half their meals fruit and vegetables, emphasising plant-based proteins and deemphasising animal-derived foods (Government of Canada, 2019).

This collection offers the best available evidence to answer wicked questions and wrestle with trade-offs. These efforts focus on illuminating principles, policies, and practices of optimal food production and consumption. The valuable knowledge and essential skills to help societies, communities, households, and individuals shift to sustainable diets comes from women, civil society, and Indigenous Peoples active in addressing food security. It comes from farmers, scholars, international and national development institutions, from enlightened actors of the private sector, and key change agents in governments at all levels. Working across sectors to create regional, nutritional, and delectable solutions for sustainable diets is the most pressing challenge facing humanity, which must be solved together, with due haste.

References

Brunori, G., Hudson, L. R., Báldi, A., Bisoffi, S., Cuhls, K., Kohl, J., Treyer, S., Ahrné, L., Aschemann Witzel, J., De Clerck, F., Dunca, J., Hansen, H. O., Ruiz, B., & Siebielec, G. (2020). *Resilience and transformation: Report of the 5th SCAR foresight exercise expert group-natural resources and food systems: Transitions towards a 'safe and just' operating space.* European Commission.

Burlingame, B., & Dernini, S. (2010). *Sustainable diets and biodiversity directions and solutions for policy, research and action.* FAO.

Clark, M. A., Domingo, N. G., Colgan, K., Thakrar, S. K., Tilman, D., Lynch, J., Azevedo, I. M. L., & Hill, J. D. (2020). Global food system emissions could preclude achieving the 1.5° and 2° C climate change targets. *Science, 370*(6517), 705–708.

Council for the Environment and Infrastructure. (2018). *Sustainable and healthy: Working together towards a sustainable food system.* https://en.rli.nl/sites/default/files/advisory_report_pdf_3_mb.pdf

Dai, Z., Kroeger, C. M., Lawrence, M., Scrinis, G., & Bero, L. (2020). Comparison of methodological quality between the 2007 and 2019 Canadian dietary guidelines. *Public Health Nutrition, 23*(16), 2879–2885.

Dury, S., Bendjebbar, P., Hainzelin, E., Giordano, T. and Bricas, N., eds. 2019. *Food Systems at risk: New trends and challenges.* FAO, CIRAD and European Commission.

Fears, R., Canales, C., Ter Meulen, V., & von Braun, J. (2019). Transforming food systems to deliver healthy, sustainable diets: The view from the world's science academies. *The Lancet Planetary Health, 3*(4), e163–e165.

Food and Agriculture Organisation of the United Nations. (2010). *The state of food insecurity in the world. Address food insecurity in protracted crises.* FAO.

Food and Agriculture Organisation of the United Nations. (2014). *The state of food and agriculture. Innovation in family farming.* FAO.

FAO, IFAD, UNICEF, WFP, & WHO. (2020). *The state of food security and nutrition in the world 2020. Transforming food systems for affordable healthy diets.* FAO.

Government of Canada. (2019). *Canada's food guide.* https://food-guide.canada.ca/en/

Gussow, J. D., & Clancy, K. L. (1986). Dietary guidelines for sustainability. *Journal of Nutrition Education, 18*(1), 1–5.

Harwatt, H., Ripple, W. J., Chaudhary, A., Betts, M. G., & Hayek, M. N. (2020). Scientists call for renewed Paris pledges to transform agriculture. *The Lancet Planetary Health, 4*(1), e9–e10.

Hawken, P. (Ed.). (2017). *Drawdown: The most comprehensive plan ever proposed to reverse global warming*. Penguin.

High Level Panel of Experts on Food Security and Nutrition. (2017). *Nutrition and food systems: A report by the high level panel of experts on food security and nutrition of the committee on world food security*. FAO.

High Level Panel of Experts on Food Security and Nutrition. (2020). *Food security and nutrition: Building a global narrative towards 2030. A report by the high level panel of experts on food Security and nutrition of the committee on world food security*. FAO.

Interacademy Partnership. (2018). *Opportunities for future research and innovation on food and nutrition security and agriculture The InterAcademy Partnership's global perspective*. https://www.interacademies.org/publication/opportunities-future-research-and-innovation food-and-nutrition-security-and

The Intergovernmental Panel on Climate Change. (2018). *Global Warming of 1.5°C*. IPCC.

The Intergovernmental Panel on Climate Change. (2019). *Special report on climate change, desertification, land degradation, sustainable land management, food security, and greenhouse gas fluxes in terrestrial ecosystems*. IPCC.

The Intergovernmental Panel on Climate Change. (2022). *Climate change 2022: Impacts, adaptation and vulnerability*. IPCC.

Mason, P., & Lang, T. (2017). *Sustainable diets: How ecological nutrition can transform consumption and the food system*. Taylor & Francis.

Puska, P. (2002). Successful prevention of non-communicable diseases: 25 Year experiences with North Karelia project in Finland. *Public Health Medicine, 4*(1), 5–7.

Searchinger, T., Waite, R., Hanson, C., Ranganathan, J., Dumas, P., Matthews, E., & Klirs, C. (2019). *Creating a sustainable food future: A menu of solutions to feed nearly 10 billion people by 2050*. World Resources Institute.

United Nations. (1987). *Report of the world commission on environment and development. Our common future*. United Nations.

Waltner-Toews, D., & Lang, T. (2000). A new conceptual base for food and agricultural policy: The emerging model of links between agriculture, food, health, environment and society. *Global Change and Human Health, 1*(2), 116–130.

Willett, W., Rockström, J., Loken, B., Springmann, M., Lang, T., Vermeulen, S., Garnett, T., Tilman, D., DeClerck, F., Wood, A., Jonell, M., Clark, M., Gordon, L. J., Fanzo, J., Hawkes, C., Zurayk, R., Rivera, J. A., De Vries, W., Majele Sibanda, L., Afshin, A., … Murray, C. (2019). Food in the anthropocene: The EAT-lancet commission on healthy diets from sustainable food systems. *The Lancet, 393*(10170), 447–492.

2

DIGNITY, JUSTICE, AND THE RIGHT TO FOOD

Francis Adams

Introduction

Over the course of the past half century, access to safe, sufficient, and nutritious food gradually became recognised as a universal human right. The process through which food emerged as a basic human right involved an extended series of international conferences, declarations, resolutions, and agreements. The United Nations (UN) and its three food-related agencies were at the forefront of this process.[1] Recognising food as a human right affirmed the intrinsic dignity of all people and helped lay the foundation for a more just and equitable global order. This chapter chronicles the emergence of the right to food before turning to international efforts to make food security a reality for millions of desperate and displaced people around the world. These efforts include the provision of food aid in response to natural disaster and civil conflict as well as longer-term food assistance to promote agricultural productivity, advance rural development, and preserve natural environments.

Right to food

The contemporary right to food regime can be traced to the mid-1930s.[2] The League of Nations identified food insecurity as an increasingly urgent concern for the global community and established a Nutrition Committee to promote improved food production and distribution. Although this initiative had little noticeable impact, it set the stage for a similar effort by the newly created United Nations a decade later.

Food security was included in the early efforts to define and codify international human rights. The *Universal Declaration of Human Rights*—adopted by the United Nations General Assembly on December 10, 1948—identifies food as a fundamental human right: "Everyone has the right to a standard of living ade-

DOI: 10.4324/9781003174417-3

quate for the health and well-being of himself and of his family, including food" (United Nations [UN], 1948, Article 25, para. 1). This principle was also included in the 1966 *International Covenant on Economic, Social and Cultural Rights* (ICESCR): "Everyone has the right to an adequate standard of living for himself and his family, including adequate food" (UN, 1966, Article 11.1). State parties to the covenant were expected to work toward the progressive realisation of this right to the maximum extent possible.

World Food Conference

The world food crisis of the early 1970s added momentum to the campaign to make food a universally recognised human right. A series of natural disasters, including devastating droughts in Africa and Asia, caused a precipitous decline in global food production. Cereal stocks dropped to their lowest levels in 20 years, and prices for many basic food items skyrocketed. Nearly a quarter of all people in the Global South were unable to meet their nutritional needs, and millions of people, especially in Sub-Saharan Africa and South Asia, were suffering from famine.

As the severity of the global food crisis escalated, the United Nations convened a World Food Conference in November 1974.[3] Delegates to the conference drafted a series of measures to address the immediate crisis and improve global food security more generally. Ten million tonnes of in-kind food aid was distributed to food-deficit countries in the Global South, and grain reserves were established to prevent future emergencies. Delegates to the World Food conference also adopted the *Universal Declaration on the Eradication of Hunger and Malnutrition*.[4] This declaration, which was a foundational document in the campaign to make food a human right, stated, "[e]very man, woman and child has the inalienable right to be free from hunger and malnutrition" (UN, 1974, para. 1). The international community was called upon to "ensure the availability at all times of adequate world supplies of basic food-stuffs" and to "co-operate in the establishment of an effective system of world food security" (UN, 1974, para. 12). Delegates to the conference also established a World Food Council (WFC) to oversee implementation of the conference mandates and coordinate the work of all UN agencies involved in global food issues.[5] A separate Committee on World Food Security (CFS) was created to formulate a strategic framework for achieving universal food security.

General Comment 12: Right to adequate food

In the 1980s and early 1990s, a series of global agreements strengthened the right to food. In 1987 the Economic and Social Council (ECOSOC) of the United Nations adopted Resolution 90, which affirmed that "the right to food is a universal human right" and "should be guaranteed to all people" (UN, 1987, p. 1). The resolution also stated that international cooperation on food and agricultural issues was necessary for food security (UN, 1987, p. 1). The following year, the World Food Council identified access to food as a "human right which must be defended by Governments, peoples and the international community" (World

Food Council, 1988, p. 1). In June 1992, leaders from nearly all United Nations member countries gathered in Rio de Janeiro for the United Nations Conference on Environment and Development. The central document of the conference, *Agenda 21*, outlined strategies for addressing the most significant challenges of the twenty-first century. Among the many recommendations advanced, countries were called to "[u]ndertake activities aimed at the promotion of food security and, where appropriate, food self-sufficiency within the context of sustainable agriculture" (UN, 1992, Section 3.8.l). At the end of 1992, an International Conference on Nutrition concluded with a pledge to substantially reduce chronic hunger, malnutrition, and micronutrient deficiencies. The conference's *World Declaration and Plan of Action for Nutrition* stated that "[h]unger and malnutrition are unacceptable in a world that has both the knowledge and the resources to end this human catastrophe" (Food and Agriculture Organization [FAO] & World Health Organization [WHO], 1992, para. 1).

The *Rome Declaration on World Food Security*, adopted at the 1996 World Summit on Food Security, reaffirmed "the right of everyone to have access to safe and nutritious food, consistent with the right to adequate food and the fundamental right of everyone to be free from hunger" (FAO, 1996b, para. 1). Signatory countries pledged to "pursue participatory and sustainable food, agriculture, fisheries, forestry and rural development policies and practises … which are essential to adequate and reliable food supplies at the household, national, regional and global levels" (FAO, 1996b, para. 10). A *World Food Summit Plan of Action* identified specific targets for governments to achieve, including a 50 percent reduction in the number of undernourished people by 2015. Food security was deemed to exist "when all people, at all times, have physical and economic access to sufficient, safe and nutritious food to meet their dietary needs and food preferences for an active and healthy life" (FAO, 1996a, para. 1).

Delegates to the summit also called for a more definitive statement on the rights identified in the *International Covenant on Economic, Social and Cultural Rights*.[6] This led to preparation of General Comment 12, entitled *The Right to Adequate Food*, by the Committee on Economic, Social and Cultural Rights in 1999.[7] The right to adequate food was defined as "[t]he availability of food in a quantity and quality sufficient to satisfy the dietary needs of individuals, free from adverse substances, and acceptable within a given culture" (United Nations Committee on Economic, Social, and Cultural Rights [UNCESCR], 1999, para. 8). It found this right to be "indivisibly linked to the inherent dignity of the human person and … indispensable for the fulfilment of other human rights enshrined in the International Bill of Human Rights … and inseparable from social justice" (UNCESCR, 1999, para. 4). The right to adequate food would be realised when "every man, woman and child, alone or in community with others, has physical and economic access at all times to adequate food or means for its procurement" (UNCESCR, 1999, para. 6).

General Comment No. 12 also identified the obligations of state parties to ICESCR. "States have a core obligation," it argued, "to take the necessary action to mitigate and alleviate hunger" (UNCESCR, 1999, para 6). Further, "Every state is obliged to ensure for everyone under its jurisdiction access to the minimum essential food which is sufficient, nutritionally adequate and safe, and to ensure their freedom from hunger." (UNCESCR, 1999, para. 14). These obligations include: *Respect* (states should refrain from any measures that prevent people from having access to food); *protect* (states should ensure that people are not deprived by others of access to food); and *fulfil* (states must directly aid people to gain access to food) (UNCESCR, 1999, para. 15). States were urged to "recognise the essential role of international cooperation and to comply with their commitment to take joint and separate action to achieve the full realisation of the right to adequate food" (UNCESCR, 1999, para. 36).

The UN Commission on Human Rights subsequently established a Special Rapporteur on the Right to Food who was responsible for monitoring the nutritional wellbeing of vulnerable groups, highlighting violations of the right to food, and working with UN agencies and other international organisations to promote food security.[8] The Special Rapporteur defined the right to food as

> the right to have regular, permanent and free access, either directly or by means of financial purchases, to quantitatively and qualitatively adequate and sufficient food corresponding to the cultural traditions of the people to which the consumer belongs, and which ensures a physical and mental, individual and collective, fulfilling and dignified life free of fear.

(United Nations Commission on Human Rights [UNCHR], 2002, para. 17; E/CN.4/2001/53, Paragraph 14)

Right to Food Guidelines

In September 2000, world leaders gathered at the UN headquarters in New York for the Millennium Summit. The purpose of this meeting was to address the major challenges facing the global community at the turn of the twenty-first century. World leaders ratified the *United Nations Millennium Declaration* and identified eight Millennium Development Goals (MDG) to be achieved by 2015. The first goal called for eradicating extreme poverty and hunger.

The Millennium Declaration and MDG created the impetus for the June 2002 World Food Summit in Rome. The declaration adopted at this summit reaffirmed the "right of everyone to have access to safe and nutritious food" and recognised the "important role of food assistance in situations of humanitarian crisis as well as an instrument for development" (FAO, 2002, p. 1). The declaration also called upon the FAO's General Council to prepare specific guidelines for implementing the right to food.

The *Right to Food Guidelines*, which were endorsed by the Committee on World Food Security and adopted by the FAO General Council in 2004, were presented as practical guidance to member states for achieving the progressive realisation of the right to adequate food as stated in Article 11 of the ICESCR.[9] The guidelines offered a wide range of recommendations, including incorporating the right to food into all poverty reduction strategies, building enabling environments for people to meet their own food needs, and establishing safety nets for the most vulnerable populations. Although the guidelines did not create new rights or obligations, and are not considered legally binding under international law, they remain among the most comprehensive set of recommendations for building food and nutritional security. As noted by CFS, the *Right to Food Guidelines* were deliberately structured to affirm *rights and duties* rather than advocate *charity and benevolence* (Committee on World Food Security, 2017, p. 13).

The *Right to Food Guidelines* also prompted the UN Human Rights Council to adopt a 2008 resolution that affirmed "the right of everyone to have access to safe and nutritious food, consistent with the right to adequate food and the fundamental right of everyone to be free from hunger" (United Nations Human Rights Council [UNHRC], 2008b, p. 3). The resolution encouraged

> all States to take steps to achieve progressively the full realisation of the right to food, including steps to promote the conditions for everyone to be free from hunger and, as soon as possible, to enjoy fully the right to food, and to create and adopt national plans to combat hunger.
>
> *(UNHRC, 2008b, p. 4)*[10]

Food Security Principles

The November 2009 World Summit on Food Security reaffirmed international responsibility for meeting global food and nutritional needs. Participating countries agreed to "collectively accelerate steps … to set the world on a path to achieving the progressive realisation of the right to adequate food in the context of national food security" (FAO, 2009, p. 1). Under the *Five Rome Principles for Sustainable Global Food Security*, which was adopted at the summit, the international community was called upon to: 1) Invest in country-led strategies aimed at channelling resources into nutrition programs; 2) foster strategic coordination at national, regional, and global levels to improve governance on food-related issues; 3) implement a twin-track approach to food security that consists of direct action to immediately tackle hunger and longer-term initiatives to promote sustainable agricultural, nutrition, and rural development; 4) improve the efficiency, responsiveness, coordination, and effectiveness of multilateral institutions in promoting food security; and, 5) ensure the sustained commitment of all partners to invest in agriculture, food security, and nutrition (FAO, 2009, pp. 3–7).

The *Five Rome Principles for Sustainable Global Food Security* helped shift the international agenda from an emphasis on emergency food aid to the broader objectives of food assistance. This transition was reflected in adoption of the 2012 *Food Assistance Convention* (FAC).[11] Rather than simply providing food to vulnerable communities, food assistance is structured to address the underlying causes of food insecurity.[12] As stated in the convention, food assistance should be structured "in a manner that protects livelihoods and strengthens the self-reliance and resilience of vulnerable populations and local communities" (UN, 2012a, Article 2.a.iii). In June 2012, the United Nations sponsored a Conference on Sustainable Development in

Rio de Janeiro to coincide with the 20th anniversary of the 1992 Conference on Environment and Development. Delegates to the conference reaffirmed their commitment to "the right of everyone to have access to safe, sufficient, and nutritious food, consistent with the right to adequate food and the fundamental right of everyone to be free from hunger" (UN, 2012b, section 108). Food assistance also was stressed at the Second International Conference on Nutrition in November 2014. The conference's *Rome Declaration on Nutrition* reaffirmed "the right of everyone to have access to safe, sufficient, and nutritious food, consistent with the right to adequate food and the fundamental right of everyone to be free from hunger" (FAO & WHO, 2014, p. 1). The declaration called upon the UN system to strengthen international collaboration and cooperation to achieve global food security and stated that "the elimination of malnutrition in all its forms is an imperative for health, ethical, political, social, and economic reasons" (FAO & WHO, 2014, p. 3). FAO was tasked with preparing a number of handbooks for incorporating the right to food into national constitutions and legislation (FAO, 2014a).

Sustainable Development Goals

With the *Millennium Development Goals* elapsing, the United Nations began work on a new set of objectives to guide the global agenda. In 2015 the UN adopted 17 Sustainable Development Goals (SDG) to be achieved by 2030. The second goal calls for ending hunger, achieving food security, improving nutrition, and promoting sustainable agriculture. Target 2.1 focuses on ending hunger and ensuring access by all people to safe, nutritious, and sufficient food; target 2.2 advocates ending all forms of malnutrition, reducing stunting and wasting in children, and addressing the nutritional needs of adolescent girls, pregnant and lactating women, infants, and the elderly; target 2.3 prescribes doubling agricultural productivity and the incomes of small-scale food producers, especially women, Indigenous Peoples, family farmers, pastoralists, and fishers; and target 2.4 recommends agricultural practises that increase productivity, help maintain ecosystems, and strengthen resilience to climate change.

Lastly, the Human Rights Council adopted a resolution in March 2019 that reaffirmed "the right of everyone to have access to safe, sufficient and nutritious food" and the "fundamental right of everyone to be free from hunger" (UNHRC, 2019, p. 3). Hunger was deemed an "outrage" and a "violation of human dignity" which must be eliminated (UNHRC, 2019, p. 3). The resolution called upon states, individually and through international cooperation, to take all measures necessary to ensure the realisation of the right to food as an "essential" human rights objective (UNHRC, 2019, p. 5). The international community was urged to provide the assistance needed to ensure universal access to food (UNHRC, 2019, pp. 4–5).

Food aid and assistance

Establishing a right to food was a significant achievement for the United Nations and its food-related agencies. At the same time, ensuring that freedom from hunger becomes a reality for all people requires the effective mobilisation of the world's resources. Since countries with the highest levels of food insecurity often face the greatest obstacles to meeting basic human needs, guaranteeing the right to food requires a collective response by the global community. The importance of food aid and assistance was continually stressed in the many global agreements, declarations, and resolutions that have been adopted to promote food security since the early 1970s.

As noted above, food aid refers to emergency nutritional relief to people with urgent needs and is typically extended at times of natural disaster or civil conflict. Such aid can be provided either in kind or through cash-based transfers and commodity vouchers. In-kind aid refers to the direct provision of food to people in need, while cash-based transfers and commodity vouchers allow people to purchase food in local markets.[13]

Although food aid is imperative when local food supplies have been destroyed or are beyond the reach of poor communities, such aid must be carefully structured to avoid causing more harm than good. Although the provision of surplus grains to poor countries may be carried out with the best of intentions, this can actually undermine the goal of achieving food security in these countries. Local farmers are adversely impacted by foreign grains being freely distributed in domestic markets, and there will be less incentive for both public and private actors to invest in domestic agriculture. With fewer farmers, less agrarian investment, and declining output, these countries could become even *more* dependent on external sources of food.

Food aid can be structured in ways that do not undermine domestic agriculture, harm local farmers, or increase external dependency. When in-kind food aid is provided, local purchases should be made whenever possible. Food is often available in a country, just not in specific areas. If severe food shortages preclude local purchases, food should be obtained from neighbouring countries in the region. Local and regional purchases (LRP) can aid small-scale farmers in the Global South, increase incentives to invest in domestic agriculture, and help build national capacities. Cash-based transfers or commodity vouchers to obtain locally produced food achieve these same objectives, provided they are carefully targeted to the most food insecure households.[14] When in-kind food aid is provided, it could be fortified with vitamins and minerals to ensure the highest nutritional content possible.

Broader food assistance should address the underlying social, economic, and environmental causes of hunger and malnutrition. Such assistance can be structured to build the local resources, knowledge, and capacities needed to ensure people can meet the nutritional needs of themselves and their families on a continuing basis. This assistance typically involves longer-term efforts to increase agricultural and aquacultural productivity in the Global South. Given the scarcity of land and water resources, the potential for horizontal expansion of food output is limited. Future increases in output will largely depend on raising productivity per unit of land and water. Emphasis should be placed on aiding smallholder farmers and fishers to improve crop, livestock, and fish productivity. Training and extension services should be designed to ensure farmers and fishers can benefit from new knowledge and technologies. Food assistance can also be structured to support the transformation of rural economies in the Global South. Programs that support income-generating opportunities in agro-processing and the wider non-farm economy are needed to increase value-addition and higher revenue opportunities to reduce rural poverty and inequality. Food assistance can also promote the sustainable management of natural resources. This includes protecting land, water, forests, and fisheries, as well as crafting effective responses to the crises posed by climate change.

Conclusion

The work of the United Nations and its affiliated agencies over the course of the past half century was critical to establishing access to safe, sufficient, and nutritious food as a universally recognised human right. International institutions are now in a stronger position to pressure governments to ensure the right to food is respected and promoted within their territories. This process also strengthened the position of societal groups working to hold governments accountable for their actions. By empowering both international institutions and civil societies, the right to food regime can be a powerful tool for promoting global food security.

Ensuring that the right to food becomes a reality for all people requires a collective effort by the community of nations. Food aid and assistance should be structured to meet both immediate needs and address the root causes of hunger and malnutrition. An integrated food policy framework is essential to building healthy and sustainable food systems (De Schutter et al., 2020). The participation of civil society organisations is especially important. When local communities are involved in the design, implementation, and monitoring of development projects, success rates increase, and the projects are more likely to continue after external assistance ends. The participation of community groups helps ensure the relevance, ownership, and sustainability of all initiatives, which create more self-reliant communities that have the resources and capabilities necessary to meet their nutritional needs on a long-term basis.

These principles are reflected in the movement for *food sovereignty*. The concept of "food sovereignty" was developed by the international movement La Via Campesina that brings together peasants, small-scale farmers, landless people, Indigenous People, migrants, and agricultural workers in an attempt to influence global agricultural policy.[15] The *Declaration of Nyéléni* defined food sovereignty as the "right of peoples to healthy and culturally appropriate food produced through ecologically sound and sustainable methods, and their right to define their own food and agriculture systems" (Forum on Food Sovereignty [FFS], 2007, p. 1). This movement works to strengthen smallholder agriculture, artisanal-fishing, pastoralist-led grazing, and food production systems that are socially and environmentally sustainable (FFS, 2007, p. 1). It "puts those who produce, distribute, and consume food at the heart of food systems and policies rather than the demands of markets and corporations" (FFS, 2007, p. 1).[16] Food sovereignty thus represents an alternative vision to the agro-industrial model of large-scale production for distant markets (Vivero-Pol et al., 2020). Emphasis is placed on locally oriented, small-scale food production for domestic consumption and ensuring lands, waters, seeds, livestock, and biodiversity are in the hands of those who produce food in local communities (FFS, 2007).

It is hard to imagine a more fundamental human right than the right to food. Beyond the moral imperative that no person should suffer from hunger or malnutrition, food insecurity imposes enormous economic, social, and environmental costs on all countries. The world clearly has the resources and capacity to construct a global food system that is economically viable, socially just, and environmentally sustainable.

Notes

1 The three food-related agencies of the United Nations are the Food and Agriculture Organisation (FAO), International Fund for Agricultural Development (IFAD), and World Food Programme (WFP).
2 There is an extended literature on the right to food regime. See especially Claeys, 2015; Riol, 2017; Food and Agriculture Organisation, 2014a; Knuth & Vidar, 2011; Margulis, 2013; Schanbacher, 2019; and Ziegler et al., 2011.
3 The 1974 World Food Conference was sponsored by the Food and Agriculture Organisation and took place in Rome.
4 *The Universal Declaration on the Eradication of Hunger and Malnutrition* was endorsed by the United Nations General Assembly through Resolution 3348 of December 17, 1974 (United Nations, 1974).
5 The World Food Council was later made a subsidiary body of the UN General Assembly.
6 This request was included in objective 7.4 of the *Plan of Action* of the World Food Summit, Food and Agriculture Organisation, 1996b.
7 General Comment 12 was adopted by the Economic and Social Council of the United Nations in 1999.
8 The appointment of the Special Rapporteur on the Right to Food was made by the UN Commission on Human Rights through Resolution 2000/10 in April 2000. When the Human Rights Commission was replaced by the Human Rights Council (HRC) in June 2006, the mandate of the Special Rapporteur was endorsed and extended through Resolution 6/2 of September 27, 2007.
9 The formal name of the Right to Food Guidelines was the *Voluntary Guidelines to Support the Progressive Realisation of the Right to Adequate Food in the Context of National Food Security*, Food and Agriculture Organisation, 2004.
10 See also United Nations Human Rights Council, 2008b.
11 United Nations, 2012a. The *Food Assistance Convention* was opened for signature on June 11, 2012 and entered into force on January 1, 2013.
12 The World Food Programme defines food assistance as "a full range of instruments, activities, and platforms that empower vulnerable and food-insecure people and communities so they can regularly have access to nutritious food" (World Food Programme, 2017, p. 8).
13 The World Food Programme (WFP) is the largest provider of food aid.
14 LRP also avoid the costs associated with storing, transporting, and distributing in-kind food aid from donor countries, which are approximately 25 percent of the total costs of such aid.
15 The World Development Movement, Food First Information and Action Network (FIAN), and the International Planning Committee on Food Sovereignty also work to promote a more just, equitable, and sustainable global food system.
16 Other important documents in the food sovereignty movement include Forum on Food Sovereignty 2002; La Via Campesina, 2003; World Development Movement, 2012; and International Land Coalition, 2011.

References

Claeys, P. (2015). *Human rights and the food sovereignty movement: Reclaiming control.* Routledge.

Committee on World Food Security. (2017). *Global strategic framework for food security and nutrition.* FAO. https://www.fao.org/3/mt648e/mt648e.pdf

De Schutter, O., Jacobs, N., & Clément, C. (2020). A 'common food policy' for Europe: How governance reforms can spark a shift to healthy diets and sustainable food systems. *Food Policy, 96,* 101849.

Food and Agriculture Organisation. (1996a). *Rome declaration on world food security.* The World Food Summit. https://www.fao.org/3/w3548e/w3548e00.htm

Food and Agriculture Organization. (1996b). *Plan of action for World Food Summit.* https://www.fao.org/WAICENT/OIS/PRESS_NE/PRESSENG/H36F.HTM#

Food and Agriculture Organization. (2002). *Declaration of the world food summit: Five years later.* https://www.fao.org/3/y1780e/y1780e00.htm#TopOfPage

Food and Agriculture Organization. (2004). *Voluntary guidelines to support the progressive realization of the right to adequate food in the context of national food sovereignty.* https://www.fao.org/3/y7937e/y7937e00.pdf

Food and Agriculture Organization. (2009). *Declaration of the world summit on food security.* https://www.mofa.go.jp/policy/economy/fishery/wsfs0911-2.pdf

Food and Agriculture Organization. (2014a). *Right to food handbooks.* https://www.fao.org/3/i3454e/i3454e.pdf

Food and Agriculture Organization. (2014b). *The right to food within the international framework of human rights and country constitutions.* https://www.fao.org/3/i3448e/i3448e.pdf

Food and Agriculture Organization, & World Health Organization. (1992). *World declaration and plan of action for nutrition.* https://apps.who.int/iris/bitstream/handle/10665/61051/a34303.pdf?sequence=1&isAllowed=y

Food and Agriculture Organization, & World Health Organization. (2014). Second International Conference on Nutrition. https://www.who.int/news-room/events/detail/2014/11/19/default-calendar/fao-who-second-international-conference-on-nutrition-(icn2)

Forum on Food Sovereignty. (2002). *Declaration of the NGO forum at the FAO summit Rome + 5.* https://www.fao.org/worldfoodsummit/english/documents.htm

Forum on Food Sovereignty. (2007). *Declaration of Nyéléni.* https://nyeleni.org/DOWNLOADS/Nyelni_EN.pdf

International Land Coalition. (2011). *Tirana declaration.* https://d3o3cb4w253x5q.cloudfront.net/media/documents/Tirana_Declaration_2011_EN.pdf

Knuth, L., & Vidar, M. (2011). *Constitutional and legal protection of the right to food around the world.* FAO. https://www.fao.org/3/ap554e/ap554e.pdf

La Via Campesina. (2003). *People's food Sovereignty: WTO out of agriculture.* https://viacampesina.org/en/food-sovereignty/

Margulis, M. E. (2013). The regime complex for food security: Implications for the global hunger challenge. *Global Governance, 19,* 53.

Riol, K. S. C. (2017). *The right to food guidelines, democracy and citizen participation: Country case studies.* Routledge.

Schanbacher, W. (2019). *Food as a human right: Combatting global hunger and forging a path to food sovereignty.* Praeger.

United Nations. (1948). *Universal declaration of human rights.* General Assembly Resolution 217A. https://www.ohchr.org/en/resources/educators/human-rights-education-training/universal-declaration-human-rights-1948

United Nations. (1966). *International covenant on economic, social and cultural rights.* General Assembly Resolution 2200A (XXI). https://www.un.org/en/development/desa/population/migration/generalassembly/docs/globalcompact/A_RES_2200A(XXI)_civil.pdf

United Nations. (1974). *Universal declaration on the eradication of hunger and malnutrition.* General Assembly Resolution 3348. https://www.ohchr.org/en/instruments-mechanisms/instruments/universal-declaration-eradication-hunger-and-malnutrition

United Nations. (1987). *Resolution 90 of the economic and social council of the United Nations.* United Nations.

United Nations. (1992). *Agenda 21: The United Nations programme of action from Rio.* https://sustainabledevelopment.un.org/content/documents/Agenda21.pdf

United Nations. (2012a). *Food assistance convention.* https://treaties.un.org/doc/Treaties/2012/04/20120425%2002-57%20PM/CTC_XIX-48.pdf

United Nations. (2012b). *Rio + 20: United Nations conference on sustainable development outcome document* (A/RES/66/288). https://www.un.org/en/development/desa/population/migration/generalassembly/docs/globalcompact/A_RES_66_288.pdf

United Nations Commission on Human Rights. (2002). *Report of the special rapporteur on the right to food.* https://digitallibrary.un.org/record/459188?ln=en

United Nations Committee on Economic, Social and Cultural Rights. (1999). *Substantive issues arising in the implementation of the international covenant on economic, social, and cultural rights: General comment 12: The right to adequate food* (E/C.12/1999/5). https://www.law.umich.edu/facultyhome/drwcasebook/Documents/Documents/Committee%20on%20Economic,%20Social,%20and%20Cultural%20Rights%20General%20Comment%2012.pdf

United Nations Human Rights Council. (2008a). *Promotion and protection of all human rights, civil, political, economic, social, and cultural rights, including the right to development. Report of the special rapporteur on the right to food.* Jean Ziegler, Human Rights Council, 7th Session, United Nations General Assembly, A/HRC/7/5. https://digitallibrary.un.org/record/616943?ln=en

United Nations Human Rights Council. (2008b). *The right to food.* Res. 7/14. https://ap.ohchr.org/documents/E/HRC/resolutions/A_HRC_RES_7_14.pdf

United Nations Human Rights Council. (2019). *The right to food: United Nations General Assembly Resolution A/HRC/40/L.12.* https://documents-dds-ny.un.org/doc/UNDOC/LTD/G19/070/30/PDF/G1907030.pdf?OpenElement

Vivero-Pol, J. L., Ferrando, T., De Schutter, O., & Mattei, U. (Eds.). (2020). *Routledge handbook of food as a commons.* Routledge.

World Development Movement. (2012). *Transforming our food system: The movement for food sovereignty.* https://www.globaljustice.org.uk/sites/default/files/files/resources/food_sovereignty_briefing_10.12_web.pdf

World Food Council. (1988). *Beijing declaration.* FAO. https://www.fao.org/3/ag415e/ag415e00.htm

World Food Programme. (2017). *World food assistance 2017: Taking stock and looking ahead.* https://docs.wfp.org/api/documents/WFP-0000019564/download/?_ga=2.46073060.1764996960.1649079476-237320691.1648617165

Ziegler, J., Golay, C., Mahon, C., & Way, S. (2011). *The fight for the right to food: Lessons learned.* Palgrave Macmillan.

3

REFRAMING THE SUSTAINABLE DIETS NARRATIVE

Shifting diets by confronting systemic racism in the U.S. food systems

Gabriel R. Valle

Introduction

A sustainable diet has the potential to do more than lower greenhouse gas emissions, improve soil health, and move humanity toward more nutritionally dense meals. It has the potential to heal and undo centuries of oppression. The industrial food system is rooted in the historical legacy of colonialism and imperialism. Industrial food has done more than alter our social relationships with each other, the land, our labour, and our food (Alkon & Agyeman, 2011; Daniel, 2013; Reese, 2019; Sbicca, 2018). It has also worked to destroy the histories, traditions, and knowledge about food and food production that sustain people and their communities (LaDuke, 2005; Mohawk, 2008; Peña et al., 2017; Penniman, 2018). When I arrived at the study of sustainable diets, I was both hopeful and critical. Could a "diet" allow oppressed, colonised, and displaced communities to heal from the traumas of industrial foods and indeed be sustainable? Could a sustainable diet also be anti-racist? The more questions like this surfaced in my research, the more I was drawn to our food systems' untold, overlooked, and hidden histories.

In this chapter, I seek to shift the narrative of sustainable diets. I argue that movements for sustainability and sustainable diets tend to silence the voices and experiences of People of Colour by assuming a white epistemology (Finney, 2014). Regardless of how good local food tastes or how nutrient-dense it may be, we know that it will not dismantle the white supremacy embedded in our food system without challenging the asymmetries of power already firmly established throughout food systems (Guthman, 2008; Slocum, 2010). By not foregrounding

DOI: 10.4324/9781003174417-4

the histories of displacement experienced by marginalised, oppressed, and colo-nised communities, the pursuit for sustainable diets may reinforce already existing systemic inequities in the food system.

The shift toward sustainable diets is about more than soil health, producing quality foods, or expanding access to nutrient-dense and culturally appropriate foods to all people at all times. It must also address the inequalities and structural racism at the heart of our food system—from production to consumption. For a diet to accomplish such lofty expectations, it must have a different starting point, because it aims to achieve more than producing good food with a minimal envi-ronmental footprint. Such a diet aims to heal our bodies, cultures, histories, and ways of being through the practices of growing, sharing, and eating food.

In this chapter, I draw on examples and ethnographic data from the United States. I begin by providing context for my anti-racist approach and how it fits into sustainable diets. My anti-racist and decolonial framework addresses three elements central to sustainable diets. I first explore access to land as a critical component in sustainable diets. For any sustainable diet to be anti-racist, we must acknowledge that most U.S. agricultural land is situated on the unceded land of Indigenous Peoples. The excellent soil health that once characterised these agricultural zones did not just happen to be there when Europeans arrived. Instead, it was cultivated and managed by centuries of intentional herding, burning, farming, and foraging by Indigenous Peoples (Anderson, 2005). Second, I propose that a sustainable diet must decolonise the practices of cooking and eating. This move is not to ignore or overlook the biopolitics of racial capitalism embedded in food production (Holt-Giménez, 2018) but rather to acknowledge the many ways Black, Indigenous, and People of Colour (BIPOC) are reimagining their labour to create and restore ways of being (Bowens, 2015). Third, I suggest that a sustainable diet allows marginal-ised, oppressed, and displaced people to heal from the trauma of industrial food and therefore must move beyond what to eat and consider the implication of how to eat.

It is clearly true that not all BIPOC foods or farming practices are or should be considered sustainable. I intend to bring awareness to how sustainable diets, regardless of their perceived good intentions, can marginalise the food and food practices of non-white folks (Burt, 2021). On the other hand, an anti-racist and decolonial sustainable diet brings awareness to this erasure in hopes of validating lifestyles, livelihoods, foods, and food traditions that have helped sustain people and places for generations.

Toward an anti-racist sustainable diet

Ibram Kendi (2019) insists that there is no such thing as a neutral policy: There are only racist and anti-racist policies. So, what if we apply this to a food system? To a diet? To how we shop, cook, eat, and grow our food? Can a diet be anti-racist? What else changes when we begin to think and act in such a way? We are at a moment when people around the world are seeking anti-racist strategies for some of the most challenging issues of our time. Our diets have always been sites of contestation. During the last five centuries of European colonialism, many traditional foods were made illegal with the intention to "civilise" native populations (Mohawk, 2008). By the nineteenth century, global capitalism began to homogenise diets with cash crops, displacing heirloom and landrace varieties (Laudan, 2013). By the twentieth century, the rise of expert nutritional advice began to universalise concepts of health and well-being (Scrinis, 2013). Today, genetically modified seeds are an extension of that expert knowledge that continues to seek control over the food system (Shiva, 2016).

The struggle of what to eat is historical and manifests in ways that produce unequal power relations. The overabundance of diet fads is just one result of this history. Diets such as Adkins, Keto, Paleo, South Beach, and even the Mediterranean diet all have one thing in common: They all overlook how a diet has the potential to make visible or render invisible the web of relationships inherent in our food and food system because they reinforce already established social norms of what is good and healthy to eat (Burt, 2021; Hayes-Conroy & Hayes-Conroy, 2013)

An anti-racist sustainable diet, on the other hand, is an active process that makes visible the plurality of ways Black, Indigenous, and People of Colour actively and intentionally choose to eat, cook, grow, and share foods that identify and challenge systems and institutions of oppression and racism. Like anti-racism, which can be defined as "the active process of identifying and eliminating racism by changing systems, organisational structures, policies and practices and attitudes, so that power is redistributed and shared equitably,"[1] this approach to a diet is an active process. For a sustainable diet to be suitable for people and the planet while also being anti-racist, it must transform processes and practices of cooking, eating, and farming as a means to cultivate liberation with the land (Penniman, 2018), find meaning in our labour (Bowens, 2015), and discover and recover ways that food can heal our bodies, communities, and histories (Peña et al., 2017).

Pamela Mason and Tim Lang (2017) insist that coming to terms with sustainable diets means understanding that a diet is "more than a meal; it's a sequence and even a lifetime of meals, a total dietary intake and pattern. In that respect, diet is about normality, habits, acculturation, what people eat conventionally across a year or their full lifespan" (p. 9). The authors define a sustainable diet as one that "optimises good sound food quality, health, environment, socio-cultural values, economy, and governance" (Mason & Lang, 2017, p. 9). In these respects, a sustainable diet appears to be about normalising sound food quality, health, environment, socio-cultural values, economy, and governance. An anti-racist diet, on the other hand, requires additional steps because it works to restructure the process of cooking, eating, growing, and sharing food in ways that challenge normality. In other words, an anti-racist diet challenges how diet-related diseases such as obesity, diabetes, and high blood pressure and outcomes of environmental racism such as asthma, food insecurity, contaminated water, and forms of cancer are normalised in communities of colour (Bullard, 1990; Cole & Foster, 2000; Hatch, 2016; Mendez, 2020; Montoya, 2011; Taylor, 2014).

From the fields

Eric Holt-Gimenez (2017) explains that "the politics of food is never far from the politics of land, water, or labour" (p. 3), which means efforts to change the food system must coincide with attempts to change systems of land access, tenure, and use. An essential first step for sustainable diets is acknowledging how accumulation and dispossession of land have always been at the heart of U.S. agriculture (Daniel, 2013). Cheryl Harris (1993) insists that property rights are "contingent on, intertwined with, and conflated with race" (p. 1714). The author points to how "whiteness as property" works to mask the "maintenance and domination of white privilege" (Harris, 1993, p. 1715). Thus, any attempt to create sustainable diets that work for both people and the planet must also undo that historical legacy.

Throughout the history of the United States, race alone has not operated to oppress communities of colour; instead, it has been the interaction between conceptions of race and ideas of land ownership (Harris, 1993). The United States agricultural industry is predicated on the removal of Indigenous Peoples and situated on their unceded traditional homelands. The Indian Removal and Homestead Acts work in concert to serve as the foundation of land tenure in the United States. The removal of Indigenous Peoples from their homelands, followed by the ability of white Americans to claim legal ownership of that land, serves as the backdrop that helps explain the intergenerational traumas and historical erasure of Indigenous communities (Dunbar-Ortiz, 2014; Gilio-Whitaker, 2019).

These forms of racialised property ownership dehumanised non-white peoples living in the United States. They assured their limited opportunity to achieve liberation within the confines of the U.S. legal system because "land is a social relation. Whether by private property, or a common, or by spiritual or normative agreements, land is an integral part of the culture and the organisation of society" (Holt-Gimenez & Williams, 2017, p. 259). We form our communities in relation to the land we live with and upon, and a sustainable diet must reflect this. We need to create more opportunities for BIPOC farmers to be successful, but more than that, we need systems of land ownership that reflect the diversity of ways in which different communities come to know and organise on the land.

A diet that reflects the many ways communities grow food to affirm their ways of knowing and being occurs when the value of food is celebrated within that community (Valle, 2017). Leah Penniman, co-founder of Soul Fire Farm, explains that the farm's Afro-ecology approach of deciding what and how to plant is "informed by an understanding of the cultural relationships between Black people and crops, as well as the relationships among the plants themselves" (Penniman, 2018, p. 104). Black farmers in the United States have a unique relationship of working the landscape that is tied to the historical legacy of slavery, Jim Crow laws, Black Codes, and the anti-Black practices of the USDA that have historically made it difficult, if not impossible, for Black farmers to succeed (Daniel, 2013).

Latinx farmers have a different but equally unique relationship to growing food in the United States. In southern Colorado and Northern New Mexico, Latinx farmers engage in the community-governed irrigation institution known as *acequia*, a gravity feed flood irrigation method that has helped the region "maintain higher water quality, promote soil regeneration, and preserve indigenous land race crops" (Peña, 2017, p. 129; also see Peña, 2003). Like Penniman, Latinx *acequeros* engage in specific farming practices to preserve ways of knowing and to nurture a culture of place (Romero, 2021). *Acequia* farmer Joe Gallegos explains the communal practice of making *chicos del horno*, a culinary cultural traditional food for *acequia* farmers in the San Luis Valley of Colorado as "from the planting of the corn, through the stripping of the roasted kernels from the cobs ... to the cleaning and packaging of the dry hard kernels—the whole process requires family, friends, neighbours, and patience" (Gallegos, 2017, p. 152).

To be anti-racist, a sustainable diet must work to create and preserve spaces like those described by Penniman and Gallegos, which encourage access to the spaces of growing foods that nourish our minds, bodies, cultures, and lands. Latinx-owned farms make up just over five percent of farms, yet Latinx farmworkers make up over 80 percent of the labour that supports U.S. agriculture (Horst & Marion, 2019). What is more, those who work in the agriculture sector of the food chain are the least likely to earn a living wage. So, while Latinx workers produce the bulk of food grown and processed in the United States, they experience some of the highest levels of food insecurity of any ethnic group working in the food chain (The Food Chain Workers Alliance, 2012). An anti-racist sustainable diet must address these discrepancies in ways that activate our commitment to social and environmental justice.

From the kitchen

Chef Marcus Samuelsson's recent cookbook confronts racism, erasure, and equity in the United States. *The Rise* (2020) celebrates chefs, foodways, and hidden histories of Black America that the author insists make up the "soul of American food." He explains that "Black food is not monolithic" (Samuelsson, 2020, p. xiii). As a chef, writer, and community organiser, Samuelsson (2020) tells the untold food stories that make up Black foodways:

> The roots of these recipes [detailed in the book] are born of major global shifts—from hundreds of African cultures, forcibly mixed during centuries of enslavement, whose regional culinary practice evolved on this land with cooking techniques and ingredients, both old and new.
>
> *(Samuelsson, 2020, p. xx)*

Similarly, the goal of chef Bryant Terry's (2020) recent cookbook, *Vegetable Kingdom*, is to "create a dish through the lens of the African Diaspora" (p. 2). He argues that "recipe creation is a praxis where I honour and bring to life the teachings, traditional knowledge, and hospitality of my blood and spiritual ancestors by making food" (ibid. p. 2). His Afro-Asian vegan creations are examples of an anti-racist sustainable diet that confronts the "violence of forgetting" (Lowe, 2006, p. 206) all too common in sustainability studies (Finney, 2014), while working to improve and restore cultural traditions, lower greenhouse gas emissions, and renew an appreciation for cooking and eating.

Chef Jocelyn Ramirez's decolonial approach to diet and cuisine challenges how colonialism and globalisation have altered traditional Mexican-origin foodways. Her work at Todo Verde, a woman of colour-owned food business, seeks to advance food equity in the Latinx communities of greater Los Angeles. She states, "I invite you to explore the intersection of Mexican cuisine that remembers the connection to the food, land and community and the dishes that have become part of the culture" (Ramirez, 2020, p. 8). The work of Luz Calvo and Catriona Esquibel expresses similar sentiments. The authors move for a liberated kitchen that challenges the colonial roots of the Standard American Diet. By re-centring Indigenous foodways, the authors "want to rework the activities of the kitchen so that they become central to the revolutionary practice of love" (Calvo & Esquibel, 2015, p. 35).

These are only a handful of BIPOC chefs who have already begun the long and challenging process of transitioning toward an anti-racist and decolonial sustainable diet. Love, conviviality, and remembering are central to such an approach because these core values remind us of who we are, where we are going, and what we will become. Food can help liberate us from oppressive systems that have erased histories of BIPOC farmers and chefs living sustainably with their landscapes and communities, but food can also be oppressive and reinforce unjust power relations regardless of how sustainable a diet may be. As the new era of intentional eating unfolds, we must constantly be reminded of the lessons our ancestors have taught us.

The healing power of food

Chayoes are what Danna Haraway might consider "situated" (1988). Not only does the *chayote* have a long history of being a staple food for groups from Mesoamerica, but it remains a vital source of situated knowledge for immigrant women. According to Hayes-Conroy and Hayes-Conroy (2013), the practice of nutrition must be informed by situated knowledge, which entails spending "more time attending to the limits and location of what we know about 'what is good to eat'" (p. 178). The situatedness of healthy eating relocates the practice of decolonisation to the scale of the body. This process is both personal and collective, and "rather than a distinct and predetermined set of core values and goals, nutrition becomes a collective, negotiated process of transformation toward healthier people and communities" (Hayes-Conroy & Hayes-Conroy, 2013, p. 178).

Box 3.1 A living example of healing with food

Lupe is a recent immigrant from Mexico. After about five years of living in the United States, Lupe developed an intestinal condition that her doctors could not diagnose. She felt chronic pain and nausea in her gut, which she insisted came from

---

relying on the Standard American Diet. Doctors prescribed her medication and told her to stick to a "simple, healthy diet." In other words, the medical advice she received was a form of hegemonic nutrition (Hayes-Conroy & Hayes-Conroy, 2013). Doctors told her to eat lots of vegetables, fruits, whole grains, lean meats, skimmed milk, and healthy fats like olive oil, fish, and nuts. They also told her to avoid saturated fats from butter and red meat while limiting salt and sugar.

As the weeks and months passed, very little had changed for Lupe, and she grew frustrated. She wanted to eat healthy food, but more importantly, she wanted to eat meaningful food. The people Lupe received advice from were both white and male, which according to Hayes-Conroy and Hayes-Conroy, matters because storytelling is an effective feminist practice. "In terms of nutrition," the authors argue, "this means that the details, differences, and discrepancies that people's life experiences reveal play a crucial role in producing effective, meaningful advice" (Hayes-Conroy & Hayes-Conroy, 2013, p. 178). Several months after being diagnosed with an undiagnosable intestinal condition, Lupe rejected the hegemonic nutritional advice she received from so-called experts and returned to her embodied knowledge of "what is good to eat." Not only did corn, beans, squash, and chillies return to her table, but she returned to her garden. *Chayotes* were eaten again—cooked, raw, steamed, and roasted. Hegemonic nutrition is nutritional advice that is served up without context and as such, has the power to erase centuries of embodied knowledge.

For Lupe to return to a healthy diet did not mean visiting the health food store like many middle-class white Americans because her embodied knowledge of "what is good to eat" could not be found there. Instead, she returned to the values, ethics, and ways of living and being that are only present in her garden. She recalled how to grow the foods she cultivated with her parents and grandparents, and today produces the majority of what she eats. Soon after Lupe's dietary shift, her body began to heal, and today she has fully recovered. Her doctor had no simple answer for her recovery but told her to keep doing what was working for her. In other words, she had identified what was "good to eat" for her.

Lupe's dietary shift to a sustainable diet that is both healthy and meaningful moves beyond the "right" versus "wrong" blame game of classical nutritional approaches. Her dietary shift coincides with the perspective of Laura Newcomer (2013), who explains nutrition as "re-teaching myself how to eat on a regular basis … [and] The authority to define and re-define what 'just right' means, on a daily basis, for my body and for me … [is] listening" (p. 196). Newcomer insists that an approach to health that prioritises nutrients over food cannot come to terms with a diet, because a diet is socially constructed and culturally reinforced. Gyorgy Scrinis (2013) argues that this "reductive understanding of food in terms of nutrients is also mirrored in the reductive ways of understanding the body and bodily health"

(p. 242).The ability to listen to our bodies is a much-needed step toward a dietary shift that embodies learning and re-learning, defining and re-defining.

Conclusion: Cooking and growing for an anti-racist, decolonial sustainable diet

How we eat is a fundamental aspect of social change. An anti-racist, decolonial approach to sustainable diets has the potential to accomplish far more than reduce greenhouse gas emissions, restore soil health, and curb food insecurity.When what we eat becomes more than food "without context," the practices of growing, preparing, and sharing food enable us to heal from oppressive histories, institutions, and practices that have shaped us and the world we inhabit. Healing is a decisive step for any approach to sustainable diets because we cannot begin to heal the planet until we heal ourselves.When healing guides how we grow and consume foods, we are helping to restore and renew the biocultural diversity that has sustained people and places for generations.

Mason and Lang (2017) insist that the general public needs to understand how their food choices have ecological impacts. Still, the question remains whether the general public has the skills to engage in a sustainable diet over the long term.The authors ask, "what skills are needed to enable people to eat sustainably?" (Mason and Lang, 2017, p. 305).The truth of the matter, the authors insist, is that a "sustainable diet means eating differently ... it applies a shift from seeing humans as consumers to seeing them as citizens" (ibid. 2017, p. 306). Seeing differently is about raising awareness and acquiring the critical thinking skills that allow one to see and act differently. An anti-racist, decolonial sustainable diet also requires eating and seeing differently, but more than that, it needs us to eat and grow in ways that challenge oppressive institutions, policies, and histories that have distanced us from our food and each other. Eating is a humble act, but eating extends deep in the cultural traditions, material realities, and ideological frameworks that make us who we are.

A sustainable diet that strives to meet "culturally appropriate" standards only describes an acceptable diet in different cultural contexts. However, an anti-racist sustainable diet is something everyone can participate in regardless of cultural contexts. As consumers, this means using our purchase power to support sustainable farming practices, farmers, and a living wage for farmworkers. As chefs and cooks, it means cooking in ways that affirm cultural values, traditions, and ways of knowing and being. It is a way of eating that challenges how food has been used as a tool to suppress and keep marginalised the peoples, cultures, and ways of living that do not conform to a universal standard. It is a way of eating that validates the histories and worldviews of underrepresented peoples so that how we grow, eat, share, and consume food can move us forward into a sustainable and just future together.

Note

1 This definition is used by the National Action Committee on the Status of Women International Perspectives:Women and Global Solidarity.

References

Alkon, A. H., & Agyeman, J. (2011). *Cultivating food justice: Race, class, and sustainability*. MIT Press.

Anderson, K. M. (2005). *Tending the wild: Native American knowledge and the management of California's natural resources*. University of California Press.

Bowens, N. (2015). *The colour of food: Stories of race, resilience, and farming*. New Society Press.

Bullard, R. D. (1990). *Dumping in dixie: Race, class, and environmental quality*. Westview Press.

Burt, K., (2021). The whiteness of the Mediterranean diet: A historical, sociopolitical, and dietary analysis using critical race theory. *Critical Dietetics, 5*(2), 41–52.

Calvo, L., & Esquibel, C. R. (2015). *Decolonize your diet: Plant-based Mexican-American recipes for health and healing*. Arsenal Pulp Press.

Cole, L. W., & Foster, S. R. (2000). *From the ground up: Environmental racism and the rise of the environmental justice movement*. New York University Press.

Daniel, P. (2013). *Dispossession: Discrimination against African American farmers in the age of civil rights*. University of North Carolina Press.

Dunbar-Ortiz, R. (2014). *An Indigenous peoples' history of the United States*. Beacon Press.

Finney, C. (2014). *Black faces, white spaces: Reimaging the relationship of African Americans to the great outdoors*. University of North Carolina Press.

Food Chain Workers Alliance. (2012). *The hands that feed us: Challenges and opportunities for food workers along the food chain*. https://foodchainworkers.org/wp-content/uploads/2012/06/Hands-That-Feed-Us-Report.pdf

Gallegos, J. (2017). Chicos del horno: A local, slow, and deep food. In D. G. Peña, L. Calvo, P. McFarland, & G. R. Valle (Eds.), *Mexican-origin foods, foodways, and social movements: Decolonial perspectives* (pp. 151–168). University of Arkansas Press.

Gilio-Whitaker, D. (2019). *As long as grass grows: The Indigenous fight for environmental justice, from colonisation to Standing Rock*. Beacon Press.

Guthman, J. (2008). "If they only knew": Colour blindness and universalism in California alternative food Institutions. *The Professional Geographer, 60*(3), 387–397.

Haraway, D. (1988). Situated knowledges: The science question in feminism and the privilege of partial perspective. *Feminist Studies, 14*(3), 575–599.

Harris, C. I. (1993). Whiteness as property. *Harvard Law Review, 106*(8), 1707–1791.

Hatch, A. R. (2016). *Blood sugar: Racial pharmacology and food justice in Black America*. University of Minnesota Press.

Hayes-Conroy, A., & Hayes-Conroy, J. (Eds.) (2013). *Doing nutrition differently: Critical approaches to diet and dietary intervention*. Taylor & Francis Group.

Holt-Giménez, E. (2017). Agrarian questions and the struggle for land justice in the United States. In J. M. Williams & E. Holt-Giménez (Eds.), *Land justice: Re-imagining land, food, and the commons in the United States* (pp. 1–14). Food First Books.

Holt-Giménez, E. (2018, Spring). *Overcoming the barrier of racism in our capitalist food system*. Institute for Food and Development Policy. https://foodfirst.org/wp-content/uploads/2018/03/Backgrounder_Spring_2018_Final.pdf

Holt-Giménez, E., & Williams, J. M. (2017). Together toward land justice. In J. M. Williams, & E. Holt-Giménez (Eds.), *Land justice: Re-imagining land, food, and the commons in the United States* (pp. 258–263). Food First Books.

Horst, M., & Marion, A. (2019). Racial, ethnic and gender inequities in farmland ownership and farming in the U.S. *Agriculture and Human Values, 36*(1), 1–16.

Kendi, I. X. (2019). *How to be an antiracist*. Penguin Random House.

LaDuke, W. (2005). *Recovering the sacred: The power of naming and claiming*. South End Press.

Laudan, R. (2013). *Cuisine & empire: Cooking in world history*. University of California Press.

Lowe, L. (2006). The intimacies of four continents. In A. L. Stoler (Ed.), *Haunted by empire: Geographies of intimacy in North American history* (pp. 191–212). Duke University Press.

Mason, P., & Lang, T. (2017). *Sustainable diets: How ecological nutrition can transform consumption and the food system*. Routledge.

Mendez, M. (2020). *Climate change from the streets: How conflict and collaboration strengthen the environmental justice movement.* Yale University Press.

Mohawk, J. (2008). From the first to the last bite: Learning from the food knowledge of our ancestors. In M. L. Nelson (Ed.), *Original instructions: Indigenous teachings for a sustainable future* (pp. 170–179). Bear & Company.

Montoya, M. J. (2011). *Making the Mexican diabetic: Race, science, and the genetics inequality.* University of California Press.

Newcomer, L. (2013). Nutrition is…. In A. Hayes-Conroy & J. Hayes-Conroy (Eds.), *Doing nutrition differently: Critical approaches to diet and dietary intervention* (pp. 191–196). Taylor & Francis Group.

Peña, D. G. (2003). The watershed commonwealth of the upper Rio Grande. In J. K. Boyce & B. G. Shelley (Eds.), *Natural assets: Democratising environmental ownership* (pp. 169–185). Island Press.

Peña, D. G. (2017). Settler colonialism and new enclosures in Colorado acequia communities. In M. Williams, & E. Holt-Giménez (Eds.), *Land justice: Re-imagining land, food, and the commons in the United States* (pp. 125–140). Food First Books.

Peña, D. G., Calvo, L., McFarland, P., & Valle, G. R. (Eds.). (2017). *Mexican-origin foods, foodways, and social movements: Decolonial perspectives.* University of Arkansas Press.

Penniman, L. (2018). *Farming while black: Soul Fire Farm's practical guide to liberation on the land.* Chelsea Green Publishing.

Ramirez, J. (2020). *La vida verde: Plant-based Mexican cooking with authentic flavour.* Page Street Publishing Co.

Reese, A. M. (2019). *Black food geographies: Race, self-reliance, and food access in Washington, D. C.* University of North Carolina Press.

Romero, E. (2021). Southwestern acequia systems and communities: Nurturing a culture of place. *Natural Resources Journal. 61*(2), 169–172.

Samuelsson, M. (2020). *The rise: Black cooks and the soul of American food.* Voracious /Little, Brown and Company.

Sbicca, J. (2018). *Food justice now: Deepening the roots of social justice.* University of Minnesota Press.

Scrinis, G. (2013). *Nutritionism: The science and politics of dietary advice.* Columbia University Press.

Shiva, V. (2016). *Seed sovereignty, food security: Women in the vanguard of the fight against GMOs and corporate agriculture.* North Atlantic Books.

Slocum, R. (2010). Race in the study of food. *Progress in Human Geography, 35*(3), 303–327.

Taylor, D. (2014). *Toxic communities: Environmental racism, industrial pollution, and residential mobility.* New York University Press.

Terry, B. (2020). *Vegetable kingdom: The abundant world of vegan recipes.* Ten Speed Press.

Valle, G. R. (2017). Food values: Urban kitchen gardens and working-class subjectivity. In D. G. Peña, L. Calvo, P. McFarland, & G. R. Valle (Eds.), *Mexican-origin foods, foodways, and social movements: Decolonial perspectives* (pp. 41–62). University of Arkansas Press.

4
WHERE SUSTAINABLE DIETS FIT IN GLOBAL GOVERNANCE

Lucy Hinton and Caitlin Scott

Introduction

With greater attention to the concept of sustainable diets throughout the 2010s and since, debates persist around what mechanisms can encourage the consumption of sustainable diets (FAO, 2012; Mason & Lang, 2017). This chapter describes the historical antecedents of global governance on food to demonstrate how sustainable food systems have become a dominant narrative. The focus of this chapter is on governance at the global level, though we recognise the need to understand how global mechanisms interact with more localised, context-specific governance (López Cifuentes et al., 2021). Drawing on literature from international organisations, we briefly trace the evolution of the United Nations (UN) agencies' attempts to define and integrate sustainable diets into existing programmes and recommendations for healthy diets (FAO & WHO, 2019). While internationally agreed documents are usually non-binding, they are intended to guide national policy implementation and are often adopted in low- and middle-income countries (LMICs) without major changes (Thow et al., 2019). We then briefly describe efforts by international civil society groups and social movements to influence domestic governments to take action on sustainable food systems (Blay-Palmer et al., 2015). We summarise the ways that corporate actors influence governance of food systems (e.g., international public policy and trade) and govern themselves through supply chains. Finally, we conclude by recommending that sustainable diet practitioners be aware of the favoured language of sustainable food systems at the global level and work towards coherent, multisectoral approaches to policy alongside food democracy.

While sustainable diets are frequently recommended (Delabre et al., 2021; Willett et al., 2019), contemporary global governance has developed around the broader umbrella idea of 'food systems.' What we eat is the product of the global food system (Clapp, 2020; Stuckler & Nestle, 2012)—"the web of relationships

DOI: 10.4324/9781003174417-5

that span the production, processing, trade, and marketing of the food we eat" (Clapp, 2020, p. 2). Systems theory suggests that "Once we see the relationship between structure and behaviour, we can begin to understand how systems work, what makes them produce poor results, and how to shift them into better behaviour patterns" (Meadows, 2008, p. 1). Often sustainable diets are focused on the consumption side of the food system; we can think of sustainable diets as being subsumed within the broader conceptualisation of food systems discourse that is currently favoured by global governance actors.

Food as a global governance problem

As food systems have become increasingly global in nature, they require increasingly global solutions. Food is now frequently grown in one location that is mentally and physically distant from the people who buy it and consume it, making the interconnections—the brokers, financiers, trade and agricultural policies, manufacturing regulations etc.—difficult to see and regulate (Clapp, 2015). Identifying responsibility in the system becomes nearly impossible. These complex systems are characterised through multiple, overlapping, and reinforcing nested systems and connections (Meadows, 2008), representing what policymakers call a "wicked" problem (Baker & Demaio, 2019). Solving wicked problems like sustainability, let alone health, equity, or other issues embedded in the food system, requires the coming together of stakeholders who frequently have "competing interests, worldviews and beliefs," to collaborate on creating change (Baker & Demaio, 2019, p. 182). Multisectoral approaches are considered the way to tackle wicked, reinforcing problems at the global scale (Bennett et al., 2018). The global food system represents a massive, wicked problem: One that requires a global governance[1] approach because it involves so many actors and jurisdictions (Weiss & Wilkinson, 2014, p. 213).

Contemporary global governance of food is highly fragmented, as many initiatives have emerged in the jurisdictionally complex global environment that aim to govern sustainability at different connection points in the food system (Biermann et al., 2009; Clapp & Scott, 2018). As is the case for many transnational wicked problems, "[t]he policy, authority, and resources necessary for tackling [them] remain vested in individual states rather than collectively in universal institutions" (Weiss & Wilkinson, 2014, p. 213). Food is regulated largely at the national level, resulting in a mismatch between the global scale of the problem and the level at which it is often governed. Initiatives like agricultural production policies and dietary guidelines usually happen at the national level, but food can also be governed at sub-national and municipal levels. For example, backyard chicken by-laws and urban agriculture policies govern where and how people raise and access food (Miller, 2011; Samanta et al., 2018). In Brazil, local procurement policies have increased demand for locally grown food (The Food Foundation & Institute of Development Studies, 2017). Urban planning rules dictate where farmers' markets or supermarkets are located, leaving parts of cities with less access to food, often in poor and/or racialised areas (Wagner et al., 2019).

Global jurisdictional complexity and fragmented governance means there are a range of actors governing food in ways that overlap, reinforce, and sometimes contradict each other. International food trade largely falls under the legal jurisdiction of the World Trade Organisation (WTO), while the decisions of the UN's Committee for World Food Security (CFS) are considered voluntary. Different governance bodies make rules that may or may not have to be adhered to. While governments or international agencies and organisations are frequently seen as the main "governance actors," this overlooks the impacts that civil society can have on governance through social movements and the ways that corporations govern through certifications and supply chain requirements.

This chapter offers an introductory overview of who governs across the global jurisdictional spectrum. A positive outcome of the increasing demand for sustainable diets is that it has become a shared policy goal, bringing together policy makers that previously may not have been in close conversation. Governance actors in the public health realm have come into conversation with those focused on agriculture and environmental policy, which we examine first through the UN system.

Trends in global governance of sustainable diets—sustainable food systems

The UN system: Sustainable diets as part of sustainable food systems

The bodies that make up the UN system have guided the development of food systems since the Second World War in ways that were the product of their time: Policy guidance was frequently scientifically reductive and siloed. The UN Food and Agriculture Organisation (FAO) was primarily concerned with improving agricultural yields in the post-war period, using scientific approaches and technical assistance to decrease global hunger (Shaw, 2007). The productivity approach was reinforced at the 1974 Food Conference, after devastating droughts decimated global wheat yields, spiking food prices around the world. The 1974 Food Conference introduced food security as an idea, and the FAO reaffirmed its strategy of increasing yields to ensure sufficient food supplies. At the same time, the World Health Organisation (WHO) was responsible for nutrition and also used reductive approaches to isolate and deal with individual nutritional elements (Scrinis, 2013).

Food remained separately siloed in the UN system until the 1990s, when more coordinated efforts and principles of inclusivity became priorities. A 1992 conference was the first to formally recognise the need for joint efforts of the FAO and WHO (Shaw, 2007)—in other words, for both "agriculture" and "nutrition" to be together at the table. And while the Committee for World Food Security (CFS) had been officially convened out of the 1974 Food Conference to tackle food in a more holistic way (Shaw, 2007), it was underutilised until its major reform in 2009 (Duncan & Barling, 2012; McKeon, 2009). It is now considered the primary body responsible for food systems issues in the UN, as well as the "foremost inclusive body" in the UN system (CFS, n.d.), signalling the rise of multisectoral and multistakeholder approaches to food systems issues.

Sustainable diets were an even more recent addition to UN programming but were quickly swept into more multisectoral approaches. The idea was officially recognised at the International Scientific Symposium on Biodiversity and Sustainable Diets United Against Hunger, convened by the FAO in Rome in 2010. The Second International Conference on Nutrition (ICN2) was held in 2014, a follow-up from the 1992 conference where the FAO, WHO, and member-states[2] came together to discuss diets in new, more interdisciplinary ways. The FAO continued to move towards more sustainability-focused approaches to food at ICN2, in comparison to its traditional productivity focus. Unlike the 1992 conference, ICN2 propelled a Decade of Action on Nutrition (2016–2025). Like the Millennium Development Goals and the Sustainable Development Goals,[3] Decades of Action aim to galvanise the international community around a particular topic area. At the same time, the 17 Sustainable Development Goals were being developed, signalling a shift in international thinking towards the need for overlapping and reinforcing goals and strategies to target complex global problems.

Contemporary treatments of food in the UN system seem to prefer the language of sustainable food systems rather than focusing more narrowly on sustainable diets. Events and flagship campaigns are good milestone markers that help us to see how ideas in the UN system are shaped over time. In 2019 the *Sustainable Healthy Diets Guiding Principles* were jointly published by FAO and WHO, and publicised at the CFS. The guidelines focused on the role of diets in sustainable food systems, signalling the broader conceptualisation of sustainability in both production and consumption sides of the food system. In February 2021, the CFS Member-States endorsed Voluntary Guidelines on Food Systems and Nutrition. While these guidelines failed to adopt the word *Sustainable* in their title, the systems-approach tried to cover sustainable and multisectoral approaches to food. The guidelines have been described as "a comprehensive, systemic, and science and evidence-based approach to achieving healthy diets through sustainable food systems" (CFS, 2021, p. 6).[4] Perhaps the most concrete example of this focus on systems was the United Nations Food Systems Summit (FSS) held in September 2021, that promised to bring together diverse stakeholders to identify solutions for the future of food systems (United Nations, 2020). Of course, there are actors who continue to focus on the consumption side within the broader conceptualisations of sustainable food systems, as exemplified in the UN FSS Action Track 2 "Shift to sustainable consumption patterns." The UN's focus on food as systems recognises the interconnected and complex nature of food production and consumption, but it also represents a fundamentally different and evolved idea than hunger or even food security which was the original goal of the UN system's treatment of food.

Global civil society: Social movements to inform sustainable diets

At the same time that food systems discourse was becoming dominant in the UN system, a broader range of stakeholders, including civil society, were becoming influential in informing food systems work. In 2009, the CFS developed mechanisms to include new voices. The Civil Society and Indigenous Peoples' Mechanism (CSM) intends to give voice to fisherfolk, Indigenous People, peasant farmers, women, and others who have been traditionally marginalised on food issues. The Private Sector Mechanism welcomes input from private enterprises across the agri-food value chain, from farmers to input providers, cooperatives, processors, SMEs, and food companies. Neither mechanism, for civil society nor for the private sector, had voting rights. Formalised mechanisms that include marginalised voices, like those at the CFS, help people access spaces where governance decisions are made, enabling civil society to push food systems in more sustainable directions (CSM, 2021). Research shows that civil society's engagement through this mechanism can impact decisions that are made, including sustainability of food systems (Pictou, 2018).

Multi-stakeholder platforms are frequently at risk of power imbalances and conflicts of interest that may potentially minimise civil society's effect (McKeon, 2015). Formal mechanisms do not guarantee influence. Even though the CSM was present and included, it aimed criticism at the new CFS Voluntary Guidelines on Food Systems and Nutrition for not doing more to promote or protect sustainable diets: "The complete missed opportunity to include the concept of sustainable healthy diets and the subsequent failure to adequately and consistently recognize even the link between environmental and human health throughout the document is a main limitation in our view" (CSM, 2021).

While civil society movements are frequently the strongest voices for progressive and sustainable action on food systems, the characteristics that make up different groups and the strategies they use vary considerably. Since national governments are still considered the locus of most food systems policies, some civil society actors use a 'boomerang' strategy to influence national action by first securing commitments at the global level (Keck & Sikkink, 1998). The rural peasant movement, La Via Campesina (LVC), has notably shaped discussions around food sustainability through a sovereignty lens (Lang & Heasman, 2015) by successfully mobilising hundreds of thousands of rural, poor, and/or landless farmers and demanding a shift in the global conversation around food away from conventional, industrial agricultural systems. This is the 'food sovereignty' movement—one that puts humans and the planet at the centre of decision making (Altieri & Toledo, 2011). Outside the UN system, reports can highlight problems and opportunities that target both national and global governance. These reports can give civil society the tools to push governments into action on viable levels; two examples are the Global Panel on Agriculture and Food Systems for Nutrition (GloPan)'s Foresight Report, and the expert-driven consensus reports of the International Panel of Experts on Sustainable Food (IPES-Food). Civil society, in the form of charities or non-government organisations (NGOs), can also sometimes impact global supply chains of food and consumer buying habits through certification schemes such as Fair Trade or Marine Stewardship Council (Bernstein & Cashore, 2007; Dubuisson-Quellier & Lamine, n.d.; Jaffee & Howard, 2010). These types of certifications can be NGO-led or partnerships between NGOs and private actors. Critics suggest both are at risk of co-optation by corporate interests (see, for example, Jaffee & Howard, 2010), while proponents suggest that as long as the initiatives encourage progress they achieve the goal.

Finally, the EAT Foundation is a relatively new entrant to global governance on sustainable food and perhaps one more focused specifically on diets. It operates in a new, public-private way. Privately founded by a vegan supermodel-turned-doctor, the EAT Foundation translates planetary limitations into actual, recommended diets. While grassroots movements suggest that diets must be local and culturally appropriate, the EAT foundation gathered climate scientists and public health researchers to recommend regional diets that would meet local needs (EAT-Lancet Commission, 2019). The EAT Foundation acts similarly to the Gates Foundation in other spheres of global governance: It is a private foundation with star and convening power that can bring both public and private decision makers together. This version of civil society influencing governance is starkly different from the grassroots collective action movements of LVC or the Civil Society Mechanism at the CFS, and each may articulate differing versions of sustainable diets and food systems.

Civil society is a diverse collection of actors and movements, but they have in common their desire to support progressive movements in food systems. They do this through formal governance mechanisms, grassroots social movements, big global reports, bold new foundations, or labelling and certification schemes. For the most part, civil society governs sustainable diets by aiming to influence other governance actors. However, when they participate in certification schemes they are governing as rule-makers. Civil society is neither government actors nor private actors, though it frequently, and often controversially, works with both.

Corporate actors: Influencing or making rules

As with civil society, there are two overarching ways to think about corporate actors exerting power over sustainable food systems and diets: 1) Corporate actors use different strategies to influence governments or intergovernmental organisations to make decisions; and 2) sometimes corporate actors govern by acting as rule-makers or decision makers themselves. In the first example, corporate actors might influence governments to make sustainable policy changes—a company that specialises in low-carbon catering might lobby national governments to limit school food procurement policies to 100-mile foodsheds. Or, they lobby governments to maintain the status quo and prevent transformation: Such as corporations' resistance to the recent development of front-of-pack nutrition labelling in Colombia, or aiming to prevent reductions in recommended intakes of meat and dairy in national dietary guidelines (Mialon et al., 2020; Rose et al., 2021; S et al., 2014).

In the second case, corporate actors directly control supply chains or consumer choices through their market power and rule-making abilities. If Walmart demands that producers change production practises to be more sustainable, the sheer scale of the company makes changes far more widespread than if some countries regulated food in the same way (Elder & Dauvergne, 2015). Similarly, retailers can introduce their own certification schemes to signal virtue to consumers, sometimes working with charities or NGOs (Busch, 2011; Fuchs et al., 2009). Big food companies frequently use sustainability claims to increase legitimacy (Scott, 2019), and though the scale can be meaningful, the substance is not always.

Food corporations, whether retailers, manufacturers, or producers, can all influence the sustainability of customer diets: The chosen inputs control individual consumers' carbon or sustainability outputs. Contrary to this view, consumers are often held personally responsible for choosing sustainable diets (IPES-Food, forthcoming), even though public health research shows that external food environments are the main determinants for food choices. The built environment, advertising, and food costs are all choices that corporate actors are largely responsible for, yet consumers are usually expected to demonstrate willpower and responsibility for their health and the planet's (Lang, 2020).

What are governments' roles?

National governments are still usually considered the main actor in food systems governance, since they control most of the policies, legislation, and regulations that govern domestic sectors. Some national government examples include: Banning of plastic straws; lowering waste associated with fast food purchases; limiting use of pesticides and fertilisers in agriculture; subsidising agroecological forms of growing instead of conventional, industrial agriculture; or re-orienting public procurement around local foodsheds. One way that some governments have tried to create momentum towards sustainable diets is through national dietary guidelines (Gonzalez Fischer & Garnett, 2016; Lang, 2017). Efforts here are directed at creating lower carbon dietary patterns through recommendations that include reduced consumption of animal products. However, there has been considerable debate over the effectiveness of this strategy, given that most citizens do not closely follow dietary guidance and other legal, and normative mechanisms may contradict the dietary guidance provided (Deckha, 2020).

Different levels of governments can set policies for sustainability as well. Municipal governments can play key roles in the built environment by making it easier for farmer's markets to operate or creating programs like mobile fresh food markets to offer fresh foods in areas that are typically underserved (Mattioni, 2021; Young et al., 2011). State or provincial governments can regulate public procurement, and national governments can change tax incentives and subsidies (Vonthron et al., 2020). In general, we can think about governments as being able to use four forms of influence to shift food systems in a more sustainable direction: 1) They can "nudge" consumers towards "better" choices (e.g., through dietary guidelines or labelling schemes) (Scrinis & Parker, 2016); 2) they can regulate (e.g., by setting limits on packaging or pesticide use) (Benbrook, 2018); 3) they can disincentivise unsustainable practises (e.g., by taxing sugar-sweetened beverages or removing existing subsidies for conventional industrial agriculture) (Franck et al., 2013; George, 2018); or 4) they can incentivise desired practises (e.g., providing grants for sustainable food production or access to land). Governments are perhaps the best positioned to institute wide-scale changes towards sustainable diets and food systems, but without an engaged and supportive citizenry, lack of political will can be an impediment. Their willingness to adopt a more systemic approach may also hold the difference between meaningful change and piecemeal action.

Recommendations and conclusions

This chapter was a brief introduction to the main categories of international actors and how they govern sustainable diets and sustainable food systems. We have demonstrated that sustainable food systems have contemporary discursive sway at the global level, especially in the UN system. This focus on systems is important: It recognises the interconnected and mutually reinforcing conditions of food in the world today, but it also accounts for multiple, overlapping, and fragmented governance. In particular, this discourse recognises that food choices are the result of social, economic, and environmental factors that individuals do not necessarily have control over (Garnett et al., 2015). Consumers must navigate food systems through jurisdictional, political, and corporate-level policies. Implementing isolated policies is not enough to ensure a sustainable, healthy, and equitable food future. Intervention points in the food system must be targeted through multi-sectoral and multi-level initiatives. In this sense, practitioners of sustainable diets must target key players appropriately. The relative weight of actions and influence by different actors should be considered when targeting interventions. While civil society might be able to register concern or influence intergovernmental or national-level action, corporations can dictate demands directly to supply chains, and national governments can regulate or legislate. Different actors have different levels of power.

Simply organising multi-stakeholder forums to receive input is not enough. The pushback towards the UN Food Systems Summit suggests that although the intention was to gather perspectives and solutions from all stakeholders, the process privileges stakeholders with power and financial resources. Since the Summit aimed to analyse input by theme and frequency, those with power and motivation can produce an outsized influence. This approach does not suggest the radical re-orientation required for sustainable, healthy, and equitable food futures (Canfield et al., 2021). In fact, it suggests a locking in of status-quo, corporate co-opted power dynamics.

In conclusion, rather than recommending specific policies, programs, or mechanisms to encourage sustainable diets,[5] we make two main suggestions. First, we suggest that practitioners of sustainable diets be aware of the language *du jour* around transformation towards a sustainable food system. This systematic approach to food sustainability includes the more focused approach of sustainable diets but also allows for multisectoral approaches to food sustainability, both in terms of a common narrative towards food systems change (HLPE, 2020) and resulting calls for policy coherence (Lang, 2021; OECD, 2021). Second, we suggest that sustainable diet practitioners recognise relative power in different actors' ability to transform food systems. While consumers are often seen as responsible for dietary choices, corporate and regulatory decisions are often more consequential in scale. By approaching food systems and sustainable diets with power in mind, practitioners can help shift policies and governance towards food democracy, where new forms of engagement could challenge existing power relations and path dependencies (De Schutter et al., 2020). A food future with more inclusive governance could take advantage of the scaled action available to governments and corporate actors, while allowing civil society the space to truly impact these decisions in meaningful ways.

Notes

1 Weiss & Wilkinson (2014) write that global governance "remains notoriously slippery" (p. 207). Here we use one of the more imprecise definitions to simply illustrate the global landscape on governing sustainable diets for readers of this handbook: We see global governance here as the "growing complexity in the way the world is organised and authority is exercised as well as a shorthand for referring to a collection of institutions with planetary reach" (Weiss & Wilkinson, 2014, p. 207).
2 Including the High-Level Task Force on the Global Food Security Crisis (HLTF), IFAD, IFPRI, UNESCO, UNICEF, World Bank, WFP, and the WTO, alongside other major international actors and agencies.
3 The widely known Millennium Development Goals and Sustainable Development Goals attempt to create coherent agendas within global development circles to drive attention and action towards specific targets.
4 See more criticism on this below.
5 Many others have done this before us and with great quality. See, for example:
  World Resources Institute Shift Wheel report; Chatham House report (Bailey and Harper 2015); GloPan Reports; IPES-Food; Garnett and Finch 2019; SDGs/FSS; Making Better Policies for Food Systems (OECD) www.oecd-ilibrary.org/agriculture-and-food/making-better-policies-for-food-systems_ddfba4de-en.

References

Altieri, M. A., & Toledo, V. M. (2011). The agroecological revolution in Latin America: Rescuing nature, ensuring food sovereignty and empowering peasants. *Journal of Peasant Studies, 38*(3), 587–612.
Bailey, R., & Harper, D. R. (2015). *Reviewing interventions for healthy and sustainable diets.* Chatham House.
Baker, P., & Demaio, A. (2019). The political economy of healthy and sustainable food systems. In M. Lawrence & S. Friel (Eds.), *Healthy and sustainable food systems* (1st ed., pp. 181–192). Routledge.
Benbrook, C. M. (2018). Why regulators lost track and control of pesticide risks: Lessons from the case of glyphosate-based herbicides and genetically engineered-crop technology. *Current Environmental Health Reports, 5*(3), 387–395.
Bennett, S., Glandon, D., & Rasanathan, K. (2018). Governing multisectoral action for health in low-income and middle-income countries: Unpacking the problem and rising to the challenge. *BMJ Global Health,* 3(Suppl 4), e000880. https://doi.org/10.1136/bmjgh-2018-000880
Bernstein, S., & Cashore, B. (2007). Can non-state global governance be legitimate? An analytical framework. *Regulation & Governance, 1*(4), 347–371.
Biermann, F., Siebenhüner, B., & Schreyögg, A. (2009). *International organizations in global environmental governance.* Routledge.
Blay-Palmer, A., Sonnino, R., & Custot, J. (2015). A food politics of the possible? Growing sustainable food systems through networks of knowledge. *Agriculture and Human Values, 33*, 1–17.

Busch, L. (2011). The private governance of food: Equitable exchange or bizarre bazaar? *Agriculture and Human Values, 28*(3), 345–352.

Canfield, M., Anderson, M. D., & McMichael, P. (2021). UN Food Systems Summit 2021: Dismantling democracy and resetting corporate control of food systems. *Frontiers in Sustainable Food Systems, 5*, 1–15.

CFS. (n.d.). *CFS homepage. Committee on world food security.* Committee on World Food Security: Making a Difference in Food Security and Nutrition. Retrieved August 31, 2020, from http://www.fao.org/cfs

CFS. (2021). *Voluntary guidelines on food systems and nutrition.* http://www.fao.org/fileadmin/templates/cfs/Docs2021/Documents/CFS_VGs_Food_Systems_and_Nutrition_Strategy_EN.pdf

Clapp, J. (2015). Distant agricultural landscapes. *Sustainability Science, 10*(2), 305–316.

Clapp, J. (2020). *Food* (3rd ed.). Polity.

Clapp, J., & Scott, C. (2018). The global environmental politics of food. *Global Environmental Politics, 18*(2), 1–11.

CSM. (2021, January 25). *CSM statements on the negotiations of the guidelines on food systems and nutrition.* Csm4Cfs. http://www.csm4cfs.org/csm-opening-statement-negotiations-voluntary-guidelines-food-systems-nutrition/

De Schutter, O., Jacobs, N., & Clément, C. (2020). A 'Common Food Policy' for Europe: How governance reforms can spark a shift to healthy diets and sustainable food systems. *Food Policy, 96*, 101849.

Deckha, M. (2020). Something to celebrate?: Demoting dairy in Canada's national food guide. *Journal of Food Law and Policy, 16*, 11.

Delabre, I., Rodriguez, L. O., Smallwood, J. M., Scharlemann, J. P. W., Alcamo, J., Antonarakis, A. S., Rowhani, P., Hazell, R. J., Aksnes, D. L., Balvanera, P., Lundquist, C. J., Gresham, C., Alexander, A. E., & Stenseth, N. C. (2021). Actions on sustainable food production and consumption for the post-2020 global biodiversity framework. *Science Advances, 7*(12), eabc8259.

Dubuisson-Quellier, S., & Lamine, C. (2008). Consumer involvement in fair trade and local food systems: Delegation and empowerment regimes. *GeoJournal, 73*(1), 55–65. https://doi.org/10.1007/s10708-008-9178-0

Duncan, J., & Barling, D. (2012). Renewal through participation in global food security governance: Implementing the international food security and nutrition civil society mechanism to the committee on world food security. *International Journal of the Sociology of Agriculture and Food, 19*(2), 143–161.

EAT-Lancet Commission. (2019). *Healthy diets from sustainable food systems: Food, planet, health.* Wellcome Trust.

Elder, S. D., & Dauvergne, P. (2015). Farming for Walmart: The politics of corporate control and responsibility in the global South. *The Journal of Peasant Studies, 42*(5), 1029–1046.

FAO. (2012). Sustainable diets and biodiversity—Directions and solutions for policy research and action. In Proceedings of the International Scientific Symposium Biodiversity and Sustainable Diets United Against Hunger, 3–5 November 2010. Rome: Nutrition and Consumer Protection Division FAO.

FAO, & WHO. (2019). *Sustainable healthy diets: Guiding principles.* FAO.

Franck, C., Grandi, S. M., & Eisenberg, M. J. (2013). Agricultural subsidies and the american obesity epidemic. *American Journal of Preventive Medicine, 45*(3), 327–333.

Friel, S., Barosh, L. J., & Lawrence, M. (2014). Towards healthy and sustainable food consumption: An Australian case study. *Public Health Nutrition, 17*(5), 1156–1166. https://doi.org/10.1017/S1368980013001523

Fuchs, D. A., Kalfagianni, A., & Arentsen, M. (2009). Retail power, private standards, and sustainability in the global food system. In J. Clapp & D. A. Fuchs (Eds.), *Corporate power in global agrifood governance* (pp. 29–60). MIT Press.

Garnett, T., Mathewson, S., Angelides, P., & Borthwick, F. (2015). *Policies and actions to shift eating patterns: What works?* (Food Climate Research Network; p. 85). Chatham House.

George, A. (2019). Not so sweet refrain: Sugar-sweetened beverage taxes, industry opposition and harnessing the lessons learned from tobacco control legal challenges. *Health Economics, Policy, and Law, 14*(4), 509–535. https://doi.org/10.1017/S1744133118000178

Gonzalez Fischer, C., & Garnett, T. (2016). *Plates, pyramids, planet. Developments in national healthy and sustainable dietary guidelines: A state of play assessment.* FAO and the Environmental Change Institute & The Oxford Martin Programme on the Future of Food, The University of Oxford. www.fao.org/3/a-i5640e.pdf

HLPE. (2020). *Food security and nutrition: Building a global narrative towards 2030* (p. 112). CFS.

Jaffee, D., & Howard, P. H. (2010). Corporate cooptation of organic and fair trade standards. *Agriculture and Human Values, 27*(4), 387–399.

Keck, M. E., & Sikkink, K. (1998). *Activists beyond borders: Advocacy networks in international politics.* Cornell University Press.

Lang, T. (2017). *Re-fashioning food systems with sustainable diet guidelines.* Food Research Collaboration | Friends of the Earth. http://foodresearch.org.uk/publications/re-fashioning-food-systems-with-sustainable-diet-guidelines/

Lang, T. (2020). Policy to promote healthy and sustainable food systems. In M. Lawrence & S. Friel (Eds.), *Healthy and sustainable food systems* (pp. 193–204). Routledge.

Lang, T. (2021). The sustainable diet question: Reasserting societal dynamics into the debate about a good diet. *The International Journal of Sociology of Agriculture and Food, 27*(1), 12–34.

Lang, T., & Heasman, M. (2015). *Food wars: The global battle for minds, mouths, and markets* (2nd ed.). Earthscan.

López Cifuentes, M., Freyer, B., Sonnino, R., & Fiala, V. (2021). Embedding sustainable diets into urban food strategies: A multi-actor approach. *Geoforum, 122*, 11–21.

Mason, P., & Lang, T. (2017). *Sustainable diets: How ecological nutrition can transform consumption and the food system.* Routledge.

Mattioni, D. (2021). Constructing a food retail environment that encourages healthy diets in cities: The contribution of local-level policy makers and civil society. *The International Journal of Sociology of Agriculture and Food, 27*(1), 87–101.

McKeon, N. (2009). *The United Nations and civil society: Legitimating global governance: Whose voice?* Palgrave Macmillan.

McKeon, N. (2015). GFG: Global food governance in an era of crisis: Lessons from the United Nations Committee on World Food Security. *Canadian Food Studies / La Revue Canadienne Des Études Sur l'alimentation, 2*(2), 328–334.

Meadows, D. (2008). *Thinking in systems: A primer.* Chelsea Green Publishing. https://www.overdrive.com/search?q=C6858C87-FE44-4231-B42B-2EA3AD20F10B

Mialon, M., Gaitan Charry, D. A., Cediel, G., Crosbie, E., Baeza Scagliusi, F., & Pérez Tamayo, E. M. (2020). "The architecture of the state was transformed in favour of the interests of companies": Corporate political activity of the food industry in Colombia. *Globalization and Health, 16*(1), 97.

Miller, K. (2011). *Backyard chicken policy: Lessons from Vancouver, Seattle and Niagara Falls.* School of Urban and Regional Planning. Queens University. https://qspace.library.queensu.ca/handle/1974/6521

OECD. (2021). *Making better policies for food systems.* OECD.

Pictou, S. (2018). The origins and politics, campaigns and demands by the international fisher peoples' movement: An Indigenous perspective. *Third World Quarterly, 39*(7), 1411–1420.

Rose, D., Vance, C., & Lopez, M. A. (2021). Livestock industry practices that impact sustainable diets in the United States. *The International Journal of Sociology of Agriculture and Food, 27*(1), Article 1.

Samanta, I., Joardar, S. N., & Das, P. K. (2018). Biosecurity strategies for backyard poultry: A controlled way for safe food production. *Food Control and Biosecurity,* 481–517. https://doi.org/10.1016/B978-0-12-811445-2.00014-3

Scott, C. M. (2019). *Challenging big food sustainability: Dietary change and corporate legitimacy in the agrifood landscape.* University of Waterloo.

Scrinis, G. (2013). *Nutritionism: The science and politics of dietary advice*. Columbia University Press.

Scrinis, G., & Parker, C. (2016). Front-of-pack food labelling and the politics of nutritional nudges. *Law & Policy*, *38*(3), 234–249.

Shaw, D. J. (2007). *World food security: A history since 1945*. Palgrave Macmillan US.

Stuckler, D., & Nestle, M. (2012). Big food, food systems, and global health. *PLoS Medicine*, *9*(6), e1001242.

The Food Foundation, & Institute of Development Studies. (2017). *Brazil's policies to guarantee food rights (No. 5; International learning series)*. The Food Foundation. https://foodfoundation.org.uk/wp-content/uploads/2017/07/5-Briefing-Brazil_vF.pdf

Thow, A. M., Jones, A., Schneider, C. H., & Labonté, R. (2019). Global governance of front-of-pack nutrition labelling: A qualitative analysis. *Nutrients*, *11*(2), 268.

United Nations. (2020). *Action tracks*. United Nations. https://www.un.org/en/food-systems-summit/action-tracks

Vonthron, S., Perrin, C., & Soulard, C.-T. (2020). Foodscape: A scoping review and a research agenda for food security-related studies. *PLoS ONE*, *15*(5). https://doi.org/10.1371/journal.pone.0233218

Wagner, J., Hinton, L., McCordic, C., Owuor, S., Capron, G., & Arellano, S. G. (2019). Do urban food deserts exist in the Global South? An analysis of Nairobi and Mexico City. *Sustainability*, *11*(7), 1963.

Weiss, T. G., & Wilkinson, R. (2014). Rethinking global governance? Complexity, authority, power, change. *International Studies Quarterly*, *58*(1), 207–215.

Willett, W., Rockström, J., Loken, B., Springmann, M., Lang, T., Vermeulen, S., Garnett, T., Tilman, D., DeClerck, F., Wood, A., Jonell, M., Clark, M., Gordon, L. J., Fanzo, J., Hawkes, C., Zurayk, R., Rivera, J. A., Vries, W. D., Sibanda, L. M., … Murray, C. J. L. (2019). Food in the Anthropocene: The EAT–lancet commission on healthy diets from sustainable food systems. *The Lancet*, *393*(10170), 447–492.

Young, C., Karpyn, A., Uy, N., Wich, K., & Glyn, J. (2011). Farmers' markets in low income communities: Impact of community environment, food programs and public policy. *Community Development*, *42*(2), 208–220.

PART 2

Environmental strategies

5

CLIMATE CHANGE, FOOD SECURITY, AND SUSTAINABLE DIETS

Francis Adams

Introduction

Climate change threatens global food security. Although there are multiple causes of hunger and malnutrition, there is little doubt that climate change is disrupting food production, distribution, and consumption worldwide, especially in the Global South.[1] Reduced food availability and higher food prices pose the greatest risk to households and communities that are already the most food insecure.[2] At present, nearly two billion people do not have reliable access to safe, sufficient, and nutritious food.[3]

Climate change refers to long term alterations in average weather patterns due to increases in the earth's air, surface, and ocean temperatures. Average global temperatures are one degree Celsius above pre-industrial levels and are projected to rise by another 1.5 degrees Celsius by 2050 (Intergovernmental Panel on Climate Change, 2018). The warming of the earth is largely due to higher levels of atmospheric carbon dioxide and other greenhouse gasses that are produced by the burning of fossil fuels.

This chapter outlines the ways in which climate change is disrupting global food production and consumption. Although the impact of climate change varies considerably between and within countries, a warmer earth is adversely impacting the natural environments and resources needed to maintain sustainable food systems in the world's poorest countries.[4] This is evident with respect to land degradation, water scarcity, extreme weather events, insect infestations, the spread of crop diseases, and the decline of fish populations. This chapter also identifies measures to mitigate the greenhouse gas emissions that originate from food systems and better manage the adverse impacts of climate change on agricultural and rural communities.

Land degradation

Climate change is reducing the land available for agricultural cultivation (Intergovernmental Panel on Climate Change, 2019a). Higher temperatures, greater evaporation, less precipitation, and the decomposition of organic carbon lower the water content of soils. Soils are eroding faster than they are being formed in many parts of the world. Nearly 40 percent of all agricultural lands are now degraded, and an additional 12 million hectares are lost to soil erosion and desertification each year.[5] Nearly half of the world's people live in areas marked by high levels of land degradation. In some parts of Central America and South Asia soil erosion is so widespread that agricultural production is no longer possible. Land degradation is most extensive in Africa where two-thirds of the region's most productive agricultural land is either moderately or severely degraded. As plants die off, the soil is more easily eroded, beginning a new cycle of land degradation that is exceedingly difficult to reverse (Food and Agriculture Organisation, 2018, p. 53; Intergovernmental Panel on Climate Change, 2019a).

Saline intrusion, where saltwater overruns the land or seeps in from below, is also causing soil erosion. Warmer ocean temperatures cause the thermal expansion of seawater while warmer atmospheric temperatures cause the melting of polar ice caps, ice sheets, and glaciers. Together, these twin processes are causing sea level rise and the flooding of coastal areas. In Latin America, saline intrusion is damaging agricultural lands in northeast Argentina, southern Brazil, Ecuador, northwest Mexico, northwest Peru, and Uruguay. Higher sea levels in the Mediterranean are causing saline intrusion into the fragile agricultural lands of North Africa. In the Middle East, sea level rise is causing saline intrusion in low lying agricultural lands of Kuwait, Oman, Qatar, and the United Arab Emirates. In South and Southeast Asia, rising sea levels are damaging fertile agricultural lands, especially in the coastal parts of the Bay of Bengal and Mekong Delta.

Water scarcity

Climate change is also causing greater water scarcity. Higher temperatures and increased evaporation are drying up lakes, rivers, and other inland water bodies. This is reducing the supply of renewable freshwater available for agriculture and livestock. A quarter of the world's population currently resides in regions where water resources are insufficient for the level of food production required.[6] The decline in water resources coincides with greater needs for water in agriculture. Because higher temperatures increase plant transpiration, the water requirements of crops have also increased in many parts of the Global South.

Water scarcities are evident in most world regions. Dry weather conditions in Latin America have contributed to lower levels in a number of inland water bodies. Lake Poopó in the Bolivian Altiplano, once the second largest lake in this country, is now completely dried up without any chance of recovery. Latin American countries have some of the fastest rates of aquifer depletion in the world. A decline in freshwater resources is also occurring in Africa, which has the greatest number of water-stressed countries in the world. Water levels in the Niger River have decreased by nearly half and Lake Chad has shrunk by 90 percent. Lower water levels in the Nile River have reduced the silt needed to replenish soils in the delta region. In the Middle East, the world's most arid region, the Euphrates, Jordan, and Tigris Rivers have declined dramatically and aquifers are being drained faster than they are refilling.[7] Per capita freshwater resources have decreased by two-thirds in the past forty years and are currently just one-sixth of the global average. More than 60 percent of all people in the Middle East live in areas with high surface water stress and ten of the seventeen most water scarce countries in the world are in this region.

Variations in precipitation

Climate change is also reducing rainfall in large parts of the Global South. The vast majority of the crops in these regions are dependent on precipitation. Warmer temperatures in the Pacific Ocean have fuelled the El Niño Southern Oscillation (ENSO) and corresponding changes in the hydrological cycle. ENSO is altering precipitation patterns in a number of areas. In Latin America, droughts have been especially severe in western Argentina, the Bolivian-Peruvian Altiplano, north-eastern Brazil, and central Chile. The Dry Corridor of Central America, which runs along the Pacific coast from Guatemala to Panama, has also experienced severe drought in recent years. This drought has shortened crop cycles and reduced food production, especially the cultivation of barley, maize, rice, and wheat. In El Salvador, Guatemala, and Honduras, crop yields have declined between 50 and 90 percent from pre-drought levels.

Many parts of Africa and Asia are also experiencing irregular rainfall patterns and less precipitation. Declining precipitation in East Africa is shortening the length of growing seasons and reducing the availability of the pasture available for livestock. Prolonged drought has caused consecutive poor harvests in a number of southern African countries that are already facing high levels of food insecurity. In South Asia, reduced precipitation has disrupted agricultural production and food supplies, especially in more arid and semi-arid areas. Drought has been especially intense in Afghanistan and parts of Bangladesh, India, Pakistan, and Sri Lanka. Crops in these countries were already being grown with limited water resources. Southeast Asian countries have endured unusually dry monsoon seasons.

The melting of inland glaciers is also limiting the water available for agricultural cultivation. Farmers often rely on snowmelt from mountain glaciers for seasonal renewal of water supplies. However, warmer temperatures at higher elevations are contributing to glacier loss. Some glaciers in the Andes have receded to levels not seen in a thousand years and many have disappeared altogether.[8] The decreased runoff from glacier snowmelt is reducing the availability of water to irrigate farmlands and maintain livestock throughout western Argentina, Bolivia, southern Chile, central Peru, and parts of Colombia and Ecuador. Glaciers on the Tibetan plateau are also critical for seasonal renewal of water supplies throughout much of South and Southeast Asia. Glacier melt due to warmer temperatures is reducing water supplies in Bangladesh, northern India, Nepal, and Pakistan. Water levels in the Ganges and Indus Rivers have significantly declined. Similarly, the Mekong River, which stretches over four thousand kilometres from its source in the Tibetan highlands through China, Myanmar, Thailand, Lao PDR, Cambodia, and Vietnam before emptying into the South China Sea, is at its lowest level in a century.

Extreme weather events

Climate change is also increasing the frequency, intensity, size, and duration of extreme weather events. Despite a decline in rainfall in many areas, highly destructive torrential rains are occurring in others. A warmer climate, which increases evaporation from water bodies, soils, and plants, creates the conditions for more extreme precipitation events. When stored water vapour is rapidly released, the heavier rains cause extensive flooding, crop loss, and the destruction of rural infrastructure. Heavier rain storms also increase surface run-off, which intensifies soil erosion. Because degraded soils are less able to hold water, flooding becomes even more extensive when the next storm occurs.

Heavier rains have been especially destructive in Africa and Asia. The storms that ravaged Malawi, Mozambique, Tanzania, and Zimbabwe in the Spring of 2019 were among the strongest on record. Thousands of hectares of farmland were ruined, along with roads, bridges, and other rural infrastructure (Congressional Research Service, 2019). Much larger rainfalls than normal have also caused record flooding in the Ganges Delta in South Asia and in the Irrawaddy and Mekong deltas in Southeast Asia. Severe flooding in these areas has destroyed crops and damaged rural infrastructure. Heavy rains have been especially destructive in Bangladesh where nearly two-thirds of all land is less than 15 feet above sea level. Heavy rains have submerged large parts of the country for extended periods. Similar flooding has caused catastrophic damage to rice crops in low-lying parts of India, Myanmar, Nepal, Pakistan, and Vietnam.

Increased ocean temperatures and elevated humidity have also fuelled more frequent cyclones, hurricanes, and typhoons.[9] The number of extreme weather-related disasters each year has more than doubled since the early 1990s. In recent years, severe storms, especially in the Caribbean, Central America, and Southeast Asia, have caused enormous destruction to crops, arable lands, rural infrastructure, and community assets. The disruption of agricultural production and post-production processes jeopardises the quantity, quality, and safety of food supplies. People who are already food insecure are the least able to endure the shocks caused by natural disasters.

Insects and disease

Climate change is also causing the spread of insect infestations and disease pathogens, both of which are highly destructive to food crops. Warmer temperatures and greater moisture create more favourable conditions for the reproduction and survival of many of the insect species that constitute the greatest threat to food crops.

Moreover, because warmer temperatures increase the metabolism of insects, they need to consume even more plants to meet their biological needs. The increased variety and number of invasive species, and their spread into new geographic areas, are contributing to the large-scale loss of food crops.

African countries have been especially impacted by insect infestations. East African countries have recently endured the largest invasion of desert locusts on record.[10] The locusts, which rapidly reproduce after unusually heavy rains, cause massive damage to crops and rangelands. Locust swarms can travel 150 kilometres and consume 400 million pounds of vegetation in a single day. Fall armyworms, an invasive caterpillar that devastates maize crops, also have an expanded presence throughout the continent. The increased incidence of Rift Valley fever, a mosquito-borne viral disease that infects livestock, is a direct consequence of climate change.

The spread of crop diseases is reducing the size and quality of agricultural yields. Rising temperatures, moisture, humidity, and concentrations of atmospheric carbon dioxide all contribute to the growth of pathogens, especially fungal infections. Warmer temperatures and changes in rainfall patterns also lead to the spread of some of the more destructive plant pathogens. Extreme weather events can disperse plant pathogens into new areas. Virulent strains of wheat rust, for example, have spread from Africa to the Middle East and South Asia. Higher humidity and moisture also cause stored grains to be more susceptible to fungal infections.

Climate change also impacts the nutritional value of food crops. As carbon dioxide levels rise, the openings in plant shoots and leaves shrink causing them to lose less water. Because this slows the circulation of plants, they draw fewer minerals from the soil. Higher atmospheric carbon dioxide has been linked to reduced concentrations of iron, zinc, and other essential minerals in a number of plant species (Smith & Myers, 2018).

Decline of fish populations

Climate change is also reducing fish populations (Food and Agriculture Organisation, 2020e). The global food system depends to a considerable extent on healthy freshwater and marine ecosystems. Fish and seafood are especially important to the diets of people in the Global South and provide approximately half of all protein consumed. Inland fisheries and aquaculture are rapidly declining as a result of changes in water temperature, precipitation, and evaporation.

Marine resources are also threatened by changes in the global climate. Oceans absorb the vast majority of the excess heat that is trapped in the atmosphere, as well as a quarter of the carbon dioxide released from the burning of fossil fuels. Higher temperatures and carbon dioxide concentrations cause ocean acidification, oxygen depletion, and the spread of aquatic pathogens, as well as the degradation, bleaching, and loss of coral reefs, which provide habitat for 25 percent of all marine species (Intergovernmental Panel on Climate Change, 2019b). These changes disrupt the ecosystems, food supply, and breeding habitats upon which marine life depend. Warmer temperatures have caused the extinction of some marine species, shifted the habitats of others, and spread disease throughout the food chain.

Climate change mitigation

Climate change constitutes a major threat to global food security. Rising atmospheric, surface, and ocean temperatures are damaging the natural environments and resources needed for sustainable food production. To increase global security around food, especially in the world's poorest countries, will require collective efforts to lessen the rate and magnitude of climate change. Climate change mitigation, which works to reduce the greenhouse gasses that are produced by food systems and increase the absorption of those gasses already emitted, will be critically important. At present, the global food system contributes approximately a third of all anthropogenic greenhouse gas emissions (Tubiello et al., 2021).

Agroecology

Agroecology, which incorporates ecological principles into the design and management of food systems, can significantly reduce greenhouse gas emissions and contribute to carbon sequestration.[11] Agroecology is based on a number of core principles. Minimal soil disturbance is especially important. The tilling of fields, which involves turning over and breaking up the soil, exposes carbon that was locked in the soil. Rather than tilling fields, seeds can be directly sown into undisturbed soil. Farmers are also encouraged to diversify their crop varieties through intercropping and multiyear crop rotations. Planting cover crops during the offseason helps build soil health by reducing carbon loss, preventing erosion, and replenishing nutrients. Composts that incorporate organic matter from crop residues help maintain the fertility and water-holding capacity of soils. Natural fertilisers, such as peat, manure, and guano, and biological pest control techniques that incorporate living organisms to suppress pest populations, can be utilised in place of greenhouse gas-intensive chemical inputs.

Agroecology ensures a long-term balance between food production and the sustainability of natural resources. The organic structure, nutrient content, and overall fertility of soils improve. In addition to nitrogen, phosphorus, and potassium, healthy soils require the restoration of depleted carbon. Higher organic carbon in soils increases nutrient and water intake by plants, reduces soil erosion, and better conserves land and water resources (Food and Agriculture Organisation, 2019a). Increasing carbon sequestration in soils helps mitigate climate change while simultaneously increasing crop yields. By adapting to natural conditions and cycles, agroecology creates synergies between agricultural production and ecological processes. For detailed review of the environmental benefits of agroecology see Altieri and Nicholls, 2021; Giraldo, 2019; Mossi et al., 2020; and Pimbert, 2018.

Since forests have an enormous capacity to sequester carbon, sustainable forest management is also imperative. Agroforestry, where trees and shrubs are grown around or among crops and livestock, creates additional carbon sinks on farms. Agroforestry also enhances the sustainability of farming systems. Trees provide shade to protect crops, improve water retention in soils, reduce erosion, and protect fields from insects and disease.

Climate change adaptation

It also will be important to aid farmers, fishers, and rural communities to better adapt to climate change. A number of tools, technologies, and services can be utilised to help manage the adverse impacts of a warmer earth and build resilience to extreme weather events. While adaptation measures will depend on the specific characteristics of local environments, they should be introduced before rather than after climate change impacts occur. Proactive adaptations are far more cost-effective and significantly reduce the long-term effects of climate change.

Disaster risk management

Disaster risk management should be mainstreamed into all development planning and programs.[12] Vulnerability mapping can identify areas that are at the highest risk from changes in climatic conditions. Geographic information systems and remote sensing technologies can be utilised to strengthen hazard risk monitoring. Investing in disaster risk management, including early warning systems and emergency planning, typically pays off many times over in the prevention of loss and damage (Food and Agriculture Organisation, 2020b; World Food Programme, 2019).

Crop breeding

Crop breeding can also lessen vulnerability to climate change. Although highly controversial, especially among agroecologists, seeds can be genetically modified and edited to increase resilience to heat, drought, saline, insects, and disease. Improved seed varieties account for at least half of all gains in agricultural productivity in the past half century (World Resources Institute, 2018, p. 24). The micronutrient content of crops can also be increased through the biofortification of seeds. Flour can be fortified with folic acid, iron, niacin, riboflavin, thiamine, and zinc; sweet potatoes, rice, and legumes can be fortified with iron and zinc; maize can be fortified with vitamin A and zinc; and cassava and sorghum can be fortified with amino acids. Genetically modified or edited crops reduce production costs, ensure more consistent food supplies, and require fewer chemical inputs. Of course, these benefits will only be fully realised if small scale farmers in the Global South are guaranteed access and ownership rights to the improved seed varieties.

Erosion prevention

Erosion-prevention measures should also be emphasised. Adjusting tilling and irrigation practices can lessen runoff and help retain more nutrients in soils. Alternate cropping systems, composts, and agroforestry all help keep soil in place. Combining deep-rooted and shallow-rooted crops improves soil structure and lessens erosion. Terraces and windbreaks also reduce soil erosion and allow more water to flow to crops. Restoring coastal mangroves, watersheds, and wetlands helps reduce storm surges and saline intrusion caused by sea level rise. Wetlands provide a range of essential ecosystem services, including water filtration, storm protection, flood control, and biodiversity preservation.

Preserving freshwater

The preservation of freshwater resources is also critical for climate change adaptation. New methods can be introduced to increase water availability, including rainwater harvesting, wastewater recycling, and desalination. Reservoirs, farm ponds, and check dams can store water in rainy seasons for use during dry seasons. A number of low-cost, environmentally safe methods can be utilised to expand irrigation systems. Drip irrigation simply requires a network of perforated tubes linked to a water source; treadle pumps can draw water through a hose from a well or stream; and shallow tube wells and boreholes with submersible pumps can irrigate sizable areas at minimal cost. Improved water management can help restore water tables and groundwater aquifers that are under intense pressure in many parts of the Global South.

Increasing fish production

Increasing fish production can help offset declines in agricultural output. Freshwater and marine aquaculture, where fish are raised in tanks, ponds, estuaries, or ocean enclosures, can increase food supplies and expand income opportunities for rural communities. Aquaculture already accounts for approximately half of all fish consumption worldwide. When sustainable practices are employed, aquaculture has less environmental impact and emits fewer greenhouse gasses than the production of other animal source foods, such as beef or poultry. Coastal fishers should also have access to the equipment and supplies needed to succeed in the face of declining marine fish populations.

Reduce food loss and waste

Lastly, it will be important to ensure that the food produced is fully utilised. At present, nearly a fifth of all food commodities are lost or wasted worldwide (United Nations Environmental Programme, 2021, p. 8). Improved processing, preservation, and storage technologies can offset declining agricultural output and reduce stress on natural environments. Farmers and fishers can reduce post-harvest loss through enhanced drying and threshing equipment, processing facilities, and refrigeration units. Increased availability of hermetic storage containers, metallic silos, and polyethylene packaging bags will help prevent the spread of insects, mould, and moisture (International Fund for Agricultural Development, 2019a).

Conclusion

Climate change clearly threatens sustainable diets worldwide. This chapter outlined the multiple ways in which a warmer earth is adversely impacting food production and consumption, especially in the world's poorest countries. There is little question that climate change is increasing land degradation, water scarcity, extreme weather events, insect infestations, the spread of crop diseases, and the decline of fish populations. The world's poor, who contribute the least to global greenhouse gas emissions, endure the greatest impacts of climate change.

It will be important to reduce the greenhouse gas emissions that are produced by food systems and increase the absorption of gasses already emitted. Agroecology and agroforestry can significantly reduce emissions from agriculture and contribute to carbon sequestration. It will also be important to ensure agricultural systems and rural communities can manage the adverse impacts of climate change. Adaptation measures should include disaster risk management, the introduction of more resilient seed varieties, the protection of land and water resources, support for aquaculture and coastal fishing communities, and improved food preservation.

If current population and consumption trends continue, food production will need to increase by 60 percent by the mid-point of this century. It will not be possible to meet global nutritional needs without major investments in climate change mitigation and adaptation. Priority should be placed on aiding those countries and communities in the Global South that are the most food insecure. Limited resources in the world's poorest countries, the interdependence of food systems, and the global nature of climatic threats necessitate a coordinated response by the international community.

Notes

1 The projected impact of climate change on natural resources in the Global South is outlined in the multiple reports of the Intergovernmental Panel of Experts on Climate Change, especially 2019a and 2019b. See also Food and Agriculture Organisation, 2020a and International Fund for Agricultural Development, 2019b.

2 Children are often the first to suffer from food insecurity. Nearly half of child mortality in the Global South is due to malnutrition or malnutrition related diseases (Food and Agriculture Organisation, 2020e, p. 31.) See also Mayer and Anderson (eds.), 2020.

3 Approximately 1.9 billion people in the world suffer from moderate or severe food insecurity. Food and Agriculture Organisation, 2020e, p. 9. The Food Security Information Network (FSIN) reports that at least 100 million people have such high levels of food insecurity that they require urgent humanitarian action (Food Security Information Network, 2020, p. 2).

4 The World Food Programme defines food systems as "interlocking networks of relationships that encompass the entire range of functions and activities involved in the production, processing, marketing, consumption, and disposal of goods that originate from agriculture, forestry, or fisheries" (World Food Programme, 2017, p. 23).

5 Desertification refers to a type of land degradation in which the biological productivity of soils is lost and fertile areas become increasingly arid. Desertification, which is caused by both natural processes and human activities, especially threatens dryland ecosystems.

6 At present, over two billion people live in water-stressed areas and this number is expected to grow to half of the world's population by 2030 (United Nations Educational, Scientific, and Cultural Organisation, 2021, vi).

7 Over the past decade, nearly all regions of Iran have endured some level of drought. The drought that occurred in the Levant (especially Jordan, Lebanon, the Palestinian Territories, and Syria) between 1998 and 2012 is thought to have been the worst drought in the past nine centuries in this region.

8 Glaciers in Bolivia, Chile, Colombia, Ecuador, and Peru have lost between 20 and 50 percent of their surface area. The three major ice fields in the Patagonian Andes have also declined by one-third.

9 See Food and Agriculture Organisation, 2019b, pp. 50–53 and Food and Agriculture Organisation, 2019c, p. 36.

10 The countries most severely impacted by the locust infestations include Djibouti, Eritrea, Ethiopia, Kenya, Somalia, South Sudan, Sudan, Tanzania, and Uganda.

11 The Food and Agriculture Organisation advocates Climate Smart Agriculture (CSA), which includes many of the basic principles of agroecology. FAO defines CSA as "an integrated approach to managing cropland, livestock, forests, and fisheries that addresses the interlinked challenges of food security and climate change." Food and Agriculture Organisation, 2019a, p. 55). See also Food and Agriculture Organisation, 2020d.

12 The Sendai Framework for Disaster Risk Reduction was adopted by the United Nations in 2015 to promote global cooperation in this area. The United Nations Office of Disaster Risk Reduction (UNDRR) plays a lead role in coordinating international efforts. The World Bank also established a Global Facility for Disaster Reduction and Recovery.

References

Altieri, M., & Nicholls, C. L. (2021). *Agroecology for sustainable and resilient farming*. Routledge.

Congressional Research Service. (2019). *Cyclones Idai and Kenneth in Southeastern Africa: Humanitarian and recovery response in brief*. Congressional Research Service.

Food and Agriculture Organization. (2018). *State of food security and nutrition in the world 2018: Building climate resilience for food security and nutrition*. Food and Agriculture Organization.

Food and Agriculture Organization. (2019a). *Africa: Regional overview of food security and nutrition 2018: Addressing the threat from climate variability and extremes for food security and nutrition*. Food and Agriculture Organization.

Food and Agriculture Organization. (2019b). *Asia and the Pacific: Regional overview of food security and nutrition 2019: Placing nutrition at the centre of social protection*. Food and Agriculture Organization.

Food and Agriculture Organization. (2019c). *Latin America and the Caribbean: Regional overview of food security and nutrition 2018: Inequality and food systems*. Food and Agriculture Organization.

Food and Agriculture Organization. (2020a). *Climate change and food security*. Food and Agriculture Organization.

Food and Agriculture Organization. (2020b). *Global information and early warning system*. Food and Agriculture Organization.

Food and Agriculture Organization. (2020c). *Incorporating climate considerations into agricultural investment programmes*. Food and Agriculture Organization.

Food and Agriculture Organization. (2020d). *Latin America and the Caribbean: Regional overview of food security and nutrition 2019: Towards healthier food environments that address all forms of malnutrition*. Food and Agriculture Organization.

Food and Agriculture Organization. (2020e). *State of world fisheries and aquaculture 2020*. Food and Agriculture Organization.

Food Security Information Network. (2020). *2020 Global report on the food crisis: Joint analysis for better decisions*. Food Security Information Network.

Giraldo, O. F. (2019). *Political ecology of agriculture: Agroecology and post-development*. Springer Nature.

Intergovernmental Panel on Climate Change. (2018). *Global warming of 1.5 C*. Intergovernmental Panel on Climate Change.

Intergovernmental Panel on Climate Change. (2019a). *Climate change and land*. Intergovernmental Panel on Climate Change.

Intergovernmental Panel on Climate Change. (2019b). *The Ocean and cryosphere in a changing climate*. Intergovernmental Panel on Climate Change.

International Fund for Agricultural Development. (2019a). *Climate action report 2019*. International Fund for Agricultural Development.

International Fund for Agricultural Development. (2019b). *The food loss reduction advantage: Building sustainable food systems*. International Fund for Agricultural Development.

Mayer, T., & Anderson, M. D. (Eds.). (2020). *Food insecurity: A matter of justice, sovereignty, and survival*. Routledge.

Mossi, A. J., Petry, C., & Wilson, F. (Eds.). (2020). *Agroecology: Insights, experiences and perspectives*. Nova Science Publishers.

Pimbert, M. P. (Ed.). (2018). *Food sovereignty, agroecology and biocultural diversity: Constructing and contesting knowledge*. Routledge.

Smith, M. R., & Myers, S. S. (2018). Impact of anthropogenic CO2 emissions on global human nutrition, *Nature Climate Change*, 8, 834–839.

Tubiello, F. N., Rosenzweig, C., Conchedda, G., Karl, K., Gütschow, J., Xueyao, P., Obli-Laryea, G., Wanner, N., Qiu, S. Y., De Barros, J., Flammini, A., & Sandalow, D. (2021). Greenhouse gas emissions from food systems: Building the evidence base. *Environmental Research Letters*, *16*(6), 065007.

United Nations Educational, Scientific and Cultural Organization. (2021). *The United Nations world water report 2021: Valuing water*. United Nations Educational, Scientific and Cultural Organization.

United Nations Environmental Programme. (2021). *UNEP food waste index report 2021*. United Nations Environmental Programme.

World Food Programme. (2017). *World food assistance 2017: Taking stock and looking ahead*. World Food Programme.

World Food Programme. (2019). *Climate risk financing: Early response and anticipatory actions for climate hazards*. World Food Programme.

World Resources Institute. (2018). *Creating a sustainable food future: Synthesis report*. World Resources Institute.

6

THE SIGNIFICANCE OF AGROBIODIVERSITY FOR SUSTAINABLE DIETS

Roland Ebel, Carmen Byker Shanks, Georges Félix, and José Pablo Morales-Payán

Agrobiodiversity

Agrobiodiversity is the result of evolution and thousands of years of adaptive agricultural management (Padmanabhan, 2011; Thaman, 2014). Diversified farming systems (referring to elevated bio-, landscape, management, and use diversity) mimic the biodiversity and functioning of local ecosystems (Altieri & Nicholls, 2020), generating multiple ecosystem services, including increased resource efficiency, elevated resilience to environmental disturbances, increased per-plant yield, and sequestration of atmospheric carbon (Zimmerer et al., 2019). Agrobiodiversity comprises all domesticated genetic resources used to obtain agricultural products, the species that support the functioning of agroecosystems, as well as wild foods and other products gathered within subsistence systems (Herforth et al., 2019; Howard, 2010). Agrobiodiversity is inherently linked to food production and consumption, and maintaining food production has been a core incentive for supporting this diversity (Veteto & Skarbø, 2009). Equally, agrobiodiversity sustains non-materialist values in agricultural communities, mutually enhancing agriculture, such as, traditional ecological knowledge and core elements of Indigenous languages (Zimmerer et al., 2019).

However, the expansion of global markets and industrialised farming, aiming for product homogeneity and non-seasonality, has monotonised farming and food systems with implications for diets. The expansion of industrialised food has caused the paradoxical circumstance that today, over two billion people lack food security, and, simultaneously, another two billion people contend with an over-abundance of food calories and are overweight or obese (Egal, 2019).

Farm system and diet simplification

With the onset of the third agricultural revolution (early 20th century in industrialised countries, globally exported as the "Green Revolution" after World War II),

DOI: 10.4324/9781003174417-8

characterised by mechanisation and homogenisation of farming systems, the species and intraspecific diversity of crops and agriculturally-used animals decreased considerably (Howard, 2010). While the total number of domesticated plant species is estimated at 7000, today, only 30 crop species provide more than 90% of plant-based calories and nutrients consumed by humans. A similar genetic erosion is observed for the approximately 100 domesticated vertebrate species (Hammer et al., 2003). The homogenisation of farming systems not only affects genetic diversity but also the socio-cultural resilience of farming communities, changing social structures, local culture, language, and cuisine (Ebel et al., 2021).

The expansion of a few high-yielding staples at the expense of locally adapted, diverse farming systems, and the almost global availability of (often inexpensive) industrial food led to dietary homogenisation and intensified the consumption of energy-dense and nutrient-poor food (Johns & Eyzaguirre, 2006). The ongoing loss of intraspecific agrobiodiversity is further responsible for poor nutrition of those depending on staples (Medhammar et al., 2012).

In contrast, agrobiodiversity often enhances balanced and healthy diets, primarily by providing elevated dietary diversity and food abundant in micronutrients and minerals (Arimond et al., 2010; Campbell et al., 2011; Whitmee et al., 2015) and decreasing malnutrition (Frison et al., 2011; Tucker, 2001). Studies from the United States (Conklin et al., 2016), China (Zhang et al., 2017), Vietnam (Ogle et al., 2001), the Philippines (Cabalda et al., 2011), India (Nithya & Bhavani, 2018), Australia (Livingstone & McNaughton, 2016), Italy (Fernandez et al., 2000), and Mali (Torheim et al., 2004) demonstrate a direct link between dietary diversity and nutrient intake. Asian and Mediterranean cuisines provide examples of diets that emerged from traditional farming systems and prevent chronic non-communicable diseases (Dans et al., 2011; Trichopoulou & Vasilopoulou, 2000).

Contemporary counter-movements to farm systems and consequent diet simplification have emerged: The Slow Food movement (as discussed elsewhere in this handbook) involves small farmers in efforts to save endangered varieties, breeds, and foods (Lotti, 2010). Agritourism contributes to a revival of farmer cuisine and the conservation of intraspecific crop and livestock diversity (Jäger et al., 2019). While selected traditional crops of agrobiodiverse systems have become popular "superfoods," like quinoa or chia seeds, hundreds of plants capable of contributing to agroecosystem, cultural, and dietary diversity, remain neglected and underutilised (Mayes et al., 2011).

Consequences of diet simplification

Especially in subsistence farming communities (still dominant in most low- and mid-income countries and cultivating over 80% of the world's agricultural land), traditional, agrobiodiverse farming practices, adapted to the local environment and culture, have been maintained (Lowder et al., 2016; Mazoyer & Roudart, 2006). Yet, care must be taken with categorisation: For example, conventional farmers in semi-arid West Africa are often characterised by low dependence on off-farm inputs, diversified production, and low yields, similar to traditional farms.

Additionally, numerous rural residents of low- and mid-income countries depend on hunting and gathering (Malézieux, 2012). The fact that most rural dwellers in these countries remain involved in traditional agriculture does not necessarily translate into sustainable diets (Morgan & Trubek, 2020). First, climate change, among other external factors, and the loss of traditional farming knowledge affect the diversity and productivity of traditional farming systems (Ebel, 2020). Second, the global nutrition transition towards consumption of highly- and ultra-processed food is on the rise in these countries (Popkin, 1994).

One consequence of the nutrition transition is micronutrient deficiency. Iron deficiency, especially, is a leading risk for disability and death worldwide (Zimmermann & Hurrell, 2007). In South Asia and Sahelian Africa, for example, millets are important components of traditional diets and, if correctly prepared, a good source of dietary iron but are increasingly substituted by rice (*Oryza sativa*) (Johns & Eyzaguirre, 2006). Likewise, the consumption of chaya (*Cnidoscolus aconitifolius*), a semi-perennial Mesoamerican euphorbiaceous and essential component of the traditional Maya cuisine, is being reduced at the expense of greens such as spinach, which contains less than a third of the iron (Ebel et al., 2019).

Another consequence of the nutrition transition is the global pandemic of diet-related non-communicable diseases, especially heart disease and diabetes (Forouhi & Unwin, 2019). Research points to a clear relationship between the development of non-communicable diseases and the consumption of highly processed food (Lawrence & Baker, 2019; Srour et al., 2019). Evidence from India (Misra et al., 2011) and Egypt (Galal, 2002) underlines the negative health consequences of an elevated intake of meat and staples (Kearney, 2010), especially rice, wheat (*Triticum aestivum*), corn (*Zea mays*), and soy (*Glycine max*). Common staples in traditional farming systems such as buckwheat (*Fagopyrum esculentum*) and finger millet (*Eleusine coracana*), instead, reduce the risk of obesity, hypertension, heart disease, and diabetes (Johns, 2007).

Burkina Faso is an exemplary low-income country shaped by early-stage farm system simplification and nutrition transition. Especially in their urban centres, staples, including cowpea (*Vigna unguiculata*), sorghum, millets, and maize, have been replaced by rice and bread. In rural areas of Burkina Faso instead, traditional diets still feature foraged nutrient-rich ingredients that complement the sauces of traditional dishes, such as leaves or fruits of perennials like cotton silk tree (*Ceiba pentandra*), baobab (*Adansonia digitata*), and shea nut tree (*Vitellaria paradoxa*). Typically, shea nut trees grow alongside sorghum and millet during the rainy season, providing households with cooking oil. Although the use of shea nut butter in rural diets is widespread, an ongoing decrease of trees within and around farming systems has induced a decreased availability of this source of sustainable diets in Burkina Faso (Félix et al., 2018).

Farm system simplification and nutrition transition are more advanced in a mid-income country like Puerto Rico, where a reduction in locally available and produced food, together with industrialisation and the rise of a service economy, have significantly changed the diet. For example, starchy crops such as cassava (*Manihot esculenta*) and breadfruit (*Artocarpus altilis*) played an essential role in 19th century Puerto Rican agriculture and cuisine (Álvarez-Febles & Félix, 2020). Cassava, traditionally intercropped with maize and legumes, was part of the ancestral crops of the pre-Columbian Taino people. Breadfruit is native to the South Pacific and was imported to the Caribbean in 1793, where its use spread rapidly, being a good source of (gluten-free) carbohydrates and essential amino and fatty acids (Golden & Williams, 2001). Still in the early 20th century, breadfruit was a common wild and backyard tree (Roberts-Nkrumah, 2007). Subsequently, Puerto Rican food policies and businesses promoted the consumption of potato (*Solanum tuberosum*) to the detriment of "weaker" carbohydrate sources such as cassava and breadfruit. The rise of imported potato largely replaced locally-grown and purportedly less stylish cassava and breadfruit. By the end of the 20th century, commercial cassava production nearly disappeared on the island, and today, most of the cassava consumed is imported (Pagán & Sagebien, 2009). Today, imported products, often highly processed, generally dominate the shelves of Puerto Rican supermarkets. Accordingly, none of the ingredients of "rice-and-beans" (white rice, red kidney beans), daily food for most families, is produced on the island (FAO 2021).

In high-income countries, diets have widely shifted away from biodiverse sources to an increased intake of sugar, saturated fats, sodium, and artificial ingredients in ultra-processed food (Afshin et al., 2019; Monteiro et al., 2016). For example, 71% of the packaged food supply in the United States and over 60% of calories consumed by Americans are ultra-processed, while 42% of adults are obese (Martínez-Steele et al., 2016). The global pandemic of diet-related non-communicable diseases is creating an increasing financial burden on healthcare budgets globally, affecting nations' welfare (Muka et al., 2015).

Countercultures and movements

Peasant movements

Existing imbalances between peasants and corporations over resource control have motivated farmers across the world to organise at different scales and raise their voices against food injustice, food system simplification, and consolidation (Alonso-Fradejas et al., 2015). Locally, farmer-to-farmer seed exchange initiatives have been an entry point for peasants to share knowledge, genetic material, food, and recipes. Exchanged landraces are often accompanied by the transmission of their culinary uses, for example processing of different corn landraces in Mexico (Jenatton & Morales, 2019) or use of sorghum varieties for specific porridge or beer recipes in West Africa (Le Garff, 2016). Similar informal knowledge exchange occurs at the Organización Boricuá's agroecological solidarity work brigades in Puerto Rico, where farmers meet, discuss, and share food (McCune et al., 2019). The Zero Budget Natural Farming peasant movement in the Indian state of Karnataka promotes and organises farmer-to-farmer exchanges at the state level (Khadse et al., 2018).

At the national scale, the Rural Women's Assembly in South Africa fights the loss of biodiversity and nutritious diets (RWA, 2019). Together with the Alliance for Food Sovereignty in Africa (AFSA), it has established a program to restore "orphan crops" and "seed sovereignty" (AFSA, 2021). The Agroecology Movement of the National Association of Small Farmers (ANAP) in Cuba is an example of an institutionalised, national peasant organisation (Rosset et al., 2011). At the regional scale, the Movimiento Agroecológico Latinoamericano has built on traditional diets to reinforce the need for seed conservation (Altieri, 2015). La Via Campesina (LVC) is the largest international peasant movement, emphasising relationships between food sovereignty, agroecosystem management, local cultural practices, and the importance of balanced diets (LVC, 2015).

Slow food and farm-to-table movements

Food consumption shapes food production at least as much as the reverse (Morgan & Trubek, 2020), and some of the most important dietary and farm diversity movements in high-income countries originated from the nutrition side. The most prominent example is the Slow Food Movement (Chapter 3), a global grassroots organisation that emerged in 1989 as a spontaneous reaction to the opening of a fast-food restaurant in central Rome (van der Meulen, 2004). Slow Food favours agrobiodiversity (Petrini, 2003), opposes the standardisation and homogenisation of the food system, and advocates for more time for the acquisition, preparation, consumption, and sharing of food (Wexler et al., 2017). While the movement understands itself as a networking organisation of small farmers, artisanal food producers, peasant communities, and anti-globalisation activists, critics argue that it occasionally operates more like a gourmet eating club (Carruth, 2017), ignoring the social reality of biodiverse farmers (Bowens, 2015). Other critics attest nostalgia for "the kitchen in the good old days," disregarding contemporary gender roles and the financial and time resources of low- and mid-income people (Wexler et al., 2017).

Following the idea of place-based food production and consumption, similar movements have evolved across the Western World. In North America, the farm-to-table (also farm-to-fork) movement has become popular, promoting local food at restaurants and public kitchens through direct sales, community-supported agriculture, farmers' markets, or by restaurants or schools producing their own food. These movements urge consumers to trace the origin of their food (Coplen, 2018). Movements aiming for agrobiodiversity often have a focus on food sovereignty, such as the US Food Sovereignty Alliance (USFSA) and Food First (USFSA 2021). The Earthseed Land Collective (2021) emphasises equity and ethnic diversity. The Native American Food Sovereignty Alliance (2021) defends Native American culture, cuisine, food production, and gathering. Similar movements have emerged in Europe, including the French Alternatives Agroécologiques et Solidaires (2021).

Superfoods

The term *superfood* is a construct for marketing foods of exotic origin claimed to preserve or restore health, wellness, and youthfulness. These foods are usually not exotic in their centres of origin or where they have been cultivated for a long time. The "exotic" component is more important to foreign markets, where superfoods are difficult or impossible to grow because of climate or expertise; and being exotic/imported, high prices can be commended from health-conscious high-income consumers. Also, superfoods are generally marketed with claims (real or not) that they are grown in environmentally friendly systems and that growers are being well-compensated (Loyer & Knight, 2018).

A plethora of fruits, vegetables, herbs, seeds, and algae have a place in the superfood group. Examples include quinoa (*Chenopodium quinoa*), chia (*Salvia hispanica*), maca (*Lepidium meyenii*), goji (*Lycium* spp), amaranth (*Amaranthus* spp), kale (*Brassica oleracea var. acephala*), moringa (*Moringa oleifera*) and other legumes, curcuma (*Curcuma longa*), açai (*Euterpe oleracea*), avocado (*Persea americana*), blueberry (*Vaccinium corymbosum*), pomegranate (*Punica granatum*), and breadfruit (Ashfaq et al., 2012; Cassiday, 2017; Gupta & Mishra, 2021). While some of these plants are exotic for Western culture (e.g. goji), others have been known for centuries (e.g. kale, pomegranate, avocado) or had geographically limited markets (e.g. chia, quinoa) until becoming superfoods abroad. Because of high profits, new superfoods are added to the list regularly (Peña-Lévano et al., 2021). Most superfoods contribute to diet in the form of vitamins, minerals, amino acids, unsaturated fatty acids, saponins, phenols, flavonoids, other nutrients, antioxidants, or functional compounds (Scrinis, 2012). However, there are no firm scientific criteria for the designation as a superfood (Magrach & Sanz, 2020).

In their traditional environments, most superfoods have been grown in subsistent and small commercial systems, using sustainable practices (crop rotations and associations, and reduced reliance on synthetic inputs such as pesticides) and consumed by the family, the community, or small domestic markets (Loyer & Knight, 2018). Today, superfoods such as quinoa and açai in South America and goji in Himalayan countries have become a big business (Ashfaq et al., 2012). In Puerto Rico, breadfruit, neglected for decades, is now experiencing a revival as a superfood thanks to its high nutritional value, good climatic adaptation, and attractive market value. The growing interest in breadfruit has prompted research to develop new varieties and sustainable management practices (Morales-Payan, 2014, 2016).

As superfoods become globally popular, the market pushes growers to increase their output, which is done by expanding the area under production, eliminating crop rotations and intercropping, and using synthetic fertilisers and pesticides. The result is disruption of the traditional sustainable production and a consequent negative impact on the environment and the diversity of foods available for the local population. Rising prices often detain the local population from buying superfoods (Peña-Lévano et al., 2021). Superfoods even replace nutritional diets: In India, the importation of quinoa is threatening traditional diets based on millets despite having a similar nutritional content (Dhaka & Prasad, 2020).

The commercial success of breadfruit, quinoa, chia, and others, indicates that achieving the status of superfood may help preserve their genetic diversity and contribution to local diets. Nevertheless, there is the danger that careless expansion of area under production, monoculture, and reckless use of inputs for the cultivation of species labelled as superfoods may have dire environmental effects and/or jeopardise other plant species considered less profitable, eroding local food sources for the sake of exporting superfoods.

Outlook and conclusion

Restoring agrobiodiversity is a powerful strategy for securing sustainable diets rooted in the cultural environment of their eaters. We presented several examples of how agrobiodiversity decline has caused poor diets coupled with negative health outcomes, and how engaged farmers and consumers in different parts of the world are fighting for a comeback of diversified farms and meals. Currently, these efforts are widely limited to privileged consumers in high-income countries and subsistent farmers in lower-income countries who are struggling with the political and financial power of the agri-food industry. Given the multiple benefits of diversified farming and eating—from sustainable agroecosystems to healthy diets—and making use of the achievements of the 21st century, we call for policy, science, and education efforts to make diversity what it used to be before monotonisation: The mainstreaming of the food system.

References

Afshin, A., Sur, P. J., Fay, K. A., Cornaby, L., Ferrara, G., Salama, J. S., & Mullany, E. C. (2019). Health effects of dietary risks in 195 countries, 1990–2017: A systematic analysis for the Global Burden of Disease Study 2017. *The Lancet*, *393*(10184), 1958–1972.

Alliance for Food Sovereignty in Africa [AFSA]. (2021). *Case studies of seed*. Retrieved from https://afsafrica.org/case-studies-seed/

Alonso-Fradejas, A., Borras, S. M., Holmes, T., Holt-Giménez, E., & Robbins, M. J. (2015). Food sovereignty: Convergence and contradictions, conditions and challenges. *Third World Quarterly*, *36*(3), 431–448.

Alternatives Agroécologiques et Solidaires. (2021). *Qui sommes nous?* Retrieved from https://www.sol-asso.fr/en/

Altieri, M. A. (2015). Breve reseña sobre los orígenes y evolución de la Agroecología en América Latina. *Agroecología*, *10*(2), 7–8.

Altieri, M. A., & Nicholls, C. I. (2020). Agroecology and the reconstruction of a post-COVID-19 agriculture. *The Journal of Peasant Studies*, *47*(5), 881–898.

Álvarez-Febles, N., & Félix, G. F. (2020). Hurricane María, agroecology, and climate change resiliency. In B. Tokar & T. Gilbertson (Eds.), *Climate justice and community renewal: Resistance and grassroots solutions* (pp. 131–146). Routledge.

Arimond, M., Wiesmann, D., Becquey, E., Carriquiry, A., Daniels, M. C., Deitchler, M., Fanou-Fogny, N., Joseph, M. L., Kennedy, G., Martin-Prevel, Y., & Torheim, L. E. (2010). Simple food group diversity indicators predict micronutrient adequacy of women's diets in 5 diverse, resource-poor settings. *The Journal of Nutrition*, *140*(11), 2059S–2069S.

Ashfaq, M., Basra, S. M. A., & Ashfaq, U. (2012). Moringa: A miracle plant for agro-forestry. *Journal of Agriculture and Social Sciences*, *8*(3), 115–122.

Bowens, N. (2015). *The color of food: Stories of race, resilience and farming*. New Society Publishers.

Cabalda, A. B., Rayco-Solon, P., Solon, J. A. A., & Solon, F. S. (2011). Home gardening is associated with Filipino preschool children's dietary diversity. *Journal of the American Dietetic Association*, *111*(5), 711–715.

Campbell, K., Cooper, D., Dias, B., Prieur-Richard, A.-H., Campbell-Lendrum, D., Karesh, W. B., & Daszak, P. (2011). Strengthening international cooperation for health and bio-diversity. *EcoHealth, 8*(4), 407–409.

Carruth, A. (2017). Slow food, low tech: Environmental narratives of agribusiness and its alternatives. In U. Heise, J. Christensen, & M. Niemann (Eds.), *The Routledge companion to the environmental humanities* (pp. 329–338). Routledge.

Cassiday, L. (2017). Chia: Superfood or superfad? *American Oil Chemists' Society Inform, 28*(1), 6–13.

Conklin, A. I., Monsivais, P., Khaw, K.-T., Wareham, N. J., & Forouhi, N. G. (2016). Dietary diversity, diet cost, and incidence of Type 2 Diabetes in the United Kingdom: A prospec-tive cohort study. *PLOS Medicine, 13*(7), e1002085.

Coplen, A. K. (2018). The labor between farm and table: Cultivating an urban political ecol-ogy of agrifood for the 21st century. *Geography Compass, 12*(5), e12370.

Dans, A., Ng, N., Varghese, C., Tai, E. S., Firestone, R., & Bonita, R. (2011). The rise of chronic non-communicable diseases in southeast Asia: Time for action. *The Lancet, 377*(9766), 680–689.

Dhaka, A., & Prasad, M. (2020). Imported superfood quinoa versus Indian nutricereal mil-lets. *Current Science, 118*(11), 1646–1649.

Earthseed Land Collective. (2021). *Our collective.* Retrieved from https://earthseedlandcoop .org/about/

Ebel, R. (2020). Are small farms sustainable by nature?: Review of an ongoing misunder-standing in agroecology. *Challenges in Sustainability, 8*(1), 17–29.

Ebel, R., Aguilar, M. d. J. M., Cocom, J. A. C., & Kissman, S. (2019). Genetic diversity in nutritious leafy green vegetable—Chaya (Cnidoscolus aconitifolius). In D. Nandwani (Ed.), *Genetic diversity in horticultural plants* (pp. 161–189). Springer.

Ebel, R., Menalled, F., Ahmed, S., Gingrich, S., Baldinelli, G. M., & Félix, G. (2021). How biodiversity loss affects society. In S. Harvey (Ed.), *Handbook on the human impact of agri-culture* (pp. 352–376). Edward Elgar Publishing.

Egal, F. (2019). Review of the state of food security and nutrition in the world, 2019. *World Nutrition, 10*(3), 95–97.

Félix, G. F., Diedhiou, I., Le Garff, M., Timmermann, C., Clermont-Dauphin, C., Cournac, L., Groot, J. C., & Tittonell, P. (2018). Use and management of biodiversity by small-holder farmers in semi-arid West Africa. *Global Food Security, 18*, 76–85.

Fernandez, E., Negri, E., La Vecchia, C., & Franceschi, S. (2000). Diet diversity and colorectal cancer. *Preventive Medicine, 31*(1), 11–14.

Food and Agriculture Organization [FAO]. (2021). *Puerto Rico FAOSTAT.* Retrieved from http://www.fao.org/faostat/en/#data/QC

Forouhi, N. G., & Unwin, N. (2019). Global diet and health: Old questions, fresh evidence, and new horizons. *The Lancet, 393*(10184), 1916–1918.

Frison, E. A., Cherfas, J., & Hodgkin, T. (2011). Agricultural biodiversity is essential for a sustainable improvement in food and nutrition security. *Sustainability, 3*(1), 238–253.

Galal, O. M. (2002). The nutrition transition in Egypt: Obesity, undernutrition and the food consumption context. *Public Health Nutrition, 5*(1a), 141–148.

Golden, K. D., & Williams, O. J. (2001). Amino acid, fatty acid, and carbohydrate content of Artocarpus altilis (Breadfruit). *Journal of Chromatographic Science, 39*(6), 243–250.

Gupta, E., & Mishra, P. (2021). Functional food with some health benefits, so called super-food: A review. *Current Nutrition & Food Science, 17*(2), 144–166.

Hammer, K., Arrowsmith, N., & Gladis, T. (2003). Agrobiodiversity with emphasis on plant genetic resources. *Naturwissenschaften, 90*(6), 241–250.

Herforth, A., Johns, T., Creed-Kanashiro, H. M., Jones, A. D., Khoury, C. K., Lang, T., Maundu, P., Powell, B., & Reyes-García, V. (2019). Agrobiodiversity and Feeding the World: More of the same will result in more of the same. In K. S. Zimmerer & S. de Haan (Eds.), *Agrobiodiversity: Integrating Knowledge for a Sustainable Future* (pp. 185–210). MIT Press.

Howard, P. L. (2010). Culture and agrobiodiversity: Understanding the links. In S. Pilgrim & J. N. Pretty (Eds.), *Nature and Culture: Rebuilding Lost Connections*. Routledge.

Jäger, M., van Loosen, I., & Giuliani, A. (2019). *How Have Markets Affected the Governance of Agrobiodiversity?* MIT Press.

Jenatton, M., & Morales, H. (2019). Civilized cola and peasant pozol: Young people's social representations of a traditional maize beverage and soft drinks within food systems of Chiapas, Mexico. *Agroecology and Sustainable Food Systems, 44*(8), 1–37.

Johns, T. (2007). Agrobiodiversity, diet, and human health. In D. I. Jarvis, C. Padoch, & H. D. Cooper (Eds.), *Managing biodiversity in agricultural ecosystems* (pp. 382–406). Columbia University Press.

Johns, T., & Eyzaguirre, P. B. (2006). Linking biodiversity, diet and health in policy and practice. *Proceedings of the Nutrition Society, 65*(2), 182–189.

Kearney, J. (2010). Food consumption trends and drivers. *Philosophical Transactions of the Royal Society B: Biological Sciences, 365*(1554), 2793–2807.

Khadse, A., Rosset, P. M., Morales, H., & Ferguson, B. G. (2018). Taking agroecology to scale: The Zero Budget Natural Farming peasant movement in Karnataka, India. *The Journal of Peasant Studies, 45*(1), 192–219.

La Via Campesina [LVC]. (2015). Nutrition and food sovereignty. *Nyéléni Newsletter, 22*, 1–6.

Lawrence, M. A., & Baker, P. I. (2019). Ultra-processed food and adverse health outcomes. *British Medical Journal, 365*, l2289.

Le Garff, M. (2016). *Nutritional functional diversity in farmer households: Case study from semi-arid Burkina Faso*. Wageningen University & Research. Retrieved from https://edepot.wur.nl/396180

Livingstone, K. M., & McNaughton, S. A. (2016). Diet quality is associated with obesity and hypertension in Australian adults: A cross-sectional study. *BioMed Central Public Health, 16*(1), 1037.

Lotti, A. (2010). The commoditization of products and taste: Slow food and the conservation of agrobiodiversity. *Agriculture and Human Values, 27*(1), 71–83.

Lowder, S. K., Skoet, J., & Raney, T. (2016). The number, size, and distribution of farms, smallholder farms, and family farms worldwide. *World Development, 87*, 16–29.

Loyer, J., & Knight, C. (2018). Selling the "Inca superfood": Nutritional primitivism in superfoods books and maca marketing. *Food, Culture & Society, 21*(4), 449–467.

Magrach, A., & Sanz, M. J. (2020). Environmental and social consequences of the increase in the demand for 'superfoods' worldwide. *People and Nature, 2*(2), 267–278.

Malézieux, E. (2012). Designing cropping systems from nature. *Agronomy for sustainable development, 32*(1), 15–29.

Martínez-Steele, E., Baraldi, L. G., Louzada, M. L. d. C., Moubarac, J.-C., Mozaffarian, D., & Monteiro, C. A. (2016). Ultra-processed foods and added sugars in the US diet: Evidence from a nationally representative cross-sectional study. *British Medical Journal Open, 6*(3), e009892.

Mayes, S., Massawe, F. J., Alderson, P. G., Roberts, J. A., Azam-Ali, S. N., & Hermann, M. (2011). The potential for underutilized crops to improve security of food production. *Journal of Experimental Botany, 63*(3), 1075–1079.

Mazoyer, M., & Roudart, L. (2006). *A history of world agriculture: From the neolithic age to the current crisis*. NYU Press.

McCune, N., Perfecto, I., Avilés-Vázquez, K., Vázquez-Negrón, J., & Vandermeer, J. (2019). Peasant balances and agroecological scaling in Puerto Rican coffee farming. *Agroecology and Sustainable Food Systems, 43*(7–8), 810–826.

Medhammar, E., Wijesinha-Bettoni, R., Stadlmayr, B., Nilsson, E., Charrondiere, U. R., & Burlingame, B. (2012). Composition of milk from minor dairy animals and buffalo breeds: A biodiversity perspective. *Journal of the Science of Food and Agriculture, 92*(3), 445–474.

Misra, A., Singhal, N., Sivakumar, B., Bhagat, N., Jaiswal, A., & Khurana, L. (2011). Nutrition transition in India: Secular trends in dietary intake and their relationship to diet-related non-communicable diseases. *Journal of Diabetes, 3*(4), 278–292.

Monteiro, C. A., Cannon, G., Levy, R., Moubarac, J. C., Jaime, P., Martins, A. P., Canella, D., Louzada, M., & Parra, D. (2016). NOVA. The star shines bright. *World Nutrition, 7*(1–3), 28–38.

Morales-Payan, J. P. (2014). Propagation of the fruit crop Artocarpus altilis by root cuttings of various lengths and diameters. *HortScience, 49*(9S), S330.

Morales-Payan, J. P. (2016). Effects of a phytostimulant amino acid formulation on breadfruit plants in the nursery. *Proceedings of the Caribbean Food Crops Society Annual Scientific Meeting, 52*, 154–155.

Morgan, C. B., & Trubek, A. B. (2020). Not yet at the table: The absence of food culture and tradition in agroecology literature. *Elementa: Science of the Anthropocene, 8*, 40.

Muka, T., Imo, D., Jaspers, L., Colpani, V., Chaker, L., van der Lee, S. J., Mendis, S., Chowdhury, R., Bramer, W. M., Falla, A., & Pazoki, R. (2015). The global impact of non-communicable diseases on healthcare spending and national income: A systematic review. *European Journal of Epidemiology, 30*(4), 251–277.

Native American Food Sovereignty Alliance. (2021). About us. Retrieved from https://nativefoodalliance.org/our-work-2/about-us/

Nithya, D. J., & Bhavani, R. V. (2018). Dietary diversity and its relationship with nutritional status among adolescents and adults in rural India. *Journal of Biosocial Science, 50*(3), 397–413.

Ogle, B. M., Hung, P. H., & Tuyet, H. T. (2001). Significance of wild vegetables in micronutrient intakes of women in Vietnam: An analysis of food variety. *Asia Pacific Journal of Clinical Nutrition, 10*(1), 21–30.

Padmanabhan, M. (2011). Women and men as conservers, users and managers of agrobiodiversity: A feminist social–ecological approach. *The Journal of Socio-Economics, 40*(6), 968–976.

Pagán, M. C., & Sagebien, J. (2009). Robustecer la cadena de suministro ante el cambio climático. *Harvard Business Review, 87*(10), 34–47.

Peña-Lévano, L., Adams, C., & Burney, S. (2021). Latin America's superfood economy: Producing and marketing Açaí, Chia Seeds, and Maca Root. *Choices, 35*(4), 1–6.

Petrini, C. (2003). *Slow food: The case for taste.* Columbia University Press.

Popkin, B. M. (1994). The nutrition transition in low-income countries: An emerging crisis. *Nutrition Reviews, 52*(9), 285–298.

Roberts-Nkrumah, L. B. (2007). An overview of breadfruit (Artocarpus altilis) in the Caribbean. *Acta Horticulturae, 757*, 51–60.

Rosset, P. M., Machín Sosa, B., Roque Jaime, A. M., & Ávila Lozano, D. R. (2011). The Campesino-to-Campesino agroecology movement of ANAP in Cuba: Social process methodology in the construction of sustainable peasant agriculture and food sovereignty. *The Journal of Peasant Studies, 38*(1), 161–191.

Rural Women's Assembly [RWA]. (2019). *International Rural Women's Day and World Food Day.* Retrieved from https://ruralwomensassembly.wordpress.com/2019/10/17/media-statement-international-rural-womens-day-and-world-food-day/

Scrinis, G. (2012). Nutritionism and functional foods. In D. Kaplan (Ed.), *The Philosophy of Food* (pp. 269–292). University of California Press.

Srour, B., Fezeu, L. K., Kesse-Guyot, E., Allès, B., Méjean, C., Andrianasolo, R. M., Chazelas, E., Deschasaux, M., Hercberg, S., Galan, P. and Monteiro, C. A., & Touvier, M. (2019). Ultra-processed food intake and risk of cardiovascular disease: Prospective cohort study (NutriNet-Santé). *British Medical Journal, 365*, 11451.

Thaman, R. (2014). Agrodeforestation and the loss of agrobiodiversity in the Pacific Islands: A call for conservation. *Pacific Conservation Biology, 20*(2), 180–192.

Torheim, L. E., Ouattara, F., Diarra, M. M., Thiam, F. D., Barikmo, I., Hatløy, A., & Oshaug, A. (2004). Nutrient adequacy and dietary diversity in rural Mali: Association and determinants. *European Journal of Clinical Nutrition, 58*(4), 594–604.

Trichopoulou, A., & Vasilopoulou, E. (2000). Mediterranean diet and longevity. *British Journal of Nutrition, 84*(S2), S205–S209.

Tucker, K. L. (2001). Eat a variety of healthful foods: Old advice with new support. *Nutrition Reviews, 59*(5), 156–158.

US Food Sovereignty Alliance [USFSA]. (2021). *Vision and Operating Principles*. Retrieved from http://usfoodsovereigntyalliance.org/visions-and-operating-principles/

van der Meulen, H. (2004). The slow food movement. *Low External Input Sustainable Agriculture Magazine, 20*(3), 30.

Veteto, J. R., & Skarbø, K. (2009). Sowing the seeds: Anthropological contributions to agro-biodiversity studies. *Culture & Agriculture, 31*(2), 73–87.

Wexler, M. N., Oberlander, J., & Shankar, A. (2017). The slow food movement: A 'Big Tent' ideology. *Journal of Ideology, 37*(1), 1.

Whitmee, S., Haines, A., Beyrer, C., Boltz, F., Capon, A. G., de Souza Dias, B. F., Ezeh, A., Frumkin, H., Gong, P., Head, P. and Horton, R., & Yach, D. (2015). Safeguarding human health in the Anthropocene epoch: Report of The Rockefeller Foundation–Lancet Commission on planetary health. *The Lancet, 386*(10007), 1973–2028.

Zhang, Q., Chen, X., Liu, Z., Varma, D. S., Wan, R., & Zhao, S. (2017). Diet diversity and nutritional status among adults in southwest China. *PloS One, 12*(2), e0172406.

Zimmerer, K. S., de Haan, S., Jones, A. D., Creed-Kanashiro, H., Tello, M., Carrasco, M., Meza, K., Amaya, F. P., Cruz-Garcia, G. S., Tubbeh, R., & Olivencia, Y. J. (2019). The biodiversity of food and agriculture (Agrobiodiversity) in the anthropocene: Research advances and conceptual framework. *Anthropocene, 25*, 100192.

Zimmermann, M. B., & Hurrell, R. F. (2007). Nutritional iron deficiency. *The Lancet, 370*(9586), 511–520.

7

PRACTISING AGROECOLOGY FOR SUSTAINABLE DIETS AND HEALTHY COMMUNITIES

Amanda Shankland

Introduction

Can you hear the sound
Beating from the ground
Deep inside the valley hidden in the river
Dancing in the wind upon your skin
I can hear Her calling telling me to make amends
Mother, we'll meet again

(Raye Zaragoza, used with permission)

Indigenous singer and songwriter Raye Zaragoza sung these words in a powerful song of gratitude for mother earth. The song is featured in a documentary titled *Gather*, which follows the lives of several people of Indigenous descent from Arizona, South Dakota, and Northern California as they try to reclaim their connections to the earth and to their culture through the rebirthing of traditions around food.

Agroecology is a philosophy and practice that is deeply rooted in Indigenous ways of knowing and land practices that is scientifically substantiated. This purposeful approach to the production and consumption of food is based on the values of harnessing the connections that exist between plants, animals, humans, and *all* the natural elements, grounded in community and cooperative values. It works to uncover and support the historically prevalent ecology of a given area. As a philosophy it recognises the significance of local, specific traditional knowledge while incorporating scientific insights to support and advance the existing connections that are observed within natural systems. Agroecology can be seen in contrast to dominant capitalist modes of understanding and relating to the earth

DOI: 10.4324/9781003174417-9

that emphasise human domination of the natural world, including of food systems. An agroecological approach can promote sustainable diets through an emphasis on health and wellbeing, as well as low environmental impacts and affordable, culturally appropriate foods. Central concerns for agroecologists include fostering nutrient dense soils, conserving and reusing water systems, and harnessing the power of the natural elements including the wind and the sun. Agroecology also offers a challenge to hegemonic discourses by attempting to decolonise the food system and valorise Indigenous knowledge and cultures (Figueroa-Helland et al., 2018). As an agroecologist, it is important to recognise how the current food system has been constructed through colonisation and racism, a process that has erased much of the rich historical knowledge and experience of Indigenous Peoples. As a non-Indigenous person and agroecology scholar, I seek to foreground the voices and work of Indigenous People and assert my commitment to reconciliation and decolonisation. I embrace a holistic approach that emphasises community, sustainability for future generations, connection with the land, and respect for our ancestors—values that are significantly inspired by, and often embodied in, Indigenous worldviews.

This chapter presents an overview of some core theoretical principles that define agroecology as well as a brief description of the history of the concept. The chapter then explores several innovative approaches that two groups of Indigenous farmers incorporate into developing agricultural landscapes and harnessing earth's natural cycles and energies to create ecologically vibrant agricultural environments. Applied examples are used to demonstrate agroecology and how it involves reclaiming ancient practices that were nearly destroyed by the impacts of colonisation, or the so-called "green revolution." The first example involves a farming technique called *chinampas* retained since pre-colonial times in regions near Mexico City. The second example, in Anishinaabeg territory around the Great Lakes, explores Indigenous farmers' efforts to re-establish *manoomin*, traditional varieties of rice. Agroecology is an enticing way to re-imagine how food is produced, shared, and consumed.

Agroecology as a philosophy and scientific approach

In the early 1970s, agroecology was being reclaimed as an approach to increase efficiency and sustainability in agricultural systems through respecting the local, natural ecosystems. The significance of this knowledge held by farmers, became recognised among the early innovators and adopters of agroecology. One of the first thinkers to advance the idea of agroecology was Efrain Hernández Xolocotzi, who from the 1940s through the 1970s worked to advance the understanding that cultural practices guide agroecological knowledge through an interchange of ecological science and people's knowledge (Pimbert, 2018). The focus of agroecology as a movement was first centred in Mexico as part of a broader movement to resist the green revolution spreading across the country. Influenced by developments south of the border, American scholars Miguel Altieri and Stephen Gliessman became well known proponents of agroecology and the struggles for a more sustainable vision of agriculture in the United States. At the same time, Pierre Rabhi, an Algerian born farmer, scholar, and activist, drew international attention to the movement by running agroecology training sessions in France and West Africa in the 1980s. These theorists, particularly Rabhi, an Algerian-born farmer/scholar who has worked in West Africa, asserted a worldview grounded in traditional knowledge, that advanced a life-affirming ethic for the earth (Pimbert, 2018). In essence, agroecology seeks to combine modern ecological science with experiential knowledge held by farmers and Indigenous People as holders of diverse and durable agrarian culture. Such knowledge is grounded in hundreds, even thousands of years of observation of adaptive, and dynamic, management practices.

The elements of agroecology are complex and are grounded in the notion that human, animal, and plant systems are tied together and function best when those relationships are nurtured (Barrios et al., 2020). Agroecology, has been described by the Food and Agricultural Organisation of the United Nations (FAO, 2018) as including multiple elements: 1) Diversity; 2) co-creation of knowledge; 3) synergies; 4) efficiency; 5) recycling; 6) resilience; 7) human and social values; 8) culture and food traditions; 9) responsible governance; and 10) circular and solidarity economy. Agroecologists protect these strategies as counter measures to modern approaches that undermine traditional agrarian systems and resource conserving technologies. Centred within traditions of local farming knowledge, agroecology provides the basis for an affordable, highly productive, and ecologically sound model of agriculture (Altieri, 1999).

Agroecology is a movement that also works toward improving the way humans inter-relate with the earth, particularly through means of sustainable food production (Bookchin, 1982). Similar to feminist and food sovereignty movements, agroecology questions the dominant hierarchical power structures that control the production and distribution of food and that engage in practices of environmental mismanagement driven by capitalism. Industrial food systems are destroying forests, wetlands, and wild habitats in favour of industrialised farming operations, with extensive technology and tools, largely exported from the Global North (Shiva, 1989). Agroecology is closely connected to feminist concerns that the industrial food system tends to place more control in the hands of fewer people and predominantly male farmers, when the majority of smallholder farmers in the world are women (Swinbank, 2021). Women farmers are responsible for producing a large share of food across the world, particularly in the Global South (Swinbank, 2021). Women ensure these ecosystems are nurtured to provide the food, medicines, and energy to feed and care for their families. Agroecology recognises the critical role that women play in sustainably managing ecosystems over the long term (Bruil et al., 2020). Giving voice to women is essential to create opportunities for the goals of agroecology to be realised (Gleissman et al., 2018).

There are numerous examples of agroecology around the world and in a wide array of settings. Important examples of agroecology are found in urban farming, organic agriculture, and indoor farms that seek to create an autonomous system within an enclosed environment. These approaches seek to mimic the processes of natural environments using both traditional and modern methods. Some of the most promising advances in agroecology are found in agroforestry, particularly in Central America, where there has been a move away from "slash and burn" agriculture. Agroforestry incorporates a diversity of trees to support building nutrient dense soils to create a canopy that helps retain water, reduce erosion, preserve native species of plants and animals, and to create a diversified source of income for farmers. Livestock may be used as a source of fertiliser in some agroforestry systems.

Following are two examples that exemplify the principles of agroecology—the chinampas in Mexico, which are unique because they are developed on wetlands and the land is built up using soil from the bottom of the lakebed, and Indigenous rice farming, also situated in lake environments—with both having cultural and social significance to the people involved.

Chinampas

In what was the heart of one of the most powerful empires in central Mexico, Tenochtitlan, the Aztecs developed the chinampas system of farming. A chinampa consisted of several layers of vegetation and waste built up in a wetland area so that it could be used for agricultural purposes. This design was born of necessity to make the local environment and soil conditions productive. The chinampa was secured in place by trees planted around the edges so the roots could keep the soil in place. When a significant number of chinampas were built, they formed a large network of canals that could double as irrigation routes and pathways for needed drainage (Robles et al., 2019). These networks were ideal for moving produce to the markets. Trading hubs grew up around these networks and provided a significant source of livelihood for the local people. In a bid to expand their influence and power over the valley, the Aztecs built this intricate network for trade and urban development.

When the Spanish conquistadors, led by Hernan Cortes, arrived in Tenochtitlan in 1519, the basin of the Aztec capital was characterised by this interconnected system of canals and lagoons of varying sizes, elevation, and salinity. At this time, the Aztecs ruled the Valley of Mexico, encompassing a vast empire that had a population of an estimated 200,000 people (Robles et al., 2019). The Spanish settlers decided, over time, that vast areas of rich farmlands would be made available by draining the lake. The ecological consequences of the draining turned parts of the valley into semi-arid zones and decimated local production systems. The consequences of these disruptions are still a concern today, with a lack of potable water in some parts of modern Mexico City's metropolitan area. Due to overuse, the aquifer beneath the city is rapidly depleting and Mexico City is estimated to have sunk 10 metres (33 feet) in the last century. Furthermore, due to lake sediments underlying most of Mexico City, the area has proven vulnerable to soil liquefaction during earthquakes, most notably in the 1985 earthquake when the infrastructure of the city was severely affected, causing buildings to collapse and thousands of lives to be lost (Robles et al., 2019).

While it is impossible to restore the lake, the chinampa system, combined with contemporary ecological farming practices offer a potential and viable solution to water scarcity (Ebel, 2020). Further, the cultural resurgence of traditional environmental values provides political impetus to research and measure the impact of this solution. Well documented outcomes of the centuries-old chinampas system include sustainable and productive economic, social, environmental, and cultural benefits (Altieri, 1999; Ebel, 2020; Robles, 2018). The benefits of the chinampa system to small Indigenous farming operations have been recognised and have sparked a movement to expand this system to other areas of the country and internationally (Ebel, 2020). Farmers using these techniques provide valuable insight into how agroecological methods can successfully function in modern-day Mexico and may be adapted to other regions. These approaches also benefit Mexican farmers and the local surrounding population by providing food, employment, and the conservation of natural resources, thereby advancing a circular and solidarity economy (Ebel, 2020). The system also prevents the deterioration of wetlands, which helps to conserve vital recharge to Mexico City's aquifers and protect a diverse ecosystem. The chinampas have proven to be more efficient than many modern techniques and have helped preserve valuable botanical and cultural knowledge from the past. A testament to their effectiveness is the way they have survived environmental stressors in the modern context, from urban sprawl to increasingly dry climatic conditions, to over-extraction of the springs in the area (Robles et al., 2019).

According to Altieri (1999), the level of productivity that characterises the chinampas is attributed to four central factors. First, a chinampa always has something growing in it, and three or four different crops can be grown in just one year, by germinating seeds in the beds before the older crop is harvested. Second, they have highly fertile soil. In spite of a continuous crop harvest, they maintain high quality soil because they are consistently supplied with potent organic fertiliser, as the lake where the chinampa is located serves as a basin for nutrients that are ideal for soil conditions. Further, aquatic plants concentrate nutrients that are available in lower amounts in the water and store them in their tissue. The use of these plants, along with mud and waste, ensures that an ample supply of nutrients are always available to the crops. Third, there is a large supply of water for the crops. The narrowness of the canals allows the water to reach the crops' roots and provide moisture to the soil. And even if the lakebed dries, and the water no longer reaches the roots, there is sufficient room for farmers to irrigate using canoes. Fourth, there is a significant amount of individual attention given to each plant by farmers, which results in much higher yields (Altieri, 1999). This productive system of cultivation also allows Indigenous farmers to plant traditional varieties of crops with no chemical inputs. A historically significant cropping method, the three sisters, which includes planting corn, squash, and climbing beans together, is commonly used. These healthy and culturally significant foods contribute to sustainable diets for local people.

There have been efforts by local Indigenous farmers to revitalise chinampas and adjust their functions to meet modern day production demands. Even with significant actions on the part of farmers located on the southern outskirts of Mexico City, it appears that the chinampas are in danger of disappearing. It has been estimated that by 2057 the chinampas in the area of Xochimilco (sotʃi'milko) will have been converted to housing developments (Ebel, 2020). The survival of chinampas into the future will depend on a concerted effort to demonstrate their importance for social, cultural, and economic values to Mexico City. Further, political interventions will be needed to ensure they are protected. There have been some successes in this regard. For instance, as early as 1984, the Chinampa in the Valley of Mexico was declared a Cultural Heritage of Humanity Site by the United Nations Educational, Scientific, and Cultural Organisation (UNESCO). In 2004, it was declared a protected site by the Convention on Wetlands of International Importance (RAMSAR) (Braulio Robles et al., 2019). Nonetheless, these sites are under intense threat of development and Mexico City likely will have to initiate a program to restore the chinampas to provide a series of ecosystem services to Mexico City. These services could result in benefits such as water filtration, regulating water levels, microclimate regulation, increased biodiversity, and carbon capture and storage. The potential benefit to tourism also be considerable (Ebel, 2020).

The chinampas offer an opportunity to apply sound historically-proven methods of agricultural production with modern technology to produce one of the most efficient and sustainable farming agroecological systems in the world (Robles et al., 2019). Nonetheless, there are pressures to expand and adapt these systems to the modern world. Countering urban expansion and capital-intensive forms of agriculture present significant challenges.

Manoomin

Prior to colonial settlement most of the wetlands and waterways of south-eastern Ontario were cultivated with wild rice known as *manoomin,* meaning "wild rice" in Ojibwe. Manoomin illustrates the significance of agroecology as a practice that asserts the cultural and social values of a people. Rice Lake, situated in the southern end of the Kawartha Lakes Watershed, was the hub of rice cultivation in south-eastern Ontario for centuries (Anderson & Whetung, 2018). This area was the traditional territory of the Anishinaabeg Peoples. Various types of colonisation in the early 20th century—including the development of the Trent Severn Waterway—brought about industrialised agriculture along the edge of the waterways. Further, lakefront and waterway developments to accommodate tourism led to the destruction of most of the wild rice strands in south-eastern Ontario. By the 1920s there was virtually no manoomin in Rice Lake, and so the Elders took some of the remaining seeds and planted them in Mud Lake (Mississippi River) near Ardoch (150 km away) (Anderson & Whetung, 2018). The Anishinaabeg lost much of their territory in 1923 through the Williams Treaty negotiations. They lost access to their traditional land and food system base by being confined to reserve lands that often were far too small for cultivation of wild rice. The reserve system had the effect of legislating traditional foods and local food economies out of existence and creating food deserts forcing much of the local populations to rely on the settler's foods. Foods consisting of flour, sugar, and oils, previously foreign to the

Indigenous populations, contributed to serious health-related issues like diabetes and heart disease (Anderson & Whetung, 2018).

Alongside the Anishinaabeg, the manoomin was subjected to the systematic impacts of colonisation, undermining the capacity to produce the food that was considered sacred and integral to their identity (Anderson & Whetung, 2018). This stark reality, accompanied by the removal of an entire generation of Indigenous children from their families, homes, and communities into the residential school system meant that the knowledge and skills required to maintain *manoomin* cultivation were severely undermined (Anderson & Whetung, 2018). By the 1970s, corporations like Monsanto and DuPont took an interest in the cultivation of wild rice and began to appropriate and to claim ownership of some varieties. They mapped the wild rice genome, changed the genetic structure of the rice, and patented it. There are concerns that a new variety that contains male-sterile seeds could genetically contaminate the natural varieties, by pollen drift and pollen transfer by ducks. This could potentially wipe out the capacity of other varieties to reproduce naturally. This is threatening the Anishinaabeg role and responsibility of ensuring the continued life and vitality of manoomin. The ability to maintain their traditional food systems and therefore their communities' health and well-being is also being undermined (Anderson & Whetung, 2018). This exemplifies another form of colonisation and commercialisation, as the Anishinaabeg saw manoomin as a sacred food and a gift from the Creator, one that should not be owned and controlled as a commodity (Anderson & Whetung, 2018). This respect for the manoomin was demonstrated through their practices. Though they understood the value of the manoomin for food and trade, the Anishinaabeg at a time, would not plant the seeds in the soil. They would sometimes wrap the seeds in clay and throw them in the water but they would not sow the seeds as they believed that doing so would wound their Mother, the earth (Morgan, 2016).

Like the chinampas, spreading and gathering manoomin were culturally, socially, and environmentally critical to these Indigenous peoples (Anderson & Whetung, 2018). Manoomin provided an important staple and could be easily carried and preserved for extended lengths of time. During early years of settlement, when moving westward many settlers would have faced starvation if not for the manoomin. Among the manoomin, wildfowl was a rich source of protein for the Algonquin and Ojibwe as it reproduced quickly and grew fat (Morgan, 2016). Ducks and geese, in particular, are drawn to areas where wild rice is abundant. Thus, cultivating manoomin and combining with waterfowl, were illustrations of the principles of agroecology in action. These practices harnessed the synergies in the natural world and gave the community members a reliable source of nutrition.

Today, the Anishinaabeg, for whom wild rice is of vital importance to their way of life, have been harvesting and spreading the seeds of the plants where it is still possible for it to grow. Currently, in particular locations, Manoomin can be found in lakes and streams. The Anishinaabeg, however, face a new threat as cottagers in the area, desiring an unobstructed view and easily navigable waterways, have been tearing out the plants (Krotz, 2017). In order to educate cottagers about the cultural and ecological relevance of manoomin for the Anishinaabeg, some

Anishinaabeg community members have shared vivid descriptions of the historical ecology that they are trying to restore and the deep cultural significance. Leanne Betasamosake Simpson, an Indigenous studies scholar, and a member of Alderville First Nation, explains that the process of gathering manoomin is embedded in their teachings about how to exist in the world by honouring the web of connections to each other, to the plant nations, the animal nations, the rivers and lakes, the cosmos, and honouring neighbouring Indigenous nations (as cited by Jackson, 2016). Betasamosake Simpson says that manoomin creates and sustains a sense of intimacy with the ecology (as cited by Jackson, 2016). The growth of manoomin speaks to the activation of diverse agroecological principles like synergies; cultural and social values; resilience; and circular economy (FAO, 2018). In particular, manoomin helps to preserve important cultural and food traditions. Harnessing important environmental and social principles, foster conditions for a more resilient system overall.

The cultivation of manoomin is an important way for the Anishinaabe to cope with the impacts of colonisation: The erosion of natural livelihoods and ways of life, dismantling of access to traditional foods, and the creation of food deserts. Growing manoomin is part of broader movements around the world, exemplified by the work of organisations like the Via Campesina, The Institute for Sustainable Development, Sociedad Científica Latinoamericana de Agroecologia, and the International Panel of Experts on Sustainable Food Systems, organisations working to assert the right of Indigenous peoples to reclaim their traditional diets and food sources.

Agroecology and sustainable diets

The use of agroecology has been expanded to include critical analysis of global food systems and to support more localised food production and consumption. This may include creating connections between consumers and producers by fostering more personal, or informed, relationships through building up networks of accessible, intimate, local markets (Pimbert, 2018). Farm organisations, consumer advocacy groups, and social movements cite the principles of agroecology in their efforts to advance the vision for alternative food systems beyond the industrial model (Gliessman et al., 2018; Massicotte, 2010). By increasing the visibility and accessibility of agroecology, these efforts may enable it to become a normative philosophy and advance its goals of facilitating bottom-up processes, circular economies, non-hierarchical food systems, and socially engaged communities of production and conscious consumption.

This handbook's purpose is to advance and extend the reach of the ideas and practices of sustainable diets. Agroecology provides a blueprint for this goal, as it not only advances a sustainable farming system but also emphasises the importance of farmers, Indigenous communities, consumers, animals, and plants as integral components of healthy ecosystems. Increasing familiarity and interconnectivity among stakeholders in these alternative food systems increase awareness of, respect for, and fair compensation for the work, products, and ecological services of producers. Agroecological farming systems purposefully attend to the air, water, soil, and ecosystem qualities, and protect natural space for the greater community into the future. Further, agroecology emphasises the cultural importance of foods by sustaining the unique practices to protect and strengthen diversity. Agroecology also can be appreciated as placing more attention on Indigenous values and practices and active commitments to decolonisation. In these ways, agroecology can be an approach to fundamentally change and reorient human interaction with each other and the natural world.

In a detailed study by Gliessman and colleagues (2018) of seven illustrations of transitions toward agroecology, the authors explore the ways that the limitations of industrial food systems can be exposed and agroecological systems adopted for greater ecological, economic, social, and historical sustainability. The authors identify ways to leverage the principles of agroecology to create a more sustainable food system (Gliessman et al., 2018). Table 7.1 adapts these principles and provides a valuable road map for working towards agroecological systems.

Conclusion

Across the world there are growing movements to affirm the values of Indigenous and peasant farmers and to nurture alternative models of agriculture. Agroecology plays a meaningful role in food sovereignty and affording greater sustainability in food systems (Pimbert, 2018). At the heart of agroecology is the rejection of the commodification of nature and a focus on the natural cycles of the earth essential for the sustainability of food and of humanity. Agroecology is also respected for its use of scientific principles, and engaging researchers to help inform, develop, and extend the environmental, health, and economic benefits of circular systems that honour natural ecosystems (Pimbert, 2018). To assess the efficacy of these movements and to investigate concerns across these fields of inquiry, further research is needed. While outside the scope of this study, more analysis could be applied to understand how applying such principles in cityscapes and in urban agriculture can offer the possibility of combining food and energy production with water and waste management, and reveal ways to significantly reduce carbon footprints.

Elemental to agroecology are: Biodiversity; cultural suitability; and meaningful development practices; optimising and recycling; localised production and consumption; along with ecological design to reorient agricultural landscapes to regenerativity. Broadly, agroecology can be understood as the study of the interrelationships between human, animal, and plant worlds, and the policies and practices that maintain the delicate balance of these systems as farmers cultivate the land to provide sustainable sources of food. The climate, health, and equity crises have accelerated the drive for reorienting to more durable and humane systems of food production and consumption. The principles embedded in agroecology provide invaluable tools for resilient and respected food systems. Policy makers and change agents, that is all of us, are called upon to supplant these crises with workable solutions; the principles of agroecology provide a proven roadmap for the production and consumption of sustainable diets.

Table 7.1 Leveraging the principles of agroecology. Adapted from Gliessman et al., 2018, used with permission

Principles	Examples of practices
Community-led governance	• Develop structures of engagement that reinforce a community-based approach • Create initiatives that support sharing of locally based knowledge • Initiate community led consultations with governments and other external interests • Determine that the land and resources are used for the collective good and demonstrate good governance and stewardship
Honour the roles of key stakeholders	• Support peasant organisations in growing their influence through strengthening economic justice and political influence • Engender farmer-to-farmer knowledge exchange and cooperative markets • Increase community-based activities and public education
Strengthen alliances with farmer, consumer, health, justice, and environmental groups	• Create opportunities for growing engagement and broader application of the principles of agroecology • Engage diverse actors in creative initiatives • Help multiple organisations advance multiple goals through amplifying sustainable food systems
Draw attention to cultural narratives and messaging	• Illuminate the narratives and worldviews of food producers, land, air, and water stewards, and shine light on the unique local geographies, soil types, biodiversity, along with diversity among cultural identities • Conduct detailed research and analysis of the discourse surrounding agricultural and environmental practices, their efficacy and impacts
Relocalise the food system when possible	• Foster connections to local markets, culture, and community • Support multiple, decentralised outlets for food availability through home gardens, farmers' markets, CSA schemes, direct marketing, and securing public commitments for local procurement practices • Emphasise the full use of all items, the circular economy, revalorising would-be waste, and commitments to zero waste
Farmer-to-farmer knowledge sharing	• Disseminate farmer knowledge through field schools, farmer collaborations with universities and through grass roots organisations specialising in environmental knowledge dissemination • Reject linear extension models that do not account for local contexts
Empower young people and women	• Prioritise young people and women as critical in developing strong communities and agroecological production and consumption

References

Anderson, P. & Whetung, J. (2018). *Black duck wild rice: A case study.* Wilfrid Laurier University.

Altieri, M. A. (1999). Applying agroecology to enhance the productivity of peasant farming systems in Latin America. *Environment, Development and Sustainability, 1,* 197–217.

Barrios, E., Gemmill-Herren, B., Bicksler, A., Siliprandi, E., Brathwaite, R., Moller, S., Batello, C., & Tittonell, P. (2020). The 10 elements of agroecology: Enabling transitions towards sustainable agriculture and food systems through visual narratives. *Ecosystems and People, 16*(1), 230–247.

Bookchin, M. (1982). *The ecology of freedom: The emergence and dissolution of hierarchy.* Cheshire Books.

Bruil, J., Delvaux, F., Diouf, A., Hogan, R., Milgroom, J., Petersen, P., Prado, B., & Serneels, S. (2020). Agroecology and feminist economics: New values for new times. *Farming Matters, 36*(1), 3–6.

Ebel. (2020). Chinampas: An urban farming model of the Aztecs and a potential solution for modern megalopolis. *Horticulture Technology, 30*(1), 13–19.

FAO. (2018). *The 10 elements of agroecology: Guiding the transition to sustainable food and agricultural systems.* FAO.

Figueroa-Helland, L., Thomas, C., & Aguilera, A. P. (2018). Decolonizing food systems: Food sovereignty, indigenous revitalization, and agroecology as counter-hegemonic movements. *Perspectives on Global Development and Technology, 17*(1–2), 173–201.

Gliessman, S. R., Jacobs, N., Clément, C., Grabs, J., Agarwal, B, Anderson, M., Belay, M., Li Ching, L., Frison, E., Herren, H., Rahmanian, M. & Yan, H. (2018). *Breaking away from industrial food and farming systems: Seven case studies of agroecological transition.* FAO.

Jackson, L. (2016, February 20). Canada's wild rice wars: How a conflict over wild ricing on Pigeon Lake is drawing attention to Indigenous rights and traditional foods. *Al Jazeera.* Retrieved from https://www.aljazeera.com/features/2016/2/20/canadas-wild-rice-wars

Krotz, S. (2017). The affective geography of wild rice: A literary study. *Studies in Canadian Literature/Études en littérature canadienne, 42*(1), 13–30.

Massicotte. (2010). La Via Campesina, Brazilian peasants, and the agribusiness model of agriculture: Towards and alternative model of agrarian democratic governance. *Studies in Political Economy, 85,* 69–98.

Morgan, F. (2016, December 13). Wild rice harvest: Looking back at the tradition of wild rice harvesting among the Indigenous people of the Great Lakes region. *Canada's History.* www.canadashistory.ca/explore/first-nations-inuit-metis/wild-rice-harvest

Pimbert, M. P. (2018). Global status of agroecology: A perspective on current practices, potential and challenges. *Review of Environment and Development, 53*(41), 53–57.

Shiva, V. (1989). *Staying alive: Women, ecology and development.* Zed books.

Swinbank, V. A. (2021). Women feed the world: Biodiversity and culinary diversity/food security and food sovereignty. In V. A. Swinbank (Ed.), *Women's Food Matters* (pp. 187–218). Palgrave Macmillan.

8

CONSERVING INSECT BIODIVERSITY IN AGROECOSYSTEMS IS ESSENTIAL FOR SUSTAINABLE DIETS

Paul Manning and Jennifer Marshman

An introduction to insects in agroecosystems

Approximately half of the earth's habitable area can be classified as agroeco-systems (Ritchie & Roser, 2013): Ecosystems modified by humans to produce food, fuel, fibre, or medicine. How these ecosystems are managed varies along a continuum of methods and intensity. Using regenerative practices (Box 8.1) can restore degraded ecosystems and provide humans and other members of the biotic community with numerous goods and services. In contrast, many aspects of industrial agricultural practices, such as intensive, high-input production of crops and livestock, impose major costs for society including loss of species, increased incidence of antibiotic resistance, and alteration of geochemical cycles, all incurred with the rationale of achieving greater agricultural productivity (Leighton, 2021).

Box 8.1 Regenerative agriculture defined. Source: IPCC, 2019

Regenerative agriculture: "land management practices focused on ecological functions that 'can be effective in building resilience of agro-ecosystems'"

(IPCC, 2019, p. 389).

DOI: 10.4324/9781003174417-10

Compared to natural ecosystems, conventional agroecosystems are generally characterised by low levels of biodiversity (measured at the genetic, species, and ecosystem level). By design, mono-cropping (i.e. growing a single crop species) reduces plant biodiversity and has implications for the "unplanned" components of biodiversity in agroecosystems, which are largely represented by insects: A tax on representing roughly 90% of all animal species and more than half of all the living organisms on Earth (Royal Entomological Society, 2021). Insects within agroecosystems either persist when natural ecosystems are converted to farmland or later colonise from the surrounding landscape. Decades of research show that higher levels of biodiversity support numerous beneficial functions within ecosystems, including control of herbivorous pests (Barnes et al., 2020) and improving ecosystem resilience to environmental stressors.

Harmful, benign, and beneficial Insects

Nothing biological determines whether an insect is considered a pest, but instead a pest is broadly defined as any organism that causes injury to humans or human interests. Though more than one million species of insects are known to science, relatively few act regularly as pests. Agroecosystems managed as intensive mono-cultures can regularly host hundreds of insect species, whereas diverse and regenerative agroecosystems characterised by high levels of plant biodiversity may support thousands of different species (Altieri & Nicholls, 2018).

Insects that live among agroecosystems have attracted considerable scientific attention. The mention of insects in relation to agriculture conjures up the image of a pest: The maggoty apple, a plague of locusts, or biting flies swarming the flanks of cattle as they graze on pasture. The role of insects as pests has been well established, and in many production systems insect pests represent some of the most significant threats to food security through limiting food production or quality (Jankielsohn, 2018). Much of the scientific effort dedicated to managing insect pests has focused on insecticide use, which can pose a range of serious implications for ecological health. Routine use of insecticides, particularly when used prophylactically, is not consistent with sustainable production practices nor with sustainable diets (Altieri & Nicholls, 2018).

Despite the considerable economic and social costs of insect pests to sustainable food production, most of the insects found within agroecosystems would not be considered pests. Most insects are relatively neutral to food production, although agroecologists and entomologists still have a limited understanding of all the interconnections between biotic organisms. All insects have a role within their respective food chains and broader food webs (Box 8.2), and their presence does not reflect any inherent benefits or costs to the agroecosystem (Figure 8.1).

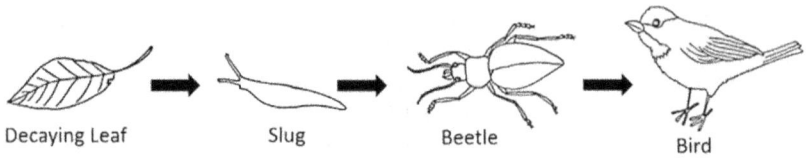

Decaying Leaf   Slug   Beetle   Bird

Figure 8.1 An example of a food chain. Working from right to left, a bird eats an insect, the insect eats a gastropod, the gastropod eats plant matter. Source: Image by co-author Paul Manning.

Box 8.2 Food chains defined. Source: The work of the authors

Food chains: A food chain is the hierarchical sequence of organisms who eat other organisms. For example, owls eat mice or toads, who eat grasshoppers, who eat plants. Interconnected food chains create **food webs** that better represent ecosystems.

In the context of agroecosystems, beneficial insects are considered agents of ecosystem services (Box 8.3). Ecosystem services are classified according to four categories of services: Provisioning, regulating, cultural, and supporting. In agroecosystems insects contribute to each of these categories.

Box 8.3 Ecosystem services defined. Source: Reid et al., 2005

Ecosystem services: The benefits that humans obtain from ecosystems (Reid et al., 2005).

Ecosystem services provided by insects are foundational to sustainable diets. In this chapter, we look at three different groups of insects that provide ecosystem services that allow production of food for sustainable diets. Focusing on the social, economic, and environmental pillars of sustainability, we describe three case studies that highlight the importance of insects in this realm.

Case-study #1 Social sustainability: The human-pollinator nexus

In this case study, we take an unconventional approach to the social pillar of sustainability—an examination of the human-pollinator relationship. We consider insect pollinators to be important members of the ecological community "with whom our livelihoods are interdependent and interrelated" (Gibson et al., 2015, p. 10), especially in the context of food production. Roughly 90% of the flowering plants on Earth, including 75% of the world's food crops, depend to some degree on animal pollination (including insects) (FAO, 2018). These percentages translate into an estimated annual market value that ranges between \$235–\$577 billion USD (IPBES, 2016), supporting farmers and other livelihoods. Medicines, biofuels, fibres, arts, recreation, and cultural practices are also directly attributable to biotic pollination (Box 8.4). In addition, pollinator-dependent plants contain significant amounts of the micronutrients that comprise healthful diets, including 100% of the lycopene and 90% of the vitamin C, as well as other antioxidants, calcium, vitamin A, and folic acid (Eilers et al., 2011).

Box 8.4 Biotic pollination defined. Source: The work of the authors

Biotic pollination: Plants can only produce seeds (and hence, fruit) if pollination occurs. Pollination can happen abiotically by wind or water, or through animal pollination (including insects) referred to as biotic pollination.

Pollinators are a diverse group of animals including bees, beetles, birds, butterflies, moths, wasps, flies, bats, and other small mammals. While there are diverse pollinators (Figure 8.2), bees are responsible for the majority of the biotic pollination of crops used by people for food, fuel, fibre, medicine, and cultural practices (IPBES, 2016).

The common slogan of "save the bees" is often accompanied by images of honey bees on flowers or in their constructed hives. The western honey bee (*Apis mellifera*)[1] dominates depictions of the pollinator, perpetuating industry-based priorities and market interests (Colla & MacIvor, 2017; Marshman & Knezevic, 2021;

Figure 8.2 The first and third images are bees (order Hymenoptera), the second and last image are flies (order Diptera) that are sometimes called "bee mimics," all of which can provide important pollination services. These insects are all black and yellow, however, pollinating insects come in many colours, shapes, and sizes. Source: Photos by co-author Jennifer Marshman.

Reilly et al., 2020). With an estimated 30,000–100,000 bees per colony, commercial beekeepers can readily transport colonies to areas where crops require pollination. Rarely do images of honey bees depict the reality of the manner or the magnitude by which many commercial bees are managed and deployed for their pollination services (i.e. hundreds of hives stacked on transport trucks for shipping cross-country). Instead, depictions of bees tend to use imagery of chubby bees happily flitting from flower to flower, leveraging the "pollinator idyll" to promote the "save the bees" narrative (Marshman & Knezevic, 2021, p. 129).

While there are several bee species that are managed in agricultural contexts (e.g. *Bombus impatiens* for blueberries, *Megachile rotundata* for alfalfa pollination, *Osmia lignaria* in orchard pollination), the western honey bee is the most widely bred and managed pollinator on the planet, with thousands of commercial and hobbyist beekeepers in Canada alone. While colony collapse disorder (CCD)—the unexplained disappearance of entire managed honey bee colonies first named in the early 2000s—is no longer a primary concern, seasonal honey bee colony losses, especially overwinter losses, are of ongoing concern for beekeepers in North America and beyond. However, a 2019 study found strong evidence of adaptation to colony losses (Rucker, Thurman and Burgett, 2019), and the western honey bee is not identified as endangered or at risk on any formalised list. Conservation biologists argue that honey bee losses are not a conservation concern (Champetier et al., 2015; Colla & MacIvor, 2017; Fürst et al., 2014) but instead represent a livestock management issue. This is an important distinction that could help refocus conservation efforts on the bees most at risk: Wild, unmanaged bee species.

Historically, and largely still, the honey bee has been the only bee species used for pesticide risk assessments, and many best management practices remain honey bee-centric, favouring their large colony sizes and easy mobility of their hives (Marshman & Knezevic, 2021). Of course, additional precautions must be considered for wild bees, who do not benefit from such convenient mobility, nesting, and foraging in or near sites of pesticide application (Chan & Raine, 2021). The important role of wild bees in crop pollination is increasingly being recognised, and in some cases, wild bees are more important for pollination than honey bees (Garibaldi et al., 2013; Nicholson & Ricketts, 2019; Reilly et al., 2020). With these combined concerns, the need for an emphasis on the protective measures for wild, unmanaged bee species becomes clear.

While our understanding of wild bees grows, one study showed that general bee knowledge among the public is low. Although the majority of bee species in Canada are solitary bees that nest in the ground, in a representative telephone survey of 2,000 Canadians, only one respondent was able to name a solitary bee species. In the same study, half of the respondents thought that managed honey bees are a wild, native bee species (van Vierssen Trip et al., 2020). In fact, honey bees are an introduced species that arrived with European colonisers in the early 17th century.

Even though general bee knowledge is low, public interest in pollinators continues to grow (Marshman & Knezevic, 2021; Wilson et al., 2017), particularly for the ecosystem services bees provide (Vierssen Trip et al., 2020). This is important because there is a reciprocal relationship required for bees to thrive in human-dominated ecosystems, such as agroecosystems. Bees are needed for their pollination services, and with human agency comes the responsibility to manage agroecosystems using environmentally sustainable practices.

The biggest threats to bees are directly linked to human activity. Direct threats include the loss of habitat from urbanisation and agricultural intensification combined with the use of pesticides and other farming practices that are harmful to bees. The introduction of managed bees into new areas for pollination services or through beekeeping or feral colonies, can create competition for floral resources, disrupt plant-pollinator networks, and spread pathogens and parasites from managed to wild bees (Graystock et al., 2016; Mallinger et al., 2017; Valido et al., 2019). Indirectly, anthropogenic climate change (Box 8.5) is increasingly creating inhospitable environments for plants that provide important food sources to bees, creating disruptions to interspecies phenological synchrony.[2]

Box 8.5 Climate change mitigation defined. Source: Adapted from Ritchie and Roser, 2013

Climate Change Mitigation: Reducing emissions of greenhouse gasses (GHGs) that trap heat in Earth's atmosphere. Agriculture, forestry and land use account for nearly 20% of global GHG emissions (Ritchie & Roser, 2013).

Human activity has large and negative impacts on the health, distribution, and abundance of many wild bee species who do not benefit from the same extensive breeding programs and management practices as honey bees. Developing a greater understanding of the relationship between humans and bees, by developing a greater awareness of the contributions of diverse bee species, is a critical step to mitigating further negative impacts to wild bee populations.

Case-study #2: Economic sustainability—free and functional pest control by natural enemies

An estimated 18–20% of global agricultural production is lost to insect pests (Sharma et al., 2017). The economic injury level model—when the density of a pest insect reaches a point whereby the cost of intervening is equal to the cost of the damage caused by the insect—can be a useful decision-making tool when determining whether to intervene. Using this model, an intervention would happen only once the injury level was reached or was deemed highly likely to occur. Many insect pests develop quickly, have high fecundity (producing many eggs), and regularly produce multiple generations per year. When an insect is presented with

Figure 8.3 Left: The larva of cabbage white butterflies (*Pieris rapae*) consume kale seedlings growing in a planter. Right: The larvae of Colorado potato beetles (*Leptinotarsa decemlineata*) feed on potato plants. Each of these pests are economically significant pests of agricultural crops at a wide range of scales and production intensities. Source: Photos by co-author Paul Manning.

an abundance of a preferred food source, pest populations can grow quickly and cause significant economic injury to farmers (Figure 8.3).

The diamondback moth (*Plutella xyllostella*) is an illuminating case study of how pest populations can cause large crop failures. Native to Asia, the diamondback moth is now found across most of the world (Sarfraz et al., 2005). Its larvae feed on plants in the mustard family (Brassicaceae) including canola, broccoli, kale, and turnip. The moth takes about 30 days to develop from egg to adult, and an average female lays 160 eggs. If each moth and its successive offspring were to successfully reach sexual maturity, mate, and lay an average number of eggs, a single mated female arriving at a canola field at the beginning of the season would result in > 1,000,000 individuals in three egg laying cycles.

Although insects like the diamondback moth are theoretically capable of exponential growth, insect pests do not exist in a vacuum. An insect must contend with numerous biophysical challenges, including temperature fluctuations, resource limitations, pathogenic infections, predation, and parasitism. Herein, we focus on the ecological processes of parasitism and predation, each of which are supported by diverse insect communities in agroecosystems. In many cropping systems, the regulating forces of these biological interactions quietly play important roles in keeping pest damage below the economic injury level.

Predation describes the process by which an organism kills and consumes another organism. Insect predators tend to be common and abundant in agroecosystems, even when agroecosystems are highly simplified (Purtauf et al., 2005). Many predatory insects are inconspicuous or are only active during nocturnal hours, and as such, may not be routinely observed by farmers. Some of the most common insect predators in agroecosystems include: Ground beetles (Coleoptera: Carabidae), shield bugs (Hemiptera: Pentatomidae), and social wasps (Hymenoptera: Vespidae). A healthy community of predatory insects can keep pest abundances below the economic injury level, thus mitigating the need for pesticides, which benefits the economic and environmental sustainability of food production.

The benefits achieved through insect predation can be further realised by providing additional on-farm habitats for predatory insects. For example, a study comparing predation rates of Colorado potato beetle (*Leptinotarsa decemlineata*) egg masses showed better control in a more diversified landscape (Werling et al., 2012). The effect was driven by predatory insects (e.g. ladybird beetles and carabids) and arthropods (e.g. spiders and harvestmen) spilling-over from prairie grass habitat into the potato plots. For highly-mobile insects, such as social wasps, conserving habitat in the surrounding landscape can be effective in supporting pest control. A study comparing predation rates of the coffee leaf miner (*Leucoptera coffeella*) showed that an increase in surrounding forest landscape enhanced biological control by promoting the abundance of predatory social wasps (Medeiros et al., 2019).

Parasitoidism is a second ecological process that supports the control of pests within agroecosystems. Most parasitoids known to science are wasps, but flies, beetles, and other insects can be parasitoids. The strategy of a parasitoid differs from a predator through the way an individual consumes its prey. Rather than locating a smaller insect and directly consuming it, parasitoids lay their eggs in, within, or near a host insect. After hatching, the parasitoid larva slowly consumes the tissues of the host—ultimately killing the host when completing development (Figure 8.4).

The effectiveness of a parasitoid in controlling its enemies is related to a number of ecological traits. Two of these traits can be influenced by a farmer's actions: How long the individual lives (longevity) and the number of eggs it produces (fecundity). To maximise the potential economic benefits supported by parasitoids, a farmer must consider how their management decisions can influence these endpoints. Though larval parasitoids consume insects, adults rely on the nectar and pollen from floral resources. By providing floral resources for the adult parasitoids, farmers can reap the benefits of enhanced pest control.

Both laboratory and field studies demonstrate that the presence of buckwheat increases pest control by parasitoid wasps (Géneau et al., 2012; Lee & Heimpel, 2005). This context dependence highlights the interconnectedness of agroecosys-

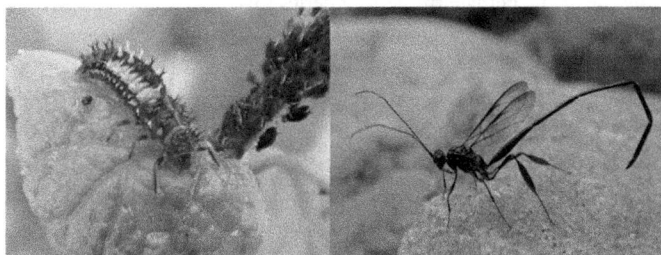

Figure 8.4 Left: A larval Asian multicoloured ladybird beetle (*Harmonia axyridis*) stalks aphid prey. This beetle is commonly encountered in agroecosystems of varying production intensities (photo by Ian Manning). Right: An American pelecinid wasp (*Pelecinus polyturator*) rests near a vegetable garden. This parasitoid uses its long ovipositor to lay eggs on larval June beetles (*Phyllophaga* spp.) feeding underground. Many species of June bugs are pests of numerous agricultural plants. Source: Photos by co-author Paul Manning.

tems to the wider landscape, and the potential benefits that habitat conservation might provide to sustainable agricultural production.

Case-study #3: Environmental sustainability—dirty work by dung beetles underpins cleaner and greener grazing practices

Although beef and dairy are foods that are understood to have large per unit costs to environmental health, both are widely enjoyed, culturally-important, and allow production of food where agronomic potential may be limited due to biophysical constraints. Across the Global North and South, production systems where cows have some access to the outdoors for exercise and grazing have strong potential to align with principles of sustainable production; farmers using grazing practices reap numerous benefits including improved herd health and wellbeing (Mee & Boyle, 2020).

On top of the benefits that grazing-based production has for animal welfare, agroecosystems where grazing is practised (spanning a continuum from highly diverse prairies to fertilised grass monocultures) represent ecosystems with considerably higher stability relative to agroecosystems regularly subject to crop rotations (i.e. arable land). The stability of these agroecosystems allows for the assembly and persistence of ecological communities that support numerous processes integral to sustainable production. One group of organisms that play an important role in the health and functioning of pasture-based agroecosystems are dung beetles (Coleoptera: Scarabaeoidea) (Nichols et al., 2008). Although many people associate dung beetles with habitats found within the Global South such as tropical rainforests and savannahs, dung beetles can be found across a wide range of terrestrial habitats.

More than 7,000 species of dung beetles are known to science. Dung beetles can be classified into three main groups based on their feeding and nesting habits: Dwellers, tunnellers, and rollers.[3] Dwellers arrive at a pile of dung, take up residence, mate, and lay eggs. Once hatched, larvae spend the entirety of their development inside the dung. Female tunneller beetles form dung into lumps, drag them into their tunnels, and lay their eggs inside the dung balls. Once there is an egg inside, the dung lump is called a brood ball. The males compete for access to a female and her tunnel, which they defend until the mated female lays her eggs. Male rolling beetles sculpt dung into a ball using their hind legs (Figure 8.5). If the dung ball is acceptable to a receptive female, he buries the dung in the soil. Once buried, females lay an egg inside the ball and the larva inside will feed on it for the entirety of its development, safely within the soil.

Regardless of the method they use for feeding and nesting, the activity of the dung beetle enhances the rate of dung decomposition which frees space for plants to grow. As beetles feed on the dung, they also mix organic matter into the soil, effectively enhancing soil carbon sequestration. Enhanced levels of soil carbon are well understood to support numerous critical functions within the soil, including improving drought resistance and mitigating climate change (Lal, 2008). The decomposition and burial of dung provides many other environmental benefits

Figure 8.5 A roller dung beetle (*Kheper* sp.) perched on top of its dung ball in Northern Tanzania. Source: Photo by Jennifer Marshman.

including enhanced plant growth, improved hydrological properties of soil, and disruption of pest fly lifecycles (Nichols et al., 2008).

A case study that clearly illustrates the benefits of dung beetles occurred during the middle part of the 20th century in Australia. Cattle represent a large component of agri-food production in Australia, and during the 1950s, serious concerns regarding the environmental sustainability of the sector were laid bare. Farmers across the country observed that grazing areas were covered in undecomposed cattle dung, which severely limited pasture productivity. Blood-feeding pest flies that use cattle dung for larval development, such as the buffalo fly (*Haematobia exigua*), were exceptionally abundant due to the quantity of undecomposed dung available for larval feeding.

George Bornemissza, an entomologist, predicted this phenomenon was attributable to the inability of the native dung beetles to utilise cattle dung. At that time more than 400 dung beetle species could be found across Australia, where the vast majority had evolved using the dung of marsupials. Because marsupial dung is typically formed of small pellets that are highly fibrous and dry, native beetles struggled to adapt to the high moisture cattle dung, and thus cattle dung remained undecomposed posing numerous barriers to production and to wider environmental health.

Led by Bornemissza, the Australian Government began importing species of dung beetles from different parts of the world through a breeding and biosecurity program that was established to ensure that the beetles that were imported were free of any parasites or pathogens (Bornemissza, 1976). At the end of the 25-year program, following 43 species introductions, more than 23 species became successfully established within Australia. As a result, many of the environmental challenges associated with cattle dung were addressed.

The success of the program spread globally, and research continues to explore how dung beetles can enhance sustainable production. An intriguing branch of research involves the potential of dung beetles to reduce the greenhouse gas

(GHG) footprint of beef and dairy production. Methane-producing microbes (methanogens) that live within digestive systems of ruminant livestock are excreted and dung pats on pasture continue to release methane. The activity of dung beetles enhances the speed at which a dung pat dries out while increasing oxygen levels (Penttilä et al., 2013). In a study in Finland, the activity of dung beetles reduced the production of GHGs by roughly 12% at the pasture level, and authors suggested this benefit could be enhanced where livestock spend a greater portion of the year on pasture (Slade et al., 2016).

Recommendations and conclusion

Insects are so integral to the social, economic, and environmental sustainability of food production—and thus sustainable diets—that it is imperative to implement agricultural practices that help to conserve, protect, and enhance insect biodiversity. However, many of our current food systems are designed and managed in a manner that is not positioned to conserve insects and the important contributions they provide to sustainable diets. Although the needs of various insect species vary widely and, in some cases, can be highly specialised, certain practices can be broadly useful in conserving insect biodiversity in agroecosystems.

First, pest control ought to align with the principles of integrated pest management (IPM) (Box 8.6). Chemical insecticides are often the *de facto* response to managing insect pests, but insecticides often have serious negative repercussions within the wider ecosystem. Within an IPM framework, use of insecticides is part of a wider suite of actions. Following the hierarchy of integrated management, non-chemical treatment options can be prioritised, such as choosing pest-resistant crop varieties, altering planting dates, and using companion planting. These strategies can be effective at reducing the overall quantity of insecticides used within the system.

Box 8.6 Integrated pest management defined. Source: FAO, 2021, para. 1

Integrated Pest Management (IPM):"an ecosystem approach to crop production and protection that combines different management strategies and practices to grow healthy crops and minimise the use of pesticides" (FAO, 2021, para. 1).

A second means of conserving insect biodiversity is conserving existing natural habitat and enhancing biodiversity in existing agroecosystems. Insects living within natural ecosystems can permeate agroecosystems where they provide a wide variety of beneficial functions. Conserving natural habitats is important for insects that cannot live *within* agroecosystems because natural habitats contain resources for nesting and provisioning (e.g. deadwood, hostplants, and thermal refuges) that might not be available within agroecosystems. Because conversion of natural habitat to agriculture is widely known to be a leading cause of global biodiversity loss, efforts must be made wherever possible to conserve natural habitats and the associated biodiversity (insect and otherwise) that these landscape features support. When agroecosystems are already established, enhancement features such as hedgerows (Box 8.7) can provide a much-needed refuge for insects (Dainese et al., 2017; Dondina et al., 2018; Garratt et al., 2017), particularly when the wider landscape is highly simplified. Most insects are mobile to some degree and can move amongst landscape features to access food or shelter. For example, the western honey bee can forage at distances of greater than eight kilometres, whereas bumble bees may travel only one kilometre, and the smaller solitary bees may only travel 100–500 metres from their nests. Because foraging ranges vary widely, it is important to consider non-crop habitat features as corridors to facilitate movement, provide sufficient suitable shelter, and allow for safe nesting and reproductive spaces.

Box 8.7 Hedgerow defined. Source: The work of the authors

Hedgerow: A habitat resource in agroecosystems with many benefits, including preventing soil erosion, decreased runoff and improved water filtration, increased carbon capture, and habitat for beneficial insects.

A third and final means of conserving insect biodiversity involves addressing the greatest challenge facing agriculture and humanity today: Anthropogenic climate change. Agriculture is a major contributor of global emissions, and sweeping changes must be implemented across all parts of the food system. Climate change has a variety of serious negative impacts on insects, for example, the phenological mismatches. Mitigation measures include pursuing lower-carbon diets (e.g. including more plant-based proteins), reducing fossil fuel inputs (e.g. scaling up renewable energy generation), and working to prevent food losses and food waste. Conserving insect biodiversity for more sustainable production systems is but one of the important endpoints of solving this pressing global challenge to create a healthier and more equitable future for the entire biosphere.

Notes

1 Also called the European honey bee, with extensive global breeding programs, commercial and hobby beekeeping practices.
2 Disruptions of this nature means that bees may emerge out of sync with the flowering plants they depend on for food sources during specific bloom times. See Kerr et al. (2015).
3 See Hanski & Cambefort (2014).

References

Altieri, M., & Nicholls, C. (2018). *Biodiversity and pest management in agroecosystems*. CRC Press.

Barnes, A. D., Scherber, C., Brose, U., Borer, E. T., Ebeling, A., Gauzens, B., Giling, D. P., Hines, J., Isbell, F., Ristok, C., Tilman, D., Weisser, W. W., & Eisenhauer, N. (2020). Biodiversity enhances the multitrophic control of arthropod herbivory. *Science Advances, 6*(45), eabb6603.

Bornemissza, G. E. (1976). Australian dung beetle project, 1965–1975. *Australian Meat Research Committee Review, 30*(1976), 1–30.

Champetier, A., Sumner, D. A., & Wilen, J. E. (2015). The bioeconomics of honey bees and pollination. *Environmental and Resource Economics, 60*(1), 143–164.

Chan, D. S. W., & Raine, N. E. (2021). Population decline in a ground-nesting solitary squash bee (Eucera pruinosa) following exposure to a neonicotinoid insecticide treated crop (Cucurbita pepo). *Scientific Reports, 11*(1), 1–11.

Colla, S., & MacIvor, J. (2017). Questioning public perception, conservation policy, and recovery actions for honeybees in North America. *Conservation Biology, 31*, 1202–1204.

Dainese, M., Montecchiari, S., Sitzia, T., Sigura, M., & Marini, L. (2017). High cover of hedgerows in the landscape supports multiple ecosystem services in Mediterranean cereal fields. *Journal of Applied Ecology, 54*(2), 380–388.

Dondina, O., Saura, S., Bani, L., & Mateo-Sánchez, M. C. (2018). Enhancing connectivity in agroecosystems: Focus on the best existing corridors or on new pathways? *Landscape Ecology, 33*(10), 1741–1756.

Eilers, E. J., Kremen, C., Greenleaf, S. S., Garber, A. K., Klein, A. M. (2011). Contribution of pollinator-mediated crops to nutrients in the human food supply. *PLoS ONE, 6*, e21363.

Food and Agriculture Organisation of the United Nations (FAO). (2018). *Why bees matter: The importance of bees and other pollinators for food and agriculture*. http://www.fao.org/publications/card/en/c/I9527EN/

FAO. (2021). *Integrated pest management*. Retrieved April, 4, from http://www.fao.org/agriculture/crops/thematic-sitemap/theme/pests/ipm/en/

Fürst, M. A., McMahon, D. P., Osborne, J. L., Paxton, R. J., & Brown, M. J. F. (2014). Disease associations between honeybees and bumblebees as a threat to wild pollinators. *Nature, 506*(7488), 364–366.

Garibaldi, L. A., Steffan-Dewenter, I., Winfree, R., Aizen, M. A., Bommarco, R., Cunningham, S. A., Kremen, C., Boreux, V., Cariveau, D., Chacoff, N. P., DudenHöffer, J. H., Freitas, B. M., Ghazoul, J., Greenleaf, S., Hipolito, J., Hozschuh, A., Howlett, B., Isaacs, R., Javorek, S. K., … Klein, A. M. (2013). Wild pollinators enhance fruit set of crops regardless of honey bee abundance. *Science, 339*(6127), 1608–1611.

Garratt, M. P., Senapathi, D., Coston, D. J., Mortimer, S. R., & Potts, S. G. (2017). The benefits of hedgerows for pollinators and natural enemies depends on hedge quality and landscape context. *Agriculture, Ecosystems & Environment, 247*, 363–370.

Géneau, C. E., Wäckers, F. L., Luka, H., Daniel, C., & Balmer, O. (2012). Selective flowers to enhance biological control of cabbage pests by parasitoids. *Basic and Applied Ecology, 13*(1), 85–93.

Gibson, K., Rose, D., & Fincher, R. (Eds.). (2015). *Manifesto for living in the Anthropocene*. Punctum Books.

Graystock, P., Blane, E. J., McFrederick, Q. S., Goulson, D., & Hughes, W. O. H. (2016). Do managed bees drive parasite spread and emergence in wild bees? *International Journal for Parasitology: Parasites and Wildlife, 5*(1), 64–75.

Hanski, I., & Cambefort, Y. (Eds.). (2014). *Dung beetle ecology* (Vol. 1195). Princeton University Press.

IPBES (Intergovernmental Science-Policy Platform on Biodiversity and Ecosystem Services). (2016). *The assessment report of the intergovernmental science-policy platform on biodiversity and ecosystem services on pollinators, pollination and food production.* https://www.ipbes.net/assessment-reports/pollinators

IPCC (Intergovernmental Panel on Climate Change). (2019). *Climate change and land: IPCC special report on climate change, desertification, land degradation, sustainable land management, food security, and greenhouse gas fluxes in terrestrial ecosystems.* https://www.ipcc.ch/srccl/download/

Jankielsohn, A. (2018). The importance of insects in agricultural ecosystems. *Advances in Entomology, 6*(2), 62–73.

Kerr, J. T., Pindar, A., Galpern, P., Packer, L., Potts, S. G., Roberts, S. M., Rasmont, P., Schweiger, O., Colla, S.R., Richardson, L. L., Wagner, D. L., Gall, L. F., Sikes, D. S., Pantoja, A. (2015). Climate change impacts on bumblebees converge across continents. *Science, 349*, 177–180.

Lal, R. (2008). Soils and sustainable agriculture. *Agronomy for Sustainable Development, 28*(1), 57–64.

Lee, J.C., & Heimpel, G.E. (2005). Impact of flowering buckwheat on Lepidopteran cabbage pests and their parasitoids at two spatial scales. *Biological Control, 34*(3), 290–301.

Leighton, P. (2021). The harms of industrial food production: How modern agriculture, livestock rearing and food processing contribute to disease, environmental degradation and worker exploitation. In Davies, P., Leighton, P., & Wyatt, T. (Eds), *The Palgrave handbook of social harm* (pp. 199–225). Palgrave Macmillan.

Mallinger, R. E., Gaines-Day, H. R., & Gratton, C. (2017). Do managed bees have negative effects on wild bees?: A systematic review of the literature. *PloS One, 12*(12), e0189268.

Marshman, J., & Knezevic, I. (2021). What's in a name? Challenging the commodification of pollination through the diverse economies of 'Bee Cities'. *Journal of Political Ecology, 28*(1), 124–145

Medeiros, H. R., Grandinete, Y. C., Manning, P., Harper, K. A., Cutler, G. C., Tyedmers, P., Righi, C.A., & Ribeiro, M. C. (2019). Forest cover enhances natural enemy diversity and biological control services in Brazilian sun coffee plantations. *Agronomy for Sustainable Development, 39*(6), 1–9.

Mee, J. F., & Boyle, L. A. (2020). Assessing whether dairy cow welfare is "better" in pasture-based than in confinement-based management systems. *New Zealand Veterinary Journal, 68*(3), 168–177.

Nichols, E., Spector, S., Louzada, J., Larsen, T., Amezquita, S., Favila, M. E., & Network, T. S. R. (2008). Ecological functions and ecosystem services provided by Scarabaeinae dung beetles. *Biological conservation, 141*(6), 1461-1474.

Nicholson, C. C., & Ricketts, T. H. (2019). Wild pollinators improve production, uniformity, and timing of blueberry crops. *Agriculture, Ecosystems & Environment, 272*, 29–37.

Penttilä, A., Slade, E. M., Simojoki, A., Riutta, T., Minkkinen, K., & Roslin, T. (2013). Quantifying beetle-mediated effects on gas fluxes from dung pats. *PLoS One, 8*(8), e71454.

Purtauf, T., Roschewitz, I., Dauber, J., Thies, C., Tscharntke, T., & Wolters, V. (2005). Landscape context of organic and conventional farms: Influences on carabid beetle diversity. *Agriculture, Ecosystems & Environment, 108*(2), 165–174.

Reid, W. V., Mooney, H. A., Cropper, A., Capistrano, D, Carpenter, S. R., Chopra, K., Dasgupta, P., Dietz, T., Duraiappah, A. K., Hassan, R., Kasperson, R., Leemans, R., May, R. M., McMicheal, A. J., Pingali, P., Samper, C., Scholes, R., Watson, R. T., Zakri, A. H., … Zurek, M. B. (2005). *Ecosystems and human well-being-synthesis: A report of the Millennium Ecosystem Assessment.* Island Press.

Reilly, J. R., Artz, D. R., Biddinger, D., Bobiwash, K., Boyle, N. K., Brittain, C., Brokaw, J. Campbell, J. W., Daniels, J., Elle, E., Ellis, J. D., Fleischer, S. J., Gibbs, J., Gillespie, R. L., Gundersen, K. B., Gut, L., Hoffman, G., Joshi, N., Lundin, O., … Winfree, R. (2020). Crop production in the USA is frequently limited by a lack of pollinators. *Proceedings of the Royal Society B, 287*(1931), 20200922.

Ritchie, H., & Roser, M. (2013, September) *Land use*. Our World in Data. https://our-worldindata.org/land-use.

Royal Entomological Society. (2021). *Entomology: Facts and figures*. Retrieved October, 2021, from https://www.royensoc.co.uk/facts-and-figures

Rucker, R. R., Thurman, W. N., & Burgett, M. (2019). Colony collapse and the consequences of bee disease: Market adaptation to environmental change. *Journal of the Association of Environmental and Resource Economists, 6*(5), 927–960.

Sarfraz, M., Keddie, A. B., & Dosdall, L. M. (2005). Biological control of the diamondback moth, *Plutella xylostella*: A review. *Biocontrol Science and Technology, 15*(8), 763–789.

Sharma, S., Kooner, R., & Arora, R. (2017). Insect pests and crop losses. In R. Arora, & S. Sandhy, (Eds.), *Breeding insect resistant crops for sustainable agriculture* (pp. 45–66). Springer.

Slade, E. M., Riutta, T., Roslin, T., & Tuomisto, H. L. (2016). The role of dung beetles in reducing greenhouse gas emissions from cattle farming. *Scientific Reports, 6*(1), 1–9.

Valido, A., Rodríguez-Rodríguez, M. C., & Jordano, P. (2019). Honeybees disrupt the structure and functionality of plant-pollinator networks. *Scientific Reports, 9*(1), 1–11.

van Vierssen Trip, N., MacPhail, V. J., Colla, S. R., & Olivastri, B. (2020). Examining the public's awareness of bee (Hymenoptera: Apoidae: Anthophila) conservation in Canada. *Conservation Science and Practice, 2*(12), e293.

Werling, B. P., Harmon, J., Straub, C., & Gratton, C., 2012. Influence of native North American prairie grasses on predation of an insect herbivore of potato. *Biological Control, 61*(1), 15–25.

Wilson, J. S., Forister, M. L., & Carril, O. M. (2017). Interest exceeds understanding in public support of bee conservation. *Frontiers in Ecology and the Environment, 15*, 460–466.

9

BREAD FROM BUGS AND THE SIGNIFICANCE OF EMPLOYING INSECT FOODS IN THE HUMAN DIET

Bruno Borsari

Introduction

Entomophagy, or the consumption of insects, has been practised since the dawn of human evolution (Lesnick, 2018; Morris, 2006; Ramos-Elorduy, 2009), and it continues to be relevant for the daily nutrition of approximately two billion people worldwide (DeFoliart, 1999; Raubenheimer et al., 2014; van Huis et al., 2013). Diets relying on insects favour their consumption at the larval or pupal stage because these forms of development lack a full exoskeleton, thus, making them more digestible. Heat treatments as applied in various cooking methods (e.g., boiling, frying, and/or smoking), give insects distinctive flavours, while getting rid of parasites and food pathogens (van Huis, 2020). Presently, in response to increasing food demands by a growing population and fast societal changes, enhanced by expanding immigrant communities in cities of industrial countries (Borsari, 2022), the idea of looking at insects as food has been getting traction worldwide.

Many studies report that edible insects can complement human diets, making entomophagy an established dietary practice that can help attenuate the ill effects of malnutrition in countries affected by recurrent food crises (Durst et al., 2010, Feng, 2018; Gahukar, 2011). Also, farming insects with its modest resource use contrasts with the environmental impacts caused by livestock agriculture, prompting this practice to be scalable to rural as well as urban areas (van Huis & Oonincx, 2017). A plethora of edible insect species has potential to fulfil people's nutritional needs (van Huis et al., 2014) and to aid in diversifying human diets. Industrial agriculture seeks to maximise the yields of certain crops and livestock, thus, homogenising and lowering food variety around the world (Vandermeer et al., 2018). Belluco and colleagues (2013) argued that when foods derived from insects are prepared correctly they should not pose adversary risks to human health. Therefore, insects'

DOI: 10.4324/9781003174417-11

processing and preparation for sale and consumption will require compliance with specific protocols, similar to more conventional animal-derived foods, to ensure their quality and safety (Dickel et al., 2020).

There are several advantages of growing insects as food, whether for human or animal nutrition. With more than a thousand edible species, some scientists have been arguing that growing insects as food provides tangible advantages toward achieving sustainability, besides optimising nutrition benefits and health. Insects' high content of macronutrients (proteins, lipids, Omega-3, fibre) is comparable and even higher than that of meat and fish (Svanberg & Berggren, 2021; van Huis et al., 2013). In addition, raising insects minimises water use and feeding costs while reducing emissions of ammonia and other greenhouse gases (Oonincx & Boer, 2012), making insects useful for producing feed employable in livestock farming and in aquaculture (Table 9.1).

For these reasons, insects are beginning to be viewed as micro-livestock species of significant interest for their potential to yield nutritious food for people and farm animals (Belluco et al., 2015; Glover & Sexton, 2015). Insects are attractive also to investors and entrepreneurs, who envision further developing this specific sector of the food supply chain (Fowles & Nansen, 2020; Gahukar, 2011) with the expectation of occupying soon-legitimate market niches (Reverberi, 2021). Yet, a reluctance toward consuming insects or insect foods in the industrialised countries of the northern hemisphere remains and acts as a constraint to insect-market expansions (Belluco et al., 2013; van Huis et al., 2013). Concerns about insect-based food consumption safety require legal consents and approvals of production protocols from governments so that insect farming, processing, and marketing may become better regulated (Collins et al., 2019). Additional obstacles to acceptability of these foods are perceptions of a general disgust for entomophagy (Chang et al., 2019; Collins et al., 2019; Menozzi et al., 2017).

Nonetheless, it remains important to recognise that besides food (e.g., honey, lerp, honeydew) insects provide a variety of valuable products (e.g., silk, carmine, wax, propolis), including medicines (Svanberg & Berggren, 2021). The ecological services of insects—such as pollination of most flowering plants, biological control of noxious insect species, and contributions to soil health through a decom-

Table 9.1 Comparisons of feed and water requirements to produce 1 kg of meat among selected farm animals and crickets. Source: Modified from the author after the work of van Huis et al., 2013. Used with permission

Animal species (1 kg) of meat	Water (l)	Feed (kg)
Cow	15.416	10
Sheep	8.763	10
Pig	5.956	4
Chicken	4.325	2
Cricket	5	2

position of soil detritus (Borsari, 2020)—continue to be vital processes to the ecological productivity of ecosystems (Borsari, 2022; van Huis et al., 2013). Thus, insect products and services legitimise the strengthening of environmental conservation strategies to avert more degradative changes to ecosystems that extirpate insect species (Wagner, 2020). Loss of natural habitats, caused by a use of land that is highly skewed toward a homogenisation of landscapes, and extensive use of biocides in intensive livestock production and industrial agriculture demand more effective policy measures to ensure better protection of global entomofauna (Cardoso et al., 2020).

The importance of this work can be seen in evaluating the suitability of insect food for consumption and its role in enhancing healthy nutrition and sustainability in the human diet. A thorough review of entomophagy considered the benefits of accepting insects as quality products when consumed in foods like bread and other baked goods, and analysed the impediments of becoming established within the food-chain-sector. Bread is an iconic food that for millennia has been sustaining human populations in vast regions around the world. Therefore, assessing the feasibility of using insect flours/powders as a novel ingredient in the baking process is another important focus of this chapter.

Trends in insects consumption and baked foods with insect flours/powders, in Europe

During the last decade, the European Union (EU) has been moving promptly to evaluate the safety of insect-derived foods, as these are becoming available through international trade agreements with producers from outside Europe. For example, insects and related food products were classified in a specific category—novel foods—and their sale is dependent upon permission granted by the EU Commission (Regulation (EU) 2015/2283, European Parliament, 2015). In addition, more recent provision has been regulating the use of insects as processed animal proteins (PAPs) in aquaculture-farming operations (Regulation (EU) 2017/893, European Parliament, 2017). The Netherlands has been leading the European insect food market; for several years they have already been offering insect products to Dutch consumers (Collins et al., 2019).

To produce high quality insect foods on a large-scale, growers need a safe feed source for raising these invertebrates, such as vegetables, fruits, or grains, and an appropriately designed space with energy availability for temperature control, light, ventilation, and clean water for maintaining hygiene and automation (Berggren et al., 2018). Organic waste may be used as food for insects grown for human consumption. However, researchers remain uncertain whether insects grown on these diets accumulate mycotoxins (Osimani & Aquilanti, 2021). In human nutrition, the most common risk to people's health posed by food is nutrient malabsorption, including chemical and microbiological contamination, which may cause allergenic reactions (Testa et al., 2016). Therefore, special attention should be paid to these issues when edible insects are processed, making their preparation phase a priority task. For these reasons, producers should implement a Hazard Analysis

Critical Control Plan (HACCP) to limit risks for consumers' health, such as the one developed by Kooh and team (2020) for flour obtained from mealworm larvae (*Tenebrio molitor*), used in bread and other baked products.

An accumulation of heavy metals is another concern for some cultivated species of insects, which requires appropriate regulations and quality control protocols. From a food safety perspective, management in raising, harvesting, and post-harvest technologies for insects remain keystone processes to guarantee food safety and quality (Testa et al., 2016). Therefore, to make the mass production of insects attractive and acceptable to consumers of industrialised countries, compared to other animal-derived foods, it is necessary to continue researching rearing and harvesting practises, as well as post-harvest processing technologies, including those for extracting proteins, fatty acids, and micronutrients (Durst et al., 2010). A review study by Belluco et al. (2013) confirmed that it is feasible for humans to eat some insect species when they are processed and prepared according to specific protocols that regulate their sale for consumption. Although the analysis of selected insect species showed a high microbial load and diversity, the research conducted by Osimani and his collaborators (2016) supported insects to be consumed on a large-scale for their richness in proteins and lipids. However, the presence of pathogenic microorganisms and their spores in bread samples reiterated the need of focusing on the safety issues posed by insect consumption, including their edible products (Osimani & Aquilanti, 2021). For these reasons, further expansions of this novel sector within the food industry in Europe will depend on achieving a better understanding of the impact of diverse rearing and processing techniques and on the nutritional and microbiological quality of insect foods (Berggren et al., 2018).

Bread-making: Its fortification and consumption

The famous motto "Fiat Panis"—"Let there be bread" of the United Nations Food & Agriculture Organisation (FAO) epitomises the importance of bread as food among several human cultures. The history of bread began in Neolithic times in western Asia, and wheat (*Triticum spp.*) remains most valued among other cereals (e.g., barley, maize, rye, sorghum, oats) for its superior nutritive value (Kourkouta et al., 2017). The gluten proteins in its seeds give wheat dough ideal traits (stickiness and the ability to rise when leavened) that optimise its baking qualities, making wheat the preferred staple among many farming communities where this species and its many cultivars can be grown (Zohary et al., 2012). However, the earliest breads were unleavened, and variations in grain type, thickness, shape, and texture diversified bread across cultures. Archaeological evidence confirmed that yeast (both as a leavening agent and for brewing ale) was used in Egypt as early as 4000 B.C., and food historians generally consider this date as a breakthrough discovery (Kourkouta et al., 2017). However, Baldoni and Giardini (1982) supported the theory that brewing began when the first cereal crops were domesticated in the Fertile Crescent region of Mesopotamia. Nonetheless, there is an overall agreement that the discovery of yeast induced fermentation was accidental and, according to the Cambridge World History of Food (Kiple & Ornelas, 2000), additional evidence suggests that barley (*Hordeum spp.*) was being fermented around 3500 B.C. in ancient Mesopotamia, or present-day Iran.

Eglite and Kunkulberga (2017) reported that people of European countries consume on average 59 kg/capita/year of bread, ranging from 104 kg/capita/year in Turkey to 32 kg/capita/year in Great Britain. Bread consumption has been stable in recent years in Greece (68 kg), Italy (52 kg), and Spain with 37 kg/capita/year. Bread made with wheat (*Triticum spp.*) flour is rich in carbohydrates, containing about 50% of these macromolecules in its dry weight, and its protein content ranges from 6% to 8% (Roncolini et al., 2019). Therefore, the idea of fortifying bread is legitimised by the goal of maintaining good nutrition and health among older Europeans to support increasing lifespans and to remediate from losses of skeletal muscle tissue (sarcopenia) as they age, which can be amplified by sub-optimal protein intake (Deer & Volpi, 2015). These are keystone reasons that have spurred enthusiasm and interest among many food scientists, whose research now includes improvements to the quality of proteins supplied with the diets of senior citizens. Therefore, an addition of essential amino acids to foods, while exploring different approaches to increase protein intake through the consumption of bread that may include insect flours/powders, has been an important focus of this research agenda (González et al., 2019), especially in Europe, where bread is consumed in large quantities on a regular basis (Osimani et al., 2018).

However, a fortification of bread and other leavened baked goods can decrease the palatability of these foods and consequently reduce the acceptance of consumers, as happened when cricket (*Acheta domesticus*) flour was mixed in muffin dough (Höglund et al. 2017). Also, a presence of bacterial spores raised safety concerns in cricket-based breads, as these may spoil the bread or make people sick, although irradiation treatments with gamma rays will eliminate spores from crickets and other insect powders (Osimani & Aquilanti, 2021). Nonetheless, a study by Osimani and his team (2018) showed that breads containing cricket powder were more nutritious than the wheat-based breads, with more proteins and essential amino acids, including fatty acids that add calories and better prevent the bread from becoming stale.

González and her collaborators (2019) found that breads containing *A. domesticus* flour had similar specific volume and texture when compared to conventional wheat bread. However, the fortified bread had a higher content of proteins and fibre, thus verifying the nutritional advantages possessed by bread containing cricket flour. Similarly, in a recent study from Thailand, researchers showed that an addition of cricket powder (10%) to the dough used for baking bread and cookies made these products rich in proteins, iron, phosphorus, low in carbohydrates, and gave these foods a darker colour, and that consumers' acceptability was comparable to control products (Bawa et al., 2020). Also, a study of biscuits showed that protein and ash contents increased while carbohydrates decreased when these were supplemented with mealworm (*T. molitor*) flour (Zielińska & Pankiewicz, 2020). The same researchers also discovered that mealworm flour increased antioxidants and other nutrients, thus suggesting that flours from mealworms can enrich baked foods by increasing nutritional value and overall healthy attributes (Zielińska & Pankiewicz, 2020). These studies substantiated the viability of improving the nutritional benefits of bread and baked foods when flours from cultivated cereal crops are enriched with modest amounts of insect powders. Finally, on January 2021, the European Food Safety Authority (EFSA) published the first report about insects as a novel food, declaring that dry mealworm larvae (*T. molitor*) are safe as food if in compliance with Regulation (EU) 2015/2283 (Turck et al., 2021). This requirement has been interpreted as a significant achievement for the insect food sector, benefitting now from greater opportunities for opening the market throughout Europe of many other insect species as food and feed. Most recently, a new provision by the European Union (EU) has been enacted to allow the use of these invertebrates as feed for chickens and pigs (starting on September 7), thus boosting consumption of insects for these industries from 10,000 tons/year to an estimated 500,000 tons/year by 2030 (Dusi, 2021).

Conclusion: Overcoming neophobia and supporting "Out of the Box Gastronomy"

Neophobia is a behaviour illustrated by hesitance and even rejection to eating a new food that has been introduced into a culture from somewhere else (Pliner & Hobden, 1992). In Europe, where there is no history of entomophagy, a reluctance for considering insects as food persists (Payne et al. 2016). According to Piha and team (2018), consumers in northern Europe are more receptive to trying insect foods than those from Central Europe, whereas young, European consumers showed sincere interest in entomophagy (Kostecka et al., 2017; Sogari et al., 2017). Yet, more studies have indicated that despite the health and environmental benefits of consuming insect foods, these are not sufficient to justify their establishment in gastronomy (House, 2016; van Huis, 2017), even when food is simply enriched with insect flours (Tan et al., 2017). Recently, Mancini and collaborators (2019) codified factors determining either an acceptance or rejection of novel foods (both personal and social factors) to predict people's intentions to consume insect foods. Among personal factors, disgust emerged as a main constraint to insect consumption (Gmuer et al., 2016; Lorenz et al., 2014), in addition to views about insects as pests and risks they may pose to public health (Hartmann et al. 2015). Social constraints have included lack of familiarity (Caparros et al., 2016; Cicatiello et al., 2016; Hartmann et al. 2015) and old misconceptions about whether it is appropriate to eat insects (Tan et al., 2016).

While there is much interest in further exploring consumers' perceptions and attitudes about insects as food, there also are more considerations, such as cooking with insects (Tan et al., 2017), including insects in recipes (van Huis et al., 2014), and with opportunities for degustation sampling that could be accomplished through outreach at fairs, symposia, and more public events. To this end, an inclusion of sensorial analyses of insect foods to study consumers' preferences is a sector of marketing research that deserves attention. Photos of insects and posters of various dishes (including recipes) could be shown while sampling insect foods, which could become the new norm of familiarising the public with the benefits and potential of this new gastronomy. Concurrently, these and similar approaches will assist to describe a more realistic scenario about insect food consumption and consumer behaviour. Insects have been foods of only ancillary importance in Europe, and their use has differed among countries, but now a renewed interest in insects as food is envisioned as benign and achievable (Svanberg & Berggren, 2021). For example, in Italy, a feasible strategy for improving acceptance of insect-based foods could include insect ingredients in iconic foods like pasta, pizza, and bread (Moruzzo et al., 2021).

However, this realisation will require policies aimed at regulating insect consumption and sound, functional infrastructures to fulfil the management and sanitary aspects of insect rearing (Berggren et al., 2018). Most of all, an inclusion of insects in Western food culture will depend on cooks' abilities to create not only nutritious but also tasty, attractive food that blends harmoniously with foods and food ingredients that are already familiar (van Huis et al., 2014). The "planetary diet" was proposed in 2019 as a route to ensure adequate, healthy food for humanity in the 21st century and beyond, while harnessing agriculture to continue operating within the limits of planetary boundaries (Rockström & Gaffney, 2021). Present models of industrial agriculture are unsustainable and threaten planetary health for their immense demand on resources (e.g., land, fossil fuels, water). Their environmental and public health impacts, like pollution caused by fertiliser, pesticide applications, and more biocides (antihelminthic drugs and antibiotics in animal farming), generate massive emissions of greenhouse gases and toxins that contaminate air, soil, and water sources. Therefore, the planetary diet was conceived as a feasible action plan for abating pollution and resource needs of agriculture, while persuading people to shift their nutrition toward a diet that is more plant-based. This societal change would initiate transformative steps toward sustainable agriculture, achievable through a decreasing consumption of animal and dairy foods (Rockström & Gaffney, 2021). This shift has potential to swerve human health toward a sustainable diet, while restoring Earth's ecology and health. Yet, entomophagy was not included among the recommendations for implementing the planetary diet. Instead, eating insects could contribute a viable role in the establishment of sustainable diets, as these nutritional tactics merge efforts directed toward transforming food systems, restoring human health, and planetary homeostasis (Figure 9.1).

Finally, the cities of industrialised countries are becoming more diverse due to an on-going immigration flow toward urban centres, and this trend will increase market demands for insect foods, thus exposing more people to entomophagy and its many benefits and opportunities. This will help steer peoples' dietary habits a step closer toward sustainability, while expanding culinary boundaries beyond established comfort zones. This "out of the box" gastronomy has potential for improving and diversifying the human diet, while enhancing health for all.

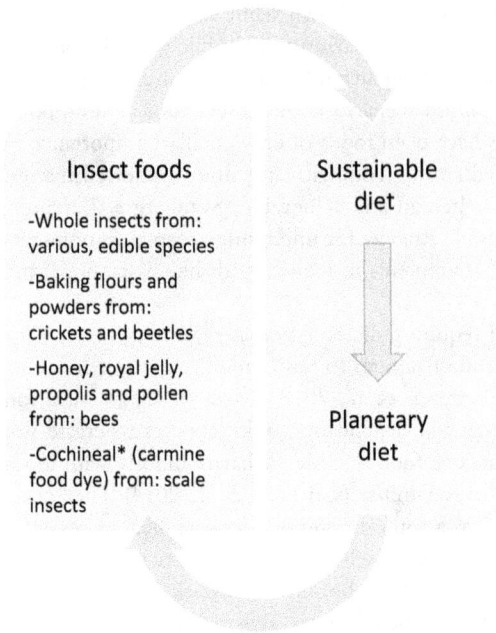

Figure 9.1 Model presentation. Source: Work of the author.

References

Baldoni, R. & Giardini, L. (1982). *Coltivazioni erbacee*. Pátron.

Bawa, M., Songsermpong, S., Kaewtapee, C., & Chanput, W. (2020). Nutritional, sensory, and texture quality of bread and cookie enriched with house cricket (*Acheta domesticus*) powder. *Journal of Food Processing & Preservation, 44,* Article e14601.

Belluco, S., Losasso, C., Maggioletti, M., Alonzi, C. C., Paoletti M. G., & Ricci, A. (2013). Edible insects in a food safety and nutritional perspective: A critical review. *Comprehensive Reviews in Food Science and Food Safety, 12,* 296–313.

Belluco, S., Losasso, C., Maggioletti, M., Alonzi, C. C., Ricci, A., & Paoletti, M. G. (2015). Edible insects: A food security solution or a food safety concern? *Animal Frontiers, 5*(2), 25–30.

Borsari, B. (2020). Soil quality and regenerative, sustainable farming systems. In W. Leal Filho, A. Azul, L. Brandli, P. Özuyar, & T. Wall (Eds.), *Zero hunger: Encyclopedia of the UN sustainable development goals*. Springer.

Borsari, B. (2022). Entomophagy as the new frontier in food systems sustainability. In W. Leal Filho & A. Machnick (Eds.), *Sustainable consumption and quality of life. Handbook of sustainability*. Springer (In press).

Berggren, Å., Jansson, A., & Low, M. (2018). Using current systems to inform rearing facility design in the insect-as-food industry. *Journal of Insects as Food and Feed, 4*(3), 167–170.

Caparros, M. R., Gierts, C., Blecker, C., Brostaux, Y., Haubruge, É., Alabi, T., Francis, F. (2016). Consumer acceptance of insect based alternative meat products in Western countries. *Food Quality and Preference, 52,* 237–243.

Cardoso, P., Barton, P. S., Birkhofer, K., Chichorro, F., Deacond, C., Fartmanne, T., Fukushima, C. S., Gaigher, R., Habel, J. C., Hallmann, C. A., Hill, M. J., Hochkirch, A., Kwak, M. L., Mammola, S., Noriega, J. A., Orfingern, A. B., Pedraza, F., Pryke, J. S., Roque, F. O., ...

Samways, M. J. (2020). Scientists' warning to humanity on insect extinctions. *Biological Conservation, 242*, 1–12.

Chang, H. P., Ma, C. C., & Chen, H. S. (2019). Climate change and consumer's attitude toward insect food. *International Journal of Environmental Research and Public Health, 16*(9), 1606.

Cicatiello, C., De Rosa, B., Franco, S., Lacetera, N. (2016). Consumer approach to insects as food: Barriers and potential for consumption in Italy. *British Food Journal, 118*, 2271–2286.

Collins, C. M., Vaskou, P., & Kountouris, Y. (2019). Insect food products in the Western world: Assessing the potential of a new 'green' market. *Annals of the Entomological Society of America, 112*(6), 518–528.

Deer, R. R., & Volpi, E. (2015). Protein intake and muscle function in older adults. *Current Opinion in Clinical Nutrition and Metabolic Care, 18*(3), 248–253.

DeFoliart, G. R. (1999). Insects as food: Why the Western attitude is important. *Annual Review of Entomology, 44*, 21–50.

Dickel, M., Eilenberg J., Falcao Salles, J., Jensen, A. B, Lecocq, A., Pijlman, G. P., van Loon, J. J. A., & van Oers, M. M. (2020). Edible insects unlikely to contribute to transmission of coronavirus SARS-CoV-2. *Journal of Insects as Food and Feed, 6*(4), 333–339.

Durst, P. B., Johnson, D. V., Leslie R. N., & Shono, K. (Eds.) (2010). *Forest insects as food: Humans bite back.* FAO. https://www.doc-developpement-durable.org/file/Elevages/Insectes/edible%20forest%20insects.pdf

Dusi, E. (2021, September 10). *Gli insetti si avvicinano alle nostre tavole: L'Ue li ha approvati per i mangimi di polli e maiali.* La Repubblica. https://www.repubblica.it/cronaca/2021/09/10/news/gli_insetti_si_avvicinano_alle_nostre_tavole_l_ue_li_ha_approvati_per_i_mangimi_di_polli_e_maiali-316901396/?ref=RHTP-VS-I287621970-P15-S3-T1&__vfz=medium%3Dsharebar&fbclid=IwAR1PJgYcSGAH9dzA9WEnSHIOkYzZ1AwYDphm7guQ1pB3yOqYGHydB5_nBQI

Eglite A., & Kunkulberga D. (2017). Bread choice and consumption trends. In [Conference presentation]. 11th Baltic Conference on Food Science and Technology: Food Science and Technology in A Changing World (pp. 178–182). Jelgava, Latvia. https://llufb.llu.lv/conference/foodbalt/2017/FoodBalt_2017_Conference_Proceedings.pdf

European Parliament (2015). *Regulation 2015/2283 on novel foods.* https://eur-lex.europa.eu/legal-content/EN/TXT/PDF/?uri=CELEX:32015R2283&from=EN

European Parliament (2017). *Regulation 2017/893 as regards the provisions on processed animal protein.* https://eur-lex.europa.eu/legal-content/EN/TXT/PDF/?uri=CELEX:32017R0893&from=IT

Feng, S. (2018). Tenebrio molitor L., entomophagy and processing into ready to use therapeutic ingredients: A review. *Journal of Nutritional Health & Food Engineering, 8*(3), 280–285.

Fowles, T. M., & Nansen, C. (2020). Insect-based bioconversion: Value from food waste. In E. Närvänen, N. Mesiranta, M. Mattila, & A. Heikkinen (Eds.), *Food waste management.* Palgrave Macmillan.

Gahukar, R. T. (2011). Entomophagy and human food security. *International Journal of Tropical Insect Science, 31*(3), 129–144.

Glover, D., & Sexton, A. (2015). *Edible insect and the future of food: A foresight scenario exercise on entomophagy and global food security.* Institute of Development Studies.

Gmuer, A., Nuessli G. J., Hartmann, C., & Siegrist, M. (2016). Effects of the degree of processing of insect ingredients in snacks on expected emotional experiences and willingness to eat. *Food Quality and Preference, 54*, 117–127.

González, C. M., Garzón, R., & Rosell, C. M. (2019). Insects as ingredients for bakery goods. A comparison study of *H. illucens, A. domestica* and *T. molitor* flours. *Innovative Food Science & Emerging Technologies, 51*, 205–210.

Hartmann, C., Shi, J., Giusto, A., & Siegrist, M. (2015). The psychology of eating insects: A cross-cultural comparison between Germany and China. *Food Quality and Preference, 44*, 148–156.

Höglund, E., Albinsson, B., Stuhr-Olsson, G., Signas, M., Karlsson, C., Rothenberg E, & Wendin, K. (2017). Protein and energy enriched muffins designed for nutritional needs of older adults. *Nutrition & Food Science*, *2*(4), 555592. DOI: 10.19080/NFSIJ.2017.02.555592.

House, J. (2016). Consumer acceptance of insect-based foods in the Netherlands: Academic and commercial implications. *Appetite*, *107*, 47–58.

Kiple K. F., & Ornelas, K. C. (Eds.). (2000). *The Cambridge world history of food*. Cambridge University Press.

Kooh, P., Jury, V., Laurent, S., Audiat-Perrin, F., Sanaa, M., Tesson, V., Federighi, M., & Boué, G. (2020). Control of biological hazards in insect processing: Application of HACCP method for yellow mealworm (*Tenebrio molitor*) powders. *Foods*, *9*, 1528.

Kostecka, J., Konieczna, K., & Cunha, L. M. (2017). Evaluation of insect-based food acceptance by representatives of polish consumers in the context of natural resources processing retardation. *Journal of Ecological Engineering*, *18*, 166–174.

Kourkouta, L., Koukourikos, K., Iliadis, C., Ouzounakis, P., Monios, A., & Tsaloglidou, A. (2017). Bread and Health. *Journal of Pharmacy and Pharmacology*, *5*, 821–826.

Lesnik, J. J. (2018). *Edible insects and human evolution*. University Press of Florida.

Lorenz, A. R., Libarkin, J. C., & Ording, G. J. (2014). Disgust in response to some arthropods aligns with disgust provoked by pathogens. *Global Ecology and Conservation*, *2*, 248–254.

Mancini, S., Sogari, G., Menozzi, D., Nuvoloni, R., Torracca, B., Moruzzo, R., & Paci, G. (2019). Factors predicting the intention of eating an insect-based product. *Foods*, *8*, 270.

Menozzi, D., Sogari, G., Veneziani, M., Simoni, E., Mora, C. (2017). Eating novel foods: An application of the theory of planned behaviour to predict the consumption of an insect-based product. *Food Quality and Preference*, *59*, 27–34.

Morris, B. (2006). Insects as food among hunter-gatherers. *Anthropology Today*, *24*(1), 6–8.

Moruzzo, R., Mancini, S., Boncinelli, F., & Riccioli, F. (2021). Exploring the acceptance of entomophagy: A survey of Italian consumers. *Insects*, *12*, 123.

Oonincx, D. G. A. B., & Boer, I. J. M. (2012). Environmental impact of the production of mealworms as a protein source for humans: A life cycle assessment. *PLoS ONE*, *7*(12), Article e51145.

Osimani, A., & Aquilanti, L. (2021). Spore-forming bacteria in insect-based foods. *Current Opinion in Food Science*, *37*, 112–117.

Osimani, A., Garofalo, C., Milanovic´, V., Taccari, M., Cardinali, F., Aquilanti, L., Pasquini, M., Mozzon, M., Raffaelli, N., Ruschioni, S., Riolo, P., Isidoro, I., & Clementi, F. (2016). Insight into the proximate composition and microbial diversity of edible insects marketed in the European Union. *European Food & Research Technology*, *243*(7), 1157–1171.

Osimani, A., Milanovic, V., Cardinali, F., Roncolini, A., Garofalo, C., Clementi, F., et al. (2018). Bread enriched with cricket powder (*Acheta domesticus*): A technological, microbiological and nutritional evaluation. *Innovative Food Science & Emerging Technologies*, *48*, 150–163.

Payne, C. L. R., Dobermann, D., Forkes, A., House, J., Josephs, J., McBride, A., Müller, A., Quilliam, R. S.& Soares, S. (2016). Insects as food and feed: European perspectives on recent research and future priorities. *Journal of Insects as Food and Feed*, *2*, 269–276.

Piha, S., Pohjanheimo, T., Lähteenmäki-Uutela, A., Křečková, Z., & Otterbring, T. (2018). The effects of consumer knowledge on the willingness to buy insect food: An exploratory cross-regional study in Northern and Central Europe. *Food Quality and Preference*, *70*, 1–10.

Pliner, P., & Hobden, K. (1992). Development of a scale to measure the trait of food neophobia in humans. *Appetite*, *19*, 105–120.

Ramos-Elorduy, J. (2009). Anthropo-entomophagy: Cultures, evolution and sustainability. *Entomological Research*, *39*(5), 271–288.

Raubenheimer D., Rothman J. M., Pontzer H., & Simpson S. J. (2014). Macronutrient contributions of insects to the diets of hunter–gatherers: A geometric analysis. *Journal of Human Evolution*, *71*, 70–76.

Reverberi, M. (2021). The new packaged food products containing insects as an ingredient. *Journal of Insects as Food and Feed, 7*(5), 901–908.

Rockström, J., & Gaffney, O. (2021). *Breaking boundaries: The science of our planet.* Dorling Kindersley Limited.

Roncolini, A., Milanović, V., Cardinali, F., Osimani, A., Garofalo, C., Sabbatini, R., Clementi, F., Pasquini, M. Mozzon, M., Foligni, R., Raffaelli, N., Zamporlini, F., Minazzato, G., Trombetta, M. F., Van Buitenen, A., Campenhout, L. V., & Aquilanti, L. (2019). Protein fortification with mealworm (*Tenebrio molitor* L.) powder: Effect on textural, microbiological, nutritional and sensory features of bread. *PLoS ONE, 14*(2), Article e0211747.

Sogari, G., Menozzi, D., & Mora, C. (2017). Exploring young foodies: Knowledge and attitude regarding entomophagy: A qualitative study in Italy. *International Journal of Gastronomy and Food Science, 7,* 16–19.

Svanberg, I., & Berggren, Å. (2021). Insects as past and future food in entomophobic Europe. *Food, Culture & Society, 24*(5), 624–638.

Tan, H. S. G., Fischer, A. R. H., van Trijp, H. C. M., & Stieger, M. (2016). Tasty but nasty? Exploring the role of sensory-liking and food appropriateness in the willingness to eat unusual novel foods like insects. *Food Quality and Preference, 48,* 293–302.

Tan, H. S. G., Verbaan, Y. T., & Stieger, M. (2017). How will better products improve the sensory-liking and willingness to buy insect-based foods? *Food Research International, 92,* 95–105.

Testa, M., Stillo, M., Maffei, G., Andriolo, V., Gardois, P., & Zotti, C. M. (2016). Ugly but tasty: A systematic review of possible human and animal health risks related to entomophagy. *Critical Reviews in Food Science and Nutrition, 57,* 17.

Turck, D., Castenmiller, J., De Henauw, S., Hirsch-Ernst, K. I., Kearney, J., Maciuk, A., Mangelsdorf, I., McArdle, H. J., Naska, A., Pelaez, C., Pentieva, K., Siani, A., Thies, F., Tsabouri, S., Vinceti, M., Cubadda, F., Frenzel, T., Heinonen, M., Marchelli, R., ... & Knutsen, H. K. (2021). Scientific opinion on the safety of dried yellow mealworm (Tenebrio molitor larva) as a novel food pursuant to Regulation (EU) 2015/2283. *EFSA Journal, 19*(1), 6343.

van Huis, A. (2017). Edible insects and research needs. *Journal of Insects as Food and Feed, 3*(1), 3–5.

van Huis, A. (2020). Nutrition and health of edible insects. *Current Opinion in Clinical Nutrition and Metabolic Care, 23*(3), 1.

van Huis, A., & Oonincx, D. G. A. B. (2017). The environmental sustainability of insects as food and feed. A review. *Agronomy for Sustainable Development, 37,* 43.

van Huis A., van Itterbeeck J., Klunder H., Mertens E., Halloran A., Muir G., & Vantomme P. (2013). *Edible insects future prospects for food and feed security.* FAO. http://www.fao.org/docrep/018/i3253e/i3253e00.htm

van Huis, A., van Gurp, H., & Dicke, M. (2014). *The insect cookbook: Food for a sustainable planet.* Columbia University Press.

Vandermeer, J., Aga, A., Allgeier, J., Badgley, C., Baucom, R., Blesh, J., Shapiro, L. F., Jones, A. D., Hoey, L., Jain, M., Perfecto, I., & Wilson, M. L. (2018). Feeding prometheus: An interdisciplinary approach for solving the global food crisis. *Frontiers in Sustainable Food Systems, 2,* 39.

Wagner, D. L. (2020). Insect declines in the anthropocene. *Annual Review of Entomology, 65*(23), 1–24.

Zielińska, E., & Pankiewicz, U. (2020). Nutritional, physiochemical, and antioxidative characteristics of shortcake biscuits enriched with *Tenebrio molitor* flour. *Molecules, 25,* 5629.

Zohary, D., Hopf, M., & Weiss, E. (2012). *Domestication of plants in the old world* (4th ed.). Oxford University Press.

10

REALISING THE POTENTIAL FOR AQUATIC FOODS TO CONTRIBUTE TO ENVIRONMENTALLY SUSTAINABLE AND HEALTHY DIETS

Anna K. Farmery and Jessica R. Bogard

Introduction

Aquatic foods, including plants and animals from fisheries and aquaculture, have been an important part of human diets for hundreds of millennia (Marean et al., 2007). However, until recently they have been largely excluded from research and policy on sustainable diets and food and nutrition security (Farmery et al., 2017; Stetkiewicz et al., 2022). Many aquatic foods are high in bio-available nutrients, and some can be harvested or produced with environmental impact and resource use that is lower than terrestrial animal sourced foods (Gephart et al., 2021) and equivalent to plant-based foods (Koehn et al., 2022). Aquatic foods also play an important role in the prevention of non-communicable diseases such as cardiovascular disease (Raatz et al., 2013), and their consumption could be increased globally to achieve greater alignment with national dietary guidelines (Springmann et al., 2020). Aquatic foods already play a key role in the diets of many people, in particular those in Indigenous and coastal communities and in many low and middle-income countries, and there is substantial potential for aquatic foods from marine and freshwater fishing and farming to make a greater contribution to sustainable diets globally. This potential will be mediated by challenges including the impacts of climate change, habitat loss and degradation, overfishing, and the provision of aquaculture feed. The ability of aquatic foods to contribute to sustainable and equitable diets will also be challenged by accessibility to resources by small-scale fishers and farmers and consumer access to affordable products. Realising the

DOI: 10.4324/9781003174417-12

opportunities to manage and utilise aquatic foods to improve nutrition and reduce environmental food impacts, as well as to support peoples' cultures and livelihoods, will determine the extent to which aquatic foods can contribute to more sustainable diets.

Availability and accessibility of aquatic foods

Aquatic foods contribute to global diets in different ways, including through direct consumption of animals and plants, indirectly as income from the sale of products, and as inputs to feed for livestock and aquaculture for human consumption. Global production and consumption of aquatic foods has increased steadily since the 1960s, predominantly due to the growth of aquaculture production, combined with improvements in supply chain infrastructure facilitating greater trade. Aquaculture now provides about half of all aquatic foods, with the annual growth of aquatic food supply (3.1%) almost twice that of annual world population growth (1.6 %) for the same period, and higher than that of all other animal protein foods (meat, dairy, poultry, etc.) (FAO, 2020). Per capita consumption of aquatic foods has increased from 9.0 kg (live weight equivalent) in 1961 to 20.5 kg in 2018, although actual consumption of aquatic foods varies between and within countries, as well as within regions and even households (FAO, 2021a), due to a mix of social, physical, environmental, and institutional factors.

There is currently limited high-quality data available on consumption patterns that are sourced through dietary surveys and are comparable across countries and over time. Production data, which are often used as a proxy for consumption data, indicate that aquatic foods are the primary source of animal-sourced foods in many countries, particularly in Melanesia and Micronesia, as well as in countries in Eastern and South-Eastern Asia and in Western Africa (Figure 10.1a). In Eastern and South-Eastern Asia, aquatic foods account for over 50% of all animal sourced foods, and in Micronesia, they account for over 70%. In contrast, in Central Asia, Southern Africa, and Western Asia, aquatic foods make a relatively small contribution to overall food supply of animal sourced foods, and in these regions, increasing consumption could significantly improve public health.

Fish from marine environments, including demersal and pelagic[1] fish, account for the greatest total volume of aquatic food supply in most regions (Figure 10.1b). The supply of freshwater fish is greater than marine fish in some regions, particularly in Asia where freshwater aquaculture is well established. The supply of aquatic plants for food is small in most regions, except in Eastern Asia, where they are an important source of aquatic food.

While the supply of aquatic foods has increased substantially over the past 50 years, accessibility remains a challenge for many people. For example, the distribution of highly perishable products to non-coastal areas can be problematic in countries with limited infrastructure and can lead to poor food hygiene and safety, as well as high rates of food loss. As aquatic foods are one of the most highly traded commodities, disruptions to supply chains can also influence accessibility, as evidenced during the COVID-19 pandemic (Love et al., 2020).

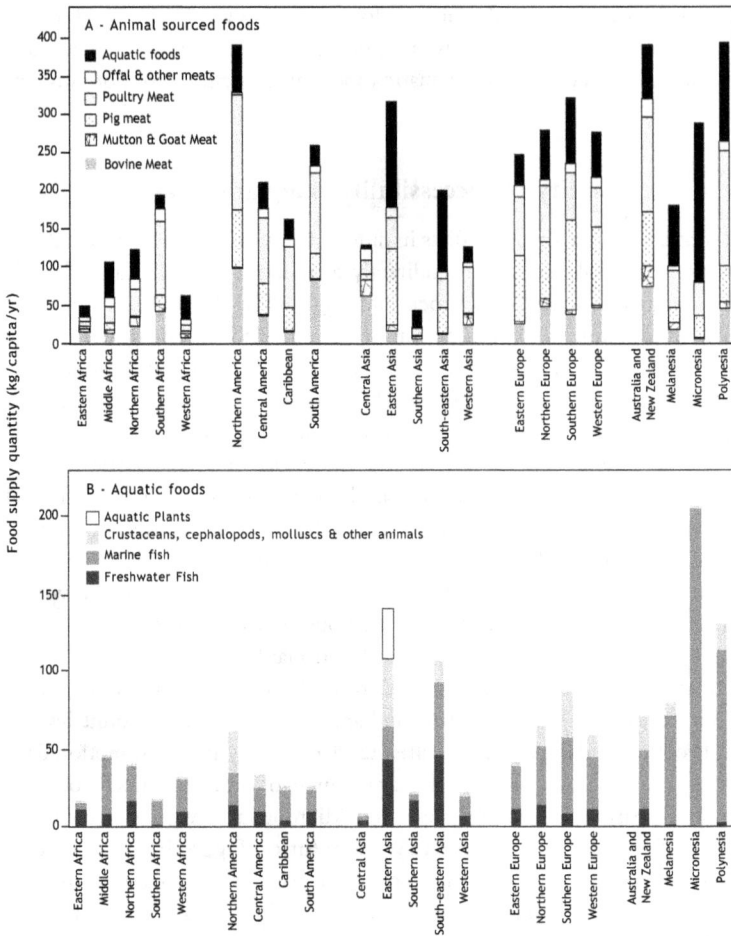

Figure 10.1   Food supply of animal sourced foods by region. A) animal-sourced foods by category; B) aquatic foods in A by category. FAOSTAT Food supply data is used as a proxy for consumption. This includes food production plus imports less exports, non-food uses, and losses during transport and storage. It does not account for losses at retail or consumer levels or non-edible parts and, therefore, likely overestimates actual consumption. Source: FAO, 2021a. Used under Creative Commons.

Affordability can also limit access to fresh aquatic foods. Studies from Europe and Australia have shown that aquatic foods are consumed in greater volumes by wealthier, more educated consumers than those who are less wealthy and educated (Cantillo et al., 2021; Farmery et al., 2018). The growth of aquaculture has enabled consumers from low- to high-income nations to benefit from year-round availability and access to aquatic foods (Naylor et al., 2021), however, ongoing efforts are required to ensure aquatic foods remain affordable and accessible (Belton et al., 2020).

Are aquatic foods sustainable?

The environmental sustainability of wild-capture and farmed aquatic foods is a topic of popular debate and promoting increased consumption for health has met with a range of environmental concerns. For wild-capture fisheries, concerns tend to focus on the potential for overfishing of target stocks and by-catch in wild capture fisheries, as well as habitat damage from destructive fishing practices (Jenkins et al., 2009). Continued overfishing will not only cause negative impacts on biodiversity and ecosystem functioning but will also reduce aquatic food production (FAO, 2020). While the number of fisheries considered to be overfished has increased since the 1970s, the majority of wild caught aquatic foods come from stocks considered to be biologically sustainable (FAO, 2020). Improving management and rebuilding of fisheries that are overfished could dramatically improve overall fish abundance while increasing food security and profits (Costello et al., 2016; Ye et al., 2013). Successful rebuilding of fisheries and ecosystems, including both large- and small-scale fisheries, has been occurring globally, and sustained effort is needed to further promote sustainable fisheries across a broad suite of management measures at local, national, and international levels (Melnychuk et al., 2021).

All food has an environmental footprint, and aquatic foods can perform well in comparison to other animal sourced foods in terms of carbon footprints, water, and land use (Gephart et al., 2021). Some aquatic foods also perform well in comparison to vegetable protein sources, particularly for indicators such as land use (Nijdam et al., 2012) and carbon footprints (Koehn et al., 2022). Aquatic foods are increasingly recognised for their potential to contribute to sustainable diets (Thilsted et al., 2016), however, identifying the most sustainable species to consume in a given context is an important consideration, as environmental footprints vary substantially between different aquatic foods. For example, fuel use by boats, which is typically the main contributor to the carbon footprint of wild caught aquatic foods, can vary with vessel type and by species. Small pelagic fish such as sardines have a very low carbon footprint (0.2 kg CO_2-eq per kg) in comparison to other wild caught aquatic foods, while crustaceans have a comparatively high carbon footprint (7.9 kg CO_2-eq per kg) (Parker et al., 2018).

Aquaculture is the world's most diverse farming practice, encompassing a range of different plant and animal species, farming methods, and environments (Metian et al., 2020) (Figure 10.2). The environmental performance of aquaculture systems varies because of this diversity. Aquaculture has been associated with a range of environmental concerns, including its impact on wild-fish supplies and eutrophication of waterways (Naylor et al., 2021). The amount and type of feed used strongly influences environmental impacts of aquaculture systems. Commercial feeds have increased their percentage of terrestrial ingredients, such as rice and chicken, in response to concerns around the use of wild fish and to increasing costs of wild fish inputs. While feeds now contain less fish, new trade-offs have emerged regarding the sustainability of terrestrial sourced feed ingredients and reduced nutritional profile of farmed fish (Troell et al., 2014).

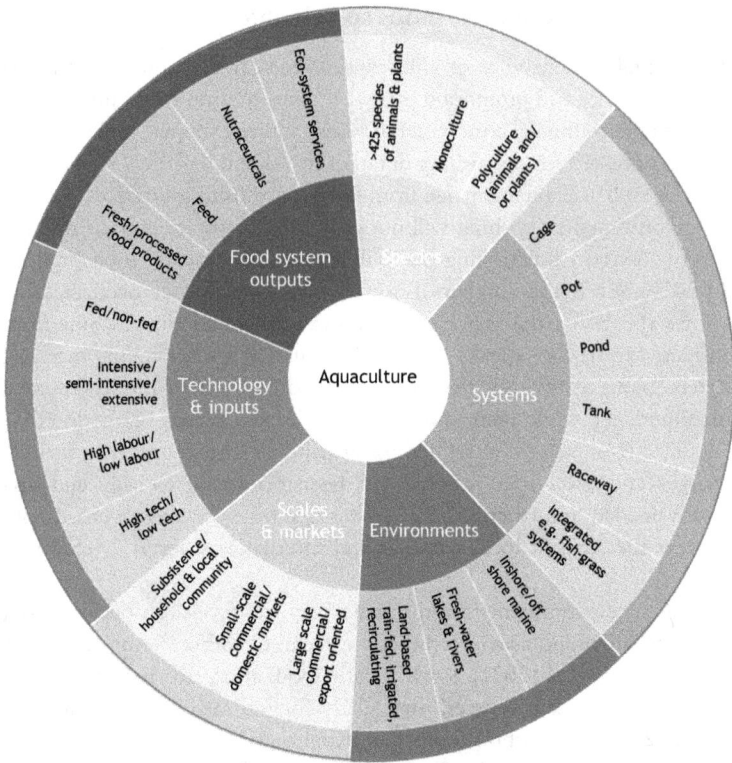

Figure 10.2 Examples of diversity in aquaculture production. Source: Created by the authors.

Non-fed aquaculture presents an opportunity to develop systems with less environmental impact. In addition, to reduced reliance on inputs, species including molluscs, filter-feeding finfish, and seaweeds can provide eco-system services such as removing nutrients from the water column. However, while farming unfed species continues to increase, the farming of fed species is increasing more rapidly (FAO, 2020).

Estimates suggest the production of food from the ocean can increase by six times the current level (Costello et al., 2020), with the bulk of this increase to come from aquaculture. It is, therefore, critical that the scaling-up of aquaculture is achieved in a sustainable manner (Fiorella et al., 2021). Unfortunately, many aquaculture systems still lack the motivation to meet sustainability criteria because their targeted markets do not reward producers through improved prices or access (Naylor et al., 2021).

Nutritional profile of aquatic foods

The nutritional benefits of aquatic foods, both plants and animals, are numerous, thereby making their contribution to nutrition and health irreplaceable in many regions. Aquatic foods from animal sources provide a rich source of high-quality protein including the full complement of essential amino acids required for human health.[2] Less well-recognised is that many aquatic species are also rich sources of several micronutrients including Omega-3 fatty acids; minerals such as iron, zinc, and calcium; and vitamins such as vitamin A and B12. Notably, it is these same nutrients which are often limited in diets of nutritionally vulnerable groups. Nutrients in aquatic foods are often highly bioavailable meaning that they are easily absorbed and utilised by the human body compared to other plant- and even some animal-sourced foods with the same nutritional profile. For example, aquatic foods have been identified as the only dietary source of a particular form of vitamin A (3,4-didehydroretinol, also known as vitamin A2) which has more than double the biological activity of the more common form of vitamin A (retinol) found in many other animal-source foods (La Frano et al., 2018). Certain aquatic foods, including types of fish, shrimp, and prawn, have also been found to have a high proportion of haem iron (the more biologically active form of iron compared to non-haem iron found predominantly in plant foods), even higher than other animal-sourced foods (Wheal et al., 2016). Fish has also been found to have an enhancing effect on the absorption of micronutrients from other foods, though the mechanisms remain unclear and require further research (Michaelsen et al., 2009). In this context, aquatic foods are uniquely placed to reduce the triple burden of malnutrition, undernutrition, and micronutrient deficiencies; overweight and obesity; in addition to related non-communicable diseases and risks (Figure 10.3).

Whilst these features of nutritional quality are common to many aquatic foods, they should not be considered a single homogenous category. There is a great diversity in the micronutrient content across and within types of aquatic foods (Bernhardt & O'Connor, 2021) (see Table 10.1). Therefore, promoting diversity of consumption is key to achieving the nutritional benefits of these foods. Existing nutrient composition data reflect only a fraction of species commonly consumed throughout the world and are biased towards species of high commercial value compared to small scale fisheries relied upon by food-insecure populations (Byrd et al., 2020). For example, from more than 2,000 fish species catalogued by the FAO, nutrient composition is available for only 25%. Recent efforts to overcome this data deficit include development of the Pacific Nutrient Database (PNDB) (SPC et al., 2020), which matches 822 food items, including 56 aquatic foods, from regional household income expenditure surveys with their respective macro- and micro-nutrient composition available from the literature (Table 10.1). Further research in this area will help support a nutrition-sensitive approach to aquaculture and fisheries to increase nutrient supplies (Thilsted et al., 2016).

Alongside the nutritional benefits, there are also health risks related to aquatic food consumption (particularly animal-based) due to the presence of contaminants from algae, bacteria, viruses, and some chemical pollutants such as heavy metals. Methylmercury accumulates up the food chain and can be found in toxic concentrations in predatory species such as shark, billfish, and tuna. Farmed aquatic foods also pose human health risks through antibiotic resistance (linked to overuse of antibiotics in production systems) and chemical contaminants (HLPE, 2014b). Whilst regulation can safeguard against these risks, enforcement remains a challenge in some settings. Despite these risks, it is broadly agreed that the positive health benefits of aquatic foods largely outweigh the risks (Hoekstra et al., 2013).

Undernutrition	Micronutrient Deficiencies	Overweight and Obesity	Non-communicable Diseases
Scale of the problem globally			
768 million people are hungry 149 million children are stunted	2 billion people are deficient in micronutrients	2 billion adults overweight or obese 39 million children overweight	874 million adults raised blood pressure 523 million suffer cardiovascular disease
Role for aquatic foods			
Rich source of high quality protein. Reduce stunting and improve child development outcomes. Crucial part of diets among poor and vulnerable.	Rich source of micronutrients such as iron, zinc and vitamin A commonly deficient in diets. Micronutrients are highly bioavailable. Enhance absorption of other nutrients in a meal.	Replacing meat with lean aquatic foods can assist weight loss.	Reduce risk factors for NCDs (including blood pressure and cholesterol). Reduce risk of and death from cardiovascular disease.

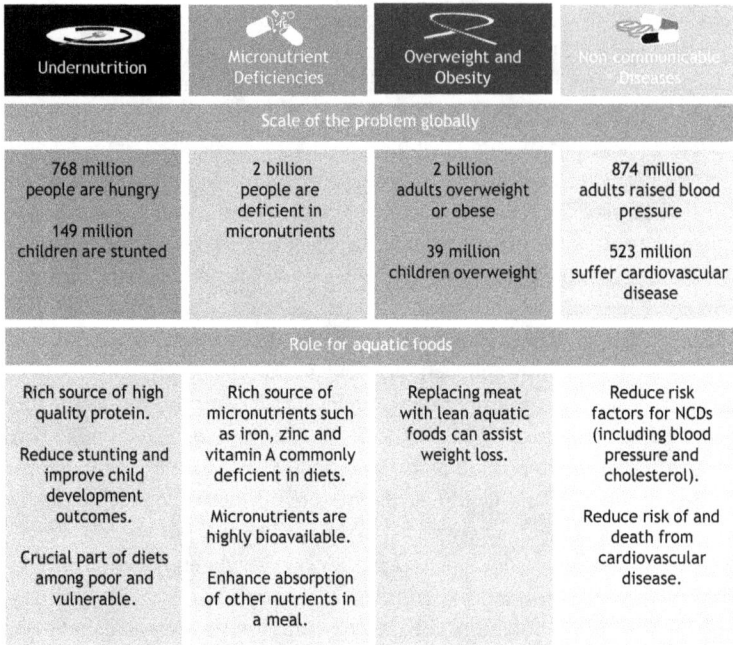

Figure 10.3 Unique potential for aquatic foods to reduce all forms of malnutrition and related non-communicable diseases and risks. Source: Adapted by the authors from Development Initiatives, 2020; FAO, 2021; FAO et al., 2021; WHO & FAO, 2006.

Managing aquatic foods for human and planetary health

Substantial potential exists for aquatic foods to make a greater contribution to sustainable diets, in particular to nutrition outcomes where consumption is low (Golden et al., 2021) and to environmental outcomes where it can replace high red meat consumption. However, simultaneously realising the nutritional and environmental potential of aquatic foods will be challenging as human health and environmental health are typically managed in isolation from each other by different government departments and across organisations. In many countries, fishery management is devolved to a local level through approaches such as community-based management, while food and nutrition security, and environment, are managed more centrally (Pomeroy, 2001). In addition, encouraging consumers to actively choose sustainable seafood is widely promoted by environmental organisations and through media (Farmery et al., 2020b), while laying the responsibility for healthy food choices on the consumer without broader structural changes to the food environment is not supported in public health and nutrition (Swinburn et al., 2011). Furthermore, fisheries and aquaculture are currently managed for economic and environmental targets, not for their contribution to food and nutrition (Bennett et al., 2021). Realising the potential of aquatic foods to contribute

Table 10.1 Nutritional composition of aggregated aquatic food groups consumed in the Pacific

Aquatic food group	Protein (g)	Fat (g)	Calcium (mg)	Iron (mg)	Zinc (mg)	Vitamin A (µg RAE)	Vitamin B12 (µg)
Tuna (n=4)	25 (3)	5.4 (8)	5.5 (6)	1.2 (0.9)	0.5 (0.23)	54.8 (58)	1.1 (1.4)
Small pelagic fish (n=1)	19.7	2.9	725	4	3.1	106	8.3
Large pelagic fish (n=2)	21	5.2	14	0.8	0.7	31	0.9
Demersal/reef fish (n=3)	19.8 (0.3)	2.2 (3.1)	25.3 (21)	0.6 (0.3)	0.6 (0.1)	30.3 (2)	1.7 (1.6)
Elasmobranchs (n=2)	22.6	0.8	11.0	1.0	0.5	12.0	1.1
Prawn/shrimp (n=2)	20.4	1.2	89	1.6	1.31	27	1.5
Crabs/crayfish (n=5)	15.7 (5.2)	1.4 (0.8)	143 (139)	1.3 (1.1)	3.5 (0.6)	16.2 (21)	2.9 (3.9)
Bivalves, gastropods (n=6)	19.1 (16)	1.9 (3)	77.5 (227)	4.0 (8)	4.4 (17)	56.3 (180)	9.4 (15)
Cephalopods (n=2)	17.2	1.1	14.5	0.8	1.4	21.5	10.2
Echinoids (n=1)	8.2	6.5	50.0	0.9	0.4	tr	0.0
Sea cucumber (n=1)	12.8	0.1	87.0	1.2	0.2	tr	2.3
Turtle (n=2)	16	1.0	100	1	1.3	5	1.1
Seaweed (n=2)	0.6	0.3	56	8	tr	50	0
Canned Fish (mixed fish, oil, brine, other) (n=12)	21.8 (14.7)	9.5 (11.1)	154 (399)	1.5 (1.7)	1.2 (2.4)	16.8 (37)	3.3 (7.6)

Source: Farmery et al. (2020a).

Nutrient values are based on 100 g edible portions. Mean value displayed with range between minimum and maximum values is provided in parentheses for categories with three or more species. Nutrient composition data from the PNDB. "tr" indicates trace amounts detected.

to sustainable diets will therefore depend on the identification of mutually sup-portive actions that can benefit both human and planetary health. As awareness of the potential for aquatic foods to contribute to sustainable diets gains momen-tum, more opportunities to manage and utilise aquatic foods for both nutrition and environment will be realised, however, many examples already exist, three of which are summarised below.

Promotion of the most nutritious and sustainable aquatic foods

Studies have identified small pelagic fish, such as sardines and anchovies; bivalves, such as mussels and oysters; and seaweed as some of the most nutritious aquatic foods as well as the most environmentally sustainable (Costello et al., 2020; Hallström et al., 2019; Koehn et al., 2022). Consuming more of these species could result in nutritional benefits, particularly for people with poor quality diets, as well as reduced environmental impacts where these foods replace more intensive animal sourced foods. However, despite these benefits, consumption remains low in most regions. Much of the global catch of pelagic fish is processed as feed for aquaculture, due to growing demand for fed-aquaculture species such as salmon. Similarly, much of the investment in seaweed is focused on its use as a sustainable feed ingredient for aqua-culture and livestock, rather than for direct human consumption.

Increasing the consumption of these more sustainable and nutritious aquatic foods is also challenged by lack of demand from consumers. Many products are not directly substitutable with other animal sourced foods like beef or chicken, and consuming comparative amounts may be unappealing. Novel processing of new foods and the inclusion of aquatic foods in public procurement policy such as low trophic and underutilised species in school feeding programs (FAO, 2021b) offer significant potential to increase demand.

Increased diversity for nutrition, sustainability, and resilience

Many wild capture fisheries target a limited range of higher value species, for which there is demand, or lower value species for feed. Similarly, aquaculture in the western hemisphere has largely developed around single- or dual-species and single-production systems (for example, Atlantic salmon in cages, Nile tilapia, and channel catfish (Ictalurus punctatus) in ponds), although aquaculture in Asia is typi-cally more diverse than other regions in terms of production systems and cultivated species (Naylor et al., 2021). The current emphasis on economic targets in fisheries and aquaculture management in many countries is exacerbating this situation.

Limiting aquaculture to only a few species makes production systems more prone to shocks, such as COVID-19, climate change, and market volatility (Naylor et al., 2021), and can reduce the nutritional quality of production (Bogard et al., 2018). Diversifying species through wild-catch and ecological aquaculture and polyculture can build more sustainable and resilient food production systems. In addition, aquatic biodiversity enhances multiple nutritional benefits to humans (Bernhardt & O'Connor, 2021). Enhancing resilience of fisheries and aquaculture is critical for these systems to continue producing food in a changing climate in which reduced fish harvests are expected with sea temperature rise and declines in fish catch, putting more than 10% of the population at risk of micronutrient deficiencies by 2050 (Golden et al., 2016).

Supply chain efforts to improve quality and avoid food waste and nutrition loss

Food loss and waste along supply chains is a feature of inefficient food systems, with implications for sustainable resource use, as well as for livelihoods, and food and nutrition security (Chen et al., 2020). Perishable nutrient-rich aquatic foods are typically at much greater risk of contamination by food-borne pathogens than products with a longer shelf-life. Product degradation from inefficient supply chains and poor handling practices can affect quality, leading to products being sold or exchanged at lower prices with reduced nutritional value and increased food safety issues (HLPE, 2014a). Reducing food loss and waste, through upgrading supply systems, has the potential to significantly improve benefits related to diet quality, food safety, and sustainability of aquatic foods. Opportunities to improve the efficiency of aquatic food production in line with the principles of the circular economy include the use of bivalve shell waste to partially replace non-renewable mineral sources in industrial applications (Alonso et al., 2021), as well as the use of animal by-products in aquaculture feed (Regueiro et al., 2021). Central to these improvements will be efforts that address gender and power imbalances in relation to access to resources experienced by women who form the majority of aquatic value-chain actors, particularly in low- and middle-income countries. The production of shelf-stable convenient aquatic products such as fish powder and fish chutney also offers significant potential to both reduce waste and loss and make significant contributions to the nutritional quality of diets, whilst also addressing issues of seasonal accessibility (Bogard et al., 2015).

Conclusion

The ability of aquatic foods to contribute to sustainable and healthy diets depends on the effectiveness of ongoing efforts by the public and private sectors to manage aquatic foods for environmental, economic, and human health outcomes. Improved understanding of the nutrient composition of aquatic foods will support adoption of a nutrition sensitive lens to fisheries and aquaculture; however, this lens needs to be compatible with approaches to sustainability. Efforts to sustain and rebuild both large- and small-scale fisheries must be strengthened at local, national, and international levels. The anticipated increase in production of aquatic foods will be necessary for some regions; however, it will not benefit sustainable diets if it is not accessible or affordable. Ensuring availability and accessibility of the most sustainable and nutritious aquatic foods is critical for people who actively choose to consume them over more resource intensive foods, as well as for people who rely on aquatic foods for food and nutrition security.

Notes

1 Demersal fish are those which live on or near to the bottom of lakes or oceans, while pelagic fish live in the water column, away from the lake or ocean floor.
2 Essential amino acids are those that cannot be synthesised by the human body and therefore must be obtained from the diet. In contrast, plant-based proteins individually do not provide the full complement of EAAs.

References

Alonso, A. A., Álvarez-Salgado, X. A., & Antelo, L. T. (2021). Assessing the impact of bivalve aquaculture on the carbon circular economy. *Journal of Cleaner Production, 279*, 123873.

Belton, B., Reardon, T., & Zilberman, D. (2020). Sustainable commoditization of seafood. *Nature Sustainability, 3*(9), 1–8.

Bennett, A., Basurto, X., Virdin, J., Lin, X., Betances, S. J., Smith, M. D., Allison, E., Best, B. A., Brownell, K. D., Campbell, L. M., Golden, C. D., Havice, E., Hicks, C. C., Jacques, P. J., Kleisner, K., Lindquist, N., Lobo, R., Murray, G. D., Nowlin, M., Patil, P. G., Rader, D. N., Roady, S. E., Thilsted, S. H., & Zoubek, S. (2021). Recognize fish as food in policy discourse and development funding. *Ambio, 50*(5), 981–989.

Bernhardt, J. R., & O'Connor, M. I. (2021). Aquatic biodiversity enhances multiple nutritional benefits to humans. *Proceedings of the National Academy of Sciences, 118*(15), e1917487118.

Bogard, J. R., Hother, A.-L., Saha, M., Bose, S., Kabir, H., Marks, G. C., & Thilsted, S. (2015). Inclusion of small indigenous fish improves nutritional quality during the first 1000 days. *Food and Nutrition Bulletin, 36*(3), 276–289.

Bogard, J. R., Marks, G. C., Wood, S., & Thilsted, S. H. (2018). Measuring nutritional quality of agricultural production systems: Application to fish production. *Global Food Security, 16*, 54–64.

Byrd, K. A., Thilsted, S. H., & Fiorella, K. J. (2020). Fish nutrient composition: a review of global data from poorly assessed inland and marine species. *Public Health Nutrition*, 1–11.

Cantillo, J., Martín, J. C., & Román, C. (2021). Determinants of fishery and aquaculture products consumption at home in the EU28. *Food Quality and Preference, 88*, 104085.

Chen, C., Chaudhary, A., & Mathys, A. (2020). Nutritional and environmental losses embedded in global food waste. *Resources, Conservation and Recycling, 160*, 104912.

Costello, C., Ovando, D., Clavelle, T., Strauss, C. K., Hilborn, R., Melnychuk, M. C., Branch, T. A., Gaines, S. D., Szuwalski, C. S., & Cabral, R. B. (2016). Global fishery prospects under contrasting management regimes. *Proceedings of the National Academy of Sciences, 113*(18), 5125–5129.

Costello, C., Cao, L., Gelcich, S., Cisneros-Mata, M. Á., Free, C. M., Froehlich, H. E., ..., & Lubchenco, J. (2020). The future of food from the sea. *Nature, 588*(7836), 95–100.

Development Initiatives (2020). *Global nutrition report: Action on equity to end malnutrition.* Development Initiatives.

FAO (2020). *The state of world fisheries and aquaculture 2020. Sustainability in action.* FAO.

FAO. (2021a). *Aquatic food supply quantity. FAOSTAT.* FAO. Retrieved July 8, 2021 from http://www.fao.org/faostat/en/#home

FAO (2021b). *The role of aquatic foods in sustainable healthy diets. Discussion paper.* FAO.

FAO, IFAD, UNICEF, WFP, & WHO. (2021). *The state of food security and nutrition in the world 2021. Transforming food systems for food security, improved nutrition and affordable healthy diets for all.* FAO.

Farmery, A. K., Gardner, C., Jennings, S., Green, B. S., & Watson, R. A. (2017). Assessing the inclusion of seafood in the sustainable diet literature. *Fish and Fisheries, 18*(3), 607–618.

Farmery, A. K., Hendrie, G., O'Kane, G., McManus, A., & Green, B. (2018). Sociodemographic variation in consumption patterns of sustainable and nutritious seafood in Australia. *Frontiers in Nutrition, 5*, 118.

Farmery, A. K., Scott, J. M., Brewer, T. D., Eriksson, H., Steenbergen, D., Albert, J., Raubani, J. Wate, T. J., Sharp, M., & Andrew, N. L. (2020a). Aquatic foods and nutrition in the Pacific. *Nutrients, 12*(12), 3705.

Farmery, A. K., van Putten, I. E., Phillipov, M., & McIlgorm, A. (2020b). Are media messages to consume more under-utilized seafood species reliable?. *Fish and Fisheries, 21*(4), 844–855.

Fiorella, K. J., Okronipa, H., Baker, K., & Heilpern, S. (2021). Contemporary aquaculture: implications for human nutrition. *Current Opinion in Biotechnology, 70*, 83–90.

Gephart, J. A., Henriksson, P. J., Parker, R. W., Shepon, A., Gorospe, K. D., Bergman, K., Eshel, G., Golden, C. D., Halpern, B. S., & Hornborg, S. (2021). Environmental performance of blue foods. *Nature, 597*(7876), 360–365.

Golden, C., Allison, E. H., Cheung, W. W., Dey, M. M., Halpern, B. S., McCauley, D. J., Smith, M., Vaitla, B., Zeller, D., & Myers, S. S. (2016). Fall in fish catch threatens human health. *Nature, 534*(7607), 317–320.

Golden, C. D., Koehn, J. Z., Shepon, A., Passarelli, S., Free, C. M., Viana, D. F., Matthey, H., Eurich, J. G., Gephart, J. A., Fluet-Chouinard, E., Nyboer, E. A., Lynch, A. J., Kjellevold, M., Bromage, S., Charlebois, P., Barange, M., Vannuccini, S., Cao, L., . . . Thilsted, S. H. (2021). Aquatic foods to nourish nations. *Nature, 598*(7880), 315–320.

Hallström, E., Bergman, K., Mifflin, K., Parker, R., Tyedmers, P., Troell, M., & Ziegler, F. (2019). Combined climate and nutritional performance of seafoods. *Journal of Cleaner Production, 230*, 402–411.

HLPE. (2014a). *Food losses and waste in the context of sustainable food systems. A report by the high level panel of experts on food security and nutrition of the committee on world food security.* FAO.

HLPE. (2014b). *Sustainable fisheries and aquaculture for food security and nutrition.* FAO.

Hoekstra, J., Hart, A., Owen, H., Zeilmaker, M., Bokkers, B., Thorgilsson, B., & Gunnlaugsdottir, H. (2013). Fish, contaminants and human health: quantifying and weighing benefits and risks. *Food and Chemical Toxicology, 54*, 18–29.

Jenkins, D. J. A., Sievenpiper, J. L., Pauly, D., Sumaila, U. R., Kendall, C. W. C., & Mowat, F. M. (2009). Are dietary recommendations for the use of fish oils sustainable? *Canadian Medical Association Journal, 180*(6), 633–637.

Koehn, J. Z., Allison, E. H., Golden, C. D., & Hilborn, R. (2022). The role of seafood in sustainable diets. *Environmental Research Letters, 17*(3), 035003.

La Frano, M. R., Cai, Y., Burri, B. J., & Thilsted, S. H. (2018). Discovery and biological relevance of 3, 4-didehydroretinol (vitamin A2) in small indigenous fish species and its potential as a dietary source for addressing vitamin A deficiency. *International Journal of Food Sciences and Nutrition, 69*(3), 253–261.

Love, D., Allison, E. H., Asche, F., Belton, B., Cottrell, R. S., Froehlich, H. E., Gephart, J. A., & Zhang, W. (2020). Emerging COVID-19 impacts, responses, and lessons for building resilience in the seafood system. *Global Food Security, 28*, 100494.

Marean, C. W., Bar-Matthews, M., Bernatchez, J., Fisher, E., Goldberg, P., Herries, A. I. R., Jacobs, Z., Jerardino, A., Karkanas, P., Minichillo, T., Nilssen, P. J., Thompson, E., Watts, I., & Williams, H. M. (2007). Early human use of marine resources and pigment in South Africa during the Middle Pleistocene. *Nature, 449*, 905.

Melnychuk, M. C., Kurota, H., Mace, P. M., Pons, M., Minto, C., Osio, G. C., Jensen, Olaf P.

Metian, M., Troell, M., Christensen, V., Steenbeek, J., & Pouil, S. (2020). Mapping diversity of species in global aquaculture. *Reviews in Aquaculture 12*(2), 1090–1100.

Michaelsen, K. F., Hoppe, C., Roos, N., Kaestel, P., Stougaard, M., Lauritzen, L., Mølgaard, C., Girma, T., & Friis, H. (2009). Choice of foods and ingredients for moderately malnourished children 6 months to 5 years of age. *Food and Nutrition Bulletin, 30*(3_suppl3), S343–S404.

de Moor, C. L., Parma, A. M., Little, R. L., Hively, D., Ashbrook, C. E., Baker, N., Amoroso, R. O., Branch, T. A., Anderson, C. M., Szuwalski, C. S., Baum, J. K., . . . Hilborn, R. (2021). Identifying management actions that promote sustainable fisheries. *Nature Sustainability, 4*(5), 440–449.

Naylor, R. L., Hardy, R. W., Buschmann, A. H., Bush, S. R., Cao, L., Klinger, D. H., Little, D. C., Lubchenco, J., Shumway, S. E., &. Troell, M. J. N. (2021). A 20-year retrospective review of global aquaculture. *Nature, 591*(7851), 551–563.

Nijdam, D., Rood, T., & Westhoek, H. (2012). The price of protein: Review of land use and carbon footprints from life cycle assessments of animal food products and their substitutes. *Food Policy, 37*(6), 760–770.

Parker, R. W., Blanchard, J. L., Gardner, C., Green, B. S., Hartmann, K., Tyedmers, P. H., & Watson, R. A. (2018). Fuel use and greenhouse gas emissions of world fisheries. *Nature Climate Change, 8*(4), 333.

Pomeroy, R. (2001). Devolution and fisheries co-management. Collective action, property rights devolution of natural resource management–exchange of knowledge implications for policy. *Zentralstelle fur Ernahrung und Landwirtschaft.* Retrieved from https://rmportal.net/framelib/devolution-pomeroy.pdf

Raatz, S., Silverstein, J., Jahns, L., & Picklo, M. (2013). Issues of fish consumption for cardiovascular disease risk reduction. *Nutrients, 5*(4), 1081.

Regueiro, L., Newton, R., Soula, M., Méndez, D., Kok, B., Little, D. C., Pastres, R., Johansen, J., & Ferreira, M. (2021). Opportunities and limitations for the introduction of circular economy principles in EU aquaculture based on the regulatory framework. *Journal of Industrial Ecology,* 1–12.

SPC, UOW, & FAO. (2020). *The Pacific Nutrient Database User Guide: A tool to facilitate the analysis of poverty, nutrition and food security in the Pacific region.* FAO & The Pacific Community Noumea.

Springmann, M., Spajic, L., Clark, M. A., Poore, J., Herforth, A., Webb, P., Rayner, M., & Scarborough, P. (2020). The healthiness and sustainability of national and global food based dietary guidelines: modelling study. *British Medical Journal, 370,* 2322.

Stetkiewicz, S., Norman, R. A., Allison, E. H., Andrew, N. L., Ara, G., Banner-Stevens, G., Belton, B., Beveridge, M., Bogard, J. R., Bush, S. R., Coffee, P., Crumlish, M., Edwards, P., Eltholth, M., Falconer, L., Ferreira, J. G., Garrett, A., Gatward, I., Islam, F. U. … Little, D. C. (2022). Seafood in Food Security: A Call for Bridging the Terrestrial-Aquatic Divide. *Frontiers in Sustainable Food Systems. 5*(703152), 1–11.

Swinburn, B. A., Sacks, G., Hall, K. D., McPherson, K., Finegood, D. T., Moodie, M. L., & Gortmaker, S. L. J. T. L. (2011). The global obesity pandemic: shaped by global drivers and local environments. *The Lancet. 378*(9793), 804–814.

Thilsted, S. H., Thorne-Lyman, A., Webb, P., Bogard, J. R., Subasinghe, R., Phillips, M. J., & Allison, E. H. (2016). Sustaining healthy diets: The role of capture fisheries and aquaculture for improving nutrition in the post-2015 era. *Food Policy, 61,* 126–131.

Troell, M., Naylor, R. L., Metian, M., Beveridge, M., Tyedmers, P. H., Folke, C., Arrow, K. J. Barrett, S., Crepin, A. S., Ehrlich, P. R., Gren, A., Kautsky, N., Levin, S. A., Nyborg, K., Osterblom, H., Polasky, S., Scheffer, M., Walker, B. H., Xepapadeas, T., & de Zeeuw, A. (2014). Does aquaculture add resilience to the global food system? *Proceedings of the National Academy of Sciences of the United States of America, 111*(37), 13257–13263.

Wheal, M. S., DeCourcy-Ireland, E., Bogard, J. R., Thilsted, S. H., & Stangoulis, J. C. (2016). Measurement of haem and total iron in fish, shrimp and prawn using ICP-MS: Implications for dietary iron intake calculations. *Food Chemistry, 201,* 222–229.

WHO, & FAO. (2006). *Guidelines on food fortification with micronutrients.* WHO & FAO.

Ye, Y., Cochrane, K., Bianchi, G., Willmann, R., Majkowski, J., Tandstad, M., & Carocci, F. (2013). Rebuilding global fisheries: The World Summit Goal, costs and benefits. *Fish and Fisheries, 14*(2), 174–185.

11

LIFE, DEATH, AND DINNER AMONG THE MOLLUSCS

Human appetites and sustainable aquaculture

L. Sasha Gora

Introduction

Not everyone eats oysters, or even shellfish for that matter. Religion, culture, science, and personal preference all inform which plants and animals we call food. Recently, because of the climate crisis, as well as growing concerns about the health of humans, plants, animals, and the planet at large, an awareness of environmental impacts also informs what ends up on the table. But whether you chew an oyster, swallow it in one slurp, or take a pass, eating is one of the most intimate ways humans interact with environments by literally digesting them. As sociologist Krishnendu Ray writes in dialogue with the work of the environmentalist Gary Snyder: "'Food is the field in which we daily explore our "harming" of the world'—the world at large [...] the animal world, and the plant world" (Ray, 2016, p. xx). Hunger is, thus, an architect of the world on land and the world at sea. It is one of the main forces behind the Anthropocene, and any quest to live sustainably must consider the impact of human appetites. Narrowing in on seafood: How do human appetites transform, harm, but perhaps also heal watery environments? What is a sustainable diet if we look at the sea, animal mariculture, and at oysters?[1]

In human history, the coast is never far away. Around 80 percent of the global population live within 60 miles of a coast, be it freshwater or salt (Mentz, 2020). If eating, as Wendell Berry claims, is an agricultural act (2007, p. 552), then it is also increasingly an aquacultural act. For centuries humans have farmed water, a practice that is only growing (FAO, 2020). Fish features regularly in the diet of more than one billion people (Hamada & Wilk, 2019), and in 2017, fish "accounted for about 17 percent of total animal protein" (FAO, 2020, p. 67) consumed globally. Fittingly, Steve Mentz, a pioneering scholar in the blue humanities, titled one of his chapters in *Ocean* "Seafood before History." "We've needed the sea since

DOI: 10.4324/9781003174417-13

prehistory," he writes. "The bitter waves provide food and transportation" (Mentz, 2020, p. 11). The sea has shaped humans, their histories, and their appetites.

This chapter—"Life, death, and dinner among the molluscs"—proposes to think about water and shellfish from a culinary perspective that prioritises the pursuit of sustainable diets.

As a category, seafood is as endless as the sea itself, it includes everything from wild to farmed animals and plants and from fresh to saltwater. In response to how seafood might contribute to sustainable diets, which is to say, what is favourable to the people who produce and consume them as well as to their environments, this chapter zooms in on oysters. It addresses their cultural, culinary, and biological lives in order to discuss human appetites and contemporary aquaculture. In dialogue with the environmental humanities, it employs a multispecies perspective in its discussion of the impact of human appetites on marine life. This chapter studies oysters and mariculture as a reflection of human appetites and hunger as an ecological architect, and considers the waters we farm and eat. This chapter unfolds in two parts: The first discusses oysters, appetites, and how humans know these bivalves through eating them; the second addresses aquaculture and sustainable appetites, before concluding with further food for thought about eating as an encounter between life and death.

The oyster, the world

Studying the history of food can reveal how everyday eating habits shape environments. In fact, "food systems account for over one-third of global greenhouse gas emissions" (Crippa et al., 2021). Although humans have farmed the sea for centuries, up until recently seafood has been one of the few wild foods—something that has been caught, hunted, or gathered rather than farmed—that many people eat.[2] But today, more and more seafood is farmed. Aquaculture now makes up nearly half of global fish production, including freshwater and oceans.[3] And some species, like Atlantic salmon, are almost all farmed (Lien, 2015). In 2018, shelled molluscs, including clams, mussels, and oysters made up around 56.3 percent of marine and coastal aquaculture production, around 17.3 million tonnes (FAO, 2020, p. 6).[4]

"Follow the chicken and find the world" (p. 274) writes Donna Haraway (2008). What happens if you follow the oyster? A type of mollusc—an animal with an outer shell but without a backbone—the oyster is a bivalve. Like the clam or the mussel, it has a shell with two parts that are held together by elastic ligament. Left alone, an oyster can spend a dozen years inhaling seawater, absorbing its nutrients, and then releasing it back into the water (Kurlansky, 2006, xvii). But when courted by human hunger, an oyster lives three or four years before becoming a snack eaten straight from the hand, from a platter, or a plate.

Members of the family Ostraeidae, the "true oyster" includes a variety of genera, but two species dominate culinary discussions: The flat oyster with its rounder shell (*Ostrea*) and the Pacific oyster with its oval shell (*Crassostrea*). Because of diseases over the last century, oyster farms around the world have largely replaced flat oysters with Pacific ones (Stott, 2004). Returning to Haraway's question, the oyster is already entangled with popular imaginations of the world. As the English idiom goes, the world is your oyster, which is another way to say that you can achieve what you want and do as you please. The origin of this saying is Shakespeare's *The Merry Wives of Windsor* from circa 1600: "Why then the world's mine oyster," the character Pistol claims, "which I with sword will open" (Act 2, scene 2, 2–5). To open the world is no different, for Pistol, than to take a knife—or a sword—to shuck an oyster.

Continuing with common—even cliché—phrases, many enthusiastic eaters describe oysters as tasting like the sea: A mouth full of flesh seasoned by saltwater. As Elspeth Probyn (2016), a cultural studies scholar, writes, "Oysters are as close as most of us get to eating the sea" (p. 52). But what does it mean to eat the sea? How are human encounters with water mediated through food? Probyn (2016) considers oysters "a marvellous sustainable and hardworking marine entity that is also delicious to eat" (p. 11). Sustainable is an adjective that frequently appears in discussions about oysters—a food that has transformed over time from a working-class staple to a luxury, and has also been both simultaneously.

"Oysters are both on the tongue and beyond the power of the tongue," writes literary scholar Rebecca Stott (2004). "They are not only slippery: they are evasive, almost beyond knowing" (p. 10). It is for this reason she claims: "The history of the oyster-human encounter is a history characterised by intimacy and distance" (p. 10). This marriage between intimacy and distance also describes human relationships to the sea. To eat the sea is one way to attempt to know it. Rachel Carson, marine biologist and writer, began her 1937 essay "Undersea" with a question: "Who has known the ocean?." She then gives an answer: "Neither you nor I, with our earth-bound senses" (p. 233). What does it mean to know the ocean? Food cultural history, for example, approaches the question of how humans know animals and plants through culinary practices: Through eating them or not.

How do we know the oyster? The American food writer M. F. K. Fisher titled her 1941 collection of recipes *Consider the Oyster*. It begins with the chapter "Love and Death among the Molluscs," from which this chapter borrows its title, though instead of love, I write about life, and add dinner to the equation to meditate on how eating connects life and death. "Life is hard we say. An oyster's life is worse," writes Fisher (1988). "She lives motionless, soundless, her own cold ugly shape her only dissipation, and if she escapes the menace of duck-slipper-mussel-Black-Drum-leech-sponge-borer-starfish, it is for man to eat, because of man's own hunger" (p. 7). An oyster's life ends in order to please human hunger but also to sustain human life, an example of the messy multispecies relations that make up eating.

Today, nearly 95 percent of oysters are farmed, which requires no fertiliser, feed, or antibiotics. For these reasons some farmed shellfish, like clams, oysters, and mussels, are considered "environmentally pristine foods" (McWilliams, 2009, p. 177). This farming reflects human appetites and hunger as landscape architecture. In the words of writer Bill Buford (2008), "you should be able to taste the place [an oyster] came from; in this still-living creature you will find the water and the food it ate—these living, fragile, handmade creatures tasting wonderfully of the health of the planet" (p. 272). Oysters work hard to filter water, but they, too, have limits. As historian Mark Kurlansky (2006) sums up, "If the water is not pure, that, too, can be tasted in the oyster" (p. 279). Oysters mirror the health of the waters they inhabit.

The oyster's geography is as vast as it is specific. Just like the land-based term terroir, the word *merroir* recognises how the particularities of a body of water flavour its fish and fruits—from the water's salinity and temperatures to the foods it hosts, and from the amount of space and degree of crowding to the seabed's structure. It is worth reflecting, however, on how imaginations of food take different forms on land rather than in water. The term pescetarianism is one example. This type of vegetarianism distinguishes between land and water-based animals; it sees animals on land one way and those in water another. Reflecting on how food politics "has been overwhelmingly focused on terrestrial animal protein" (p. 25), Probyn (2016) argues that seafood of all kinds: "compels us to understand how entangled we are as consumers in the geopolitical, economic, cultural, and structural intricacies of the fishing industries" (p. 4). This is to say that "fish-as-food requires us to go beyond a simplistic food politics" (Probyn, 2016, p. 4). It is for this reason studies of sustainable diet should look beyond the shoreline and consider the role of water, as well as seafood's "water footprints" (Farmery et al., 2016).

Sustainable appetites

Any form of agriculture, be it on land or at sea, is cultural; the word even has culture inside of it. Environmental historian Donald Worster (1994) makes this clear by stating, "People invent agriculture; that is, they choose some plants to eat, cultivating and breeding them, while ignoring others" (p. 10). The same goes for aquaculture. Humans select plants and animals that live in, around, and under water, and it is this process of selection that transforms them into food. Aquaculture is a cultural construction and what one culture selects can differ from another. To borrow from terrestrial terms, fishing is a form of hunting and aquaculture is a form of farming.

However, farming at sea can also blur these distinctions. For example, in her summary of the development of mussel farming in the Venetian Lagoon, which started in the 1950s, anthropologist Rita Vianello (2018) writes: "The farmers stress the fact that their work is not real farming, but only guiding their natural growth, that packaged feed is never used and that the mussels are left to grow in their natural environment" (p. 47). These farmers imagine "real" farming to entail more pronounced interventions than what they do to "guide" the natural growth of bivalves. Farming at sea can raise questions about farming on land, as well as reveal how the two are connected, such as through water runoff.

As the chapters in this book make clear, sustainable diets are not just about food. They are about the world as a whole and shaping its future. Humans are literally eating the planet up. They are drinking it up too. At present, the global level of "food production accounts for the use of 48 percent and 70 percent of land and fresh water resources respectively" (FAO & WHO, 2019, p. 5). In addition to being responsible for around one-third of greenhouse gas emissions, food systems have also been driving the loss of biodiversity, deforestation, and water pollution (FAO & WHO, 2019, p. 8). Therefore, there is an urgent need for sustainable diets that not only nourish people, but also heal, rather than harm, the environments on which they depend. For this reason, the FAO & WHO (2019) defines a sustainable diet, in part, as one that preserves "biodiversity, including [...] aquatic genetic resources" and avoids overfishing (p. 10). To eat is, thus, to shape the ecosystems of the world and to support or threaten their future. Water-based foods are essential in this process. However, not all seafood has a positive environmental record, and life cycle assessment (LCA) is a tool for employing an ecosystem approach to evaluate the sustainability of different animals (Ziegler et al., 2016).

The aquaculture and fisheries sector has a large role to play in advancing all the United Nation's Sustainable Development Goals, but especially SDGs 2, "Zero Hunger," and 14: "Conserve and sustainably use the oceans, seas and marine resources for sustainable development" (FAO, 2020, vi).[5] It is for this reason that the FAO (2020) recommends "blue growth" as it recognises fish and fishery products as not only some of the planet's healthiest foods "but also as some of the less impactful on the natural environment" (vii). Since the 1970s, aquaculture production has been growing at 7.5 percent per year (FAO, 2020, p. vi). People are also eating more seafood. Surpassing any other animal protein, from 1961 to 2017, global fish consumption increased annually by 3.1 percent—more than double the population growth rate for this period (FAO, 2020, p. 3). The demand for shellfish has also increased—but why? One reason is rising incomes, but another is that consumers favour bivalves because of their positive environmental impacts coupled with their nutritional benefits (FAO, 2020, p. 88).

But what makes aquaculture sustainable? As marine social scientist and environmental policy scholar Kate Barclay and Alice Miller (2018) point out: "With seafood being the most highly traded food globally and per capita consumption increasing more rapidly than other animal proteins, wild capture fisheries face real and imminent environmental limits" (p. 1). The promotion of aquaculture has been one response to these limits. However, despite the rapid growth of aquaculture, there are many critiques that it is "growing the wrong way." As environmental scientist Jennifer Jacquet (2017) writes: "Similar to factory farming, aquaculture is becoming an industrialised food system that is unsustainable and unnecessarily cruel." To avoid repeating the same mistakes that were made on land, Jacquet suggests that government policies and stakeholders, including farmers and investors, should promote bivalves. She considers bivalves not only the best option in the ocean, but the best choice if one chooses to eat animals, period (2017).

Bren Smith seconds this opinion. He calls himself a restorative ocean farmer and recounts his journey toward developing his "blue thumb" in his 2019 memoir *Eat Like a Fish: My Adventures as a Fisherman Turned Restorative Ocean Farmer*. Once a commercial fisherman—a hunter of the seas—he now farms sea greens and shellfish in 20 acres of saltwater in Long Island Sound in the eastern United States. Smith makes it clear, however, that he is not a fish farmer and is critical of the direction the US has taken in its approach to ocean agriculture. "If the nation had chosen to focus energies on growing restorative species such as seaweeds rather than jailing and feeding fish," Smith (2019) argues, "we'd have a more sensible dinner plate today. We'd be feeding the planet while breathing life back into our seas, and protecting wild fish stocks while creating middle-class jobs" (p. 7). This, he believes, is the potential of sustainable water-based farming: It can combat climate change, create jobs, and "feed the planet" (2019, p. 11).

A turning point for Smith was the summer of 1992. He was fishing black cod in Alaska, and remembers hearing the dire news from Newfoundland: The Atlantic northwest cod fishery had collapsed. With over 30,000 people in Newfoundland out of work and the number of cod stock tragically low, this is an example of what marine biologist Daniel Pauly calls an "Aquacalypse" (2009). The news killed Smith's appetite for pillage and sparked a "search for sustainability." He found his answer in farming sea vegetables and shellfish.

The history of cultivating oysters is long, and the first to do so were the Chinese. Though few written records remain, a 475 BCE Chinese treatise archives the developed state of aquaculture at that time (Stott, 2004, p. 37). The Romans also left records documenting their work in cultivating oysters; by 95 BCE Baia, located in the Gulf of Naples, hosted human-made oyster beds (Stott, 2004, p. 210). Oyster cultivation was established in Italy, France, and Britain, as well as China and Japan by CE 600 (Stott, 2004, p. 211). But Kurlansky argues that a commercial interest in oyster farming only really developed in the nineteenth century when European and North American beds were clearly exhausted (2006, p. 115). To cultivate oysters is to attempt to regulate their supply. At this point, humans learned that they "could make a better oyster than nature" (p. 136–137) something that is unusual in the world of farming animals (Kurlansky, 2006). Just like a wild oyster, a farmed one takes in the same nutrients and filters its local waters.

Restorative ocean farming is one term Smith uses. He also calls it "regenerative ocean farming" or "3D ocean farming" (2019, p. 9). Interestingly, he dislikes the term aquaculture, although does not explain why. But he does detail what kind of cuisine mariculture can produce: Climate cuisine (2019, p. 12). This, for him, is a cuisine composed of underutilised ocean vegetables and shellfish. There are other names for and interpretations of what Smith calls "climate cuisine." For example, in 2015, the *New York Times* listed "climatarian" as one of the year's top ten new food words. This list defines climatarian (n.) as:

> A diet whose primary goal is to reverse climate change. This includes eating locally produced food (to reduce energy spent in transportation), choosing pork and poultry instead of beef and lamb (to limit gas emissions), and using every part of ingredients (apple cores, cheese rinds, etc.) to limit food waste.
>
> *(Moskin, 2015)*

In 2016, the following year, the Cambridge Dictionary added climatarian to its repertoire. However, what exactly a climatarian is remains open to interpretation. Generally, it advocates consuming local and organic foods, minimising drought-prone crops and feedlot beef, cutting back on waste, and prioritising land- and water-efficient foods, the likes of sea vegetables and molluscs.

Cooking Sections—a duo of London-based cultural practitioners who focus on the systems that organise the world through food—use a similar word: Climavore. This name doubles as the title of a long-term project Cooking Sections initiated in 2015 with the aim of investigating how to eat as climate changes. In 2016 they launched a project on the Isle of Skye that promotes oysters and other regenerative seafoods while discouraging the consumption of farmed salmon, a practice that comes with ethical and environmental ills, such as escaped fish, viruses and parasites, and contaminated waters (Cooking Sections, 2019; Coates, 2006; Gora, 2020; Lien, 2015). The oyster is one food that *Climavore* champions.

Sustainability is about endurance, about long-term commitment, and so sustainable seafood must contribute to an ecosystem instead of just taking from it. One definition of sustainable seafood is: "seafood that is derived from either wild-capture or cultivated fisheries that can be maintained in the long term without detrimental effects to the structure or function of the wider ecosystem" (Koldewey et al., 2009, p. 71). The sustainable seafood movement emerged in the mid to late 1990s and prioritises environmentally sensitive production and consumption. In dialogue with the environmental movement at large, it focuses on promoting ecological stewardship and raising awareness of pressing issues in contemporary fishing practices and aquaculture, as well as the responsibility of all of those involved. It uses various methods and largely draws from market-based approaches (Barclay & Miller, 2018, p. 2.)

Navigating seafood markets and grocery store freezer aisles requires a great deal of what I call seafood fluency (Gora, 2021a, p. 90). Flesh alone reveals little about geographical and biological origins (Watson et al., 2015). Therefore, shoppers must rely on labels. However, around 30 percent of labels falsely identify fish, an example of fish fraud or even laundering (Barendse et al., 2019). This is of concern for sustainable diets because mislabelling obscures important information related to species stocks and environmental impacts. Compared to fish fillets or frozen fish sticks, shellfish, like oysters, are easier to identify.

There are also two market-based incentives that have become common aides in navigating fresh, frozen, or canned seafood: Certification and ecolabeling. Both have the aim of shifting "industry practises in commercial fisheries and aquaculture toward sustainability" (Cooke et al., 2011, p. 911). Examples of "social marketing," these strategies emerged as part of the sustainable seafood movement, which resulted from "a collaboration between industry and environmental NGOs that recognised that informed choices made by consumers could contribute to the conservation of marine biodiversity" (Cooke et al., 2011, p. 912). However, practices like ecolabeling are not always consumer-oriented and, instead, function "as business strategic tools for conventional marketing to compete in large-scale retailing" (Prosperi et al., 2020, p. 12). There are also awareness campaigns, which take an educational approach to informing consumers about what is or is not sustainably caught and any additional environmental impacts. Nonetheless, this is still knowledge that a consumer requires to feel confident about shopping for sustainable seafood. Additionally, there is a lack of evidence that initiatives like ecolabels have reduced overfishing (Barclay & Miller, 2018, p. 2).

Efforts to make the global seafood market more sensitive to environmental impacts are key. Because as much as 70 percent of seafood is consumed in restaurants, chefs play a crucial role in promoting sustainable practices (Koldewey et al., 2009). They set the bar for the industry and are also able to educate diners about what to eat and how. This is also why food writers and those working in food production need to be included in discussions around sustainable appetites and diets. Also, there is great potential in restaurants partnering with oyster farmers and recycling shells to return them to the sea or exploring their potential for other uses.

But discussions about sustainability also raise questions about limits. Kurlansky (2006) brings this up in his account of the history of New York—the world's oyster-trading capital in the nineteenth century. A city that once sold oysters from street carts like it does hot dogs today. He writes that the history of New York oysters is, in fact, the history of the city itself, a history that ends with excessive waste and the killing of its grand estuary. "The reality is," writes Kurlanksy (2006), "that millions of people produce far too much sewage to coexist with millions of oysters" (p. 269). Oysters have great potential in terms of positively contributing to the waters they call home. A mighty keystone species, a single adult oyster can filter up to 50 gallons of water each day, removing nitrogen pollution—which is responsible for dead zones in the ocean—and gobbling up dirt and phytoplankton (Probyn, 2016, p. 53). An oyster can contribute to clearer water, which allows light to spread nutrients and nourish underwater habitats. But this assumes that the water is in good condition. After all, the hardworking oyster can only do so much. Humans must also keep the waters clean and take environmental responsibility for their voracious appetites.

Conclusion

The language of seafood turns up in many English-language idioms. This chapter shared "The world is your oyster," but there are many more: "A fish out of water"; "To be a shrimp;" "Fishing for compliments;" "Something's fishy;" "A big fish in a small pond;" and, "To fish around." Clams also make an appearance: "To clam up" references how this bivalve claps together its shell when disturbed; and "Happy as a clam" is the short version of "happy as a clam at high tide," which is when a clam cannot be dug up and so is left alone and, therefore, content. This last expression circles back to Fisher's (1988) description of an oyster surviving the hunger of other predators only to be gobbled up by humans. This leaves one to wonder just how happy is a clam, and how sustainable is an oyster?

Fisher opened *Consider the Oyster* with a quote—a "proverbial cliché" (Stott, 2004, p. 101)—from Jonathan Swift's *Polite Conversation* from 1738: "He was a bold man that first ate an oyster" (1988, n.p.). What does this imply? Why bold? Eating an oyster is a confrontation between life and death, as dinner is in general. But an oyster is also an acquired taste. It is a polarising food: Either loved or loathed. When shucked with precision, "the diner is eating an animal with a working brain, a stomach, intestines, liver, and a still-beating heart" (Kurlansky, 2006, p. 52). Although not about the oyster alone, there is even a word for the fear of shellfish: Ostraconophobia. The oyster is one of the few foods that humans eat alive. As Probyn (2016) describes: "Oyster eating is a rare instance when live flesh meets live flesh" (p. 11–12). A living food and a sustainable one when farmed with care in healthy waters. Human concerns about sustainability may wax and wane, but sustainable food production must go beyond the water's edge, an example the oyster provides if we, as Fisher writes, consider it.

Notes

1 Some of the questions I raise here have emerged from an ongoing dialogue with the marine ecologist Camilla Bertolini about the history of oysters in Venice and their reintroduction to the Venetian Lagoon. I am grateful to her for further pushing my focus beyond cuisine and culture. This chapter also builds on "Self-Portrait, with Shellfish," a text based on a talk I gave at Ocean Space, Venice, in 2020 (Gora, 2021a).
2 I also address this in Gora, 2021b.
3 In 2018 "Aquaculture accounted for 46 percent of the total production and 52 percent of fish for human consumption, up from 25.7 percent in 2000" (FAO, 2020, 2). Aquaculture in freshwater, as opposed to saltwater, produces most farmed fish, but here I largely focus on shellfish mariculture (FAO, 2020, p. 6).
4 This excludes freshwater aquaculture production.
5 For the role that aquaculture plays in achieving SDGs see Troell et al., 2021.

References

Barclay, K., & Miller A. (2018). The sustainable seafood movement is a governance concert, with the audience playing a key role. *Sustainability*, *10*(180), 1–20.

Barendse, J., Roel, A., Longo, C., Andriessen, L., Webster, L. M. I., Ogden, R., & Neat F. (2019). DNA barcoding validates species labelling of certified seafood. *Current Biology*, *29*(6), 198–99.

Berry, W. (2007). The pleasures of eating. In M. O'Neill (Ed.), *American food writing: An anthology with classic recipes* (pp. 551–8). Literary Classics of the United States.

Buford, B. (2008). On the bay. In D. Remnick (Ed.), *Secret ingredients: The New Yorker book of food and drink* (pp. 259–272). The Modern Library.

Carson, R. L. (1937). Undersea. *Atlantic Monthly*, *78*, 55–67.

Coates, P. (2006). *Salmon*. Reaktion Books.

Cooke, S. J., Murchie, K. J., & Danylchuk, A. J. (2011). Sustainable 'seafood' ecolabeling and awareness initiatives in the context of inland fisheries: Increasing food security and protecting ecosystems. *BioScience*, *61*(11), 911–918.

Cooking Sections. (2019, July 9). *CLIMAVORE: On tidal zones*. http://www.cooking-sections.com/CLIMAVORE-On-Tidal-Zones

Crippa, M., Solazzo, E., Guizzardi, D., Monforti-Ferrario, F., Tubiello, F. N., & Leip, A. (2021). Food systems are responsible for a third of anthropogenic GHG emissions. *Nature Food*, *2*, 198–209.

FAO. (2020). *The state of world fisheries and aquaculture: Sustainability in action*. FAO. https://www.fao.org/3/ca9229en/CA9229EN.pdf

FAO, & WHO (2019). *Sustainable healthy diets: Guiding principles*. FAO. https://www.fao.org/3/ca6640en/ca6640en.pdf

Farmery, A., Gardner, C., Jennings, S., Green, B. S., & Watson, R. A. (2016). Assessing the inclusion of seafood in the sustainable diet literature. *Fish and Fisheries*, *18*, 607–18.

Fisher, M. F. K. (1988). *Consider the Oyster*. North Point Press. (Original work published 1941).

Gora, L. S. (2020). The sparrow that turned salmon: Unveiling the systems that govern food through Art. *Preserve Journal, 4*, 110–17.

Gora, L. S. (2021a). On ice: Life and lunch at Mercato di Rialto. *Lagoonscapes: The Venice Journal of Environmental Humanities, 1*, 79–100.

Gora, L. S. (2021b). Self-portrait, with shellfish. *Riot and Roux!, 2*. https://www.riotandroux .com/product-page/issue-2-water

Hamada, S., & Wilk, R. (2019). *Seafood: Ocean to the plate*. Routledge.

Haraway, D. J. (2008). *When species meet*. University of Minnesota Press.

Jacquet, J. (2017, January 23). Why oysters, mussels and clams could hold the key to more ethical fish farming. *The Guardian*. https://www.theguardian.com/sustainable-business/ 2017/jan/23/aquaculture-bivalves-oysters-factory-farming-environment.

Koldewey, H. J., Atkinson, J., & Debney, A. (2009). Threatened species on the menu? Towards sustainable seafood uses in zoos and aquariums. *International Zoo Yearbook*, 43, 71–81.

Kurlansky, M. (2006). *The big oyster: History on the half shell*. Random House.

Lien, M. (2015). *Becoming salmon: Aquaculture and the domestication of a fish*. University of California Press.

McWilliams, J. E. (2009). *Just food: Where locavores get it wrong and how we can truly eat responsibly*. Back Bay Books.

Mentz, S. (2020). *Ocean*. Bloomsbury.

Moskin, J. (2015, December 15). 'Hangry'? Want a slice of 'piecaken'? The top new food words for 2015. *New York Times*. https://www.nytimes.com/2015/12/16/dining/new-food-words.html.

Pauly, D. (2009, September 28). *Aquacalypse now*. New Republic. https://newrepublic.com/ article/69712/aquacalypse-now

Probyn, E. (2016). *Eating the ocean*. Duke University Press.

Prosperi, P., Vergamini, D., & Bartolini, F. (2020). Exploring institutional arrangements for local fish product labelling in Tuscany (Italy): A convention theory perspective. *Agriculture and Food Economics, 8*, Article 6.

Ray, K. (2016). *The ethnic restaurateur*. Bloomsbury.

Shakespeare, W. (2004). *The Merry wives of windsor*. Simon & Schuster. (Original work published 1600).

Smith, B. (2019). *Eat like a fish: My adventures as a fisherman turned restorative ocean farmer*. Alfred A. Knopf.

Stott, R. (2004). *Oyster*. Reaktion Books.

Troell, M., Costa-Pierce, B., Stead, S. M., Cottrell, R. S., Brugere, C., Farmery, A. K., Little, D. C., Strand, Å., Soto, D., Pullin, R., Beveridge, M., Salie, K., Dresdner, J., Valenti, P., Blanchard, J., James, P., Yossa, R., Allison, E., Devaney, C., & Barg, U. (2021). Perspectives on aquaculture's contribution to the SDGs for improved human and planetary health. *Aquaculture and the SDGs*. FAO. https://aquaculture2020.org/uploads/gca-tr3-aquaculture-sdgs.pdf

Vianello, R. (2018). The rehabilitation of lice (mussels, *Mytilus galloprovincialis* Lamark 1819) in the lagoon of Venice: An example of a change in the perception of sea resources. *Regional Studies in Marine Science, 21*, 39–49.

Watson, R. A., Green, B. S., Tracey, S. R., Farmery, A., & Pitcher, T. J. (2015). Provenance of global seafood. *Fish and Fisheries, 17*, 585–595.

Worster, D. (1994). *An unsettled country: Changing landscapes in the American West*. University of New Mexico Press.

Ziegler, F., Hornborg, S., Green, B. S., Eigaard, O. R., Farmery, A. K., Hammar, L., Hartmann, K., Molander, S., Parker, R. W. R., Hognes, E. S., Vázquez-Rowe, I., & Smith, A. D. M. (2016). Expanding the concept of sustainable seafood using life cycle assessment. *Fish and Fisheries, 17*, 1073–1093.

12

GRASS-FED LIES

The mythology of sustainable meat

Jason Hannan

Introduction

In October 2021, *The Hamilton Spectator* published an article titled, "Where does your food come from?" (Antonacci, 2021). The article tells the story of Kaitlyn Krakar and Emma Krakar, two livestock farmers determined to challenge popular conceptions about animal agriculture in Canada. Kaitlyn is introduced as a representative of the Beef Farmers of Ontario, an industry lobby group representing some 19,000 beef farmers. We learn that Kaitlyn has come to the fair with Venus, a pregnant Charolais cow, who is a major attraction for visiting children. "She loves the scratches," Kaitlyn says about Venus. We then learn about the "pig mobile," an elaborate display designed by Ontario Pork, another industry lobby group, to showcase the different stages of the life-cycle of farmed pigs. The pig mobile features a gestation crate, the notorious device long condemned by animal welfare advocates as a de facto prison cell for pregnant sows. Emma Krakar calls it a "piglet protection program." The two sisters assure us that their family-run livestock farm is a far cry from the large-scale, corporate-owned confined animal feeding operations (CAFOs), more commonly known as factory farms. Family farms, the readers are told, are by definition humane and sustainable. One of the popular misconceptions the Krakar sisters hope to dismantle concerns the ecological impact of cattle farming. Kaitlyn insists that cows grazing on "marginal land" do "a lot of good for the environment." The sisters also stress just how much they love their work. Although they express sadness over having to watch the animals they raise be sent off to slaughter, they nonetheless feel content in having made a positive difference to their lives. As Emma Krakar puts it, "They're alive because I was there" (Antonacci, 2021).

The *Spectator* article reads less like journalism and more like an advertisement for Ontario's beef and pork industries. To readers versed in the ecology and political economy of animal farming, the red flags will be many and obvious. It is an

DOI: 10.4324/9781003174417-14

awkward highwire act, for example, to present oneself as an embattled under-dog while simultaneously representing a powerful industry lobby group with the political muscle to secure massive government subsidies, even during times of eco-nomic crisis.[1] The same lobby groups that represent so-called family farms also happily represent factory farms. In fact, the distinction between the two is highly dubious. Factory farms can be, and often are, family-operated. There is nothing to prevent family-run operations from adopting the methods and principles of large-scale industrial animal agriculture. The gestation in crate in Ontario Pork's "pig mobile," for example, is perhaps the most salient symbol of factory farming. In truth, the sheer number of farm animals slaughtered each year in Canada alone would not be possible but for the vast number of factory farms spread out across the country. Another red flag in the article is the empty claim about the environ-ment. Somehow, animal welfare bleeds into sustainability, though the structural connection is not explained. The concept of "marginal land" does not come out of environmental science but rather from the dictionary of finance and invest-ment. Marginal land is land that holds little agricultural or economic potential; it does not mean that such land plays no important part in our ecosystems. The idea that cows grazing on so-called marginal land do "a lot of good for the environ-ment" carelessly conflates agricultural development with environmental renewal (Dutkiewicz & Rosenberg, 2021). As will be discussed later, the science does not support the meat industry's bloated claim about sustainability. The *Spectator* article is merely another iteration of a familiar, if self-confused and scientifically incoher-ent, narrative about the supposed virtues of animal farming. How exactly did this narrative emerge? Why does it continue to flourish?

This chapter explores the emergence and evolution of the mythology of humane and sustainable meat. It begins by revisiting one of the most critical chap-ters in the history of that mythology: Michael Pollan and Joel Salatin's promotion of the idea of "humane" and "sustainable" meat. It then identifies the catalyst for the latest reinvention of the meat industry's self-image: Climate change. After dis-cussing the role of Allan Savory in introducing the fantasy of regenerative grazing into popular agricultural lore, the chapter then examines how this fantasy has taken hold of some of the largest corporations in the world, including Walmart, McDonald's, and Cargill. The chapter concludes with some thoughts on the meat industry's strategy of chameleonism.

Michael Pollan and the dietary third way

Although the original model of animal agriculture in North America was the family-run farm, the now-popular narrative of the family-run farm as the humane and sustainable alternative to the industrial animal agriculture entered the Western social imaginary some time in the early 2000s. By far, the most influential voice for the framing and dissemination of this narrative is Michael Pollan. A promi-nent author, journalist, and professor, Pollan has written a number of bestsell-ing books about food, including *The Omnivore's Dilemma, In Defense of Food: An Eater's Manifesto, Food Rules: An Eater's Manual*, and *Cooked: A Natural History of*

Transformation. In 2002, Pollan published "An Animal's Place" in *The New York Times Magazine* (Pollan, 2002). It was later republished as a chapter in *The Omnivore's Dilemma*, a book that played a critical role in crafting the idea of ethical animal farming and bringing about the so-called slow food movement in North America (Pollan, 2006). "An Animal's Place" is a work of powerful storytelling and seductive mythmaking. In a very real sense, it laid the initial foundations for the now-popular mythology of humane and sustainable meat.

Pollan opens the essay by describing his experience reading Peter Singer's *Animal Liberation* while eating a rib-eye steak at The Palm, a high-end steakhouse with multiple locations across the United States (Singer, 2009). The Palm is not for everyday families—it epitomises bourgeois dining in North America. Rib-eye steaks cost between $70 to $120. With appetisers, drinks, desserts, and taxes, a dinner for four can easily cost an entire month's rent for working families. Pollan tells us that he consciously chose to read *Animal Liberation* at The Palm to experience cognitive dissonance. *Animal Liberation*, he notes, is the Bible of the animal rights movement. It has rightly secured a reputation for turning readers off meat and convincing them to go vegan or vegetarian. As a meat lover with an ostensible moral conscience, Pollan felt compelled to put his moral beliefs to the test.

The core of *Animal Liberation's* argument against the exploitation of animals concerns the principle of equality. In a move evocative of Hegel and Marx, Singer observes that the struggle for equality needs to be understood as part of an ongoing historical continuum. Equality isn't a fixed and timeless idea but develops and expands through the course of history. The women's movement, the abolition movement, the civil rights movement, and the 2SLGBTQIA+ movement, for example, are but chapters in the ongoing expansion of our idea of equality. No cause can be taken as the last and final fight for social justice. If anything, says Singer, we should be suspicious of all talk of having reached "the last remaining form of discrimination" (Singer, 2009, p. 23). Instead, we should stand vigilant against residual and unconscious prejudice in our society. Just because we dismantle some form of oppression today does not mean there are no more forms of oppression left to dismantle. The exploitation of animals is one such form of enduring oppression that needs to be dismantled. Defending animal exploitation is like defending racism or sexism; it is protective and reactionary.

Pollan apparently found this argument so compelling that he was at one point forced to put down his fork and knife. However, upon subsequent reflection, he was able to formulate a response that he believes definitively puts Singer's central thesis to rest. Pollan's counterargument looks something like the following. The modern world, he observes, has been cloven asunder by two untenable extremes. On the one side, there is modern industrial animal farming. As Pollan acknowledges, this is a cruel and inhumane system that reduces farm animals to machines—insentient objects without emotions or the capacity to feel pain. Drawing from Matthew Scully's *Dominion*, Pollan runs through a long list of familiar horrors that are standard practices on factory farms (Scully, 2002): Cramming thousands of chickens into filthy coops, where they are so tightly packed that they will never walk or see the light of day; snipping the beaks of chickens so they do not peck each other to death; tail-docking piglets without any anaesthetic to prevent their tails from getting chewed by other traumatised piglets; clubbing sick pigs to death; and leaving cattle to stand in knee-deep piles of their own waste. Tragically, this list could be extended. If this sort of brutality is the essence of animal farming, says Pollan, then it is little wonder that so many people should give up eating meat altogether. But that, too, he warns, is another untenable modern extreme. Giving up meat is both dangerous and unnatural: Dangerous, because a vegetarian food system would allegedly kill more animals than an omnivorous one; unnatural, because giving up meat is a denial of our own animal nature. We evolved to eat meat. Just look at our canine teeth, says Pollan.

What we need, in his view, is to rediscover a lost middle path between these two extremes of the modern world. So, what exactly is the middle ground between factory farming and abstaining from meat? Pollan's model for the dietary and agricultural Third Way is Polyface Farm. Run by Joel Salatin, a self-described "Christian libertarian environmentalist capitalist lunatic farmer," Polyface Farm has become something of a gold standard for so-called humane and eco-friendly animal farming (The lunatic farmer, n.d.). Salatin and his family raise cows, pigs, chickens, turkeys, and rabbits. They allow the animals to move about freely on open grasslands. They practice rotational grazing, a method in which animals eat and defecate on one patch of pasture before being shepherded on to another patch, allowing the nutrients to be absorbed into the soil and the grass to regrow. Salatin celebrates rotational grazing for imitating natural lifecycles, thereby promoting healthy soil, healthy grass, and healthy animals all at once. Much of the slaughtering takes place on the farm property. Polyface Farm celebrates transparency. They encourage visitors to see how the animals live and how they are killed. Visitors are even invited to participate in the slaughter directly. Salatin calls this "food with a face" (as cited in Pollan, 2002).

Pollan admires Polyface Farm so much that he has been tirelessly promoting it for two decades, effectively turning Salatin into an international celebrity. Salatin has become a hero and role model to animal farmers around the world. He has been featured in numerous documentaries, including the Oscar-nominated film *Food, Inc* (Kenner, 2009). He regularly goes on international speaking tours. Salatin has also published 12 books, which are read in farming circles like scripture. Not only is traditional, grass-fed animal farming perfectly humane, the defence goes, it is also sustainable. This line of defence, however, has undergone a critical mutation in recent years with the advent of climate activism.

Allan Savory and the salvific promise of regenerative grazing

In 2006, the United Nation's Food and Agriculture Organization released "Livestock's Long Shadow," an assessment of the environmental impact of animal farming (Steinfeld et al., 2006). The report examined the economic determinants of increasing livestock production, the extent of land use, the severity and cost of land degradation, the environmental impact of pesticides and herbicides, the role of livestock production in water depletion and pollution, the impact on biodiversity, and, most importantly, the role of livestock farming on climate change and air pollution. The report found that livestock farming was responsible for almost one-fifth of all greenhouse gas emissions, a shocking figure that put animal farming in an entirely new and unsettling light. In 2013, the FAO released a follow-up report entitled, "Tackling Climate Change Through Livestock," in which the estimate of greenhouse gas emissions from animal agriculture was revised to 14 percent (Gerber et al., 2013). However, numerous reports and peer-reviewed studies have confirmed the link between animal agriculture and climate change. Among the most significant are the comprehensive assessment reports from the UN Intergovernmental Panel on Climate Change. The IPCC's 2019 "Special Report on Climate Change and Land," for example, raised alarm about the impact of livestock farming on the Amazon rainforest (Shukla et al., 2019). The IPCC report identifies ranching as a primary driver of deforestation. It warns that, left unchecked, farming-driven deforestation threatens to convert the Amazon into a dry and ecologically desolate savannah. This nightmare scenario would release upwards of 50 billion tons of carbon into the atmosphere in as little as three decades.

Although climate scientists disagree about the precise percentage of climate emissions from animal agriculture, there is nonetheless broad consensus concerning its impact upon the climate. There is also a growing consensus that we need to change our food system if we wish to keep greenhouse gas emissions in check and preserve a habitable planet (de Boer et al., 2016; Springmann et al., 2020; Stehfest et al., 2009). The 2019 IPCC report, for example, encourages reducing the consumption of meat and making up the difference with grains, legumes, fruits, and vegetables (Steinfeld et al., 2006). The growing public realisation that the global appetite for meat is slowly choking the planet has spawned a dietary revolution. While many consumers have opted to cut down on their consumption of meat, many others have decided to give up on meat altogether. The growing number of climate vegans—those who give up animal products for the sake of the planet—has led to material changes in production and consumption (Kerschke-Risch, 2015). The explosive popularity of plant-based meat alternatives at restaurants, schools, and grocery stores can be attributed in part to anxiety over a rapidly warming planet and the demand for sustainable alternatives to meat.

Public concern over the climate impact of meat production has created a public relations crisis for the animal agriculture industry. The industry has reacted swiftly. One response is the denial of any relationship between meat production and climate change. This is a common response from industry lobby groups and trade associations, especially those representing cattle ranchers. A second type of response is to argue that cattle ranching can help fight climate change. This idea has spread like wildfire across the animal farming community in recent years. The most extreme version of this claim comes from Allan Savory, founder of the Savory Institute. Savory is another celebrity livestock farmer and a proponent of what he calls "holistic management" (Savory & Butterfield, 2016). Holistic management is indistinguishable in principle and practice from Salatin's rotational grazing. The basic idea behind holistic management is that properly managed livestock grazing can restore healthy soil and grasslands, which in turn can sequester carbon. By consuming grass instead of grains, cows produce a special kind of manure, which they then trample into the soil. This natural process nourishes the soil, enriching it with valuable nutrients, thereby creating stronger and healthier grass. This grass is not just good for cows. It also serves as a carbon sink more powerful even than forests.

Savory says he developed his method of holistic grazing to combat the problem of desertification. He later reached the conclusion that his farming method is also a powerful weapon against climate change. While the claim about carbon sequestration might not seem all that controversial, Savory goes much further than this. In a famous 2013 Ted Talk, which to date has been viewed almost five million times, Savory makes the following incredible argument:

> people who understand far more about carbon than I do calculate that for illustrative purposes, if we do what I'm showing you here, we can take enough carbon out of the atmosphere and safely store it in the grassland soils for thousands of years, and if we just do that on about half the world's grasslands that I've shown you, *we can take us back to pre-industrial levels while feeding people.* I can think of almost nothing that offers more hope for our planet, for your children, for their children and all of humanity.
>
> *(Savory, 2013)*

This isn't just an extraordinary claim; it is completely unsupported by empirical evidence. While some studies have shown that livestock grazing can, under certain conditions, enable soil to sequester limited amounts of carbon for limited amounts of time, multiple studies have disputed Savory's wildly extravagant claim that his holistic management model can reverse climate change to preindustrial levels.

The best known of these studies is *Grazed and Confused?* from the Oxford Martin School's (Garnett et al., 2017). The report identifies four main problems with Savory's holistic grazing method. The first concerns grazing intensity. The holistic grazing method is said to "shock" local flora such that they lay deeper roots in the soil, which helps create a more capacious carbon sink. The problem with this argument, however, is that root biomass does not appear to expand following intense grazing, but rather shrink. An inverse relationship arises between the grazing intensity and the ability of the forage to recover, a situation that would release more carbon into the atmosphere than it would sequester. Second, holistic grazing is said to rely upon the cows trampling the soil, which both enables seeds to grow and enables the soil to absorb carbon. But as the report notes, the evidence suggests the opposite—that trampling in fact weakens the ability of soil to absorb carbon. Third, holistic grazing is said to introduce critical nutrients into the soil through bovine manure that helps the soil absorb carbon. The problem here is that bovine manure alone is insufficient. Other nutrients, such as nitrogen, are necessary for soil to become carbon-absorptive. Fourth, even if grasslands were able to capture such massive amounts of carbon, they would need to be able to do this year-round for several decades in a row. Evidence of this is lacking. Holistic grazing thus rests on big claims, but very little evidence. As the *Grazed and Confused?* authors note, "whether or not adaptive grazing approaches offer advantages, it is clear that the extremely ambitious claims its proponents make are dangerously misleading" (Garnett et al., 2017, p. 57.) The brute reality is that reversing climate change through livestock farming is just not physically possible. Yet this belief has only grown in popularity, in effect becoming a kind of ecological mythology.

A makeover for the meat industry

Not surprisingly, the dubious science behind regenerative grazing has not stopped the meat industry from embracing the wild and scientifically meaningless claims about "sustainability" (O'Grady, 2003). In just a short time, mega-corporations, industry lobby groups, and trade associations have recognised the lucrative potential in the idea of sustainable meat and are now undertaking a full-scale campaign to reinvent the image of animal agriculture. Incredibly, this campaign has been aided by one of the largest environmental organisations in the world. In 2020, the World Wildlife Fund announced the Ranch Systems and Viability Planning project, a $6 million initiative between the WWF, The Walmart Foundation, McDonald's, and Cargill to implement regenerative grazing practices on one million acres of grasslands across the Northern Great Plains of the United States. The Ranch Systems and Viability Planning project purports to "help fight the climate crisis" by "increasing underground carbon storage, filtering clean water, and providing habitat for wildlife" (World Wildlife Fund, 2020, p.?). The paltry $6 million cost of this project is a trivial and meaningless sum considering Walmart's $560 billion in revenue and $13 billion in profits in 2021, McDonald's $19 billion in revenue and $4.7 billion in profits in 2020, and Cargill's $134 billion in revenue and $4.93 billion profits in 2021 (Kohan, 2021; McDonald's Corporation, 2022). Between Walmart, McDonald's, and Cargill, $6 million is an infinitesimal investment with a massive return in the form of a new public image. There is also something to be said about the bizarre collaboration between the world's leading environmental organisation and three of the world's largest corporations. This collaboration is perhaps not all that surprising in light of the WWF's record of partnering with private corporations and even paramilitary groups at the cost of both human rights and the environment. In 2017, Survival International gave the WWF its "Greenwashing of the Year Award" for partnering with logging companies that have devastated nearly 4 million acres of land belonging to the Indigenous Baka and Bayaka communities of central Africa (Survival International, 2017).

In September 2020, Walmart announced plans to become a "regenerative company," taking the concept of regeneration to incoherent heights. Claiming to have 15 years of "sustainability leadership," Walmart declared a commitment to "help protect, manage, or restore 50 million acres of land and one million square miles of ocean by 2030 to help combat the cascading loss of nature threatening the planet" (Walmart Inc, 2020). Walmart President and CEO Doug McMillon announced plans to transform "the world's supply chains to be regenerative." According to McMillon,

The commitments we're making today not only aim to decarbonize Walmart's global operations, they also put us on the path to becoming a regenerative company—one that works to restore, renew and replenish in addition to preserving our planet, and encourages others to do the same.

(Walmart Inc, 2020, Schwab, 2018)

How a global company like Walmart can be "regenerative" is a mystery. The term "regenerative" serves less as a precise metric for environmental standards and more as a marketing device for cultivating the image of corporate social responsibility.

McDonald's announced that it was "the first restaurant company in the world to address global climate change" (Gibbs, n.d.) The fast-food giant says it plans to reduce climate emissions from its restaurants by 36 percent and reduce emissions intensity from its global supply chain by 31 percent, all by 2030. This amounts to 150 million metric tons of carbon. McDonald's notably plans to reduce greenhouse gas emissions without any reduction in beef production or consumption. This position, of course, runs against the growing call from the scientific community for a massive reduction in both global beef production and consumption. Here, McDonald's is big on ambition, but unsurprisingly weak on science. Asked how exactly they plan to pull off the impossible, a McDonald's representative responded by saying, "We are looking for ways to incorporate soil health initiatives into our supply chain sustainability programs through managed grazing practices and regenerative agricultural practices" (Schwab, 2018). They fail to cite studies to support their ambitious plans. McDonald's thus fills the vacuum of scientific evidence with the empty and speculative mythology of regenerative grazing.

Meat industry trade associations and lobby groups have also pivoted from explicit denialism to championing regenerative agriculture as an integral part of the solution to climate change. In 2019, the Canadian Cattlemen's Association released a public statement in response to the IPCC's 2019 Special Report on Climate Change and Land. Contrary to the scientific consensus, the CCA claims that cows are "part of the climate change solution" and that Canadian beef cattle "help to preserve one of the world's most endangered ecosystems." Grazing, we are told, is necessary to preserve the habitats of "579 birds, mammals, reptiles and amphibians." Grazing is also apparently vital to "help mitigate the risk of wildfires" (Canadian Cattlemen's Association, 2019). The evidence provided for this last claim is not peer-reviewed science, but rather a guest column in *Beef in BC Magazine* "Grazing for Wildfire Prevention" (Beef in BC, 2019). While the CCA acknowledges that consumers "certainly have the right to choose the food they eat," they insist that "reducing meat consumption is not a solution to climate change" (Canadian Cattlemen's Association, 2019).

A mythology solidified

The rise of the regenerative agriculture movement has spawned a new talking point in media coverage of diet, food production, and climate change. *Forbes*, for example, published a story in 2018 titled "Going Vegan? Eating Sustainable Beef Can Be Good For The Environment." The article claims that regenerative agriculture can aid in reversing the adverse impacts, like climate change from agricultural practices. It cites two studies exploring the promise of regenerative grazing for sequestering carbon. At best, however, these studies suggest some promise in reducing the animal agriculture industry's carbon emissions. Neither study offers any indication that regenerative agriculture can "reverse the climate change impacts of agricultural practices" (Yeoh, 2018). The *Forbes* article confuses a hypothetical claim about a mere *reduction* in carbon emissions for concrete evidence of regenerative grazing's ability to *reverse* the climatic effects of meat production. Reversal appears to be another choice from the general grab bag of meaningless concepts.

A 2019 article in the CBC titled, "How 'regenerative farmers' help reduce greenhouse gases" similarly propagates the mythology of regenerative grazing (Osman & Chevalier, 2019). Although written as an objective news story, the article once again reads like an advertisement for Canada's meat industry. The article largely presents the perspective of the regenerative grazing movement, which argues that "people can still take a bite out of climate change while eating red meat, pushing back against global headlines calling for major changes to the world's farming and eating habits" (Osman & Chevalier, 2019). It paints a sympathetic portrait of animal farmers who lament the deteriorating reputation of meat. The article quotes Ottawa farmer Amber Payne, who says, "It hurts my soul to hear that we're viewing red meat as detrimental to climate change." Payne goes on to say, "I look at it as a solution to fix many global problems." The CBC article reviews the basic philosophy and general principles of regenerative grazing, which it calls an "age-old method" of farming, thereby adding to the aura and mystique. Although the article does cite one sceptical scientific opinion about the promises of regenerative grazing, this opinion is very brief and only appears near the end.

Conclusions: Regenerative grazing as meat industry chameleonism

This chapter has sought to show how the animal agriculture industry has reinvented itself in light of the terminological fashions of the day to preserve its market power. This pattern of shapeshifting according to changing political winds may rightly be described as a form of industry chameleonism, the shrewd marketing tactic of reinventing animal agriculture and redefining the very idea of meat. The industry has demonstrated a remarkable capacity for self-redefinition and reinvention while remaining fundamentally unchanged. In response to the forceful critiques of the industry's horrific treatment of animals, it has crafted the fantasy humane meat—the fiction that an industry whose principal business is the exploitation and slaughter of animals is somehow more humane than a lifestyle that shuns the exploitation and slaughter of animals. Now, in response to equally forceful critiques of the industry's role in driving up greenhouse gas emissions, it has crafted the fantasy that animal agriculture is the answer to our climate crisis (Hannan, 2020).

The effectiveness of these fantasies is a testament to the power of meat industry chameleonism, a phenomenon with material consequences both for the lives of animals and the fate of the planet. It is imperative that the meat industry's claims about sustainability be subjected to both scientific as well as ideological critique. That is, it is not enough merely to scrutinise the ecological merits of regenerative grazing, but also to examine the purpose, nature, and contours of regenerative grazing as industry mythology. Subjecting that mythology to critique can help to foster basic agricultural and political economic literacy, both of which are vital to a civically engaged public concerned about both animal justice and climate justice. That critique is best presented in the light of sustainable food systems. The evidence here strongly converges upon plant-based food systems, which take up a fraction of the land and water used by animal agriculture, produce far less greenhouse gas emissions, dramatically reduce water pollution, and do not inflict unspeakably cruel and gratuitous violence upon sentient beings (Godfray et al., 2018; Poore & Nemecek, 2018; Springmann et al., 2020).

Note

1 Government of Ontario (2020, December 18). Governments support beef farmers and protect food supply chains. Retrieved from https://news.ontario.ca/en/release/59780/governments-support-beef-farmers-and-protect-food-supply-chains

References

Antonacci, J.P. (2021, October 12). Where does your food come from? Livestock farmers share their stories at the Norfolk County Fair. *The Hamilton Spectator*. Retrieved from https://www.thespec.com/news/hamilton-region/2021/10/12/where-does-your-food-come-from-livestock-farmers-share-their-stories-at-the-norfolk-county-fair.html

Beef in BC. (2019). *Grazing for wildfire prevention.* July/August.

de Boer, J., de Witt, A., & Aiking, H. (2016). Help the climate, change your diet: A cross-sectional study on how to involve consumers in a transition to a low-carbon society. *Appetite, 98*, 19–27.

Canadian Cattlemen's Association (2019, August 8). *CCA statement on UN IPCC report, 'Climate change and land.'* Retrieved from https://www.cattle.ca/news-events/news/view/cca-statement-on-un-ipcc-report-climate-change-and-land/

Dutkiewicz, J & Rosenberg, G. (2021, September 23) The myth of regenerative ranching. *The New Republic*. Retrieved from https://newrepublic.com/article/163735/myth-regenerative-ranching

Garnett, T., Godde, C., Muller, A., Röös, E., Smith, P., de Boer, I.,zu Ermgassen, E. K. H. J & van Zanten, H. (2017). Grazed and confused. *Food Climate Research Network, 127*, 522.

Gerber, P. J., Steinfeld, H., Henderson, B., Mottet, A., Opio, C., Dijkman, J., Falcucci, A. & Tempio, G. (2013). *Tackling climate change through livestock: a global assessment of emissions and mitigation opportunities.* FAO.

Gibbs, R. (n.d.). Using our scale for good: Taking big steps to reduce our carbon footprint. Retrieved from https://corporate.mcdonalds.com/corpmcd/en-us/our-stories/article/ourstories.carbon_footprint.html

Godfray, H. C. J., Aveyard, P., Garnett, T., Hall, J. W., Key, T. J., Lorimer, J., Pierrehumbert, R. T., Scarborough, P., Springmann, M. & Jebb, S. A. (2018). Meat consumption, health, and the environment. *Science, 361*(6399). https://www.science.org/doi/abs/10.1126/science.aam5324

Hannan, J., (Ed.). (2020). *Meatsplaining: The animal agriculture industry and the rhetoric of denial.* Sydney University Press.

Kenner, R. (Director). (2009). *Food, inc.* Magnolia Pictures.

Kerschke-Risch P. (2015) Vegan diet: motives, approach and duration. Initial results of a quantitative sociological study. *Ernahrungs Umschau*, 62(6), 98–103.

Kohan, S. (2021, February 18). Walmart revenue hits $559 billion for fiscal year 2020. *Forbes.* Retrieved from https://fortune.com/company/walmart/fortune500/

McDonald's Corporation (2022, January 27). McDonald's reports fourth quarter and full year 2020 results. Retrieved from https://corporate.mcdonalds.com/corpmcd/en-us/our-stories/article/press-releases.Q4-2021-results.html

O'Grady, J. P. (2003). How sustainable is the idea of sustainability? *Interdisciplinary Studies in Literature and Environment*, 10(1), 1–10.

Osman L., & Chevalier, J. (2019). How 'regenerative farmers' help reduce greenhouse gases. Retrieved from https://www.cbc.ca/news/canada/ottawa/red-meat-local-farmer-ottawa-1.5244343

Pollan, M. (2002). An animal's place. *New York Times Magazine.* Retrieved from https://www.nytimes.com/2002/11/10/magazine/an-animal-s-place.html.

Pollan, M. (2006). *The omnivore's dilemma: A natural history of four meals.* Penguin Press.

Poore, J., & Nemecek, T. (2018). Reducing food's environmental impacts through producers and consumers. *Science, 360*(6392), 987–992.

Savory, A. (2013, March 4). How to green the world's deserts and reverse climate change. *YouTube.* Retrieved from https://youtu.be/vpTHi7O66pI

Savory, A., & Butterfield, J. (2016). *Holistic management: a commonsense revolution to restore our environment.* Island Press.

Schwab, T. (2018, October 15). If McDonald's is serious about reducing its carbon footprint, it may need to rethink the hamburger. *The Counter.* Retrieved from https://thecounter.org/mcdonalds-greenhouse-gas-emissions-reduction-pledge-beef/

Scully, M. (2002). *Dominion: The power of man, The suffering of animals, and the call to mercy.* St. Martin's Press.

Shukla, P. R., Skea, J., Calvo Buendia, E., Masson-Delmotte, V., Pörtner, H. O., Roberts, D. C., Zhai, P. & Malley, J. (2019). *IPCC, 2019: Climate Change and Land: an IPCC special report on climate change, desertification, land degradation, sustainable land management, food security, and greenhouse gas fluxes in terrestrial ecosystems.* IPCC.

Singer, P. (2009). *Animal liberation: The definitive classic of the animal movement.* Harper Perennial.

Springmann, M., Spajic, L., Clark, M. A., Poore, J., Herforth, A., Webb, P., Rayner, M. & Scarborough, P. (2020). The healthiness and sustainability of national and global food based dietary guidelines: Modelling study. *British Medical Journal, 370*, 1–16.

Stehfest, E., Bouwman, L., Van Vuuren, D. P., Den Elzen, M. G., Eickhout, B., & Kabat, P. (2009). Climate benefits of changing diet. *Climatic Change, 95*(1), 83–102.

Steinfeld, H., Gerber, P., Wassenaar, T. D., Castel, V., Rosales, M., Rosales, M., & de Haan, C. (2006). *Livestock's long shadow: Environmental issues and options.* Food & Agriculture Org.

Survival International (2017). *WWF wins Survival's "Greenwashing of the year" award.* Retrieved from https://survivalinternational.org/news/11677

The Lunatic Farmer (n.d.). Retrieved from https://www.thelunaticfarmer.com/

Walmart Inc. (2020). Walmart sets goal to become a regenerative company. Retrieved from https://corporate.walmart.com/newsroom/2020/09/21/walmart-sets-goal-to-become-a-regenerative-company

World Wildlife Fund (2020). New project will help improve 1 million acres of grasslands to help fight the climate crisis. Retrieved from https://www.worldwildlife.org/stories/new-project-will-help-improve-1-million-acres-of-grasslands-to-help-fight-the-climate-crisis

Yeoh, N. (2018). Going vegan? Eating sustainable beef can be good for the environment. *Forbes.* Retrieved from https://www.forbes.com/sites/neilyeoh/2018/04/14/going-vegan-eating-sustainable-beef-can-be-good-for-the-environment/?sh=330c80f867ea

13

RE-MEATIFICATION

The potential of plant-based alternatives to animal foods in transitions towards more sustainable and humane diets

Tony Weis and Allison Gray

Introduction: Problematising meatification

The meatification of diets describes the steady, though highly uneven, rise in per capita meat consumption. Animal flesh has moved from the margins to the centre of human diets in many parts of the world—a course that has been led by high-income countries (Weis, 2013). From 1961 to 2020, per capita meat consumption nearly doubled on a world scale, from 23 kg to 43 kg, at the same time as the human population nearly tripled from around 3 billion to almost 8 billion. Per capita egg consumption also rose at a similar pace on a world scale, while per capita dairy consumption has held relatively steady amid rapid human population growth (FAOSTATS, 2022). There are vast disparities in per capita consumption of meat and other animal products between high-income and low-income countries, while middle-income countries lie at the forefront of recent and projected increases.

The meatification of diets has been driven by the industrialisation of livestock production, especially the explosive growth of pigs and poultry, which together account for 72 percent of the total volume of animal flesh produced annually (FAOSTATS, 2022). Industrial livestock production commands close to one-third of the world's arable land, with large shares of annual coarse grain (especially corn) and oilseed (especially soybeans) harvests used to feed rising populations of concentrated animals, principally pigs, poultry birds (mainly chickens), and cattle. Pasture is by far the world's largest anthropogenic land use and is a major factor in both tropical deforestation (especially in Amazonia) and desertification. The extent of livestock production and consumption features heavily in environmental assessments of agro-food systems due to the enormity of land use and its impacts on climate change, biodiversity loss, freshwater consumption and pollution, and

DOI: 10.4324/9781003174417-15

other problems (IPCC, 2019; Machovina et al., 2015; Poore & Nemecek, 2018; Springmann et al., 2016, 2018; Willett et al., 2019; WWF, 2020).

The industrialisation of livestock production entails fundamentally different land and interspecies relationships than those that prevailed over agrarian history. For most of the past 10,000 years, livestock tended to live at relatively low densities in farming landscapes, grazing on fallowed land, eating crop stubble and household food scraps, and consuming fodder produced in rotations, as well as being moved over large areas of land by migratory herders. In addition to their flesh, livestock animals were priced for their labouring power, fertilisation role, and generation of milk, eggs, and fibre, which meant that killing for meat tended to be a sporadic event for farmers and herders (Fagan, 2015; Harris, 1974; Lappe, 1991).

One of the most basic goals in producing animals at high densities in warehouses, sheds, and feedlots is that it greatly increases labour productivity, or output per farmer or farm worker. One of the most basic implications of these enclosures is that they necessitate different ways of feeding animals from those that prevailed in mixed farming systems, establishing a far greater reliance on cultivating crops directly for animals. This reliance on concentrated feed to produce rising volumes of animal flesh, milk, and eggs is an inherently wasteful process, as much of the usable nutrition from grains and oilseeds gets burned in the "unproductive" metabolic processes of animals. Further, the land devoted to feed crop monocultures could conceivably be used in ways that generate more plant nutrition per land area. This intractable inefficiency necessarily expands the land area needed for agriculture, with negative reverberations for biodiversity, and is a big part of the highly regressive character of meatification. It means that people with meat-heavy diets command a lot more grains and oilseeds on a per capita basis than people who eat little to no meat, which connects them to the additional resources, carbon dioxide and nitrous oxide emissions, and pollution associated with the machinery, fertilisers, pesticides, and irrigation used in this production (Poore & Nemecek, 2018; Springmann et al., 2018; Weis, 2013).

Additionally, there are the multidimensional costs associated with industrial livestock operations and processing plants. A great deal of energy is required for: Ventilation, monitoring, and automated feeding systems in industrial livestock operations; the slaughter and packing lines in processing plants; the refrigeration units from processing to retail; and in shipping animals over greater distances between sites of breeding, growing, and slaughter as operations become more consolidated. This energy budget entails further carbon dioxide emissions, which add to the atmospheric burden from ruminant livestock animals—the leading source of global methane emissions. Industrial livestock operations and processing plants also generate localised airborne pollution, as the stench and particulate matter that emanates from these spaces reverberate negatively on the physical health and psychosocial well-being in local communities. They are also water intensive spaces in the course of hydrating large populations of thirsty animals, flushing out immense aggregations of faeces and urine from enclosures, and cleaning slaughter and packing lines from the blood and corporeal wastes (Blanchette, 2020; Imhoff, 2011; Lynbery, 2014; Neubert, 2020; PEW Commission, 2008; Weis, 2013). The heavy water use in livestock operations and processing plants combined with the water embedded in feed crops explains why animal-based foods tend to have much higher water budgets than plant-based foods, as well as disproportionately contributing to nutrient loading and other sorts of water pollution (Hoekstra, 2013; Mekonnen & Hoekstra, 2012).

The resource budgets and pollution loads associated with livestock production are a central reason why, if the current trajectory of meatification continues (along with expected human population growth), the damaging environmental impacts from agro-food systems are expected to intensify by as much as 50 to 90 percent by 2050 (IPCC, 2019; Springmann et al., 2018). The continuing meatification of diets runs directly counter to the prospect of containing further warming to potentially safe levels. Global average surface temperatures have already warmed by 1 °C above pre-industrial levels, and annual emissions must be cut by at least half by 2030 for there to be any hope of containing further warming to 1.5 °C (IPCC, 2018). It is also clear that responding to this planetary burden can simultaneously benefit public health, as heavy meat consumption is well-established as a major contributing factor in the proliferation of a number of non-communicable diseases (Springmann et al., 2016, 2018; Willett et al., 2019).

Beyond questions of sustainability lie other untold costs and risks. The growing populations and densities of animals, rapid turnover times, weakened immune systems, and heavy antibiotic use together magnify the risks of zoonotic disease evolution, namely that dangerous new variants of infectious diseases like avian and swine influenza could emerge and eventually jump the species barrier. While it is impossible to know when or how the next dangerous variant of avian or swine flu will emerge, the COVID-19 pandemic illustrates the sort of impact that a new highly virulent type of influenza could bring (Davis, 2020; Economou & Gousia, 2015; Wallace, 2020). Industrial livestock operations also create prime conditions for antibiotic-resistant bacteria to develop and spread within animal populations (due to the scale of antibiotic use, the great densities of animals with weakened immune systems, and the accelerated turnover times of herds and flocks). Potentially, these can be transmitted to humans in a variety of ways, such as through residues on food, the use of faecal slurries as fertiliser, and ground- and surface water supplies. The inadvertent consumption of residual antibiotics in animal foods can also exacerbate the risks of antibiotic-resistant bacteria developing within human populations along with those posed by the scale of drug use in modern societies. The problem of antibiotic resistance already adversely affects millions of people every year with little popular recognition, while public health researchers warn that its scale could begin to spike issues at any moment (Morehead & Scarbrough, 2018; O'Neill, 2016; Ventola, 2015).

Finally, it is also important to recognise that the explosive growth and industrialisation of livestock production now shapes the living conditions for a large and growing share of all mammals and birds (Weis, 2016). The number of animals killed annually for food increased approximately ten-fold from 1961 to 2019, from around 8 to 80 billion (FAOSTATS, 2022); if meat production and consumption continue rising on the current course, led by the number of chickens, this is poised to reach 120 billion by 2050. Industrialisation entails far more than quantitative growth; it is also entwined with a qualitative transformation in the control over and suffering inflicted upon animals. This transformation includes: Accelerating genetic changes to increase rates of weight gain (shortening the time, birth to slaughter), lactation, and egg laying that strain animal bodies; specialisation of breeding operations and reliance on artificial insemination; immobilisation of bodies to enable automation and reduce the feed lost to metabolic processes; routine physical mutilations geared to inhibiting animals from harming neighbours and themselves; and fears associated with shipment and loading onto large, fast-moving slaughter lines (Imhoff, 2011; Lymbery, 2014; Singer, 2020; Weis, 2013). These miserable conditions also translate into harsh working conditions for a labour force that tends to be precarious and poorly paid, with jobs marked by intense psychological strain and high risks of both sudden and repetitive stress injuries, as well as a proneness for violence to spill over into homes and communities (Blanchette, 2020; Eisnitz, 2006; Fitzgerald et al., 2009; Genoways, 2015; Pachirat, 2011).

In sum, the need to problematise and reverse the course of meatification is increasingly acknowledged on environmental, health, and ethical grounds, starting in high-income countries where the per capita consumption of animal-based foods is the greatest. The remainder of this chapter considers the key reasons why the surging development and marketing of plant-based meats, dairy products, and egg replacers can potentially contribute to more sustainable and humane diets.

De-meatification and the surge of plant-based alternatives to animal foods

Plant-based livestock alternatives derive their nutritional content from varying combinations of protein-rich pulses like beans, lentil, or peas (the most common foundation), as well as nuts, mushrooms, grains (e.g., wheat, oats), tubers (e.g., potatoes), and vegetable oils. A range of processing techniques are used to manufacture plant-based alternatives to animal foods, guided by the goal of mimicking some combination of key characteristics, including taste, appearance, texture, and cooking and baking properties. These processes have some old antecedents, as a range of food cultures have long histories of processing plants into protein-dense foods akin to some meat and dairy products, such as varieties of seitan, tempeh, tofu, and soy-based beverages developed in various parts of East and Southeast Asia. However, there is a dramatically different speed at which plant-based alternatives to livestock alternatives are now being developed, resulting in a fast-widening diversity of products (Reece, 2018). To understand this surge of innovation in production, along with the fast-growing presence in retail environments, from fast food chains to supermarkets, it helps to position it within a brief overview of the three basic de-meatification pathways.

Reducing animal consumption

One pathway acknowledges the problems with industrial livestock production and the importance of reducing animal consumption to some degree, whether for environmental, health, or ethical considerations, often while striving to support smaller-scale mixed farms producing meat, milk, and eggs (Fairlee, 2010; Pollan, 2008). This trend is reflected in the rise of so-called "flexitarian" diets and campaigns that encourage people to reduce but not eliminate meat consumption. The second pathway, vegetarianism, abstains from consuming animal flesh, often premised on a desire to avoid direct responsibility for killing, while participating in the consumption of milk and eggs from livestock (Fox, 1999). Whether its adherents think of it in these terms or not, vegetarianism is premised on an acceptance of the functional role that livestock play in mixed farming systems, especially with respect to nutrient cycling and conceivably with respect to labour. The third pathway, veganism, rejects the consumption of all animal products, which is often rooted in ethical and environmental arguments and a desire to end all animal use in agriculture (Singer, 2020), though health concerns are occasionally at the forefront. Veganism rejects a premise of vegetarianism, arguing that it is impossible for animals to have functional roles in mixed farming systems in practice without killing some (or even most) individuals short of their natural lifespans. For instance, if the goal in rearing livestock is only to extract milk and eggs, there will be a lot of unproductive or "excess" males and potentially some excess females (should the population grow beyond the replacement rates of productive animals), and hence it is not realistic to assume that all of these animals could be allowed to live out full lifespans in peace. Thus, vegans reject the possibility that one can consume milk and eggs from animals without some level of complicity in killing, regardless of whether or not animal flesh is consumed.

Plant-based alternatives

The increasing array of plant-based meats, milks, and egg replacers both responds to the momentum of de-meatification and growing consumer demand (Macdiarmid et al., 2016; Pohjolainen et al., 2016; Reece, 2018) and has the potential to significantly augment demand and the popular consciousness that surrounds it. Plant-based livestock alternatives directly confront some of the key perceptual and practical barriers that could otherwise inhibit people from moving towards flexitarian, vegetarian, or vegan diets, or to stick with this decision. At the forefront of these barriers is the strength of prevailing palate preferences for animal-based foods, from bacon and eggs at breakfast, to chicken sandwiches and hot dogs at lunch, to pepperoni and cheese pizzas, hamburgers, and chicken breasts at dinner. Animal-heavy consumption patterns are perceived with a deep sense of entitlement in many places and powerfully entwined with food preparation and cooking skills, especially in high-income countries (Adams, 2010; Chiles & Fitzgerald, 2018).

A central motivation in the development of plant-based meats, dairy products, and egg replacers is to respond to a combination of cravings, convenience, and entrenched modes of thinking and cooking associated with livestock-heavy diets. One clear reflection of this is the proliferation of products like veggie burgers, dogs, and sausages, and plant-based milks, cheeses, and yogurts in mainstream food environments. Plant-based product development is increasing quickly in poultry meat alternatives and more recently began to diversify into fish and egg alternatives. In observing this dynamism, a central question to consider is whether these products have the potential to reach a level of broad-based consumer acceptance where they begin to substitute for animal-based foods on a scale that contributes to the de-industrialisation and de-growth of livestock production and consumption.

Re-meatification: The prospects of substitution

A core premise fuelling the rise of plant-based livestock alternatives is that the taste preferences, eating habits, and cooking practices associated with meats, dairy products, and eggs are so deeply entrenched in modern societies that it is difficult for many people to imagine mealtime without them. To put it another way, there is a recognition that conceptual and practical barriers are bound to reduce the sway of ethical and environmental arguments to a considerable degree. Rather than simply running head-first into these barriers and challenging everyone to radically alter their palates and develop new food preparation and cooking skills, we can expect that some people will be more likely to consider dietary transitions towards flexitarian, vegetarian, and vegan diets if disruptions are minimised through more directly substitutable plant-based products. Direct substitution hinges on a close resemblance to the taste and food preparation characteristics of animal products, in forms that resonate with the dominant food culture and can be consumed in familiar ways, such as between buns, on sandwiches and cereals, and at fast-food restaurants and social events like barbeques (Twine, 2018). We suggest that efforts to increase this direct substitution can be understood to be aiming towards a "re-meatification" of diets, where conceptions of meat, dairy products, and eggs continue to occupy a central place in modern cuisines but get increasingly composed of processed plants rather than animals' flesh and reproductive outputs.

While it is impossible to quantify, there have undoubtedly been immense improvements in taste, texture, cooking properties, and overall quality of plant-based alternatives. Quality improvements are augmented by increasingly competitive prices, growing availability in a range of retailers, and frequent placement near animal-based meat and dairy products in supermarket refrigerators, rather than being confined to less visible spaces. As a condition for retailers to sell plant-based products made by larger plant-based corporations, greater visibility within retail environments is sometimes a requirement and this placement has the potential to broaden the exposure of these products to consumers and increases the likelihood that more people will try them as a substitute.

Environmental benefits from animal alternatives

The re-meatification of diets has the potential to bring considerable environmental benefits. While there has been a great deal of research focused on technologies to reduce the emissions and other pollution loads from livestock production—from the genetic modification of animals to the development of wearable contraptions to capture ruminant methane—there are insurmountable biophysical barriers that limit what this can achieve. One part of these barriers relates to the energy, water, and pollution intensity associated with livestock operations and processing plants and with shipping animals over greater distances in the course of production, and the impossibility of capturing all the methane associated with enteric fermentation in ruminants and manure. Another barrier is the inherent wastage of usable nutrition in feed crops in the unproductive metabolic processes of animals, which inevitably greatly expands the land and resources in cultivation (Poore & Nemecek, 2018; Springmann et al., 2018; Weis, 2013). Even if more efficient, emerging technologies were widely deployed in livestock operations around the world, these could only reduce the total GHG emissions from production by a modest amount (Cederberg et al., 2013).

In contrast, there is overwhelming scientific research that indicates the vastly superior efficiency of deriving nutrition directly from plants, as plant-based foods tend to require far less land, fossil energy, water, and other resources to produce, than animal-based foods relative to a given unit of nutrition. They are responsible for far less greenhouse gas emissions and other pollution loads, such that a global-scale transition towards plant-based diets could potentially reduce the GHG emissions associated with agriculture and food by up to 80 percent (IPCC, 2019; Poore & Nemecek, 2018; Springmann et al., 2016, 2018; Vermeulen et al., 2012). This case is further augmented by evidence that eating mostly or entirely plant-based diets greatly reduces incidence of non-communicable diseases (Willett et al., 2019).

Research indicates that plant-based meats tend to require very similar amounts of land as pulse production on average (relative to the same level of nutritional content) and generate similar GHG emissions, regardless of the primary ingredient, which is typically a pulse (Dettling et al., 2016; Fresán et al., 2019; Goldstein et al., 2017; Keoleian & Heller, 2018; Mejia et al., 2019; Smetana et al. 2015). In addition to cultivation, plant-based meats also demand less energy in the food processing facilities (Mejia et al., 2019), with the associated GHG emissions largely contingent on the nature of the electricity grid where they are located (which means that their overall atmospheric impact will shrink considerably if the composition of electricity grids increasingly moves towards renewables like solar and wind). A life-cycle assessment that compared the Beyond Burger to a similarly-sized burger composed of cattle flesh found that it generated 90 percent less GHG emissions and 93 percent less land and water (Keoleian & Heller, 2018).

While improved resemblance and the ease of substitution are important to the growth of plant-based alternatives and can make it easier for some people to act on their desire to reduce environmental impacts and interspecies violence, ultimately the prospects for plant-based re-meatification rest on a growing awareness of the problems of meatification. Here, it is possible that the increasing range and visibility of plant-based alternatives—especially in cities in high-income countries—might not just be something that helps some people respond to their convictions but could also trigger new reflections on established behaviours, including thinking more about the socio-ecological and interspecies dimensions of previously taken-for-granted food choices and destabilising perceptions about the superior protein density of animal foods. In other words, it is possible, though far from inevitable. See the related chapter on potential pitfalls with the surge of plant-based alternatives to animal foods, and how the mainstreaming of plant-based alternatives can help challenge some of the powerful illusions that prevail in meat-heavy food cultures that have helped to normalise the dramatically changing scale and nature of livestock production.

Conclusion

Reducing the production and consumption of animal foods are essential to the prospect of building more equitable, sustainable, and humane agro-food systems. The de-meatification imperative starts in high-income countries, where consumption is greatest, and is key to potentially shrinking the total land area devoted to cultivation and pasture, which would enable more ecological restoration (and in turn more carbon sequestration and habitat for wild animals) and reduce the fossil energy use, greenhouse gas emissions, freshwater consumption and pollution, and scale of animal suffering associated with modern agro-food systems, all while improving human health outcomes and food security in both the near and long term (IPCC, 2019; Machovina et al., 2015; Poore & Nemecek, 2018; Springmann et al., 2016, 2018; Willett et al., 2019).

However, dietary change cannot rely on compelling scientific or moral arguments alone. This chapter has considered how efforts to promote the increasing substitution of animal foods with plant-based alternatives—what we refer to as re-meatification—can augment de-meatification pathways. The fast-improving array, quality, and accessibility of plant-based alternatives to animal foods cannot resolve tensions surrounding de-meatification pathways, but they do have the potential to: Make it easier for flexitarians, vegetarians, and vegans alike to navigate the dominant food culture; reduce the likelihood that engrained eating habits, palate preferences, and food preparation skills will trump environmental or ethical considerations; and help people take the first step towards reducing their consumption of animal foods, which research has shown is often the hardest (Gray, 2020; Clark & Bogdan, 2019). It is easy to be cynical about the mainstreaming of products like veggie burgers in fast-food restaurants and supermarkets that are overflowing with unhealthy and unsustainable foods. Yet there are also many good reasons to believe that continuing advances in food science that turn plant inputs directly into protein-dense foods with mass appeal can play a significant part in transitions towards more sustainable and humane diets.

References

Adams, C. J. (2010). *The sexual politics of meat: A feminist-vegetarian critical theory*. Bloomsbury.
Blanchette, A. (2020). *Porkopolis American animality, standardised life, and the factory farm*. Duke University Press.
Cederberg, C., Hedenus, F., Wirsenius, S., & Sonesson, U. (2013). Trends in greenhouse gas emissions from consumption and production of animal food products: Implications for long-term climate targets. *Animal*, 7(2), 330–340.

Chiles, R. M. & Fitzgerald, A. (2018). Why is meat so important in Western history and culture? A genealogical critique of biophysical and political- economic explanations. *Agriculture and Human Values, 35*(1), 1–17.

Clark, L. F. & Bogdan, A. (2019). Plant-based foods in Canada: Information, trust and closing the commercialization gap. *British Food Journal, 121*(10), 2535–2550.

Davis, M. (2020). *The monster enters: Covid-19, Avian Flu, and the plagues of Capitalism*. OR Books.

Dettling, J., Qingshi, T., Faist, M., DelDuce, A., & Mandlebaum, S. (2016). *A comparative life cycle assessment of plant-based foods and meat foods*. Quantis.

Economou, V., & Gousia, P. (2015). Agriculture and food animals as a source of antimicrobial-resistant bacteria. *Infection and Drug Resistance, 8*, 49–61.

Eisnitz, G. (2006) *Slaughterhouse: The shocking story of greed, neglect, and inhuman treatment inside the US meat industry*. Prometheus.

Fagan, B. (2015). *The intimate bond: How animals shaped human history*. Bloomsbury.

Fairlie, S. (2010). *Meat: A benign extravagance*. Chelsea Green.

Fitzgerald, A., Kalof, L., & Dietz, T. (2009). Slaughterhouses and increased crime rates: An empirical analysis of spillover from 'The Jungle' into the surrounding community. *Organisation & Environment, 22*(2), 158–184.

Food and Agriculture Organisation Statistics Division (FAOSTATS). (2022). *Production & resource STAT calculators*. FAO. http://www.fao.org/faostat/en/

Fox, M. (1999). *Deep vegetarianism*. Temple University Press.

Fresán, U., Maximino-Alfredo, M., Craig, W. J., Jaceldo-Siegl, K., & Sabaté, J. (2019). Meat analogs from different protein sources: A comparison of their sustainability and nutritional content. *Sustainability, 11*(1), 3231–3241.

Genoways, T. (2015). *The chain: Farm, factory, and the fate of our food*. Harper.

Goldstein, B., Moses, R., Sammons, N., & Birkved, M. (2017). Potential to curb the environmental burdens of America beef consumption using a novel plant-based beef substitute. *PLos ONE, 12*(12), e0189029.

Gray, A. (2020). *Eating in the Anthropocene: Perceptions of dietary-based environmental harm and the role of plant-meat consumption* (Unpublished doctoral dissertation, University of Windsor, Windsor, Ontario).

Harris, M. (1974). *Cows, pigs, wars, and witches: The riddles of culture*. Vintage.

Hoekstra, A.Y. (2013). *The water footprint of modern consumer society*. Routledge.

Imhoff, D., ed. (2011). *The CAFO reader: The tragedy of industrial animal factories*. University of California Press.

Intergovernmental Panel on Climate Change (IPCC). (2018). *Global Warming of 1.5°C*. IPCC.

Intergovernmental Panel on Climate Change (IPCC). (2019). *Special report on climate change, desertification, land degradation, sustainable land management, food security, and greenhouse gas fluxes in terrestrial ecosystems*. IPCC.

Keoleian, G. A. & Heller, M. C. (2018). *Beyond Meat's Beyond Burger life cycle assessment: A detailed comparison between a plant-based and an animal-based protein source*. University of Michigan. http://css.umich.edu/sites/default/files/publication/CSS18-10.pdf

Lappé, F. M. (1991[1971]). *Diet for a small planet*. Ballantine.

Lymbery, P. (2014). *Farmageddon: The true cost of cheap meat*. Bloomsbury.

Macdiarmid, J. I., Douglas, F., & Campbell, J. (2016). Eating like there's no tomorrow: Public awareness of the environmental impact of food and reluctance to eat less meat as part of a sustainable diet. *Appetite, 96*, 487–493.

Machovina, B., Feeley, K. J., & Ripple, W. J. (2015). Biodiversity conservation: The key is reducing meat consumption. *Science of the Total Environment, 536*, 419–31.

Mejia, M. A., Fresán, U., Harwatt, H., Oda, K., Uriegas-Mejia, G., & Sabaté, J. (2019). Life cycle assessment of the production of a large variety of meat analogs by three diverse factories. *Journal of Hunger & Environmental Nutrition, 15*(5), 699–711.

Mekonnen, M. M. & Hoekstra, A. Y. (2012). A global assessment of the water footprint of farm animal products. *Ecosystems*, *15*(3), 401–415.

Morehead, M. S. & Scarbrough, C. (2018). Emergence of global antibiotic resistance. *Primary Care*, *45*(3), 467–484.

Neubert, C. (2020). The Anthropocene stinks! Odour, affect, and the entangled politics of livestock waste in a rural Iowa watershed. *Society and Space*, *38*(4), 736–752.

O'Neill, J. (2016). *Tackling drug-resistant infections globally: Final report and recommendations*. The Review on Antimicrobial Resistance.

Pachirat, T. (2011). *Every twelve seconds: Industrialised slaughter and the politics of sight*. Yale University Press.

Pew Commission on Industrial Farm Animal Production (2008). *Putting meat on the table: Industrial farm animal production in America*. The Pew Charitable Trusts and The John Hopkins Bloomberg School of Public Health.

Pohjolainen, P., Tapio, P., Vinnari, M., Jokinen, P., Räsänen, P. (2016). Consumer consciousness on meat and the environment: Exploring differences. *Appetite*, *101*, 37–45.

Pollan, M. (2008). *In defence of food: An eater's manifesto*. Penguin.

Poore, J., & Nemecek, T. (2018). Reducing food's environmental impacts through producers and consumers. *Science*, *360*(6392), 987–992.

Reece, J. (2018). *The end of animal farming: How scientists, entrepreneurs, and activists are building an animal-free food system*. Beacon.

Singer, P. (2020). *Why vegan? Eating ethically*. WW Norton.

Smetana, S., Mathys, A., Knoch, A., & Heinz, V. (2015). Meat alternatives: Life cycle assessment of most known meat substitutes. *The International Journal of Life Cycle Assessment*, *20*(9), 1254–1267.

Springmann, M., Godfray, H. C. J., Rayner, M., & Scarborough, P. (2016). Analysis and valuation of the health and climate change co-benefits of dietary change. *Proceedings of the National Academy of Sciences*, *113*(15), 4146–4151.

Springmann, M., et al. (2018). Options for keeping the food system within environmental limits. *Nature*, *562*(7728), 519–525.

Twine, R. (2018). Materially constituting a sustainable food transition: The case of vegan eating practice. *Sociology*, *52*(1), 166–181.

Ventola, C. L. (2015). The antibiotic resistance crisis. Part 1: Causes and threats. *Pharmacy and Therapeutics*, *40*(4), 277–283.

Vermeulen, S. J., Campbell, B.M, & Ingram J. S. I. (2012). Climate change and food systems. *Annual Review of Environment and Resources*, *37*, 195–222.

Wallace (2020). Dead epidemiologists: On the origins of Covid-19. *Monthly Review*.

Weis, T. (2013). *The ecological hoofprint: The global burden of industrial livestock*. Zed Books.

Weis, T. (2016). Towards 120 billion: Dietary change and animal lives. *Radical Philosophy*, *199* (5), 8–13.

Willett, W., et al. (2019). Food in the anthropocene: The EAT-lancet commission on healthy diets from sustainable food systems. *The Lancet*, *393*(10170), 447–492.

WWF (2020). *Living planet report 2020: Bending the curve of biodiversity loss*. WWF.

PART 3

Health and well-being

14

HEALTH, WELL-BEING, AND BURDEN OF DISEASE

Gabriella Luongo, Catherine L. Mah, and Sara F. L. Kirk

Introduction

In this chapter, we present the current state of under, over and malnutrition globally, alongside the health outcomes associated with existing dietary patterns. We then explore nutritional requirements relevant to sustainable diets and the status of current consumer lifestyle, consumption patterns, and food environments.

Paucity in the face of plenty: The triple burden of over, under, and malnutrition

The conclusion of the 20th century marked the first time in human history when the number of people with excess adiposity in the world was equivalent to the number of people who were underfed (Gardner et al., 2000). That some population groups have access to unlimited food supplies while others do not have enough is an inequity that is persistent and punishing. After some initial progress, the number of people affected by hunger globally has been slowly rising since 2014 (FAO et al., 2020). This means that "zero hunger" and "good health and well-being," two of the 17 UN Sustainable Development Goals, are not even close to being achieved (United Nations Development Programme, 2020). As countries across the world deal with the challenges and aftermath of the COVID-19 pandemic, the food security of their most vulnerable populations continues to be threatened. The prevalence of both moderate and severe levels of food insecurity was estimated at 25.9% in 2019 globally, equivalent to 2 billion people (FAO et al., 2020). Food insecurity is not confined to countries of the Global South; even wealthy countries in North America and Europe have populations that experience high rates of food insecurity (FAO et al., 2020). The prevalence of moderate and severe food insecurity is also higher among women than men, threatening achievement of another of the SDGs: "gender equality" (United Nations Development Programme, 2020).

DOI: 10.4324/9781003174417-17

Diet-related disease

Data from the Non-Communicable Disease Risk Factor Collaboration (NCD-RisC) suggest that as the prevalence of obesity has increased, the prevalence of underweight has decreased, primarily driven by a shift in the population distribution of BMI ((NCD-RisC, 2021). This global study provides evidence of both excess obesity and persistent underweight as representing growing social inequalities that restrict access to healthy foods in those at highest risk of undernutrition and requiring policy actions that support equitable access to nutritious food ((NCD-RisC), 2021). Malnutrition is not just a consequence of under-nutrition but is also associated with obesity, an apparent paradox that is now well established (Hawkes et al., 2020; Popkin et al., 2020). Inadequate nutrition is a risk factor for non-communicable diseases (NCD) like heart disease, stroke, and diabetes (Vos et al., 2020). These NCDs typically have shared risk factors; actions taken to address one can also address others (Hawkes et al., 2020). Therefore, obesity and undernutrition are considered together as "malnutrition in all its forms," recognising that they are driven by the same underlying food systems and are the global leading cause of ill health (Swinburn et al., 2019).

National food based dietary guidelines and sustainable diets

Many countries have developed and published food-based dietary guidelines to support healthier diets (Food and Agriculture Organisation of the United Nations, 2021). These guidelines have been used as the basis of public food and nutrition information, and have informed the development of policies, programs, interventions, and monitoring and surveillance measures to improve population health and reduce diet-related disease risk (Food and Agriculture Organisation of the United Nations, 2021). Current food-based dietary guidelines typically encourage the intake of fruits, vegetables, whole grains, fish, plant-based proteins, and healthy fats and oils, and limiting the consumption of sugar, processed foods, red meats, and foods high in sodium (Food and Agriculture Organisation of the United Nations, 2021). These guidelines are consistent with the 2017 Global Burden of Disease Study, which indicated that high intake of sodium, and low intake of whole grains and fruits, were the leading dietary risk factors for death and years lost to disability, globally (Afshin et al., 2019).

Recent national food based dietary guidelines in Canada, the United Kingdom (UK), Australia, and Brazil, have included guidance around dietary practices within the context of ecological and social sustainability, including issues on the affordability, accessibility, and social and ecological context for a healthy diet. Guidelines might refer to *how* food choices can be made, as well as *what* food or dietary patterns to select. For example, Canada's Food Guide and the UK Eatwell Guide promote the increased consumption of pulses and legumes relative to red meat and processed meat as sustainable diet choices that are also nutritious (Health Canada, 2019; Public Health England, 2016). Further, Canadian, Australian, and Brazilian food guide documents have recognised that healthy diets include foods that are both nutritious and sustainably produced and distributed (i.e., considerations for production, packaging, processing, and transportation of foods) (Health Canada, 2019; Ministry of Health of Brazil, 2015; National Health and Medical Research Council, 2013; Public Health England, 2016). An analysis of the UK Eatwell Guide identified, diets improvements can be made to reduce carbon emissions, water consumption, and land use to create more sustainable diets (Carbon Trust, 2016).

Dietary reference intakes for food based dietary guidelines

Food based dietary guidelines are informed by national dietary reference intakes (DRIs) (Government of Canada, 2013; U.S. Department of Health and Human Services, 2020). DRIs identify the recommended value of key nutrients (elements, vitamins, and macronutrients) required for healthy populations (Government of Canada, 2013; U.S. Department of Health and Human Services, 2020). The DRI amounts are set by age-sex groupings and are established to prevent nutrient deficiencies and reduce non-communicable disease risk (Government of Canada, 2013; U.S. Department of Health and Human Services, 2020). DRIs include four types of reference values: Estimated average requirement (EAR), recommended dietary allowance (RD), adequate intake (AI), and tolerable upper intake level (UL) (Government of Canada, 2013).

In 2019, the National Academies of Sciences, Engineering, and Medicine in the USA updated the sodium DRI from 2005 and included a new DRI reference value called the Chronic Disease Risk Reduction Intake (CDRR) (The National Academies of Sciences, Engineering, Medicine, 2021). The CDRR defines the nutrient intakes that are recommended to reduce the risk of chronic disease development when there is at least moderate strength of evidence for a casual and intake-response relationship between the nutrient intake and disease risk (The National Academies of Sciences, Engineering, Medicine, 2021). Since 1995, scientists in Canada and the United States have collaborated to review the evidence for DRIs based on an expert review of the nutrition literature relating to nutrient intake and disease risk (Government of Canada, 2013). The DRI process is overseen by the US Food and Nutrition Board of the National Academies of Sciences, Engineering, and Medicine (Government of Canada, 2013; The National Academies of Sciences, Engineering, Medicine, 2021).

National dietary intake and population nutritional monitoring and surveillance

To examine the dietary intakes of a population, nutrition monitoring and surveillance are required. While monitoring and surveillance are often conflated, in actuality, they serve two unique purposes (Byers & Sedjo, 2012). Monitoring includes the collection and analysis of precise population-representative nutrition data for the purposes of measuring trends at the population level (Byers & Sedjo, 2012). Conversely, surveillance involves the collection and analysis of data that is less representative of the entire population and may be used to study sub-populations in response to monitoring data or to efficiently evaluate policies or interventions (Byers & Sedjo, 2012). Together, monitoring and surveillance measures can examine nutrition trends over time, identify the dietary intake and nutritional status of a population, and compare the dietary intakes of different sub-populations of interest (Byers & Sedjo, 2012).

The complexity of the human diet introduces a unique challenge to quantifying dietary intake and its association to diet-related disease (Willett & Sampson, 2012). Dietary intake can be defined by chemical composition (essential nutrients, energy sources, additives, chemicals, and toxins) or as foods, food groups, or dietary patterns (Willett & Sampson, 2012). Traditionally, dietary intake has tended to be described in relation to nutrients (major energy sources and essential nutrients) and foods (and food groups) (Willett & Sampson, 2012). The advantage of describing dietary intake in terms of nutrients is that the nutrients can be directly linked to biological disease risk and are a useful measure, particularly if the nutrient is only in small amounts within food items (Willett & Sampson, 2012). Conversely, to understand underlying nutrients contributing to disease risk, exploratory analyses of foods or food groups may be advantageous (Willett & Sampson, 2012). Food and food groups also form the basis of public health dietary recommendations, thus describing dietary intakes of foods and food groups may provide more useful nutrition information for the public (Willett & Sampson, 2012). Examining both nutrients and foods when describing dietary intake provides a complete understanding of the effects of dietary intake on disease risk (Willett & Sampson, 2012). In recent years, food-based dietary guidelines and nutrition knowledge translation activities have shifted to an emphasis on healthier dietary patterns, in contrast to individual nutrients, foods, or food groups (Tapsell et al., 2016). This shift acknowledges that the important features of healthy diets can be obtained through many different healthy dietary patterns, with consideration for individual needs and sociocultural preferences (Tapsell et al., 2016).

Dietary intake data can be used to quantify the relationships between sociodemographic factors, dietary intake, and disease risk (Willett, 2012b). An important difficulty in measuring the substantial day-to-day variation in individuals' dietary intake (i.e., reported versus usual intake); in addition, no single diet exposure variable (i.e., food or nutrient) can entirely capture or quantify dietary risk, and the interaction of different dietary components may differentially influence an individual's disease risk within dietary patterns (Willett, 2012b). Regardless of the dietary assessment measure selected, it is important to be aware of their relative strengths, limitations, and utility to answer specific questions (Willett, 2012b).

Four main dietary assessment methods are used in population nutritional monitoring and surveillance: Biomarkers, diet records, 24-hour dietary recalls (24HR), and Food Frequency Questionnaires (FFQs). Dietary biomarkers are a retrospective objective tool used to measure the dietary intake of energy and nutrients, such as potassium, protein, and sodium, typically through blood or urine collection (Van Dam & Hunter, 2012). Diet records are a prospective, self-report dietary assessment method wherein participants report the type and quantity of food consumed on a given day at the time of consumption (Baranowski, 2012). 24HRs are a retrospective dietary assessment method that collects the type and quantity of food consumed over the previous 24-hours or the previous day (Baranowski, 2012). Finally, FFQs retrospectively ask participants to report their usual frequency of consumption of a pre-created list of food items over a specific period of time, such as 12 months (Willett, 2012a).

Biomarkers are regarded as the closest estimator of unbiased dietary intake due to their objective nature (Van Dam & Hunter, 2012). Biomarkers have the most utility for validating and calibrating other self-report dietary assessment measures for energy, protein, sodium, and potassium (Van Dam & Hunter, 2012). Diet records and 24HRs are particularly useful because of their ability to collect detailed dietary intake information, unlike biomarkers and FFQs (Baranowski, 2012). One single administration of a diet record or 24HR at the population level can be used to estimate mean intakes of a population (Baranowski, 2012). If at least a representative proportion of the population completes more than one diet record or 24HR, usual dietary intakes can be used to estimate the distribution of population level intakes and examine the diet-disease relationship (Baranowski, 2012). Due to potential reactivity bias in diet records, 24HRs are preferred for evaluating interventions (Baranowski, 2012). Finally, FFQs can be used to esti-mate usual dietary intake due to the longer recall period. However, due to the larger systematic error introduced in FFQs, it is recommended that other dietary assessment methods (preferably biomarkers) are used to calibrate the responses (Willett, 2012a).

The application of appropriate dietary intake methods in nutrition research can further elucidate the relationship between dietary intake and disease risk and pro-vide insight into the utility of public policy interventions to improve population health. Increasingly, proxy indicators for diet, such as purchasing or sales data, have been used to capture the relationship between consumption behaviours, dietary intakes, and disease risk and health outcomes, as discussed below.

Food environments as key drivers of dietary intakes

Food environments can be defined as the accessibility, availability, and affordability of food in a community or region (Health Canada, 2013). In recent decades, epi-demiological evidence has accumulated to show that the food environments are a key contributor to dietary intakes and diet quality (Caspi et al., 2012). Although psychosocial research has shown that proximally, individual food selection behav-iour is largely affected by factors such as taste, convenience, cost, and health con-cerns (Glanz et al., 1998), at the population level, it is now widely recognised that food environments play an important influence in shaping diet-related risk for non-communicable diseases and obesity.

Food choices are made by individuals and households, but food selection behaviour is not carried out in a void. Food choices are social. They are modi-fied by influences of family, friends, and social networks, and purchased within the constraints of a household budget (Health Canada, 2019). Food choices are made at an increasingly vast array of purchasing locations: Grocery stores, restaurants, workplaces, convenience stores, gas stations, and even hardware stores. Food choices occur within larger institutional settings where individuals work, learn, and play.

COVID-19 has further shaped the food environment landscape, also known as the foodscape (Lake et al., 2010). With public health requirements for physical distancing, many households have turned to online food environments to acquire food, which could have both positive and negative effects on dietary quality and nutrition (Jilcott Pitts et al., 2018). Dietary choices are hence an excellent example of an economic behaviour affecting health. Yet in recent decades, advances in behavioural economics have demonstrated that *homo economicus* is widely recognised as neither entirely rational nor utility-maximising; rather, consumption decisions are made within conditions of relative uncertainty and a given choice architecture (Kahneman & Tversky, 1979; Thaler & Sunstein, 2008). Consumer behaviour affects diet, and consumer behaviour is rational, multifaceted, and social, and shaped by the common economic and environmental drivers that comprise communities, nations, and society (Cummins & Macintyre, 2006; World Health Organisation, 2017).

Many researchers have attempted to characterise the nature of food environments' effect on consumption behaviour, and consequently diet, both conceptually and in practice. Conceptual models for food environments distinguish among various features of the food environment that can shape dietary choices. Some food environment models highlight the location where food is consumed, such as food at home versus eating-out venues (food away from home, or FAFH). Other models distinguish between public and private sector facilities where food is provided, highlighting the governance of food procurement and government responsibility for overseeing food environments (Swinburn et al., 2013). An increasing concern in many countries is the retail food environment, with the dominance of large national and multinational supermarkets as the main location where food is purchased (Hawkes, 2008).

The reliance on market-based or store food to access nourishment is of particular concern for Indigenous communities, where such nutrition transitions have often been forced as a result of centuries of colonisation and violence. Research on retail food environments and diet and health in Indigenous communities has shown that store foods tend to contribute a large proportion of dietary energy and increase non-communicable disease risk as compared to non-Indigenous populations in the same regions (Luongo et al., 2020).

Measuring the food environment and its relationship to diet

As research has elaborated on the role of food environments in diet, a growing range of methodological research has attempted to strengthen and refine measurement of the food environment (Lytle, 2009; Minaker et al., 2011) to better describe those food environment exposures that may shape diet and health. Measurement efforts have largely centred on four components of the food environment: *Community* food environment (distribution of food sources within a geographic area); *consumer* food environment (retailer features directly experienced by shoppers, such as in-store); *information* environments (informational aspects encountered such as labelling and marketing); and *organisational* food environments (parameters of the schools, hospitals, etc., where food is available to defined membership groups rather than the general population) (Glanz, 2009). To measure exposure to the food environment and its correlation with dietary outcomes, researchers have used a variety of instruments and methodologies (Lytle & Sokol, 2017). Methodologies used to measure the food environment have spanned a wide range of geography, health, and marketing disciplines, including geographic analysis, food supply analysis, menu analysis, nutrient analysis, receipt analysis, sales analysis, and Universal Product Code scanning (Lytle & Sokol, 2017; McKinnon et al., 2009).

To date, the majority of research measuring food environment exposures has come from disciplines of health and economic geography, and on the community food environment (Lytle & Sokol, 2017). Community food environments are sometimes referred to as geographic food access, and can include the number, proximity, and density of retail food sources in a given geographic area. Geographic analyses collect and analyse data from a specific spatial area including store density, store accessibility (i.e., distance to the store), and store type (e.g., supermarket, grocery store, fast food outlet, gas station) (Lytle & Sokol, 2017; McKinnon et al., 2009).

Of course, food environments are not static, and neither are the people who purchase food in them. Businesses are open at only select hours, and in both urban and rural communities, substantial travel of different kinds is often required by residents to access food. These dynamic features of the food environment have prompted other geographic and spatio-temporal measurement efforts to better account for mobility and transportation in people's exposure to the food environment. Such research focuses on how exposure to environments are shaped by individuals' active everyday use of their neighbourhoods and communities, also referred to as *activity spaces* (Kestens et al., 2010). Other research has highlighted the importance of temporal features in mobile use of space (Widener et al., 2011). Recent qualitative research in health and social sciences has expanded to integrate greater focus on the socio-cultural aspects of space (place) and how it affects diet (Clary et al., 2017).

More recently, interest in consumer environments and merchandising (sometimes referred to as the 4Ps of marketing: Price, product, placement, and promotion) is growing in public health nutrition, including on how the commercial determinants of health affect diet (Chavez-Ugalde et al., 2021). Consumer food environments include in-store food shopping exposures and experiences related to the availability, quality, and pricing of food products, including merchandising activities by retailers (Liberato et al., 2014). Several kinds of consumer food environment instruments have been developed, intended to be completed by subjects or trained observers to measure the food environment (Lytle & Sokol, 2017).

Check-list and market basket measures are the two most commonly used instruments to measure the consumer food environment (Lytle & Sokol, 2017). Both instruments collect information on food availability (including variety/ diversity), price, and/or quality. A check-list tool is a list of preselected indicator foods that represent current dietary recommendations (Lytle & Sokol, 2017;

McKinnon et al., 2009). The tool is used within and between food retail stores to compare the availability, price, and quality of food items, commonly comparing a "healthy" food item to the "less healthy" option (Lytle & Sokol, 2017; McKinnon et al., 2009). The market basket measure is a list of food items (a food basket) that reflect the total overall diet for an individual or family based on current dietary recommendations (Lytle & Sokol, 2017; McKinnon et al., 2009). For example, a food basket may be composed of food items that meet the recommended dietary intake of a four-person household over a month (Health Canada, 2020). The market basket is most used to compare the price of the food basket over time to examine food affordability (Health Canada, 2020). The foods included, and methods used to develop the tools, are specific to a particular geographic location that the tool will be used to measure the consumer nutrition environment (Lytle & Sokol, 2017; McKinnon et al., 2009). Therefore, while pre-existing checklists and market basket instruments have been modified to fit within diverse settings to measure the consumer nutrition environment, the proportion of the diet and influence of the food environment captured by these tools are not known. Over the last decade, the popularity of check-list tools for research purposes has increased while that of market baskets have declined, although food-based market-baskets tend to be a staple of consumer and economic monitoring (Statistics Canada, 2019).

The development of the Nutrition Environment Measures Survey (NEMS), an easy to use and adaptable check-list tool, made check-lists popular among academics (Glanz et al., 2007; Jani et al., 2018; Lo et al., 2016). Despite this, the market basket remains the most used tool among government organisations to measure food environments. For instance, the Canadian Government developed the *National Nutritious Food Basket*, which is a surveillance tool used as an indicator of food affordability, regionally and over time, in Canada (Health Canada, 2020). The food basket items have been updated to reflect the foods in the 2019 Canadian Food Guide (Health Canada, 2020). Other common instruments include interviews or questionnaires that collect information in food stores using respondent self-report and inventory collection of all food items available in that store (Lytle & Sokol, 2017; McKinnon et al., 2009).

Newer methodologies have attempted to further characterise purchasing patterns, food quality, cost, and their complexities (Lytle & Sokol, 2017; McKinnon et al., 2009). Of growing interest to public health research on food environments data is sales data, originally used for marketing research or economic and administrative monitoring (Lytle & Sokol, 2017; McKinnon et al., 2009). Sales data refers to the collection of the type, quantity, and price of foods purchased from sales, cashier receipts, and food service reporting forms (Lytle & Sokol, 2017; McKinnon et al., 2009). At the population level, sales data is utilised under the assumption that individuals consume what they purchase and that a vast majority of dietary intake comes from store purchases (i.e., in contrast to eating out) (Eyles et al., 2010; Ransley et al., 2001). Sales data is of growing interest in research on food environment interventions to modify diet, as discussed below. For example, sales data in the USA and Barbados have been utilised to examine the impact of beverage taxes (Alvarado et al., 2019; Roberto et al., 2019; Silver et al., 2017).

Conclusions: Food environment interventions to modify diet

Diet is modifiable and food selection is a modifiable health behaviour; from a prevention standpoint, diet amenable to population-level actions. In its landmark declaration for a Decade of Action on Nutrition (2016–2025), only the second such public health declaration on non-communicable diseases besides that on tobacco control, the United Nations General Assembly highlighted that coordinated action, including policy and program intervention on common economic, social, and environmental drivers, would be needed to address the burden of diet-related risk globally (United Nations General Assembly, 2016).

Interventions aimed at improving diets have been designed, implemented, and evaluated in many contexts globally, with a predominance in urban food environments (Mah et al., 2019). The World Cancer Research Fund International's NOURISHING database reports on policies that promote healthier diets and reduce obesity (World Cancer Research Fund International, 2018). The NOURISHING Framework highlights ten areas within the food environment, food system, and behaviour change and communication domains, where governments can act to promote healthy diets and reduce overweight and obesity (World Cancer Research Fund International, 2018). Within the food environment domains, such interventions include: Nutrition labelling standards and regulations; food advertising and commercial promotion restrictions; retailer incentives to create healthier food environments; and economic tools (e.g., taxation and subsidies) to encourage the purchase and consumption of more nutritious foods (World Cancer Research Fund International, 2018).

To date, reviews of population nutrition interventions within food environments have shown that existing interventions continue to emphasise behaviour change and communication (information based) interventions, compared to food environment and food system policies, despite the widespread recognition of the role of the food environment in diet, and with mixed results (Mah et al., 2019). While food environment interventions have largely used sales or purchasing data to quantify the effects of the intervention, few have used robust dietary assessment methods to examine the impact of the food environment on dietary intake (Mah et al., 2019). More recent food environment interventions have incorporated environmental and social planning aspects, fiscal strategies, and multiple components to strengthen the intervention effect (Escaron et al., 2013; Gittelsohn et al., 2012; Mah et al., 2019). However, there remains more work to do to achieve global dietary patterns that address both health and sustainability.

References

Afshin, A., Sur, P. J., Fay, K. A., Cornaby, L., Ferrara, G., Salama, J.S., Mullany, E.C., Abate, K.H., Abbafati, C., Abebe, Z., Afarideh, M., Aggarwal, A., Agrawal, S., Akinyemiju, T., Alahdab, F., Bacha, U., Bachman, V. F., Badali, H., Badawi, A. … & Murray, C. J. L. (2019). Health effects of dietary risks in 195 countries, 1990–2017: A systematic analysis for the Global Burden of Disease Study 2017. *The Lancet, 393*(10184), 1958–1972.

Alvarado, M., Unwin, N., Sharp, S. J., Hambleton, I., Murphy, M. M., Samuels, T., Suhrcke, M., & Adams, J. (2019). Assessing the impact of the Barbados sugar-sweetened beverage tax on beverage sales: An observational study. *International Journal of Behavioural Nutrition and Physical Activity, 16*(1), 13.

Baranowski, T. (2012). 24-Hour recall and diet record methods. In W. Willett (Ed.), *Nutritional epidemiology* (3rd ed.). Oxford University Press.

Byers, T., & Sedjo, R. L. (2012). Nutrition monitoring and surveillance. In W. Willett (Ed.), *Nutritional Epidemiology* (3rd ed.). Oxford University Press.

Carbon Trust (2016). *The eatwell guide: A more sustainable diet.* Carbon Trust. Retrieved from https://prod-drupal files.storage.googleapis.com/documents/resource/public/The%2 0Eatwell%20Guide%20a%20More%20Sustainable%20Diet%20-%20REPORT.pdf

Caspi, C. E., Sorensen, G., Subramanian, S. V., & Kawachi, I. (2012). The local food environment and diet: A systematic review. *Health and Place, 18*(5), 1172–1187.

Chavez-Ugalde, Y., Jago, R., Toumpakari, Z., Egan, M., Cummins, S., White, M., Hulls, P., & De Vocht, F. (2021). Conceptualising the commercial determinants of dietary behaviours associated with obesity: A systematic review using principles from critical interpretive synthesis. *Obesity Science & Practice, 7*(4), 473–486.

Clary, C., Matthews, S. A., & Kestens, Y. (2017). Between exposure, access and use: Reconsidering foodscape influences on dietary behaviours. *Health & Place, 44,* 1–7.

Cummins, S., & Macintyre, S. (2006). Food environments and obesity: Neighbourhood or nation? *International Journal of Epidemiology, 35*(1), 100–104.

Escaron, A. L., Meinen, A. M., Nitzke, S. A., & Martinez-Donate, A. P. (2013). Supermarket and grocery store-based interventions to promote healthful food choices and eating practices: A systematic review. *Preventing Chronic Disease, 10*(4), 1–20.

Eyles, H., Jiang, Y., & Ni Mhurchu, C. (2010). Use of household supermarket sales data to estimate nutrient intakes: A comparison with repeat 24-Hour dietary recalls. *Journal of the American Dietetic Association, 110*(1), 106–110.

FAO, IFAD, UNICEF, WFP, & WHO (2020). *The state of food security and nutrition in the world 2020: Transforming food systems for affordable healthy diets.* FAO.

Food and Agriculture Organisation of the United Nations (2021). *Food-based dietary guidelines.* FAO.

Gardner, G., Halweil, B., & Peterson, J. A. (2000). *Underfed and overfed: The global epidemic of malnutrition.* Worldwatch Institute.

Gittelsohn, J., Rowan, M., & Gadhoke, P. (2012). Interventions in small food stores to change the food environment, improve diet, and reduce risk of chronic disease. *Preventing Chronic Disease, 9,* E59.

Glanz, K. (2009). Measuring food environments: A historical perspective. *American Journal of Preventive Medicine, 36*(4) Supplement, S93–S98.

Glanz, K., Basil, M., Maibach, E., Goldberg, J., & Snyder, D. (1998). Why Americans eat what they do: Taste, nutrition, cost, convenience, and weight control concerns as influences on food consumption. *Journal of the American Dietetic Association, 98*(10), 1118–1126.

Glanz, K., Sallis, J. F., Saelens, B. E., & Frank, L. D. (2007). Nutrition environment measures survey in stores (NEMS-S): Development and evaluation. *American Journal of Preventive Medicine, 32*(4), 282–289.

Government of Canada (2013, April 23). *Dietary reference intakes.* Health Canada. Retrieved from https://www.canada.ca/en/health-canada/services/food-nutrition/healthy-eating/dietary-reference-intakes.html

Hawkes, C. (2008). Dietary implications of supermarket development: A global perspective. *Development Policy Review, 26*(6), 657–692.

Hawkes, C., Ruel, M. T., Salm, L., Sinclair, B., & Branca, F. (2020). Double-duty actions: Seizing programme and policy opportunities to address malnutrition in all its forms. *The Lancet, 395*(10218), 142–155.

Health Canada (2013). *Measuring food environments in Canada.* Government of Canada. Retrieved from https://www.canada.ca/en/health-canada/services/food-nutrition/healthy-eating/nutrition-policy-reports/measuring-food-environment-canada.html

Health Canada (2019). *Canada's dietary guidelines for health professionals and policy makers.* Government of Canada. Retrieved from https://food-guide.canada.ca/sites/default/files/artifact-pdf/CDG-EN-2018.pdf

Health Canada (2020). *National nutritious food basket.* Government of Canada. Retrieved from https://www.canada.ca/en/health-canada/services/food-nutrition/food-nutrition-surveillance/national-nutritious-food-basket.html

Jani, R., Rush, E., Crook, N., & Simmons, D. (2018). Availability and price of healthier food choices and association with obesity prevalence in New Zealand Māori. *Asia Pacific Journal of Clinical Nutrition, 27*(6), 1357–1365.

Jilcott Pitts, S. B., Ng, S. W., Blitstein, J. L., Gustafson, A., & Niculescu, M. (2018). Online grocery shopping: Promise and pitfalls for healthier food and beverage purchases. *Public Health Nutrition, 21*(18), 3360–3376.

Kahneman, D., & Tversky, A. (1979). Prospect theory: An analysis of decision under risk. *Econometrica, 47*(2), 263–291.

Kestens, Y., Lebel, A., Daniel, M., Thériault, M., & Pampalon, R. (2010). Using experienced activity spaces to measure foodscape exposure. *Health & Place, 16*(6), 1094–1103.

Lake, A. A., Burgoine, T., Greenhalgh, F., Stamp, E., & Tyrrell, R. (2010). The foodscape: Classification and field validation of secondary data sources. *Health & Place, 16*(4), 666–673.

Liberato, S. C., Bailie, R., & Brimblecombe, J. (2014). Nutrition interventions at point-of-sale to encourage healthier food purchasing: A systematic review. *BioMed Central Public Health, 14*(1), 919.

Lo, B. K., Minaker, L. M., Mah, C. L., & Cook, B. (2016). Development and testing of the Toronto nutrition environment measures survey–store (ToNEMS-S). *Journal of Nutrition Education and Behaviour, 48*(10), 723–729.

Luongo, G., Skinner, K., Phillipps, B., Yu, Z., Martin, D., & Mah, C. L. (2020). The retail food environment, store foods, and diet and health among indigenous populations: A scoping review. *Current Obesity Reports, 9*(3), 288–306.

Lytle, L. A. (2009). Measuring the food environment: State of the science. *American Journal of Preventive Medicine, 36*(4 Suppl) Suppl, S134–S144.

Lytle, L. A., & Sokol, R. L. (2017). Measures of the food environment: A systematic review of the field, 2007–2015. *Health and Place, 44*, 18–34.

Mah, C. L., Luongo G., Hasdell R., Taylor N., Lo, B. (2019). Systematic review of the effect of retail food environment interventions on diet and health with a focus on the enabling role of public policies. *Current Nutrition Reports, 8*, 411–428.

McKinnon, R. A., Reedy, J., Morrissette, M. A., Lytle, L. A., & Yaroch, A. L. (2009). Measures of the food environment. A compilation of the literature, 1990–2007. *American Journal of Preventive Medicine, 36*(4 Suppl.), S124–S133.

Minaker, L., Fisher, P., Raine, K., & Frank, L. (2011). Measuring the food environment: From theory to planning practice. *Journal of Agriculture, Food Systems, and Community Development, 2*, 65–82.

Ministry of Health of Brazil (2015). *Dietary Guidelines for the Brazilian Population.* Secretariat of Health Care, Primary Health Care Department. Retrieved from http://bvsms.saude.gov.br/bvs/publicacoes/dietary_guidelines_brazilian_population.pdf

National Health and Medical Research Council (2013). *Australian dietary guidelines.* National Health and Medical Research Council, Retrieved from https://www.eatforhealth.gov.au/sites/default/files/content/n55_australian_dietary_guidelines.pdf

NCD Risk Factor Collaboration (NCD-RisC) (2021). Heterogeneous contributions of change in population distribution of body mass index to change in obesity and underweight. *ELife, 10*, e60060.

Popkin, B. M., Corvalan, C., & Grummer-Strawn, L. M. (2020). Dynamics of the double burden of malnutrition and the changing nutrition reality. *The Lancet, 395*(10217), 65–74.

Public Health England (2016). *From plate to guide: What, why and how for the eatwell model.* PHE. Retrieved from https://assets.publishing.service.gov.uk/government/uploads/system/uploads/attachment_data/file/579388/eatwell_model_guide_report.pdf

Ransley, J. K., Donnelly, J. K., Khara, T. N., Botham, H., Arnot, H., Greenwood, D. C., & Cade, J. E. (2001). The use of supermarket till receipts to determine the fat and energy intake in a UK population. *Public Health Nutrition, 4*(6), 1279–1286.

Roberto, C. A., Lawman, H. G., LeVasseur, M. T., Mitra, N., Peterhans, A., Herring, B., & Bleich, S. N. (2019). Association of a beverage tax on sugar-sweetened and artificially sweetened beverages with changes in beverage prices and sales at chain retailers in a large urban setting. *Journal of the American Medical Association, 321*(18), 1799–1810.

Silver, L. D., Ng, S. W., Ryan-Ibarra, S., Taillie, L. S., Induni, M., Miles, D. R., Poti, J. M., & Popkin, B. M. (2017). Changes in prices, sales, consumer spending, and beverage consumption one year after a tax on sugar-sweetened beverages in Berkeley, California, US: A before-and-after study. *PLoS Medicine, 14*(4), e1002283.

Statistics Canada (2019). *The Canadian consumer price index reference paper.* Statistics Canada. Retrieved from https://www150.statcan.gc.ca/n1/pub/62-553-x/62-553-x2019001-eng.pdf

Swinburn, B., Sacks, G., Vandevijvere, S., Kumanyika, S., Lobstein, T., Neal, B., Barquera, S., Friel, S., Hawkes, C., Kelly, B.,. L'Abbé, M., Lee, A. Ma, J., Macmullan, J., Mohan, S., Monteiro, C., Rayner, M., Sanders, D., Snowdon, W., & Walker, C. (2013). INFORMAS (international network for food and obesity/non-communicable diseases research, monitoring and action support): Overview and key principles. *Obesity Reviews: An Official Journal of the International Association for the Study of Obesity, 14*(1 Suppl.), 1–12.

Swinburn, B. A., Kraak, V. I., Allender, S., Atkins, V. J., Baker, P. I., Bogard, J. R. Brinsden, H., Calvillo, A., De Schutter, O., Devarajan, R. Ezzati, M., & Dietz, W. H. (2019). The global syndemic of obesity, undernutrition, and climate change: The Lancet Commission report. *The Lancet, 393*(10173), 791–846.

Tapsell, L. C., Neale, E. P., Satija, A., & Hu, F. B. (2016). Foods, nutrients, and dietary patterns: Interconnections and implications for dietary guidelines. *Advances in Nutrition, 7*(3), 445–454.

Thaler, R. H., & Sunstein, C. R. (2008). *Nudge: Improving decisions about health, wealth, and happiness.* Yale University Press.

The National Academies of Sciences, Engineering, Medicine (2021). *Expansion of the dietary reference intake model: Learning from sodium and potassium.* Retrieved from https://www.nap.edu/resource/25353/interactive/

U.S. Department of Health and Human Services (2020). *About the dietary guidelines.* Retrieved from https://health.gov/our-work/food-nutrition/about-dietary-guidelines

United Nations Development Programme (2020). *Sustainable development goals.* Retrieved from https://sdgs.un.org/goals

United Nations General Assembly (2016). *United Nations decade of action on nutrition (2016–2025). Seventieth session of the United Nations General Assembly.* United Nations General Assembly. Retrieved from https://www.un.org/ga/search/view_doc.asp?symbol=A/70/L.42

Van Dam, R. M., & Hunter, D. (2012). Biochemical indicators of dietary intake. In W. Willett (Ed.), *Nutritional Epidemiology* (3rd ed.). Oxford University Press.

Vos, T., Lim, S. S., Abbafati, C., Abbas, K. M., Abbasi, M., Abbasifard, M., Abbasi-Kangevari, M., Abbastabar, H., Abd-Allah, F., Abdelalim, A., Abdollahi, M., Abdollahpour, I., Abolhassani, H., Aboyans, V., Abrams, E. M., Abreu, L. G., Abrigo, M. R. M., Abu-Raddad, L. J., Abushouk, A. I. ... & Murray, C. J. L. (2020). Global burden of 369 diseases and injuries in 204 countries and territories, 1990–2019: A systematic analysis for the Global Burden of Disease Study 2019. *The Lancet, 396*(10258), 1204–1222.

Widener, M. J., Metcalf, S. S., & Bar-Yam, Y. (2011). Dynamic urban food environments a temporal analysis of access to healthy foods. *American Journal of Preventive Medicine, 41*(4), 439–441.

Willett, W. (2012a). Food frequency methods. In W. Willett (Ed.), *Nutritional Epidemiology* (3rd ed.). Oxford University Press.

Willett, W. (Ed.) (2012b). *Nature of variation in diet* (3rd ed.). Oxford University Press.

Willett, W., & Sampson, L. (2012). Foods and nutrients. In W. Willett (Ed.), *Nutritional Epidemiology* (3rd ed.). Oxford University Press.

World Cancer Research Fund International (2018). *Nourishing database.* Retrieved from https://www.wcrf.org/int/policy/nourishing-database

World Health Organisation (2017). *Double-duty actions for nutrition: Policy brief.* WHO. Retrieved from https://apps.who.int/iris/rest/bitstreams/1084413/retrieve

15

GLOBAL BURDEN OF ZOONOTIC DISEASE, PANDEMICS, COVID-19, AND SUSTAINABLE DIETS

Anna Okello, Jessica E. Raneri, Delia Grace Randolph, and Hung Nguyen-Viet

Introduction: Zoonotic diseases and their relevance to food systems

Zoonotic diseases are infectious diseases that transmit between humans and animals. Whilst zoonotic diseases are well-known causes of global human and animal morbidity and mortality, determination of the global burden of zoonoses is difficult. This difficulty is largely due to the complexity of human-animal interactions and the significant number of context-specific risk pathways depending on the pathogens involved. The zoonotic diseases that generate the most attention, politically and financially are those termed "Emerging Infectious Diseases" (EIDs). These are "new" diseases that emerge as a result of a spillover event between humans and animals, with the potential to cause a global pandemic. Examples of EID events in the 21st century include Sudden Acute Respiratory Syndrome (SARS), H5N1 Highly Pathogenic Avian Influenza (HPAI), and Middle Eastern Respiratory Syndrome (MERS). COVID-19 is also thought to have emerged from animals, although the exact pathway is not yet known (Cohen, 2021).

However, in addition to the growing number of newly emerging zoonoses, there are a multitude of zoonoses that are endemic within many populations in the world. These endemic zoonoses include the Neglected Tropical Diseases, as classified by the World Health Organisation, which include diseases such as rabies, brucellosis, *Taenia solium* neurocysticercosis, and zoonotic tuberculosis (WHO, 2021).

Outside the WHO-defined neglected tropical diseases, there are also a large number of foodborne diseases, many of which are also zoonotic. Examples of these diseases include bacterial pathogens such as *Salmonella*, *E. coli*, and *Campylobacter*, parasitic diseases (e.g., *Echinococcus granulosus* and *Taenia solium*) and some viruses (e.g., hepatitis A and norovirus). Foodborne pathogens, such as *Campylobacter*, are

DOI: 10.4324/9781003174417-18

common environmental contaminants of soil and water. Other foodborne diseases are carried by both wild and farmed animals, such as pigs and poultry, or transmitted through cross-contamination of raw ingredients and poor food storage and cooking practices.

Issues, trends, debates

Impact of COVID-19 on food systems resilience

The COVID-19 pandemic has brought the critical role of food production and supply to the forefront. In Southeast Asia alone, reduced agricultural productivity in the first quarter of 2020 is estimated to have resulted in a 1.4% reduction in regional Gross Domestic Product (GDP), valued at 3.76 billion USD and impacting over 100 million farm labour jobs (Gregorio & Ancog, 2020). The International Food Policy Research Institute (IFPRI) estimated an aggregated decline in GDP of up to 3.6% for developing countries as a group, with sub-Saharan Africa, Southeast Asia, and Latin America more affected than China and East Asia due to their relatively high dependence on primary commodity exports (Laborde et al., 2020). The recent concern raised by FAO and others regarding the likelihood of achieving SDG2 "Zero Hunger" by 2030 (FAO, IFAD, UNICEF, WFP, & WHO, 2020) has been exacerbated by the economic impacts of COVID-19, with estimations that the number of people living in poverty could increase by 20% from current levels, or by 420–580 million people, relative to the latest official recorded figures for 2018 (Sumner et al., 2020). Food insecurity was also impacted, with one in three people (an increase of almost 320 million people) not having access to adequate food and the prevalence of undernourishment increasing by 1.5% in 2020 compared to 2019 (FAO, IFAD, UNICEF, WFP & WHO, 2021).

Resilient food systems that ensure sustainable and nutritious diets will, therefore, be critical in underpinning economic and social recovery from the COVID-19-induced food and economic crises in various parts of the world. The important role of animal-sourced foods and livestock production in national COVID-19 recovery strategies has already been clear in some cases; for example, asset distribution programs provided chickens and goats to overcome the food security and economic impacts of disease (World Vision Australia, 2020). In the longer term, an acknowledgement of livestock's capacity to mitigate nutritional and economic risks faced by low- and middle-income country (LMIC) households in the aftermath of COVID-19 is an opportunity to further advocate for sustainable and improved food production systems within the Sustainable Development Goals, the Universal Health Coverage Declaration, and the Sendai Disaster Risk Reduction Framework.

In contrast, the decline in household incomes may lead to a rebound decline in affordability of animal-sourced food (ASF); the correlation between incomes and consumption is well documented. For example, Rabobank estimated an annual decline in beef (9–13%), pork (4–17%), and poultry (1–4%) consumption in Southeast Asia in 2020, noting, however, the intra-regional differences to these figures and consumption increasing in some cases (e.g., poultry in Vietnam) (RaboResearch, 2020). Given ASF are critical sources of dietary nutrients in many LMICs, especially for women and children, there will likely be negative effects on public health and nutrition if dietary quality in many of the world's poorest regions is further constrained. These effects of the pandemic on global agricultural production, health, and nutrition are likely to continue for the next decade, with the agricultural sector gradually responding to lower consumer demand by reducing agricultural output, particularly for high-value livestock products (OECD, 2020).

Complex relationships between animal, human, and environmental health

The increased demand for ASF has resulted in increasing intensification of agricultural practices in almost all regions of the world. This increase also includes intensification of livestock production, by which large numbers of animals—and in some cases animal species—mix together. Accompanying the myriad nutritional, welfare, and environmental considerations of intensive animal production is the propensity for disease spread—and in some cases the crossing of species barriers—where large numbers of animals congregate. The latter is of particular concern if there is an accompanying wild or feral animal dimension, for example, where clearing land for livestock intensification encroaches on wildlife habitats.

Practices associated with livestock intensification have been linked to several high-profile zoonotic disease outbreaks, including bovine spongiform encephalopathy (BSE), highly-pathogenic avian influenza, and enteric foodborne disease from the contamination of human food and water sources with animal manure (Hrudey et al., 2002; Stevik et al., 2004). The increased size and intensity of many livestock production enterprises, particularly in high-income countries (HICs), has been associated with an increased risk of chronic obstructive pulmonary disease in farm workers (Sigsgaard & Balmes, 2017). Other studies note that livestock workers are exposed to greater amounts of allergens, a small percentage develop allergies, some exhibiting significant respiratory concerns (Ngajilo, 2016; Taluja et al., 2018). Conversely, zoonotic parasitic diseases, such as *Echinococcus granulosus* and *Taenia solium*, have largely been controlled or eradicated as a result of intensified livestock systems in HICs. Viral diseases, largely as a result of wildlife vectors (e.g., bats, waterfowl, rats, etc.) can impact both smallholder and intensive systems, depending on the pathogen and how close the interface between domestic livestock and wildlife. For example, the emergence of Nipah virus in Malaysia in 1998 has been linked to agricultural expansion and intensification (Epstein et al., 2006). Threats of disease spread remain.

The capacity for commercial livestock enterprises to maintain stringent on-farm biosecurity is often greater than for smallholder settings where animals are often free range, or biosecurity practices such as vaccination and quarantine are more difficult to enforce. However, transmission of avian influenza in Europe and other countries demonstrates that even with highly biosecure commercial poultry operations, wildlife vectors can still transmit disease (WHO, 2020). In this way, both intensive and free-range production systems have their own unique sets of opportunities and challenges for disease control, depending on the broader environmental context, species involved, and risk factors of various pathogens. In summary, the safety of animal source foods is a dynamic situation heavily influenced by multiple factors along both formal and informal production chains.

The hunting and farming of wild animals for food is a common practice in many parts of the world. For example, estimates of the total volume of wildlife products consumed and exported in Vietnam have been around 3,050 tonnes per year, around half of which is consumed domestically (Van Song, 2003). Human contact with wildlife may represent a risk of zoonotic disease transmission and spill-over events, and global efforts to reduce disease risk have increased since the emergence of COVID-19 in late 2019. For example, China banned all trade and consumption of wildlife for food on 26 January 2020. Vietnam followed with a ban on wildlife imports in July 2020, ordering the closure of illegal wildlife markets to protect human health and ecological balance. In 2021, the WHO, in conjunction with the World Organisation for Animal Health (OIE) and the United Nations Environment Program (UNEP), released interim guidance notes for reducing public health risks associated with the sale of live wild animals of mammalian species in traditional food markets (WHO, OIE, UNEP, 2021).

The broader societal role of livestock beyond food and nutrition security

While many of the narratives around the consumption of ASF in HIC focus on overconsumption and the resulting human health and environmental concerns, ASFs result in positive impacts on nutrition and livelihoods in many LMICs (Kavle et al., 2015; Murphy & Allen, 2003; Neumann et al., 2003). However, livestock also has a vital role in the traditional economies of exchange, in trade and other social activities, such as the payment of dowries or bride price (Forkuor et al., 2018; Shava & Masuku, 2019). The role of livestock as a bank or asset source also is well documented, with livestock losses or deaths resulting in far-reaching consequences for a range of social structures, including education, health, and funeral costs. In many parts of the world, the symbolism provided by livestock is just as, if not more, important than the monetary value of the animal itself (Shava & Masuku, 2019). For example, pigs have a significant role in the cultural fabric of Timor-Leste (Smith et al., 2019; Ximenes & Rose, 2021), whilst the monetary value of cattle and buffalo is often intertwined with complex social and cultural value-additions including insurance, prestige, and other locally relevant determinants of social capital (Hänke & Barkmann, 2017; Mayala et al., 2019; Turner et al., 2015)

The gendered dimension of livestock keeping also has a considerable effect on household income, nutrition, and division of labour. For example, the differing roles of men and women in pastoralist communities result in differing opportunities for income generation and status within these systems (Hertkorn et al., 2015; Okello et al., 2014). The importance of horses, donkeys, and oxen in reducing human workloads through the provision of draft traction is significantly under-recognised and under-valued (Fazili & Kirmani, 2011;Valette, 2014).

Strategies for greater accessibility to safe, nutritious foods
Recognising the global burden of foodborne disease

Food safety is part of the definition of food security (FAO, 1996). However, food safety is often neglected, with the majority of current agricultural discourse focused on food sufficiency and nutrition. However, this is starting to change as new evidence emerges on the health and economic burdens of foodborne disease. For example, in 2015, the World Health Organisation's Foodborne Disease (FBD) Burden Epidemiology Reference Group (FERG) provided the first estimates of the incidence, mortality, and burden of 31 foodborne hazards. The findings indicated the global burden of FBD was comparable to that of HIV/AIDS, malaria, and tuberculosis; and 33 million disability adjusted life years (DALYs), with 40% of that burden borne by children under five years of age (Havelaar et al., 2015). The majority of illnesses were caused by infectious agents, most of which were zoonotic, and LMICs were disproportionately affected (Gibb et al., 2019; Havelaar et al., 2015).

Diarrhoeal diseases such as norovirus and *Campylobacter spp.* were found to be the most frequent cause of disease and death. Also notable is the variance of FBD burden across the world: Some hazards, such as non-typhoidal *S. enterica*, is important to all regions of the world, whereas others, such as certain parasitic helminths, are highly localised. Therefore, risk analysis methodologies need to take into consideration the broader social, cultural, and economic contexts. Research on the attribution of health burdens to different foods is less advanced, but strongly suggests that most illnesses are due to fresh foods sold in traditional, informal, or wet markets, primarily livestock products, aquatic products, and vegetables (Grace, 2015).

There are a multitude of secondary societal and development effects of poor food safety. The World Bank estimated that the economic cost of foodborne disease was more than 100 billion USD a year in LMICs (Jaffee et al., 2018), nearly all due to the loss of income. Other well-known development impacts include the impact of childhood diarrhoea on development goals such as literacy and numeracy. Amassing evidence that illuminates suspected causal linkages between enteric bacterial infections and chronic malnutrition and stunting is well underway.

Promoting a holistic understanding of informal food supply systems

Traditional markets provide an affordable daily source of fresh, nutritious food for billions of people globally, particularly low-income consumers, and also play important socioeconomic and cultural roles. However, the COVID-19 pandemic has brought traditional markets into the spotlight due to concerns around food safety (WHO, OIE, & UNEP, 2021). Moreover, the historical focus of agricultural development policy on the promotion of formal markets through the "supermar-ketisation" of traditional food products (Raneri & Wertheim-Heck, 2020) has led to systemic neglect of traditional and informal food systems. Evidence suggests that interventions to suppress the informal sector may even decrease food safety and quality in some cases (Leksmono et al., 2006; Vorley, 2013).

A recent review of interventions to improve the value and quality of infor-mal food supply chains has demonstrated that participatory methods to facilitate moderate, step-wise approaches to food safety in informal settings may lead to substantial nutritional and economic benefits (Robinson & Yoshida, 2016). Gender also has been found to have a major influence on food safety outcomes, given the different roles undertaken by men and women in food safety risk manage-ment. For example, evidence from Nigeria highlighted that women are mainly responsible for buying and preparing food, whereas men are involved more in food production and slaughtering (Johnson et al., 2015). Evidence from Southeast Asia also demonstrates that women are more involved in food selling, purchasing, and preparation (Nguyen-Viet et al., 2019). Finally, it is important to leverage exist-ing consumer knowledge and practices, such as thorough cooking and good hand hygiene, shown to contribute to improved human health and nutrition outcomes (Robinson & Yoshida, 2016).

Future opportunities to reduce risks in food safety and zoonotic disease in global food systems: Incentive-based approaches and emerging tools

There has been a concerted effort over the last two decades to promote more incentive-based approaches to support and improve the economic and food safety performance of informal food systems. Evidence from risk analyses and other stud-ies of livestock value chains have found that product aggregators—for example, traders, processors, chilling plants, slaughterhouses—can play a key role in main-taining and improving food quality; product aggregators are often easier to engage with, given they are significantly fewer in number compared to producers or con-sumers (Grace et al., 2012; Kouamé-Sina et al., 2012; Makita et al., 2010).

The International Livestock Research Institute (ILRI) and partners have been working since 2006 on assessing and improving food safety in traditional markets. Key to this approach are: The concept of light-touch interventions that are sus-tainable and scalable; changing practice through capacity building for food safety actors such as farmers, slaughtering workers, and butchers; and providing incen-tives and an enabling policy environment (Nguyen-Viet et al., 2017; Nguyen-Viet et al., 2019). ILRI has pioneered a *triple pathway* approach that professionalises—rather than penalises—the informal sector, with the aims of supporting smallholder market access, safeguarding the supply of cheap nutritious food to the poor and reducing the burden of foodborne disease (Nguyen-Viet et al., 2021).

An early example of applying this triple pathway was a training and certifica-tion scheme launched in Kenya in the early 2000s to improve the quality and safety of informal dairy markets. Traders were trained in hygienic milk handling and business practices, and at the end of their training could apply for a certificate from the Kenya Dairy Board that entitled them to legally sell milk, which was a strong incentive. This approach has since been extended and adapted to different value chains and countries, such as milk traders in India and Tanzania, and butchers in Nigeria. It is based on the hypothesis that professionalising—rather than criminalising—informal-market actors improves food safety outcomes whilst at the same time improving nutrition, and protecting and enhancing important sources of income and employment for the poor (Johnson et al., 2015). For example, an ILRI project trained butchers from butcher associations in Nigeria to improve hygiene practices, increase their knowledge and understanding of risk, and hear stakeholders' views on their own experiences of food safety and hygiene (Grace et al., 2019).

Similar examples of methods and approaches to improve food safety exist in Southeast Asia. To address the issues of food safety and consumer concern in Vietnam, ILRI and partners developed a comparative tool to consistently assess food safety performance across different value chains. Initially focused on pork, the Food Safety Performance Tool provided a rapid but holistic assessment of food safety outcomes across retail supermarkets, traditional markets, and street food vendors through three pillars: Safety, scalability, and societal concerns (Thinh et al., 2020). Initial assessments have highlighted not just the risks in terms of quantitative measurements of food safety, but how these different risks and hazards are perceived by various actors along the pork value chains (Thinh et al., 2020).

Recommendations and conclusion

This chapter has highlighted the urgent need to better consider the central role of food safety and zoonotic disease mitigation in sustainable food systems discourse. Even though the true global burden of foodborne and zoonotic disease is only just starting to be realised, it is undeniable that insufficient attention and action to address these burdens—in both informal and formal settings—will continue to have a wide-ranging impact on livelihoods, human and animal health, and the environment.

However, it's also important to re-consider the way we think about disease risk, and mitigation of this risk, in global food systems. Traditional approaches to risk assessment have often failed to consider the sociocultural, economic, and gender impacts of mitigation measures, resulting in poor compliance with regulations and, in some cases, more serious consequences such as increased food insecurity or loss of women's economic empowerment. In this way, consideration of the broader political economy of food safety and zoonotic disease management—including power dynamics, risk perception, cost-benefit, and motivations for behavioural change—all need to be incorporated into more robust methodologies for how we "assess risk." While there has been some good progress in this area over the last decade, a fundamental shift in our understanding of the value and contribution of safe food to global food and nutrition security, as well as the broader health of humans, animals and our environment under One Health, is still required.

References

Cohen, J. (2021, September 1). Call of the Wild: Why many scientists say it's unlikely that SARS-COV-2 originated from a "lab leak". *Science*. https://www.science.org/content/article/why-many-scientists-say-unlikely-sars-cov-2-originated-lab-leak?s=08&

Epstein, J. H., Field, H. E., Luby, S., Pulliam, J. R., & Daszak, P. (2006). Nipah virus: Impact, origins, and causes of emergence. *Current Infectious Disease Reports, 8*(1), 59–65.

FAO (1996). *Rome declaration on world food security and world food summit plan of action. World food summit*. FAO.

FAO (2018). *Sustainable food systems: Concept and framework*. FAO.

FAO, IFAD, UNICEF, WFP, & WHO (2020). *The state of food security and nutrition in the world 2020. Transforming food systems for affordable healthy diets*. FAO.

FAO, IFAD, UNICEF, WFP, & WHO (2021). *The state of food security and nutrition in the world 2021. Transforming food systems for food security, improved nutrition and affordable healthy diets for all*. FAO.

Fazili, M. R., & Kirmani, M. A. (2011). Equine: The ignored working animal of Kashmir: Status, constraints, research areas and ways for improvement. *Asian Journal of Animal Sciences, 5*(2), 91–101.

Forkuor, J. B., de Paul Kanwetuu, V., Ganee, E. M., & Ndemole, I. K. (2018). Bride price and the state of marriage in North-West Ghana. *International Journal of Social Science Studies, 6*, 34.

Gibb, H. J., Barchowsky, A., Bellinger, D., Bolger, P. M., Carrington, C., Havelaar, A. H., Oberoi, S., Zang, Y., O'Leary, K. & Devleesschauwer, B. (2019). Estimates of the 2015 global and regional disease burden from four foodborne metals—arsenic, cadmium, lead and methylmercury. *Environmental Research, 174*, 188–194.

Grace, D. (2015). Food safety in low and middle income countries. *International Journal of Environmental Research and Public Health, 12*(9), 10490–10507.

Grace, D., Dipeolu, M., & Alonso, S. (2019). Improving food safety in the informal sector: Nine years later. *Infection Ecology & Epidemiology, 9*(1), 1579613.

Grace, D., Mutua, F., Ochungo, P., Kruska, R. L., Jones, K., Brierley, L., Lapar, M., Said, M. Y., Herrero, M. T., Phuc, P. M., Thao, N. B., Akuku, I., & Ogutu, F. (2012). *Mapping of poverty and likely zoonoses hotspots. A final report to the Department for International Development, UK*. International Livestock Research Institute.

Gregorioa, G. B., & Ancog, R. C. (2020). Assessing the impact of the covid-19 pandemic on agricultural production in Southeast Asia: Toward transformative change in agricultural food systems. *Asian Journal of Agriculture and Development, 17*(1362-2020-1097), 1–13.

Hänke, H., & Barkmann, J. (2017). Insurance function of livestock, farmers coping capacity with crop failure in southwestern Madagascar. *World Development, 96*, 264–275.

Havelaar, A. H., Kirk, M. D., Torgerson, P. R., Gibb, H. J., Hald, T., Lake, R. J., Praet, N., Bellinger, D. C., de Silva, N. R., Gargouri, N., Speybroeck, N., Cawthorne, A., Mathers, C., Stein, C., Angulo, F. J., Devleesschauwer, B., on behalf of World Health Organization Foodborne Disease Burden Epidemiology Reference Group (2015). World health organization global estimates and regional comparisons of the burden of foodborne disease in 2010. *PLOS Medicine 12*, e1001923.

Hertkorn, M. L., Roba, H., & Kaufmann, B. (2015). Caring for livestock. Borana women's perceptions of their changing role in livestock management in southern Ethiopia. *Nomadic Peoples, 19*(1), 30–52.

Hrudey, S. E., Huck, P. M., Payment, P., Gillham, R. W., & Hrudey, E. J. (2002). Walkerton: Lessons learned in comparison with waterborne outbreaks in the developed world. *Journal of Environmental Engineering and Science, 1*(6), 397–407.

Jaffee, S., Henson, S., Unnevehr, L., Grace, D., & Cassou, E. (2018). *The safe food imperative: Accelerating progress in low-and middle-income countries*. World Bank.

Johnson, N., Mayne, J. R., Grace, D., & Wyatt, A. J. (2015). *How will training traders contribute to improved food safety in informal markets for meat and milk? A theory of change analysis*. IFPRI Discussion Paper 1451. IFPRI.

Kavle, J. A., El-Zanaty, F., Landry, M., & Galloway, R. (2015). The rise in stunting in relation to avian influenza and food consumption patterns in Lower Egypt in comparison
to Upper Egypt: Results from 2005 and 2008 Demographic and Health Surveys. *British
Medical Journal Public Health, 15*(1), 1–18.

Kouamé-Sina, S. M., Makita, K., Costard, S., Grace, D., Dadié, A., Dje, M., & Bonfoh, B.
(2012). Hazard identification and exposure assessment for bacterial risk assessment of
informally marketed milk in Abidjan, Côte d'Ivoire. *Food and Nutrition Bulletin, 33*(4),
223–234.

Laborde, D., Martin, W., & Vos, R. (2020). Poverty and food insecurity could grow dramatically as COVID-19 spreads. In J., Swinnen, & J., McDermott (Eds.), *Covid-19 and global
food security*. IFPRI.

Leksmono, C., Young, J., Hooton, N., Muriuki, H., & Romney, D. L. (2006). *Informal trad-
ers lock horns with the formal milk industry: The role of research in pro-poor dairy policy shift in
Kenya*. Overseas Development Institute and International Livestock Research Institute
Working Paper No. 266. Overseas Development Institute; International Livestock
Research Institute.

Makita, K., Fèvre, E. M., Waiswa, C., Eisler, M. C., & Welburn, S. C. (2010). How human
brucellosis incidence in urban Kampala can be reduced most efficiently? A stochastic
risk assessment of informally-marketed milk. *PLoS One, 5*(12), e14188.

Mayala, N. M., Katundu, M. A., & Msuya, E. E. (2019). Socio-cultural factors influencing
livestock investment decisions among smallholder farmers in Mbulu and Bariadi districts, Tanzania. *Global Business Review, 20*(5), 1214–1230.

Murphy, S. P., & Allen, L. H. (2003). Nutritional importance of animal source foods. *The
Journal of Nutrition, 133*(11) Supplement 2, 3932S–3935S.

Neumann, C. G., Bwibo, N. O., Murphy, S. P., Sigman, M., Whaley, S., Allen, L. H., Guthrie,
D., Weiss, R. E., & Demment, M. W. (2003). Animal source foods improve dietary
quality, micronutrient status, growth and cognitive function in Kenyan school children: Background, study design and baseline findings. *The Journal of Nutrition, 133*(11)
Supplement 2, 3941S–3949S.

Ngajilo, D. (2016). *Allergic sensitization and work related asthma among poultry workers in South
Africa* [Master's thesis, University of Cape Town].

Nguyen-Viet, H., Tuyet-Hanh, T. T., Unger, F., Dang-Xuan, S., & Grace, D. (2017). Food
safety in Vietnam: Where we are at and what we can learn from international experiences. *Infectious Diseases of Poverty, 6*(1), 1–6.

Nguyen-Viet, H., Dang-Xuan, S., Pham-Duc, P., Roesel, K., Huong, N. M., Luu-Quoc, T.,
… & Grace, D. (2019). Rapid integrated assessment of food safety and nutrition related to
pork consumption of regular consumers and mothers with young children in Vietnam.
Global Food Security, 20, 37–44.

Nguyen-Viet, H., Grace, D., Unger, F., Lindahl, J. F., Tum, S., Dang-Xuan, S., Chea, R.,
Chhay, T., Srey, T. Nguyen, C. & Young, M. (2021). Pork and poultry safety in traditional
markets in Cambodia: Understanding complexities and scaling up good interventions.
ILRI Policy Brief. ILRI. https://hdl.handle.net/10568/114472

OECD (2020). The impact of COVID-19 on agricultural markets and GHG emissions.
OECD Policy Responses to Coronavirus (COVID-19). https://www.oecd.org/coro-
navirus/policy-responses/the-impact-of-covid-19-on-agricultural-markets-and-ghg-
emissions-57e5eb53/#biblio-d1e563

Okello, A. L., Majekodunmi, A. O., Malala, A., Welburn, S. C., & Smith, J. (2014). Identifying
motivators for state-pastoralist dialogue: Exploring the relationships between livestock
services, self-organisation and conflict in Nigeria's pastoralist Fulani. *Pastoralism, 4*(1),
1–14.

RaboResearch. (2020, April). Impact of coronavirus on Southeast Asian food & agribusiness. https://research.rabobank.com/far/en/sectors/regional-food-agri/coronavirus-
impact-on-sea-fa.html

Raneri, J. & Wertheim-Heck, S. (2020). 'Supermarketization,' food environments, and the urban poor. International Food Policy Research Institute. Retrieved from https://www.ifpri.org/blog/supermarketization-food-environments-and-urban-poor

Robinson, E., & Yoshida, N. (2016). *Improving the nutritional quality of food markets through the informal sector: Lessons from case studies in other sectors* (No. IDS Evidence Report; 171). IDS.

Shava, S., & Masuku, S. (2019). Living currency: The multiple roles of livestock in livelihood sustenance and exchange in the context of rural indigenous communities in southern Africa. *Southern African Journal of Environmental Education, 35*, 1–13.

Sigsgaard, T., & Balmes, J. (2017). Environmental effects of intensive livestock farming. *American Journal of Respiratory and Critical Care Medicine, 196*(9), 1092–1093.

Smith, D., Cooper, T., Pereira, A., & da Costa Jong, J. B. (2019). Counting the cost: The potential impact of African Swine Fever on smallholders in Timor-Leste. *One Health, 8*, 100109.

Stevik, T. K., Aa, K., Ausland, G., & Hanssen, J. F. (2004). Retention and removal of pathogenic bacteria in wastewater percolating through porous media: A review. *Water Research, 38*(6), 1355–1367.

Sumner, A., Hoy, C. & Ortiz-Juarez, E. (2020). *Estimates of the impact of COVID-19 on global poverty.* WIDER Working Paper 2020/43. UNU-WIDER.

Taluja, M. K., Gupta, V., Sharma, G., & Arora, J. S. (2018). Prevalence of symptoms (respiratory and non-respiratory) among poultry farm workers in India. *American Journal of Physiology, Biochemistry and Pharmacology, 8*(2), 1–7.

Thinh, N.T., Grace, D., Van Hung, P., Nguyen-Viet, H., Dang-Xuan, S., Nga, N.T. D., Luong, N.T., Huyen, N.T.T., Ngoc, T.T.B., Phuc, P.D. & Unger, F. (2020). Food safety performance in key pork value chains in Vietnam. *ILRI Research Brief.* 94. ILRI. Retrieved from https://hdl.handle.net/10568/108320

Turner, S., Bonnin, C., & Michaud, J. (2015). *Frontier livelihoods: Hmong in the Sino-Vietnamese borderlands.* University of Washington Press.

Valette, D. (2014). Invisible helpers: Women's views on the contributions of working donkeys, horses, and mules to their lives. Key findings from research in Ethiopia, Kenya, India and Pakistan. *The Brooke.* Retrieved from https://www.thebrooke.org/sites/default/files/Advocacy-and-policy/Invisible-helpers-voices-from-women.pdf

Van Song, N. (2003). *Wildlife trading in Vietnam: Why it flourishes. EEPSEA research reports.* IDRC Regional Office for Southeast and East Asia.

Vorley, B. (2013). *Meeting small-scale farmers in their markets: Understanding and improving the institutions and governance of informal agrifood trade.* International Institute for Environment and Development (IIED).

WHO (2020). Increase in 'bird flu' outbreaks: WHO/Europe advice for handling dead or sick birds. *WHO Health Topics.* https://www.euro.who.int/en/health-topics/communicable-diseases/influenza/news/news/2020/01/increase-in-bird-flu-outbreaks-whoeurope-advice-for-handling-dead-or-sick-birds

WHO (2021). Neglected zoonotic disease. *WHO Control of Neglected Tropical Diseases, Section 4.* https://www.who.int/teams/control-of-neglected-tropical-diseases/neglected-zoonotic-diseases

WHO, OIE, & UNEP (2021, April 12). Reducing public health risks associated with the sale of live wild animals of mammalian species in traditional food markets. *Interim Guidance.* https://www.who.int/publications/i/item/WHO-2019-nCoV-Food-safety-traditional-markets-2021.1

World Vision Australia (2020, July 2). *COVID-19 emergency response Asia Pacific regional situation report.* https://asiapacific.unfpa.org/sites/default/files/pub-pdf/apro_covid-19_regional_sitrep_5_june_2020.pdf

Ximenes, A. & Rose, M. (2021, July 20). COVID-19: The view from a Timorese village. *DevPolicy Blog,* The Development Policy Centre. https://devpolicy.org/covid-19-the-view-from-a-timorese-village-20210720/

16

EATING FOR HEALTH AND THE ENVIRONMENT

Food systems analysis and the ecological determinants of health

Sarah Elton and Donald Cole

Introduction

In the popular sitcom about the afterlife, *The Good Place*, the character named Chidi believes he's in the so-called "Bad Place" because he enjoyed almond milk despite knowing that almond production is bad for the environment. Chidi is a moral philosopher and understands that one's personal choices have implications beyond the self—including which foods we decide to eat. That this joke about almond milk is a plot point in a mainstream American sitcom is testimony to the contested nature of choosing what foods to eat in this era of climate change and environmental crises. So when the new Canada Food Guide (Health Canada, 2019) moved from a meat- and dairy-centred diet to a plant-based one, the range of responses was to be expected: Some heralded the guide as a triumph of science over the food industry lobby; while others decried it (Vogel, 2018). Contrary to Chidi's belief that almonds would get you to a bad place, plant-based dietary guidelines appear to lead eaters to almond milk in their morning coffee.

The new guide urges Canadians to: "have plenty of vegetables and fruits;" cut down on animal proteins, especially fatty meats; reduce their intake of processed foods; drink water over sweet beverages; and be aware of the persuasive effects of food marketing. Foregrounding plant-based foods over the kinds of edible products, often ultra-processed, that the global industrial food system is good at producing (Tempels et al., 2017) is in line with other recent dietary guidelines, including the much-quoted EAT Lancet Commission (Willett et al., 2019), as well as insight papers on sustainable food systems released by the Nordic Countries (Stockholm Resilience Centre, 2020). All these new guidelines support not only individual health but also planetary health. As the EAT Lancet group writes: "Our vision is a fair and sustainable global food system for healthy people and the planet"

DOI: 10.4324/9781003174417-19

(EAT Forum, n.d.). However, in focusing on plant-based foods, the dietary recommendations provide a blunt tool for assessing whether a diet is sustainable or not. We ask: What are better ways to navigate the complexities of what to eat and what to avoid when thinking of the health of ourselves, other species, and the planet?

We suggest that the ecological determinants of health (EDoH) is a dextrous concept, adaptable to different dietary contexts, which may help scholars, practitioners, and the public to assess the healthiness of their food from both a personal and planetary perspective. Articulated by the Canadian Public Health Association (CPHA, 2015) to describe the various systems and lifeforms of the biosphere upon which human health depends, EDoH include air, water, and food, as well as soil systems and a climate favourable to human survival. The EDoH are the ecological equivalent of the social determinants of health that together shape health and interact with biomedical determinants.

In this chapter, we start with relevant theory and the origins and meaning of the EDoH. We then argue that the EDoH applied to eating and diets will allow for a shift in discourse in agri-food policy and practice in three ways: 1) Offering a clear system for appraisal of the impact of food production systems on ecosystems; 2) shifting the focus of agri-food policy to root food systems in their foodsheds, replacing a one-diet-fits all approach; and 3) making way for a new relationship with nonhuman nature that understands human health and nutrition to be intrinsically connected to all lifeforms. In settler colonial states, this approach may even support decolonisation of agri-food systems and push thinking towards more equitable eco-social relationships (Parkes et al., 2020).

Relevant knowledge systems and literature

It is important for us to start by recognising that Indigenous Peoples on Turtle Island have long conceived of the connection between human health, food, and ecosystems (for one Anishinabek example, see Geniusz, 2015). Such relationships are recognised in Indigenous law of the territory where the authors of this chapter live and work, the Dish with One Spoon Wampum (Six Nations Polytechnic, 2016). This legal agreement pre-exists the settler colonial written record in the Great Lakes Region—known as the dish (Simpson, 2008). Indigenous scholar Rick Hill summarises its philosophy in his video lecture: "Nature provides everything that we need to be happy and healthy. A dish full of all the animals, the plants, the birds, the medicine plants, the crops, waters. It's like nature's prescription for a whole health plan" (Six Nations Polytechnic, 2016). Indigenous food sovereignty and Indigenous resurgence social movements are working to ensure the longevity of this "whole health plan" in sovereign food systems (Wendimu et al., 2018).

As recognised by the Food and Agriculture Organisation's (FAO) Globally Important Heritage Agricultural Systems, Indigenous Peoples around the world have managed complex agroecosystems with an implicit understanding that human health and wellbeing are tied to intergenerational food production. One example is the rice-fish-duck agroecosystem in southern China where farmers have managed mountain rice paddies, practising intensive agriculture over generations

(FAO, n.d.). On the steep slopes that they or their forebears have terraced, they produce grain, duck, and fish in a way that integrates culture with ecosystem function, supports biodiversity, produces a range of foods, and supports health.

Such longstanding worldviews and knowledge systems set the context for Euro-Western ideas around environments, human health, and food consumption. Carlsson and colleagues (2019) described the late 19th century work of "ecological nutritionist" Ellen Swallow Richards, who brought an ecological mindset to home economics. In the 1960s, a counterculture promoted the idea that "natural foods" and vegetarianism could help you be healthy and also save the planet (Belasco, 2014).[1] Perhaps best captured in the best-selling book, *Diet for a Small Planet*, Francis Moore Lappé (1971) argued that meat production damaged the earth and that vegetarianism was a solution for global food scarcity.

A growing interest in ecology has given rise to a number of relevant academic subfields: Political ecology, that theorises relationships between politics, economics, and nature (Blaikie & Brookfield, 1987); One Health, that considers human health as connected to that of animals and other aspects of nature (Zinsstag, 2020); and ecological public health, bringing public health concerns to the forefront of food system thinking (Lang, 2009). Leitzmann (2003) has proposed a system called "nutrition ecology"; Seed and Rocha (2018) argue that dietary guidelines are a good way to connect food systems with human and ecological health; and FAO has proposed a "forest, human health and nutrition nexus" (FAO, 2020), which connects food security and nutrition to the presence of forests in people's lives. Most recently, Beacham (2021) has argued for planetary approaches to food regimes in the Anthropocene.

The ecological determinants of health (EDoH)

The Canadian Public Health Association (CPHA) had long considered the linkages between humans and ecosystems (CPHA, 1992). Yet there has been little systemic change, despite growing evidence that many aspects of Canadian life, including the economy, were contributing to environmental degradation and climate change, with consequences for health. The EDoH framework was developed when Dr. Trevor Hancock recruited colleagues to update the 1992 CPHA framework and generate proposals for action for public health practitioners, scholars, and policy makers. Their extensive review, meetings, and conversations led to the discussion paper that defined the EDoH as a new concept (CPHA, 2015). The first three EDoH are the founding pillars of life on earth: Oxygen, water, and food. The other systems categorised as EDoH (central to food production, though not the focus of the original document) are: Soil systems, including the circulation of phosphorus and nitrogen, elements integral to soil fertility; water systems that can provide clean water and support plant growth as well as marine life; a stable climate and an ozone layer that provide the appropriate environment for plant and animal growth and survival; "abundant energy," which includes energy needed to fuel food production and distribution systems; and the material humans require to build tools and structures. It was suggested in 2021 that the earth microbiome, including the human microbiome, be also recognised as an EDoH (Elton, 2021a).

The document also positioned the EDoH within the current-day context of the Anthropocene and the unfolding crises of environmental destruction: What the authors call global ecological change. In listing these threats to the EDoH—climate change, ecotoxicity, and loss of biodiversity and species extinction—the document identifies major threats to food systems today. Further, it flags the social production of these problems, including the industrial and agricultural activities that are referred to as economic development, and the potential threat they constitute for the health and wellbeing of humans, other species, and their habitats. In this way, the concept both identifies the physical connection of human and non-human nature, and the social-political context in which such relations are either made or disrupted.

Applying the EDoH

Assessing sustainability

In the fields of dietetics, nutrition, and public health, the health of food is typically assessed in one of three ways. *First*, a biomedical framework considers how healthy the food is for the human body: Does it provide important nutrients or have deleterious effects, such as leading to diabetes? *Second*, a food security/insecurity lens, at individual, community, and population levels, associates health with who has access to nutritious food and how much, and who does not. *Third*, the healthiness of a food system can be assessed by characterising the retail foodscape. For example, does the foodscape present an *obesogenic environment*, defined by Egger and Swinburn (1997) as a "pathological environment" (p. 480) with limited opportunities for physical movement and poor quality of food that together increase BMI?[2]

Applying an EDoH framework to assessing the healthiness of food widens the lens to an analysis of the entire food system, from food production through to disposal. If the way a food is produced damages and kills the biodiverse lifeforms in the soil with pesticides and fossil-fuel derived fertiliser, then the production of this food endangers the EDoH—even if the food is nutrient-rich. Taking almond milk as an example, the most purchased non-dairy milk in the United States (Winans et al., 2020) we apply the EDoH framework to assess the healthfulness of this product. Although the EAT Lancet commission suggested boosting our consumption of plant-based foods like nuts by 200%, with reduction of animal-based food intake by half (Willett et al., 2019), a food system analysis using the EDoH complicates the assumption that a nut-based food is a healthy choice.

According to one life-cycle assessment of commercial almond milk made from nuts grown in irrigated orchards in California, which is the site of 80% of the world's almond production, one 1.42 litre bottle is equal to 175 kgs of freshwater—that's about six cubic feet (0.17 cubic metres) of water used to produce one bottle of almond milk. One bottle also equates to 0.71 kg of CO_2 produced. According to the Environmental Protection Association in the United States' emissions calculator, a coniferous tree planted in a city takes ten years to sequester 10.5 kgs of

CO_2, or the carbon dioxide equivalent of 14.8 bottles of almond milk (US EPA, 2015). When one includes packaging in the calculation, the embedded CO_2 rises by 55% and water by 5% (Winans et al., 2020). In the San Joaquin Valley, where many orchards are located, groundwater (typically extracted from deep water aquifers) is being depleted without adequate replenishment (Fleischer, 2018). Irrigation systems for California almonds[3] are so intensive that one kernel alone requires 12 litres to grow (Fulton et al., 2019). Winans et al. (2020) also found that energy-intensive pumps are often needed for increasingly deep groundwater wells, which increase embedded energy use. Further, almond production relies on the commercial pollination industry, as the tree's fruit flowers require pollination by bees and other insects to fruit. The industrial beekeeping operations that pollinate California orchards are described in the popular press, industry (Watson, 2020) and the documentary film *The Pollinators*. The environmental toll of these pollination services contribute to the ecological cost of almond production (Marvinney & Kendall, 2021).

Such material realities of California almond milk production, when assessed through the EDoH framework, problematise the idea that plant foods are implicitly healthy choices. Water, a stable climate, an environment free from fungicides and insecticides (Wade et al., 2019) are all EDoH impacted by the dominant Californian approach to almond production. Of course, dairy production also contributes substantially to climate change—cattle belch methane, particularly those fed corn and soy mixtures in intensive livestock operations; in addition along the food production path significant amounts of carbon dioxide and nitrous oxide are emitted. To grow feed for such dairy cattle requires CO_2 intensive, hydrocarbon inputs. We are not advocating for one type of milk over the other, as agroecological options exist in Spanish almond production (Reisman, 2020) and grass-fed dairy ranches. What we are suggesting is that the binary between plant- and animal-based foods simplifies a complicated question. To unpack which food is preferable, one must assess the production systems, their technology, and the range of their negative and positive impacts, as has been done for different diets and beef consumption in particular (Rangananthan et al., 2016). An EDoH approach also invites a more fine-tuned analysis that considers a person's relationship to particular places, including those ecosystems from which they eat.

Ecosystem-responsive diets

A shift in agri-food policy to focus on EDoH would help to root food systems in their more proximal ecosystems. Increased reliance on food produced in a regional foodshed could lead to less dependency on global supply chains and their associated environmental costs. The EDoH in a more proximal regional ecosystem—soil, water, weather, biota, and peoples— are more visible to consumers, the way something thousands of kilometres away, through multiple intermediaries, is not. This idea of rooting food systems in a more proximal foodshed has been articulated from different scholarly perspectives. Working from a nutrition ecology perspective, Leitzman (2003) called for "regionally produced" food. Lang (2009, p. 332) flagged:

> An immediate area for research is how to define a sustainable diet in locally appropriate ways. It is not likely that a sustainable diet would be the same in uplands of the United States as in Africa or China, but the principles might be.

In the following, we provide three examples of eco-regional approaches.

In their edited collection, *Diet for a sustainable ecosystem*, Cuker (2020) provided a detailed exploration of what an ecologically-responsive food system might look like in one American region. Working with an interdisciplinary team, Cuker conceived of a human diet that would foster a sustainable ecosystem in Chesapeake Bay and its watershed. An early chapter describes the Algonquin food system centred on fishing, foraging, and hunting in and around the Bay, which sustained the Indigenous Peoples before the invasion of colonists. Then a set of chapters document impacts of settler colonialism, describing how settlers took advantage of local ecosystems, displaced its peoples, and, over the following centuries, caused great damage to the land, the estuary, and its species. The authors then put forward a plan to restore the health of the bay and its peoples together, guided by a sustainability spectrum involving different food system approaches. They suggest a shift to a plant foods diet, aiming to reduce greenhouse gas released into the air and the associated sea level rise and loss of land. They also suggest seafood consumption from local bivalve production, providing not only food but filtered water, which is also health supporting. They envisage these land and aquatic production systems as organic so that they reduce water contamination and promote healthy soils (Kaufman, 2020). The authors explicitly connect each part of this food system to the EDoH as well as the recommended diet. At the same time, they pay important attention to governance, equity for workers and families facing food insecurity, and other social determinants of health (Cuker & Davis, 2020).

A second ecosystem-responsive diet paradigm comes from the Andean highlands where an agroecology movement flourishes.

> Agroecology is as a dynamic, locally adapted concept [which] applies regenerative ecological principles to agricultural practices—such as by optimising production diversity, eliminating harmful inputs, and leveraging beneficial biotic relationships; and integrates these practices with environmental, social and economic priorities to contribute to sustainable food systems.
>
> *(Deaconu et al., 2021a, p. 1)*

Agroecology seeks to preserve the EDoH in practice. One example would be the use of animal manures to enrich soil health. Mixed methods demonstrated that agroecological farmers outperform[ed] reference farming neighbours on both nutrient adequacy (i.e. meeting key nutrient needs) and dietary moderation (i.e. avoiding dangerous excesses). Stronger nutrient adequacy is likely related to agroecological farmers' higher production diversity as well as the social and human capital developed within their networks, while stronger dietary moderation is likely related to their greater consumption of foods obtained through own-production and the social economy (e.g., barter).

(Deaconu et al., 2021a, p. 102033)

Third, a promoter of regional agroecosystem responsive diets is Eat Local Grey Bruce (ELGB), a bi-county, not-for-profit, producer-eater cooperative committed to "local food production, ecological practices, healthy eating and meaningful employment." (Kralt & Cole, 2021, p. 14). The agricultural context in Grey County, Ontario, Canada is one where cash crops superseded natural and seeded pasture land, predominantly for the production of corn and soybeans to feed increased beef steers and heifers (Grey County, 2017). Consumption of fossil fuels for mechanised cash crop production, methane production by feed fed livestock, and breach of intensive livestock operation sewage lagoons (Weis, 2013) all compromise EDoH. In fact, a prime example of the combined effects of disregarding EDoH (Ali, 2004) was the major enteric illness outbreak due to runoff from livestock operations contaminating a town's wells during a heavy rainfall event in Bruce County in 2000. ELGB involves producers practising regenerative agriculture (about 1% of all producers) and bolsters social inclusion among eaters (about 1% of households in the two counties), both exemplifying local food charter principles (Cole et al., 2022). ELGB has initiated collaborations with Indigenous communities on shared products for sale, developed support memberships that help pay down amounts owed to producers, and created *solidarity tokens* that eaters can contribute to subsidise lower income members' purchases from member producers.

Box 16.1 Living within and contributing to food systems

Fair Fields Organics

Fair Fields Organics is one Eat Local Grey Bruce producer farm, of which Cole is a family member. It is a certified organic polyculture, producing rhubarb, asparagus, strawberries, corn, and honey for sale, and eggs, poultry meat, and pork for home consumption. Situating our work in our ecosystem involves understanding the hardiness zone (4a) and soil type (Harrison loam) and recognising what grows best in this climate and tilth. To control unwanted plants that compete with our crops, we use tarps or compost to block these plants' access to sunlight, cutting off their food supply, as opposed to herbicides that kill plants along with soil microbes and water fauna. We create buffers between our production and nearby farms to reduce pesticide drift onto our crops and bees. We have located the apiary and orchard in a protected area, where they are less likely to suffer the effects of stronger winds. To mitigate against the increasingly frequent extended droughts due to climate change, we irrigate from spring-fed ponds, relying on surface water rather than drilling wells to deplete groundwater. Finally, we work on building soil organic matter through green manures, cover crops, and turkey manure application. Together, these practices aim to retain and build EDoH to ensure the sustainability of the farm and our health.

Source: Personal example from one co-author

Ontological shift for decolonising food systems

The preamble to the World Health Organisation constitution, adopted at the International Health Conference in New York, defined health to be "a state of complete physical, mental and social well-being and not merely the absence of disease or infirmity" (WHO, 1946, p. 1). This definition has influenced policy and practice globally. Missing from this definition is a recognition of the ways that human health is interdependent with the biosphere. During subsequent decades, the subfields of One Health and Ecohealth, amongst others, have theorised humanity's interconnections with nonhuman nature, though with marginal impact (Butler & Friel, 2006; Hancock, 2015; Lang & Rayner, 2015).

Thus, another benefit of an EDoH framework is that it makes way, among dominant Euro-Western paradigms, for the possibility of a new relationship with nonhuman nature. By understanding human health to be intrinsically connected to the biosphere, earth systems, and all the planet's lifeforms, it challenges the Euro-Western notions of the self-sufficient individual (Whatmore, 1997). This idea that a person is an independent individual is rooted in Enlightenment thinking and creates a dualism between humans and nature (Plumwood, 1993). This idea of *man* is also bound up in settler-colonialism and patriarchy (see Sylvia Wynter's (2003) monumental essay for insight), ideologies that render some humans exceptional—white, European men—while casting other groups as lesser. The idea that any person is self-sufficient has been picked apart by feminist theory, critical race theory, and disability and crip theory. In short: We all survive and thrive as a result of relationships—the often invisible labour performed by women, disabled, and racialised people. Health is dependent on these relationships, this labour, and this biosphere.

When health is defined as the physical and mental state possessed by an individual, it plays into an Enlightenment narrative of some humans as an exceptional species. Recognition that an individual's health is mediated by social structures, social relations, and human institutions, as per the social determinants of health, stops short of including the health-supporting role of non-human nature.[4] Adoption of an EDoH framework recognises that humans are interdependent with nonhuman nature, implicitly critiquing the Enlightenment ontology. It invites a profound shift in perspective, as the original EDoH document recognises: "we need to view humans as part of the web of life, and understand that human health depends on the effective functioning of ecosystems, the health of other species and the sustainable use of available resources" (CPHA, 2015, p. 18).

The potential benefits of such an ontological shift extend beyond health discourses. Canada, where we live and work, is a settler-colonial state. Here, settler-colonialism helped to birth the food system as we know it, oppressing non-white humans and nonhumans alike. The food system is implicated here—as it is globally—as a product and instrument of colonial violence (Daschuk, 2019), continuing to function in many instances by perpetuating structural violence on those working in it, to the detriment of their food security (e.g., migrant workers) (Rotz, 2017; Weiler et al., 2017). A recognition of the interdependence of humans and nonhuman nature in terms of health and food systems may help to dispel a colonial binary and invite an opportunity to build on more holistic knowledge systems and philosophies.

Conclusion

We have sketched out an approach, building on the EDoH framework, to respond to our initial question: What are better ways to navigate the complexities of what to eat and what to avoid when thinking of the health of our bodies, other species, and the planet? The EDoH can assist in balancing individual human's health with the planet's health, region-by-region. Implementation of the EDoH framework is possible at various scales, starting from the individual deciding about what kind of milk to add to their morning coffee. It is important that policymakers, working at municipal and regional scales, apply the EDoH framework. In such situations, the EDoH can complement social determinants of health—equity frameworks that consider the intersection of different ethnic, income, origin, ability, and other group experiences. The corporate commercial determinants of health (Mialon, 2020) must also be considered. So while EDoH does not offer quick answers to the complex question of what to eat, it does provide tools to navigate these eco-social complexities (Parkes et al., 2020). The task is multisectoral, with roles for scholars, public health practitioners, and farmers, as well as bureaucrats, politicians, and civil society organisations working together for more sustainable food systems.

Notes

1 In the book, Belasco critiques these conclusions and argues that food did not spur a counterculture revolution but rather mass consumerism.
2 For a critique of this discourse from an EDoH perspective, see Elton (2021).
3 Marvinney and Kendall (2021) note in their life-cycle assessment of California almonds that water used in nut production generally is similar to that of almonds. This means switching from almond milk to milk made from another nut would not necessarily improve the environmental cost of one's plant-based diet.
4 This is why the original CPHA (2015) document, which lays out the ecological determinants of health, suggests that a health equity lens put forward by a social determinant of health analysis be integrated with the ecological determinants of a health lens. Also, a subsequent CPHA endeavour used the term eco-social (Parkes et al., 2020).

References

Ali, S. H. (2004). A socio-ecological autopsy of the E.coli 0157:h7 outbreak in Walkerton, Ontario, Canada. *Social Science & Medicine, 58*(12), 2601–2612.

Beacham, J. (2021). Planetary food regimes: Understanding the entanglement between human and planetary health in the Anthropocene. *Geographic Journal, 188*(3) 1–10.

Belasco, W. J. (2014). *Appetite for change: How the counterculture took on the food industry.* Cornell University Press.

Blaikie, P. M., & Brookfield, H. C. (1987). *Land degradation and society.* Methuen.

Butler, C. D., & Friel, S. (2006). Time to regenerate: Ecosystems and health promotion. *PLOS Medicine, 3*(10), Article e394.

Canadian Public Health Association (CPHA) (1992). *Human & ecosystem health. Canadian perspectives, Canadian action.* https://www.cpha.ca/sites/default/files/assets/policy/eco-system_health_e.pdf

Carlsson, L., Mehta, K., & Pettinger, C. (2019). Critical dietetics and sustainable food systems. In J. Coveney, & S. Booth. (Eds.), *Critical dietetics and critical nutrition studies* (pp. 97–115). Springer Science+Business Media.

Cole, D. C., Needham, L., Markowitz, P. 2022. A food charter as a critical food guidance tool in rural areas: The case of Bruce & Grey Counties in Southwestern Ontario. *Canadian Food Studies. 9*(1), 12–36

CPHA (2015). *Global change and public health: Addressing the ecological determinants of health.* https://www.cpha.ca/sites/default/files/assets/policy/edh-discussion_e.pdf

Cuker, B. E. (2020). *Diet for a sustainable ecosystem: The science for recovering the health of the Chesapeake Bay and its people.* Springer Nature.

Cuker, B. E., & Davis, K. (2020). Ethics and economics of building a food system to recover the health of the Chesapeake Bay and its people. In B. E. Cuker (Ed.), *Diet for a sustainable ecosystem: The science for recovering the health of the Chesapeake Bay and its people* (pp. 407–430). Springer Nature.

Daschuk, J. W. (2019). *Clearing the plains: Disease, politics of starvation, and the loss of Indigenous life.* University of Regina Press.

Deaconu, A., Berti, P.R., Cole, D.C., Mercille, G., & Batal, M. (2021a). Agroecology and nutritional health: A comparison of agroecological farmers and their neighbours in the Ecuadorian highlands. *Food Policy, 101*, 1–14.

Deaconu, A., Berti, P. R., Cole, D. C., Mercille, G., & Batal, M. (2021b). Market foods, own production and social economy: How food acquisition sources influence nutrient intake among Ecuadorian farmers and the role of agroecology in supporting healthy diets. *Sustainability, 13*(8), 4410.

EAT Forum (n.d.). *Our vision, mission and values.* Retrieved April, 4, from https://eatforum.org/about/who-we-are/our-vision-mission-and-values/

Egger, G., Swinburn, B. (1997). An ecological approach to the pandemic. *British Medical Journal, 315*, 477–80.

Elton, S. (2021a). Intimate ecosystems: The microbiome and the ecological determinants of health. *Canadian Journal of Public Health, 112*, 1004–1007.

Elton, S. (2021b). Is the 'obesity crisis' really the health crisis of the food system? The ecological determinants of health for food system change. *Canadian Food Studies / La Revue Canadienne Des Études Sur l'alimentation, 8*(1), Article 1.

Fleischer, D. (2018, January). *Almond milk is taking a toll on the environment. UCSF office of Sustainability.* https://sustainability.ucsf.edu/1.713

Food and Agriculture Organisation of the United Nations (2020). *Forests for human health and well-being.* FAO. https://www.fao.org/3/cb1468en/cb1468en.pdf

Food and Agriculture Organisation of the United Nations (n.d.). *Dong's rice fish duck system.* Globally Important Agricultural Heritage Systems (GIAHS). Retrieved April, 4, from http://www.fao.org/giahs/giahsaroundtheworld/designated-sites/asia-and-the-pacific/dongs-rice-fish-duck-system/en/

Fulton, J., Norton, M., & Shilling, F. (2019). Water-indexed benefits and impacts of California almonds. *Ecological Indicators, 96*, 711–717.

Geniusz, M. S. (2015). *Plants have so much to give us, all we have to do is ask: Anishinaabe botanical teachings*. University of Minnesota Press.

Grey County (2017, May 16). *Agriculture in Grey County by the numbers*. County of Grey - Colour It Your Way. https://www.grey.ca/news/agriculture-grey-county-numbers

Hancock, T. (2015). Population health promotion 2.0: An eco-social approach to public health in the Anthropocene. *Canadian Journal of Public Health, 106*(4), e252–e255.

Health Canada (2019, January). *Canada's dietary guidelines for health professionals and policy Makers*. Canada.ca/Food Guide. https://food-guide.canada.ca/sites/default/files/artifact -pdf/CDG-EN-2018.pdf

Kaufman, M. (2020). An organic-based food system: A voyage back and forward in time. In B. E. Cuker (Ed.), *Diet for a sustainable ecosystem: The science for recovering the health of the Chesapeake Bay and its people* (pp. 375–395). Springer Nature.

Kralt, J., & Cole, D. (2021). Awakening to the need for community resilience: Experiences within a rural food distribution cooperative. *Journal of Critical Dietetics, 5*(3), 12–19.

Lang, T. (2009). Reshaping the food system for ecological public health. *Journal of Hunger & Environmental Nutrition, 4*(3–4), 315–335.

Lang, T., & Rayner, G. (2015). Beyond the golden era of public health: Charting a path from sanitarianism to ecological public health. *Public Health, 129*(10), 1369–1382.

Leitzmann, C. (2003). Nutrition ecology: The contribution of vegetarian diets. *The American Journal of Clinical Nutrition, 78*(3) Supplement, 657S–659S.

Marvinney, E., & Kendall, A. (2021). A scalable and spatiotemporally resolved agricultural life cycle assessment of California almonds. *The International Journal of Life Cycle Assessment, 26*, 1123–1145.

Mialon, M. (2020). An overview of the commercial determinants of health. *Globalisation and Health, 16*(1), 1–7.

Parkes, M. W., Poland, B., Allison, S., Cole, D. C., Culbert, I., Gislason, M. K., Hancock, T., Howard, C., Papadopoulos, A., & Waheed, F. (2020). Preparing for the future of public health: Ecological determinants of health and the call for an eco-social approach to public health education. *Canadian Journal of Public Health, 111*(1), 60–64.

Plumwood, V. (1993). *Feminism and the mastery of nature*. Routledge.

Ranganathan, J., Vennard, D., Waite, R., Searchinger, T., Dumas, P., Lipinski, B. (2016). Shifting diets: Toward a sustainable food future. In *Global food policy report* (pp. 66–79). IFPRI. https://ebrary.ifpri.org/utils/getfile/collection/p15738coll2/id/130216/file-name/130427.pdf

Reisman, E.D. (2020). *Orchard entanglements: Political ecologies of almond production in California and Spain* [Doctoral dissertation, University of California Santa Cruz]. https://eschol-arship.org/content/qt8vc1k18h/qt8vc1k18h_noSplash_bac321faea966acdb1fee5e 73f44937a.pdf

Rotz, S. (2017). "They took our beads, it was a fair trade, get over it": Settler colonial log-ics, racial hierarchies and material dominance in Canadian agriculture. *Geoforum, 82*, 158–169.

Seed, B., & Rocha, C. (2018). Can we eat our way to a healthy and ecologically sustainable food system? *Canadian Food Studies / La Revue Canadienne Des Études Sur l'alimentation, 5*(3), 182–207.

Simpson, L. (2008). Looking after Gdoo-naaganinaa: Precolonial Nishnaabeg diplomatic and treaty relationships. *Wicazo Sa Review, 23*(2), 29–42.

Six Nations Polytechnic (2016). Food security & three sisters sustainability: Conversations in cultural fluency no.3 [Video]. *Youtube*. https://www.youtube.com/watch?v=39 y11jrfHjg

Stockholm Resilience Centre (2020). *Towards sustainable Nordic food systems* (Insight Paper, no.1). Nordic Food System Transformation Series.

Tempels, T., Verweij, M., & Blok, V. (2017). Big food's ambivalence: Seeking profit and responsibility for health. *American Journal of Public Health, 107*(3), 402–406.

US EPA, O (2015, August 10). *Greenhouse gases equivalencies calculator: Calculations and references [Data and Tools]*. US EPA. https://www.epa.gov/energy/greenhouse-gases-equivalencies-calculator-calculations-and-references

Vogel, L. (2018). Meat and dairy supporters seek industry-friendly changes to food guide. *Canadian Medical Association Journal, 190*(7), E201–E202.

Wade, A., Lin, C.-H., Kurkul, C., Regan, E. R., & Johnson, R. M. (2019). Combined toxicity of insecticides and fungicides applied to California almond orchards to honey bee larvae and adults. *Insects, 10*(1) 1–11.

Watson, E. (2020, February 7). *Special report: Bee friendly? Pollinating California's almond crop*. Foodnavigator-Usa. https://www.foodnavigator-usa.com/Article/2020/02/07/SPECIAL-REPORT-Bee-friendly-Pollinating-California-s-almond-crop

Weiler, A. M., McLaughlin, J., & Cole, D. C. (2017). Food security at whose expense? A critique of the Canadian temporary farm labour migration regime and proposals for change. *International Migration, 55*(4), 48–63.

Weis, T. (2013). The meat of the global food crisis. *Journal of Peasant Studies, 40*(1), 65–85.

Wendimu, M. A., Desmarais, A. A., & Martens, T. R. (2018). Access and affordability of "healthy" foods in northern Manitoba? The need for Indigenous food sovereignty. *Canadian Food Studies / La Revue Canadienne Des Études Sur l'alimentation, 5*(2), 44–72.

Whatmore, S. (1997). Dissecting the autonomous self: Hybrid cartographies for a relational ethics. *Environment and Planning D: Society and Space, 15*(1), 37–53.

Willett, W., Rockström, J., Loken, B., et al. (2019). Food in the anthropocene: The EAT–lancet commission on healthy diets from sustainable food systems. *The Lancet, 393*(10170), 447–492.

Winans, K. S., Macadam-Somer, I., Kendall, A., Geyer, R., & Marvinney, E. (2020). Life cycle assessment of California unsweetened almond milk. *The International Journal of Life Cycle Assessment, 25*(3), 577–587.

World Health Organisation (WHO) (1946). Preamble to the Constitution of WHO as adopted by the International Health Conference. https://treaties.un.org/doc/Treaties/1948/04/19480407%2010-51%20PM/Ch_IX_01p.pdf

Wynter, S. (2003). Unsettling the coloniality of being/power/truth/freedom: Towards the human, after man, its overrepresentation: An argument. *The New Centennial Review, 3*(3), 257–337.

Zinsstag, J. (Ed.). (2020). *One health: The theory and practice of integrated health approaches* (2nd ed.). CAB International.

17

BREASTFEEDING

A foundational strategy to strengthen sustainability in infant nutrition and development

Joseph W. Dorsey and Marian E. Davidove

Introduction

This chapter illuminates how breastfeeding can contribute to sustainable diets for infants and children. We offer insights with parent(s) building a holistic approach and strategies for introducing solid food. Breastfeeding, more than using formula, can be a substantial foundation as part of sustainable diets as it enables lactating mothers, those who are able to use this method, to provide bonding time with the child, along with a cost-effective source of nutrition. This inquiry considers ways to promote and sustain breastfeeding over time. This chapter considers the complexity and nuances that can be involved with breastfeeding. We include the biological, economic, societal, cultural, and environmental aspects of breastfeeding. Ways to foster social arrangements and community norms for experienced breast-feeding women and parents to offer coaching for new parents are encouraged. We conclude with descriptions of the long-term implications of sustainable diets in early childhood nutrition and human development.

Food sustainability is often regarded as an environmental or ecological prob-lem but has become essentially a human problem, particularly as driven by life-styles fuelling the Anthropocene. People are an integral part of the food system, both producing and consuming nutrients. Anthropocene activities are degrading, destroying, and/or disrupting our food supplies, however, the way people eat can also conserve and regenerate Earth's natural systems. The holistic approach to a sustainable diet is confirming that "Food is the single strongest lever to opti-mise human health and environmental sustainability" (EAT-Lancet Commission, 2019, p. 5). At each life stage, food intake and production can be optimised for human and planetary health. For infants and young children an optimal source of food is breast milk. Breastfeeding comprises both the bio-physiological process

DOI: 10.4324/9781003174417-20

that happens post-pregnancy and the product of lactation, which is breast milk. Breastfeeding is a unique food system component since it is a natural, renewable, and a safe source of nutrients and protections supplied by the lactating woman; it is the gold standard for developing sustainable dietary patterns for nursing mothers, infants, and children. In this chapter, we examine breastfeeding's vital role as both a sustainable diet and within sustainable diets.

According to Mason and Lang (2017), "a sustainable diet is one that optimises good sound food quality, health, environment, socio-cultural values, economy and governance" (p. 9). Sustainable diets contain all the essential nutrients and quality control needed for the optimal well-being of consumers. To understand what constitutes a sustainable diet, we need to look at different populations, life stages, and contexts. Conditions that support sustainable diets can differ in low-income and high-income environments. Still, there are three commonalities in building a sustainable diet that cut across all communities: First, a sustainable diet must be *appealing*; sustainable food needs to be visually attractive and taste delicious. Second, a sustainable diet should feel *normal and culturally acceptable*. Third, a sustainable diet should be *easy* to obtain and affordable. These three tenets, when applied to infant and early childhood nutrition, support the core argument of this chapter that breastfeeding constitutes a sustainable diet and a promising way of life for mothers, infants, and children.

For many, a sustainable diet would be a major shift in their eating pattern, but it is also a perspective and practice that can start at an early age and be normalised through one's lifetime. Promoting breastfeeding as a natural alternative to formula feeding can help strengthen economic stability, food security, and climate change adaptation. Promoting and supporting breastfeeding is a necessary foundation for a sustainable diet plan for the mother and the baby. Dietary shifts that entail improving breastfeeding rates and duration could initiate sustainable diets for infants and children. Breastfeeding early in a child's life is the cornerstone for initiating a sustainable diet with long-term positive effects.

The practice of breastfeeding as a key to long term sustainability in nutrition and health may not be evident at first glance. Lactation is a natural biological function of post-pregnancy, for most, but not all, women. We acknowledge that there are instances where breastfeeding is not possible for an array of reasons. As a mammalian species, humans are naturally endowed with nipples for lactation, although they are only functional in pubescent and post-puberty females after bearing a child. While breasts have been sexualised in some societies, it is valuable to appreciate the nutritional role breasts can play in feeding new-born babies and infants for the first few months or years of their lives. When new mothers are able to breastfeed, breast milk can be the first form of nutrients, enzymes, antibodies, and pre- and probiotics that new-borns receive after birth (Newburg & Walker, 2007). Breast milk contains essential nutrition to keep a baby healthy, protected, comfortable, and perhaps content during the early stages of infancy and possibly beyond (Walters et al., 2016). However, changes in choice architecture, economic roles, social perceptions, and cultural norms by modern and post-modern society have complicated the practice of breastfeeding for women and their families around the world.

Exclusive breastfeeding is the practice of only giving breast milk to the infant. The introduction of powdered and liquid baby formulas over the last century has provided a valid but environmentally, economically, and socially costly alternative to natural breast milk and has jeopardised the value of mother's milk as the standard form of infant nutrition. For decades,[1] the promotion of infant formula has been pushed by corporations as the best alternative to breastmilk, and in some cases touted as superior, thus undermining breastfeeding rates and duration globally. Weaning early from breastmilk can push recent mothers to introduce cow's milk to their children's diet to compensate (Couto et al., 2020). Like many other aspects of the current food system, the industrialisation of infant feeding through formulas has provided some levels of convenience, time management, and physical freedom for young mothers, career women, and those unable to or wishing not to breastfeed. The absence of substantial breastfeeding has been shown to have short and long-term repercussions for the baby as well as local economies, social cohesion, environmental quality, and human population growth (Smith, 2019). Therefore, instilling the basics of a sustainable diet through breastfeeding is a critical factor in the early phases of human development.

Biological

Health benefits to breastfeeding infants

The first few years of a child's life are crucial to forming eating habits that can lead to a sustainable diet as they grow older. The decision to breastfeed or not to breastfeed is an important step in setting the stage for more natural and holistic patterns of feeding the child through its formative years. Of course, a person can grow up healthy and happy without being breastfed as an infant, but studies show that breastfeeding provides certain biological, social, and environmental advantages that are universal and irreplaceable (Binns et al., 2016; Brahm & Valdes, 2017; Moore, 2018; Ogunba et al., 2019; Pugh et al., 2002).

There is an abundance of research on the short-term and long-term health benefits of breastfeeding for the mother and the infant, and recently, there has been much written about additional societal benefits in relation to culture, economics, policies, and the environment. Worldwide, breastfeeding has been endorsed as the optimal method of nourishing human infants as compared to other forms of infant feeding. According to the World Health Organisation (WHO), the United Nations International Children's Emergency Fund (UNICEF), the Academy of Nutrition and Dietetics (AND), and the American Academy of Pediatrics (AAP), there are major health benefits for both the mother and child when exclusive breastfeeding is practised for at least six months and continued for a time once solid foods and juices are introduced. During the global pandemic of 2020–2022 (the date of this publication), WHO recommended that a new mother found to have COVID-19 should be supported to breastfeed her baby, hold her new-born skin to skin, and share a room with her baby (WHO, 2020). However, nearly two out of three infants are not exclusively breastfed for the recommended six months—a rate that has not improved in two decades (Binns et al., 2016; Couto et al., 2020; Guttman & Zimmerman, 2000; Kornides & Kitsantas, 2013; Pugh et al., 2002).

Promoting breastfeeding at least during the child's first six months of life reduces the likelihood that a child will ingest contaminated food or water during infancy, and beneficial elements in mother's milk can strengthen the child's immune system (Hanson, 2000). Breast milk also has a favourable effect on establishing a healthy and medicinal human microbiome in the infant's intestine for better digestion (Newburg & Walker, 2007).

Additional evidence shows that breastfeeding protects infants against a wide range of infectious diseases such as bacterial meningitis, bacteraemia, diarrhoea, respiratory tract infections, pneumonia, ear infections, gastrointestinal illness, otitis media, urinary infections, and necrotizing enterocolitis during the nursing period (Allen & Hector, 2005; Kornides & Kitsantas, 2013; Rollins et al., 2016). Other positive health outcomes from exclusive breastfeeding include decreased rates of sudden infant death syndrome in the first year of life, a reduction in the incidence of insulin-dependent (type 1) and non-insulin-dependent (type 2) diabetes mellitus, lymphoma, leukaemia, and Hodgkin disease, overweightness and obesity, hypercholesterolemia, and asthma in older children and adults, compared to those who were not breastfed (Stuebe, 2009; Victora et al., 2016). WHO (2020) recommends that breastfeeding be provided to children up to two years or more to protect them from premature mortality and morbidity from infectious diseases.

Another benefit for breastfed children is that they learn "self-regulated appetite control" as they begin to self-feed solid food (Moss & Yeaton, 2014). Breastfeeding may acclimatise children to control food intake and develop the ability to identify internal cues signifying hunger gratification in their minds and stomachs (Brahm & Valdes, 2017). In contrast, bottle-fed infants may over-consume when their milk intake is regulated by the caretaker. The caregiver decides how much liquid is in the bottle and when to stop. This conditioned feeding tendency may lead to a pattern of over-eating later in life (Allen & Hector, 2005; Victora et al., 2016). Obesity is of growing concern worldwide, and if the current growth rate continues, it is estimated that 2.7 billion adults will be overweight, over 1 billion affected by obesity, and 177 million adults severely affected by obesity by 2025 (World Obesity, n.d.).

According to the WHO, breastfeeding provides "up to half or more of a child's nutritional needs during the second half of the first year and up to one third during the second year of life" (n.d., para. 2). The American Academy of Pediatrics has stated that there is no upper limit to the duration of breastfeeding, as there is no evidence of psychological or developmental harm from breastfeeding into the third year of life or longer (Gartner et al., 2005). One possible harmful aspect of breastfeeding beyond one year is the higher prevalence of tooth decay. A suggested strategy is not to stop breastfeeding but to practice better oral hygiene with the child. Discontinuing breastfeeding prior to six months often leads to the earlier introduction of solid foods. Early introduction of solid foods may pose a risk to choking when the infant has not developed the coordination or ability to refuse, chew, or swallow food. When unhealthy eating patterns are established before two years of age, older children develop eating habits that are prone to consuming foods that are low in nutritional value and high in calories (Moss & Yeaton, 2014). So, the timing of the introduction to solid food can be linked to future adiposity in the child, particularly if they are infant formula fed. Childhood obesity can be avoided if the mother breastfeeds beyond six months and delays the introduction of solid food prior to six months. The evidence shows that obesity in childhood can lead to obesity in adulthood (Allen & Hector, 2005; Victora et al., 2016). Breastfeeding is associated with a desirable body weight status for children between 5 and 18 years old (Moss & Yeaton, 2014). Breastfeeding provides an early foundation for healthy and sustainable weight status development past childhood and, perhaps, into adulthood.

Cognitively, there are numerous benefits for exclusively breastfed infants and long-term benefits for children as well. Higher cognitive development in exclusively breastfed infants, as compared to other food sources, has been observed in several studies focused on various skills in learning, reasoning, decision making, visual memory, auditory memory, reading and writing ability, and mathematical skill in children (Victora et al., 2016). However, these studies conclude that it is not exclusive breastfeeding that influences the child's behaviour, as much as it is the parenting behaviour and family's socioeconomic characteristics (Couto et al., 2020).

Health benefits for breastfeeding mothers

There are physiological benefits for the breastfeeding mother in regard to rapid weight loss of the extra body fat accumulated during pregnancy. Although percentage of body fat and weight gained during pregnancy plays a role, when women breastfeed for three or more months it seems to accelerate weight loss through fat depletion. While breastfeeding does cost energy, diet and physical activity help remove body fat. Fortunately, activity and gradual weight loss (1 pound per week) is safe and does not affect the mother's ability to produce breast milk or inhibit her infant's weight gain (Behan, 2019).

Additional benefits of breastfeeding afford the uterus to contract; it signals hormonal changes that delay the return of the regular ovulation cycle thus lengthening birth intervals (though not a guaranteed method of birth control); it conserves iron stored in the blood; and it may protect women against breast and ovarian cancer, Type 2 diabetes, and high blood pressure (CDC, n.d.a.). But the extent of these protections and changes depend on the frequency, intensity, and duration of breastfeeding.

Societal

The practices of a mother-infant dyad establish widespread dietary patterns. If the present population growth trend continues, over nine billion people will inhabit the earth by the mid-21st century. If the projections hold, two billion additional inhabitants are expected in the Global South between now and 2050. Such a large global population dynamic puts a tremendous pressure on our natural resource base, food supply, and energy requirements. Searchinger and colleagues (2019) predict that the world faces a 69% gap between crop calories produced in 2006 and those likely required in 2050, with sub-Saharan Africa being the most vulnerable region. Also, industry production and waste, transportation systems, and agricultural yields all impose stress on the environment and contribute to air, water, and land pollution on a global scale. A possible contributor to root causes of this growth trend is decreasing breastfeeding rates, as breastfeeding delays onset of ovulation cycles and decreases fertility. According to Becker and colleagues (2003), not breastfeeding increases fertility rates and could boost the number of expected infants born by as much as 50% in countries with very high breastfeeding rates.

Breastfeeding rates differ from low- to middle- and high-income countries. With few exceptions, breastfeeding duration is shorter in high-income countries (Victora et al., 2016). Individually, breastfeeding initiation and duration depend on a plethora of factors including region, culture, religion, maternal knowledge, attitude, social support, and perceptions of breastfeeding. Socio-cultural factors outside and within the home may also influence these choices. As a key component of food security, promotion of breastfeeding can be considered in light of the multi-factorial dynamics (Davidove & Dorsey, 2019). Breastfeeding could be promoted through helpful programming and social supports like a network of families that enable nursing mothers to reach out for help when and where needed (CDC, n.d.b.; Smith, 2019).

Cultural

As noted, breastfeeding can aid in supporting more sustainable population growth, strengthen mother-infant dyad, enhance food security, and foster positive health outcomes (Davidove & Dorsey, 2019). Not being able to breastfeed can mean the loss of these desirable qualities. Although there are many good reasons to breastfeed, there are notable reasons why many women choose not to breastfeed. Women may not be able to breastfeed due to physical pains, child capacity, lactation challenges, and an array of other challenges. Ability to breastfeed is not uniform and can be influenced by a woman's age, marital status, employment status, social network, and social influences. For example, many mothers need to return to work after maternity leave. When returning to work, many mothers encounter fatigue, difficulty expressing their milk, and worrying about having a long-term strategy for keeping an available milk supply. Not breastfeeding allows the mother to assess the amount of food needed by the baby, choose who feeds the baby, and control feeding times. Formula feeding allows working mothers and other busy mothers to have more freedom and helps to address constraints such as inflexible working hours, setting working hours, and job insecurity (Weimer, 2001). If a woman who wants to breastfeed faces discrimination, harassment, or ridicule at work, in public and by family members, these realities can create barriers to exclusive breastfeeding. Social support is necessary at all stages of motherhood but even more so in the early phase of feeding infants. In societies where a woman's breast is considered erotic and if having to breastfeed among strangers, particularly male observers, there may be a sexual connotation even when there is a baby nursing at the nipple. And, in some places, breast exposure is prohibited by local obscenity laws. Understandably, many new mothers may become discouraged to breastfeed privately and/or publicly.

Social support is an important factor in the decision to breastfeed but is manifested differently according to ethnicity, income, and social status. According to Guttman and Zimmerman (2000), friend support was important among African Americans in the decision to breastfeed, among Hispanics their mothers' support was a determinate, and with Anglo-Americans it was the male partner that had the most influence to breastfeed. With lower participation in breastfeeding by African American and Hispanic women, this could be seen as a function of social injustice and health inequity in the United States (Guttman & Zimmerman, 2000). The authors acknowledge the value of adding to this research and acknowledging that others are researching to better understand cultural norms and social influences that may be operating. Mothers who are married and who have more than one child are more likely to exclusively breastfeed (Kornides & Kitsantas, 2013). These researchers also note that one's health condition can also influence breastfeeding, as women who are obese or smoke are less likely to breastfeed exclusively (Kornides & Kitsantas, 2013). Often the mothers who quit breastfeeding within two months either felt that they were not producing enough milk, their infant preferred the bottle to their breastmilk, or they believed that their personal lifestyles were being constrained (Moore, 2018).

Prenatal health care and postnatal education by hospital staff, clinical nutritionists, medical doctors, nurses, dietitians, and other health professions can be strong influencers on a woman's decision to breastfeed or not. Breastfeeding interventions could target not just the new mothers but also the mother's immediate family members so that their knowledge, opinions, and support could influence her to initiate breastfeeding and continue the practice for the better part of her child's infancy. Moreover, a woman who breastfeeds successfully is capable of influencing and supporting her children in the decision to breastfeed. In this way, breastfeeding becomes an enduring family tradition and part of a sustainable lifestyle ensuring food sustainability and food security for future generations.

Economics

Breastfeeding provides the mother-infant dyad a large degree of self-sufficiency, a goal of sustainable living, by decreasing reliance on processed food, medicine, and clean drinking water. Conversely, breastfeeding also leads to an economic gender equality issue, as women pay a price when they must take leave of absence from work, make sacrifices at work, and choose different career paths to enable breastfeeding successfully (Davidove & Dorsey, 2019). Globally, not breastfeeding is a costly diversion of funds. According to Walters and colleagues (2016), the cost of not breastfeeding in Southeast Asia is an additional contributor to economic losses by a noticeable decrease in cognitive abilities. It was estimated that Thailand had a significant 0.5% loss of Gross National Income in 2014 due to suboptimal breastfeeding (Walters et al., 2016). Rollins et al. (2016) state that "cognitive losses associated with not being breastfed to 6 months are estimated globally as 0.49% of world Gross National Income". Yet, "interventions aimed at promoting breast-feeding rates are among the most effective possible health policies available, with an estimated return of US \$35 per dollar invested" (Quesada et al., 2020, para. 1).

In many communities of the Global South, women with lower incomes may have less access to healthy nutritious foods and fewer resources to manage their household budgets. Having fixed expenses such as housing, electricity, water, and transportation makes it difficult to control the household budget if these costs increase. For women in lower economic brackets, dependence on infant formula may divert food dollars away from fresh, wholesome foods and increase the risks of food insecurity, which threatens their health status. In Scandinavia and other countries of the Global North, with longer, publicly supported maternity/parental leave, exclusive breastfeeding and breastfeeding duration increased through insti-tutionalised financial support of these families (Linnecar et al., 2014). More efforts are needed to find cost-effective approaches to increase breastfeeding rates while lowering the financial burden on health care systems and society.

Environmental

Failing to improve breastfeeding rates has many implications for environmental protection. Compared to the social and economic impacts of not breastfeed-ing, environmental harm is the most difficult to mitigate and therefore the most pressing concern. Carbon and water footprints are insightful tools for quantifying the toll of infant formula production on the environment. Breast milk is optimal within the spectrum of sustainable food sources with a near-zero ecological and carbon footprint, as only 500–640 additional calories per day are needed for lacta-tion. Consequently, arguments can be made of the reduced environmental impact from breastfeeding as more inclusive analyses shine light on the water, energy, and fuel embedded in food systems, including in processed foods, like baby formula. Breastfeeding is essentially a provisioning ecological service. It may be difficult to effectively evaluate and monetise; an alternate method to demonstrate the value of breastfeeding is an estimation of the environmental cost or negative externalities of formula feeding.

The Codex Alimentarius (CA) is one way to measure the energy and water needed to synthesise infant formula. CA is a complete registry of ingredients for standard food products. In regards to cow's milk infant formula, the average global green-house gas (GHG) emissions from milk production, processing, and transport are estimated to be 2.4 kilograms (kg) of carbon dioxide-equivalent (CO_2-eq.) per kilogram of fat and protein corrected milk (FPCM) (Burlingame & Dernini, 2012). Linnecar and colleagues (2014) report that infant formula production was derived from 553 million tonnes of milk generating 1,328 million tonnes of CO_2-eq. of GHG. The production of one kilo of liquid cow's milk requires 940 litres of water. Producing 1 kg of milk powder uses 4,700 litres of water and emits 21.8 kg CO_2-eq. of GHG (p. 13). Using powdered infant formula is an ecologically counter-productive process of taking liquid cows' milk, turning it into a powder, and then rehydrating it with water back into liquid (Davidove & Dorsey, 2019). The enormous, intensive water consumption, extensive use of materials for packaging, high-demand use of energy resources in manufacturing, GHG emissions from food miles' transportation, and widespread generation of household waste make infant formula production a major environmental concern (Becker et. al, 2003).

The environmental cost of not breastfeeding is more than just the carbon and water footprint of cow's milk: It includes the deforestation of land and loss of biodiversity required for growing cattle, livestock feed, and oil palms. Waste reduction considers formula feeding as a potential source of food wastage in the baby food industry due to the use of unrecyclable plastics, excessive packaging, and extensive water use. In addition, land use, nitrogen and phosphorus cycling, freshwater management, and carbon emission need to be taken into account as aspects of planetary boundaries impacts (Rockström et al., 2009). To achieve major improvements in global food production, breastfeeding should be considered part of the food production sector.

Conclusion

The purpose of this chapter is to demonstrate that breastfeeding can be understood as a sustainable diet for the first six months of life and then a major component into early childhood as it is naturally provided by the lactating mother. Biological, social, economic, and environmental benefits of breastfeeding provide sufficient evidence of the importance to promote this behaviour to lactating mothers and their social support systems. There is a planetary urgency to mitigate the negative impacts of industrialised formula feeding and to promote breastfeeding widely. An unchecked formula industry will divert food production, energy, and freshwater to cows for infant formula production; these drivers can add to the food gap, population growth, and climate change. Promoting the adoption of a sustainable diet that includes breastfeeding is a step towards achieving better nutrition and health, as well as pursuing food justice by protecting the natural ecosystems that support equitable food production and distribution. Health promotion programs may consider three human behavioural components—individual, social, and material— within strategies to bolster more breastfeeding along with sustainable household eating patterns. The Behavioural Insights Team (2020) suggests this multi-prong approach. The *individual* component consists of the "inner" psychological drivers of consumer behaviour, both conscious and unconscious. These drivers include personal tastes and preferences, values, and beliefs, as well as ingrained habits, emotions, heuristics (mental shortcuts), and cognitive biases. The *social* component considers external drivers on behaviour, including public policies, like placing public signs for "breastfeeding spaces," cultural norms, and narratives, peer pressure, and social identity. And the *material* component is the wider physical and economic context. These drivers include the physical environment and the manner in which options are made available and presented to us. Pricing, mass media, advertising, and technology also are factors to be incorporated into shaping public education campaigns and educating on the power of food choices. A healthy combination of psycho-social approaches may aid in supporting more breastfeeding. Respectful promotion strategies will pay attention to cultural diversity, economic barriers, and the active involvement of nursing mothers in developing sensitive and contextualised approaches to breastfeeding. Breastfeeding can be understood as contributing to global food security, as reducing greenhouse gas emissions and biodiversity loss, and as an integral component of sustainable diets and as a contributor to good health and longevity.

Note

1 Justus von Liebig, a chemist, "developed, patented, and marketed an infant food" (Stevens et al., 2009) in 1865. He first developed a liquid formula consisting of cows milk, wheat, and potassium bicarbonate. Later he processed the formula into a powder for preservation (Radbill, 1981; Stevens et al., 2009).

References

Allen, J., & Hector, D. (2005). Benefits of breastfeeding. *New South Wales Public Health Bulletin, 16*(3–4), 42–46.

Becker, S., Rutstein, S., & Labbok, M. H. (2003). Estimation of births averted due to breast-feeding and increases in levels of contraception needed to substitute for breast-feeding. *Journal of Biosocial Science, 35*(4), 559–574.

Behan, E. (2019, June). *Losing weight while breastfeeding.* Academy of Nutrition and Dietetics. https://www.eatright.org/health/pregnancy/breast-feeding/losing-weight-while-breastfeeding

Binns, C., Lee, M., & Low, W.Y. (2016). The long-term public health benefits of breastfeeding. *Asia-Pacific Journal of Public Health, 28*(1), 7–14.

Brahm, P., & Vades, V. (2017). Benefits of breastfeeding and risks associated with not breast-feeding. *Revista Chilena De Pediatria, 88*(1), 7–14.

Burlingame, B., & Dernini, S. (2012). *Sustainable diets and biodiversity directions and solutions for policy, research and action.* FAO & Bioversity International. https://hdl.handle.net/10568/104606

CDC (n.d.a). *Frequently asked questions (FAQs) | breastfeeding.* U.S. Department of Health and Human Services. Retrieved April, 4, from https://www.cdc.gov/breastfeeding/faq/index.htm

CDC (n.d.b). *The surgeon general's call to action to support breastfeeding.* U.S. Department of Health and Human Services. Retrieved April, 4, from https://www.cdc.gov/breastfeeding/pdf/actionguides/Communities_in_Action.pdf

Couto, G. R., Dias, V., & de Jesus Oliveira, I. (2020). Benefits of exclusive breastfeeding: An integrative review. *Nursing Practice Today, 7*(4), 245–254.

Davidove, M. E., & Dorsey, J. W. (2019). Breastfeeding: A Cornerstone of healthy sustainable diets. *Sustainability, 11*(18), 4958.

EAT-Lancet Commission on Food, Planet, Health (2019). *Healthy diets from sustainable food systems: Food, Planet, Health.* https://eatforum.org/content/uploads/2019/07/EAT-Lancet_Commission_Summary_Report.pdf

Gartner, L. M., Morton, J., Lawrence, R. A., Naylor, A. J.., O'Hare, F., Schanler, R. J., Eidelman, A. I., & American Academy of Pediatrics Section on Breastfeeding (2005). Breastfeeding and the use of human milk. *Pediatrics, 115*(2), 496–506.

Guttman, N., & Zimmerman, D. R. (2000). Low-income mothers' views on breastfeeding. *Social Science & Medicine, 50*(10), 1457–1473.

Hanson L. A. (2000). The mother-offspring dyad and the immune system. *Acta Paediatrica, 89*(3), 252–258.

Kornides, M., & Kitsantas, P. (2013). Evaluation of breastfeeding promotion, support, and knowledge of benefits on breastfeeding outcomes. *Journal of Child Health Care: For Professionals Working with Children in the Hospital and Community, 17*(3), 264–273.

Linnecar, A., Gupta, A., Dadhich, J. P., & Bidla, N. (2014). *Formula for disaster: Weighing the impact of formula feeding vs. breastfeeding on environment.* BPNI & IBFAN. http://www.gifa .org/publication/formula-for-disaster-weighing-the   -impact-of-formula-feeding-vs-breastfeeding-on-environment/

Mason, P., & Lang, T. (2017). *Sustainable diets: How ecological nutrition can transform consumption and the food system.* Routledge.

Moore, M. L. (2018). Breastfeeding benefits support-research. *Scientific Journal of Gynecology and Obstetrics, 1*(1), 2.

Moss, B. G., & Yeaton, W. H. (2014). Early childhood healthy and obese weight status: Potentially protective benefits of breastfeeding and delaying solid foods. *Maternal and Child Health Journal, 18* (5), 1224–1232.

Newburg, D., & Walker. W. (2007). Protection of the neonate by the innate immune system of developing gut and of human milk. *Pediatric Research, 61*, 2–8.

Ogunba, B., Idemudia, S., & Omikunle, I. (2019). Breastfeeding: The environmentally friendly and ideal method of infant feeding. *Nigerian Journal of Environment and Health, 2*, 30–36.

Pugh, L. C., Milligan, R. A., Frick, K. D., Spatz, D., & Bronner, Y. (2002). Breastfeeding duration, costs, and benefits of a support program for low-income breastfeeding women. *Birth, 29*(2), 95–100.

Quesada, J. A., Méndez I., & Martín-Gil, R. (2020). The economic benefits of increasing breastfeeding rates in Spain. *International Breastfeeding Journal, 15*(1), 1–7.

Radbill, S. (1981). Infant feeding through the ages. *Clinical Pediatrics, 20*(10), 613–621.

Rockström, J., Steffen, W., Noone, K., Persson, Å., Chapin, F. S., Lambin, E., Lenton, T. M., Scheffer, M., Folke, C., Schellnhuber, H. J., Nykvist, B., de Wit, C. A., Hughes, T., van der Leeuw, S., Rodhe, H., Sörlin, S., Snyder, P. K., Costanza, R., Svedin, U., … Foley, J. (2009). Planetary boundaries: Exploring the safe operating space for humanity. *Ecology and Society, 14*(2), 1–33.

Rollins, N. C., Bhandari, N., Hajeebhoy, N., Horton, S., Lutter, C. K., Martines, J. C., Piwoz, E. G., Richter, L. M., Victora, C. G., & Lancet Breastfeeding Series Group. (2016). Why invest, and what it will take to improve breastfeeding practices? *The Lancet, 387*(10017), 491–504.

Searchinger, T., Waite, R., Hanson, C., & Ranganathan, J. (2019). *Creating a sustainable food future: A menu of solutions to feed nearly 10 billion people by 2050* (Final Report). World Resources Institute. https://research.wri.org/sites/default/files/2019-07/creating-sustainable-food-future_2_5.pdf

Smith, J. P. (2019). A commentary on the carbon footprint of milk formula: harms to planetary health and policy implications. *International Breastfeed Journal, 14*, Article 49.

Stevens, E. E., Patrick, T. E., & Pickler, R. (2009). A history of infant feeding. *The Journal of Perinatal Education, 18*(2), 32–39.

Stuebe, A. (2009). The risks of not breastfeeding for mothers and infants. *Reviews in Obstetrics and Gynecology, 2*(4), 222–231.

The Behavioural Insights Team. (2020). *A menu for change: Using behavioural science to promote sustainable diets around the world.* http://www.bi.team/publications/a-menu-for-change/

Victora, C. G., Bahl, R., Barros, A. J., Franca, G. V., Horton, S., Krasevec, J., Murch, S., Sanhar, M. J., Walker, N., & Rollins, N.C., & Lancet Breastfeeding Series Group. (2016). Breastfeeding in the 21st century: epidemiology, mechanism, and lifelong effect. *The Lancet, 387*(10017), 475–490.

Walters, D., Horton, S., Siregar, A.Y., Pitriyan, P., Hajeebhoy N., Mathisen, R., Phan, L.T., & Rudert, C. (2016). The cost of not breastfeeding in Southeast Asia. *Health Policy and Planning, 31*(8), 1107–1116.

Weimer, J. P. (2001). *The economic benefits of breastfeeding: A review and analysis.* Food and Rural Economics Division, Economic Research Service, United States Department of Agriculture. https://www.ers.usda.gov/publications/pub-details/?pubid=46472

WHO (World Health Organisation) (n.d.). *Breastfeeding.* Retrieved April, 2, from https://www.who.int/health-topics/breastfeeding#tab=tab_1

World Obesity (n.d.). *Prevalence of obesity.* Retrieved April, 2, from https://www.worldobesity.org/about/about-obesity/prevalence-of-obesity

WHO (2020, June 23). Breastfeeding and COVID-19 [Scientific brief]. https://www.who.int/news-room/commentaries/detail/breastfeeding-and-covid-19

18

TOWARDS MORE COMPREHENSIVE ANALYSES OF THE NUTRITION TRANSITION AMONG ADOLESCENTS IN THE RURAL SOUTH

An empirical contribution

Fiorella Picchioni, Giacomo Zanello, Mondira Bhattacharya, Nithya Gowdru, and Chittur Srinivasan

Introduction

Contextual and historically embedded rural transformation processes, that include extensive societal and economic changes, can drive rapid dietary shifts and alterations in physical activity patterns (Kelly, 2011). Complex interactions between diets and physical activity are shaping increases of diet-related non-communicable diseases (DR-NCDs) (Popkin, 2006). Therefore, analytical frameworks and practices to achieve sustainable diets in transforming agrarian contexts will benefit from incorporating the complex linkages between what people eat and how they work, travel, and spend their leisure time in research, policy, and interventions.

These considerations are particularly important for interventions targeting adolescents, who are, especially in the Global South, at the forefront of dramatic shifts in diets and lifestyles caused by rural transformation (Aurino et al., 2017; Guthold et al., 2020). Since adolescence is a critical window of opportunity to recover from previous nutritional impairments and to enhance

DOI: 10.4324/9781003174417-21

the health of the next generation (Prentice et al., 2013), identifying what is contributing to dietary and physical activity changes is key to containing the growing rates in the prevalence of DR-NCDs, food allergies, and other chronic conditions (Bixby et al., 2019). For example, with food allergies growing in the Global South, often arising in childhood and persisting (or worsening) later in life, holistic research and policies to support human and environmental health through a food systems approach are critical. The emergence of food allergies has been associated with several possible factors, with some being environmental exposure, genetic variants, and nutrient intake, including nutrition transitions (Cai, 2014; Mazzocchi et al., 2017). While this chapter does not directly focus on food allergies, we recognise the urgent need for more research in the area of sustainable diets and allergens. However, as the remainder of this chapter and book illustrate, better results can be achieved by embedding research of specific issues within the multiple facets that shape human and environmental health.

This chapter describes an empirical approach that aims to contribute to the nascent research on the interplay between changing diets and physical activity among adolescents in the context of rural transformation in rural India, with implications for other places in the Global South. Through our contribution we aim to emphasise the importance of integrating considerations on *sustainable lifestyles* when designing interventions and policies to achieve sustainable diets.

We focus this chapter on rural India, where overweight and obesity rates have increased dramatically in the last decade (Aiyar et al., 2021; Barker et al., 2020). Our case study in rural Telangana (southern India) illustrates how to combine data derived from new technologies (i.e., wearable activity trackers), quantitative surveys, and qualitative interviews. Together with reflecting on ethical and practical considerations when engaging with adolescents in research, our chapter aims to contribute to: 1) The research on the health challenges adolescents face in the context of rural transformation and 2) raise questions that can guide the design of interventions and policies that, in tandem, aim to target diets and physical activity levels.

Blind-spots in sustainable nutrition research with adolescents: A contribution from mixed-methods

Conventionally, research on adolescent nutrition and health in the Global South has focused on the analysis of anthropometric indices and dietary intake at the individual level (Bassett et al., 2008; Stevano et al., 2020). Our *mixed-method sequential explanatory* [1] study design, as shown in Figure 18.1, addresses two broad gaps. Firstly, there is a lack of systematic data collection that links energy expenditure and physical activity patterns among teenagers in the agrarian South (Swaminathan et al., 2011). In turn, this limits the understanding of

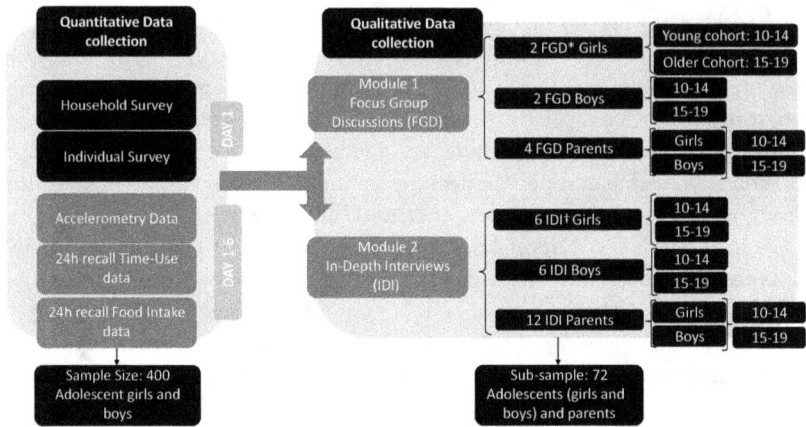

Figure 18.1 Mixed-method study design. Source: Work by authors. Notes: *Each focus group discussion included six members, one facilitator, and one note taker. †Three in-depth interviews were conducted with the younger cohort of adolescent girls and three with the older cohort. The same sex/age structure was replicated for adolescent boys and respondent's parents.

health risks that emerge from the interplay between variations in physical activity and dietary intake patterns among this important age group. A more comprehensive picture of the dietary and physical activity challenges that adolescents face in transitioning (rural) societies can be gained if dietary assessments go hand in hand with evaluations of physical activity patterns. To this end, 24-hour recall food intake data from 400 adolescent boys and girls in rural Telangana was combined with energy expenditure data from accelerometer devices. Data were collected across six consecutive days and complemented with 24-hour recall time-use surveys during the same period. A household and individual survey administered before the above-mentioned instruments provided contextual information.

The second gap we aim to address relates to the missing links between individual-level dietary practices and physical activity, on the one hand, and household and community dynamics and enforced norms, on the other (Bassett et al., 2008). There is a growing recognition that researchers and public health practitioners should move beyond individual-level approaches to design and test interventions (Hunter et al., 2019). Importantly, adolescent boys and girls should be positioned within their households and communities where gender, age, and carer-child relations of power shape intra-household food consumption and work allocation. To capture these dynamics and complement quantitative findings, we deployed a series of focus group discussions and in-depth interviews with a sub-sample of adolescents and their parents/carers (Table 18.1).

Table 18.1 Presents the mixed-methods and survey instruments used together with some examples of insights and types of outputs they can produce

Research gaps	Methods/Instruments' characteristics	Examples of insights and type of outputs
Lack of systematic and integrated data on physical activity and dietary behaviours in the agrarian South	**Accelerometer devices** detect and record both speed and direction of movements. Algorithms are used to translate data into aggregate measures of activity intensity (light, moderate, vigorous and very vigorous activity) and energy expenditure in kcal. This study employed the research-graded tri-axial ActiGraph GT3X+ accelerometers. Its validity and reliability have been extensively assessed (Santos-Lozano et al., 2013; Sasaki et al., 2011) and used in multiple studies involving free-living humans (Keino et al., 2014; Pawlowski et al., 2016) including adolescents (Robusto & Trost, 2012). Data was collected at 30 Hz/second and movements translated to energy expenditure using validated age-based cut-points (Troiano et al. 2008; Trost et al., 2011). ActiGraph GT3X produced several indicators, described in Box 1. More information on fieldwork protocol and data management can be found in Zanello et al. (2018).	• Total energy expenditure★ (daily) • % of activity energy expenditure★ across different types of activities • physical activity level★ (daily)
	24 h food intake surveys provide retrospective assessments of food and beverage consumption during the preceding 24 hours (Gibson & Ferguson, 2008). Administered by trained enumerators, questions are asked in chronological order of consumption over different context specific meals. The more days/periods of recall, the better capture of between-day and seasonal variability.	• Daily Kcal intake • % of calories by food source/group • Micronutrient adequacy

(Continued)

Table 18.1 (Continued)

Research gaps	Methods/Instruments' characteristics	Examples of insights and type of outputs
	24 h recall time-use surveys capture the use of time in relation to different categories of activities (e.g., paid and unpaid work, market and non-market activities, education, leisure and personal time) (Antonopoulos, 2008). Time-use interviews have previously been used with adolescents (Fisher et al., 2015) and provide a daily snapshot of the activities conducted during the day before the interview. We used a 30-minute activity interval and a piloted context-specific pre-defined activities list and allowed respondents to include missing options.	% of time spent on different types of activities (classified in macro and micro-groups)
Missing links between individual-level behaviours (dietary practices, physical activity) and broader household and community norms/values.	**FGDs** are a particular form of group interview intended to exploit group dynamics which is conducive to gather information of socially accepted knowledge and beliefs (Kitzinger, 1994; Morgan, 1996). The selection of participants was carefully considered to mitigate power imbalances; we interviewed groups of adolescents and their parents/carers divided by gender and age and avoided mixing groups. **IDIs** deepens the generation of understanding (Guest et al., 2017). The rationale to use IDIs and complement surveys and FGDs include: 1) To explore conforming patterns and characteristics, and 2) to investigate peculiarity and differences to document between and within-group heterogeneity. In addition to gender and age group, sampling criteria included household socio-economic status (i.e., wealth index) and school attendance, and among older adolescents, marital and parental status.	• Picture of how family context influences teens' diets and physical activities • Identification of gender and age norms that shape behaviours • Map parent-teen interactions that define dietary choices/ physical activity opportunities

Source: Work by authors.

Notes: *Full definition of the terminology is provided in Box 18.1.

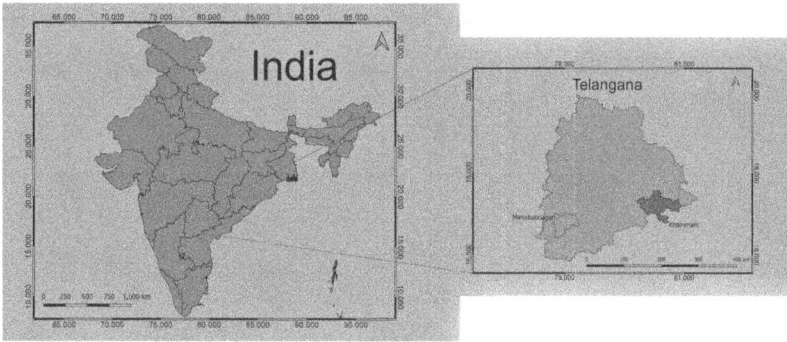

Figure 18.2 Study sites: The state of Telangana and the districts of Mahabubnagar and Khammam. Source: Work by authors.

Box 18.1   Presents indicators produced using Actigraph accelerometer devices

Activity energy expenditure (AEE): Calories used to perform different forms of physical activities. AEE is directly generated by accelerometer devices and is a function of the intensity of activities and of body weight. AEE represents approximately 30% of TEE.

Basal metabolic rate (BMR): Amount of calories required to support basal physical functions. BMR is a function of age, sex, body size, and body composition, and represents approximately 60% of TEE. BMR estimation requires the use of a formula like the Harris-Benedict (Harris & Benedict, 1918).

Total energy expenditure (TEE): Sum of BMR, AEE, and of the thermic effect of food (TEF—or the energy needed to digest and metabolise food).

Physical activity level (PAL): Ratio of TEE to BMR and a measurement of the intensity of physical activity corrected for age, sex, and body size. This feature makes PAL a suitable index to compare the intensity of work across populations.

Source: Work by authors

India case study: Settings and data

The study was conducted between 2019–2020 in Khammam and Mahbubnagar districts, located in the central and south agro-climatic zone of Telangana (Figure 18.2).

Quantitative data

Sample description

The quantitative sample included 400 adolescents (11–19 years old) from 347 households. Table 18.2 reports household and individual characteristics. Physical

Table 18.2 Descriptive statistics of households and individual respondents by age group and sex

	Early Adolescent (10–14 yrs)			Late Adolescent (15–19 yrs)		
	Boys	Girls	t-test	Boys	Girls	t-test
Household characteristics						
Head of the household age (years)	44.7	43.4	1.3	45.1	46.9	–1.8
Household size (number of HH)	4.7	4.5	0.2	4.2	4.4	–0.1
Wealth Index (based on assets ownership)	0.2	–0.1	0.3	0.2	–0.1	0.3
Individual characteristics						
Height (in cm)	145.9	146.1	–0.2	164.4	153.8	10.6★★★
Weight (in kg)	33.5	34.8	–1.3	49.2	43.8	5.4★★★
BMI Status (WHO-1SD)						
Underweight	0.71	0.57	0.14★	0.64	0.52	0.12
Normal	0.24	0.40	–0.16★	0.31	0.46	–0.15★
Overweight	0.04	0.03	0.01	0.05	0.02	0.03
Daily energy consumption and intake						
Basal metabolic rate (BMR)	1180.4	1095.9	84.5★★★	1493.1	1198.5	294.6★★★
Activity energy expenditure (AEE)	475.0	392.8	82.2★★★	430.9	346.3	84.5★★★
Total energy expenditure (TEE)	1655.4	1488.7	166.6★★★	1924.0	1544.9	379.1★★★
Physical activity level (PAL)	1.41	1.36	0.05★★★	1.29	1.29	0.00
Energy intake (kcal)	1596.6	1462.6	133.9★★★	1705.8	1486.2	219.6★★★
Percent of energy intake from processed foods	0.4	0.4	–0.0★	0.4	0.4	–0.01
Compliance of accelerometer wear						
Avg. number of days per participant[†]	4.7	4.8	–0.1★	4.6	4.7	–0.1
Total number of days[‡]	177.9	196.4	–18.5★★★	228.9	227.2	1.7
Observations (participant/ day)	421	461		472	454	

Source: Work by authors.

Notes:

[†] Average number of days with valid data (non-wearing time less than 3 h throughout the day) out of the five days of the survey.

[‡] Total number of distinct day-level observations (individuals × valid days surveyed). Asterisks show level of significance ★★★= significant at 0.1% level, ★★=significant at 1% level, and ★=significant at 5% level.

activity parameters are significantly higher for boys compared to girls. In terms of energy intake, boys exceed girls by 9% and 15% for younger and older cohorts, respectively. Average energy intake is lower than energy expenditure across all groups. The share of energy intake from processed foods is approximately 4% across the sample.[2]

Combining food intake and physical activity data: Two examples

Figure 18.3 plots the proportion of calories from processed foods to total energy expenditure (TEE), by sex and age groups. For ratios > (<) 1, energy intake from processed foods exceeds (is lower than) TEE. As rural transformation is taking place in many rural areas, diets and patterns of physical activities change. Convenient but unhealthy processed foods can become widely available, and livelihoods that were physically demanding shift to more sedentary lifestyles.

The majority in the sample has a processed food to TEE ratio <1. Across both age groups, girls have a relatively smaller proportion of energy intake from processed foods to TEE in comparison with boys. The relatively small ratio of processed foods derived calories to TEE ratio that is >1 (approximately 3% of the observations) is not surprising given our context of analysis. However, we should also question whether any amount of processed foods consumption among adolescents is acceptable at all. Are some types of processed foods qualitatively more or less harmful than others? Do higher levels of processed food intakes influence the intensity of activity patterns (i.e., PAL)? Are these patterns somehow linked with new findings that illustrate how increased ultra-processed food consumption is associated with increased appetite and development of food allergies (Hall et al., 2019)?

Figure 18.4 reports the contribution (i.e., coefficients) of time spent across four main activities (i.e., education, economic, domestic, leisure) on PAL, by sex and age groups. The data show the marginal effects on PAL of a one-unit increase in the explanatory variables. For example, one-unit change of economic and domestic activities tends to have the largest marginal effects on PAL in both

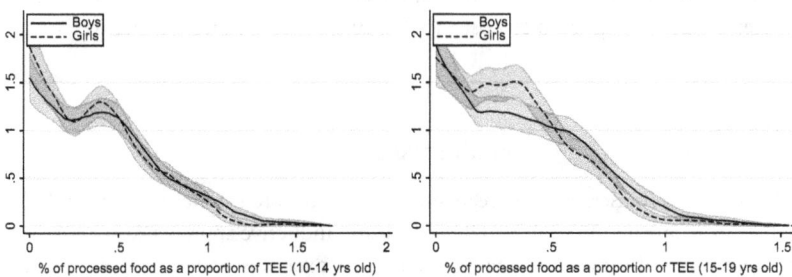

Figure 18.3 Calories from processed foods as a proportion of total energy expenditure. Source: Work by authors. Notes: 95% confidence intervals computed over 1,000 bootstrapped repetitions. Two-sample Kolmogorov-Smirnov test for equality of distribution: Younger cohort (p=0.198) and older cohort (p=0.014).

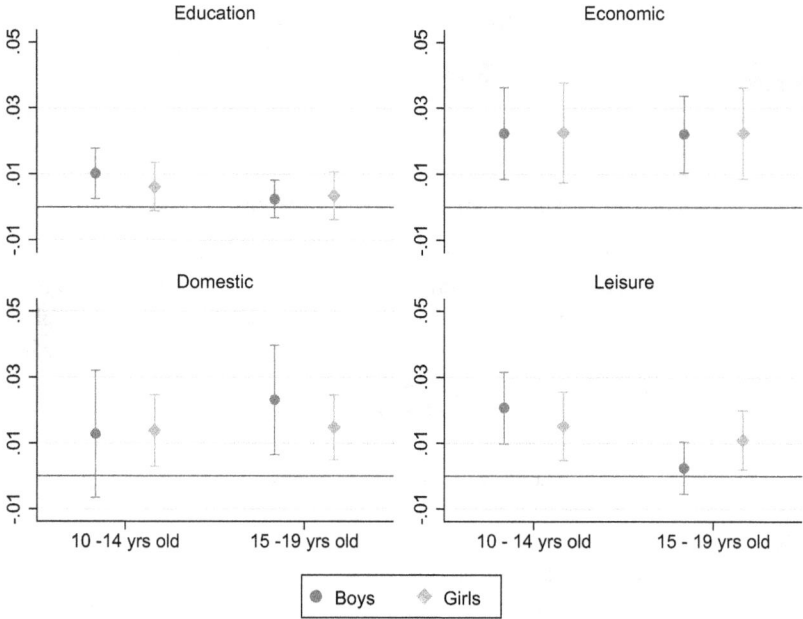

Figure 18.4 Contribution of activities on PAL, by sex and age groups. Source: Work by authors. Note: Graph plots coefficients from a regression model of PAL on a set of activities (education, economic, domestic, and leisure). Sleeping and resting as baseline. The model controls for day of the week and missing hours. Standard errors clustered at individual level. Full regressions of determinants of physical activity level (PAL) are presented in Table 18.A at the end of the chapter.

age cohorts of boys and girls. In the case of leisure and education, the marginal effect on PAL is higher in the younger cohort than the older cohorts. A unit-increase of leisure activities of boys has a larger marginal effect on PAL compared to girls. The data raises questions of the duration and nature of economic and domestic activities, how they change in relation to age and gender, and whether they represent an opportunity cost in relation to exercise and "formative" occupations.

Qualitative Themes[3]

FGDs and IDIs transcripts were read and analysed by a researcher familiar with the context of Telangana. A combined deductive and inductive approach was employed for theme identification and analysis. Once all data was coded, two researchers revisited and revised categories, which were in turn discussed and developed into main themes. Below we report a sample of themes, with some direct quotes from participants. Coding was conducted on Nvivo12. These themes enhance the evidence provided by quantitative data illustrated above.

Nutrition knowledge

Younger and older adolescents demonstrate good nutrition knowledge and a detailed understanding of how nutrition influences health outcomes. This information is mainly obtained from their families, school, media, and friends. Children listed a comprehensive range of vitamin-rich foods and distinguished them from foods that provide energy and strength. This knowledge was reinforced by interviews conducted with parents. "At home, mother, father and elders say if you drink milk, you will have more strength, and if you eat eggs, you will get energy" (young adolescent girl FGD). "At school, our physics madam (teacher), through experiments she showed us as well—in what foods, what nutrition is there" (young adolescent girl FGD).

While food consumption at home tends to be more supervised, eating out during recreational activities may be unmonitored. The options were regarded generally as "less healthy" and sometimes unsafe but more enticing. As the ability to gather outdoors is more accepted among boys, this group tended to report higher levels of processed foods consumption than girls. "Outside food is good due to its colour and taste" (young adolescent boy, FGD). "Do not eat outside food—mother says. Asks me to eat good foods—potato, ladies finger, bitter gourd, all these, she asks me to eat" (young adolescent boy, FGD)

Gendered food allocation

Virtually all parents recognised that girls and boys should have the same diets. However, in practice diets were different between them and across age groups: Leafy vegetables were prioritised for girls while animal protein was generally allocated to boys. Some of the main reasons reported included lesser body absorption of animal protein by girls, while boys needed protein and more calorific foods because they were more active. Reproductive health considerations for girls were also mentioned. These beliefs were reiterated by adolescents themselves.

> If it's girls, for them chicken, mutton, the energy to assimilate it is not sufficient. If it's boys, as they can assimilate it, that is why they mostly eat non-veg, and so girls generally eat pulses, vegetables, leafy vegetables, eggs, to remain active.
>
> *(FGD, parent of young adolescent boy)*

As [girls] grow older, non-veg [options] should be reduced, and as the age comes, even if you eat it, [they] will not digest (IDI, parent of older adolescent girl).[4]

Changing food environments

During the IDIs, parents were asked to reflect on diets during their adolescence and compare them to that of their children. Adults talked about food shortage, both in terms of variety and quantity, and often cited that skipping meals was common during their childhood. Most of the focus was given on the availability and affordability of increased varieties of vegetables, fruits, animal source proteins, and sweet options. However, interviews also highlighted a sense that some types of foods were no longer available.

> Before, one meal would be there, another would not be there, it used to be like that—and [we went] to school like that only. Now there is no such deficit, however hard it might be, for food there is not a struggle.
>
> *(IDI, parent of older adolescent boy)*

"Now compared [to before], there is a lot more food. We're eating much more, sweetly, nicely" (IDIs, parent of older adolescent boy).

Opportunities and challenges to physical activity

Physical activity and mobility opportunities were heavily influenced by gender. Boys were more likely to participate in outdoor events and economic work. Girls' physical activity, on the other hand, was primarily domestic and agricultural work on family land. "My experience is that, as girls keep doing the cooking work—while we boys have food and go and roam around here and there. Boys have more freedom. But for girls, they will not have that freedom" (FGD, older adolescent boy).

Parents' fears for their daughters' safety and norms around good behaviour were among factors for restricting their outdoor physical activity opportunities. Parents would also articulate that lack of courage of girls would justify limited mobility. These narratives could often collide with adolescent's own voices, and girls would share a sense of uneven treatment compared to boys.

> In our villages, if it is boys, they go around here and there. Now if the girls are going and coming, then they say "see how she is roaming around." Like that we shouldn't get a bad name, we should tell our girls to be disciplined, and keep them at home.
>
> *(IDI, parent of older adolescent girl)*

"Now, if it will be the boys, means at whatever time they may go, they say—he is a boy, wherever, whenever, anywhere he can go and come" (IDI, older adolescent girl).

Lessons from conducting research with adolescents

The mixed-method study presented in this chapter is an adaptation of a methodology developed to study the rural livelihoods-nutrition-energy expenditure nexus among farmers in the Global South (Zanello et al., 2017, 2020). Hence, various steps were taken to adapt the approach with adolescents in rural Telangana. Ethical and safeguarding protocols were developed following the guidelines of Santelli et al. (1995) and Brady and Graham (2019) to conduct health and social research with children and young people. Conducting research with adolescent respondents represents a valuable opportunity in agri-health research, as it enables us to incorporate the experiences of a demographic group that was previously over-looked. This section presents few core ethical and practical considerations applied during the preparation and execution of this study.

When conducting research with children and young people *no harm and providing benefits* is the key principle that shapes the study's lifecycle. The principle refers to the balance between the inclusion of children and young people's views and experiences while considering common risks of partaking in research activities (Brady & Graham, 2019). Inherent and unexpected risks should be anticipated, assessed, and mitigated, and distressing topics avoided or carefully planned. As well as benefiting society at large or improving policies, it is worth exploring what benefits participants may enjoy in return for providing time and data for research. Planning enjoyable activities while providing learning opportunities from research outputs should be considered.

Informed consent addresses the questions of ensuring that young respondents fully appreciate the consequences of the research while assessing their ability to provide consent. The following principles should be integral to the design, planning, and acquisition of informed consent (Brady & Graham, 2019; Santelli et al., 1995):

- *Working with gatekeepers* (people with parental responsibilities and/or work with minors) is common when conducting research with the under-age population. Their priority is to protect individuals under their responsibility, and they can have a better grasp of the commitment that the research would entail.
- *Autonomy* refers to the person's ability and rights to make their own decisions. Therefore, even if gatekeepers provide consent, the child's individual autonomy should remain intact and repeatedly checked. It is therefore important to use validated protocols to gain multiple level informed consents. Treating consent as a live and ongoing dynamic is critical. For young respondents, refusing consent may be expressed indirectly and/or non-verbally.
- *Capacity, age, cognitive ability* will determine whether a child or young adult is able to provide valid consent. To this end, technical jargon should be avoided, while information and consent processes should be accessible and age-appropriate.
- *Opt-out options* should be available and repeatedly reminded, even when data collection is terminated within a predefined time frame.

Researchers need to anticipate and explore potential *confidentiality issues* and protect them together with *anonymity and privacy*, for both respondents as well as others impacted transversally by research activities. Confidentiality procedures should be explained at the beginning of each session, especially when running group discussions. Finally, *balancing safeguarding concerns with confidentiality* means that the welfare and safety of young respondents involved in the study override research interests. It is therefore best practice to anticipate and plan for safeguarding policies and providing training on how to use it to all involved in fieldwork.[5]

Conclusions

The cycle of sedentarisation and nutrition transition unfolding in the Global South among adolescents is a complex and multifaceted process. The dominant narrative on rural areas is one that depicts daily and constant physical exertion in "traditional" livelihood activities and consumption of traditional diets. While these considerations are true, the transition of rural lifestyles and diets is ongoing in growing economies.

To help design and implement timely and comprehensive policies that address overweight and obesity among adolescents in rural contexts, conceptualisations and practices on how to achieve sustainable diets will benefit from integrating considerations on sustainable lifestyles. The concept of the sustainable diets in rural areas should be linked with 1) the availability and access to nutritious foods; 2) environmentally sustainable food systems; and 3) physically sustainable agricultural production in the absence of mechanisation (for example, farmers will still require energy-dense foods if agricultural work continues to be physically demanding). Additionally, in the context of rapid societal transformation, the availability of unhealthy diets can be compounded by a decline in physical activity. Debates and interventions on how to render food systems more respectful of consumers and enhancing for ecosystems should go hand in hand with promoting inclusive built environments that encourage outdoors activities and safe access to recreational facilities for girls and boys of all ages.

We trust this chapter contributes to the conversation on extending the use of mixed-methods to holistically assess the quantity and quality of diets and physical activity when designing research and interventions to achieve production and consumption of sustainable diets. Structural transformations intersect with social networks, parental guidance, school-based practices, and historical and socio-economic factors, gender, and age norms. In turn, these factors shape the interplay between food consumption and physical activity behaviours in youths. To understand these connections, context-specific and mixed-methods insights on food intakes and physical activity are central to inform the development of comprehensive interventions to address structural and environmental issues, as well as influences within households, schools, and communities.

Notes

1 Mixed-methods sequential explanatory study design implies collecting and analysing first quantitative and then qualitative data in two consecutive phases within one study.
2 The selection of processed foods was based on Monteiro et al. 2010. The ration includes consumption of both processed and ultra-processed foods (Monteiro et al., 2010).
3 Information about the sample of the qualitative module is provided in Figure 18.1.
4 Wording included in the squared brackets are added by the authors to provide more context.
5 Institutions working with under-age individuals will normally have pre-existing safeguarding procedures. Alternatively, the NSPCC provides guidance on writing adequate safeguarding policies (https://learning.nspcc.org.uk/safeguarding-child-protection/writing-a-safeguarding-policy-statement). This research was reviewed and approved by the Ethics Committee at the University of Reading and by the National Institute of Rural Development (India).

References

Aiyar, A., Rahman, A., & Pingali, P. (2021). India's rural transformation and rising obesity burden. *World Development, 138*, 105258.

Antonopoulos, R. (2008). *The unpaid care work-paid work connection.* Levy Economics Institute, Working Papers Series.

Aurino, E., Fernandes, M., & Penny, M. E. (2017). The nutrition transition and adolescents' diets in low- and middle-income countries: A cross-cohort comparison. *Public Health Nutrition, 20*(1), 72–81.

Barker, M. E., Hardy-Johnson, P., Weller, S., Haileamalak, A., Jarju, L., Jesson, J., Krishnaveni, G.V., Kumaran, K., Leroy, V., Moore, S. E., Norris, S. A., Patil, S., Sahariah, S. A., Ward, K., Yajnik, C. S., & Fall, C. H. D. (2020). How do we improve adolescent diet and physical activity in India and sub-Saharan Africa? Findings from the transforming adolescent lives through nutrition (TALENT) consortium. *Public Health Nutrition, 24*(16), 5309–5317.

Bassett, R., Chapman, G. E., & Beagan, B. L. (2008). Autonomy and control: The co-construction of adolescent food choice. *Appetite, 50*(2–3), 325–332.

Bixby, H., Bentham, J., Zhou, B., Di Cesare, M., Paciorek, C. J., Bennett, J. E., Taddei, C., Stevens, G. A., Rodriguez-Martinez, A., Carrillo-Larco, R. M., Khang, Y. H., Sorić, M., Gregg, E. W., Miranda, J. J., Bhutta, Z. A., Savin, S., Sophiea, M. K., Iurilli, M. L. C., Solomon, B. D., ... Ezzati, M. (2019). Rising rural body-mass index is the main driver of the global obesity epidemic in adults. *Nature, 569*(7755), 260–264.

Brady, L.-M., & Graham, B. (2019). *Social research with children and young people. A practical Guide.* Social Research Association Shorts.

Cai, W. (2014). Nutritional challenges for children in societies in transition. *Current Opinion in Clinical Nutrition and Metabolic Care, 17*(3), 278–284.

Fisher, K., Chatzitheochari, S., Gilbert, E., Calderwood, L., Fitzsimons, E., Cleary, A., Huskinson, T., & Gershuny, J. (2015) A mixed-mode approach to measuring young peoples time use in the UK Millennium cohort study. *Electronic International Journal of Time Use Research, 12*, 174–180.

Gibson, R.S., & Ferguson, E. L. (2008). *An interactive 24-hour recall for assessing the adequacy of iron and zinc intakes in developing countries.* International Food Policy Research Institute (IFPRI) and International Center for Tropical Agriculture (CIAT).

Guest, G., Namey, E. E., & Mitchell, M. L. (2017). In-Depth Interviews. In *Collecting Qualitative Data: A Field Manual for Applied Research.* SAGE Publications, Ltd.

Guthold, R., Stevens, G. A., Riley, L. M., & Bull, F. C. (2020). Global trends in insufficient physical activity among adolescents: A pooled analysis of 298 population-based surveys with 1·6 million participants. *The Lancet Child and Adolescent Health, 4*(1), 23–35.

Hall, K. D., Ayuketah, A., Brychta, R., Cai, H., Cassimatis, T., Chen, K.Y., Chung, S.T., Costa, E., Courville, A., Darcey, V., Fletcher, L. A., Forde, C. G., Gharib, A. M., Guo, J., Howard, R., Joseph, P.V., McGehee, S., Ouwerkerk, R., Raisinger, K., ... Zhou, M. (2019). Ultra-processed diets cause excess calorie intake and weight gain: An inpatient randomised controlled trial of ad libitum food intake. *Cell Metabolism, 30*(1), 67–77.e3.

Harris, J. A., & Benedict, F. G. (1918). A Biometric Study of Human Basal Metabolism. *Proceedings of the National Academy of Sciences, 4*, 370–373. doi:10.1073/pnas.4.12.370.

Hunter, R. F., de la Haye, K., Murray, J. M., Badham, J., Valente, T. W., Clarke, M., & Kee, F. (2019). Social network interventions for health behaviours and outcomes: A systematic review and meta-analysis. *PLoS Medicine, 16*(9), e1002890.

Keino, S., Van Den Borne, B., & Plasqui, G. (2014). Body composition, water turnover and physical activity among women in Narok County, Kenya. *BMC Public Health, 14*, 1–7. doi:10.1186/1471-2458-14-1212.

Kelly, P. F. (2011). Migration, agrarian transition, and rural change in Southeast Asia: Introduction. *Critical Asian Studies, 43*(4), 479–506.

Kitzinger, J. (1994). The methodology of Focus Groups: the importance of interaction between research participants. *Sociology of Health and Illness, 16*, 103–121. doi:10.1111/1467-9566. ep11347023.

Mazzocchi, A., Venter, C., Maslin, K., & Agostoni, C. (2017). The role of nutritional aspects in food allergy: Prevention and management. *Nutrients, 9*(8), 850–862.

Monteiro, C. A., Levy, R. B., Claro, R. M., de Castro, I. R. R., & Cannon, G. (2010). A new classification of foods based on the extent and purpose of their processing. *Cadernos de Saude Publica, 26*(11), 2039–2049.

Morgan, D. L. (1996). Focus groups. *Annual Review of Sociology, 22*, 129–152. doi:10.1146/annurev.soc.22.1.129.

Pawlowski, C. S., Andersen, H. B., Troelsen, J., & Schipperijn, J. (2016) Children's physical activity behavior during school recess: A pilot study using GPS, accelerometer, participant observation, and go-along interview. *PLoS One, 11*, e0148786, doi:10.1371/journal.pone.0148786.

Popkin, B. M. (2006). Technology, transport, globalisation and the nutrition transition food policy. *Food Policy, 31*(6), 554–569.

Prentice, A. M., Ward, K. A., Goldberg, G. R., Jarjou, L. M., Moore, S. E., Fulford, A. J., & Prentice, A. (2013). Critical windows for nutritional interventions against stunting. *The American of Clinical Nutrition, 97*(5), 911–918.

Robusto, K. M., & Trost, S. G. (2012). Comparison of three generations of ActiGraphTM activity monitors in children and adolescents. *Journal of Sports Science, 30*, 1429–1435, doi: 10.1080/02640414.2012.710761.

Santelli, J. S., Rosenfeld, W. D., DuRant, R. H., Dubler, N., Morreale, M., English, A., & Rogers, A. S. (1995). Guidelines for adolescent health research: A position paper of the society for adolescent medicine. *Journal of Adolescent Health, 17*(5), 270–276.

Santos-Lozano, A., Santín-Medeiros, F., Cardon, G., Torres-Luque, G., Bailón, R., Bergmeir, C., Ruiz, J. R., Lucia, A., & Garatachea, N. (2013). Actigraph GT3X: Validation and determination of physical activity intensity cut points. *International Journal of Sports Medicine, 34*, 975–982, doi:10.1055/s-0033-1337945.

Sasaki, J. E., John, D., & Freedson, P. S. (2011) Validation and comparison of ActiGraph activity monitors. *Journal of Science and Medicine in Sport, 14*, 411–416, doi:10.1016/j.jsams.2011.04.003.

Stevano, S., Johnston, D., & Codjoe, E. (2020). The urban food question in the context of inequality and dietary change: A study of schoolchildren in Accra. *The Journal of Development Studies, 56*(6), 1177–1189.

Swaminathan, S., Selvam, S., Thomas, T., Kurpad, A. V., & Vaz, M. (2011). Longitudinal trends in physical activity patterns in selected urban south Indian school children. *Indian Journal of Medical Research, 134*(8), 174–180.

Troiano, R. P., Berrigan, D., Dodd, K. W., Mâsse, L. C., Tilert, T., & Mcdowell, M. (2008). Physical activity in the United States measured by accelerometer. *Medicine & Science in Sports & Exercise, 40*, 181–188, doi:10.1249/mss.0b013e31815a51b3.

Trost, S. G., Loprinzi, P. D., Moore, R., & Pfeiffer, K. A. (2011) Comparison of accelerometer cut points for predicting activity intensity in youth. *Medicine & Science in Sports & Exercise 43*, 1360–1368, doi:10.1249/MSS.0b013e318206476e.

Zanello, G., Srinivasan, C. S., & Nkegbe, P. (2017). Piloting the use of accelerometry devices to capture energy expenditure in agricultural and rural livelihoods: Protocols and findings from northern Ghana. *Development Engineering, 2*, 114–131.

Zanello, G., Srinivasan, C. S., Picchioni, F., Webb, P., Nkegbe, P., Cherukuri, R., Neupane, S., Ustarz, Y., Gowdru, N., Neupane, S., et al. (2018) *Using Accelerometers in Low- and Middle-Income Countries: A field manual for practitioners.* Reading (UK).

Zanello, G., Srinivasan, C. S., Picchioni, F., Webb, P., Nkegbe, P., Cherukuri, R., & Neupane, S. (2020). Physical activity, time use, and food intakes of rural households in Ghana, India, and Nepal. *Scientific Data, 7*(1), 1–10.

Appendix

Table 18.A Full regressions of determinants of physical activity level (PAL)

	Younger adolescents (11–14 years old)		Older adolescents (15–19 years old	
	Males	*Females*	*Males*	*Females*
Education activities	0.010★★★	0.006★	0.003	0.003
	(0.004)	(0.004)	(0.003)	(0.004)
Economic activities	0.022★★★	0.023★★★	0.022★★★	0.023★★★
	(0.007)	(0.008)	(0.006)	(0.007)
Domestic activities	0.013	0.014★★	0.023★★★	0.015★★★
	(0.010)	(0.005)	(0.008)	(0.005)
Leisure activities	0.021★★★	0.015★★★	0.002	0.011★★
	(0.005)	(0.005)	(0.004)	(0.005)
Non-wearing time	–0.001★★★	–0.001★★★	–0.001★★★	–0.000★★★
	(0.000)	(0.000)	(0.000)	(0.000)
Constant	1.288★★★	1.283★★★	1.255★★★	1.222★★★
	(0.044)	(0.041)	(0.031)	(0.034)
Wald chi2	10.765★★★	18.905★★★	10.798★★★	9.793★★★
R-squared	0.171	0.170	0.217	0.159
Sample size	421	461	472	454

Notes: Regressions estimated at day/level. Sleeping and resting activities as baselines. Robust standard errors (in brackets) are clustered at individual level. ★★★Denotes statistical significance at the 1% level, ★★ at 5% level, and ★ at 10% level. Source: Work by authors

PART 4

Education and public engagement

19

BEYOND WHAT NOT TO EAT

Supporting communities to know sustainable diets

Liesel Carlsson

Introduction: Interrelated challenges

This chapter focuses on the role of nutrition and dietetic practitioners in contributing to uptake of sustainable diets. Three interrelated barriers are first introduced and then discussed in parallel with examples from the literature that help address such challenges. These begin with, but move beyond, telling people what to eat or not eat, to share insights for more systemic solutions that support communities to know sustainable diets. The insights are broadly relevant across disciplinary boundaries. As noted in chapters throughout this handbook, sustainable diets (SD) consider low environmental impacts while boosting nutrition security and healthy living for present and future generations (Burlingame & Dernini, 2012). Diets can be considered sustainable when they also are "protective and respectful of biodiversity and ecosystems, culturally acceptable, accessible, economically fair and affordable; nutritionally adequate, safe and healthy; while optimising natural and human resources" (Burlingame & Dernini, 2012, p. 7).

The title NDP is used to describe a group of practitioners and scholars in the field of nutrition and dietetics, whose role varies by jurisdiction, but generally "is a professional who applies the science of food and nutrition to promote health, prevent and treat disease to optimise the health of individuals, groups, communities and populations" (International Confederation of Dietetic Associations, n.d.). NDPs have historically been influencing dietary patterns at many levels and are extremely well positioned to do so, given their roles throughout food systems. However, there are three inter-related challenges that NDPs face in facilitating transitions to sustainable diets: Disciplinary paradigms, the complexity of behaviour change, and the need for context-specific solutions.

Disciplinary paradigms present systemic barriers. Recent decades have seen the dominant use of an individualised, biomedical paradigm within the profession, which has propelled NDPs forward as well-respected members of the allied health

DOI: 10.4324/9781003174417-23

professionals and resulted in significant advances in clinical nutrition sciences, where a focus on individualised care is critical. Examining nutritional challenges at the individual level, however, has limited use in supporting NDPs to understand and advance practice in solving nutritional problems with roots in social structures and ecological systems (Carlsson, Pettinger & Mehta, 2019b).

The second related challenge lies within the complexity of behaviour change. Scholars across many disciplines have identified key recommendations for sustainable diets, such as moderate consumption and plant-dominant diets, which are helpful in guiding concrete decisions about what dietary patterns at a population level are more sustainable. However, NDPs have long understood that knowledge does not equal behaviour. In this context, there is a need for more sophisticated approaches to applying such recommendations to support uptake of sustainable dietary patterns.

The third challenge is related to context. The validity and appropriateness of such broad-scale recommendations are highly contextual to the community in question. What is more sustainable on a global scale may be highly unsustainable on a local scale; what is most sustainable will vary between communities and geographic regions. Thus, recommendations require advanced knowledge of food and nutrition, sustainability science, and the contextual realities, as well as interdisciplinary thinking and collaboration skills. Each of these three challenges are discussed in the following sections, along with solutions drawn from the fields of nutrition education, community development, and systems thinking. These present opportunities for NDP engagement in contributing to uptake of sustainable diets.

NDPs are seasoned and well-positioned— professional paradigms slow to progress

The concept of sustainable diets is not new to this field but rather a current catalyst for renewed action. What is now known as nutrition and dietetics is a field of scholarship and practice that weaves its history back through that of home economics and human ecology. Very early leaders of these fields were engaged in work that conceptualised human nutritional health as inseparable from the health of socioecological systems in which we are embedded (Dyball & Carlsson, 2017). Since then, there have been efforts to recapture a socioecological systems approach, such as eco-dietetics (Rebrovick, 2015) and the new nutrition science project (Beauman et al., 2005; Cannon & Leitzmann, 2006), as well as early efforts to create sustainable dietary guidelines (Gussow & Clancy, 1986).

Across food system sectors, NDPs are strategically placed to leverage significant and concerted action in this field. They play many roles: In health systems, working with clients to improve health outcomes through dietary patterns; in economic systems, managing the large scale menus of food service providers for hospitals, care facilities, and universities; in consumer demand systems, working with community organisations on programs that promote and support the availability, accessibility, and adoption of healthy food environments; in food production systems, working with commodity marketing boards to promote product development that supports health; and in political systems, working with government agencies to shape national food policy. While NDPs alone do not own this problem space, they are well positioned in part because of diverse leverage points across many subsystems that are part of food systems. These roles are captured in Figure 19.1.

While clear opportunities exist, there remain significant barriers within the profession. A minority of NDPs are actively and confidently working toward facilitating the promotion and adoption of sustainable diets. One central barrier is the cultural norms of the profession. Earlier work in dialogue with over 70 NDPs from 30 different countries highlights examples where managers and fellow NDPs discouraged engagement in sustainable diets, deprioritised this work, or identified it as out of their scope-of-practice (Carlsson, Callaghan & Laycock Pedersen, 2019a; Carlsson & Callaghan, 2022). In this same research, NDPs suggested that engaging with sustainable diets might be an irresponsible dilution of the role. This is both paradoxical and self-limiting: Paradoxical because the bread-and-butter of NDPs' work, promoting dietary patterns that are nutritionally sound and health

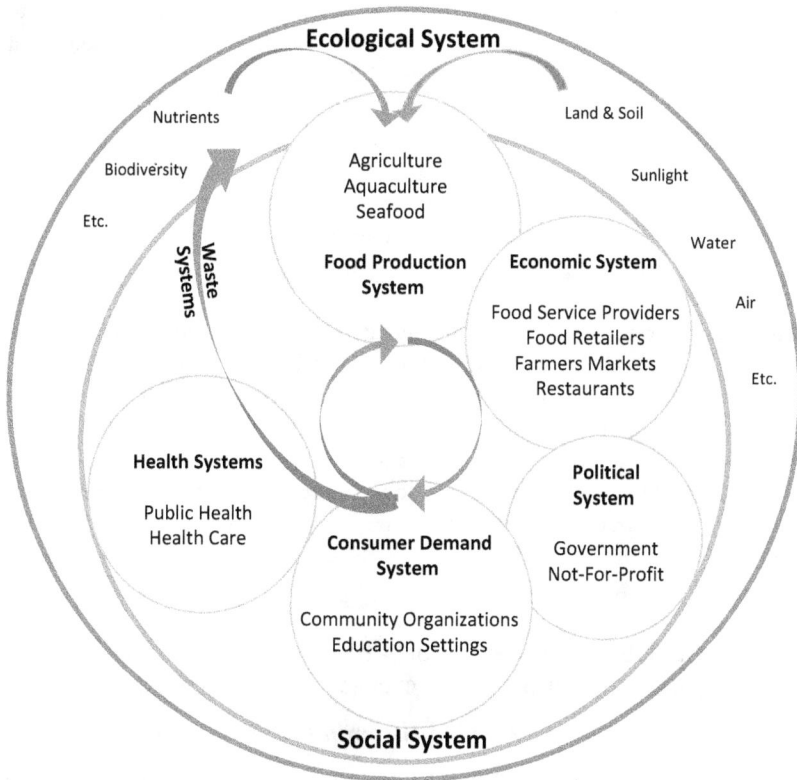

Figure 19.1 Mapping the roles of NDPs in the food systems landscape. Adapted by the author with permission from Nourish (2013), and originally published by Goodridge et al. (2022; used with permission).

promoting, is emerging as also more sustainable (though this is highly dependent on regional context) (Aleksandrowicz et al., 2016; Blackstone et al., 2018; Clark et al., 2020; Macdiarmid, 2013; Nelson et al., 2016; Springmann et al., 2016, 2018; Vanham et al., 2013); and, self-limiting because NDPs are the most trusted professionals the public looks to for food and nutrition advice (Eat Well Global, 2020). Research with other health professionals (including NDPs) indicates that dietitians are not alone; professional-level barriers to engagement with sustainability exist (Alberdi & Begiristain-Zubillaga, 2021; Guillaumie et al., 2020; Kotcher et al., 2021), further deprioritizing integration into health settings.

Cultural barriers in the profession are not unrelated to the absence of sustainability concepts in NDP training. Preliminary research (Higgins et al., in progress) examining the standards that govern training curriculum internationally show that most countries made general reference to "sustainable food systems" without more granular direction as to what this meant, and most required only a limited depth of knowledge. Theories and models that operationalise training are needed (Carlsson, Pettinger & Mehta, 2019b).

Given such barriers, where do NDPs begin? As Marsden and Morley (2014) point out, there is no template for facilitating sustainable food systems and diets. Sustainability is a lens through which one looks; it is not "tacked on" as an additional responsibility, but a way of thinking. Thus, all decisions that NDPs are making, or supporting others to make, are considered against the backdrop of nutritional and health outcomes (which is always the lens) *in the context* of the socioecological system. NDPs can reflect on whether there are ways in which more sustainable diets would help address nutritional challenges they are working on. There are resources for NDPs to support such activities (Food + Planet, 2021; International Confederation of Dietetics Associations, 2020; One Blue Dot, n.d.; Vogliano, et. al, 2021).

Few professional associations formally recognise sustainable diets as within scope through position and policy documents, but those who do (British Dietetic Association, 2017; Dietitians of Canada, 2020; Italian Association of Dietitians (ANDID), 2010; Spiker et al., 2020) help to inform dialogue among practitioners. Well defined position statements can help translate general training standards into operational guidance and support practitioners to adopt a sustainability lens in their work. In the context of the exceptionally strategic position of NDPs in the food systems, integration of this lens into practice can give broad leverage across many roles and activities for comprehensive shifts.

Guidelines that define sustainable dietary patterns abound—community-level adoption is complex

Knowledge dissemination

To guide integration of sustainable diets into practice, NDPs first need an understanding of the issues. Much research has been dedicated to understanding and defining sustainable diets (Burlingame & Dernini, 2012; Garnett, 2014; Johnston et al., 2014; Jones et al., 2016), and translating what is known into food guidelines

(EAT-Lancet Commission, 2019; Fischer & Garnett, 2016; Seed, 2015). Seed and Rocha (2018) synthesise seven emerging *environmental* sustainability messages, which are helpful to guide practice:

- Consuming a primarily plant-based diet
- Limiting/avoiding consumption of processed foods
- Reducing waste
- Reducing meat consumption (especially ruminant meat)
- Consuming seasonal, field grown fruit and vegetables
- Choosing certified food (e.g., sustainably certified fish)
- Breastfeeding

Lawrence and colleagues (2019) add *moderating consumption* to this list of key messages, a common message in general healthy eating guidance (e.g., Health Canada, 2019); however, NDPs are hesitant about messages that imply weigh-based discrimination (e.g., that people with high body weights are both unhealthy and bad for the environment, and that it can be fixed by simply eating "moderately") or fuel ethically motivated disordered eating.

Dietitians of Canada suggests similarly broad principles to guide food-related decision making that supports social sustainability (Dietitians of Canada, 2020). These include attention to the following:

- Sustainable livelihoods
- Social justice
- Human dignity
- Participation of all stakeholders

From a knowledge development perspective, NDPs can apply a sustainability lens by, for example, using the key messages or principles as a checklist. When working with a client to attain health goals, the NDP and client can ask: Can it be done in the context of reducing meat consumption, ultra-processed foods, and food waste? Can it be achieved through food choices that do not undermine the livelihoods of food producers and workers in processing facilities? For clients empowered to make such choices, information may be adequate. For example, sustainable guidelines to fuel athletes (Meyer & Reguant-Closa, 2017; Reguant-Closa et al., 2020) are one example of innovative work led by RDs to support knowledge development and adoption of sustainable dietary patterns.

Yet dietary change is extremely complex, and for many, everyday life makes such checklists irrelevant; dietary guidelines, principles, and key messages themselves do not facilitate change. Focusing on knowledge development for sustainable dietary choices unloads the burden of sustainable diets on the individual. Two concepts drawn from the field of nutrition education can help facilitate the voluntary adoption of sustainable dietary patterns: Creating more sustainable food environments from which everyone can choose, and facilitating systemic shifts in food systems (Contento, 2010).

Create sustainable food environments

In institutional settings, such as schools, hospitals, and long-term care facilities, NDPs can and do create food environments that support making sustainable dietary choices. Examples include menu design with the inclusion of field-grown, seasonal foods and procurement choices favouring foods certified as meeting environmental or social justice standards (e.g., Fair Trade and Organic certifications). Examples of well-organised efforts in hospitals exist (Health Care Without Harm, 2021), and NDP-led research show that in hospitals, engagement with local food producers and efforts to reduce food waste are most common (Carino et al., 2020), though difficult to evaluate. Institutional procurement is valuable as it can facilitate larger scale purchasing (Morgan & Morley, 2014) for otherwise niche producers, provide educational opportunities for consumers, and reduce food waste among food system actors where waste is concentrated. Examples of public policies that support institutional contributions to sustainable food environments exist, such as Sweden's food waste action plan, which proposes measures for food waste reduction in public institutions (Livsmedelsverket, 2018), and Acadia University's Food Plan, which guides food services for the campus (Acadia Food Services Advisory Committee, 2018). As with most choices, such efforts to create more sustainable food environments are vulnerable to trade-offs. For example, field-grown, seasonal vegetables may result in lower greenhouse gas emissions than imported or greenhouse grown alternatives, yet exacerbate economic inequalities if the field workers earn extremely low wages.

Facilitate systemic shifts in food systems

To facilitate systemic shifts in food systems, NDPs can work to establish processes that normalise and ensure access to sustainable diets. As suggested by Dietitians of Canada (2020), those working in public health can advocate for policies that support sustainable livelihoods and decrease inequities between actors in the system (i.e., producers, labourers, retailers). These actions contribute to a more socially sustainable system. An example of an equity-based approach, which some NDPs are advocating strongly for, is a guaranteed minimum income (Brady, 2019, 2021). Gaining buy-in from policy makers in other disciplines and sectors may be difficult, and public support may wane if policies translate to higher food prices or perceived market interference. For these types of efforts, change may come long after.

As NDPs build on direct knowledge dissemination to include affecting food environments and facilitating systemic shifts, they begin to move beyond telling people what to eat, or not eat, to generate more systemic change for sustainable diets. Yet, a final challenge remains: What constitutes a more, or less, sustainable dietary pattern is highly dependent on the social and ecological context.

Sustainable Diets are highly contextual—invite community participation and ownership

There will be times when broad guidelines are neither clinically nor culturally relevant. Context is critical. For example, a global reduction in average animal-based foods consumption, in particular ruminant meats, is strongly supported by research to be more sustainable (Garnett et al., 2017) and reflected in several food based dietary guidelines (Health Canada, 2019; Seed, 2015). This recommendation is not nutritionally sound in many low-resource, food insecure settings where consumption of animal-based foods is very low already, and micronutrient and calorie deficiencies could be directly addressed through increasing animal-based food intake—especially red meats. Diets are not sustainable if they undermine nutritional adequacy.

Through research examining community-level engagement with sustainable food systems, including the dietetics community, several concepts informed by systems thinking and community development surface as helpful to this challenge (Carlsson, 2019) and help individuals and communities *know* sustainable diets in a way that is contextually relevant.

Recognise context and multiple forms of knowledge

A deep understanding of the specific context in which sustainable dietary guidelines are being applied requires looking beyond technical knowledge in the guidelines and literature to include knowledge held by other professionals, organisations, and community members (Brown et al., 2010). Building on the example above, the sustainability-based recommendation to reduce meat consumption is also not appropriate in some remote communities in Northern Canada where community access to wild game is culturally meaningful, more affordable, and more nutritious than foods bought at grocery stores (Thompson et al., 2020). Diets are not sustainable if they are culturally inappropriate and inaccessible.

Furthermore, as Vogliano et al. (2021) suggest, people need to see themselves reflected in sustainable dietary guidelines, and one way to support this is through practising cultural humility. Cultural humility invites NDPs to reflect on how their own assumptions shape their work on sustainable diets and invites the inclusion of other cultures and ways of knowing to shape the sustainable diets discourse. Such knowledge enriches the solutions with practical ideas about what will work, values the knowledge of the community members in solution building, helps communities "see" themselves in sustainability solutions, and supports food sovereignty. For examples exploring this approach in the context of Indigenous food systems, see Settee and Shukla (2020), and in addressing income-related food insecurity, see Travers (1997), Williams et al. (2012), and Monteith et al. (2020).

Translating sustainable dietary guidelines at the community level may require the NDP to have specialised knowledge in both nutrition and sustainability, but this work invites NDPs to acknowledge the limits to their knowledge. Inviting diverse members of the relevant communities to activate their agency in sustainable diets is more likely to result in sustainable solutions and community-level uptake.

Build a common language

Critical to facilitating uptake of sustainable diets is ensuring a common language (Carlsson, Callaghan, & Laycock Pederson, 2019a). Ideas of what is sustainable and healthy differs between disciplines and communities, exacerbated by the fact that the concepts are complex and interdisciplinary. This complexity can create barriers to alignment and adoption, create confusion, and invite competing voices with vested, conflicts of interest to shape the discourse (Clapp, 2014). At a practical level, building a common language could include facilitating reflexive conversations with a client—an individual, an organisation, a community—to discuss and document the goals and issues, to develop common understanding (for an example of one process used, see Carlsson et al., 2017). Capturing and formalising such language can take many forms, such as food charters and declarations (*Toronto Food Charter*, n.d.) and food plans (Acadia Food Services Advisory Committee, 2018).

Collaborate broadly

Systemic advancement of the sustainable diets agenda requires complex, interdisciplinary thinking, and advanced knowledge of food, nutrition, sustainability, and the local context. It is not surprising that many NDPs question their ability or mandate to do this work. Current ways of working steer professionals to divide up issues by disciplinary or sectoral areas of practice to achieve a common goal. For example, while NDPs focus on nutritional aspects of dietary patterns, economic developers focus on livelihoods, and environmental consultants on the effects on water quality. Systems thinking advances the notion that these are interconnected, and working *together* using transdisciplinary techniques can maximise goals and minimise trade-offs. The Canadian Food Policy Advisory Council, whose role is to advise governments on policies to support sustainable food systems, was called together by the Government of Canada in recognition that "greater collaboration is essential to make meaningful progress on complex and systemic food issues" (Agriculture and Agri-Food Canada (AAFC), 2021). It is one example of such broad collaboration (AAFC, 2021).

Foster community ownership

Some Community Development theorists might propose that RDNs are obligated to work toward food systems that self-perpetuate, in the sense that knowledge and power in food systems be returned to community ownership through involvement, learning, and knowledge sharing (McKnight, 1994). Attention to multiple forms of knowledge, building a common language, and collaboration, will help foster such ownership. Theories on cultural adaptation (Dyball & Newell, 2015) suggest that the state of the cultural paradigms of a community have significant influence on the state of the community, including physical and cultural infrastructure that guides day to day actions, which, in turn, influence the state of the cultural paradigms. Inviting communities to be part of knowledge creation and solutions for adopting sustainable diets can influence paradigmatic shifts toward sustainability-informed dietary decisions, and therefore toward self-perpetuating, sustainable systems. Table 19.1 provides a summary of each of the three challenges and potential solutions explored to this point, and highlights some additional.

Table 19.1 Summary of barriers NDPs face, potential solutions, and examples from research and practice. Source: Work of the author

Barriers	Solutions	Examples (See references for full citation)
Disciplinary paradigms	Consider nutritional and health outcomes *in the context* of the socioecological system.	One Blue Dot—The BDA's Environmentally Sustainable Diet Project. Reguant-Closa et al. (2020). The environmental impact of the Athlete's Plate nutrition education tool.
	Initiate reflection and dialogue with colleagues in the profession and externally.	Beauman, C. et al (2005). The principles, definition and dimensions of the New Nutrition Science. Carlsson, L., & Callaghan, E. (in press). The social license to practice sustainability: Concepts, barriers and actions to support nutrition and dietetics practitioners in contributing to sustainable food systems.
	Adopt socioeconomically informed theories and models in education and training.	Spiker, M. et al (2020). Academy of Nutrition and Dietetics: Revised 2020 standards of professional performance for registered dietitian nutritionists (competent, proficient, and expert) in sustainable, resilient, and healthy food and water systems.
Complexity of behaviour change	Disseminate knowledge applying evidence-based sustainability principles and messages.	Meyer, N., & Reguant-Closa, A. (2017). "Eat as if you could save the planet and win!" Sustainability integration into nutrition for exercise and export.
	Create sustainable food environments from which everyone can choose.	Acadia Food Services Advisory Committee. (2018). Acadia University Food Services Plan.
	Facilitate systemic shifts in food systems that normalize access to sustainable diets.	Brady, J. (2019). Social justice, health equity, and advocacy: What are our roles? Brady, J. (2021). Basic income and the future of food studies.

(Continued)

Table 19.1 (Continued)

Barriers	Solutions	Examples (See references for full citation)
Need for context specific solutions	Acknowledge the specific context and multiple forms of relevant knowledge.	Settee, P., & Shukla, S. (Eds.). (2020). Indigenous food systems: Concepts, cases and conversations. Thompson, S. et al. (2020). Eco-carnivorism in Garden Hill First Nation.
	Build a common language around sustainable diets; facilitate meaning and understanding through dialogue.	Carlsson, L. et al. (2019). Building common ground for sustainable food systems in nutrition and dietetics. ICDA. Toronto Food Charter. (n.d.). City of Toronto.
	Collaborate broadly with colleagues both within the profession, and beyond.	Agriculture and Agri-Food Canada. (2021). The Canadian Food Policy Advisory Council.
	Foster community ownership of sustainable diets through involvement, learning and knowledge sharing to influence paradigmatic shifts.	Williams, P. et al. (2012). A participatory food costing model in Nova Scotia.

Conclusions: Support communities to know sustainable diets

This chapter is a call to action to move beyond telling people what not to eat, and instead to support communities to *know* sustainable diets. NDPs are strategically positioned to support transitions to sustainable diets. They are well-supported by national guidelines that clarify *what* dietary patterns at a population level are more sustainable; however, there is no template for *how* to facilitate uptake. At the community level, this is especially complex and elusive. NDP training is framed by paradigms ill-suited to understand and address challenges of this nature, behaviour change is hindered by the socioecological realities of people's lives, and what is sustainable is highly contextual. A combination of tools and concepts drawn mainly from the fields of nutrition education, systems thinking, and community development can lend support to NDPs looking to facilitate meaningful transition to sustainable diets among their clients, whether they be individuals, organisations, or communities. The work most likely to create systemic, paradigmatic changes for sustainable diets are those that: 1) Acknowledge contextual realities and personal assumptions, and value multiple types of knowledge in both understanding and solution building; 2) work to develop a common language among all stakeholders such that the issues are clear and not muddied conflicts of interest; 3) foster collaboration between actors of multiple disciplines and knowledge types who work *together* on problems rather than dividing it up into component parts; and 4) facilitate community ownership over sustainable diets in a way that supports paradigmatic shifts. While this chapter focuses on generalised insights for NDPs, their relevance may extend beyond this community to others doing important work in facilitating transitions to sustainable diets.

References

Acadia Food Services Advisory Committee (2018). *Acadia university food services plan.* Acadia University. https://sustainability.acadiau.ca/tl_files/sites/aiae/Food/Acadia%20Food%20Plan%20December%202018_Final.pdf

Agriculture and Agri-Food Canada (2021). *The Canadian food policy advisory council.* https://agriculture.canada.ca/en/about-our-department/key-departmental-initiatives/food-policy/canadian-food-policy-advisory-council

Alberdi, G., & Begiristain-Zubillaga, M. (2021). The promotion of sustainable diets in the healthcare system and implications for health professionals: A scoping review. *Nutrients, 13*(3), 747.

Aleksandrowicz, L., Green, R., Joy, E. J. M., Smith, P., & Haines, A. (2016). The impacts of dietary change on greenhouse gas emissions, land use, water use, and health: A systematic review. *Plos One,* 11(11), e0165797.

Beauman, C., Cannon, G., Elmadfa, I., Glausauer, P., Hoffmann, I., Keller, M., Krawinkel, M., Lang, T., Leitzmann, C., Lotsch, B., Margetts, B., McMichael, A., Meyer-Abich, K., Oltersdorf, U., Pettoello-Mantovani, M., Sabaté, J., Shetty, P., Soria, M., Spiekermann, U., … Zerilli-Marimo, M. (2005). The principles, definition and dimensions of the new nutrition science. *Public Health Nutrition, 8*(6A), 695–698.

Blackstone, N. T., El-Abbadi, N. H., McCabe, M. S., Griffin, T. S., & Nelson, M. E. (2018). Linking sustainability to the healthy eating patterns of the dietary guidelines for Americans: A modelling study. *The Lancet Planetary Health, 2*(8), e344–e352.

Brady, J. (2019). Social justice, health equity, and advocacy: What are our roles. In J. Coveney & S. Booth (Eds.), *Critical dietetics and critical nutrition studies,* (pp. 143–159). Springer.

Brady, J. (2021, June 9). Basic Income 101: What is it? What does it mean for food studies?. Presentation at the conference "Just Food: Because it is never Just Food". The Culinary Institute of America - New York University.

British Dietetic Association (2017). *British dietetic association policy statement: Sustainable diets.* British Dietetic Association. https://www.bda.uk.com/improvinghealth/healthprofessionals/policy_statement-_sustainable_food

Brown, V. A., Harris, J. A., & Russell, J. Y. (2010). *Tackling wicked problems through the transdisciplinary imagination.* Earthscan.

Burlingame, B., & Dernini, S. (2012). *Sustainable diets and biodiversity: Directions and solutions for policy, research and action.* Nutrition and Consumer Protection Division, Food and Agriculture Organization.

Cannon, G., & Leitzmann, C. (2006). The new nutrition science project. *Scandinavian Journal of Food & Nutrition, 50*(1), 5–12.

Carino, S., Porter, J., Malekpour, S., & Collins, J. (2020). Environmental sustainability of hospital foodservices across the food supply chain: A systematic review. *Journal of the Academy of Nutrition and Dietetics, 120*(5), 825–873.

Carlsson, L. (2019). *Inviting community into the development of globally sustainable food systems* [PhD Thesis, Blekinge Tekniska Högskola]. http://urn.kb.se/resolve?urn=urn:nbn:se:bth-18803

Carlsson, L., & Callaghan, E. (2022). The social license to practice sustainability: Concepts, barriers and actions to support nutrition and dietetics practitioners in contributing to sustainable food systems. *Journal of Hunger & Environmental Nutrition,* 1–19.

Carlsson, L., Callaghan, E., Morley, A., & Broman, G. (2017). Food system sustainability across scales: A proposed local-to-global approach to community planning and assessment. *Sustainability, 9*(6), 1061–1075.

Carlsson, L., Callaghan, E., & Laycock Pederson, B. (2019a). *Building common ground for sustainable food systems in nutrition and dietetics.* Blekinge Tekniska Högskola, ICDA. https://icdasustainability.org/about/

Carlsson, L., Pettinger, C., & Mehta, K. (2019b). Critical dietetics and sustainable food systems. In J. Coveney & S. Booth (Eds.), *Critical Dietetics and Critical Nutrition Studies* (pp. 97–115). Springer. https://www.springer.com/us/book/9783030031121

Clapp, J. (2014). Distant agricultural landscapes. *Sustainability Science, 10*(2), 305–316.

Clark, M., Macdiarmid, J., Jones, A. D., Ranganathan, J., Herrero, M., & Fanzo, J. (2020). The role of healthy diets in environmentally sustainable food systems. *Food and Nutrition Bulletin, 41*(2), 31S–58S.

Contento, I. (2010). *Nutrition education: Linking research, theory, and practice.* Jones & Bartlett Learning, LLC.

Dietitians of Canada (2020). The role of dietitians in sustainable food systems and sustainable diets: A role paper. https://www.dietitians.ca/Advocacy/Toolkits-and-Resources?n=The%20Role%20of%20Dietitians%20in%20Sustainable%20Food%20Systems%20and%20Sustainable%20Diets%20(role%20paper)&Page=1#

Dyball, R., & Carlsson, L. (2017). Ellen Swallow Richards: Mother of human ecology? *Human Ecology Review, 23*(2), 17–28.

Dyball, R., & Newell, B. (2015). *Understanding human ecology: A systems approach to sustainability.* Routledge.

EAT-Lancet Commission (2019). *EAT Lancet Commission summary: Healthy diets from sustainable food systems.* https://eatforum.org/content/uploads/2019/01/EAT-Lancet_Commission_Summary_Report.pdf

Eat Well Global (2020). *The consumer voice: Global insights on food, nutrition, trust and influence.* Eat Well Global.

Fischer, C. G., & Garnett, T. (2016). *Plates, pyramids, planet.* Food Climate Research Network. http://www.fao.org/documents/card/en/c/d8dfeaf1-f859-4191-954f-e8e1388cd0b7

Food + Planet (2021). *Promoting sustainable and climate-friendly food systems.* https://foodandplanet.org/

Garnett, T. (2014). *What is a sustainable healthy diet: A discussion paper.* Food Climate Research Network. http://www.fcrn.org.uk/sites/default/files/fcrn_what_is_a_sustainable_healthy_diet_final.pdf

Garnett, T., Godde, C., Muller, A., Roos, E., de Boer, I., zu Ermgassen, E., Herrero, M., van Middelaar, C., Schader, C., & van Zanten, H. (2017). *Grazed and confused.* Food Climate Research Network. https://www.fcrn.org.uk/projects/grazed-and-confused

Goodridge, L., Carlsson, L., & Callaghan, E. (2022). Mapping the roles of nutrition and dietetics professionals in sustainable food systems and exploring opportunities for strategic collaboration. *Canadian Journal of Dietetic Practice and Research.* (e-first pub).

Guillaumie, L., Boiral, O., Baghdadli, A., & Mercille, G. (2020). Integrating sustainable nutrition into health-related institutions: A systematic review of the literature. *Canadian Journal of Public Health, 111*(6), 845–861.

Gussow, J. D., & Clancy, K. L. (1986). Dietary guidelines for sustainability. *Journal of Nutrition Education, 18*(1), 1–5.

Health Care Without Harm (2021). *Global green and healthy hospitals.* Health Care Without Harm. https://noharm-global.org/issues/global/global-green-and-healthy-hospitals

Health Canada (2019). *Canada's dietary guidelines for health professionals and policy makers.* Health Canada. http://epe.lac-bac.gc.ca/100/201/301/weekly_acquisitions_list-ef/2019/19-04/publications.gc.ca/collections/collection_2019/sc-hc/H164-231-2019-eng.pdf

Higgins, M., Strother. H, Burkhart, S., Carlsson, L., Meyer, N., Spiker, M., & Wegener, J. (in progress). *Sustainable food systems and diets in dietetic training standards: An international content analysis study.*

International Confederation of Dietetic Associations (n.d.). *About ICDA*. https://www.int ernationaldietetics.org/About-ICDA.aspx

International Confederation of Dietetics Associations (2020). *ICDA sustainability toolkit: Supporting sustainability in nutrition.* https://icdasustainability.org/

Italian Association of Dietitians (2010). *Role of dietitians in food sustainability: Position of the Italian Association of Dietitians.* https://icdasustainability.org/wp-content/uploads/2020 /05/ANDIDPositionStatementFoodSustainability.pdf

Johnston, J. L., Fanzo, J. C., & Cogill, B. (2014). Understanding sustainable diets: A descriptive analysis of the determinants and processes that influence diets and their impact on health, food security, and environmental sustainability. *Advances in Nutrition, 5*(4), 418–429.

Jones, A. D., Hoey, L., Blesh, J., Miller, L., Green, A., & Shapiro, L. F. (2016). A systematic review of the measurement of sustainable diets. *Advances in Nutrition, 7*(4):641–64. doi: 10.3945/an.115.011015. PMID: 27422501; PMCID: PMC4942861.

Kotcher, J., Maibach, E., Miller, J., Campbell, E., Alqodmani, L., Maiero, M., & Wyns, A. (2021). Views of health professionals on climate change and health: A multinational survey study. *The Lancet Planetary Health, 5*(5), e316–e323.

Lawrence, M., Baker, P., Wingrove, K., & Lindberg, R. (2019). Sustainable diets: The public health perspective. In B. Burlingame & S. Dernini (Eds.), *Sustainable diets: Linking nutrition and food systems* (pp. 13–21). CAB International.

Livsmedelsverket. (2018). *Fler gör mer! Handlingsplan matsvinn 2030: Åtgärdsförslagen i korthet.* https://www.livsmedelsverket.se/bestall-ladda-ner-material/sok-publikationer/artiklar /2018/2018-kortversion-fler-gor-mer-handlingsplan-for-minskat-matsvinn-2030

Macdiarmid, J. I. (2013). Is a healthy diet an environmentally sustainable diet? *The Proceedings of the Nutrition Society, 72*(1), 13–20.

Marsden, T., & Morley, A. (2014). Current food questions and their scholarly challenges: Creating and framing a sustainable food paradigm. In T. Marsden & A. Morley (Eds.), *Sustainable food systems: Building a new paradigm* (pp. 1–29). Routledge.

McKnight, J. (1994). *Community and its counterfeits.* CBC Radio Works.

Meyer, N., & Reguant-Closa, A. (2017). "Eat as if you could save the planet and win!" Sustainability integration into nutrition for exercise and sport. *Nutrients, 9*(4), 412.

Monteith, H., Anderson, B., & Williams, P. L. (2020). Capacity building and personal empowerment: Participatory food costing in Nova Scotia, Canada. *Health Promotion International, 35*(2), 321–330.

Morgan, K. J., & Morley, A. S. (2014). The public plate: Harnessing the power of purchase. In T. Marsden & A. Morley (Eds.), *Sustainable food systems: Building a new paradigm* (pp. 84–102). Routledge.

Nelson, M., Hamm, E., Michael, W., Abrams, S. A., & Griffin, T. S. (2016). Alignment of healthy dietary patterns and environmental sustainability: A systematic review. *Advances in Nutrition, 7*, 1005–1025.

Nourish (2013). *Food system tools: Food and community.* https://www.nourishlife.org/teach/ food-system-tools/

One Blue Dot (n.d.). *The BDA's environmentally sustainable diet project.* British Dietetic Association. Retrieved July 14, 2020, from https://www.bda.uk.com/resource/one -blue-dot.html

Rebrovick, T. (2015). The politics of diet: "Eco-dietetics," neoliberalism, and the history of dietetic discourses. *Political Research Quarterly; Salt Lake City, 68*(4), 678–689.

Reguant-Closa, A., Roesch, A., Lansche, J., Nemecek, T., Lohman, T. G., & Meyer, N. L. (2020). The environmental impact of the athlete's plate nutrition education tool. *Nutrients, 12*(8), 2484.

Seed, B. (2015). Sustainability in the Qatar national dietary guidelines, among the first to incorporate sustainability principles. *Public Health Nutrition, 18*(13), 2303–2310.

Seed, B., & Rocha, C. (2018). Can we eat our way to a healthy and ecologically sustainable food system? *Canadian Food Studies / La Revue Canadienne Des Études Sur l'alimentation, 5*(3), 182–207.

Settee, P., & Shukla, S. (2020). *Indigenous food systems: Concepts, cases and conversations*. Canadian Scholars' Press.

Spiker, M., Reinhardt, S., & Bruening, M. (2020). Academy of Nutrition and Dietetics: Revised 2020 standards of professional performance for Registered Dietitian Nutritionists (competent, proficient, and expert) in sustainable, resilient, and healthy food and water systems. *Journal of the Academy of Nutrition & Dietetics*, *120*(9), 1568–1568.

Springmann, M., Godfray, H. C. J., Rayner, M., & Scarborough, P. (2016). Analysis and valuation of the health and climate change cobenefits of dietary change. *Proceedings of the National Academy of Sciences of the United States of America*, 113(15), 4146–4151.

Springmann, M., Wiebe, K., Mason-D'Croz, D., Sulser, T. B., Rayner, M., & Scarborough, P. (2018). Health and nutritional aspects of sustainable diet strategies and their association with environmental impacts: A global modelling analysis with country-level detail. *The Lancet Planetary Health*, *2*(10), e451–e461.

Thompson, S., Pritty, P., & Thapa, K. (2020). Eco-carnivorism in garden hill first nation. In R. M. Katz-Rosene & S. J. Martin (Eds), *Green meat? sustaining eaters, animals, and the planet*. McGill-Queen's University Press. https://www.mqup.ca/green-meat--products-9780228001331.php

Toronto Food Charter (n.d.). City of Toronto. http://tfpc.to/to-food-policy-archive/toronto-food-charter

Travers, K. D. (1997). Nutrition education for social change: A critical perspective. *Journal of Nutrition Education*, *29*(2), 57–62.

Vanham, D., Mekonnen, M. M., & Hoekstra, A. Y. (2013). The water footprint of the EU for different diets. *Ecological Indicators*, *32*, 1–8.

Vogliano, C., Geagan, K., Chou, S., & Palmer, S. (2021). *Empowering nutrition professionals to advance sustainable food systems*. Food + Planet. https://foodandplanet.org/

Williams, P., Amero, M., Anderson, B., Gillis, D., Green-LaPierre, R., Johnson, C., & Reimer, D. (2012). A participatory food costing model in Nova Scotia. *Canadian Journal of Dietetic Practice and Research*, *73*(4), 181–188.

20
FOOD LITERACY, PEDAGOGIES, AND DIETARY GUIDELINES

Converging approaches for health and sustainability

Alicia Martin, Katherine Eckert, Jess Haines, and Evan Fraser

Introduction

While the globalisation of food systems has fostered efficiencies that have led to increasing availability and variety of food for some, it has also contributed to food insecurity (Magnan, 2016), environmental crises (Willett et al., 2019), the Nutrition Transition (i.e., the shift in dietary intake and energy expenditure in response to changes in food environments and population-level socio-economic status) (Popkin, 2001), and a rapid rise in non-communicable diseases (Hawkes et al., 2013). Due to a decline in the need to prepare and cook meals, due to the increasing availability of ready to eat meals, citizen consumers have also become "deskilled" (Slater, 2013) through a culinary skills transition (Lang & Caraher, 2001). To understand and address these interconnected issues, a broader understanding of food, nutrition, and food systems is vital among citizens. As such, there is an increasing interest in food literacy and pedagogies to promote healthy and sustainable diets.

Although food literacy has been a topic of interest over the last decade, a common definition has not been established (Truman et al., 2017). Initially approached from a health perspective, conceptualisations of food literacy now consist of multiple dimensions that incorporate food systems (Classens & Sytsma, 2020; Cullen et al., 2015; Hernandez, 2019; Hernandez et al., 2021; Martin & Massicotte, 2021; Rosas et al., 2021). Cullen and colleagues (2015) offer a broad definition:

> Food literacy is the ability of an individual to understand food in a way that they develop a positive relationship with it, including food skills and

DOI: 10.4324/9781003174417-24

practices across the lifespan in order to navigate, engage, and participate within a complex food system. It's the ability to make decisions to support the achievement of personal health and a sustainable food system considering environmental, social, economic, cultural, and political components.

(p. 143)

This definition recognises ongoing learning "across the lifespan," and outlines personal health and well-being and the sustainability of food systems as outcomes of food literacy. While previous approaches to food literacy and pedagogies have been siloed within disciplines, this definition recognises health, socio-environmental, economic, cultural, and political structures.

Given such a broad framing of food literacy, in this chapter we explore how food systems literacy and dietary guidelines are converging to promote health and sustainability outcomes. We highlight that moving forward, further research is required to evaluate the effects of food systems literacy on sustainable diets and food systems transformation.

We begin the chapter by discussing the convergence of health and sustainability in food literacy and pedagogies. We then present dietary guidelines as a tool to promote food literacy and sustainability. Next, we provide considerations for the incorporation of food systems literacy and sustainability in dietary guidelines. We conclude by highlighting key gaps in the research, notably the need to comprehensively evaluate food literacy programming and the impacts of multidimensional dietary guidelines on citizens' food literacy.

The convergence of health and sustainability in food literacy and pedagogies

The increased availability and consumption of ultra-processed foods and greater mental and physical distance between farm and fork created during the Nutrition Transition (Dixon, 2009) have posed challenges for citizens' food literacy. This increased distance has also allowed for the perpetuation of injustices (i.e., labour exploitation) and environmental degradation within food systems (Clapp, 2020). During the Nutrition Transitions, diets shifted as convenient pre-prepared foods were widely available and consumed (Popkin, 2001). Unfortunately, many of these foods were ultra-processed, high in calories, and offered minimal nutritional benefits (Dixon, 2009). The Nutrition Transition also consisted of increased consumption of meat (Popkin et al., 2012) and reduced intake of traditional foods, contributing to the loss of culture, food security, and increased rates of noncommunicable disease, especially among Indigenous communities (Martin, 2012). These dietary shifts were often not by "choice" and were largely influenced by the social determinants of health such as income, food security, and availability (Marcone et al., 2020).

In response, some countries developed dietary guidelines to help consumers in making food "choices" (Lang & Heasman, 2004, p. 5) and to improve their food literacy. However, as nutrition is a young science, dietary guidance continues to evolve and be debated (Mozaffarian et al., 2018). Current dietary guidelines are beginning to move towards a more holistic perspective, shifting from a focus on single nutrients towards guidance that includes both *what* to eat (i.e., foods and nutrients), *how* to eat (i.e., the context of eating and food), and how this relates to food systems and sustainability (more details can be found in other chapters offering analysis of national food guides, like the Framework for Integrating Sustainability in International Food Based Dietary Guidelines). As shown in Table 20.1, this shift aligns with food literacy frameworks that incorporate functional ("knowing of/ about"), interactive ("knowing how to"), and critical (ability to contextualise and critique) knowledge. Dixon's (2009) analysis of the Nutrition Transition demonstrated the linkages between global food systems and nutritional recommendations (p. 327). Notably, nutritional recommendations form a basis of information not only for health and education programs but also for agricultural and food systems policies (Gonzalez Fischer & Garnett, 2016). Dietary guidelines are therefore essential for educational, agricultural, and food systems policy coherence and may also be a food literacy tool to promote sustainable diets.

Shifts in conceptualisations of food literacy, especially for pedagogies and pedagogical tools, are needed to empower citizens with food systems knowledge and critical thinking skills (Yeatman, 2016). The use of an interdisciplinary food system lens to inform food literacies and pedagogies fosters an integrated and more comprehensive approach to promote sustainable diets and food systems.

The evolution of food-related literacies and pedagogies

Food literacy is rooted in the concept of health literacy (Poelman et al., 2018), which first emerged during the 1970s (Nutbeam, 2000). Nutbeam (2000) provides a conceptual framework for health literacy using three levels: Functional health literacy ("communication of information"); interactive health literacy ("development of personal skills"); and critical health literacy ("personal and community empowerment") (Nutbeam, 2000). This health literacy framework was modified for food literacy by Slater (2013) to be food and health outcomes focused, and more recently by Martin and Massicotte (2021) to also incorporate food systems, as shown in Table 20.1.

The concept of food literacy has gained traction over the last decade to educate individuals about healthy eating and provide them with food skills and the confidence to apply them (Poelman et al., 2018; Vidgen & Gallgoes, 2014) primarily for health outcomes. Although still debated (Truman et al., 2017), current efforts to conceptualise and define food literacy are increasing (Hernandez, 2019; Hernandez et al., 2021; Martin & Massicotte, 2021; Rosas et al., 2021). Within the literature, food literacy may be composed of multiple dimensions including food, nutrition, agricultural, food systems, food justice and sovereignty, and environment, as described in the following sections.

Beyond literacies related to food and nutrition, there are also a broad range of literacies and pedagogies focused on the environmental impact of food systems.

Table 20.1 A framework of food literacy levels. Source: Modified by authors from Martin & Massicotte (2021), adapted from Slater (2013, p. 623), used with permission

Food literacy level	Definition	Examples
Functional food literacy	Basic knowledge and communication of credible, evidence-based nutrition, food, and food systems information, involving accessing, understanding, and evaluating information	• "Awareness of the type and/or varieties of foods (e.g., grain, vegetable and/or fruit, milk, meat)" (Azevedo-Perry et al., 2017, p. 2410) • Awareness of food systems activities and their implications on health, society and the environment; knowledge for making sustainable food choices (Azevedo-Perry et al., 2017, p. 2411)
Interactive food literacy	Development of personal skills regarding food and nutrition issues, and food systems, involving informed decision-making, goal setting, and practices to enhance nutritional health and well-being, and food systems sustainability	• "Ability to perform basic kitchen skills like chop/mix/stir/measure ingredients and prepare meals" (Azevedo-Perry et al., 2017, p. 2410) • Ability to perform basic gardening/farming skills like planting seeds, pulling weeds, caring for plants, and harvesting fruits and vegetables
Critical food literacy	Respect for different cultural, family, and religious beliefs in terms of food and nutrition. Understanding the wider context of food systems (production, processing, distribution, consumption, and waste) and nutritional health, and advocating for individual, community, and institutional changes that enhance nutritional and food systems health at the local, regional, national, and global scales	• Advocating for the availability of culturally relevant foods where one purchases food; advocating for the ability to hunt and/or forage locally for foods • Advocating for sustainable food production and systems; advocating for community growing space and acceptability policies (i.e., front yard gardening, etc.)

Food systems pedagogy is "an approach to teaching and learning about food that is interdisciplinary, that embraces complexity, and that includes experiential learning opportunities" (Classens & Sytsma, 2020, p. 8). Agricultural literacy has a longer standing interest and became more prominent during the rise of corporate farming in the 1980s as calls grew for all citizens to have basic agricultural knowledge

(Kovar & Ball, 2013) so that they would be able to participate in decision-making related to agriculture (Hess & Trexler, 2011). These calls aligned with major changes in the World Trade Organization that broadened the global integration of agriculture and food systems (Magnan, 2016). Like other approaches in food literacies, the focus in agricultural literacy shifted from basic knowledge and understanding to one that also incorporates critical thinking, values (Powell et al., 2008), attitudes, and behaviours (Brune et al., 2020). These approaches are more likely to influence behaviours, which is ideal to foster changes in public perception and promote involvement in agriculture (Brune et al., 2020).

Food systems literacy takes a broader systems approach compared to agricultural literacy (Martin & Massicotte, 2021; Widener & Karides, 2014). Food systems education and pedagogies have sometimes arisen through an environmental literacy lens to educate consumers to promote sustainable diets. These approaches have been critiqued for being elitist and for lacking a critical literacy lens that recognises inequities across food systems (Meek & Tarlau, 2016). The more critical approaches seeking food systems transformation are often for food justice (Widener & Karides, 2014), transformative agroecology learning (Anderson et al., 2018), and critical food systems education for food sovereignty (Meek & Tarlau, 2016).

Considering the critiques of the burden of healthy and sustainable diets being placed on the consumer, many approaches recognise the need to emphasise the "critical" level of food literacy to empower citizens to advocate for changes in their communities (Sumner, 2013). Critical food literacy aligns with the emphasis on agency in the six-dimension conceptualisation of food security by the High-Level Panel of Experts on Food and Nutrition Security for the UN Committee on World Food Security (HLPE, 2020). They define agency as:

> Individuals or groups having the capacity to act independently to make choices about what they eat, the foods they produce, how that food is produced, processed, and distributed, and to engage in policy processes that shape food systems. The protection of agency requires socio-political systems that uphold governance structures that enable the achievement of FSN [food security and nutrition] for all. (p. 10)

Therefore, with enhanced food literacy, citizens might be empowered to use their agency to "engage in policy processes that shape food systems" and their resulting health and well-being. A shift in conceptualisations of food literacies needs to continue to go beyond individual behaviour change (Kimura, 2011) to also include the promotion of agency for broader systems engagement. Since food systems are a significant contributor to environmental degradation, analysts have been calling for more sustainable consumption behaviours given the "influence" of the consumer in the current market-based economy (Lang & Heasman, 2004). Like with the individualisation of the approaches in the other food literacies, this has been seen as elitist because many citizens may not have a socio-economic status that affords them the ability to "vote with their dollars." Thus the importance of an emphasis on improving agency for systems change is growing.

The compartmentalisation of food-related literacies and pedagogies has made it challenging to understand how they could be integrated to promote a more comprehensive understanding (Classens & Sytsma, 2020). This reductionism may have negative implications for policymakers and food-related educators (i.e., classroom teachers and dietitians). As such, scholars call for systems-based, critical, and interdisciplinary approaches to food literacy and pedagogy to improve health and sustainability outcomes (Ingram et al., 2020; Rose & Lourival, 2019; Yeatman, 2016).

Dietary guidelines as a tool to promote food literacy and sustainability

Dietary guidelines are a key tool that have been underexplored regarding their connection with food literacy for sustainable diets. While dietary guidelines are a tool for individual food literacy, they are also a policy tool used in the development of best practices for pedagogical approaches. As dietary guidelines are inherently a type of food-related policy, they provide a basis for the development of educational curricula (Seed & Rocha, 2018). Dietary guidelines, therefore, influence policies at all levels. Implementation of national dietary guidelines influences regional and local jurisdictions (Seed & Rocha, 2018) and, consequently, have significant implications for a country's overall food policy directions (Gonzalez Fischer & Garnett, 2016). A broad conceptualisation including healthy and sustainable diets for dietary recommendations are essential. Figure 20.1 depicts some of these interactions, and while it is not comprehensive, in the context of policy jurisdictions in Canada, it shows how national dietary guidelines can influence policies. For example, dietary guidelines might influence pedagogy in the formal education system. They can also influence informal educational initiatives such as those led by public health organisations or initiatives on a local farm.

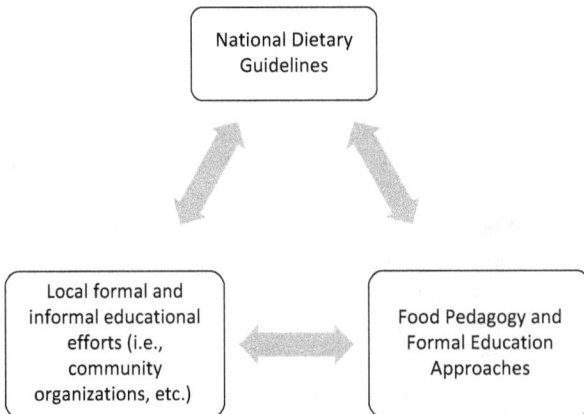

Figure 20.1 Bidirectional influences between policy jurisdictions, food literacy, and pedagogies. Source: Created by the authors.

Emerging policies and pedagogical approaches aim to reflect the comprehensive inclusion of health and well-being, and food systems and sustainability. For example, the Government of Ontario in Canada considered the implementation of a *Food Literacy for Students Act* (Bill 216) that would have included "experiential or hands-on skills learned in gardens and kitchens" (Kramp, 2020). This bill considered making food literacy education mandatory for students from Grades 1 to 12 (ages 6–18). Unfortunately, the bill did not pass before government was dissolved for an election in June 2022. Nonetheless, the momentum of this bill led the Government of Ontario to include some new food literacy expectations across grades (1–8 and 9) in the updated 2022 Science and Technology curriculum (Ontario Ministry of Education, 2022). The curriculum indicates that "skills and knowledge related to food literacy are wide-ranging, from students developing an understanding of where food comes from, including the importance of locally sourced food and how it is grown and prepared, to students investigating the importance of biodiversity in agriculture" (Ontario Ministry of Education, 2022, p. 85). Furthermore, specific learning expectations reflect both health and well-being and food systems lens: "students describe various plants used for food; explain how food literacy can support decisions related to physical and mental health; describe the purpose, inputs, and outputs of systems related to food processing; identify food as a source of energy for living things; and describe how different soils are suited to growing different types of food, including crops" (Ontario Ministry of Education, 2022, p. 85). These changes and this momentum may lead to further curricula-wide changes and possible resubmission of a *Food Literacy for Students Act*, making food literacy mandatory across grades. Given the development of food literacy across the lifespan (Cullen et al., 2015), early and long-term exposure to food literacy is necessary. Effective food literacy education among children and youth may be highly impactful, as research indicates that dietary habits formed as children often continue as kids grow older (Scaglioni et al., 2018).

In Brazil, critical food pedagogies, with an emphasis on place, have been institutionalised in the formal education system through the National Program of Agrarian Reform Education (PRONERA) (Meek, 2015). This institutionalisation has been beneficial for teaching about agroecology in rural lives and contexts and was developed in response to advocacy by the Landless Workers Movement (MST) (Meek, 2015). Despite these benefits, there have also been concerns around industry co-optation from this institutionalisation (Tarlau, 2015). This type of food systems education discourse is also present in the Brazilian food guide, which emphasises that healthy diets must come from socially and environmentally sustainable food systems, ensuring that policies "take into account the impact of the means of production and distribution of food on social justice and environmental integrity" (MHB, 2014, p. 18). For social sustainability, The Brazilian Food Guide notes that this includes:

> the size and use of farms, the freedom of farmers to choose seeds, fertilisers and ways to control pests and diseases, working conditions and exposure to occupational hazards, the nature and number of intermediaries

between farmers and consumers, the fairness of the trading system, employment generation and the sharing of profit between capital and labour.

(MHB, 2014, pp. 18–19)

This recognition of modes of production and farmer agency is reflective of the MST movement and food sovereignty discourses that are critical of agribusiness and demand more agency and justice for food and farm workers (Meek, 2015).

For food literacy pedagogical initiatives to be successful, there are basic components that must be considered. First, food literacy education policies need to be co-constructed with educators to ensure that they will be realistic and effective (Stevenson, 2007). Second, an effective pedagogical framework for food literacy should include knowledge "about/of" (declarative), knowledge "how to" (procedural), and the ability to be critical of information and behaviours, as shown in Figure 20.2.[1] These components can be consistent across exposure levels/age groups with the difficulty and content changing according to the specific context. Pedagogy for food literacy also needs to include both health and well-being, and food systems and sustainability dimensions. Third, the conceptualisation within these dimensions should be broad and interconnected, recognising cultural, political, environmental, and economic factors (Cullen et al., 2015). There are some frameworks that highlight additional dimensions (see Hernandez, 2019), however, we have kept to two core dimensions, for the sake of keeping this framework simple in a policy context. Each dimension could be broken down further to address the different competencies and indicators. A good example of such work has been done by Azevedo-Perry and colleagues (2017) for health and well-being. However, this work is lacking a detailed food systems and sustainability dimension

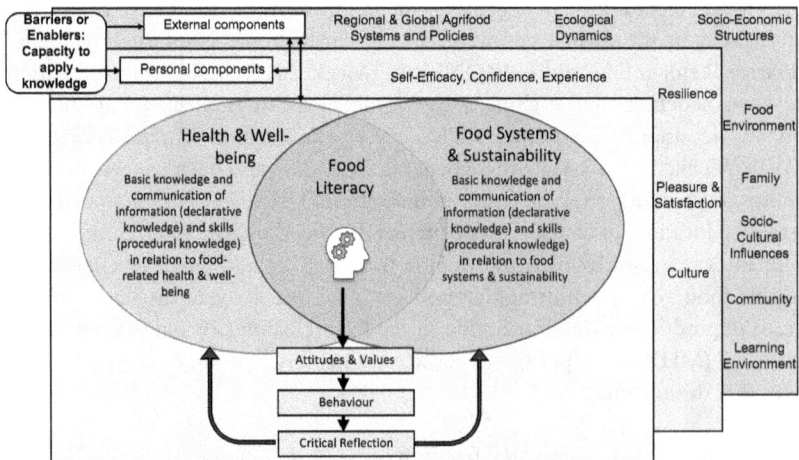

Figure 20.2 Food literacy conceptual framework Source: Adapted from Martin & Massicotte, 2021, used with permission.

(Slater et al., 2018), a common issue that is critical for the implementation of the recommended broad approach. Fourth, acknowledging the barriers and enablers to food literacy is also important (Kimura, 2011). Although food literacy encompasses a behavioural component, which indeed has its own complexities (i.e., attitudes, values, mood, etc.), knowledge does not always translate into action. Figure 20.2 builds upon the approaches of previous frameworks by highlighting barriers and enablers to actualising food literacy, incorporating the different levels of food literacy, and adding critical reflection below the core dimensions (see Anderson et al., 2018; Azevedo-Perry et al., 2017; Block et al., 2011; Colatrugio & Slater, 2016; Cullen et al., 2015; Desjardins et al., 2013; Meek & Tarlau, 2016; Ronto et al., 2016; Sumner, 2013). The framework presented is unique in its focus on the two core dimensions of food literacy.

There are many ways that food literacy and food-related pedagogies have been conceptualised and approached: Some for health, some for food systems, and some for sustainable diets. All of these silos make it challenging to understand the concept in addition to developing pedagogical tools and approaches that are more holistic. There is merit in developing frameworks that contain multiple literacies. Outlining two core dimensions of food literacy—1) health and well-being, and 2) food systems and sustainability—may make this interdisciplinary approach more approachable. Given the overlap between the social, economic, cultural, and political components of health and food systems, further subdivision might lead to more fragmented approaches that make interdisciplinary work more challenging. Given the significance of dietary guidelines as a tool for food literacy, the debates and advancements in food literacy and other food pedagogies should be broadly considered when updating guidelines. These considerations include going beyond providing basic information for declarative knowledge/functional food literacy to providing information for procedural and critical knowledge, including for interactive and critical food literacy. Going forward, incorporating these views and perspectives will help provide holistic pedagogical approaches and guidelines for healthy and sustainable diets. Empirical studies should be conducted to assess the effectiveness of these approaches, as current knowledge is mostly theoretical. In the next section, we provide a summary of considerations for food literacy in the development of dietary guidelines.

Considerations for the incorporation of food systems literacy and sustainability in dietary guidelines

To be an effective food literacy tool, dietary guidelines should:

1) Use comprehensive, evidence-based information (FAO & WHO, 2019)
2) Be easy to find and use clear, simple formatting
3) Explicitly incorporate all levels of food literacy for health and well-being, and food systems and sustainability
4) Be developed by interdisciplinary teams (Gonzalez Fischer & Garnett, 2016)
5) Address cultural and family food traditions, especially in contexts where oppression and colonisation are pronounced (Katz-Rosene, 2020)

6) Avoid perpetuating neo-colonial ideals about what is "healthy." Indigenous and diverse groups should be involved in the development of guidelines to ensure the guidelines do not perpetuate harms (Katz-Rosene, 2020)

Table 20.2 brings together elements from the food literacy frameworks discussed in Table 20.1 and Figure 20.2. In doing so, we have indicated ways that functional, interactive, and critical food literacy under health and well-being and food systems and sustainability can be comprehensively incorporated into dietary guidelines. This conceptualisation of food literacy could be used in policy development to ensure a broad incorporation of food literacy in dietary guidelines, since it's important that dietary guidelines go beyond provision of basic nutritional information to address complex societal issues and promote more sustainable diets. The critical level of food literacy is essential for understanding one's food environment, food systems, and promoting citizens' connection to food and sense of empowerment. The agency of citizens is an important aspect in recent conceptualisations of food security (HLPE, 2020) and is important to dismantle inequalities in the access to and availability of food. The more that dietary guidelines incorporate information that contribute to all three levels of citizens' food literacy, the more we may engage citizens for food security, personal health and well-being, and sustainability. Citizens also need to be empowered with food skills to enact their food literacy with confidence and efficacy (Azevedo-Perry et al., 2017). Incorporating interactive food literacy in dietary guidelines is challenging, as it does not lead to the actual development of skills; however, it is an important element to be included for food educators.

While this may seem like a tall order, the Brazilian and Swedish dietary guidelines do an exemplary job of covering most of these areas. For example, in the Swedish dietary guidelines, pros and cons of various food "choices" are provided to inform citizens and enable them to make decisions by also considering health and environmental outcomes (SNFA, 2015). The Brazilian guidelines discussed the relevance of eating cultural, familial, or traditional foods. They also put highly or ultra-processed foods into context and discuss how they can damage food culture (MHB, 2014). This contextualisation of information may enable citizens to reflect on their food choices, fostering critical food literacy.

To date, few countries have incorporated environmental sustainability and well-being within official dietary guidelines (Gonzalez Fischer & Garnett, 2016). Thus, the consideration of food literacy within dietary guidelines is timely given they may be updated to be multi-dimensional and multi-outcome. The level of interdisciplinarity for tools like dietary guidelines affects approaches for food literacy and pedagogies for policies in other jurisdictions (Seed & Rocha, 2018). As research is converging for healthy and sustainable diets, we need to continue to consider how dietary guidelines can be developed to reflect interconnections between health and sustainability (Willet et al., 2019).

Table 20.2 Framework for incorporating food literacy levels in dietary guidelines. Source: The work of the authors

Health and well-being			Food systems and sustainability		
Functional	*Interactive*	*Critical*	*Functional*	*Interactive*	*Critical*
Provision of credible, evidence-based food and nutrition information	Recommendations to promote and support health and well-being through food and nutrition skills (i.e., cooking, shopping for groceries, etc.).	Acknowledges the complexity of health and well-being and the need to respect cultural, social, family, and religious traditions in terms of food and nutrition Provides broad context of nutritional health to weigh pros and cons of food "choices" and to advocate for individual, community, and institutional changes that enhance nutritional health at the local, regional, national, and global scales.	Provision of credible, evidence-based information about food systems sustainability.	Recommendations to promote and support sustainability through food and food systems skills (i.e., farming/ growing, reducing food waste).	Acknowledges the complexity of food systems and sustainability issues and the need to respect cultural, social, family, and religious traditions in terms of foods produced and available for consumption. Provides broad context of food systems to weigh pros and cons of food "choices" and to advocate for individual, community and institutional changes that enhance food systems health at the local, regional, national, and global scales.

Conclusion

Addressing global environmental and health-related problems is now more critical than ever before (Willett et al., 2019). The increasing impacts of climate change, non-communicable disease, and food insecurity are complicated and arduous issues that will require international and interdisciplinary collaboration. Although solving these problems may seem overwhelming, the benefits for society—healthier people and a healthier planet—will be considerable.

Some claim that fostering food systems literacy among citizens may facilitate solutions to address these complex issues (Ingram et al., 2020; Rose & Lourival, 2019). However, empirical evidence that examines the development of indicators is necessary to assess the effects of food literacy strategies (Slater et al., 2018). To address these gaps, interdisciplinary approaches to education and pedagogy for sustainable diets, and food literacy policies and interventions, need to be better integrated.

Academic and governmental discussion around individual health and sustainability originated as separate discussions, siloed by disciplinary boundaries. Today, these issues are recognised as complex interrelated problems best solved through interdisciplinary collaboration among academics, the agrifood industry, policymakers, educators, healthcare practitioners, and (most importantly) citizens. Among this collaboration, governments and policymakers play key roles in acknowledging the interconnected nature of food systems and sustainability, human health, and well-being to bolster better outcomes (Ingram et al., 2020). A broad understanding of food literacy and food systems is needed among citizens but first needs to be reflected in influential documents such as national dietary guidelines, which may be used as a tool to promote food literacy, health, and sustainability.

Note

1 Declarative, procedural, and critical knowledges are used synonymously in this case with functional, interactive, and critical food literacies. However, the former are depicted in the framework since they are more common terms in the field of education.

References

Anderson, C. R., Maughan, C., & Pimbert, M. P. (2018). Transformative agroecology learning in Europe: building consciousness, skills and collective capacity for food sovereignty. *Agriculture and Human Values, 36*(3), 531–547.

Azevedo Perry, E., Thomas, H., Samra, H. R., Edmonstone, S., Davidson, L., Faulkner, A., Petermann, L., Manafo, E., & Kirkpatrick, S. I. (2017). Identifying attributes of food literacy: A scoping review. *Public Health Nutrition, 20*(13), 2406–2415.

Block, L. G., Grier, S. A., Childers, T. L., Davis, B., Ebert, J. E. J., Kumanyika, S., Laczniak, R. N, Machin, J. E., Motley, C. M., Peracchio, L., Pettigrew, S., Scott, M., & van Ginkel Bieshaar, M. N. G. (2011). From nutrients to nurturance: A conceptual introduction to food well-being. *Journal of Public Policy & Marketing, 30*(1), 5–13.

Brune, S., Stevenson, K. T., Knollenberg, W., & Barbieri, C. (2020). Development and validation of a children's agricultural literacy instrument for local food. *Journal of Agricultural Education, 61*(3), 233–260.

Clapp, J. (2020). Unpacking the world food economy. In J. Clapp (Ed.), *Food* (3rd ed., pp. 1–28). Polity.

Classens, J. & Sytsma, E. (2020). Student food literacy, critical food systems pedagogy, and the responsibility of postsecondary institutions. *Canadian Food Studies*, 7(1), 8–19.

Colatruglio, S., & Slater, J. (2016). Challenges to acquiring and utilizing food literacy: Perceptions of young Canadian adults. *Canadian Food Studies*, 3(1), 96–118.

Cullen, T., Hatch, J., Martin, W., Higgins, J. W., & Sheppard, R. (2015). Food literacy: Definition and framework for action. *Canadian Journal of Dietetic Practice and Research*, 76(3), 140–145.

Desjardins, E., Davidson, L., Samra, R., MacDonald, A., Dunbar, J., Thomas, H., Munoz, M., King, B., Maxwell, T., Wong-McGraw, P., & Shukla, R. (2013). Making something out of nothing: Food literacy among youth, young pregnant women and young parents who are at risk for poor health. Food Literacy for Life. https://foodsecurecanada.org/sites/foodsecurecanada.org/files/food_literacy_study_technical_report_web_final.pdf

Dixon, J. (2009). From the imperial to the empty calorie: how nutrition relations underpin food regime transitions. *Agriculture and Human Values*, 26(4), 321–333.

FAO, & WHO (2019). *Sustainable healthy diets: Guiding principles.* https://document-cloud.adobe.com/link/review?uri=urn:aaid:scds:US:631911c0-56ee-48cc-aa8a-8ddbc1ef16f8

Gonzalez Fischer, C., & Garnett, T. (2016). *Plates, pyramids, planet: Developments in national health and sustainable dietary guidelines: A state of play assessment.* FAO and The Food Climate Research Network at The University of Oxford. https://www.fao.org/3/I5640E/i5640e.pdf

Hawkes, C., Thow, A. M., Downs, S., Ghosh-Jerath, S., Snowdon, W., Morgan, E., Thiam, I., & Jewell, J. (2013). Leveraging agriculture and food systems for healthier diets and noncommunicable disease prevention: The need for policy coherence. In *Expert paper for the second international conference on nutrition.* FAO. http://www.fao.org/fileadmin/user_upload/agn/pdf/HawkesICN2paper_Jul1.pdf

Hernandez, K. (2019). *Exploring food literacy in nova scotia public schools: A critical analysis* [Doctoral dissertation, Dalhousie University].

Hernandez, K. J., Gillis, D., Kevany, K., & Kirk, S. (2021). Towards a common understanding of food literacy: a pedagogical framework. *Canadian Food Studies / La Revue Canadienne des Études sur l'Alimentation*, 8(4), 8–25.

Hess, A. J., & Trexler, C. J. (2011). A qualitative study of agricultural literacy in urban youth: What do elementary students understand about the agri-food system? *Journal of Agricultural Education*, 52(4), 1–12.

HLPE (2020). *Food security and nutrition: Building a global narrative towards 2030.* https://www.fao.org/3/ca9733en/ca9733en.pdf

Ingram, J., Ajates, R., Arnall, A., Blake, L., Borrelli, R., Collier, R., de Frece, A., Häsler, B., Lang, T., Pope, H., Reed, K., Sykes, R., Wells, R., & White, R. (2020). A future workforce of food-system analysts. *Nature Food*, 1(1), 9–10.

Katz-Rosene, R. (2020). "Ditch red meat and dairy, and don't bother with local food": The problem with universal dietary advice aiming to save the planet (and your health). *Canadian Food Studies / La Revue Canadienne des Études sur l'Alimentation*, 7(2), 5–19.

Kimura, A. H. (2011). Food education as food literacy: Privatized and gendered food knowledge in contemporary Japan. *Agriculture and Human Values*, 28(4), 465–482.

Kovar, K., & Ball, A. (2013). Two decades of agricultural literacy research: A synthesis of the literature. *Journal of Agricultural Education*, 54(1), 167–178.

Kramp, D. (2020). *Food literacy for students act 2020, bill 216.* https://www.ola.org/sites/default/files/node-files/bill/document/pdf/2020/2020-10/b216_e.pdf

Lang, T., & Caraher, M. (2001). Is there a culinary skills transition? Data and debate from the UK about changes in cooking culture. *Journal of the HEIA*, 8(2), 2–14.

Lang T., & Heasman M. (2004). *Food wars: The global battle for mouths, minds and markets.* Earthscan.

Magnan, A. (2016). *When wheat was king: The rise and fall of the Canadian-UK grain trade*. UBC Press.

Marcone, M. F., Madan, P., & Grodzinski, B. (2020). An overview of the sociological and environmental factors influencing eating food behavior in Canada. *Frontiers in Nutrition, 7:77.* doi: 10.3389/fnut.2020.00077

Martin, A. & Massicotte, M.-J. (2021). Agrifood systems and food literacy: Insights from two high schools' programs in Ontario. *Canadian Food Studies / La Revue Canadienne des Études sur l'Alimentation, 8*(4), 135-160.

Martin, D. (2012). Nutrition transition and the public-health crisis: Aboriginal perspectives on food and eating. *Critical Perspectives in Food Studies*, 228–241.

Meek, D. (2015). Learning as territoriality: The political ecology of education in the Brazilian landless workers' movement. *The Journal of Peasant Studies, 42*(6), 1179–1200.

Meek, D., & Tarlau, R. (2016). Critical food systems education (CFSE): Educating for food sovereignty. *Agroecology and Sustainable Food Systems, 40*(3), 237–260.

Ministry of Health of Brazil (MHB) (2014). *Dietary guidelines for the Brazilian population.* http://bvsms.saude.gov.br/bvs/publicacoes/dietary_guidelines_brazilian_population .pdf

Mozaffarian, D., Rosenberg, I., & Uauy, R. (2018). History of modern nutrition science—implications for current research, dietary guidelines, and food policy. *British Medical Journal*, 361.

Nutbeam, D. (2000). Health literacy as a public health goal: a challenge for contemporary health education and communication strategies into the 21st century. *Health Promotion International, 15*(3), 259–267.

Ontario Ministry of Education (2022). *Science and Technology Curriculum.* https://www.dcp .edu.gov.on.ca/en/curriculum/science-technology/context

Poelman, M. P., Dijkstra, S. C., Sponselee, H., Kamphuis, C. B. M., Battjes-Fries, M. C. E., Gillebaart, M., & Seidell, J. C. (2018). Towards the measurement of food literacy with respect to healthy eating: The development and validation of the self-perceived food literacy scale among an adult sample in the Netherlands. *The International Journal of Behavioral Nutrition and Physical Activity, 15*(1), 54–54.

Popkin, B. M. (2001). The nutrition transition and obesity in the developing world. *The Journal of Nutrition, 131*(3), 871S–873S.

Popkin, B. M., Adair, L. S., & Ng, S. W. (2012). Global nutrition transition and the pandemic of obesity in developing countries. *Nutrition Reviews, 70*(1), 3–21.

Powell, D., Agnew, D., & Trexler, C. (2008). Agricultural literacy: Clarifying a vision for practical application. *Journal of Agricultural Education, 49*(1), 85–98.

Ronto, R., Ball, L., Pendergast, D., & Harris, N. (2016). Adolescents' perspectives on food literacy and its impact on their dietary behaviours. *Appetite, 107*, 549–557.

Rosas, R., Pimenta, F., Leal, I., & Schwarzer, R. (2021). FOODLIT-PRO: Conceptual and empirical development of the food literacy wheel. *International Journal of Food Sciences and Nutrition, 72*(1), 99–111.

Rose, N., & Lourival, I. (2019). Hegemony, counter-hegemony and food systems literacy: Transforming the global industrial food system. *Australian Journal of Environmental Education, 35*(2), 110–122.

Scaglioni, S., De Cosmi, V., Ciappolino, V., Parazzini, F., Brambilla, P., & Agostoni, C. (2018). Factors influencing children's eating behaviours. *Nutrients, 10*(6), 706. https://www. mdpi.com/2072-6643/10/6/706/htm

Seed, B., & Rocha, C. (2018). Can we eat our way to a healthy and ecologically sustainable food system? *Canadian Food Studies/La Revue Canadienne des Études sur l'Alimentation, 5*(3), 182–207.

Slater, J. (2013). Is cooking dead? The state of Home Economics Food and Nutrition education in a Canadian province. *International Journal of Consumer Studies, 37*, 617–624.

Slater, J., Falkenberg, T., Rutherford, J., & Colatruglio, S. (2018). Food literacy competencies: A conceptual framework for youth transitioning to adulthood. *International Journal of Consumer Studies*, *42*(5), 547–556.

Stevenson, R. B. (2007). Schooling and environmental/sustainability education: From discourses of policy and practice to discourses of professional learning. *Environmental Education Research*, *13*(2), 265–285.

Sumner, J. (2013). Food literacy and adult education: Learning to read the world by eating. *The Canadian Journal for the Study of Adult Education*, *25*(2), 79–92.

Swedish National Food Agency (SNFA) (2015). *Find your way to eat greener, not too much and be active*. https://www.livsmedelsverket.se/globalassets/publikationsdatabas/andra-sprak/kostraden/kostrad-eng.pdf

Tarlau, R. (2015). Education of the countryside at a crossroads: rural social movements and national policy reform in Brazil. *The Journal of Peasant Studies*, *42*(6), 1157–1177.

Truman, E., Lane, D., & Elliott, C. (2017). Defining food literacy: A scoping review. *Appetite*, *116*, 365–371.

Vidgen, H. A., & Gallegos, D. (2014). Defining food literacy and its components. *Appetite*, *76*, 50–59.

Widener, P., & Karides, M. (2014). Food system literacy: Empowering citizens and consumers beyond farm-to-fork pathways. *Food, Culture & Society*, *17*(4), 665–687.

Willett, W., Rockstrom, J., Loken, B., Springmann, M., Lang, T., Vermeulen, S., Garnett, T., Tilman, D., DeClerck, F., Wood, A., Jonell, M., Clark, M., Gordon, L. J., Fanzo, J., Hawkes, C., Zurayk, R., Rivera, J. A., DeVries, W., Sibanda, L. M., Murray, C. J. L. (2019). Food in the Anthropocene: The EAT-lancet commission on healthy diets from sustainable food systems. *The Lancet*, *393*(10170), 447–492.

Yeatman, H. (2016). Developing food literacy through food production. In H. Vidgen (Ed.), *Food literacy* (pp. 205–220). Routledge.

21

FOOD SYSTEMS LITERACY AND CRITIQUE

Liz Nix and Chris Fink

Introduction

The current food system has been criticised for fostering inequality, environmental strain, and deleterious impacts on human health. These impacts vary in their scope across the Global North and South, yet calls for action have grown in frequency and come from various sectors. In recent decades, especially (but not exclusively) in the Global North, the concept of food literacy has been leveraged as a tool to aid consumers in making food choices that could serve to enhance their health. However, it has not historically accounted for systemic barriers to food choices (Covic, 2019). This chapter explores the evolution of food literacy and how it might be used as a tool for systemic action in improving environmental, social, and health-based outcomes from food production and consumption in both the Global North and Global South.

Food system literacy included in food literacy

The concept of food literacy has gained traction in recent years as a way to bridge the gap between dietary health and consumption patterns (Vigden, 2016). Food literacy focuses on the acquisition of food-related knowledge and skills, and the subsequent translation into behaviour and action, all aimed at positive outcomes related to food (Vidgen & Gallegos, 2014). While this operationalised definition appeared only recently, the underlying principles of food literacy have been of great importance for millennia. Had food literacy been defined throughout human existence it may have indicated the knowledge among prehistoric ancestors about which foods were edible or which geographical areas offered the best game, berries, or seeds. As food environments and acquisition have changed, food literacy might have indicated people's abilities to grow their own food or acquire and prepare appropriate and safe foods. In more recent times with increased industri-

DOI: 10.4324/9781003174417-25

alisation, those in the Global North may need the abilities to analyse ingredients and nutrients included in pre-packaged and highly processed foods. In short, food literacy adapts with the concerns of a society and in response to threats to the environment and population.

As food systems become increasingly globalised, definitions of food literacy would need to incorporate more of this complexity (Anand et al., 2015; Popkin, 2004). Food systems, as defined by Neff (2015), encompass "all the activities and resources that go into producing, distributing, and consuming food; the drivers and outcomes of those processes; and, the extensive and complex relationships between system participants and components" (p. 2). Climate change, social injustice, famine and diet-related chronic disease are major worldwide concerns largely influenced by the food system. In an effort to incorporate food system concerns into food literacy, Cullen et al. (2015) proposed the following definition of food literacy:

> Food literacy is the ability of an individual to understand food in a way that they develop a positive relationship with it, including food skills and practises across the lifespan in order to navigate, engage, and participate within a complex food system. It's the ability to make decisions to support the achievement of personal health and a sustainable food system considering environmental, social, economic, cultural, and political components.
>
> *(p. 145)*

This definition includes elements of the food system in individual food literacy. However, creating a conceptual framework for food literacy, with the inclusion of elements of food systems, depends, in large part, on the context in which these systems are found. Measurements and assessment of food systems literacy must therefore be defined by the target population's social, cultural, and economic settings (Amouzandeh et al., 2019; Palumbo et al., 2019; Park et al., 2020; Thomas et al., 2019). Also missing from this definition is an emphasis on the disparities that are explicit in and promoted by the historic context and current manifestation of the globalised food system, including colonialism, displacement, and oppression. Efforts from an individual and systemic perspective should also be focused on building equity within the current food system, recognising the negative impacts of colonialism and systemic racism in diet-related disparities. In addition to addressing social inequities within our food system, food system literacy is an opportunity to elevate the voices, perspectives, and experiences of those populations that have been marginalised and colonised and to elevate food literacy and food system literacy as a form of food sovereignty. As co-authors we acknowledge our position in the Global North, and particularly within the United States. Our intention with the chapter is to engage in discussion, give space for frequently marginalised voices, and illuminate important work by others. We respectfully acknowledge the limitations of our worldview.

In the next section, we will address how food systems literacy might differ between populations in the Global North and Global South. We will also explore how food systems literacy, as part of food literacy, is not only essential to consumers but also to those who work either directly (e.g. food producers, processors, distributors) or peripherally (environmentalists, nutrition specialists, and policymakers) with the food systems.

Consumer food system literacy in the Global North

As humans evolved, they have become increasingly disconnected from the sources and processes that are involved in distributing food to consumers (Popkin, 2004; Tiffen, 2003). This disconnect between food and the broader food systems appears to have both individual physiological effects, as well as widespread societal impacts on social justice, food insecurity, and the environment (Anand et al., 2015). There are many platforms where consumers can seek nutrition and food information that may provide opportunities to promote food systems literacy.

One such platform is from worldwide food-based dietary guidelines (FBDG). Only a few national FBDGs include language regarding sustainability, and some have even been associated with negative environmental impacts, such as increased greenhouse gases (Bergman et al., 2020; Lei & Shimokawa, 2020; Monteiro et al., 2015; Seed, 2015; Springmann et al., 2020). As FBDGs are the basis of advice from health professionals and educational programs, clear recommendations of improving health while also protecting the environment and promoting food justice can impact a variety of programs and perspectives.

While nutrition education that emphasises nutrients may lead to increased interest in supplemental nutrition, an emphasis on whole foods supports broader dietary pattern changes or, better yet, an evaluation and critique of food systems (Krause et al., 2018; Renwick & Smith, 2020; Scrinis, 2016). Education and awareness may provide consumers with an opportunity to improve food system outcomes and sustainability by "voting with their fork." The reciprocal effect of how the food system aids or inhibits dietary behaviours, leading to increased chronic disease and nutrient disparities, should also be considered.

Significant evidence has demonstrated that the general population have little or misconstrued knowledge of agri-food systems (Bellotti, 2010), particularly in urbanised environments (Hess & Trexler, 2011; Pense et al., 2006). Increasingly in the Global North, there are programs intended to reconnect individuals with farm-to-fork processes, in the form of school and urban garden projects, farm tours, or agricultural courses in primary and secondary schools (Powell et al., 2008; Widener & Karides, 2014; Yamashita & Robinson, 2016). Popular farm-to-fork programs have many positive impacts on society by bringing awareness to the food supply chain and often improving dietary health and affinity for fruits and vegetables (Robinson-O'Brien et al., 2009). However, these programs may still be inadequate in fully connecting people to the sources and processes of food acquisition, as many of these programs focus on historical and traditional methods of farming and gardening, whereas in the Global North, most food is produced through industrialised mono-cropping (Specht et al., 2014). Thus, agricultural literacy is just one small component of broader food systems literacy. These raw food commodities often go through extensive food processing and distribution, with each step along the way being highly regulated by food policies. Small urban gardens, therefore, do little to educate individuals about the origins of their food and the various factors that might influence the food industry such as culture, economics, and food policies (Colbath & Morrish, 2010).

Increasing food systems literacy among consumers should be done thoughtfully to avoid creating a shallow understanding of food systems, which could lead to food industry exploitation of consumers' limited knowledge (Specht et al., 2014). Having some level of knowledge, without understanding the broader complexities of the current food system, may lead to reductionist "sustainabilitism" just as early nutritional knowledge helped to build "nutritionism." Scrinis (2016) argues that nutritionism is used by the food industry as a way to promote highly processed food by including phrasing around "good" and "bad" nutrients (Scrinis, 2016). There is early evidence that this same approach could be applied to sustainabilitism in marketing, wherein terms that are popularised but not fully understood by consumers could lead to choices that do not promote health and may have limited effects on environmental sustainability (Torma & Thøgersen, 2021). Food labels that include "organic" or "GMO-free" may be appealing to many consumers, but it is not clear that consumers fully understand the meaning of this labelling (Boccia et al., 2018; Campbell et al., 2013; Wunderlich & Gatto, 2015).

There is evidence of a population eager for food system literacy in the Global North, but there are no standardised methods or resources for providing this literacy (Rainbolt et al., 2012; Wezel et al., 2018). What programs are available are limited in scope and may not address the complexities of the current globalised food system. Education and interventions focused on nutrition or food literacy might expand their scope to extend beyond factors influencing human chronic disease, to chronic conditions of the environment, such as global warming, decreasing biodiversity, and deforestation. However, this individual knowledge and empowerment must be coupled with institutional and political support in promoting sustainable diets. In fact, relying on the consumer to drive sustainable diets doesn't differ significantly from relying on the consumer to make the "right" choices for healthy foods in an environment where food is available in such energy dense forms that the easiest choice is rarely the healthiest choice. It is clear that changes to promote sustainable diets need to be made by other food systems stakeholders with the most power to implement change.

Food systems literacy in food systems stakeholders of the Global North

Influencing environmental, political, and structural factors of food systems takes communication and commitment between various actors and stakeholders of the systems. While food systems literacy is important for the consumer, it is even more crucial for actors within food systems, such as food producers, processors, distributors, dietitians, environmentalists, social activists, and politicians. Stakeholders in the food systems can be primary or peripheral and, often, somewhere along a continuum of involvement. Primary stakeholders are those responsible for producing, processing, and distributing our food. While many of these stakeholders are involved in ongoing discussions of sustainability, it may often be leveraged as a near-term way of enhancing profits. Peripheral stakeholders are those that have significant influence over food systems but are not directly involved in the food supply chains, such as health professionals, dietitians and nutritionists, environmentalists, social justice advocates, and local and national policymakers. In this section, we explore how a systemic understanding and approach can help to address the wicked problems of food systems, potentially leading to solutions that benefit all stakeholders. We also explore common misperceptions, conflicting interests, or restrictive policies for these stakeholders that limit their capacity to make decisions to enhance environmental sustainability, public health, and social justice.

Food producers are one of the primary food supply actors to benefit from the promotion of sustainable diets. Unsustainable food production and greenhouse gas emissions can lead to increased crop difficulties related to weather pattern changes, soil erosion, and increased natural disasters (Giannini et al., 2017; Masud et al., 2017; Sultan et al., 2019; Wiréhn, 2018). However, even as these effects of climate change affect their everyday life and livelihood, some farmers are slow to adopt adaptation and mitigation strategies (Arbuckle et al., 2015; Haden et al., 2012; Mase et al., 2017). This slow adaptation is surprising, given the vast knowledge farmers have of the food supply chains. One possible explanation is a difference in beliefs and attitudes toward the potential effect of food production on climate change (Masud et al., 2017). In addition, immediate cost benefits and policies that reinforce industrial and commodity crops may outweigh long term cost benefits when the payoff of sustainable farming is uncertain (Bonisoli et al., 2018). Lastly, family farms historically have been passed down through generations, leaving the farmer to ensure that the land is in a healthy and productive state. With increasing corporate farms and decreasing family farms, there may be less emphasis on the long-term sustainability of farm land (Magnan, 2015). Also, farmers are not functioning on their own but are part of a larger system of consumers that also have to buy in. Foods need to be produced and consumers need to buy in keeping with the costs of production.

While farmers may be the most likely to benefit from sustainable food practices, moving too quickly to impose policy changes and mitigation strategies on farmers without input from this group may have further economic impact on individuals that are already experiencing financial strain. Farmers across the Global North may receive inadequate payment for crops, leaving them at increased risk of debt and food insecurity (Birthal et al., 2017; Clapp, 2009). Farmers should be key players and decision-makers in the discussion of promoting sustainable diets in the Global North.

Regularly missing from the conversation of sustainability are food processors. Food processing in itself is a viable way to reduce waste from seasonal crops and prevent spoilage. In contrast however, the resources needed for large-scale food processing and the added waste of food packaging may outweigh the environmental benefits of this manner of food waste reduction. Another challenge of large-scale food processing is the prolific consumption of value-added and ultra-processed foods (Juul et al., 2018). Intake of ultra-processed food appears to grow along with the economy of a given country and is associated with increased risk of chronic disease (Baker et al., 2020; Steele et al., 2021). In some ways, the adoption of sustainable and nutritious diets may have major economic impacts on the food processing industry, which may limit their incentive to engage in sustainable food conversations. Large scale food processors, however, do have the capacity and capital to engage in sustainability efforts and may do so in either perfunctory or meaningful ways. These interest groups could be a powerful asset to sustainability conversations that move beyond the use of sustainability buzzwords for marketing to consumers.

The world has seen increases in food production and preservation, and yet we continue to see widespread hunger and malnutrition. In addition to this, 17% of food production globally is wasted, with rates estimated to be over 30% in the Global North (Forbes et al., 2021). Clearly, feeding the world isn't only about production; it is also about distribution and who has access to safe, nutritious, and sustainable food. In the Global North, many neighbourhoods and cities may have limited access to nutritious, let alone sustainable food sources. Lack of quality food in these areas may be attributed to a variety of factors and could be influenced by collaborative efforts and policies of national and international food distributors, as well as local health departments and city planners.

Additionally, healthcare workers in the Global North have become increasingly interested in the impact of food and nutrition in relation to chronic disease, as evidenced by the increase in culinary medicine programs and the integration of sustainability into nutrition and dietetic programs (Carlsson et al., 2019; Parks & Polak, 2020). The incorporation of food literacy into healthcare settings could have an important impact on food choices and consumption. Therefore, it is crucial that these programs also incorporate food system literacy to prevent nutritional promotion at the expense of the environment or the workers within the food system. Furthermore, rooting these efforts in anti-racist, anti-colonial tenets provides an opportunity to recognise and address the structural and systemic forces that reinforce social inequities. These concepts are expounded upon in the chapter.

Food system literacy in the Global South

In the Global South, there have been a variety of solutions proposed by industrialised countries in the Global North as well as grassroots movements to address food access, security, and nutritional deficiencies. These programs have had moderate success at decreasing world hunger and nutrition-related disease. However, more work is needed, particularly as climate change continues to influence production and some societies within the Global South are still being exploited in neo-colonial food production systems, leaving many suffering from extreme poverty, hunger, and death (Fasona & Ogunkunle, 2018; Mugambiwa & Tirivangasi, 2017). Early approaches to food systems literacy often have used similar reductionist "nutritionism" logic, "by prioritising the delivery of individual molecular components of food to those lacking them" (Patel et al., 2015), rather than addressing the underlying oppression, colonialism, poverty, and exploitation that many suffer from in regions of extreme food insecurity. Recent reductionist approaches have included fortification or biotechnology, such as genetically modified rice and bananas to decrease vitamin A deficiency, rather than improving access to naturally rich vitamin A sources, such as red and orange vegetables, leafy greens, or eggs (Beyer, 2010). While these solutions can help ease deficiency burdens in the short term, long-term systemic solutions are needed.

New solutions should be aimed at increasing food literacy and food systems literacy among both consumers and other food systems stakeholders. These efforts should be targeted at reversing inequities brought about by colonialist and exploitative acts of the Global North. Continuing to facilitate the work toward food sovereignty (Nyéléni, 2014) can aid in achieving these outcomes. Food sovereignty, or "the right of communities at all scales, from a village to a whole nation, to democratic self-determination about what food is grown, imported, distributed, and eaten" (Parasecoli, 2019, p. 187), requires food systems literacy as a springboard to activism and the self-determination of communities in relation to food.

In the Global South, there has been an increase in agricultural programs that have a primary objective of increasing food access and security, imparting knowledge and skills of sustainable agriculture, and increasing variety and access in low-income areas (Martin-Guay et al., 2018; Morvaridi, 2012). While these programs are more efficient at addressing some of the root causes of food insecurity, they may still be inadequate if we do not assess the food system in all its complexities (Holt-Giménez & Altieri, 2013). Looking at only the food supply chain as a form of food system literacy could prove inadequate in complex political, economic, and cultural contexts. For instance, many humanitarian agricultural programs have failed in rural Africa due to a lack of understanding of structural gender inequalities (Ogunlela & Mukhtar, 2009; Tibesigwa & Visser, 2016). Similar cultural misunderstandings could lead to failed attempts at promoting food security and sovereignty. Therefore, it is important that these efforts come from the communities that are affected. Grassroots projects such as La Via Campesina and the Network of West African Peasant Organisations have been met with some success (ROPPA) (Ouedraogo, 2010; Rosset, 2013). These organisations have developed successful frameworks for promoting food sovereignty while giving power to those that are most affected by these programs (Shilomboleni, 2017). Other marginalised communities within the Global North that have been also greatly affected by colonialism have become increasingly important voices for food sovereignty movements. Indigenous populations within Canada and the United States have been catalysts for change to promote environmental sustainability and nutritional adequacy while maintaining traditional food practices (Coté, 2016).

Conclusion

Many people may feel powerless to change food systems, particularly if they have limited related knowledge. Promoting more widespread food systems literacy may provide empowerment to consumers and stakeholders to increase food security and sovereignty in local, regional, and international settings. Food systems transformations have the power to create a more just food system that provides adequate nutritional quality, environmental sustainability, and food security.

References

Amouzandeh, C., Fingland, D., & Vidgen, H. A. (2019). A scoping review of the validity, reliability and conceptual alignment of food literacy measures for adults. *Nutrients*, *11*(4), 801.

Anand, S. S., Hawkes, C., de Souza, R. J., Mente, A., Dehghan, M., Nugent, R., Zulyniak, M. A., Weis, T., Bernstein, A. M., Krauss, R. M., Kromhout, D., Jenkins, D. J. A., Malik, V., Martinez-Gonzalez, M. A., Mozaffarian, D., Yusuf, S., Willett, W. C., & Popkin, B. M. (2015). Food consumption and its impact on cardiovascular disease: Importance of solutions focused on the globalised food system. *Journal of the American College of Cardiology*, *66*(14), 1590–1614.

Arbuckle, J. G., Morton, L. W., & Hobbs, J. (2015). Understanding farmer perspectives on climate change adaptation and mitigation: The roles of trust in sources of climate information, climate change beliefs, and Perceived Risk. *Environment and Behaviour*, *47*(2), 205–234.

Baker, P., Machado, P., Santos, T., Sievert, K., Backholer, K., Hadjikakou, M., Russell, C., Huse, O., Bell, C., Scrinis, G., Worsley, A., Friel, S., & Lawrence, M. (2020). Ultra-processed foods and the nutrition transition: Global, regional and national trends, food systems transformations and political economy drivers. *Obesity Reviews*, *21*(12), e13126.

Bellotti, B. (2010). Food literacy: Reconnecting the city with the country. *Agricultural Science*, *22*(3), 29–34.

Bergman, K., Lövestam, E., Nowicka, P., & Eli, K. (2020). 'A holistic approach': Incorporating sustainability into biopedagogies of healthy eating in Sweden's dietary guidelines. *Sociology of Health & Illness*, *42*(8), 1785–1800.

Beyer, P. (2010). Golden Rice and 'Golden' crops for human nutrition. *New Biotechnology*, *27*(5), 478–481.

Birthal, P. S., Negi, D. S., & Roy, D. (2017). *Enhancing farmers' income: Who to target and how?* Policy Paper, 30. National Centre for Agricultural Economics and Policy Research.

Boccia, F., Covino, D., & Sarnacchiaro, P. (2018). Genetically modified food versus knowledge and fear: A Noumenic approach for consumer behaviour. *Food Research International*, *111*, 682–688.

Bonisoli, L., Galdeano-Gómez, E., & Piedra-Muñoz, L. (2018). Deconstructing criteria and assessment tools to build agri-sustainability indicators and support farmers' decision-making process. *Journal of Cleaner Production*, *182*, 1080–1094.

Campbell, B. L., Mhlanga, S., & Lesschaeve, I. (2013). Perception versus Reality: Canadian consumer views of local and organic. *Canadian Journal of Agricultural Economics/Revue Canadienne d'Agroeconomie*, *61*(4), 531–558.

Carlsson, L., Mehta, K., & Pettinger, C. (2019). Critical dietetics and sustainable food systems. In Coveney, J. & Booth, S. (Eds.), *Critical dietetics and critical nutrition studies*, (pp. 97–115). Springer.

Clapp, J. (2009). Food price volatility and vulnerability in the Global South: Considering the global economic context. *Third World Quarterly*, *30*(6), 1183–1196.

Colbath, S. A., & Morrish, D. G. (2010). What do college freshmen know about agriculture? An evaluation of agricultural literacy. *North American Colleges and Teachers of Agriculture Journal*, *54*(3), 14–17.

Coté, C. (2016). "Indigenizing" food sovereignty. Revitalising indigenous food practices and ecological knowledges in Canada and the United States. *Humanities*, *5*(3), 57.

Covic, N. (2019, October 14). *The THE EAT-Lancet Healthy Reference Diet: Perspectives from the Global South*. CGIAR Research Program on Agriculture for Nutrition and Health. https://a4nh.cgiar.org/2019/10/14/the-eat-lancet-healthy-reference-diet-perspectives-from-the-global-south/

Cullen, T., Hatch, J., Martin, W., Higgins, J. W., & Sheppard, R. (2015). Food literacy: definition and framework for action. *Canadian Journal of Dietetic Practice and Research*, *76*(3), 140–145.

Fasona, M., & Ogunkunle, O. J. (2018). Some aspects of natural resource exploitation and management in Nigeria. In Nduka Eneanya, A. (Ed.), *Handbook of Research on Environmental Policies for Emergency Management and Public Safety*, (pp. 22–39). IGI Global.

Forbes, H., Quested, T., & O'Connor, C. (2021). *UNEP Food Waste Index Report 2021*. United Nations Environment Programme.

Giannini, T. C., Costa, W. F., Cordeiro, G. D., Imperatriz-Fonseca, V. L., Saraiva, A. M., Biesmeijer, J., & Garibaldi, L. A. (2017). Projected climate change threatens pollinators and crop production in Brazil. *Plos One*, *12*(8), e0182274.

Haden, V. R., Niles, M. T., Lubell, M., Perlman, J., & Jackson, L. E. (2012). Global and local concerns: What attitudes and beliefs motivate farmers to mitigate and adapt to climate change? *Plos One*, *7*(12), e52882.

Hess, A., & Trexler, C. (2011). A qualitative study of agricultural literacy in urban youth: What do elementary students understand about the agri–food system? *Journal of Agricultural Education*, *52*(4), 1–12.

Holt-Giménez, E., & Altieri, M. A. (2013). Agroecology, food sovereignty, and the new Green Revolution. *Agroecology and Sustainable Food Systems*, *37*(1), 90–102.

Juul, F., Martinez-Steele, E., Parekh, N., Monteiro, C. A., & Chang, V. W. (2018). Ultra-processed food consumption and excess weight among US adults. *British Journal of Nutrition*, *120*(1), 90–100.

Krause, C., Sommerhalder, K., Beer-Borst, S., & Abel, T. (2018). Just a subtle difference? Findings from a systematic review on definitions of nutrition literacy and food literacy. *Health Promotion International*, *33*(3), 378–389.

Lei, L., & Shimokawa, S. (2020). Promoting dietary guidelines and environmental sustainability in China. *China Economic Review*, *59*, 101087.

Magnan, A. (2015). The financialization of agri-food in Canada and Australia: Corporate farmland and farm ownership in the grains and oilseed sector. *Journal of Rural Studies*, *41*, 1–12.

Martin-Guay, M.-O., Paquette, A., Dupras, J., & Rivest, D. (2018). The new Green Revolution: Sustainable intensification of agriculture by intercropping. *Science of The Total Environment*, *615*, 767–772.

Mase, A. S., Gramig, B. M., & Prokopy, L. S. (2017). Climate change beliefs, risk perceptions, and adaptation behaviour among Midwestern U.S. crop farmers. *Climate Risk Management*, *15*, 8–17.

Masud, M. M., Azam, M. N., Mohiuddin, M., Banna, H., Akhtar, R., Alam, A. S. A. F., & Begum, H. (2017). Adaptation barriers and strategies towards climate change: Challenges in the agricultural sector. *Journal of Cleaner Production*, *156*, 698–706.

Monteiro, C. A., Cannon, G., Moubarac, J.-C., Martins, A. P. B., Martins, C. A., Garzillo, J., Canella, D. S., Baraldi, L. G., Barciotte, M., Louzada, M. L. da C., Levy, R. B., Claro, R. M., & Jaime, P. C. (2015). Dietary guidelines to nourish humanity and the planet in the twenty-first century. A blueprint from Brazil. *Public Health Nutrition*, *18*(13), 2311–2322.

Morvaridi, B. (2012). Capitalist philanthropy and the new Green Revolution for food security. *The International Journal of Sociology of Agriculture and Food*, *19*(2), 243–256.

Mugambiwa, S. S., & Tirivangasi, H. M. (2017). Climate change: A threat towards achieving "Sustainable Development Goal number two" (end hunger, achieve food security and improved nutrition and promote sustainable agriculture) in South Africa. *Jàmbá: Journal of Disaster Risk Studies*, *9*(1), 1–6.

Neff, R. (2015). *Introduction to the U.S. food system: Public health, environment, and equity.* John Wiley & Sons.

Nyéléni (2014, April 2). *Newsletter No. 13: Food sovereignty.* Nyéléni. https://nyeleni.org/spip .php?article409

Ogunlela, Y. I., & Mukhtar, A. A. (2009). Gender issues in agriculture and rural development in Nigeria: The role of women. *Humanity & social sciences Journal, 4*(1), 19–30.

Ouedraogo, M. (2010). The network of peasant organizations and producers in West Africa (ROPPA): The role of peasant organisations. *SCN News, 38*(Suppl.), 22–25.

Palumbo, R., Adinolfi, P., Annarumma, C., Catinello, G., Tonelli, M., Troiano, E., Vezzosi, S., & Manna, R. (2019). Unravelling the food literacy puzzle: Evidence from Italy. *Food Policy, 83*, 104–115.

Parasecoli, F. (2019). *Food.* Cambridge. MIT Press.

Park, D., Park, Y. K., Park, C. Y., Choi, M. K., & Shin, M. J. (2020). Development of a comprehensive food literacy measurement tool integrating the food system and sustainability. *Nutrients, 12*(11), 3300.

Parks, K., & Polak, R. (2020). Culinary medicine: Paving the way to health through our forks. *American Journal of Lifestyle Medicine, 14*(1), 51–53.

Patel, R., Bezner Kerr, R., Shumba, L., & Dakishoni, L. (2015). Cook, eat, man, woman: Understanding the new alliance for food security and nutrition, nutritionism and its alternatives from Malawi. *The Journal of Peasant Studies, 42*(1), 21–44.

Pense, S. L., Beebe, J. D., Leising, J. G., Wakefield, D. B., & Steffen, R. W. (2006). The agricultural literacy of urban/suburban and rural twelfth grade students in five Illinois high schools: An ex post facto study. *Journal of Southern Agricultural Education Research, 56*(1), 5–17.

Popkin, B. M. (2004). The nutrition transition: An overview of world patterns of change. *Nutrition Reviews, 62*(suppl_2), S140–S143.

Powell, D., Agnew, D., & Trexler, C. (2008). Agricultural literacy: Clarifying a vision for practical application. *Journal of Agricultural Education, 49*(1), 85–98.

Rainbolt, G. N., Onozaka, Y., & McFadden, D. T. (2012). Consumer motivations and buying behaviour: The case of the local food system movement. *Journal of Food Products Marketing, 18*(5), 385–396.

Renwick, K., & Smith, M. G. (2020). The political action of food literacy: A scoping review. *Journal of Family & Consumer Sciences, 112*(1), 14–22.

Robinson-O'Brien, R., Story, M., & Heim, S. (2009). Impact of garden-based youth nutrition intervention programs: A review. *Journal of the American Dietetic Association, 109*(2), 273–280.

Rosset, P. (2013). Re-thinking agrarian reform, land and territory in La Via Campesina. *The Journal of Peasant Studies, 40*(4), 721–775.

Scrinis, G. (2016). Reformulation, fortification and functionalization: Big Food corporations' nutritional engineering and marketing strategies. *The Journal of Peasant Studies, 43*(1), 17–37.

Seed, B. (2015). Sustainability in the Qatar national dietary guidelines, among the first to incorporate sustainability principles. *Public Health Nutrition, 18*(13), 2303–2310.

Shilomboleni, H. (2017). A sustainability assessment framework for the African green revolution and food sovereignty models in southern Africa. *Cogent Food & Agriculture, 3*(1), 1328150.

Specht, A. R., McKim, B. R., & Rutherford, T. (2014). A little learning is dangerous: The influence of agricultural literacy and experience on young people's perceptions of agricultural imagery. *Journal of Applied Communications, 98*(3), 63–74.

Springmann, M., Spajic, L., Clark, M. A., Poore, J., Herforth, A., Webb, P., Rayner, M., & Scarborough, P. (2020). The healthiness and sustainability of national and global food based dietary guidelines: Modelling study. *BMJ, 370*, m2322.

Steele, E. M., Batis, C., Cediel, G., Louzada, M. L. da C., Khandpur, N., Machado, P., Moubarac, J.-C., Rauber, F., Jedlicki, M. R., Levy, R. B., & Monteiro, C. A. (2021). The burden of excessive saturated fatty acid intake attributed to ultra-processed food consumption: A study conducted with nationally representative cross-sectional studies from eight countries. *Journal of Nutritional Science, 10*, E43.

Sultan, B., Defrance, D., & Iizumi, T. (2019). Evidence of crop production losses in West Africa due to historical global warming in two crop models. *Scientific Reports, 9*(1), 12834.

Thomas, H., Azevedo Perry, E., Slack, J., Samra, H. R., Manowiec, E., Petermann, L., Manafò, E., & Kirkpatrick, S. I. (2019). Complexities in conceptualising and measuring food literacy. *Journal of the Academy of Nutrition and Dietetics, 119*(4), 563–573.

Tibesigwa, B., & Visser, M. (2016). Assessing gender inequality in food security among smallholder farm households in urban and rural South Africa. *World Development, 88*, 33–49.

Tiffen, M. (2003). Transition in Sub-Saharan Africa: Agriculture, urbanisation and income growth. *World Development, 31*(8), 1343–1366.

Torma, G., & Thøgersen, J. (2021). A systematic literature review on meta sustainability labelling: What do we (not) know? *Journal of Cleaner Production, 293*, 126194.

Vidgen, H. A., & Gallegos, D. (2014). Defining food literacy and its components. *Appetite, 76*, 50–59.

Vidgen, H. (Ed.). (2016). *Food literacy: key concepts for health and education.* Routledge.

Wezel, A., Goette, J., Lagneaux, E., Passuello, G., Reisman, E., Rodier, C., & Turpin, G. (2018). Agroecology in Europe: Research, education, collective action networks, and alternative food systems. *Sustainability, 10*(4), 1214.

Widener, P., & Karides, M. (2014). Food system literacy. *Food, Culture & Society, 17*(4), 665–687.

Wiréhn, L. (2018). Nordic agriculture under climate change: A systematic review of challenges, opportunities and adaptation strategies for crop production. *Land Use Policy, 77*, 63–74.

Wunderlich, S., & Gatto, K. A. (2015). Consumer perception of genetically modified organisms and sources of information. *Advances in Nutrition, 6*(6), 842–851.

Yamashita, L., & Robinson, D. (2016). Making visible the people who feed us: Educating for critical food literacy through multicultural texts. *Journal of Agriculture, Food Systems, and Community Development, 6*(2), 269–281.

22

COLLECTIVE ACTION IN UNDERGRADUATE FOOD SYSTEMS EDUCATION TO ENHANCE SUSTAINABLE DIETS

Roland Ebel, Colin Dring, and Mary Stein

Introduction: Collective action and food systems education

Food systems are complex and adaptive political-economic-cultural-socio-ecolog ical-geographical systems, involving multiple feedback loops, actors, and institutions. They are commonly characterised by a lack of knowledge and data, heterogeneity, nonlinearity, interdependence, and self-organisation (Finegood, 2011). There is an urgent need for focused action by food system professionals capable of addressing this uncertainty (Ingram et al., 2020), associated with complex challenges such as the causes and consequences of consumption of energy-dense, ultra-processed foods (Elizabeth et al., 2020), carbon-intensive food production (Marsden, 2013), food system inequities (Cadieux & Slocum, 2015), and food sovereignty (Meek et al., 2019). This effort requires the identification of drivers and determinants of food systems, their political economy, and power relationships across scales (Brouwer et al., 2020).

Given the magnitude of these tasks, the demand for food systems in higher education (FSHE) has been consistently growing (Hilimire et al., 2014). Simultaneously, undergraduate students are increasingly compelled to address the negative human influences on global environmental change and related social inequities (Stuart & Gunderson, 2019). In addition, there is high demand for nutrition and dietetics post-secondary programs to incorporate food systems and sustainability (Harmon et al., 2011). As Hartle and colleagues (2017) note, FSHE pedagogy aims to educate learners in interdisciplinary and systems-level approaches, encompassing food system issues such as adequate nutritional intake, sustainable diets, and food security.

DOI: 10.4324/9781003174417-26

FSHE approaches are well-suited to address the multi-dimensional food system issues that sustainable diets require (Lairon, 2012).

The established pillars of FSHE include transdisciplinarity to develop students' understanding of the agri-food sector as a complex socio-ecological system (Cadieux et al., 2016) and systems thinking to understand the object of an inquiry as a larger set of relationships and factors (Hilimire et al., 2014). In addition, FSHE encourages critical reflection, self-awareness, and positionality in its students, strengthening their awareness that knowledge is socially constructed (Roy et al., 2019). Undergraduate food system professionals are further expected to develop advanced communication, team, and practical skills (Ebel et al., 2020). These skill sets serve as the basis for developing collective agency in students (Valley et al., 2020), preparing them for undertaking food system analyses and becoming involved stakeholders (Ingram et al., 2020).

Collective action and its facilitation in higher education

Emerging from assumptions around individuals' ability to effect transformative and structural change, collective action refers to voluntary action implemented collectively by a group of people, who are sufficiently disturbed by "business as usual" and care enough about an issue to incur significant individual effort (Anderson, 2019).

Collective action scholarship arises from heterogeneous, broad, and sometimes conflicting fields including political science, history, sociology, psychology, and economics. Collective action competence is the capability of a group (of students) with a collective need to direct their behaviour toward a common goal based on a collective literacy, understanding of a group's learning processes, and applying a collective skill set (Clark, 2016). For individuals to effectively work within a collective, organisational, strategic, and team skills are crucial (Ebel et al., 2020). Critical reflection to illuminate coordination among stakeholders from diverse worldviews, as well as relational and communicational skills (for cultivating trust and care among the collective), are equally important (Curnow & Jurow, 2021). The Association of American Colleges and Universities identifies a similar set of competencies linked to civic engagement, including the ability and commitment to collaboratively work across and within community contexts and structures to achieve a civic aim (AACU, 2014).

In their development of equity competence domains, Valley et al. (2020) list the domain *strategies and tactics for dismantling inequity*, which involve: (1) Awareness of historical and current strategies and projects of resistance; (2) knowledge of the effects of oppression, discrimination, and historical trauma; and (3) strategies and tactics with affected individuals and groups to set goals, generate program ideas, make organisational decisions, respect differences in communication and conflict styles, and take steps for collective action.

Drawing on this tradition, collective action in higher education is defined as a theme demonstrated when students are empowered and motivated to act together to achieve a common objective, address critical societal issues, and contribute to the public good (Valley et al., 2017). These actions can be collective or individual (Klandermans, 1997) but hold to a common intention (e.g., food justice (Alkon and Guthman, 2017) or farmworker-led social movements (Minkoff-Zern, 2017)).

One effective way for FSHE to generate solutions to complex challenges is to draw on multiple perspectives, different disciplines, and various sectors of society and impacted groups, through collective action approaches (Valley et al., 2017). We argue that collective action pedagogies support students in developing these skills and competencies for food system problems that continue to include ongoing oppression, systematic violence, and material deprivation for specific social groups.

Although a growing number of higher education institutions have increased collective action and civic engagement activities within their organisational structures and curricula (Murray, 2018; Ostrom, 2009), questions remain regarding the possibility of higher education being a site for collective action. First, current higher education still emphasises pedagogical approaches that focus on individual achievements, competition, and declarative forms of knowledge (Frisk & Larson, 2011) rather than group competencies, collective learning, agency, and action. A second challenge, especially in FSHE, is interrupting a neoliberal educational paradigm where individuals are expected to implicitly know what constitutes the "common good" (Rose, 1999) and where the society governance and individual choices and freedoms interact and collide. Thus, "good citizenship" in a neoliberal regime emphasises self-responsibility, individualism, and entrepreneurialism (Bondi & Laurie, 2005; Dean, 1999). Students are (implicitly or explicitly) taught that this logic of choice, centring on consumerism and materialism, is the baseline for addressing social injustices. Consequently, in FSHE, market-based "alternative" food projects (e.g., farmers markets) have been uncritically characterised as civic engagement, without examining their effect on social justice and environmental outcomes (Guthman, 2008; Johnston, 2008). Third, declarative knowledge approaches assume a causal link between knowledge and behaviour change (McKenzie-Mohr, 2000), generating the risk of oversimplifying complex issues and developing "one-size-fits-all" approaches to solve them.

Additional to the institutional context, FSHE is value-laden itself and requires the educational field to engage with morality, ethics, and norms (Sipos et al., 2008). This requirement becomes most notable when collective action is practising politics. Although politics does not necessarily refer to party politics, collective agency thrives in cultural environments linked to political movements. For example, in the US, several collective action projects (involving students) have emerged related to the Black Lives Matter movement (Murray, 2018). Political topics commonly featured across FSHE include equity, diversity, and inclusion (Valley et al., 2020). Importantly, FSHE educators agree that developing collective agency and systems thinking in students (including perspectives outside Western science) supports minoritised students in seeing themselves as powerful political actors and majoritarian students in developing cultural sensitivity (Mirra & Garcia, 2017; Klein, 2013). Such an understanding of collective agency is an expansion of the capability approach to justice, extending beyond economic metrics of development (Nussbaum, 2003; Sen, 2011).

Suggested curricular activities

Universities can foster collective agency in students by encouraging and supporting them to engage in civic, collective projects, on or outside campus, with or without formal curricular integration (Barnhardt, 2015; Clark, 2016). Such strategies are complementary and enhance each other. In this chapter, we describe pedagogies that are part of curricular activities. A variety of pedagogies, both within and beyond the classroom, and integrated throughout the curriculum, allow FSHE students to practise and develop collective action knowledge and skills. We distinguish between four levels of collective engagement training for students, from mostly preparational pedagogies (with little to no connection to real problems) to actual collective action projects: (1) Classroom activities (dry training) that help students develop collective agency skills without being related to a specific project; (2) classroom projects, where the collective is limited to the students themselves and the perceptions of other stakeholders are, for example, simulated in the form of roleplays; (3) experiential student projects, where the students engage with stakeholders but their role is widely limited by the course instructor or the external participants (e.g., capstone projects or participation in NGO work); and (4) community-based and participatory research projects, where the issue identification, research questions, approaches, agenda, and normativity of a project are equally determined by a collective of students, external partners, and stakeholders.

Classroom training

Diverse deep-learning activities, interactive group techniques, as well as public narratives help students develop a basis for engaging in collective agency (Ebel et al., 2020). Deep-learning activities can be integrated into almost any type of class and enhance the students' critical reflection, communication, and problem-solving skills. Examples include one-sentence summaries of food system issues; advice letters to fictitious future students; and, tabloid titles, where students develop a headline to illustrate a topic (VanGundy, 2008). Interactive group techniques (e.g., debates, role playing) enhance teamwork, facilitate problem-solving, encourage self-reflection and systems thinking, and stimulate creativity and empathy (Szitar, 2014). Jigsaws—when the class is split into different groups, where each become "experts" on one issue and learn from peer "experts" about others—is one of the most appropriate pedagogies for deepening complex diet-related issues and their connection to the wider food system (Carpenter, 2006). Peer reviews of writing assignments represent another common technique. Debates (which are more structured than traditional classroom discussions) cultivate the student's compassion and cultural sensitivity, in addition to sharpening communication skills. A common debate format is "four corners," where students are encouraged to develop a predetermined viewpoint on a topic, independent from their perception on it (Kennedy, 2009). Roleplays and other simulations provide a possibility for experiential learning inside the classroom (Wurdinger & Marlow, 2005), which are usually based on an open-ended case scenario (Szitar, 2014) (e.g., access to healthy nutrition in a food desert).

Practising a public narrative prepares students for engaging in collective action. One such narrative exercise is comprised of three short monologues performed with little preparation: The story of self, the story of us, and the story of now. In the story of self, the narrator shares biographically crucial experiences around choice points. The *story of self* should be given by the instructor. The *story of us* is based on common choice points of a collective (e.g., a group of students), and *stories of now* feature current and upcoming challenges and choices for a collective. Through public narratives, students reflect on their experiences and choices, develop empathy for the experiences of others, and begin to identify bridges for collective project opportunities (Ebel et al., 2020; Ganz, 2011).

Images and symbols, including maps, are tools for practising analysis and effective communication with food system stakeholders. A *rich picture* is a graphic image portraying key elements, events, relationships, forces, ideas, and values of diverse stakeholders in a complex system (Cadieux et al., 2016). Rich pictures can be incorporated into different classroom activities, such as storylines (Avison et al., 1992). *Maps* are pictures that help students identify and abstract the dynamics of complex food system issues. A *concept or affinity map* is a graphical representation of the relationship among concepts, developments, and actors of a food system along with their interactions (Deaton et al., 2016).

Power mapping is the process of creating a concept map that represents all stakeholders that touch a problem and the power dynamics at play. Power maps allow students to critically reflect on where power resides in a food system and how authority, privilege, and oppression play out in the daily lives of its stakeholders (Hildreth, 2014). To create a power map, students commonly interview diverse actors of a real case (Boyte & Finders, 2016; Schiffer, 2007). For example, at Montana State University (MSU), students were presented with a case study of an agricultural community in rural Montana that faces low availability of healthy and diverse food. Students were tasked to identify stakeholders capable of impacting this problem, assess who is most likely to mitigate it, and create positionality statements for each stakeholder. Findings were translated into a power matrix (range of possible actions of the diverse stakeholders), which evolved to a power map characterising stakeholders and their power interactions (Ebel & Thornton, 2019). Depending on the degree of interaction with real stakeholders, co-developed power maps can also serve as a tool for community engagement and participatory appraisal.

Classroom projects

Where consistent community engagement is not possible, students can gain insights into food system issues through classroom projects, where they do not regularly interact with stakeholders or engage with them in a joint project but learn about real cases through literature and online research, as well as sporadic interactions (e.g., interviews) with stakeholders (Ebel et al., 2020).

Multiple factors such as culture, age, and personal experience influence how an individual perceives their positionality relative to others and can influence their engagement in collective action. Students often perceive themselves as having little power, outside of an "in-group," or lacking authority or agency to influence change. Embedding opportunities for students to engage in one-to-one relational encounters with food system stakeholders in conferences, networking events, and informal conversations can improve students' perception of themselves as agents of change. One-to-one relational encounters, different from interviews or team projects, occur between a single student and a stakeholder and consist of an equal, unstructured exchange of (topic-relevant) crucial experiences. Preliminary research at MSU has found students who engaged in one-on-one relational activities to be in greater agreement with the statement "I can foster change," compared to peers that did not participate in such encounters (Ebel et al., 2020).

Case studies link theoretical learning to practical challenges, make abstract concepts concrete, provide a sense of feasibility of community projects, and enhance students' systems thinking, critical reflection, and analytical skills (Francis et al., 2017; Herreid & Schiller, 2013). They deal with specific issues challenging a particular community, and typically centre on a difficult decision (real or fictitious) faced by stakeholders with different perceptions and interests (Barnes et al., 1994). Case studies can be integrated in class examples, in outdoor research, online, and in a flipped classroom. In inductive case studies, the topic is introduced, but problems and patterns are left undefined. In deductive case studies, multiple cases are considered for addressing a broader food system question (Herreid & Schiller, 2013; Wurdinger & Marlow, 2005). In diet-related topics, scenario-based socio-environmental case studies have been proven to be effective. In these case studies, students are confronted with an actual or simulated scenario in which a problem is presented to them in the form of existing knowledge, data, methods from diverse disciplines, and opinions of stakeholders with different points of view. Commonly, students are then required to argue for their preferred solution to the problem (Herreid, 2011; Hilimire et al., 2014; Wei, 2015).

Experiential student projects

Experiential learning augments theoretical classroom learning, enabling students to make critical connections between course material and real-world challenges (Green, 2021). Experiential learning is essential in most FSHE programs, intending to integrate cognitive, psychomotor, and affective learning within the curriculum and provide a mechanism for unveiling worldviews and frames of reference (Valley et al., 2017). It takes many forms in FSHE, including campus farm courses and practicums, internships, capstone course projects, and extracurricular (e.g., student club) affiliated service projects.

Campus farms and gardens present a natural opportunity for student engagement in experiential education that can strengthen practical skills around sustainable food production, knowledge of local food system distribution models and channels, and engagement in social justice efforts (Green, 2021). Internships, often embedded as required components of FSHE programs, allow students to avail themselves of real-world experience and mentorship in roles across the food system. Internships most often have embedded reflection components to better allow for building bridges between practical experiences and theoretical concepts. Furthermore, capstone courses provide an opportunity for team-based engagement and can result in students making important connections with food system stakeholders, exploring different points of view, practising written and oral communications skills, and developing agency as a change-maker.

Service-learning projects can provide relevant, real-world opportunities for students to develop collective action skills while also contributing to the creation of networks and ongoing campus-community connections (Swords & Kiely, 2010). Service-learning projects can foster important community engagement in FSHE programs while providing real-world context through which students can expand their civic agency skills. Finally, extracurricular service projects, such as those taken on by student clubs with sustainable food system-related missions provide an additional, though not well researched, opportunity for experiential learning for FSHE students.

Community-based and participatory research projects

Community-based and participatory action inquiry projects connect learners, researchers, and community organisations. The projects' topics and issues are framed and generated by community organisations and their priorities. Through partnerships with the university, learners and researchers participate in conducting research (data gathering, analysis, dissemination) and/or in applying previously conducted research in a variety of interventions, including workshops, event organising, political campaign, and policy development (Greenberg et al., 2020). The aim of bridging collective agency and community-based and participatory research projects is to have a group of learners co-design, implement, and evaluate civic participation and social action through their joint actions. For example, community-based experiential learning, employing a group project-based approach, was instituted in the Faculty of Land and Food Systems, University of British Columbia, where student teams worked in partnership with the teaching team and community organisations on community-identified priorities (Rojas et al., 2012). Berg and colleagues (2009), utilising community ethnographic approaches with high school students, employed participatory research for social action at multiple scales across a local community intervening in early-onset sex and drug use. Similarly, Driskell et al. (2008) found during their participatory action research project with youth working on emergent and shifting patterns of demographics and settlement, that youth citizenship gained through collective practices resulted in expanded notions of citizenship.

Examples of collective action projects

Since its foundation in 2009, the Sustainable Food and Bioenergy Systems (SFBS) at MSU has involved collective action pedagogies in numerous courses, including classroom training, classroom projects, internships, and community projects, all with a strong emphasis on experiential learning. One example is the SFBS capstone course in the senior year, which involves an annual collective action food system project. In this course, students work in teams with local stakeholders, typically beginning with an in-depth examination of a specific problem conducted through a needs assessment or affinity mapping exercise. They then research an existing food system challenge and propose potential solutions. Examples of past projects include: Measuring food waste on campus and designing a food waste reduction intervention; developing an implementation plan for an edible forest trail in a new park; and, conducting a needs assessment and then developing an implementation plan for a campus food pantry with a dual mission of improving food security and reducing food waste.

Conclusion

Human nature is a complex mixture of the pursuit of self-interest combined with the ability to acquire internal norms of behaviour and to follow rules when understood and perceived as legitimate (Ostrom, 2000). While most students (not only in FSHE) have an interest in civic engagement, many of them worry about committing to anything they perceive as distracting them from meeting "purposeful" academic goals (Hollister et al., 2008). Given the complex challenges future food system professionals will face, FSHE has to (1) affirm that understanding complex challenges cannot be obtained by individual learning only, but demands a collective approach; and (2) provide the conditions and opportunities to prepare students to engage in collective action.

Sustainable diets require sustainable food systems. Sustainable food systems require change agents capable of transforming the dominant industrial food system against the power dynamics that maintain it, by effectively working with diverse stakeholders with varying levels of power and potentially conflicting interests. Such an effort demands educational approaches that can contribute to the creation of these change agents. We argue, therefore, that facilitating collective agency is an essential component of FSHE. We present pedagogies that can help develop this agency in students and invite FSHE educators to implement, study, and enhance these pedagogies.

References

Alkon, A., & Guthman, J. (2017). *The new food activism. Opposition, cooperation and collective action.* University of California Press Books.

Anderson, M. (2019). The importance of vision in food system transformation. *Journal of Agriculture, Food Systems, and Community Development, 9*, 55–60.

Association of American Colleges & Universities [AACU]. (2014). *Intercultural knowledge and competence VALUE rubric.* https://www.aacu.org/value/rubrics/intercultural-knowledge

Avison, D. E., Golder, P. A., & Shah, H. U. (1992). Towards an SSM toolkit: rich picture diagramming. *European Journal of Information Systems, 1*, 397–408.

Barnes, L. B., Barnes, L. B., Barnes, L. B., Christensen, C. R., Hansen, A. J., & Hansen, T. L. (1994). *Teaching and the case method: Text, cases, and readings.* Harvard Business Press.

Barnhardt, C. L. (2015). Campus educational contexts and civic participation: Organizational links to collective action. *The Journal of Higher Education, 86*, 38–70.

Berg, M., Coman, E., & Schensul, J. J. (2009). Youth action research for prevention: A multi-level intervention designed to increase efficacy and empowerment among urban youth. *American Journal of Community Psychology, 43*, 345–359.

Bondi, L., & Laurie, N. (2005). Working the spaces of neoliberalism: Activism, professionalisation and incorporation-Introduction. *Antipode, 37*, 394–401.

Boyte, H. C., & Finders, M. J. (2016). "A liberation of powers": Agency and education for democracy. *Educational Theory, 66*, 127–145.

Brouwer, I. D., Mcdermott, J., & Ruben, R. (2020). Food systems everywhere: Improving relevance in practice. *Global Food Security, 26*, 100398.

Cadieux, K.V., & Slocum, R. (2015). What does it mean to do food justice? *Journal of Political Ecology, 22*, 1–26.

Cadieux, K.V., Levkoe, C., Mount, P., & Szanto, D. (2016). *Visual methods for collaborative food system work*. Hamline University Press.

Carpenter, J. M. (2006). Effective teaching methods for large classes. *Journal of Family & Consumer Sciences Education, 24*, 13–23.

Clark, C. R. (2016). Collective action competence: an asset to campus sustainability. *International Journal of Sustainability in Higher Education, 17*, 559–578.

Curnow, J., & Jurow, A. S. (2021). Learning in and for collective action. *Journal of the Learning Sciences, 30*, 14–26.

Dean, M. (1999). *Governmentality: Power and rule in modern society*. SAGE.

Deaton, M., Wei, C., & Weng, Y.-C. (2016). Concept mapping: A technique for teaching about systems and complex problems. In SESYNC (Ed.), *Best practices for teaching S-E synthesis with case studies*. SESYNC.

Driskell, D., Fox, C., & Kudva, N. (2008). Growing up in the New New York: Youth space, citizenship, and community change in a hyperglobal city. *Environment and Planning A: Economy and Space, 40*, 2831–2844.

Ebel, R., & Thornton, A. (2019, October 16). Big sandy, Montana: Built on sand or food? (Module 1). SESYNC. https://www.sesync.org/big-sandy-montana-built-on-sand-or-food-module-1

Ebel, R., Ahmed, S., Valley, W., Jordan, N., Grossman, J., Byker Shanks, C., Stein, M., Rogers, M., & Dring, C. (2020). Co-design of adaptable learning outcomes for sustainable food systems undergraduate education. *Frontiers in Sustainable Food Systems, 4*, 568743.

Elizabeth, L., Machado, P., Zinöcker, M., Baker, P., & Lawrence, M. (2020). Ultra-processed foods and health outcomes: A narrative review. *Nutrients, 12*, 1955.

Finegood, D. T. (2011). The complex systems science of obesity. In J. Cawley (Ed.), *The Oxford handbook of the social science of obesity*. Oxford University Press.

Francis, C., Jensen, E., Lieblein, G., & Breland, T. (2017). Agroecologist education for sustainable development of farming and food systems. *Agronomy Journal, 109*, 23–32.

Frisk, E., & Larson, K. L. (2011). Educating for sustainability: Competencies & practices for transformative action. *Journal of Sustainability Education, 2*, 1–20.

Ganz, M. (2011). Public narrative, collective action, and power. In S. Odugbemi, & T. Lee, (Eds.), *Accountability through public opinion: From inertia to public action*. World Bank.

Green, A. S. (2021). A new understanding and appreciation for the marvel of growing things: exploring the college farm's contribution to transformative learning. *Food, Culture & Society*, 1–18.

Greenberg, M., London, R. A., & McKay, S. C. (2020). Community-initiated student-engaged research: Expanding undergraduate teaching and learning through public sociology. *Teaching Sociology, 48*, 13–27.

Guthman, J. (2008). Neoliberalism and the making of food politics in California. *Geoforum, 39*, 1171–1183.

Harmon, A., Lapp, J. L., Blair, D., & Hauck-Lawson, A. (2011). Teaching food system sustainability in dietetic programs: Need, conceptualization, and practical approaches. *Journal of Hunger & Environmental Nutrition, 6*, 114–124.

Hartle, J. C., Cole, S., Trepman, P., Chrisinger, B.W., & Gardner, C. D. (2017). Interdisciplinary food-related academic programs: A 2015 snapshot of the United States landscape. *Journal of Agriculture, Food Systems, and Community Development, 7*, 35.

Herreid, C. F. (2011). Case study teaching. *New directions for teaching and learning, 128*, 31–40.

Herreid, C. F., & Schiller, N. A. (2013). Case studies and the flipped classroom. *Journal of College Science Teaching, 42*, 62–66.

Hildreth, R. (2014). *A Coach's Guide to Public Achievement*. Center for Democracy and Citizenship.

Hilimire, K., Gillon, S., Mclaughlin, B. C., Dowd-Uribe, B., & Monsen, K. L. (2014). Food for thought: Developing curricula for sustainable food systems education programs. *Agroecology and Sustainable Food Systems, 38*, 722–743.

Hollister, R. M., Wilson, N., & Levine, P. (2008). Educating students to foster active citizenship. *Peer Review, 10*(2/3), 18–21.

Ingram, J., Ajates, R., Arnall, A., Blake, L., Borrelli, R., Collier, R., De Frece, A., Häsler, B., Lang, T., Pope, H., Reed, K., Sykes, R., Wells, R., & White, R. (2020). A future workforce of food-system analysts. *Nature Food, 1*, 9–10.

Johnston, J. (2008). The citizen-consumer hybrid: ideological tensions and the case of Whole Foods Market. *Theory and Society, 37*, 229–270.

Kennedy, R. R. (2009). The power of in-class debates. *Active Learning in Higher Education, 10*, 225.

Klandermans, P. G. (1997). *The social psychology of protest*. Blackwell.

Klein, J. T. (2013). The transdisciplinary moment(um). *Integral Review, 9*, 189–199.

Lairon, D. (2012). Sustainable diets and biodiversity. In B. Burlingame, & S. Dernini, (Eds.), *Sustainable diets and biodiversity- Directions and solutions for policy, research and action*. FAO.

Marsden, T. (2013). From post-productionism to reflexive governance: Contested transitions in securing more sustainable food futures. *Journal of Rural Studies, 29*, 123–134.

Mckenzie-Mohr, D. (2000). Promoting sustainable behavior: An introduction to community-based social marketing. *Journal of Social Issues, 56*, 543–554.

Meek, D., Bradley, K., Ferguson, B., Hoey, L., Morales, H., Rosset, P., & Tarlau, R. (2019). Food sovereignty education across the Americas: multiple origins, converging movements. *Agriculture and Human Values, 36*, 611–626.

Minkoff-Zern, L.-A. (2017). Farmworker-led food movements then and now. In J. Guthman, & A. H. Alkon, (Eds.), *The new food activism: Opposition, cooperation, and collective action*. University of California Press.

Mirra, N., & Garcia, A. (2017). Civic participation reimagined: Youth interrogation and innovation in the multimodal public sphere. *Review of Research in Education, 41*, 136–158.

Murray, J. (2018). Student-led action for sustainability in higher education: A literature review. *International Journal of Sustainability in Higher Education, 19*, 1095–1110.

Nussbaum, M. (2003). Capabilities as fundamental entitlements: Sen and social justice. *Feminist Economics, 9*, 33–59.

Ostrom, E. (2000). Collective Action and the Evolution of Social Norms. *Journal of Economic Perspectives, 14*, 137–158.

Ostrom, E. (2009). Collective action theory. In C. Boix, & S. C. Stokes, (Eds.), *Oxford handbook of comparative politics*. Oxford University Press.

Rojas, A., Sipos, Y., & Valley, W. (2012). Reflection on 10 years of community-engaged scholarship in the faculty of land and food systems at the University of British Columbia-Vancouver. *Journal of Higher Education Outreach and Engagement, 16*, 195–214.

Rose, N. (1999). *Powers of freedom: Reframing political thought*. Cambridge University Press.

Roy, S. G., De Souza, S. P., Mcgreavy, B., Druschke, C. G., Hart, D. D., & Gardner, K. (2019). Evaluating core competencies and learning outcomes for training the next generation of sustainability researchers. *Sustainability Science, 15*, 1–13.

Schiffer, E. (2007). *The power mapping tool: a method for the empirical research of power relations* (IFPRI Discussion paper 00703). IFPRI. https://ebrary.ifpri.org/digital/collection/p15738coll2/id/38994

Sen, A. (2011). *The idea of justice.* Belknap Press.

Sipos, Y., Battisti, B., & Grimm, K. (2008). Achieving transformative sustainability learning: engaging head, hands and heart. *International Journal of Sustainability in Higher Education, 9,* 68–86.

Stuart, D., & Gunderson, R. (2019, May 3). Toward climate-catalyzed social transformation? *ROAR Magazine.* https://roarmag.org/essays/toward-climate-catalyzed-social -transformation

Swords, A. C. S., & Kiely, R. (2010). Beyond pedagogy: Service learning as movement building in higher education. *Journal of Community Practice, 18,* 148–170.

Szitar, M.-A. (2014). Learning about sustainable community development. *Procedia - Social and Behavioral Sciences, 116,* 3462–3466.

Valley, W., Wittman, H., Jordan, N., Ahmed, S., & Galt, R. (2017). An emerging signature pedagogy for sustainable food systems education. *Renewable Agriculture and Food Systems, 33,* 467–480.

Valley, W., Anderson, M., Blackstone, N. T., Sterling, E., Betley, E., Akabas, S., Koch, P., Dring, C., Burke, J., & Spiller, K. (2020). Towards an equity competency model for sustainable food systems education programs. *Elementa: Science of the Anthropocene, 8,* 33.

Vangundy, A. B. (2008). *101 activities for teaching creativity and problem solving.* John Wiley & Sons.

Wei, C. (2015). *Overview of socio-environmental synthesis.* SESYNC. https://www.sesync.org/ system/tdf/resources/tutorial_1_overview_of_socio-environmental_synthesis_0.pdf ?file=1&type=node&id=967&force=

Wurdinger, S. D., & Marlow, L. (2005). *Using experiential learning in the classroom: Practical ideas for all educators.* Rowman & Littlefield Education.

23

PREFIGURATIVE SPACES OF CRITICAL FOOD LITERACY

The case for campus food growing spaces

Michael Classens and Nicole Burton

Introduction

There is a growing consensus among critical food scholars, activists, and practitioners that the industrialised, capital-intensive food system[1] must be transformed (Altieri & Nicholls, 2020; Michail, 2019.) As Kevany and Prosperi outline in the opening chapter of this collection, the accumulating socio-ecological devastation wrought by this system is staggering. Continuing on with the capital-intensive, industrialised food system—one that profoundly undervalues biophysical nature and the wellbeing of humans and nonhumans (e.g., plants, animals)—will exacerbate climate change and perpetuate social inequality. Put plainly, the operational imperatives of the dominant food system threaten our ability to continue feeding ourselves and will ultimately bring into question the prospect for planetary survival (see, for example, Gerten et al., 2020; Wallace, 2020).

Encouragingly, alternatives to the capital-intensive industrialised food system have proliferated in recent years and signal pathways toward more socially just and ecological rational food systems (Altieri & Nicholls, 2020; Mendez et al., 2015). Advocates, activists, scholars, and allies are prefiguring solutions all along the food chain. Within this context, pedagogy has emerged as an urgent focal point in acknowledgement of the fact that (re)education, broadly speaking, will be central to any putative efforts to scale up a pluriverse of alternative food systems (La Vía Campesina, 2017; Meek, 2020). However, to meet present challenges, a paradigmatic shift in food systems pedagogy is required—"fundamental changes…in both *what* and *how* we teach" (Galt et al., 2012, p. 43, emphasis original). And yet, as so clearly articulated by the co-editors of this collection (see the introductory chapter to this volume), the compounding socio-ecological crises we face leave us little time to act. The gravity of the moment calls for both a reimagining of socio-ecological relations and a radical pragmatism rooted in place. J. K. Gibson-Graham's (2006) political strategy, equal parts transformative and practical, of seizing "the

DOI: 10.4324/9781003174417-27

ubiquitous starting place of *here* and *now*" (p. 194, emphasis original) serves as a rallying cry compelling us to search the everyday for spaces and opportunities for transformation.

The postsecondary campus, inherently political, is increasingly a politicised space—a site that drives and incubates socio-ecological change. Researchers in the United States have noted a 50-year high in university and college student participation in demonstrations. Furthermore, students at more than 84 campuses across the U.S. have made explicit socio-political demands of their institutions' administrations (Morgan & Davis, 2019, p. xvi). Within this institutional ferment, Campus Food Growing Spaces (CFGS) are proliferating on campuses across North America, providing a counterpoint to the traditional agricultural pedagogical paradigm of the land grant university system. We draw on free space literature (Evans and Boyte, 1992; Polletta, 1999) on learning on campus, beyond the classroom (Buddel, 2015; Roberts-Stahlbrand, 2020), to conceptually frame CFGS as spaces of pedagogical innovation that can ignite food systems shifts and transformation.

This chapter expands on this claim, directing specific attention to the role of CFGS as opportunities to advance *critical* food literacy in support of food systems transformation. In as much as the project of animating, and ultimately achieving, substantive food systems change will require an informed and politicised citizenry, widespread critical food literacy is an essential component of food systems transformation. We make the case that CFGS provide ideal pedagogical conditions to enable teaching and learning for critical food literacy. We draw on Classens and Systsma's (2020) definition of critical food literacy, which fuses the pragmatic (planning, managing, growing, and preparing food) with the political (being mobilised to incite food systems change) as a framework to understand the actual and potential pedagogical value of CFGS.

We scaffold our argument empirically, conceptually, and normatively. The chapter draws empirically on the preliminary results from a Canada-wide survey of students (N=65) engaged in CFGS across Canada; a detailed scan of postsecondary institution websites for mention of various CFGS (N=~80); participation in an emerging network of those involved with CFGS across Canada; and our embodied experiences in working with students (and working *as* a student, second author) to advocate for, and learn within, CFGS. Our research is ongoing, responding to a conspicuous gap in critical food pedagogy in North America (and particularly in Canada). So, we combine this exploratory data related to *what is* with conceptual and normative commitments to develop a heuristic exploration of *what could be*. Ultimately, as activist-scholars we are interested in how informal critical food systems literacy learning within CFGS can contribute to transforming the unequal socio-ecological relations of the industrialised, capital-intensive food system.

Issues, trends, and debates

Until relatively recently, the many intersections between food systems transformation and pedagogy have gone largely unexamined. As Jennifer Sumner (2016) aptly states, "Those who study learning have not often turned their gaze toward food, while those who study food have generally overlooked the learning associated with it" (p. xix). Encouragingly, in recent years scholars and practitioners have begun to address these lacunae and in the process have assembled a compelling body of work demonstrating the importance of reckoning with this pedagogical "edible dynamic" (Belasco, 2007, p. 5).

This emerging body of work has a number of characteristics that calibrate it to directly contend with the interconnected social and ecological unsustainability of the contemporary industrialised food system (see, for example, Anderson et al., 2019). This scholarship is interdisciplinary and values diverse epistemologies and ontologies (Classens & Sumner, 2021; Valley et al., 2018). Relatedly, it is values-based and thus directly aimed at addressing socio-ecological inequities resulting from the contemporary food system (Flowers & Swan, 2012; Galt et al., 2012; Sumner, 2016).

Importantly, *critical* food systems pedagogy is motivated by an impulse to unpack and expose various power dynamics that permeate food systems and; more importantly, how we teach about food systems. For example, scholars have demonstrated how garden and farm-based education can operate to reinforce unequitable power relations that cleave along categories of class and racialisation (Flowers & Swan, 2012; Sumner, 2013). That is to say, there is nothing inherently liberatory about garden-based learning. Indeed, the invocation of local food and terroir may reproduce neoliberal logics that valorise particular production and consumption choices of largely upper-middle class white consumers (see Holt-Gimenez & Shattuck, 2011), masking (and reproducing) the "unbearable whiteness of alternative food" (Guthman, 2011, p. 263). In contrast to this, David Meek (2020) poses a provocative question: "How can food systems education move away from producing consumer subjects and toward the production of collective agents who are transgressive, using food knowledge and agricultural practice to systematically dismantle the system itself?" (p. 10).

Within this context, critical food literacy is a key pedagogical objective of teaching for food citizenship, not simply the reproduction of food consumers (however "ethical" their choices) or a technocratic class trained to reproduce the fundamental unsustainability of the industrialised food system (Slater, 2017). While it remains a "concept under construction" (Sumner, 2013, p. 82), food literacy provides a broad framework adaptable to diverse food-learning contexts. Truman et al. (2017), for example, found 39 different definitions of food literacy with applicability across a wide array of settings. They note that two common themes emerge from these diverse approaches: Functional knowledge, related to basic abilities to identify, procure, and prepare food; and, critical knowledge, related to people's ability to assess their relationships to food systems (Truman et al., 2017, p. 213). Yamashita and Robinson (2016) draw on critical literacy scholarship to underpin the notion of a specifically *critical* food literacy that promotes self-reflection regarding one's own positionality within food systems, scrutiny of the broader structural inequities of food systems, and action toward food systems transformation. Classens and Sytsma (2020) combine the "practical with the political" (p. 3), in defining critical food literacy as a set of skills, knowledge, and understandings that (1) equip individuals to plan, manage, prepare, and eat food that is healthy, culturally appropriate, and sustainable, while (2) enabling them to understand the broader socio-political and ecological dynamics of the food system, and (3) empowering them to incite socioecological change within the food system.

(Classens and Sytsma, 202, p. 3)

There is a robust body of scholarship that explores questions of how to teach food literacy from a variety of conceptual and practical perspectives (Brooks & Begley, 2014; Winslow, 2012; Yamashita & Robinson, 2016). However, so far, there has been virtually no work exploring how CFGS can contribute to critical food literacy.

We are particularly interested in how CFGS function as spaces of informal student learning within the broader campus context. Roberts-Stahlbrand (2020) has recently developed a useful definition of informal learning as "any activity or process—physical, mental, emotional, cultural, and/or spiritual—that leads to new understandings, knowledges, skills, values, beliefs and/or tastes" (p. 2). Her work makes a compelling case that informal food systems learning is an essential element operating within campus food dining halls. This definition builds on the notion that the campus environment, outside of the classroom, is constituted by a variety of essential informal learning spaces (see Buddel, 2015; Keeling, 2006), including on-campus wellness centres (Mirwaldt, 2010) and residences (Vetere, 2010). We situate CFGS within this context and argue below that they provide an ideal pedagogical context to teach and learn critical food literacy.

CFGS—a new pedagogical paradigm?

While historically the purview of land grant universities (particularly in the United States), agriculture-based education has proliferated in recent years. In the United States, campuses with active food growing projects increased from 23 in 1992 to an estimated 300 by 2016 (LaCharite, 2016). In Canada, there are upwards of 80 post-secondary campuses with some kind of food growing space (Classens et al., 2021). The size of these spaces can vary from several dozen square metres through to 100 acres or more. In fact, there is significant variability in most aspects of CFGS—in the governance structures, core objectives, the extent of formal integration into program curriculum, and community outreach. While little research on CFGS exists presently, some typological characteristics are beginning to emerge.

A first and crucial distinction of the emerging CFGS is that they differ from the centralised, curriculum, and research-based farms typically associated with departments of agriculture or formal food or agriculture programs. In Canada, for example, the Ontario Agricultural College, founded in 1874 and affiliated with the University of Guelph since 1964, has operated research and teaching farmland for well over a century (Ross & Crowley, 1999). Similarly, in the American context, the land grant university system established a robust network of agricultural research, teaching, and learning across the country. However, the bulk of these have been identified as being in alignment with the dominant model of capital intensive, productivist agriculture and are firmly rooted within the formal course curriculum (La Charite, 2016, p. 522).

In contrast, the emerging paradigm for CFGS provides an explicit alternative to the teaching of conventional agriculture (Barlett, 2011; Sayre & Clark, 2011). These spaces are not regulated solely by positivist and productivist commitments but instead embrace different ways of knowing the world and concerns beyond immediate-term yields. Classens and colleagues (2021) demonstrate that CFGS promote sustainable production (employing agroecological and allied techniques), provide education opportunities that incorporate social and environmental justice, strive to create inclusive community and student leadership, and support student food security on campus. Furthermore, these spaces provide opportunities for experiential and informal learning through social organising and agroecological experimentation: They serve as "insulated spaces for the growth of new nodes, actors and institutions in the food chain" (Barlett, 2011, p. 103).

Relatedly, these spaces are not bounded disciplinarily nor operated only within the context of formal food studies programs. As an example, despite upwards of 80 campus food growing projects in Canada, there are only approximately seven food studies programs across the country (Stephens & Hinton, 2021). This gap suggests that students from a wide array of programs, departments, and faculties—beyond those primarily concerned with food and agriculture—are engaged in CFGS. Indeed, respondents to our survey call a variety of departments and faculties home, from international development studies and Indigenous environmental science to environmental studies and education.

Thomas Eatmon and colleagues (2015) elaborate on this characteristic of CFGS, referring to them as an "integrating context" (p. 326) on campuses. Learning projects that attempt to address complex social, ecological, socio-political, and economic phenomena are challenging, exacerbated by disciplinary boundaries, as well as the often functional divisions between teaching, research, community engagement, and campus operations and administration. In their work, Eatmon et al. (2015) demonstrate that CFGS,

> [E]ncourage the use of transdisciplinary efforts around a single theme… by capitalising on distinct and often opposing perspectives, amalgamating skills and knowledge from multiple sources and experiences, demanding that issues and positions are framed contextually, and applying theory to practice in multiple environments.
>
> *(p. 327)*

As the teaching of productivist agriculture comes under increasing scrutiny for its contributions to the socio-ecological conditions of the dominant food system, CFGS are well positioned to offer a robust and interdisciplinary educational alternative in support of equitable and ecologically rational food systems.

A final distinction worth noting here is that CFGS encourage informal and student-directed action learning: Students play a central role in creating, facilitating, and maintaining these spaces (Gardener, 2012; LaCharite, 2016; Parr & Trexler, 2011). In some cases, the spaces are fully student-led and directed; in others, students have some ability to impact the governance of the space. Our survey found that 48% of respondents had some ability to contribute to the governance of their CFGS. Students help advocate for growing spaces on campus, develop operational and business models, organise community outreach events, and become politically engaged on campus through their involvement with CFGS (see also Sayre & Clark, 2011). These spaces, then, serve as incubation sites for students to develop a range of skills and capacities crucial to prefiguring more just and ecologically rational food systems.

Forging just and sustainable diets through student food activism

Morgan and Davis (2019) reject the cynical sentiment that student activism is divorced from, and inconsequential to, broader political participation and social change. From civil rights and fair trade to global garment workers' rights and anti-apartheid politics, postsecondary students have been a key catalyst driving global socio-ecological change beyond the campus for decades (Barlett, 2011; Morgan & Davis, 2019). The proliferation of CFGS beyond the conventional land grant university model demonstrates, in part, an appetite among students to learn about and prefigure new food systems beyond campus (Gardener, 2012; LaCharite, 2016).

Gibson-Graham (2006) highlights the role of place-based experiential learning in forging post-capitalist configurations. They urge us to locate sundry "openings for learning. Not learning in the sense of increasing a store of knowledge but in the sense of becoming other, creating connections and encountering possibilities that render us newly constituted beings in a newly constituted world" (Gibson-Graham & Roelvink, 2009, p. 322). As Barlett demonstrates, alternative food configurations fit within Gibson-Grahams' (2006) call for "an open and experimental orientation to action" (p. 196, quoted in Barlett, 2011, 104). CFGS are crucibles within which informal and experiential learning opportunities occur and, crucially, from which socio-ecological change can spill. Different ways of relating to Earth and each other, outside of the formal trappings of capitalist relations, are, according to Gibson-Graham and Roelvink (2009), symptoms of, not simply precursors to, socio-ecological change.

Adopting Gibson-Graham's (2006) perspective of "reading for difference rather than domination" (pp. xxxi–xxxii) provides a way of destabilising a tendency within food scholarship to either too hastily celebrate alternative garden projects as unmitigatedly good or too hastily dismiss them as simply reproducing neoliberal logics (see Classens, 2014). Through their feminist and queer critique of political economy, Gibson-Graham (2006) demonstrates the folly in collapsing all activity into that which simply reproduces neoliberal capitalism. They "bring into visibility the great variety of noncapitalist practices that languish in the margins of economic representation" (Gibson-Graham, 2006, p. xxxii) by revealing the lattice work of everyday interventions—such as non-waged labour, care work, and sharing—that exist within the dominant capitalist system. For Gibson-Graham, these represent fissures of possibility, not to be uncritically celebrated but to be understood as (perhaps imperfect) evidence of the practices and socio-natural relations of an emerging economic ethics for the Anthropocene (Gibson-Graham and Roelvink, 2009).

Drawing on free space literature further illuminates ways to conceptualise the pedagogical and transformational potential of CFGS. Evans and Boyte (1992) provide a seminal definition of free spaces:

> Particular sorts of public places in the community, what we call free spaces, are the environments in which people are able to learn a new self-respect, a deeper and more assertive group identity, public skills, and values of cooperation and civic virtue. Put simply, free spaces are settings between private lives and large scale institutions where ordinary citizens can act with dignity, independence and vision.

(Evans and Boyte, 1992, p. 17)

Polletta (1999) articulates the crucial pedagogical role free spaces play in developing political subjectivity and animating social movement politics. For Polletta (1999), free spaces provide movement actors opportunities to "supply leaders, recruit participants, craft mobilising frames, and fashion new identities" (p. 8). Futrell and Simi (2004) identify an important multi-spatial element of free spaces. These spaces, while bound territorially, are not insular. They are in fact a counterpoint to the so-called local trap (Purcell, 2006) and defensive localisms (DuPuis & Goodman, 2005) of NIMBYism. Within the context of food, free spaces enable informal learning experiences that resonate from the materiality of corporeal biopolitics—food that is literally ingested—through to the global issues of climate change, white supremacy, and the political economy of capital-intensive agriculture.

At the same time, free space literature acknowledges the existent power relations of any specific locality. While students have autonomy to act within CFGS, these spaces are (by definition) embedded within the institutional, material, and administrative matrix of their host universities. In some cases, this institutional context can exert an antagonistic force, threatening the viability of the CFGS through, for example, land management and tenure disputes (see, for example, Bomford, 2011; Francis, 2021). Within this context, students are furnished with a breadth of opportunities to learn about and engage in the realpolitik of local organising both on and beyond campus.

Indeed, there are many examples of activities on CFGS leading to concrete action beyond campus. Respondents to our survey overwhelmingly (mean of 4.12 out of 5) said that their experience with CFGS inspired them to become more active in broader food systems issues. What specifically this looks like in practice varies across contexts, though a couple of examples will illustrate the point. The Langside Learning Garden at the University of Winnipeg was developed in collaboration with the Spence Neighbourhood Association to turn a derelict urban lot into a collaborative learning space focused on "food security, local food production and community building" (University of Winnipeg, 2019). The Trent Vegetable Garden donates food to a variety of local front-line food insecurity organisations, including Food Not Bombs and the One Roof Community Centre (Trent Vegetable Gardens, n.d.). In short, CFGS equip and inspire students to incite change beyond their respective campuses.

Conclusion

Education is central to addressing the existing pathologies of the dominant food system while confronting the forthcoming challenges in the age of climate chaos. Success in the face of this turmoil will lie in our collective ability to imagine new paradigms for teaching and learning—to facilitate conditions that nurture engaged, politicised citizens and enable the prefiguration of food systems solutions in real time. CFGS are an example of pedagogical spaces conducive to informal and integrative modes of teaching and learning for a more sustainable and equitable future. These spaces, largely unexamined within food systems scholarship, are already inspiring the prefiguration of more equitable and ecologically sound food systems. However, much work remains to be done to better understand how these spaces as sites of informal learning contribute to student learning outcomes and how they might be more thoroughly integrated into formal curriculum and program goals. There also needs to be a clearer picture of how CFGS can collaborate with community partners to animate socio-ecological change beyond the campus.

It is also worth underscoring that additional research and analysis is required to better understand whether and to what extent the "unbearable whiteness of alternative food" (Guthman, 2011, p. 263) is reproduced within CFGS. While we see vast potential within CFGS as spaces of beneficial socio-ecological change, we do not mean to be uncritically celebratory of them. We agree with Classens et al. (2021) that CFGS "should centre Black, Indigenous, and racialised leadership and look for alternative methods of decision-making and distribution that rely on collectivity and mutuality" (p. 4). To realise the full potential of these spaces, CFGS must be rooted in anti-oppression and liberatory politics, and supported through allied pedagogical approaches, for example the equity competency framework (Valley et al., 2020). Ultimately, there is hope in these spaces and the students, staff, faculty, and allied community partners who make them possible. And in the face of current and forthcoming global challenges, there is pragmatism in the prefigurative food futures enacted within CFGS.

Note

1 We use the terms 'capital-intensive industrialised food system' and 'dominant system' interchangeably throughout, and counterpose these terms against food systems alternatives, in the plural. We do so as a short-hand distinction, cautiously, without meaning to reify the simple binary between 'alternative' and 'conventional' (see for example Harris, 2010; Holloway et al., 2007; and Watts et al., 2005).

References

Altieri, A. M., & Nicholls, C. N. (2020). Agroecology and the emergence of a post COVID-19 agriculture. *Agriculture and Human Values, 37*(3), 525–526.

Anderson, C. R., Binimelis, R., Pimbert, M. P., & Rivera-Ferre, M. G. (2019). Introduction to the symposium on critical adult education in food movements: Learning for transformation in and beyond food movements—the why, where, how and the what next? *Agriculture and Human Values, 36*(3), 521–529.

Barlett, P. F. (2011). Campus Sustainable Food Projects: Critique and Engagement. *American Anthropologist 113*(1), 101–115.

Belasco, W. (2007). *Appetite for change: How the counterculture took on the food industry.* Ithaca: Cornell University Press.

Bomford, M. (2011). University of British Columbia (2000): The improbable farm in the world city. In L. Sayre & S. Clark (Eds.), *Fields of learning: The student farm movement in North America* (pp. 249–268): University of Kentucky Press.

Brooks, N., & Begley, A. (2014). Adolescent food literacy programmes: A review of the literature. *Nutrition & Dietetics, 71*(3), 158–171.

Buddel, N. (2015). Student Affairs and Services Stream: College Quarterly. *College Quarterly, 18*(1), 1–12.

Classens, M. (2014). The nature of urban gardens: Toward a political ecology of urban agriculture. *Agriculture and Human Values, 32*(2), 229–239.

Classens, M., & Sytsma, E. (2020). Student food literacy, critical food systems pedagogy, and the responsibility of postsecondary institutions. *Canadian Food Studies / La Revue Canadienne Des Études Sur l'alimentation, 7*(1), 8–19.

Classens, M., & Sumner, J. (2021). Reflecting on food pedagogies in Canada. *Canadian Food Studies / La Revue Canadienne Des Études Sur l'alimentation, 8*(4), 1–7.

Classens, M., Adam, K., Crouthers, S. D., Sheward, N., & Lee, R. (2021). Campus food provision as radical pedagogy? Following students on the path to equitable food systems. *Frontiers in Sustainable Food Systems, 5,* 1–5.

DuPuis, E. M., & Goodman, D. (2005). Should we go "home" to eat?: Toward a reflexive politics of localism. *Journal of Rural Studies, 21*(3), 359–371.

Eatmon, T., Pallant, E., & Laurence, S. (2015). Food production as in integrating context for campus sustainability. In, W. Leal Filho, N. Muthu, G. Edwin, & M. Sima, (Eds.), *Implementing campus greening initiatives,* (pp. 325–336). World Sustainability Series, Springer International.

Evans, S. M., & Boyte, H. C. (1992). *Free Spaces: The Sources of Democratic Change in America.* University of Chicago Press.

Flowers, R., & Swan, E. (2012). Introduction: Why food?: Why pedagogy?: Why adult education? *Australian Journal of Adult Learning, 52*(3), 419–433.

Francis, F. (2021, April 15). On trent land: An alternative plan. Arthur. Retrieved from https://www.trentarthur.ca/news/an-alternative-lands-plan

Futrell, R., & Simi, P. (2004). Free spaces, collective identity, and the persistence of U.S. White power activism. *Social Problems, 51*(1), 16–42.

Galt, R. E., Clark, S. F., & Parr, D. (2012). Engaging Values in Sustainable Agriculture and Food Systems Education: Toward an Explicitly Values-Based Pedagogical Approach. *Journal of Agriculture, Food Systems, and Community Development, 2*(3), 43–54.

Gardner, L. D. (2012). *Down on the farm: A qualitative study of sustainable agriculture and food systems education at liberal arts colleges and universities* [Masters Thesis. Department of CARRS, Michigan State University].

Gerten, D., Heck, V., Jägermeyr, J., Bodirsky, B. L., Fetzer, I., Jalava, M., Kummu, M., Lucht, W., Rockström, J., Schaphoff, S., & Schellnhuber, H. J. (2020). Feeding ten billion people is possible within four terrestrial planetary boundaries. *Nature Sustainability, 3*(3), 200–208.

Gibson-Graham, J. K. (2006). *A postcapitalist politics.* University of Minnesota Press.

Gibson-Graham, J.K., & Roelvink, G. (2009). An economic ethics for the Anthropocene. *Antipode 41*(S1), 320–346.

Guthman, J. (2011). "If they only knew": The unbearable whiteness of alternative food. In A. Alkon & J. Agyeman (Eds.), *Cultivating food justice: Race, class, and sustainability* (pp. 263–282). The MIT Press.

Harris, E. M. (2010). Eat local? Constructions of place in alternative food politics. *Geography Compass, 4*(4), 355–369.

Holloway, L., Dowler, E., Toumainen, H., Kneafsey, M., Venn, L., & Cox, R. (2007). Beyond the 'alternative' – 'conventional' divide? Thinking differently about food production consumption relationships. In D. Maye, L. Holloway, & M Kneafsey (Eds.), *Alternative food geographies: representation and practice* (pp. 77–93). Elsevier.

Holt-Gimenez, R., & Shattuck. A. (2011). Food crises, food regimes, and food movements: Rumblings of reform or tides of transformation? *Journal of Peasant Studies 38*(1): 109–44.

Keeling, R. P. (2006). *Learning reconsidered 2: Implementing a campus-wide focus on the student experience.* Human Kinetics Publishers.

La Vía Campesina (2017). Peasant agroecology schools and the peasant-to-peasant method of horizontal learning. Retrieved from https://foodfirst.org/wp-content/uploads/2017/06/TOOLKIT_agroecology_Via-Campesina-1.pdf

LaCharite, K. (2016). Re-visioning agriculture in higher education: The role of campus agriculture initiatives in sustainability education. *Agriculture and Human Values, 33*(3), 521–535.

Meek, D. (2020). *The political ecology of education.* West Virginia University Press.

Méndez, V. E., Bacon, C. M., Cohen, R., & Gliessman, S. R. (2015). *Agroecology: A transdisciplinary, participatory and action-oriented approach.* Routledge, CRC Press.

Michail, N. (2019, December 19). *What's the future of farming? It can only be agroecology, says Farms of the Future.* Food Navigator. Online, Retrieved from https://www.foodnavigator.com/Article/2017/08/01/What-s-the-future-of-farming-It-can-only-be-agroecology-says-Farms-of-the-Future#

Mirwaldt, P. (2010). Health and wellness services. In D.H. Cox & C. Strange (Eds.). *Achieving student success: Effective student services in Canadian higher education* (pp. 89–99). McGill-Queen's University Press.

Morgan, D., & Davis, III C. H. F. (2019). *Student activism, politics, and campus climate in higher education.* Routledge.

Parr, D.M., & Trexler. C. J. (2011). Students' experiential learning and use of student farms in sustainable agriculture education. *Journal of Natural Resources and Life Sciences Education 40*(1), 172–180.

Polletta, F. (1999). 'Free spaces' in collective action. *Theory and Society, 28*(1), 1–38.

Purcell, M. (2006). Urban democracy and the local trap. *Urban Studies 43*, 1921–1941.

Roberts-Stahlbrand, A. (2020). *Is the meal hall part of the campus learning system? Investigating informal learning in a university residence meal hall.* [Unpublished Master's thesis, Ontario Institute for Studies in Education. University of Toronto].

Ross, A.M., & Crowley, T. (1999). *The college on a hill: A new history of the Ontario Agricultural College, 1874–1999.* Dundurn Press.

Sayre, L., & Clark, S. (2011). *Fields of learning: The student farm movement in North America.* University Press of Kentucky.

Slater, J. (2017). Food literacy: A critical tool in a complex foodscape. *Journal of Family and Consumer Science, 109*(2), 14–20.

Stephens, P., & Hinton, L. (2021). The state of postsecondary food studies pedagogy in Canada: An exploration of philosophical and normative underpinnings. *Canadian Food Studies, 8*(4), 298–325.

Sumner, J. (2013). Food literacy and adult education: Learning to read the world by eating. *Canadian Journal for the Study of Adult Education, 25*(2), 79–92.

Sumner, J. (2016). *Learning, food and sustainability.* Palgrave MacMillan.

Trent Vegetable Gardens (n.d.). *Who we are*. Retrieved from https://www.trentgardens.org/our-vision

Truman, E., Lane, D., & Elliot, C. (2017). Defining food literacy: A scoping review. *Appetite, 116*, 365–371.

University of Winnipeg (2019, September 4). *Langside learning garden opening*. Retrieved from https://news.uwinnipeg.ca/langside-learning-garden-opening/

Valley, W., Wittman, H., Jordan, N., Ahmed, S., & Galt, R. (2018). An emerging signature pedagogy for sustainable food systems education. *Renewable Agriculture and Food Systems, 33*(5), 467–480.

Valley, W., Anderson, M., Blackstone, N. T., Sterling, E., Betley, E., Akabas, S., Koch, P., Dring, C., Burke, J., & Spiller, K. (2020). Towards an equity competency model for sustainable food systems education programs. *Elementa: Science of the Anthropocene, 8*.

Veter, H. (2010). Housing and residence life. In D.H. Cox & C. Strange (Eds.). *Achieving student success: Effective students services in Canadian higher education* (pp. 77–88). McGill-Queen's University Press.

Wallace, R. (2020). *Dead epidemiologists: On the origins of COVID-19*. Monthly Review Press.

Watts, D. C. H., Ilbery, B., & Maye, D. (2005). Making reconnections in agro-food geography: Alternative systems of food provision. *Progress in Human Geography, 29*(1), 22–40.

Winslow, D. (2012). *Food for thought: Sustainability, community-engaged teaching and research, and critical food literacy* [Unpublished dissertation, Syracuse University]. Retrieved from https://surface.syr.edu/wp_etd/33

Yamashita, L., & Robinson, D. (2016). Making visible the people who feed us: Educating for critical food literacy through multicultural texts. *Journal of Agriculture, Food Systems, and Community Development, 6*(2), 269–281.

24

FOOD GARDENS FOR SUSTAINABLE DIETS IN THE ANTHROPOCENE

Daniela Soleri, David A. Cleveland, and Steven E. Smith

Introduction[1]

Food gardens with a mixture of plant species have likely been an important human subsistence strategy in many places for a long time, (e.g., 1500 years ago in Mesoamerica) (Slotten et al., 2020). So, it's not surprising that they have been a persistent component of programs and projects to improve household and community food production, nutrition, income and savings, and gender and social equity (Cleveland & Soleri, 1987). They have been a featured part of national emergency programs, such as Victory Gardens in the US during WW I, WW II, and the 1930s depression (Lawson, 2005). Since the 1950s, gardens have been part of many "development" programs in the Global South (GS), with some being sponsored by the Global North (GN) and by international organisations like the UN Food and Agriculture Organisation (FAO).

We define food garden diets as sustainable when they contribute positively—to human nutrition and health, community functioning, social equity, biodiversity, animal welfare, and environmental and climate stability—relative to other ways of obtaining food (Soleri et al., 2019). While data show that food garden diets can be sustainable, they can also be relatively unsustainable: It depends on how gardens function within their specific biophysical and social contexts.

The Anthropocene and food garden diets

Human impact on the Earth became so significant by the end of the 20th century that scientists proposed a new geological age, the Anthropocene epoch (Steffen et al., 2015). The Anthropocene is widely conceptualised to include the social, cultural, and economic consequences of increasing human population and per capita consumption, resulting in a public and planetary health crisis (IPCC, 2019; Nugent & Fottrell, 2019; Ripple et al., 2020; Steffen et al., 2018;

DOI: 10.4324/9781003174417-28

Swinburn et al., 2019), including the anthropogenic climate crisis (ACC). As a result of these Anthropocene global trends, gardens face new challenges, and new strategies may be needed for them to be sustainable.

Based on our personal and professional experiences gardening and working with gardeners in different locations in the US and elsewhere in the GS and GN, and our research and review of the literature, we have found that successful Anthropocene food gardens depend on understanding their biophysical contexts through ecological and evolutionary thinking, and their social contexts including the roles of individual behaviours, social organisation, and knowledge systems. We have summarised these five key ideas in Figure 24.1 (Soleri et al., 2019).

In this chapter, we answer the following questions, with a focus on the US and GN.

- What are the potential contributions of food gardens to sustainable diets?
- What are the challenges for achieving this?
- How can food gardens successfully respond to these challenges?

Potential contributions of food gardens to sustainable diets

Food systems are increasingly dominated by large corporations whose main goal is profit, resulting in diets dominated by ultra-processed foods (UPFs) and animal-based foods, which are environmentally, socially, and economically unsustainable.

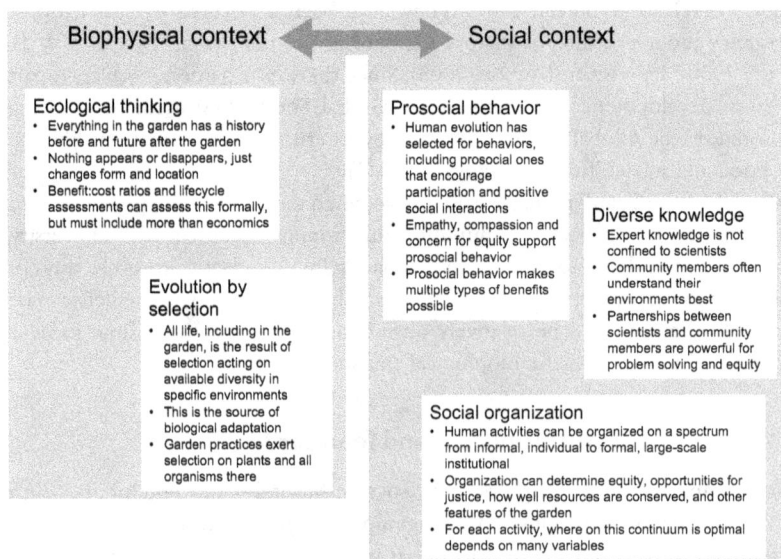

Figure 24.1 Five key ideas for fostering sustainable diets through food gardens. Adapted by the authors from *Food Gardens for a Changing World* (Soleri et al., 2019), used with permission.

These diets dominate the GN and are spreading rapidly to the GS, resulting in a pandemic of non-communicable diseases, wide scale degradation of resources, exacerbated climate change, and increasing inequity and inequality (Swinburn et al., 2019). Food gardens today can support sustainability by contributing to physical and mental health, social and economic benefits, and environmental and climate stability.

Nourishing food for physical health

Fresh vegetables and fruit from the garden are good sources of many nutrients important for human health, including vitamins, minerals, and amino acids, as well as valuable non-nutrients like antioxidants and dietary fibre (Soleri et al., 2019). These garden foods can supplement diets, or replace UPFs, either with the same number of calories while providing more nutrients and healthy compounds, or with the same volume, reducing calories and also increasing healthfulness. Garden crops like basil, chilis, fenugreek, garlic chives, *papaloquelite*, and rose, feijoa, and calendula petals, can provide flavours and colours with cultural significance and help replace less healthy ingredients in the diet, like added sugars and salt. In the GN, working in gardens can also increase vegetable and fruit consumption, for example, by children and youth (Savoie-Roskos et al., 2017), and by community gardeners, compared with home gardeners or non-gardeners (Alaimo et al., 2008; Litt et al., 2011).

Mental health and social benefits

Food gardens can be "therapeutic landscapes" (Hale et al., 2011), and gardeners appreciate the physical, emotional, and social benefits of gardening (Egerer et al., 2018b; Waliczek et al., 2005). A study in Denver found that community gardeners' perception of how pleasant their neighbourhood is was positively associated with fruit and vegetable consumption, and with community gardening compared to home gardening or no gardening (Litt et al., 2011). Gardens provided members the opportunity to experience a diversity of social roles, including being leaders, followers, and learners (Teig et al., 2009).

Working in food gardens can increase awareness of biological and ecological processes, which improve gardeners' understanding of human and environmental health (Hale et al., 2011) and facilitate interactions with natural and social environments, especially in the poorest neighbourhoods (Voicu & Been, 2008). Participating in community gardens increases face to face interactions and can create mutual trust and social cohesion. In New York City, for example, gardens were the source of the social cohesion that was the basis of organising and mutual support following Hurricane Sandy in 2012 (Chan et al., 2017). Food gardens can also help immigrants and refugees feel connected through familiar activities (Wen Li et al., 2010). When controlled by gardeners themselves, gardens growing traditional crops and varieties can be a part of healing responses to the historical trauma that affects many people (e.g., African, Asian, Mexican, and Native Americans in the US) (Ramírez, 2015).

Community gardens may also be organised to increase access and social equity. For example, in Seattle, the P-Patch Community Garden Program comprises 89 gardens throughout the city, worked by hundreds of gardeners, who in 2020 contributed 1628 kg (3583 lb) of produce/month to local food assistance programs (SDN, 2021). In the same year, this program declared an antiracism focus, with commitments to revising plot applications to encourage and support BIPOC participation.

Economic sustainability

In the US, saving money is a major motivation for food gardening (NGA, 2014). The few studies available found garden harvest values to be from \$12–\$35 per m^2, but these did not include the cost of inputs (Soleri et al., 2019). With resources becoming increasingly scarce and many growing environments becoming more stressful, the cost of inputs such as piped water or imported compost will become more important in calculating the economic sustainability of gardens; thinking in terms of ecological networks and life cycles of what comes into and out of the garden is essential.

Environmental sustainability

Compared with the industrial food system, food gardens can produce food in more environmentally sustainable ways, using less energy and resources per unit of food production, storage, cooling, and packaging (Cleveland et al., 2017). Fruit trees and other large perennials can capture and store carbon, absorb some soil pollutants, block the movement of airborne pollutants, and provide cooling and shade (Pataki et al., 2011). Gardens can attract and protect pollinators and other wildlife through their plant diversity and planting complexity (e.g., with tall and short plants and different growth habits and life cycles, Goddard et al., 2010).

Still, garden management determines whether they are providing positive ecosystem services or negative "disservices" (Cameron et al., 2012). Ecological thinking and consideration of the benefit:cost ratio is needed to distinguish between services and disservices. For example, it may take years for a transplanted fruit tree to sequester as much carbon in the plant as was released as a result of growing the tree, transporting it to the garden, and disturbing the soil during planting (Cameron et al., 2012).

Yields of harvested food per unit area are often greater from smaller areas, like gardens, primarily because small areas are more carefully managed. Gardens can also be more efficient in harvest per unit of other inputs like water or compost but are less efficient in harvest per unit of labour because of the time required. However, time spent in the garden is often a benefit rather than a cost. In addition, the environmental and social costs of the fossil fuels that increase labour efficiency in industrial agriculture would need to be calculated for an accurate estimate of their true efficiency.

The need to be vigilant

The many potential benefits of gardens for sustainable diets are not automatic. Even contributions from food gardens that are positive in some ways can have negative effects in others (Soleri et al., 2019) because they are embedded within larger inequitable, unjust, environmentally destructive food systems dominated by neoliberal economic policies that don't prioritise individual, community, or planetary wellbeing (McClintock, 2014). That is, there are trade-offs. For example, the gardening industry promotes the use of toxic pesticides and herbicides, and crops with high water, nutrient, and management requirements, which can increase short term yields and improve diets. Yet these practices also have negative effects on the environment, biodiversity, and long-term human health.

An approach that integrates biophysical and social perspectives is needed, which can help us identify and weigh the trade-offs that will often be required. For example, garden projects that increase social equity in some ways can increase social inequity in others (McClintock, 2014). Garden space in housing projects can be a source of fresh vegetables and fruits but can promote gentrification and is not a substitute for structural changes that would result in better wages and living conditions, play spaces for children, or affordable grocery stores (Wolch et al., 2014).

Challenges to food garden sustainability in the Anthropocene

The need for the benefits of food gardens is greater than ever, but so too are the challenges to realising these benefits. These challenges go beyond the environmental and social variation gardeners are familiar with, to the novel challenges brought by the accelerating directional change that is the hallmark of the Anthropocene.

Familiar variation v. trends

To respond to these Anthropocene challenges in ways that will support food gardens for sustainable diets, we need to distinguish between familiar variation and trends. Experienced gardeners are aware of a range of familiar variation that affects them and their gardens (e.g., variation in the date of first frost, annual rainfall, or the size of pest populations), and they often have effective strategies for dealing with the range of this variation. In contrast, trends are changes that are directional and cumulative, lead to conditions not previously experienced, and present new challenges that may require new strategies.

The rates of change in trends over time can be accelerating. For example, the average rate of global temperature increase went from 0.04 °C (0.07 °F) per decade over the 70 years from 1880–1950, to 0.13 °C (0.23 °F) per decade over the 64 years from 1951–2015 (NOAA NCEI, 2016). Accelerating rates of change characterise trends in many indicators of the Anthropocene (Ripple et al., 2020). For all these reasons, trends can be especially challenging.

Trends challenging food garden contributions to sustainable diets

To support sustainable diets, food gardens must respond to four major Anthropocene trends: 1) Declining quantity and quality of garden resources like water and land; 2) the ACC; 3) rising social and economic inequity; and 4) an ageing, urbanising population.

Anthropocene trends can interact synergistically in ways that increase their combined impact. For example, the ACC is decreasing the quality and quantity of water in many areas, and existing inequitable structures of resource distribution mean that people most in need of food gardens have a harder time accessing water for irrigation.

However, the results of interacting variables are often difficult to predict. For example, because of the ACC, food gardens everywhere will experience increased atmospheric CO_2 concentrations, but some areas will have higher temperatures and less precipitation, while others may receive the same amount of precipitation as in the past but it will arrive in more extreme weather events that diminish the water storage capacity. With so many interacting variables, the net effect of ACC on plants is difficult to predict in detail, but in general, growing food gardens will become more challenging. The effects of ACC also vary spatially. Parts of northern and southern Africa, southern Europe, and much of the western US will receive less precipitation (IPCC, 2014), meaning gardeners will have to supply more of the water their gardens need through irrigation. In other areas, like south Asia, northern Europe, and the north-eastern US, gardeners will be challenged by increasingly intense seasonal rainfall and flooding.

How food gardens can be a response to Anthropocene challenges

For food gardens to support sustainable diets now and in the future, they must be able to respond effectively to Anthropocene trends. Critical to success is understanding how these trends differ from familiar variation, and then responding by using observation, experimentation, and collaboration among gardeners and others across wide social and physical spaces.

Conceptualising change and response

The probability that trends will have a negative impact on food gardens depends on exposure to the change (how much and how frequently) and how sensitive the garden is to the change (Fellmann, 2012). This relationship can be quantitatively estimated, but gardeners can also use it qualitatively to see how to minimise exposure to a change by escaping it and reduce sensitivity by avoiding or tolerating potential harm.

Our ability to respond to change by reducing exposure and/or sensitivity to the effects of change, is our response capacity, as shown in Figure 24.2. Responding successfully leads to resilient, sustainable gardens—but resilience does not always mean returning to the way things were before. When a trend results in large changes, and increased uncertainty, resilience can also require successful transformation, such as replacing heat- and drought-sensitive crops that have been a good

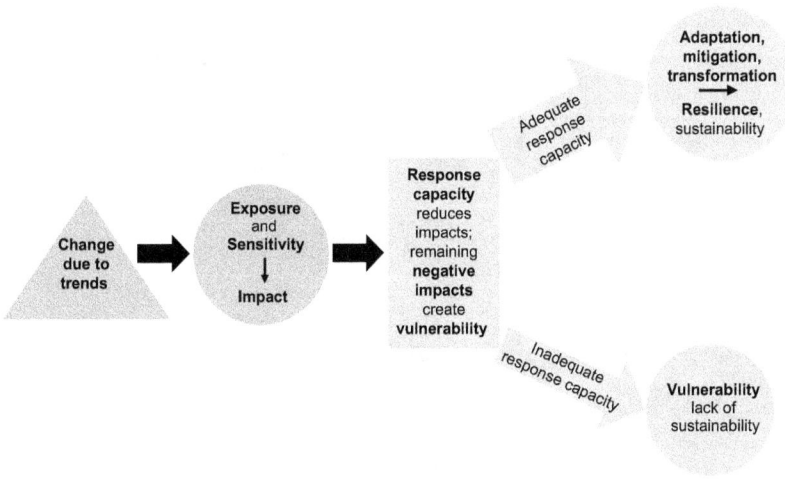

Figure 24.2 Responding to the changes of Anthropocene trends. Adapted by the authors from figure in Soleri et al., 2019, used with permission.

source of iron and vitamin C, like spinach, with new ones that are not so sensitive, like chard or nopales. Vulnerability is the negative impact experienced when our response capacity is inadequate.

The greater our response capacity, the larger the range of conditions we can effectively respond to, remaining resilient, sustainable, and not be vulnerable to the negative impact of the changes (Fellmann, 2012). Coping is our capacity to respond to the recurring, temporary changes of familiar variation. But trends can result in new conditions with which gardeners have no experience. Therefore, responding requires more than coping—it requires behavioural adaptation to the changes we are experiencing now and preparation for the future (Fellmann, 2012), which frequently requires new strategies. It's important to keep the specific goals of our garden in mind so our responses to change support sustainable food garden diets. For example, if the main goal of a community garden is growing vegetables for members, responding to change would be very different than if the primary goal is to provide youth with outdoor activities.

Some trends have been predicted with certainty. For example, increasing temperature due to ACC means that in some locations the usual coping strategies won't be sufficient, and new strategies to adapt or transform will be needed, as shown in Figure 24.3. Gardeners can reduce exposure to heat by shading and mulching, reduce sensitivity by changing to more heat-tolerant crop varieties, and ask for ideas from gardeners who have experience working with high temperatures.

Other examples of responding to trends in order to maintain sustainable food gardens and diets are reducing sensitivity to increasing urbanization and population ageing by developing infrastructure and institutions that

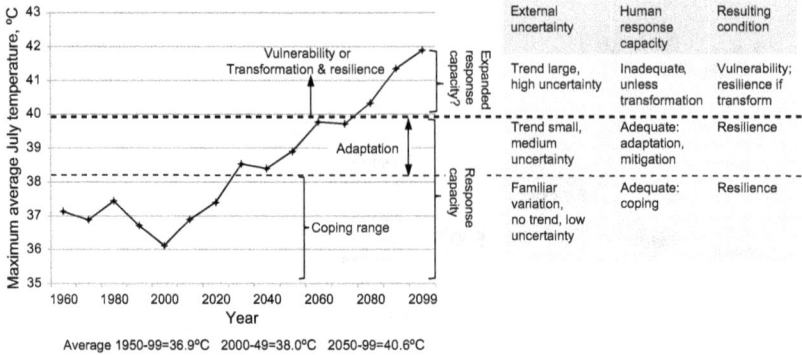

Average 1950-99=36.9°C   2000-49=38.0°C   2050-99=40.6°C

Figure 24.3   Hypothesised scenario involving coping, adaptation, and vulnerability related to average daily maximum temperature in July (°C), 1950–2099, Bakersfield, California. Concept based in part on Fellmann (2012) and sources therein. Data (decadal means) from GIF (2018), observed data 1950–2015; 2016–2099 projections based on high emissions scenario [A2] and CNRM CM3 model; station GHCND: USW00023155. Adapted by the authors from figure in *Food Gardens for a Changing World* (Soleri et al., 2019), used with permission.

make gardening easier for older city dwellers; reducing sensitivity to rising social inequity by developing partnerships and policies supporting community gardens for food sovereignty in disenfranchised communities; and reducing exposure to, and mitigating, the diet-related noncommunicable disease pandemic by expanding community and school gardens for growing and eating fruits and vegetables, while providing physical activity and social interaction.

Observing, experimenting, and collaborating

Gardeners can improve response capacity, reduce vulnerability, and support mitigation through observation and experimentation to find effective practices, and by working together to connect and organize through networks and institutions. A key to effective observation and experimentation is the complementarity of local and formal scientific knowledge. It is increasingly recognized that the local knowledge of gardeners and farmers is similar to scientists' knowledge in many ways, since both are based on empirical observations over time of the same reality, and on testing relationships between different variables (Soleri & Cleveland, 2009).

Observing—Observant gardeners can understand a lot by simply looking carefully and consistently (e.g., at pest and beneficial organism activity). Observations can help gardeners determine which garden practises work best or compare ideas about improvements that can be made right away. For example, a gardener may compare a crop's response to amounts of irrigation or added compost. However, informal experiments don't control for variables other than those being experimented with, and therefore the results can sometimes be unclear, or not supported by subsequent experience. If this is the case, more formal experiments that minimise confounding variation can be used to make the results more reliable. These experiments will require more work, time, and often space, so are most appropriate for school and community gardens (Soleri et al., 2019).

Experiments—Formal garden experiments typically involve identification of a hypothesis to be tested, including dependent and independent variables. Formal experiments can quantify responses to those variables, giving gardeners a more certain, generalisable understanding of the benefits and costs of gardening practices than is possible with single-year informal experiments. With this understanding, gardeners can adjust to increase the benefit:cost ratio by modifying or eliminating a crop, practise, or output with a low benefit:cost, or increasing those with high benefit:cost.

When estimating benefit:cost, it is important to include inputs as well as outputs and measure variables that are good indicators of the garden's goals. Harvests are a key output and can be measured in different ways (Soleri et al., 2019) but may not be appropriate for many gardeners' goals. For example, if the goal is improved nutrition, indicators should include the amount harvested, its nutrient content, and how much is eaten and by whom. Garden impacts on social networks can be measured with short surveys, asking how many and what type of new acquaintances and friendships gardeners developed through the garden (Grewell, 2015). Inputs, or costs, also have to be measured in order to estimate benefit:cost ratios. In locations where droughts and water scarcity are increasing due to ACC, documenting garden outputs per unit of water used is especially important (see Cleveland et al., 1985).

Working together—Anthropocene trends have increased awareness of the need to consider and manage food gardens on broader community, regional, and even global scales. For example, a three-year study of urban community gardens in California documented how regional social and environmental change affects garden regulations and gardeners' management decisions, which in turn affect biodiversity and ecological functioning of the entire community garden and the larger surrounding area (Lin & Egerer, 2020). Working together in formal and informal institutions with shared rules of interaction regarding a resource or process (Soleri et al., 2019) increases response capacity by expanding the social and biophysical resources available to achieve goals. For example, in gardens with more rules and regulations around water use, gardeners tend to use less water (Egerer et al., 2018a), thus conserving more of that resource, and enabling more to have access to it.

The five key ideas outlined earlier (see Figure 24.1) support working together effectively, such as *ecological thinking* for garden sustainability in larger contexts. Quantifying the impacts of different forms of composting by household gardeners (none, household, municipal) we found that waste management organised at a larger municipal *scale*, including methane capture and energy generation, would result in lower greenhouse gas emissions than household composting (Cleveland et al., 2017). This information could be used by gardeners to encourage *prosocial* action by local governments to create these facilities.

Some gardeners are developing and maintaining their own, locally grown seed stocks, and identifying new varieties to adapt to and mitigate ACC, based on an understanding of *evolution by selection*. Gardeners also do this by working together to create seed libraries, *prosocial* institutions supporting semi-formal, non-commercial seed systems with free access to seeds and information (Soleri, 2017). Seed libraries and similar institutions are organised at a community or regional *scale*, often encourage *diverse forms of knowledge*, and have been increasing rapidly in response to the Anthropocene trends of biodiversity loss and increasing neoliberal domination of the food system. In a few cases, working together in these kinds of semi-formal seed institutions expanded the response capacity when the COVID-19 pandemic occurred (Soleri et al., 2022).

Responding to the COVID-19 pandemic with social innovation

The beginning of the COVID-19 pandemic coincided with the northern hemisphere spring planting season and led to a dramatic increase in the demand for garden seed in the US, as people looked to gardens for fresh food, outdoor exercise, and improved mental health. Commercial seed suppliers could not keep up with that demand due to lack of staff and new public health requirements; many suppliers shut down for a time. Recognising the need, some small, semi-formal garden seed organisations built on their *prosocial* practices, expanding their response capacity. Two such US-based institutions are Richmond Grows Seed Lending Library (RGSLL), which was established in 2010 in Richmond, CA; and the Experimental Farm Network (EFN), founded in 2013, which started the Community Garden Commission (CGC) in March 2020 (Soleri et al., 2022). Both RGSLL and EFN have biological investigations into community-scale biodiversity conservation and crop adaptation, and implicit social investigations into prosocial processes and practises including common pool resource management and mutual aid. In March 2020, using different strategies, RGSLL and EFN, through the CGC, pivoted quickly from their biological investigations of local seed and built on their social investigations to provide emergency seed and gardening support. With 60 volunteers, RGSLL created 12 tiny free seed libraries across Richmond, distributing 20,000 seed packets in 2020. Hundreds of people joined CGC and established 257 seed distribution hubs in 41 states, providing seeds to approximately 12,000 gardeners in 2020, as well as establishing working groups to discuss and take action in areas such as food systems policy. As seen during the COVID crisis, both organisations' investigations of social processes offered pathways to more just and effective responses by food gardeners to Anthropocene crises.

Recommendations and conclusions

Food gardens at the level of household, community, school, or workplace can be a key part of sustainable diets in the Anthropocene. They can contribute to a diversity of benefits, including increased availability and consumption of healthy foods, physical and mental wellbeing, social interactions, community organisation, resource sharing, and ecological stability. However, attaining these benefits is not inevitable; it requires ongoing engagement, inquiry, and evaluation.

A major cause of the Anthropocene crises is the emphasis on the human traits of materialism, selfishness, short-sightedness, and individuality, embodied by the assumptions of neoliberalism. To successfully respond to the challenges of Anthropocene trends, we need to emphasise different human traits—creativity, compassion, generosity, and prosociality. The empirically-based knowledge of both gardeners and scientists also is needed to counter the unsupported assumptions of neoliberalism.

The five key ideas described in Figure 24.1 summarise the core of such a successful response. The combination of prosocial values and concepts from scientific research that underpin these ideas provide an empirically-based counter to neoliberal assumptions. These key ideas can help gardeners and their allies develop responses to Anthropocene challenges that support goals of equity and justice, healthy people, communities and environments, a more stable climate, and sustainable diets.

Note

1 *Acknowledgements.* This chapter is based in part on our book, *Food Gardens for a Changing World* (Soleri, Cleveland, & Smith, 2019).

References

Alaimo, K., Packnett, E., Miles, R. A., & Kruger, D. J. (2008). Fruit and vegetable intake among urban community gardeners. *Journal of Nutrition Education and Behavior, 40*, 94–101.

Cameron, R. W. F., Blanuša, T., Taylor, J. E., Salisbury, A., Halstead, A. J., Henricot, B., & Thompson, K. (2012). The domestic garden – Its contribution to urban green infrastructure. *Urban Forestry & Urban Greening, 11*(2), 129–137.

Chan, J., DuBois, B. B., Nemec, K. T., Francis, C. A., & Hoagland, K. D. (2017). Community gardens as urban social–ecological refuges in the global north. In A. M. G. A. WinklerPrins (Ed.), *Global urban agriculture* (pp. 229–241). CABI.

Cleveland, D. A., & Soleri, D. (1987). Household gardens as a development strategy. *Human Organisation, 46*, 259–270.

Cleveland, D. A., Orum, T. V., & Ferguson, N. F. (1985). Economic value of home vegetable gardens in an urban desert environment. *Hortscience, 20*, 694–696.

Cleveland, D. A., Phares, N., Nightingale, K. D., Weatherby, R. L., Radis, W., Ballard, J. Campagna, M., Kurtz, D., Livingston, K., Riechers, G., & Wilkins, K. (2017). The potential for urban household vegetable gardens to reduce greenhouse gas emissions. *Landscape and Urban Planning, 157*, 365–374.

Egerer, M. H., Lin, B. B., & Philpott, S. M. (2018a). Water use behavior, learning, and adaptation to future change in urban gardens. *Frontiers in Sustainable Food Systems, 2*(71).

Egerer, M. H., Philpott, S. M., Bichier, P., Jha, S., Liere, H., & Lin, B. B. (2018b). Gardener well-being along social and biophysical landscape gradients. *Sustainability, 10*(1), 96.

Fellmann, T. (2012). The assessment of climate change related vulnerability in the agricultural sector: Reviewing conceptual frameworks. In A. Meybeck, J. Lankoski, S. Redfern, N. Azzu, & V. Gitz (Eds.), *Building resilience for adaptation to climate change in the agriculture sector* (pp. 37–61). FAO/OECD.

GIF (2018). *Cal-adapt*. Retrieved from http://cal-adapt.org/

Goddard, M. A., Dougill, A. J., & Benton, T. G. (2010). Scaling up from gardens: Biodiversity conservation in urban environments. *Trends in Ecology & Evolution, 25*(2), 90–98.

Grewell, R. (2015). *Urban farm and garden alliance nelson report UPDATE* (No. 1405). KNCBR. Retrieved from https://conservancy.umn.edu/bitstream/handle/11299/178552/KNCBR%201405.pdf?sequence=1&isAllowed=y

Hale, J., Knapp, C., Bardwell, L., Buchenau, M., Marshall, J., Sancar, F., & Litt, J. S. (2011). Connecting food environments and health through the relational nature of aesthetics: Gaining insight through the community gardening experience. *Social Science & Medicine, 72*(11), 1853–1863.

IPCC (2014). *Climate change 2014: Impacts, adaptation, and vulnerability. Part B: Regional aspects. Contribution of working group II to the fifth assessment report of the intergovernmental panel on climate change.* Cambridge University Press. Retrieved from https://www.ipcc.ch/site/assets/uploads/2018/02/WGIIAR5-PartB_FINAL.pdf

IPCC (2019). *Climate change and land: An IPCC special report on climate change, desertification, land degradation, sustainable land management, food security, and greenhouse gas fluxes in terrestrial ecosystems: Summary for policymakers.* IPCC. Retrieved from https://www.ipcc.ch/report/srccl/

Lawson, L. J. (2005). *City bountiful: A century of community gardening in America.* University of California Press.

Lin, B. B., & Egerer, M. H. (2020). Global social and environmental change drives the management and delivery of ecosystem services from urban gardens: A case study from Central Coast, California. *Global Environmental Change, 60,* 102006.

Litt, J. S., Soobader, M. J., Turbin, M. S., Hale, J. W., Buchenau, M., & Marshall, J. A. (2011). The influence of social involvement, neighbourhood aesthetics, and community garden participation on fruit and vegetable consumption. *American Journal of Public Health, 101*(8), 1466–1473.

McClintock, N. (2014). Radical, reformist, and garden-variety neoliberal: Coming to terms with urban agriculture's contradictions. *Local Environment, 19*(2), 147–171.

NGA (2014). *Garden to Table: A 5-year look at food gardening in America.* NGA. Retrieved from http://www.hagstromreport.com/assets/2014/2014_0402_NGA-Garden-to-Table.pdf

NOAA NCEI (2016). *State of the climate: Global analysis—Annual 2015.* Retrieved from https://www.ncdc.noaa.gov/sotc/global/201513

Nugent, R., & Fottrell, E. (2019). Non-communicable diseases and climate change: Linked global emergencies. *The Lancet, 394*(10199), 622–623.

Pataki, D. E., Carreiro, M. M., Cherrier, J., Grulke, N. E., Jennings, V., Pincetl, S. Pouyat, R. V., Whitlow, T. H., & Zipperer, W. C. (2011). Coupling biogeochemical cycles in urban environments: Ecosystem services, green solutions, and misconceptions. *Frontiers in Ecology and the Environment, 9*(1), 27–36.

Ramírez, M. M. (2015). The elusive inclusive: Black food geographies and racialized food spaces. *Antipode, 47*(3), 748–769.

Ripple, W. J., Wolf, C., Newsome, T. M., Barnard, P., & Moomaw, W. R. (2020). World scientists' warning of a climate emergency. *BioScience, 70*(1), 8–12.

Savoie-Roskos, M. R., Wengreen, H., & Durward, C. (2017). Increasing fruit and vegetable intake among children and youth through gardening-based interventions: A systematic review. *Journal of the Academy of Nutrition and Dietetics, 117*(2), 240–250.

SDN (2021). *P-Patch community gardening 2020 year end report.* SDN. Retrieved from http://www.seattle.gov/Documents/Departments/Neighborhoods/PPatch/2020YearEndReport.pdf

Slotten, V., Lentz, D., & Sheets, P. (2020). Landscape management and polyculture in the ancient gardens and fields at Joya de Cerén, El Salvador. *Journal of Anthropological Archaeology, 59,* 101191.

Soleri, D. (2017). Civic seeds: New institutions for seed systems and communities: A 2016 survey of California seed libraries. *Agriculture and Human Values, 35,* 331–347.

Soleri, D., & Cleveland, D. A. (2009). Breeding for quantitative variables. Part 1: Farmers' and scientists' knowledge and practice in variety choice and plant selection. In S. Ceccarelli, E. Weltzien, & E. Guimares (Eds.), *Participatory plant breeding* (pp. 323–366). FAO, ICARDA & ICRISAT.

Soleri, D., Cleveland, D. A., & Smith, S. E. (2019). *Food gardens for a changing world.* Wallingford, England: CABI.

Soleri, D., Kleinmen, N., & Newburn, R. (2022). Community seed groups: Biological and social investigations building capacity for garden scale responses to crises. *Citizen Science: Theory and Practice.* SocArXiv. Retrieved from https://osf.io/preprints/socarxiv/swv47/.

Steffen, W., Broadgate, W., Deutsch, L., Gaffney, O., & Ludwig, C. (2015). The trajectory of the Anthropocene: The Great Acceleration. *The Anthropocene Review, 2*(1), 81–98.

Steffen, W., Rockström, J., Richardson, K., Lenton, T. M., Folke, C., Liverman, D., Summerhayes, C.P., Barnosky, A.D., Cornell, S.E., Crucifix, M., Donges, J.F., Fetzer, I., Lade, S.J., Scheffer, M., Winkelmann, R., & Schellnhuber, H. J. (2018). Trajectories of the Earth system in the Anthropocene. *Proceedings of the National Academy of Sciences, 115*(33), 8252–8259.

Swinburn, B. A., Kraak, V. I., Allender, S., Atkins, V. J., Baker, P. I., Bogard, J. R. Brinsden, H., Calvillo, A., De Schutter, O., Devarajan, R., Ezzati, M., Friel, S., Goenka, S., Hammond, R.A., Hastings, G., Hawkes, C., Herrero, M., Hovmand, P.S., Howden, M., & Dietz, W. H. (2019). The global syndemic of obesity, undernutrition, and climate change. *The Lancet Commissions, 393*(10173), 791–846.

Teig, E., Amulya, J., Bardwell, L., Buchenau, M., Marshall, J.A., & Litt, J. S. (2009). Collective efficacy in Denver, Colorado: Strengthening neighbourhoods and health through community gardens. *Health & Place, 15*(4), 1115–1122.

Voicu, I., & Been, V. (2008). The effect of community gardens on neighbouring property values. *Real Estate Economics, 36*(2), 241–283.

Waliczek, T. M., Zajicek, J. M., & Lineberger, R. D. (2005). The influence of gardening activities on consumer perceptions of life satisfaction. *HortScience, 40*(5), 1360–1365.

Wen Li, W., Hodgetts, D., & Ho, E. (2010). Gardens, transitions and identity reconstruction among older Chinese immigrants to New Zealand. *Journal of Health Psychology, 15*(5), 786–796.

Wolch, J. R., Byrne, J., & Newell, J. P. (2014). Urban green space, public health, and environmental justice: The challenge of making cities 'just green enough'. *Landscape and Urban Planning, 125*, 234–244.

25

BROADENING OUR DEFINITION OF SUSTAINABLE FOOD

Shifting perception, policy, and practice to include nonhuman animals

Terry Gibbs and Tracey Harris

Introduction

"How can we give voice to other sentient beings within the broader discourse of sustainability?" This is one of the key questions that have informed the preparation of this chapter. As advocates for social justice, we are compelled to ask pressing questions. How do we expand the parameters of inclusivity and diversity to include other species and the nonhuman world? What strategies will be helpful in making it possible to see nonhuman animals as individuals in their own right within the discourse and advocacy work, rather than seeing them only in relation to what is best for humans and/or the environment? Our experience suggests that we may seek to address these questions by expanding the ways we make visible how non-human animals are objectified, commodified, and face systematised cruelty in our food production systems. This can be done through education, advocacy, and cross-movement solidarity with an orientation towards radical compassion.

Structural violence and the animal-industrial complex

Radical compassion, or as it is termed elsewhere, "big C" compassion, goes beyond the daily gestures of kindness we usually associate with that word, to challenging and transforming broader societal structures of oppression (Gibbs, 2017). In effect, it is about changing our institutional structures to better reflect the values of social and ecological justice. The concept of structural violence is a useful framework for understanding where to begin this inquiry. Developed in the late 1960s as part of his research on peace building, Johan Galtung's concept expands the notion

 DOI: 10.4324/9781003174417-29

of violence beyond direct physical violence to include harms caused by social structures that disproportionately benefit certain groups while diminishing the ability of others to meet their fundamental needs (Galtung, 1969). It demonstrates how economic, political, and cultural systems, institutions, and processes can be inherently violent, containing daily microaggressions reinforced and perpetuated because of the invisibility of the power structures at play and the special interests protected by these systems. He notes that while there may not be a single person harming another individual in the system, "The violence is built into the structure and shows up as unequal power and consequently as unequal life chances" (ibid., p. 171). We can easily relate this concept of structural violence to the systems and structures of colonialism, patriarchy, the military industrial complex and so on. While we do not want to reduce the problems of the Animal Industrial Complex (A-IC) or any system of oppression to capitalism alone, this is the global economic system that we function in and understanding its language and "logic" can help explain how we got to where we are and how we may get out of this mess.

In his historical exploration of the human rights and environmental consequences of capitalism, Garry Leech (2012) emphasises the "logic of capital," suggesting that the system running on this logic is actually genocidal and has, since its inception, been based upon violence from the brutality of colonialism to current day neo-colonial and neoliberal systems. He draws on the work of David Harvey who frames this process as "accumulation by dispossession" whereby capitalism constantly expands its spheres of production and circulation while in the process dispossessing people from their land, livelihoods, and resources (Harvey, 2005). Other authors such as David Nibert (2013) have shown how the abuse and destruction of animals has been inextricably linked with this genocidal logic. The constant need to expand the parameters of production and consumption in search of profit has always resulted in violence towards humans and nonhuman animals and the degradation of nature and biodiversity. This logic of capital is based upon the untenable notion that "resources" are infinite and that pretty much anything, including the very seeds of life, only gain value when they become commodities (Shiva, 1993). This framework has been perpetuated by the discourse of economic growth, rooted in the almost fundamentalist drive to continually increase GDP, a discourse that still guides world governments despite the resulting environmental crisis (Klein, 2014; Picketty, 2014). Explicit in this logic is the industrialised animal-based food system. In addition to commodifying nonhuman animals and many workers, a number of credible science-based researchers such as World Bank affiliated Goodland and Anhang argue that the livestock production system is the largest single contributor to GHG (greenhouse gas) emissions globally (Goodland & Anhang, 2009; Harwatt, 2018).

Barbara Noske (1997) first discussed the animal-industrial complex (A-IC) as a system that encompasses all aspects of producing nonhuman animals for food, including hatching facilities, factory farms, transportation, and slaughter facilities. Under this system virtually no area of these animals' lives goes untouched by humans. They are artificially inseminated, fed diets that they would never consume if left to their own devices, and their life cycles are sped-up so that they grow larger, faster, and are sent to slaughter in a much-reduced timeframe (Fitzgerald, 2015; Noske, 1997). Within the logic of capital, animals become commodities as their nature and individuality as sentient emotional beings is completely erased. As Adams (1991), Arcari (2017), Joy (2011) and others have made clear, understanding and confronting the language we use to talk about animals can be a critical starting point in confronting the structural violence they face. As the struggles of other historically marginalised groups make clear (e.g., racism, sexism, homophobia, and transphobia), language in the A-IC can contain microaggressions that play out daily. Those marginalised in the A-IC suffer on a daily basis and, like those who experience these other systems of violence, the harms inherent in the A-IC are often invisible and, in many cases, deliberately made so, through opaque systems and practises, behaviours, and language that normalise oppression and make it very difficult to illuminate and combat (Arcari, 2017; Broad, 2016). This is blatantly obvious in "ag-gag" laws in several Canadian provinces and US states that prohibit undercover investigation of factory farms and meat processing plants (Fitzgerald, 2015; Sorenson, 2016).[1] As has been noted elsewhere, transparency is absolutely necessary if we are going to confront the structural violence of our current animal-intensive food system, because this requires that consumers understand both how their food was produced and the impacts that this production creates (Fitzgerald, 2015; Singer & Mason, 2006).

Paula Arcari has noted how debates within the policy literature on food and sustainability issues have effectively "de-animated" animals where they are "either aggregated—as livestock, units of production and resources, or materialised—as meat and protein" (Arcari, 2017, p. 69). This commodification of animals, further illustrated by the discourse of "standard industry practices," makes it possible to normalise systematised cruelty so that average consumers are completely freed of any moral or ethical consideration of the beings whose flesh they eat or whose bodies have been harmed. These standard industry practises, which include debeaking chickens, castration of baby pigs without anaesthetic, forcing cows to be repeatedly impregnated (with related problems of mastitis and systematic use of antibiotics), removing new born calves from their mothers on dairy farms, are just a few examples that will be elaborated on below.

The experience and discourse of alienation, directly related to the logic of capital, are important to understand here as we see how commodification, or "thingifying" animals, through language and practice, allows us to do harm and separate ourselves existentially from their suffering and from their capacity for joy (Gibbs & Harris, 2020; Noske, 1997). Under the factory farming system, animals are alienated from their own bodies, their productive output, others of their species, nature, and species life in general (Noske, 1997). The significant question of joy, living one's best life or living to one's full potential, in any discussion of future just societies is increasingly being explored in the literature (von Gall & Gjerris, 2017). Whether we are talking about humans or nonhuman animals and rights, the question is not just how to remove suffering but also to look at the bigger picture of the kind of society we are looking to build and what values may inform it.

On a practical level, looking specifically at nonhuman animals, authors such as von Gall and Gjerris (2017) have explored the role of joy in farm animal welfare legislation. They refer to the focus on the "ethical primacy of negative over positive duties" and the emphasis on suffering and distress in scientific and animal welfare discussions. Part of the problem, they note, is the difficulty of placing "economic value" on joy, once again reflecting the inherent biases in a mindset/system run on the logic of capital. They also note that there are "economic benefits" to disregarding joy, including remaining competitive and increasing profit. But, as they correctly point out, there is no sense in which this reasoning can be justified ethically.

Radical compassion and interdependence: A compassionate way forward

As noted above, radical compassion implies challenging systems of structural violence and the creation of new systems—better ways of doing things that include not only avoiding harm but that lead to positive outcomes, such as greater fulfilment and joy in life. Continuing to extend the lens of intersectionality to include nonhuman animals seems an important step that not only helps animals but is ultimately the best way forward for humans and the ecosystems we depend upon. So how might a framework of radical compassion allow us to adjust our approaches? As compared to everyday compassion (raising money for a local youth group, giving to the local food bank, volunteering for a community group, helping our neighbour), radical compassion is a way of directing these positive human instincts towards systems and reflects an underlying understanding of human interdependence with all living beings.

The point of radical compassion is to recognise and stop harm currently happening, to build the avoidance of harm into our planning and policy for the future, and to go forward with an intention to not only avoid harm but to allow humans and other animals to live their best lives on an ecologically healthy planet. Radical compassion is rooted in a view that until our economic and political systems reflect an understanding of our interdependence, we will continue to do harm and engage in processes that marginalise certain groups and undermine planetary integrity. The practises of marginalisation have created the conditions for this world view to be held. Interdependence is not a fluffy concept but, as Indigenous perspectives clearly remind us, it is the reality of our relationship to the natural world (Kimmerer, 2013). Within the debates on voice, subjectivity, and agency, exciting new research is emerging that attempts to situate land and other animals as potential bearers of rights, and animals in particular as co-participants rather than objects in processes of research and knowledge creation (Bastian et al., 2017; Marshall 2017).

Ultimately, and as Einstein noted, when we see ourselves as separate from the world around us and from other living beings, it is an *optical delusion of our consciousness* (Einstein, 2015). But this world view that starts with an acknowledgement of our interdependence is thousands of years old, and its contemporary manifestations in cultural and political debates are extensive. While it is beyond the scope of this chapter to deeply explore all of the new narratives emerging around, trying to give "voice"/political space to land and other animals, it is important to acknowledge the political implications of these shifting debates. On a practical level, we would argue for a much broader application of the *precautionary principle* in our approaches to institutional practise and policy—one that extends beyond environmental considerations and ensures that this lens of interdependence is applied to exploring the potential harm of any decision, not only to humans, but to all species as integral to the natural world. This lens will require the input of compassionate individuals, knowledgeable and skilled in holistic approaches to social change. Systems that are already causing harm, and policies and practises that have already been enacted or are in continual motion (such as those perpetuating racism, speciesism, human exceptionalism, and environmental destruction), can be stopped by making visible and confronting harms head on. Where institutions and established processes are failing marginalised groups, political protest has been a key element in this ongoing struggle, as Occupy, Black Lives Matter, Idle No More, Via Campesina, and other movements have made clear. In terms of the struggle for water, land, and species protection in particular, Indigenous communities across the globe have been on the front lines. For this reason, it is our belief that engagement with Indigenous knowledge holders and land and water protectors should be a key element of any food justice or sustainability efforts moving forward. As we work towards genuine engagement and true reconciliation, our approach requires asking and listening to Indigenous Peoples for the necessary steps and actions to build relationships of trust and sharing amongst equals. True partnerships with Indigenous Peoples necessitate addressing the power imbalances in past and present relationships.

In addition to political protest and cross movements of solidarity and engagement, traditional policy channels are also an important avenue for creating openings for broader social change. The debate on sustainability labelling in Canada is an interesting and important example of attempts to bring these debates into practical policy circles with the aim to ensure consumer agency and accountability. Nova Scotia MP Jaime Battiste, the first Mi'kmaw Member of Parliament in Canada, managed to get his private member's motion for a green grading system passed through the House of Commons in February 2021. While it is yet unclear how this system will work in practice, its broad acceptance as a framework at the national political level is an important step. To the best of our ability, we should avoid engaging in activities that may bring harm to either humans or nonhumans while respecting the ecological integrity of the earth. This is no easy feat and we do not suggest that implementation would be straightforward, but it is already clear that this approach is gaining traction in communities and is being reflected in creative ways through practice and policy.

Given what we know about our current animal-intensive food system and the implications for nonhuman animals, humans, and the environment, how then do we challenge these serious problems and advocate for positive social change? Such complex issues require a multitude of creative solutions and will involve confronting structural violence and implementing compassionate solutions individually and as communities. As has been argued elsewhere:

We find ourselves at a time in history when we cannot afford to simply sit around wondering what needs to be done to sustain ourselves, other animals, and the planet, but instead we need to be brave by jumping into potential solutions … and by making personal change to facilitate larger structural transformations.

(Harris, 2018, p. 94)

Addressing and finding meaningful solutions for such complex problems will require collaborative change to create both practical strategies and policy shifts that will facilitate food equity, environmental sustainability, and dignity and justice for both human and nonhuman animals alike. We will explore some of these examples below by highlighting how the ways we talk, eat, and act as citizens are key pillars to strengthening and expanding this agenda.

How do we talk about other animals?

In thinking through privilege and power, it becomes difficult to contemplate change if we do not have the words to easily discuss the problem. Johnson argues that "once you name it, you can think, talk, and write about it, make sense of it by seeing how it's connected to other things that explain it and point toward solutions" (Johnson, 2006, p. 9.). This certainly holds true for the way we talk about "food" animals, as Carol Adams (1991) argues,

> Through butchering, animals become absent referents. Animals in name and body are made absent *as animals* for meat to exist … Without animals there would be no meat eating, yet they are absent from the act of eating meat because they have been transformed into food.
>
> *(p. 40)*

While many may shy away from thinking about how we personally benefit from privilege and contribute to inequality generally, this also holds true when asking citizens to consider their role in the everyday oppression of other animals constructed as food. The language we use to describe intensive agriculture (e.g., "livestock," "Confined Animal Feeding Operations," "the veal industry"), the procedures that are deemed necessary to turn nonhuman animals' bodies into economically feasible products (e.g., "debeaking" egg-laying chickens," "tail docking" pigs), and the language we use to describe their actual body parts and secretions (e.g., "hamburger," "pork chops," "omelette," "cheese"), all obscure that actual nonhuman animals are individuals that were commodified in the creation of these products (Stibbe, 2001). We also use language to differentiate ourselves from nonhuman animals: "humans have 'hair' while other animals have 'fur,' humans have 'skin' while other animals have 'hide,' and, deceased humans are 'corpses' while deceased others are 'carcasses'" (Nibert, 2002, p. 219). As so-called livestock or food animals, they are commodified during their short lives, embodying a strange in-between category that while they are alive, the end product (becoming meat, eggs, dairy products) justifies all sorts of conditions and treatments that are not seen as acceptable (or legal) for other categories of animals that we hold dear to our hearts, such as companion animals.

Facing such inequality, and the power gained from it, people may "get rid of the words that name it by discrediting them or twisting their meaning or turning them into a phobia or making them invisible" (Johnson, 2006, p. 9) because "awareness requires effort and commitment" (ibid., p. 22). Although it may be tricky to understand the significance of the words and phrases we use, it's important to remember that everyday language is used to "share and teach our cultural values and norms and allows us to socially construct the world around us, including our representations and constructions of other animals" (Smith-Harris, 2004, p. 12). But, there is also the possibility of "reclaiming" language and doing so allows us to clearly bring privilege and power to the forefront (Johnson, 2006).

The way we eat: How, what, and whom we eat

In moving forward, we need to think structurally about what we eat and how it was produced (Greger, 2020), and this means that from a public safety standpoint (both in terms of epidemiology and climate change) we need to act urgently to de-intensify and drastically reduce animal agriculture (Berry, 2009; Greger, 2020; Gunderson et al., 2016). As the United Nations Framework Convention on Climate Change (1992) states, we "should take the precautionary measures to anticipate, prevent or minimise the causes of climate change and mitigate its adverse effects." We would argue that this precautionary principle needs to guide our interactions, and actions, with humans, nonhuman animals, as well as our environmental concern.

Helen Harwatt (2018) proposed a three-step strategy, as one example of what the precautionary principle might look like in action. She argued that we need to acknowledge that we are facing "peak livestock," as animal agriculture has escalated to billions of animals worldwide, and the initial step is to urgently work towards a decrease in this number (Harwatt, 2018). The second step is to move away from animal-based products, eliminating the "worst first," which are associated with the highest greenhouse gas (GHG) emissions (ibid.). This step would mean that at the global level, beef, followed by cow's milk would be the first to be replaced with "best available" plant protein (ibid). While this model focuses on worst first from an environmental standpoint, we might also consider this logic when examining animal suffering (i.e., mutilating egg-laying hens through "debeaking") (Gunderson et al., 2016), continuously taking newborn calves away from their mothers in the "dairy" industry (Berreville, 2014), or death (i.e., there are an estimated 23 billion farmed chickens worldwide (Harwatt, 2018)) and those practices that pose the most serious health risks to human populations (i.e., those that have the highest potential of creating the pathway for a serious animal-to-human pandemic) (Greger, 2020).

How do we engage as citizens in strengthening (food) justice?

Beginning to engage as citizens advocating for food justice requires not only increased concern and understanding of the animal-industrial complex but also advocating for regulatory and legislative changes. This advocacy must include rethinking what constitutes humane treatment of nonhuman animals, living wages for farmers and food processing workers, and moving away from the "bigger is better" systems of production. Deintensification of animal agriculture requires rethinking the concentration of power that exists in food production, as intensive processing of farmed animals

> raise issues of public health, of soil and water and air pollution, of the quality of human work, of the humane treatment of animals, of the proper ordering and conduct of agriculture, and of the longevity and healthfulness of food production.
>
> *(Berry, 2009, p. 13)*

In the US, the proposed Farm System Reform Act (2019) is "legislation that aims to transition animal agriculture away from factory farming by banning the opening of new large-scale concentrated feeding operations (CAFOs) [Concentrated Animal Feeding Operations] and limiting the growth of existing CAFOs in the meat and dairy sector" with CAFOs phased out entirely by 2040 (Booker, 2020). This Act can be seen as having short-term goals, like banning the opening of new CAFOs, mid-term goals may include efforts like limiting the growth of existing CAFOs, and long-term goals, for example, phasing CAFOs out entirely by 2040. Banning factory farms should help create less corporate concentration and ultimately should lead to more localisation in our food system. This goal would serve several important purposes. First, it would increase farming and food processing jobs, as more farms and local processors would be needed with less corporate concentration and reduced farm sizes. Second, it would help people reconnect with local growers and producers of food. Knowing the farming and food production process would, we believe, facilitate more concern for the workers and animals in this system. In addition to these shifts, from a consumer health, animal welfare, and environmental standpoint, a shift towards increased consumption of plant-based food, along with a serious reduction in ones that are animal-based is essential.

Imagine a food system in which fruits and vegetables were cheaper than fast-food hamburgers, chips, and soda. This shift is the case already in some places such as Cuba, where policy has intentionally confronted the logic of capital and has led to Cuba being recognised by the World Wildlife Fund in 2006 and 2016 as having the most sustainable model of development on earth (Gibbs, 2017). We may also imagine a market in which plant-based "meat," "eggs," "fish," and "dairy" options were cheaper than animal-based ones. We currently have a system where the default is food produced with milk, eggs, fish, and/or meat. Imagine turning that on its head, where the default is plant-based. There also could be a condition where the externalities associated with the production and consumption (Pieper et al., 2020) of animal-based products is met with the appropriate tax (Nordgren, 2012) or surcharge. This change would normalise plant-based foods and make them more accessible and affordable.

Conclusion: Cross-species solidarity and compassion

In thinking through the problems this chapter has examined, as well as the potential solutions offered, we are reminded of Carol Adams' (2007) piece, "The War on Compassion." In it she reminds us of the danger of erecting artificial boundaries in our compassionate framework towards others, whether human or nonhuman animals. She states:

> When the first response to animal advocacy is, how can we care about animals when humans are suffering? We encounter an argument that is self-enclosing: it re-erects the species barrier and places a boundary on compassion while enforcing a conservative economy of compassion: it splits caring at the human-animal border, presuming that there is not enough to go around.
>
> *(p. 22)*

It can become all too easy to move the threshold for compassion to suit one's interests. Instead, we need to recognise the complicated intersections of oppression (Nibert, 2002). To build a compassionate food system, calls for the recognition of a myriad of ways that human and nonhuman lives intersect. Taylor and Taylor (2020) suggest a "solidarity across species" as we work together to create a food system that is both sustainable and compassionate.

Creating this system requires that we confront, head-on, the structural violence of our economic system from the places and spaces we find ourselves able to make change. This confrontation may imply everything from protest to education, advocacy, and policy making. Working together to shift to a more humane and sustainable food system would initially lessen, and eventually eliminate, many social injustices, and would allow us to demonstrate that we value sustainability and compassion in our relationships with human and nonhuman animals alike.

Note

1 "Ag-gag" laws are typically state or provincial laws, originally in the USA, and adopted also in other Western nations, to make it unlawful to take part in undercover investigations of animal industries. Advocates indicate they are a form of threat to whistle-blowers and to illuminating any abuse of animals, or related food safety issues for consumers, in agricultural facilities.

References

Adams, C. J. (1991). *The sexual politics of meat: A feminist-vegetarian critical theory*. The Continuum Publishing Company.

Adams, C. J. (2007). The war on compassion. In J. Donovan & C. J. Adams (Eds.), *The feminist care tradition in animal ethics* (pp. 21–36). Columbia University Press.

Arcari, P. (2017). Normalized, human-centric discourses of meat and animals in climate change, sustainability and food security literature. *Agriculture and Human Values, 34*, 69–86.

Bastian, M., Jones, O., Moore, N., & Roe, E. (2017). Introduction to more-than-human participatory research contexts, challenges, possibilities. In M. Bastian, O. Jones, N. Moore, & E. Roe (Eds.), *Participatory research in more-than-human worlds* (pp. 1–15). Routledge.

Berreville, O. (2014). Animal welfare issues in the Canadian dairy industry. In J. Sorenson (Ed.), *Critical animal studies: Thinking the unthinkable* (pp. 186–207). Canadian Scholars' Press.

Berry, W. (2009). *Bringing it to the table: On farming and food.* Counterpoint.

Booker, C. (2020, July 28). America's food system is broken [Keynote address]. National Food Policy Conference, Virtual. Retrieved from https://consumerfed.org/virtual-food-policy-conference-2020/.

Broad, G. M. (2016). Animal production, ag-gag laws, and the social production of ignorance: Exploring the role of storytelling. Environmental Communication, *10*(1), 43–61.

Einstein, A. (1950) accessed from The Liberator Magazine, 21 December 2015.

Farm System Reform Act (2019). S.3221. 116th Cong. https://www.congress.gov/bill/116th-congress/senate-bill/3221/text.

Fitzgerald, A. J. (2015). *Animals as food: (Re)connecting production, processing, consumption, and impacts.* Michigan State University Press.

Galtung, J. (1969). Violence, peace and peace research. *Journal of Peace Research, 6*(3), 167–191.

Gibbs, T. (2017). *Why the Dalai Lama is a socialist: Buddhism and the compassionate society.* Zed Books.

Gibbs, T., & Harris, T. (2020). The vegan challenge is a democracy issue: Citizenship and the living world. In K. M. Kevany (Ed.), *Plant-based diets for succulence and sustainability* (pp. 139–154). Routledge.

Goodland, R., & Anhang, J. (2009). *Livestock and climate change: What if the key actors are…cows, pigs, and chickens?* WorldWatch Institute.

Greger, M. (2020). *How to survive a pandemic.* Flatiron Books.

Gunderson, R., Stuart D., & Peterson, B. (2016). Factory farming: Impacts and potential solutions. In G. Muschert, B. Klocke, R. Perrucci, & J. Shefner (Eds.), *Agenda for social justice: Solutions for 2016* (pp. 27–37). Policy Press.

Harris, T. (2018). *The tiny house movement: Challenging our consumer culture.* Lexington Books.

Harvey, D. (2005). *The new imperialism.* Oxford University Press.

Harwatt, H. (2018). Including animal to plant protein shifts in climate change mitigation policy: A proposed three-step strategy. *Climate Policy, 19*(5), 533–541.

Johnson, A. G. (2006). *Privilege, power, and difference* (2nd ed.). McGraw-Hill.

Joy, M. (2011). *Why we love dogs, eat pigs and wear cows: An introduction to carnism.* Conari Press.

Kimmerer, R. (2013). *Braiding sweetgrass: Indigenous wisdom, scientific knowledge, and the teachings of plants.* Milkweed Editions.

Klein, N. (2014). *This changes everything.* Vintage Canada.

Leech, G. (2012). *Capitalism: A structural genocide.* Zed Books.

Marshall, A. Talk given at Thinkers Lodge, 28 September – 1 October 2017, Pugwash, NS: Climate Change, Drawdown & the Human Prospect: A Retreat for Empowering our Climate Future for Rural Communities.

Nibert, D. (2002). *Animal rights human rights: Entanglements of oppression.* Rowman & Littlefield Publishers.

Nibert, D. (2013). *Animal oppression and human violence: Domesecration, capitalism, and global conflict.* Columbia University Press.

Nordgren, A. (2012). Ethical issues in mitigation of climate change: The option of reduced meat production and consumption. *Journal of Agricultural and Environmental Ethics, 25*, 563–584.

Noske, B. (1997). *Beyond boundaries: Humans and animals.* Black Rose Books.

Picketty, T. (2014). *Capital in the twenty-first century.* Harvard University Press.

Pieper, M., Michalke, A., & Gaugler, T. (2020). Calculation of external climate costs for food highlights inadequate pricing of animal products. *Nature Communications, 11*(6117), 1–13.

Shiva, V. (1993). *Monocultures of the mind.* Zed Books.

Singer, P. & Mason, J. (2006). *The ethics of what we eat: Why our food choices matter.* Rodale.

Smith-Harris, T. (2004). There's not enough room to swing a cat and there's no use flogging a dead horse: Language usage and human perceptions of other animals. *Revision, 27,* 12–15.

Sorenson, J. (2016). *Constructing ecoterrorism: Capitalism, speciesism and animal rights.* Fernwood Publishing.

Stibbe, A. (2001). Language, power, and the social construction of animals. *Society & Animals, 9*(2), 145–161.

Taylor, A., & Taylor, S. (2020, Summer). Solidarity across species. *Dissent.* Retrieved from https://www.dissentmagazine.org/article/solidarity-across-species

UNFCCC. (1992). *Framework convention on climate change.* UNFCCC. Retrieved from https://unfccc.int/resource/docs/convkp/conveng.pdf

von Gall, P., & Gjerris, M. (2017). Role of joy in farm animal welfare legislation. *Society & Animals, 25,* 163–179.

PART 5

Social policies and food environments

26

(RE)BUILDING SUSTAINABLE CITY REGION FOOD SYSTEMS AFTER COVID-19

The role of local governments and food initiatives

Francesco Cirone, Bernd Pölling, Simone Amadori, and Matteo Vittuari

Introduction

The COVID-19 pandemic emphasised both the limits of conventional food systems and opportunities for the development of more localised alternatives. The negative impact on food systems greatly varied according to the different design and features of a city, as did the coping strategies adopted by local governments and food initiatives. For example, the dimensions of the urban area determined the severity of impacts, with larger and more densely populated cities being generally more affected by restrictions than small villages (FAO, 2020a). Indeed, smaller settlements feature shorter local food chains and stronger social ties that positively influence their resilience (Zollet et al., 2021). Differences were also observed among the Global South and the Global North, as evidenced in Table 26.1.

Despite this great variability, some overarching trends could still be observed. Globally, agricultural production was hampered by the restrictions on movement of workers and products across borders and by the tightening of safety regulations (Aday & Aday, 2020), which resulted in a greater number of food losses in the field (Blay-Palmer et al., 2021). Moreover, even if the agricultural products were successfully harvested, the farmers had to contend with logistical issues and many did not have the means to store or process food which hurt their bottom line (Vittuari et al., 2021). Indeed, early on during COVID-19, the "just-in-time" approach evolving long-distance supply chains, faced faltering conditions and disruptions to deliveries (Hobbs, 2020).

DOI: 10.4324/9781003174417-31

Table 26.1 Impact of COVID-29 on food systems

Global North	Global South
Crisis of HoReCa and food services in general	Interruption in local food logistics and markets (FAO, 2020a)
Labour shortages due to structural dependency on migrant workforce (Vittuari et al., 2021)	Difficulties in accessing agricultural inputs and marketing the products (Blay Palmer et al., 2021)
Consumers mostly impacted by income losses (Béné, 2020)	Consumers mostly impacted by an increase in food prices (FAO, 2020a)

Source: Work of the authors.

Meanwhile, distribution changed significantly with the need to stay at home: Delivery of both fresh and ready-to-eat food boomed, sustained by a rapid development of online channels (Farcas et al., 2021). This shift was observed in retailers, in the food service sector, and in alternative food networks (AFN) (Zollet et al., 2021). As a general trend, however, larger scale operations implemented the required innovations more easily (FAO, 2020a). AFN showed a high degree of responsiveness (Hobbs, 2020) with many food initiatives implementing tailored coping strategies to comply with national COVID-19 directives while continuing their operations (H2020 FoodE, 2020).

Food consumption patterns were also greatly influenced by the pandemic: At the very beginning of the pandemic, authors reported an increase in the purchase of ready-to-eat meals (Aday & Aday, 2020; Farcas et al., 2021), but then consumers reoriented their food shopping towards healthy and nutrient-rich foods to prepare at home, with an increasing interest in local products (Vittuari et al., 2021; Zollet et al., 2021). Baking products and goods with long-shelf life were also bought more frequently (Laguna et al., 2020).

The exploration of these trends emphasises several key hotspots that should be addressed while rethinking the paths towards food systems' sustainability. Even if food initiatives faced greater difficulties in implementing innovations, such as the conversion towards delivery or the development of online channels, they were more resilient: They proved to be prepared to deal with sudden adversity in a way that could be sustained in the long term (Béné, 2020). The importance of their role in ensuring the basic functioning of food system operations was recognised by citizens, who actively engaged in food initiatives (Vittuari et al., 2021) and supported them by increasing the share of expenditures. Some of these initiatives emerged spontaneously (Zollett et al., 2021), and others were fuelled by local governments who sought to address the drawbacks of the pandemic (FAO, 2020b).

This chapter provides an overview of the main resilience strategies adopted at the micro level—by food initiatives—and macro level—by local governments—to look at the interplay between these two dimensions and at the opportunities for sustaining and replicating local or grassroots solutions. City-regions were chosen as the focus of this chapter because of their centrality for food consumption—

according to FAO, more than 70% of the global food supply is destined to cities—but also for their particular strategies to address the drawbacks unveiled by the pandemic. Hence, a City Region Food System (CRFS) approach is proposed as a robust framework to explore these challenges. Territorial and holistic perspectives allow the inclusion of the main trends to encourage building of sustainable CRFS. Two different approaches to CRFS are presented in the next section: A macro approach, which focuses on system level links, and a micro approach concerned with food initiatives.

Understanding City Region Food Systems (CRFS)

The City Region Food System approach has been introduced by Jennings et al. (2015) who defined it as, "the complex network of actors, processes, relationships that has to do with food production, processing, marketing, and consumption in a given geographical region which includes a more or less concentrated urban centre and its surrounding and rural hinterland" (p. 17). This definition is especially interesting because it addresses food system governance from a precise geospatial perspective including not only cities but also the closely intertwined rural space. Moreover, it emphasises distinctive socio-economic interconnections that define food systems. The CRFS theoretical framework allows researchers to examine the linkages between the solutions implemented by local governments and food initiatives in addressing the drawbacks of the pandemic and to understand how good practices could be better coordinated and replicated.

While being a broad and comprehensive framework, the CRFS approach can be tackled on different levels according to the scope of the analysis that it is supporting. The perspective adopted by FAO & RUAF Global Partnership on Sustainable Urban Agriculture and Food Systems stresses the system-level links among urban, peri-urban, and rural areas (FAO & RUAF, 2015). The focus on the policies adopted by local governments provides a complement to an overarching assessment of the resources of the food system and the analysis of its most relevant weaknesses and opportunities. The toolkit elaborated by FAO & RUAF supports this perspective by suggesting a set of indicators to explore the main linkages and assets of a CRFS and to suggest the direction of its development towards resilience and sustainability (Carey & Dubbeling, 2018).

A different approach to CRFS builds on this system level perspective to zoom in on food initiatives, framed as City Region Food System initiatives (CRFSi). This approach is currently under development in the framework of the European H2020 project FoodE, thanks to a collaboration among 24 partners from eight European countries (France, Germany, Italy, the Netherlands, Norway, Romania, Slovenia, and Spain) coming from academia, public administration, and private sectors. The project wishes to engage food initiatives, framed as city-region food systems initiatives (CRFSi), in the design, implementation, and monitoring of environmentally, economically, and socially sustainable CRFS (Vittuari et al., 2021). The emphasis on food initiatives drives the setup of a bottom-up approach aimed at collecting good practices, in terms of resilience and sustainability, that can subsequently be replicated by designing better tailored policy interventions. The alignment of policy priorities and grassroots solutions is key to build resilient and sustainable CRFS.

Exploring resilience strategies in CRFS

The pandemic highlighted the need to consider a functional distinction between resilience and sustainability: Planning for abrupt shocks cannot be the same as planning for slow-occurring changes because different degrees and pace of responsiveness are needed. A resilient CRFS should stand out as the precondition and breeding-ground for building sustainability by encouraging good practices. The importance of having a defined and well-structured plan for resilience has been demonstrated by the convincing performances of cities that have already implemented some types of resilience strategy due to other epidemic- or climate-related shocks. Cities that were traditionally characterised by a strong social fabric and the presence of socially and locally embedded AFNs—such as Rome (Italy)—demonstrated a greater capacity to interpret and address the weaknesses connected with the pandemic (Zollet et al., 2021).

Whatever the triggering forces behind a successful response to the stressors induced by COVID-19—a virtuous institutional strategy or an efficient network of grassroots initiatives—some distinct features at the macro and micro level clearly made the difference. By exploring good practices implemented by local governments and by CRFSi, it will be possible to suggest some common trends to retain as key lessons for resilience and as starting points to achieve sustainability at the macro and micro level. The two levels are closely intertwined: Local governments have a pivotal role in promoting top-down resilience plans but also in collaborating with CRFSi towards the identification and implementation of bottom-up innovative strategies. The main sources of data are two surveys, complemented with relevant insights taken from both scientific articles and blogs published on the RUAF foundation website. The survey "Urban Food Systems and COVID-19" by FAO will allow us to extract macro level examples from some selected local governments' strategies in thwarting the collapse of the food systems and to protect the most vulnerable (FAO, 2020a). This research has been conducted on a global scale with more than 800 respondents from both the Global North and Global South, taken mostly from local governments, but also from academia, NGOs, and national governments. Cities and towns of different dimensions were included, ranging from villages of less than 5,000 inhabitants to cities hosting more than five million people. A micro level perspective on food initiatives in the European context will be elaborated from a survey originally administered for the H2020 FoodE project, featuring more than 200 interviewees selected according to their unique involvement in CRFS sustainability. The field of action of the CRFSi included in the study ranges from agriculture and fishing, food processing, food distribution, food services, food waste management, and education services. However, the greatest majority is involved with some type of crops: Vegetables, fruits, cereals, or mushrooms.

Some overarching themes are discussed below as starting points for understanding resilience strategies at the macro and micro level. The recognition of common key features could help define a set of good practices to be jointly addressed by local governments and CRFSi. As pointed out by Zollet et al. (2021), an integrated city region food system defined around some key priority areas and participated by a diverse set of stakeholders will be more resilient to shocks. These designed food systems could rely on a network of public and private, urban and rural, profit and non-profit organisations able to interpret and address issues at the most appropriate level. Moreover, CRFSi could operate in a supportive policy environment aligned around shared priorities.

Food distribution

Disruptions in food distribution were tackled on many different levels: Both macro and micro solutions entailed a rethinking of logistics, both through centralisation and decentralisation strategies, with the creation of new storage and marketing facilities.

Macro level

The local government in Lima (Peru) actively supported a decentralisation of the wholesale market by repurposing public infrastructures (FAO, 2020a). Similarly, in Rome (Italy) measures were taken to allow the continuation in food market operations and the development of more logistic centres (Zollet et al., 2021). Adopting a different approach, Antananarivo (Madagascar), Toronto (Canada), and Tirana (Albania) chose to further centralise their wholesale markets to better plan food distribution (Blay-Palmer et al., 2021). Local governments in Riga (Latvia) and Quilmes (Argentina) also were engaged in food distribution reorganisation to provide school meals through alternative channels (FAO, 2020a). Cities in high-income countries mostly focused on strengthening delivery services to reach vulnerable citizens and on the provision of logistical support for food distribution (FAO, 2020a).

Micro level

Many CRFSi envisioned the need for logistical reorganisation, expansion of storing facilities and selling spaces, and process optimisation (H2020 FoodE, 2020). Another point related to distribution is the rapid development of food delivery mechanisms, both for commercial purposes and to meet the needs of the disadvantaged. The disruptions in Ho.Re.Ca.[1] pushed the development of new delivery and distribution channels, which many European CRFSi reported as potentially lasting innovations (H2020 FoodE, 2020). When related to food aid, the delivery services were usually implemented through multi-stakeholder food networks participated in by public and private actors (H2020 FoodE, 2020).

Food availability

The redistribution of food surplus, the provision of school meal alternatives, and different types of financial assistance—both to producers and consumers—are some interventions consistent with the aim of securing access to food for all.

Macro level

As revealed in the FAO survey (2020a), cities in lower-to-middle-income countries focused on public procurement from local producers to ensure food availability, as is the case of Davao City (Philippines), where the local government bought food from local small farmers to distribute to the most vulnerable. Other cities in low- and lower-middle income countries put an emphasis on ensuring stability in food prices (FAO, 2020a). For instance, Curitiba (Brazil) was able to keep prices 30–35% lower than average by coordinating with ten municipalities in its region (Blay-Palmer et al., 2021). In Mboumba (Senegal), the local government teamed up with migrant associations abroad to boost remittances to purchase food (FAO, 2020a).

Micro level

One fifth of the CRFSi interviewed throughout Europe claimed that their main aim was to ensure food security. Indeed, many of them decided to distribute food surplus to the people in need; as with the new distribution channels, most of the initiatives envisioned continuing with these activities (H2020 FoodE, 2020).

Short food chains

The focus on local short food chains was also observed both for local government policies and in the actions of CRFSi: Public procurement of local foods, restructuring of food sales towards business to consumer relationships, and the participation in local AFNs are all activities in line with the promotion of local production.

Macro level

Riga (Latvia) and Quilmes (Argentina) encouraged the maintenance of short food chains by buying directly from local producers or by promoting hyper-local food production through backyard community gardening (Blay-Palmer et al., 2021). As already noted, in Davao City (Philippines) the local government bought food from local small farmers with the double aim to sustain food production and provide the most vulnerable with food baskets (FAO, 2020a).

Micro level

One third of the initiatives included in interviews in the FoodE survey reported that they participate in food networks aimed at promoting the territory and supporting the local market. Interestingly, the initiatives indicated local governments as their most common administrative partners, followed by regional governments (H2020 FoodE, 2020). Moreover, many initiatives chose to innovate their business by implementing business to consumer structures and by relying more on short supply chains. For instance, among food producers, almost half of the initiatives employ on-site marketing activities or participate in AFNs (H2020 FoodE, 2020).

Collaborative food chains

The need for collaborative food chains and the development of food networks involving a wide range of stakeholders were perceived as priorities by both local governments and CRFSi, who recognised the value of embedded CRFS.

Macro level
Often cities collaborate with local organisations to set priorities and successfully implement their policies: This is the case of Dhaka (Bangladesh), where the partnership between the local government and community initiatives succeeded in scouting the most in need and in targeting them with food aid (FAO, 2020a). Similarly, in Quito (Ecuador) social movements cooperated with the local government for the delivery of food baskets (FAO, 2020a).

Micro level
Most of the CRFSi are managed by cooperatives or by other forms of associated ownership. Half of the interviewees collaborated with and/or initiated a food network aimed at mutually supporting each other and promoting social inclusion. One third of the food networks involves multi-stakeholder engagement with at least two different types of actors among NGOs, grassroot associations, or public bodies. Food networks, including more than three different stakeholders, have been initiated by public bodies or by NGOs, and just twice by associations or private firms (FAO, 2020a).

Education

Food education is another theme that emerged both in local governments' campaigns and in the activities envisioned by the CRFSi to spread knowledge on food related topics, namely on local and sustainable production.

Macro level
Many different cities decided to focus on food education to help in better managing food: Both Riga (Latvia) and Quilmes (Argentina) promoted campaigns for responsible food purchase behaviour, also with a concern for reducing food waste (FAO, 2020a).

Micro level
Some CRFSi reoriented their activities on training and education: As a general trend, the theme of education emerged as a preferred direction for further developments. For example, spaces were repurposed for educational activities and online learning along with live seminars were offered on food-related topics.

Sustaining good practices towards CRFS sustainability

The agenda stemming from the analysis of resilience strategies calls for different types of innovation to sustain long-term positive change. For example, the reorganisation of distribution entails a set of technological advancements to be implemented in the local governments' organisational structure and in the food initiatives' business models. However, a more wide-ranging process of social innovation is also needed to allow more horizontal power structures towards participatory governance (Zollet et al., 2021). As for food initiatives, a restructuring of their business models should also be considered to fit the new activities in their sustainable value proposition (Bocken et al., 2019). Another case for seeking innovation could emerge from wanting to rely more on local food production. New methods and supports for urban food production, and key innovation encourage multifaceted perspectives to identify the tools available for sustaining and evolving resilient local practices.

Technological innovation can fuel food system transformation in many ways. As demonstrated by the examples reviewed above, it could have economic and social impacts on food security, and also on the ways in which agricultural production and distribution systems are designed and operated (Klerkx et al., 2019). Examples of the latter are farming technologies employed in urban and peri-urban areas, such as vertical farming, rooftop farming, digital agriculture, circular agriculture, and aquaponics (Appolloni et al., 2020). These types of technologies are, and will be, fundamental for boosting local food production and contributing to food security strategies in case of new disruptions in the food chains. Other technologies that are currently being developed will have the potential to transform not only food production but also food processing, and most importantly, food distribution; between them, the internet of things (IoT), artificial intelligence (AI), machine learning, and blockchain are some of the most promising (Klerkx et al., 2019). The implementation of technological innovation in food systems is usually an incremental process and requires great investments. Hence, the collaboration of the private and public sector is needed to develop new technologies and to contribute to their dissemination (Pigford et al., 2018).

Business model innovation is another key element to sustain the transformation towards sustainable CRFS, as it is a fundamental element of risk mitigation and resilience strategies (Geissdoerfer et al., 2018). As evidenced in the FoodE survey, CRFSi often felt a need to reorganise their business activities to address new challenges and grasp new opportunities. While business models (BM) have been initially used to address production efficiency and economic performance (Donner et al., 2021; Ulvenblad et al., 2019), resilience and sustainability have now become key drivers for business model innovation. In this context, sustainable business models (SBM) are emerging as fundamental strategic management tools for increasing profits, while also addressing social and environmental concerns. Sustainable business model innovation in the food sector mostly deals with the inclusion of a wider range of stakeholders in business agenda setting and activities (Bellucci et al., 2019) and with new solutions for increasing resource use efficiency or managing waste and by-products (Ulvenblad et al., 2019). Circular business models (CBM) are frequently proposed as suitable solutions to strengthen collaborative food networks while making a more efficient use of resources (Moggi & Dameri, 2021; Zucchella & Previtali, 2019). Fortunati and colleagues (2020) exemplify the benefits of a CBM by claiming that a company could enjoy greater productivity and safer working conditions for employees, while consumers would have access to higher quality products and the whole system will be more resilient to crises, thereby being better positioned to support sustainable diets.

Social innovation also is needed to build more inclusive governance structures for CRFS. An integrated approach that, on one hand, creates or reinforces food networks utilised by CRFSi and local governments, and on the other hand, allows joint planning by opening the process of policy priorities' definition, is a key precondition for the well-functioning of sustainable CRFS. Examples of these innovative governance architectures are Food Councils, Urban Food Committees (FAO, 2020a), biodistricts, and agricultural parks. While Food Councils are well-structured and cross-sectoral organisations aimed at co-planning food system policy with consumers and food initiatives (Macrae and Donahue, 2013), Urban Food Committees are grassroot issue-oriented coalitions of a variety of actors (FAO, 2020a). Urban Food Committees could explore and be the vanguard towards more complex and inclusive structures. Biodistricts and agricultural parks take a leap further, by embedding the city region stakeholders in an even more complex structure aimed at collectively managing the local territorial resources while agreeing on some shared values. These forms of social innovation aim at pushing the threshold of sustainability further, by conceiving sustainable food system development as a holistic concept encompassing lifestyle, nutrition, human relations, and nature (FAO, 2017).

Conclusion

A need for a greater integration between the priorities of local governments and CRFSi has emerged from this review. The exploration of resilience strategies at the macro and micro levels has helped to illuminate the many directions in this process. The definition of local and collaborative food chains, the focus on food distribution and availability, and the pledge for food education are the main themes that stemmed from this work; clearly, they represent only a subset of the features of a resilient and sustainable food system. However, a bottom-up process of learning from the strategies implemented in facing a shock like COVID-19 allows for the setting of priorities and identifying contact points between local government policy and the actions of food initiatives.

A multidimensional process of change is needed to ensure that resilience strategies will not remain single-issue solutions but will eventually become the foundations for CRFS sustainability. The exploration of some of the main directions of innovation in food systems helped in demonstrating the available opportunities. However, the implementation of these innovations entails many policy challenges. A lack of adequate resilience planning can be seen in recent developments; these should be addressed by an open and inclusive process aimed at devising strategies to quickly mobilise local resources in the case of shocks (FAO, 2020a). Then, such food governance architectures should be replicated through a wider inclusion of stakeholders from different sectors and regions, allowing multi-level horizontal and vertical cross-pollination (Vittuari et al., 2021): The promotion of PPAs, CBMs, and AFNs goes in this direction. To support this process, assessment tools could be beneficial in framing good practices and shaping a more informed policy environment. Ideally, this promotion of local food production and short chains would ignite policy arrangements to encourage small-scale production systems to rely on local resources, protect local food consumption (Vittuari et al., 2021), and strengthen and harness rural-urban linkages (FAO, 2020b).

Note

1 Ho.Re.Ca, is a short form for Hotel, Restaurant, café—an abbreviation integrated into Dutch, German, Italian, Romanian, and French languages.

References

Aday, S., & Aday, M. S. (2020). Impact of COVID-19 on the food supply chain. *Food Quality and Safety*, *4*(4), 167–180.

Appolloni, E., Orsini, F., Michelon, N., Pistillo, A., Paucek, I., Pennisi, G., Bazzocchi, G., & Gianquinto, G. (2020). From microgarden technologies to vertical farms: Innovative growing solutions for multifunctional urban agriculture. *Acta Horticulturae*, *1298*, 59–70.

Bellucci, M., Bini, L., & Giunta, F. (2019). Implementing environmental sustainability engagement into business: Sustainability management, innovation, and sustainable business models. In Galanakis, C. M. (Ed.), *Innovation strategies in environmental science* (pp. 107–143). Elsevier.

Béné, C. (2020). Resilience of local food systems and links to food security – A review of some important concepts in the context of COVID-19 and other shocks. *Food Security*, 805–822.

Blay-Palmer, A., Santini, G., Halliday, J., Malec, R., Carey, J., Keller, L., Ni, J., Taguchi, M., & van Veenhuizen, R. (2021). City region food systems: Building resilience to COVID-19 and other shocks. *Sustainability*, *13*(3), 1–19.

Bocken, N., Boons, F., & Baldassarre, B. (2019). Sustainable business model experimentation by understanding ecologies of business models. *Journal of Cleaner Production*, *208*, 1498–1512.

Carey, J., & Dubbeling, M. (2018). *City region food system toolkit*. RUAF. http://www.ruaf.org/projects/developing-tools-mapping-and-assessing-sustainable-city-

Donner, M., Verniquet, A., Broeze, J., Kayser, K., & De Vries, H. (2021). Critical success and risk factors for circular business models valorising agricultural waste and by-products. *Resources, Conservation and Recycling*, *165*, 105236.

FAO (2017). *The experience of Bio-districts in Italy*. FAO.

FAO (2020a). *Cities and local governments at the forefront in building inclusive and resilient food systems*. https://bit.ly/2x1B7yI

FAO (2020). *Urban food systems and COVID-19: The role of cities and local governments in responding to the emergency*. http://www.foodpolicymilano.org/en/food-aid-system/

FAO, & RUAF (2015). *A vision for city region food systems: Building sustainable and resilient city regions*. www.cityregionfoodsystems.org

Farcas, A. C., Galanakis, C. M., Socaciu, C., Pop, O. L., Tibulca, D., Paucean, A., Jimborean, M. A., Fogarasi, M., Salanta, L. C., Tofana, M., Tofana, M., & Socaci, S. A. (2021). Food security during the pandemic and the importance of the bioeconomy in the new era. *Sustainability*, *13*(1), 1–11.

Fortunati, S., Morea, D., & Mosconi, E. M. (2020). Circular economy and corporate social responsibility in the agricultural system: Cases study of the Italian agri-food industry. *Agricultural Economics, 66*(11), 489–498.

Geissdoerfer, M., Vladimirova, D., & Evans, S. (2018). Sustainable business model innovation: A review. *Journal of Cleaner Production, 198*, 401–416.

Hobbs, J. E. (2020). Food supply chains during the COVID-19 pandemic. *Canadian Journal of Agricultural Economics, 68*(2), 171–176.

Jennings, S., Cottee, J., Curtis, T., & Miller, S. (2015). *Food in an urbanised world: The role of city region food systems in resilience and sustainable development.* https://www.alnap.org/system/files/content/resource/files/main/food-in-an-urbanised-world-report-draft-february-2015.pdf

Klerkx, L., Jakku, E., & Labarthe, P. (2019). A review of social science on digital agriculture, smart farming and agriculture 4.0: New contributions and a future research agenda. *NJAS-Wageningen Journal of Life Sciences, 90*, 100315.

Laguna, L., Fiszman, S., Puerta, P., Chaya, C., & Tárrega, A. (2020). The impact of COVID-19 lockdown on food priorities. Results from a preliminary study using social media and an online survey with Spanish consumers. *Food Quality and Preference, 86*, 104028.

Macrae, R., Donahue, K. (2013). *Municipal food policy entrepreneurs: A preliminary analysis of how Canadian cities and regional districts are involved in food system change.* https://ensser.org/wp-content/uploads/fileadmin/files/2013_MacRae&Donahue.pdf

Moggi, S., & Dameri, R. P. (2021). Circular business model evolution: Stakeholder matters for a self-sufficient ecosystem. *Business Strategy and the Environment, 30*(6), 2830–2842.

Pigford, A. A. E., Hickey, G. M., & Klerkx, L. (2018). Beyond agricultural innovation systems? Exploring an agricultural innovation ecosystems approach for niche design and development in sustainability transitions. *Agricultural Systems, 164*, 116–121.

FoodE, Horizon 2020 project: https://foode.eu/

Ulvenblad, P. O., Ulvenblad, P., & Tell, J. (2019). An overview of sustainable business models for innovation in Swedish agri-food production. *Journal of Integrative Environmental Sciences, 16*(1), 1–22.

Vittuari, M., Bazzocchi, G., Blasioli, S., Cirone, F., Maggio, A., Orsini, F., Penca, J., Petruzzelli, M., Specht, K., Amghar, S., Villalba, G., & de Menna, F. (2021). Envisioning the future of European food systems: Approaches and research priorities after COVID-19. *Frontiers in Sustainable Food Systems, 5*, 58.

Zollet, S., Colombo, L., de Meo, P., Marino, D., McGreevy, S. R., McKeon, N., & Tarra, S. (2021). Towards territorially embedded, equitable and resilient food systems? Insights from grassroots responses to covid-19 in Italy and the city region of Rome. *Sustainability, 13*(5), 1–25.

Zucchella, A., & Previtali, P. (2019). Circular business models for sustainable development: A "waste is food" restorative ecosystem. *Business Strategy and the Environment, 28*(2), 274–285

27

REORIENTING FOOD ENVIRONMENTS TO SUPPORT SUSTAINABLE DIETS

Selena Ahmed, Shauna M. Downs, and Anna Herforth

Introduction

There is a critical need to transform food systems to sustainably support human and planetary health. On the production side of food systems, human activities are having unprecedented impacts on Earth and its ecosystems, threatening the resource base upon which global food security depends (Steffen et al., 2015; Tilman & Clark, 2014). Global agriculture is the leading burden for the transgression of multiple planetary boundaries (the environmental thresholds identified as safe operating spaces for Earth systems and humanity) including biosphere integrity, biodiversity, freshwater use, land use, nutrient flows, and soil resources (Campbell et al., 2017; Steffen et al., 2015). Climate change, resulting from greenhouse gas emissions across the food system and other sectors of society, is placing pressure on agriculture, thereby exacerbating malnutrition, food insecurity, and hunger (FAO, 2018; Swinburn et al., 2019). Concurrently, industrialisation, development, and globalisation of food systems have enhanced the interdependence of food supply chains and thus reduced resilience in the face of shocks (Davis et al., 2021). These supply chain challenges are linked to issues of food quality, food safety, transparency, and resource inequities (Davis et al., 2021). On the consumption side of food systems, malnutrition in all its forms, including overweight, obesity, undernutrition, and their coexistence, is among the leading causes of death globally (Afshin et al., 2019; Swinburn et al., 2017). Most countries are experiencing a transition towards diets high in saturated fat, sugar, highly processed foods, and meat, while being low in fibre, fruits, and vegetables (Bordirsky et al., 2020).

Sustainable diets, or healthy diets that are acquired from food systems that support planetary health in ways that are culturally appropriate and acceptable, are

DOI: 10.4324/9781003174417-32

increasingly being promoted in recognition of the interconnected challenges in the way we produce, procure, prepare, consume, and waste food (Burlingame and Dernini, 2012; Gussow & Clancy, 1986; Springmann et al., 2018; Willet et al., 2019). In addition to shifts toward consumption of healthy foods, sustainable diets also consider the ecological and socioeconomic dimensions linked to dietary choices, including biodiversity, ecosystem services, greenhouse gas emissions, equity, food traditions, and food sovereignty. Multiple interventions are being implemented globally to promote sustainable diets (Bailey and Harper, 2015; Garnett et al., 2015) based on key dimensions, including ecological, economic, human health, and sociocultural and political dimensions (Ahmed et al., 2019; Downs et al., 2017; Jones et al., 2016).

Sustainable diets interventions may target different entry points in the food system (Herforth et al., 2017) as well as sub-groups within populations. Food environments are critical places in the food system to implement interventions to advance sustainable diets because they contain the total scope of factors that influence consumer food procurement (Downs et al., 2020). In this chapter, we present an overview of food environments as well as its different types and elements. We emphasise the need to focus on sustainability attributes of foods as a key entry point to reorient food environments to support sustainable diets. Specifically, there is an urgent need for interventions, programs, and policies to ensure that foods with multiple sustainability attributes are available, affordable, convenient, and desirable for consumers in their food environments. This chapter concludes with challenges and opportunities for reorienting food environments to support sustainable diets.

What are food environments?

Food environments are the consumer interface with the food system. The earliest definition of food environment is based in ecology (Palo & Robbin, 1991) and expanded to include the built environment in human societies (Bronfenbrenner, 1979), such as community neighbourhoods, schools, workplaces, and institutional settings (Cummins, 2003; Morland et al., 2002). The concept of food environments was embraced by the nutrition and public health communities in high-income countries as they embraced a socio-ecological model of examining the multi-faceted and interactive factors that influence food choices (Story et al., 2008; Townsend & Foster, 2013). More recently, the concept of a food environment began being applied in low- and middle-income countries (LMICs) (Herforth & Ahmed, 2015).

In a recent paper, we updated previous definitions of food environments to be applicable in diverse countries including LMICs. Our updated *definition of a food environment* is:

> the consumer interface with the food system that encompasses the availability, affordability, convenience, promotion and quality, and sustainability of foods and beverages in wild, cultivated, and built spaces that are

influenced by the socio-cultural and political environment and ecosystems within which they are embedded.

(Downs et al., 2021)

While the majority of food environment research has examined built food environments, equally important are natural food environments, comprised of wild and cultivated areas where communities access food for direct consumption for food security, nutrition, and health (Ahmed & Herforth, 2017). Each type of food environment comes with its own opportunities and challenges. It is thus important to acknowledge multiple types of food environments. *Natural food environments* function largely as subsistence areas of food procurement. *Wild food environments* include a range of ecosystems with relatively low-intensity management, including forests and jungles, disturbed habitats, open pastures, and aquatic areas. *Cultivated food environments* include fields, orchards, closed pastures, gardens, and aquaculture from which consumers directly procure food for household consumption (rather than for sales). *Built food environments*, or market food environments, rely on food supply chains and include both informal and formal markets.

Entry point in the food environment to support sustainable diets

Interventions, programs, and policies to support sustainable diets target different entry points in the food system, from production through consumption and waste. Within the food environment, a key entry point for supporting sustainable diets is procurement and promotion of food with multiple sustainability attributes. Without foods with sustainability properties being available, affordable, convenient, and desirable, consumers cannot make dietary choices that support sustainable diets.

First, there is a need to assess the sustainability of foods and beverages based on ecological, economic, human health, and sociocultural and political dimensions (Ahmed et al., 2019) in order for enterprises that lead food procurement to source such foods. However, currently there is a lack of assessment tools for measuring sustainability properties within food environments, of both the foods themselves and the availability of such foods. Table 27.1 presents a framework to assess the sustainability attributes of foods and beverages. Next, there is a need to acknowledge how the sustainability attributes of foods vary based on the type of food environment. For example, a key sustainability attribute of foods from wild food environments is supporting biodiversity and ecosystem well-being. Lastly, there is a need to promote foods that have diverse sustainability attributes for diverse consumers situated in different types of contexts with varying access to the different types of food environments.

Efforts to promote the availability, affordability, convenience, and desirability of foods with sustainability attributes in the food environment should involve multiple stakeholder groups as well as types of interventions. Policy makers can be involved in providing incentives to the private sector as well as to producers to provide foods with sustainability attributes that are available, affordable,

Table 27.1 Framework to assess sustainability attributes of foods in the food environment

Sustainability Dimension	Sustainability attribute of foods and beverages found in the food environment
Ecological Dimension	**Production quality:** The food environment contains food that supports production systems that cultivate for nutritional quality (crop quality).
	Biodiversity, agrobiodiversity, and ecosystem services: The food environment contains food that supports conservation and maintenance of biodiversity and agrobiodiversity as well as associated ecosystem services.
	Sustainable agriculture: The food environment contains food that supports sustainable agricultural practices and sustainable intensification that limit pesticide, herbicide and fertilizer use.
	Local and seasonal foods: The food environment contains food that are in season and are local.
	Clean energy: The food environment contains food produced through the use of clean energy and green or sustainable technologies
	Soil, land, and water conservation and protection: The food environment contains food produced and/or procured in ways that prevent contamination of soil, land, and water resources such as protecting watersheds from pollutants.
	Low GHGE and climate resilience: The food environment contains food produced and/or procured using methods with relatively low GHG emissions; cultivated in agricultural systems that manage for climate change / climate resilience.
Economic Dimension	**Distribution, supply chains, and transport:** The food environment contains food that supports direct sales between producers and consumers.
	Food loss and waste: The food environment minimizes loss of food waste across the food system from farm through fork.
	Food packaging: The food environment contains food that has minimum food packaging and/or encourages recycling.
	Food system livelihoods: The food environment contains food of which the production promotes livelihoods to support stakeholders in the food system from on farm and throughout food value chains.
	Farmers' markets and local food systems: The food environment includes farmers' markets, community supported agriculture (CSA), food cooperatives, and food hubs.
	Food storage and preparation: The food environment contains food of which the production and preparation avoids resource-intensive food storage of cold chain items and high-energy preparation such as the use of a microwave.
Human Health Dimension	**Food safety:** The food environment contains food that is safe and prevents foodborne illness, contamination, negative health influence of agriculture and diseases linked to chemicals and pesticide use.
	Plant-based and nutrient-dense foods: The food environment contains food that is plant-based and nutrient dense foods such as fruits, vegetables, and legumes.

(Continued)

Table 27.1 (Continued)

Sustainability Dimension	Sustainability attribute of foods and beverages found in the food environment
Socio-Cultural and Political Dimension	**Equity issues:** The food environment contains food of which the production supports equity in the food system including on-farm, in market, trade, distribution, food service, and policy sectors. **Labor:** The food environment contains food that supports safe labor conditions and standards for workers in the food system. **Animal welfare:** The food environment contains food that supports healthy, comfortable, well nourished, and safe conditions for animals raised for livestock.

Source: The work of the authors, adapted from Ahmed et al., 2019

and desirable for consumers. Policy makers can further provide disincentives to support sustainable diets, which may involve taxation of food with unsustainable attributes. Policies may further focus on promoting product labelling with sustainability attributes based on multiple dimensions, such as labels with environmental information on food production practices that support biodiversity and mitigating greenhouse gas emissions or socio-cultural information on labour practices Interventions and programs may further seek to inform, guide, and empower consumers to select foods with sustainability attributes. Examples of these interventions include positioning of food and signage in retail settings as well as food and nutrition education to motivate behavioural change for making more sustainable food choices, such as increasing dietary diversity, eating more locally-sourced plant-based and nutrient-dense food, and reducing consumption of ultra-processed foods (Bailey & Harper, 2015).

Challenges and opportunities for reorienting food environments to support sustainable diets

Most food environment frameworks do not explicitly integrate sustainability, which presents a major challenge for reorienting food environments to support sustainable diets. We call for the integration of sustainability as a key element of the food environment in order to enable the design and implementation of interventions, programs, and policies that support both human and planetary health. This framework should acknowledge the different types of food environments and how sustainability can best be enhanced in each type of food environment. Research and development are further called to validate metrics and tools for evaluating the availability, affordability, and desirability of foods with sustainability attributes in the food environment.

Beyond the way we approach food environments, major challenges for reorienting food environments to support sustainable diets occur across the food system, including challenges associated with the production, processing, distribution, marketing, regulation, demand, and consumption of foods that support sustainability. Within the built food environment, there is a lack of transparency for retailers and consumers with regards to sustainability of foods.

Typically, retailers and consumers are provided with limited information regarding how foods and beverages are produced beyond limited labels such as organic and/or Fairtrade certifications and types of packaging. In addition, retailers and consumers are faced with the complexity of the multi-dimensional nature of sustainability. The multi-dimensional attributes of sustainability, which may entail trade-offs between various sustainability dimensions, make it difficult for consumers to decide which products are more or less sustainable. For example, there are often trade-offs between various aspects of sustainability between food choices in formal markets, such as organic bananas wrapped in plastic, vs. conventional bananas that are free of packaging. The decision is thus left to retailers and consumers regarding which attributes they most value and how that interacts with other elements of food environments such as affordability and desirability. A lack of comprehensive understanding of sustainability by both retailers and consumers further exacerbates challenges for selecting foods with sustainability attributes.

Two additional challenges at the consumer level for reorienting food environments to support sustainable diets are the lack of consumer demand for foods based on sustainability attributes and the lack of consumer demand for information on sustainability attributes. Many consumers globally are struggling to meet food security requirements and focusing on sustainability may not be a priority. In addition, consumers may be overloaded with information and find it inconvenient to sort through additional labelling information. Thus, in many cases there is a need to overcome food access barriers in creating healthy food environments and addressing additional determinants of food insecurity at the personal and socioecological levels including inequity, food distribution, income, and livelihood opportunities. To overcome barriers of knowledge gaps and overload, a cultural shift is called for regarding the way we value food.

Consumer education that synthesises food knowledge for consumers in accessible ways along with campaigns to motivate consumers regarding the importance of supporting sustainability, can facilitate a cultural shift for empowering food choices that support sustainable diets.

Beyond challenges that occur with regards to procurement of foods with sustainability attributes by retailers and consumers, there are also challenges related to the limited availability of food that is produced and processed using sustainable practices within the food environments that consumers interact with. For example, less than 50% of commercial farms in the United States follow multiple sustainable agricultural practices. Factors linked to the limited availability of food that is produced using sustainable practices are economic limits, lack of knowledge of sustainable practices. and associated technology, farmers' characteristics, lack of incentives to adopt more sustainable agricultural practices, inflexibility and lack of adaptive capacity of producers to adopt more sustainable agricultural practices, infrastructure conditions, and reluctance to change (Rodriguez et al., 2009).

Ensuring that food environments can support sustainable diets further relies on the complementary functioning and multi-scale interactions of local, regional, national, and international supply chains. However, food supply chains are exposed to wide-ranging shocks including climate change, emerging pests, policy and market shifts, and pandemics that threaten the sustainability of food and agriculture, with heightened threats for the most marginalised. For example, the COVID-19 pandemic disrupted farmworker safety, on-farm management, access to farm inputs, and labour schedules that cascaded along food supply chains, impacting the availability and affordability of foods with sustainability attributes in food environments.

Conclusion

The food environment is a critical place in the food system to implement interventions to advance sustainable diets. However, most food environment frameworks do not explicitly integrate sustainability and equity, which limits the potential for reorienting food environments to support human and planetary health. We call for the integration of sustainability as a key element of the food environment to enable the design and implementation of interventions, programs, and policies that support both human and planetary health. More specifically, the focus on sustainability attributes of foods is a key entry point to reorient food environments to support sustainable diets. Thus, consumers' food environments must empower dietary choices of foods with sustainability attributes that are available, affordable, convenient, and desirable to advance sustainable diets. Efforts to foster enabling food environments should occur alongside efforts to address the range of socio-ecological and personal determinants of dietary choices, including systemic inequities, in the context of global change.

Acknowledgments

This work was supported by United States National Science Foundation (NSF EPSCoR Research Infrastructure Improvement Program Track-2 FEC 1632810), National Institutes of Health NIGMS Montana IDeA Network for Biomedical Research Excellence (NIH NIGMS 5P20GM103474-19), and National Institutes of Health NIGMS Center for American Indian and Rural Health Equity, Montana State University (NIH NIGMS 5P20GM104417-05).

References

Afshin, A., Sur, P. J., Fay, K. A., Cornaby, L., Ferrara, G., Salama, J. S., Mullany, E. C., Hassen Abate, K., Abbafati, C., Abebe, Z., Afarideh, M., Aggarwal, A., Agrawal, S., Akinyemiju, T., Alahdab, F., Bacha, U., Bachman, V. F., Badali, H., Badawi, A., … Murray, C. J. L. (2019). Health effects of dietary risks in 195 countries, 1990–2017: A systematic analysis for the Global Burden of Disease Study 2017. *The Lancet, 393*(10184), 1958–1972.

Ahmed, S., Downs, S., & Fanzo, J. (2019). Advancing an integrative framework to evaluate sustainability in national dietary guidelines. *Frontiers in Sustainable Food Systems, 3*(76), 1–20.

Ahmed, S., & Herforth, A. (2017). Missing wild and cultivated environments in food environment measures. Agriculture. *Nutrition, and Health Academy Blog, 30*(8), 2017. Available online: https://anh-academy.org/academy-news-events/blog/2017/08/30/missing-wild-and-cultivated-environments-food-environment

Bailey, R., & Harper, D. R. (2015). *Reviewing interventions for healthy and sustainable diets*. London: Chatham House. Retrieved from https://www.chathamhouse.org/sites/default/files/field/field_document/20150529HealthySustainableDietsBaileyHarperFinal.pdf

Bodirsky, B. L., Dietrich, J. P., Martinelli, E., Stenstad, A., Pradhan, P., Gabrysch, S., Mishra, A., Weindl, I., Le Mouël, C., Rolinski, S., Baumstark, L., (2020). The ongoing nutrition transition thwarts long-term targets for food security, public health and environmental protection. *Scientific Reports, 10*(1), 19778, 1–14.

Bronfenbrenner, U. (1979). *The ecology of human development: Experiments by nature and design*. Harvard University Press.

Burlingame, B., & Dernini, S. (2012). Sustainable diets and biodiversity directions and solutions for policy, research and action. In Proceedings of the International Scientific Symposium Biodiversity and Sustainable Diets United against Hunger. Rome: FAO.

Campbell, B. M., Beare, D. J., Bennett, E. M., Hall-Spencer, J. M., Ingram, J. S., Jaramillo, F., Ortiz, R., Ramankutty, N., Sayer, J. A., & Shindell, D. (2017). Agriculture production as a major driver of the Earth system exceeding planetary boundaries. *Ecology and Society, 22*(4), 1–11.

Cummins, S. C. (2003). The local food environment and health: Some reflections from the United Kingdom. *American Journal of Public Health, 93*(4), 521–521.

Davis, K. F., Downs, S., & Gephart, J. A. (2021). Towards food supply chain resilience to environmental shocks. *Nature Food, 2*(1), 54–65.

Downs, S. M., Payne, A., & Fanzo, J. (2017). The development and application of a sustainable diets framework for policy analysis: A case study of Nepal. *Food Policy, 70*, 40–49.

Downs, S. M., Ahmed, S., Fanzo, J., & Herforth, A. (2020). Food environment typology: Advancing an expanded definition, framework, and methodological approach for improved characterization of wild, cultivated, and built food environments toward sustainable diets. *Foods, 9*(4), 532.

FAO, IFAD, UNICEF, WFP, & WHO (2018). *The state of food security and nutrition in the world: Building climate resilience for food security and nutrition*. FAO.

Garnett, T., Mathewson, S., Angelides, P., & Borthwick, F. (2015). Policies and actions to shift eating patterns: What works. *Foresight, 515*(7528), 518–22.

Gussow, J. D., & Clancy, K. L. (1986). Dietary guidelines for sustainability. *Journal of Nutrition Education, 18*(1), 1–5.

Herforth, A., & Ahmed, S. (2015). The food environment, its effects on dietary consumption, and potential for measurement within agriculture-nutrition interventions. *Food Security, 7*(3), 505–520.

Herforth, A., Ahmed, S., Declerck, F., Fanzo, J., & Remans, R. (2017). Creating sustainable, resilient food systems for healthy diets. In S. Oenema, C. Campeau, & D. Costa Coitinho Delmuè (Eds.). *UNSCN News 42: A spotlight on the nutrition decade* (pp. 15–22). UNSCN. Retrieved from https://www.unscn.org/uploads/web/news/UNSCN-News42-2017.pdf

Jones, A. D., Hoey, L., Blesh, J., Miller, L., Green, A., & Shapiro, L. F. (2016). A systematic review of the measurement of sustainable diets. *Advances in Nutrition, 7*(4), 641–664.

Morland, K., Wing, S., Roux, A. D., & Poole, C. (2002). Neighbourhood characteristics associated with the location of food stores and food service places. *American Journal of Preventive Medicine, 22*(1), 23–29.

Palo, R. T., & Robbins, C. T. (1991). *Plant defences against mammalian herbivory*. CRC Press.

Rodriguez, J. M., Molnar, J. J., Fazio, R. A., Sydnor, E., & Lowe, M. J. (2009). Barriers to adoption of sustainable agriculture practices: Change agent perspectives. *Renewable Agriculture and Food Systems, 24*(1), 60–71.

Springmann, M., Clark, M., Mason-D'Croz, D., Wiebe, K., Bodirsky, B. L., Lassaletta, L., de Vries, W., Vermeulen, S. J., Herrero, M., Carlson, K. M., Jonell, M., Troell, M., DeClerck, F., Gordon, L. J., Zurayk, J., Scarborough, P., Rayner, M., Loken, B., Fanzo, J., Godfray, H. C. J., Tilman, D., Rockström, J., & Willett, W. (2018). Options for keeping the food system within environmental limits. *Nature, 562*(7728), 519–525.

Steffen, W., Richardson, K., Rockström, J., Cornell, S. E., Fetzer, I., Bennett, E. M., Biggs, R., Carpenter, S. R., De Vries, W., De Wit, C. A., Folke, C., Gerten, D., Heinke, J., Mace, G. M., Persson, L. M., Ramanathan, V., Reyers, B., & Sörlin, S. (2015). Planetary boundaries: Guiding human development on a changing planet. *Science, 347*(6223), 1259855.

Story, M., Kaphingst, K. M., Robinson-O'Brien, R., & Glanz, K. (2008). Creating healthy food and eating environments: Policy and environmental approaches. *Annual Review of Public Health, 29,* 253–272.

Swinburn, B. A., Kraak, V. I., Allender, S., Atkins, V. J., Baker, P. I., Bogard, J. R., Brinsden, H., Calvillo, A., De Schutter, O., Devarajan, R., Ezzati, M., Friel, S., Goenka, S., Hammond, R. A., Hastings, G., Hawkes, C., Herrero, M., Hovmand, P. S., Howden, M., ... Dietz, W. H. (2019). The global syndemic of obesity, undernutrition, and climate change: The Lancet Commission report. *The Lancet, 393*(10173), 791–846.

Tilman, D., & Clark, M. (2014). Global diets link environmental sustainability and human health. *Nature, 515*(7528), 518–522.

Townsend, N., & Foster, C. (2013). Developing and applying a socio-ecological model to the promotion of healthy eating in the school. *Public Health Nutrition, 16*(6), 1101–1108.

Willett, W., Rockström, J., Loken, B., Springmann, M., Lang, T., Vermeulen, S., Garnett, T., Tilman, D., DeClerck, F., Wood, A., Jonell, M., Clark, M., Gordon, L. J., Fanzo, J., Hawkes, C., Zurayk, R., Rivera, J. A., de Vries, W., Sibanda, L. M., & Murray, C. J. (2019). Food in the anthropocene: The EAT–lancet commission on healthy diets from sustainable food systems. *The Lancet, 393*(10170), 447–492.

28

A FRAMEWORK FOR INTEGRATING SUSTAINABILITY IN INTERNATIONAL FOOD-BASED DIETARY GUIDELINES

Rachel Mazac, Barbara Seed,
Jennifer L. Black, and Kerry Renwick

Introduction

The need for healthy and sustainable food-based dietary guidelines

Human dietary practices impact the environment and have implications for food system sustainability (Clark & Tilman, 2017; Willett et al., 2019). Trends in global food demand are estimated to increase by 100–110% by 2050 to keep up with predicted population growth and shifting consumption toward more animal products associated with increased wealth (Myers et al., 2017; Springmann et al., 2018a). Increased food demand for crops could require over 150% of current cropland and blue water use and emit greenhouse gas (GHG) equivalent levels near 200% more than current levels by 2050 (Springmann et al., 2018). Reducing agricultural crop demand through sustainable dietary practices could substantially reduce land clearing, water use, and associated biodiversity loss (Tilman & Clark, 2014).

Dietary practices with low environmental impact have been associated with beneficial health outcomes (Aleksandrowicz et al., 2016; Clark & Tilman, 2017; Gonzalez Fischer & Garnett, 2016; Soret et al., 2014; van Dooren et al., 2014). For example, diets lower in animal-based food products are linked to reduced all-cause mortality risk, as well as lower GHG emissions and less water and land use, (Aleksandrowicz et al., 2016; Soret et al., 2014; Tilman & Clark, 2014). A healthy diet, however, does not necessarily mean it is also environmentally sustainable (Macdiarmid, 2013). For example, higher diet quality is associated with lower agricultural land use, but not always associated with less fertiliser, pesticides, and water

DOI: 10.4324/9781003174417-33

use, depending on how diet quality is measured (Conrad et al., 2020). Further, diets with lower environmental impacts (e.g. restricting total food intake and cutting out intake of all animal-products) may not always meet nutritional needs (Vieux et al., 2013, 2020), particularly in developing countries (e.g. iron deficiency).

Some nutritious dietary practices, however, are consistently more environmentally sustainable. For example, nutritionally balanced semi-vegetarian—less than one animal-based food consumed in a week—and vegetarian diets are associated with prevention of noncommunicable, chronic diseases compared to nonvegetarian diets (Aleksandrowicz et al., 2016; Soret et al., 2014). Vegan and vegetarian diets are lower in total emissions estimated through food product life cycle analysis of diets, which includes food production, processing, transportation, storage, retail, consumption, and disposal (Tilman & Clark, 2014; Willett et al., 2019). Ambitious dietary changes toward plant-based and flexitarian diets are required for staying within planetary boundaries for GHG emissions, cropland and blue water use, and nitrogen and phosphorus application (Springmann et al., 2018). These efforts must further be combined with individual and systemic reductions in food waste while ensuring social and economic access to sufficient, safe, and nutritious foods for the global population.

One step that governments and non-governmental food system actors can take in shifting consumption and supporting sustainable, healthy futures is to create and disseminate food based dietary guidelines (FBDGs). These could include sustainability considerations while promoting diets that meet population dietary needs. Countries develop FBDGs to assist populations and industries in improving national public health outcomes and meeting international goals of preventing population-wide nutrient deficiencies (Mozaffarian & Ludwig, 2010). Given the mounting evidence of the need for health- and environment-related changes in food and nutrition practises (Johnston et al., 2014; Popkin, 2006; Popkin et al., 2012; Soret et al., 2014; Willett et al., 2019), the United Nations (UN) and Food and Agriculture Organisation (FAO) urge international action on the part of governments to publish FBDGs that incorporate sustainability (FAO and WHO, 2019; Gonzalez Fischer & Garnett, 2016).

Many governments already create dietary guidelines that can be the basis for action related to healthy eating and the foundation for policy and institutional change. The UN identified an estimated 100 of the 215 (47%) countries worldwide as having FBDGs, with the absence of FBDGs being most conspicuous in low- and low-middle income countries (UN FAO, 2019). The aim of existing FBDGs to improve or maintain health is consistent, yet FBDGs may be further employed to benefit environmental and sustainability concerns (Gonzalez Fischer & Garnett, 2016; Gussow & Clancy, 1986; Willett et al., 2019). The UN FAO has long promoted sustainability in dietary guidelines (UN FAO, 2010). Additionally, the recent consensus of the global EAT-Lancet Commission is that "dietary guidelines that integrate health and environmental sustainability considerations could be one tool for nutrition education" and recommend that "relevant national bodies should implement guidelines for healthy diets from sustainable food systems" (Willett et al., 2019, p. 34).

FBDGs are one factor influencing population health and, potentially, sustainability in dietary practices globally. Though evidence is limited on FBDG use in places like Canada, some level of consumer awareness and understanding of FBDGs exists (Vanderlee et al., 2015). Yet, due to a lack of evaluation and metrics for effectiveness, little is known about if/how the awareness and understanding of FBDGs translates into use/adherence. Researchers acknowledge the possibility that consumers may either feel it is not necessary to follow FBDGs or may follow guidelines without conscious realisation (Brown et al., 2011). FBDGs are the foundation for nutrition and food policy at many levels: national (e.g. food labelling), regional (e.g. residential or child care licensing regulations), local (e.g. vending machines in municipalities) and organisational (e.g. foods sold at universities). FBDGs are also used in the development of resources, such as school food guidelines and nutrition education curricula, and are used for monitoring food consumption and dietary intake of the population (Seed, 2015).

FBDGs that centre sustainability could publicly signal official policy shifts and enact future behaviours that have health and sustainability synergies (FAO & WHO, 2019; Willett et al., 2019). Rizvi et al. (2018) assert that FBDGs without sustainability do not go far enough in considering the externalities of dietary choices. If, for example, the population of each country in the world ate according to the 2015 US FBDGs, we would dramatically exceed the planet's current capacity to sustain life and need an additional giga hectare of farmland—the size of Canada—to feed the current population (Rizvi et al., 2018).

Tools that articulate and translate a comprehensive interpretation of sustainable diets could guide FBDG developers in both narrowing in on salient evidence and broadening definitions for sustainability consideration in FBDGs. Recent work has developed and adapted frameworks for analysis of sustainability in FBDGs (Ahmed et al., 2019; Downs et al., 2017; Mazac et al., 2021). Such frameworks have been applied to developing food environment typology tools to advance the understanding of socio-ecological factors influencing access to and action for sustainable diets (Downs et al., 2020). Such tools that measure properties of food environments have informed recent studies aimed at assessing policy and educational interventions related to teaching dietary guidance and examining the efficacy of health care professionals in applying dietary guidelines in practice (Lígia & Patricia, 2021; Tramontt et al., 2021). Yet, previous work does not provide specific, procedural guidelines for how to include sustainability concepts or how to use such frameworks to develop new iterations and publications for national food-based nutrition recommendations. Guidance on how to implement such frameworks for sustainability consideration in FBDG development processes is needed to incorporate specific evidence from expert input for sustainability in FBDGs.

The main objective of this chapter is to provide practical guidance on how to use a framework on Sustainability in FBDGs previously published, and now adapted, by the authors (Mazac et al., 2021). First, we aim to describe key steps for enhancing sustainability in FBDGs recommended by recent publications. Then we will delve into our main goal, which is to describe how to use the framework with examples from recent FBDGs which incorporate sustainability considerations.

Sustainability in FBDGs

International inclusion of sustainability in FBDGs

The UN FAO asserts that biodiversity, natural resource, and ecosystem impacts should be included in dietary recommendations (FAO & WHO, 2019). Internationally, inclusion of sustainability principles in FBDGs has gained momentum and broader adoption. Moving beyond a previously narrow focus on nutrition, recent FBDGs now include other sustainability considerations, such as the social context of eating (Ministry of Health of Brazil, 2015) and the national-level economic incentives of eating a healthy diet (U.S. Department of Health and Human Services & U.S. Department of Agriculture, 2015).

Very few countries have included explicit sustainability considerations in their FBDGs. Recent work by the authors found that only five countries included explicit consideration of sustainability in their FBDGs (Brazil, Germany, Qatar, Sweden, and Canada) (Gonzalez Fischer & Garnett, 2016; Mazac et al., 2021). The United States, Australia, and China have attempted to incorporate sustainability and environmental considerations into their official guidelines, but have failed to achieve sustained government support for integration in FBDGs. Though not explicitly identified in their official FBDGs, the United Kingdom, France, the Nordic countries (Estonia, Finland, Denmark, Iceland, and Norway), Belgium, and the Netherlands have supporting dietary guideline documents with environmental considerations (Mazac et al., 2021). Additionally, a recent international review of ninety-three FBDGs identified little explicit reference to environmental or sustainability concerns (de Boer & Aiking, 2021). These findings suggest explicit references to sustainability are needed to acknowledge the urgency to stimulate dietary shifts.

Guidelines for including sustainability in FBDGs

The UN FAO directs developers to include sustainability in their FBDGs and to define healthy and sustainable diets for their respective national and political contexts. A 2019 report from UN FAO and the World Health Organisation (WHO) sets out 16 "Guiding Principles for Sustainable Healthy Diets," which first consider conditions for health such as the roles of breastfeeding, dietary diversity, food types/amounts, and food safety followed by the environmental impacts of dietary choice (e.g. keeping impacts within boundaries, preventing biodiversity loss, limiting antibiotics, plastics/packaging, and food waste). Sociocultural aspects such as cultural appropriateness, food security, and gender-related impacts of dietary practices are also asserted as important in FBDG development (FAO & WHO, 2019). From evidence in guidelines and reviews, Seed and Rocha (2018) outline key sustainability messages and give advice for incorporating sustainability into dietary guidelines. Key messages for promoting sustainable diets in FBDGs include: consuming primarily plant-based diets where meat consumption is reduced; consuming in-season and organic/free-trade certified foods; breastfeeding; and, reducing food waste and limiting the consumption of processed and ultra-processed foods (Seed & Rocha, 2018).

Suggestions for developing FBDGs that explicitly incorporate sustainability are outlined in Gonzalez-Fischer and Garnett's (2016) UN FAO and Food Climate Research Network (FCRN) report. They recommend that government support is particularly important, along with a multi-stakeholder development process and in promoting FBDGs in collaboration with diverse actors, like public, health professionals, industry, and food organisations. Diverse stakeholders also are instrumental in supporting ongoing monitoring of the implementation and impacts of such guidance. According to the FCRN report, FBDGs should have clear links to other food policies implemented in the country, such as school feeding policies, public procurement, industry standards, or advertising regulations. Seed and Rocha (2018) also recommend policy coherence across other areas of government such as trade agreements. When developing FBDGs, inclusion of sustainability within the publication requires multiple government and sectoral champions, integrated with a variety of academic and professional expertise. Gonzalez Fischer and Garnett (2016) further suggest that two separate development groups are needed: one for the experts in health and environmental fields, and another for the public, society, and industry. Procedurally, Seed and Rocha propose involving the food industry later in the process, while earlier intra-government and cross-sector alliances—including civil society—can support governments who may feel constrained by industry pressure. They also assert that sustainability considerations in FBDGs can create/reinforce "win-win" messages that can meet multiple, interconnected sustainability goals (Seed & Rocha, 2018).

Finally, to be efficacious in lessening environmental impacts of diets, FBDGs would be strengthened by drawing on evidence that emphasises the interconnected nature of diets, health, and the environment. This evidence includes the externalised costs of the food system to health care and the ecosystem (Seed & Rocha, 2018). Additionally, good practice includes developing FBDGs with clear language and messages, which are accessible and ambitious, while highlighting the benefits of limiting overconsumption of any foods, reducing food waste, guiding principles for vegan/vegetarian diets, and re-asserting the importance of valuing food and the land (Gonzalez Fischer & Garnett, 2016).

Framework for sustainability in FBDGs

Through an analysis of sustainability concepts in FBDGs, we adapted previous frameworks to develop a tool for the analysis of sustainability in FBDGs (Mazac et al., 2021). Figure 28.1 depicts the Sustainability in FBDGs Framework from Mazac et al.'s (2021) analysis, proposed here as an instrument for assessing and guiding the inclusion of sustainability consideration in FBDGs. We analysed 16 countries represented in 12 FBDGs and supporting documents. Five sustainability domains, based on frameworks by Downs et al. (2017) and Ahmed et al. (2019), form the basis of the analysis, and each domain includes 10–13 sustainability concepts (Ahmed et al., 2019; Downs et al., 2017).

In the documents reviewed, sustainability concepts were interconnected and included in a diversity of ways. Irrespective of the length of the FBDG document,

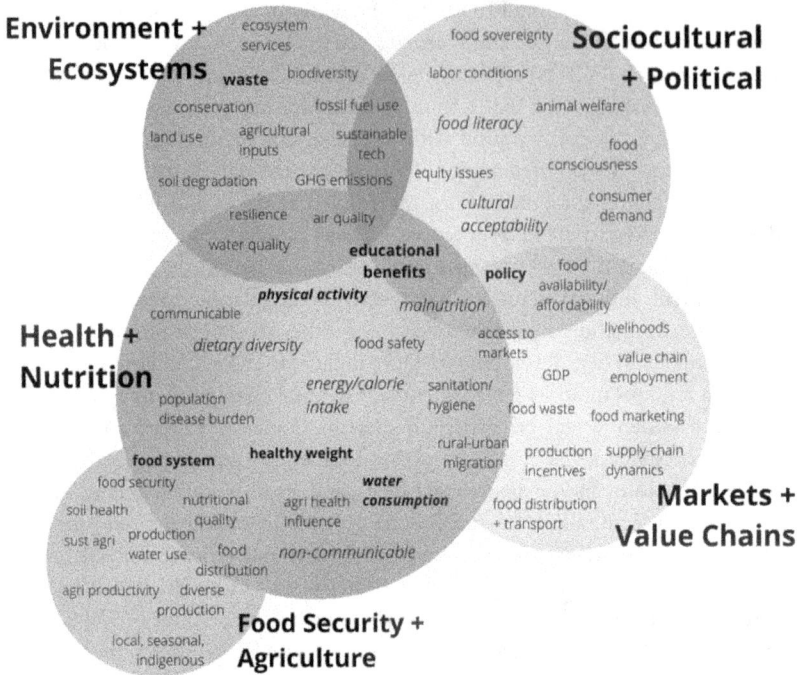

Figure 28.1 Sustainability in food-based dietary guidelines (FBDGs) framework. Domains and concepts included in FBDGs with sustainability considerations; concepts in italics included in all documents reviewed (n = 12), concepts in bold were inductively added by the author from their analysis. Reproduced from Mazac et al. (2021), with permission. Note: agriculture/al (agri), greenhouse gases (GHG), Gross Domestic Product (GDP).

ranging from 2 to 100 pages, all documents included four of the five domains, often with many concepts included in each domain. The framework is presented here as a Venn diagram with overlapping circles (Figure 28.1) to give some indication of the complexity of representing the different domains in food guides. Main concepts that consistently intersect with multiple domains (e.g. *air quality* in Environment, Health, and Sociocultural domains) are illustrated in areas of overlap between the circles/domains. To see more examples of coded excerpts from the analysed FBDGs that depict simultaneous inclusion of concepts from several of the domains, see Table 2 in Mazac et al. (2021).

How to use the framework

To use our proposed framework for inclusion of sustainability consideration in FBDGs, it is recommended to apply, to at least some degree, concepts from each of the five domains. The 10–13 concepts in each domain (as illustrated in Figure 28.1) are further elucidated in numerous sub-concepts. For further clarification on the sub-concepts within each of the sustainability domains and examples from recent FBDGs, see the Supplementary Material in Mazac et al. (2021).

To begin, FBDG developers can start by selecting domains that best suit their context. For example, if waste and climate change have been identified as priorities by a nation, Environment and Ecosystems will be an important domain to include. Next, choose the concepts to include and consider how they interconnect with those in other desired domains, as this will lead into the selection of other, connected concepts. Then, review the sub-concepts that correspond to those concepts. Consider which sustainability concepts would bring the most understanding and traction for the populace. Seed (2015) recommends identifying and targeting national priorities and regional interests to integrate relevant sustainability principles, such as resource scarcity, developing an environmentally-aware society, and ensuring food security.

If a primary emphasis of the FBDG is to be Sociocultural and Political, several of the concepts connect to the Environment and Ecosystems domain. For example, Sociocultural and Political concepts *cultural acceptability* (selected sub-concepts: convenience, preferences, share meals, eat together, food culture, eating environment) and *food literacy* (selected sub-concepts: cooking, food preparation, food education, food skills, freshly prepared, food/nutrition labels) connect with the Environment and Ecosystems domain. In this case, *land use* (agroecological production, multi-functional landscapes, productive lands) would need to be considered when cultivating *culturally appropriate* food, and a certain level of *food literacy* would be required to know how these relate. An explicit example of these concepts connected in practice could be: the provisioning of local food, which honours food culture and provides local employment and resilience, requires the on-going maintenance of ecosystem services and the sustainable management of productive and multi-functional landscapes. The understanding of this connection to, and the impacts on, the local ecosystems is an essential part of food education and preparation. Such information on sustainability should be considered for inclusion on packaging and nutrition labels.

The consideration of Markets and Value Chains could be connected to *food literacy* if consumers are considered as co-producers who also shape production through demand-side consumption patterns and who both influence and are influenced by *food marketing* (advertising, brands, food packaging, food promotion, media outreach, social marketing). Additionally, *food availability/affordability* (e.g. food prices, distribution of wealth) are connected to Food Security and Agriculture through *food security* and *nutritional quality* (e.g. food assistance, food poverty; quality of food produced, nutrient-rich foods), which reconnect with the Health and Nutrition domain. Therein, the interconnected nature of the concepts (and sub-concepts) could lead to the inclusion of multiple domains and sustainability considerations.

There are a number of possibilities for where to integrate sustainability principles in FBDGs. They can be included in overview statements, in the title, in the recommendations themselves, and in accompanying documents. Sweden illustrates an example of sustainability placed directly in the title of the FBDGs (i.e. "Find your way to eat greener, not too much and be active!"), while Australia relegated the sustainability verbiage to the appendices (i.e. Appendix G. "Food, nutrition and environmental sustainability"). Brazil's FBDGs illustrate an example of inclusion in overview statements that acknowledge and integrate sustainability at the outset of dietary guidelines; one of the key principles of their FBDG is: "Healthy diets derive from socially and environmentally sustainable food systems" (Ministry of Health of Brazil, 2015, p. 18). Examples of excerpts that integrate interconnected sustainability directly into the recommendations in FBDGs can be found in Mazac et al. (2021).

Not all FBDGs need or should be the same. For example, in some cases, limitations of meat consumption should be recommended but refined after considering the national nutritional context, and cultural and economic practises related to animal product consumption. As Gonzalez Fischer and Garnett (2016) and Willett et al. (2019) outline, high meat-consuming countries—often high income—should reduce all types of meat consumption and over-consumption of sugar-sweetened and processed foods. In middle-income countries, where population numbers, income, and demand for meat is increasing, guidelines for moderate, "less and better" meat intake are more appropriate. However, in low meat-consuming countries—often low income—recommendations should be made for meeting nutritional requirements, which may include consumption of animal products for dietary diversity and for sufficient intake of protein, essential amino acids, iron, and other micronutrients. Balancing the nutritional needs of a population with environmental impacts and other sustainability implications of diets are required for developing FBDGs with sustainability considerations (Conrad et al., 2020).

Challenges and opportunities of sustainability consideration in FBDGs

Challenges and example cases

Using the framework and process presented here may help to provide a visual display to those who are interested but have previously lacked a roadmap for the inclusion of sustainability considerations in FBDGs. Illustrating and expanding upon the connections among domains and concepts can facilitate surmounting barriers. Particularities of political, sociocultural, environmental, and economic contexts can be barriers to the integration of sustainability in FBDGs. Australia and the United States, for example, have substantive agri-business industries and these outweighed other stakeholders in the final attempts to include sustainability in their FBDGs development process. In the US process of including sustainability considerations, observers asserted that the lack of sustainability considerations in the US guidelines resulted from strong lobbying by the food industry, meat producers in particular (Gonzalez Fischer & Garnett, 2016; Jelsøe, 2015; Merrigan et al., 2015). Australia had calls from sectors of the country to include sustainability considerations in food policy, but government changes and industry input hampered suggested sustainability integration. The official 2013 Australian Dietary Guidelines did not make explicit sustainability references in the recommendations in the body of the text, but were relegated to the appendices (National Health and Medical Research Council, 2013).

Lang and Mason (2018) discuss sustainable diet policy development at an international level between 2008 and 2017. Their article examines what sustainability means in food policy and FBDGs, how it has been addressed in complex ways, and draws out emergent themes in development. Focusing on nutrition alone is possible, but they recommend a multi-criteria approach that engages a myriad of agencies for full integration of sustainability in FBDGs (Lang & Mason, 2017). From soft (e.g. emphasising consumer choice), to hard (e.g. choice-editing by retailers with marketing campaigns) messaging, FBDGs can integrate primary messages on health and nutrition with parallel inclusion of other sustainability considerations. Lang and Mason recommend acknowledgement of the interconnected nature of health and sustainability goals, akin to the framework we have presented. State leadership and support are needed to moderate industry voices and misinformation around consumption practices (Lang & Mason, 2017).

Opportunities and example cases

There are many opportunities and methods through which sustainability can be included in FBDGs, and our framework may help leverage connections among domains to integrate sustainability considerations more comprehensively. Despite context-dependent challenges, both top-down and bottom-up approaches can work to include explicit sustainability messaging food guides. The Brazilian recommendations, which focus on food processing as a guiding framework instead of individual nutrients, allowed for the 2015 FBDG to retain acceptance after overcoming the policy hurdle of a change in government leadership (Monteiro et al., 2015). Using everyday language and cultural messaging, took a broader food literacy approach (e.g. the level of processing of foods) instead of extolling or vilifying particular nutrients. Qatar's top-down approach and environmental concerns facilitated the inclusion of sustainability considerations, as did their relatively small agri-business industry and public interest in reducing overconsumption and waste (Seed, 2015). Qatar's process differs from Brazil's, where a more bottom-up approach included industry and also many others in a democratic process (Monteiro et al., 2015).

The framework presented in this chapter advises food systems governance actors on whether and how to engage with the agri-food and beverage industry to address malnutrition within the context of healthy and sustainable diets (Kraak, 2021), and informs policy-making that engages industry in the sustainable diet conversation. FBDG stakeholders, including government staff or industry lobbyists, may push back on using FBDGs to promote considerations outside of bodily or nutritional health. However, having a policy position that "stick[s only] to the health message" (Lang, 2017, p. 22) does not go far enough in signalling needed environmental, policy, and food system improvements (Rizvi et al., 2018). Excluding topics outside of health assumes that food consumption choices are driven solely by nutrient content; such arguments disregard the socio-cultural dimensions of eating. Also adhering to nutrition recommendations alone may not lead to lower environmental impacts (Lang, 2017). There is consensus in the scientific literature on the fact that humans—and their dietary practices—are responsible, in part, for global climate change (Campbell et al., 2017; Vermeulen et al., 2012; Willett et al., 2019). Conversely, climate change will also impact health (e.g. increase of extreme weather events, loss of water and air quality, droughts, and famine) (Myers et al., 2017). FBDGs must reflect such realities and help guide eaters through complex and multi-layered choices. This framework is a theoretical model that needs further testing; developers should work through the methods outlined here, think about these domains, which ones would work in their context, and get feedback from key stakeholders to both validate the tool and at the same time accomplish their goals of comprehensive FBDGs that include health and sustainability principles.

Conclusion

In this chapter, we presented a process for applying a holistic, evidence-based framework to integrate sustainability in FBDGs. With the leadership of experts and advocates, and the inclusion of scientific evidence, each of the five domains can be included in various ways and to varying degrees, based on the needs, context, and emphasis of the country. To do so, multi-objective decision-making tools, such as our framework, can be useful. This chapter outlines the process of how to apply a framework for FBDG developers to explicitly integrate sustainability considerations and create FBDGs that make strides towards actively embracing health and sustainability goals. Future sustainability advocates and policy makers can draw on this framework and examples to consider sustainability explicitly in dietary recommendations, in order to simultaneously guide nutritionally adequate diets and meet global sustainability goals. Dietary guidelines with sustainability considerations, if understood and implemented in both population-level policy and by individuals making food choices, can contribute to the prevention of nutrition-related chronic diseases and under- and malnutrition, while boosting global sustainability objectives. FBDGs can be a useful and meaningful avenue to shift towards sustainable diets at a population-level. The time for FBDGs to include sustainability is now long overdue.

Reference list

Ahmed, S., Downs, S., & Fanzo, J. (2019). Advancing an integrative framework to evaluate sustainability in national dietary guidelines. *Frontiers in Sustainable Food Systems, 3*, 76.

Aleksandrowicz, L., Green, R., Joy, E. J. M., Smith, P., & Haines, A. (2016). The impacts of dietary change on greenhouse gas emissions, land use, water use, and health: A systematic review. *PloS One, 11*(11), e0165797.

Brown, K. A., Timotijevic, L., Barnett, J., Shepherd, R., Lähteenmäki, L., & Raats, M. M. (2011). A review of consumer awareness, understanding and use of food-based dietary guidelines. *British Journal of Nutrition, 106*(1), 15–26.

Burlingame, B., & Dernini, S. (2010). Sustainable diets and biodiversity. Proceedings of the International Scientific Symposium BIODIVERSITY AND SUSTAINABLE DIETS UNITED AGAINST HUNGER 3–5 November 2010 FAO Headquarters, Rome

Campbell, B. M., Beare, D. J., Bennett, E. M., Hall-Spencer, J. M., Ingram, J. S. I., Jaramillo, F., Ortiz, R., Ramankutty, N., Sayer, J. A., & Shindell, D. (2017). Agriculture production as a major driver of the Earth system exceeding planetary boundaries. *Ecology and Society*, *22*(4), 8. https://www.ecologyandsociety.org/vol22/iss4/art8/

Clark, M., & Tilman, D. (2017). Comparative analysis of environmental impacts of agricultural production systems, agricultural input efficiency, and food choice. *Environmental Research Letters*, *12*(6), 064016.

Conrad, Z., Blackstone, N. T., & Roy, E. D. (2020). Healthy diets can create environmental trade-offs, depending on how diet quality is measured. *Nutrition Journal*, *19*(1), 1–15.

de Boer, J., & Aiking, H. (2021). Limiting vs. diversifying patterns of recommendations for key protein sources emerging: A study on national food guides worldwide from a health and sustainability perspective. *British Food Journal*, *123*(7), 2414–2429.

Downs, S. M., Ahmed, S., Fanzo, J., & Herforth, A. (2020). Food environment typology: Advancing an expanded definition, framework, and methodological approach for improved characterization of wild, cultivated, and built food environments toward sustainable diets. *Foods*, *9*(4), 532.

Downs, S. M., Payne, A., & Fanzo, J. (2017). The development and application of a sustainable diets framework for policy analysis: A case study of Nepal. *Food Policy*, *70*, 40–49.

FAO, & WHO (2019). *Sustainable healthy diets: Guiding principles*. FAO.

Gonzalez Fischer, C., & Garnett, T. (2016). *Plates, pyramids, planet Developments in national healthy and sustainable dietary guidelines: A state of play assessment*. FAO and the Environmental Change Institute & The Oxford Martin Programme on the Future of Food, The University of Oxford. www.fao.org/3/a-i5640e.pdf

Gussow, J. D., & Clancy, K. L. (1986). Dietary guidelines for sustainability. *Journal of Nutrition Education*, *18*(1), 1–5.

Jelsøe, E. (2015). Dietary guidelines: Nutritional health communication versus sustainable food policy. *Journal of Transdisciplinary Environmental Studies*, *14*(2), 36–51.

Johnston, J. L., Fanzo, J. C., & Cogill, B. (2014). Understanding sustainable diets: A descriptive analysis of the determinants and processes that influence diets and their impact on health, food security, and environmental sustainability. *Advances in Nutrition*, *5*(4), 418–429.

Kraak, V. I. (2021). Advice for food systems governance actors to decide whether and how to engage with the agri-food and beverage industry to address malnutrition within the context of healthy and sustainable food systems comment on "challenges to establish effective public-private partnerships to address malnutrition in all its forms." *International Journal of Health Policy and Management*. https://vtechworks.lib.vt.edu/handle/10919/107011

Lang, T. (2017). *Re-fashioning food systems with sustainable diet guidelines: Towards a SDG2 strategy*. Friends of the Earth.

Lang, T., & Mason, P. (2018). Sustainable diet policy development: Implications of multi-criteria and other approaches, 2008–2017. *Proceedings of the Nutrition Society*, *77*(3), 331–346.

Lígia, R., & Patricia, J. (2021). Measuring professional self and collective efficacy for dietary advice in Primary Health Care. *Nutrition and Health*, *27*(1), 49–57.

Macdiarmid, J. I. (2013). Is a healthy diet an environmentally sustainable diet? *Proceedings of the Nutrition Society*, *72*(1), 13–20.

Mazac, R., Renwick, K., Seed, B., & Black, J. L. (2021). An approach for integrating and analyzing sustainability in food-based dietary guidelines. *Frontiers in Sustainable Food Systems*, *5*, 84.

Merrigan, K., Griffin, T., Wilde, P., Robien, K., Goldberg, J., & Dietz, W. (2015). Designing a sustainable diet. *Science*, *350*(6257), 165–166.

Ministry of Health of Brazil. (2015). *Dietary guidelines for the Brazilian population*. Secretariat of Health Care Primary Health Care Department. http://bvsms.saude.gov.br/bvs/publicacoes/dietary_guidelines_brazilian_population.pdf

Monteiro, C. A., Cannon, G., Moubarac, J.-C., Martins, A. P. B., Martins, C. A., Garzillo, J., Canella, D. S., Baraldi, L. G., Barciotte, M., & da Costa Louzada, M. L. (2015). Dietary guidelines to nourish humanity and the planet in the twenty-first century. A blueprint from Brazil. *Public Health Nutrition, 18*(13), 2311–2322.

Mozaffarian, D., & Ludwig, D. S. (2010). Dietary guidelines in the 21st century: A time for food. *JAMA, 304*(6), 681–682.

Myers, S. S., Smith, M. R., Guth, S., Golden, C. D., Vaitla, B., Mueller, N. D., Dangour, A. D., & Huybers, P. (2017). Climate change and global food systems: Potential impacts on food security and undernutrition. *Annual Review of Public Health, 38*(1), 259–277.

National Health and Medical Research Council. (2013). *Australian dietary guidelines*. National Health and Medical Research Council. https://www.nhmrc.gov.au/_files_nhmrc/file/publications/n55_australian_dietary_guidelines1.pdf

Popkin, B. M. (2006). Global nutrition dynamics: The world is shifting rapidly toward a diet linked with noncommunicable diseases. *The American Journal of Clinical Nutrition, 84*(2), 289–298.

Popkin, B. M., Adair, L. S., & Ng, S. W. (2012). Global nutrition transition and the pandemic of obesity in developing countries. *Nutrition Reviews, 70*(1), 3–21.

Rizvi, S., Pagnutti, C., Fraser, E., Bauch, C., & Anand, M. (2018). Global land use implications of dietary trends. *PLoS One, 13*(8), e0200781.

Seed, B. (2015). Sustainability in the Qatar national dietary guidelines, among the first to incorporate sustainability principles. *Public Health Nutrition, 18*(13), 2303–2310.

Seed, B., & Rocha, C. (2018). Can we eat our way to a healthy and ecologically sustainable food system? *Canadian Food Studies, 5*(3), 182–207.

Soret, S., Mejia, A., Batech, M., Jaceldo-Siegl, K., Harwatt, H., & Sabaté, J. (2014). Climate change mitigation and health effects of varied dietary patterns in real-life settings throughout North America. *The American Journal of Clinical Nutrition, 100*(suppl_1), 490S–495S.

Springmann, M., Clark, M., Mason-D'TMCroz, D., Wiebe, K., Bodirsky, B. L., Lassaletta, L., de Vries, W., Vermeulen, S.J., Herrero, M., & Carlson, K. M.. (2018a). Options for keeping the food system within environmental limits. *Nature, 562*(7728), 519–525.

Springmann, M., Wiebe, K., Mason-D'Croz, D., Sulser, T. B., Rayner, M., & Scarborough, P. (2018). Health and nutritional aspects of sustainable diet strategies and their association with environmental impacts: A global modelling analysis with country-level detail. *The Lancet Planetary Health, 2*(10), e451–e461.

Tilman, D., & Clark, M. (2014). Global diets link environmental sustainability and human health. *Nature, 515*(7528), 518–522.

Tramontt, C. R., Maia, T. D. M., Baraldi, L. G., & Jaime, P. C. (2021). Dietary guidelines training may improve health promotion practice: Results of a controlled trial in Brazil. *Nutrition and Health, 27*(3), 347–356.

UN FAO (2019). *Food-based dietary guidelines*. http://www.fao.org/nutrition/education/food-dietary-guidelines/home/en/

U.S. Department of Health and Human Services, & U.S. Department of Agriculture (2015). 2015–2020 *Dietary guidelines for Americans* (8th ed.). https://health.gov/dietaryguidelines/2015/resources/2015-2020_Dietary_Guidelines.pdf

van Dooren, C., Marinussen, M., Blonk, H., Aiking, H., & Vellinga, P. (2014). Exploring dietary guidelines based on ecological and nutritional values: A comparison of six dietary patterns. *Food Policy, 44*, 36–46.

Vanderlee, L., McCrory, C., & Hammond, D. (2015). Awareness and knowledge of recommendations from Canada's food guide. *Canadian Journal of Dietetic Practice and Research: A Publication of Dietitians of Canada = Revue Canadienne De La Pratique Et De La Recherche En Dietetique: Une Publication Des Dietetistes Du Canada, 76*(3), 146–149.

Vermeulen, S. J., Campbell, B. M., & Ingram, J. S. (2012). Climate change and food systems. *Annual Review of Environment and Resources, 37*, 195–222.

Vieux, F., Soler, L.-G., Touazi, D., & Darmon, N. (2013). High nutritional quality is not associated with low greenhouse gas emissions in self-selected diets of French adults. *The American Journal of Clinical Nutrition, 97*(3), 569–583.

Vieux, F., Privet, L., Soler, L. G., Irz, X., Ferrari, M., Sette, S., Raulio, S., Tapanainen, H., Hoffmann, R., & Surry, Y. (2020). More sustainable European diets based on self-selection do not require exclusion of entire categories of food. *Journal of Cleaner Production, 248,* 119298.

Willett, W., Rockström, J., Loken, B., Springmann, M., Lang, T., Vermeulen, S., Garnett, T., Tilman, D., DeClerck, F., & Wood, A. (2019). Food in the anthropocene: The EAT–lancet commission on healthy diets from sustainable food systems. *The Lancet, 393*(10170), 447–492.

29

ODISHA MILLET MISSION

A transformative food system for mainstreaming sustainable diets

Saurabh Garg, M. Muthukumar,
Dinesh Balam, and Bindu Mohanty

Introduction

The Green Revolution re-defined food systems all over the world. While significantly increasing the number of calories produced per acre of agriculture, it resulted in negative externalities such as reducing the diversity in our diets, increasing the triple burden of malnutrition (obesity, undernourishment, micronutrient deficiency), and adversely impacting natural resources, particularly soil health, water, and biodiversity. Mainstream food systems are currently governed centrally, leading to longer supply chains and greater carbon emissions. With predicted lower yields of major staples like rice and wheat due to climate change (IPCC, 2019), mainstream food systems will be unable to sustainably nourish a growing population on a hotter planet. A business-as-usual scenario would exacerbate nutrition insecurity, especially among the most vulnerable populations. Thus, there is a growing recognition for the urgent need to move towards sustainable diets.

Sustainable diets are climate-resilient, nutritious diets, grown agro-ecologically that are also culturally appropriate, economically accessible, and promote social equity (Burlingame & Dernini, 2012; FAO, 2014; Gussow & Clancy, 1986; Mason & Lang, 2017). To sustainably feed an estimated 9 billion people in 2050,[1] it is imperative that our food systems be re-imagined to protect the environment, ensure nutritional security, and enhance livelihoods of marginalised farmers. Equally importantly, neglected and underutilised crops, like millets, agro-ecologically produced, should again become part of our diets.

In the context of South Asia, especially among India's drylands, the role of millets is indispensable in healthy diets to cope with current and future challenges. Despite the state-supported Green Revolution and its tendency for homogenisation and mono-cropping, Indigenous tribes in South Asia still grow at least 160 varieties of small millets (DHAN Foundation & WASSAN, 2012). It is vital that

DOI: 10.4324/9781003174417-34

our food systems value traditional knowledge systems by including farmers as key stakeholders in research and governance of food systems. Such an approach, as noted in the Cordoba Declaration on Promising Crops for the XXI Century (International Seminar on Traditional and New Crops to Meet the Challenges of the XXI Century, 2012), will contribute to the Universal Right to Food and Health and the Sustainable Development Goals of zero hunger, good health and well-being, and decent work and economic growth.

This chapter examines the properties of millets, their decline in production and consumption in India, and documents a novel public programme of the Government of Odisha, the Odisha Millet Mission (OMM), to re-introduce millets into the state's welfare schemes by actively engaging farmers as stakeholders. The case study of OMM, a transformative food system, gives the developmental context of Odisha and documents the innovations in the food system including challenges faced in implementation, how they were overcome, and how this transformative food system has positive effects on the environment, public health, and rural economy.

Millets as a staple of healthy diets

Millets are small seeded cereals, or nutria-cereals, as they are a very good source of carbohydrates, micronutrients, and phytochemicals with nutraceutical properties. Millets contain 7–12% protein, 2–5% fat, 65–75% carbohydrates, and 15–20% dietary fibre. Nutritional content varies in different millet species and varieties, but in general, millets, compared to the staples, rice, wheat, and maize, have a higher content of micronutrients, such as niacin, B6 and folic acid, and calcium, iron, potassium, magnesium, and zinc (Malleshi et al., 2021). Millets are easy to digest, contain a high amount of lecithin, and are excellent for strengthening the nervous system. Millets are good for people who are gluten intolerant. Their fibre content also helps to increase elimination of toxins, prevent constipation, and other bowel disorders. Millets have anti-diabetic properties due to their low glycaemic index and a high content of slow digestible starch (Rao et al., 2018). Thus millets, as a staple of healthy diets, can play a significant role in combating malnutrition, particularly anaemia and other non-communicable diseases, globally.

Millets as climate smart crops

Millets grow in harsh environments with limited water resources, where other crops do not grow easily or yield poorly. Some varieties can withstand high humidity levels or can be exposed to high temperatures. For instance, pearl millet can grow in critical drought conditions, and others can be grown on very acidic soil conditions. Millets are adaptable almost everywhere in dry regions, on clay soils, in wet lowlands, or in alluvial plains (Malleshi et al., 2021). Their powerful root system can penetrate soil, sometimes even up to two metres, to extract water and minerals. Due to their powerful root systems, millets are hardy, drought resistant, and also act as a natural soil and subsoil conditioner, which is important in no tillage systems. The high carbon content of the crop residues and the high amount of those residues makes them particularly important for maintaining and increasing soil carbon levels, maintaining a good soil cover, important for sustainable cropping systems, and providing, where needed, forage for livestock (FAO, 2018).

Millets possess several morpho-physiological, molecular, and biochemical characteristics that confer better tolerance to environmental stresses than major cereals. They can also yield well with little to no synthetic inputs. As millets are C4 crops, they produce at enhanced photosynthetic rates in warm conditions, and use water and nitrogen more efficiently, that is at 1.5 to four times higher levels of efficiency than C3 crops, such as the major cereals. In addition, C4 photosynthesis associated with millets include improved growth and ecological enactment in warm temperatures, enhanced flexible allocation patterns of biomass, and reduced hydraulic conductivity per unit leaf area (Sage & Zhu, 2011; Vivitha & Vijayalakshmi, 2015). Given all these advantages, IPCC (2019) specifically recommends a shift to coarse grains such as millets for climate-resilient diets. This recommendation is even more pertinent for rainfed districts of India where the increasing inter-and intra-annual (seasonal) variability will lead to greater vulnerability in the livelihoods of the small-scale and marginal farmers who rely on natural resources (Gangaiah, n.d.).

In spite of all the advantages, millet cultivation has seen a considerable decline in terms of total cropped area. In India, between 1952 and 1954, millets constituted 20% of national food grain production, which declined to 6% by 2018 (Rao et al., 2018). Historical policy neglect of these crops is the primary reason for this decline. Dubbed as "poor man's [sic] food," "coarse cereals," and relegated to a "minor" status unworthy of public attention, millets, a dryland crop, are neglected and underutilised species of South Asia. The Green Revolution of the 1960s favoured irrigated hybrid varieties of wheat and rice as high-yielding and high-input utilisation crops to meet the demands of national food security. The resulting national policy framework in India, that persists still today, favoured such commodified food grains, and millets did not receive policy support and research attention for their cultivation and consumption. Consequently, supply of millets declined as cropping patterns shifted in favour of rice or wheat cultivation (Nelson et al., 2019). The demand side of the equation was similarly shaped by policy support, with assured procurement prices for rice and wheat and its distribution through the Public Distribution System (PDS)[2] at subsidised prices. Other underlying barriers that negatively influence the production and consumption of millets are: Negligible production support to the farmers; absence of capacity-building to suggest improved methods of production and technologies; lack of appropriate post-harvest processing technologies for small millets; competition from other remunerative crops; induced behavioural changes in consumption due to active public promotion of rice and wheat; lack of public support for millets in the nutritious food category; and, lack of strong industrial demand and poor farm gate prices (DHAN Foundation & WASSAN, 2012).

Even among India's Indigenous tribes, for whom millets were once traditional staples, there is a decline in household consumption. This shift away from traditional sustainable diets, institutionally-engineered, is true not only for India, but for all countries in South Asia. It is increasingly recognised that while rice and wheat provided necessary food security to India's poor, it resulted in a negative external cost: Malnutrition. Today, millets are internationally regarded as being key to nutrition security (IPCC, 2019).

The developmental context of Odisha

Odisha ranks third in rainfed agriculture among the Indian states with 71.3% of its total cropped area under rainfed conditions. Sixty percent of the total population, the majority of them belonging to Indigenous tribes, depend on rainfed agriculture for their livelihoods (Agricultural Census 2010–2011, 2015). In addition to the demographic and economic aspects of rainfed agriculture in the state, there is a deepening ecological crisis: 99% of the agricultural land undergoing degradation in the state is rainfed (Space Application Centre, 2016), and the state has low to very low adaptive capacity to climate emergency (Rao et al., 2013).

Based on socio-economic indicators, Odisha has been regarded as a bit behind other states in India, particularly with a persistent problem of malnutrition. Moreover, malnutrition challenges—namely anaemia, stunting, and wastage—are particularly prevalent among the state's Indigenous tribes (Meher, 2007). To cite one example, 80% of the chronically undernourished tribal children belong to only eight states, with Odisha being one of them (UNICEF, n.d.). Despite several government initiatives, a longitudinal data analysis of National Family Health Survey (NFHS) data shows that while Odisha has made commendable strides in addressing malnutrition, the tribal population continues to be at high risk (as cited in Bharat Rural Livelihoods Foundation, 2021).

In India, in general, despite constitutionally-enshrined affirmative action, scheduled tribes (ST) tend to figure at the bottom of various development indicators (Béteille, 1991). Odisha has one of the highest populations of Indigenous tribes in the country: Spread over its 30 districts, Odisha has 62 ST communities and 13 primitive tribes, constituting 22.43 per cent of the total state population (Meher, 2007). Millets, particularly finger millet (*Eleusine coracana*), were traditionally grown by these Indigenous communities and their cultures were intermeshed with millet food systems. Nevertheless, in the past three decades, the area cultivating finger millet (*Eleusine coracana*) declined by more than 50%.

The food system design of OMM

A multi-stakeholder participatory approach in designing OMM

In this context of decline of millet production, the Government of Odisha and Nabakrushna Choudhury Center for Development Studies (NCDS) in partnership with other civil society organisations and networks, jointly organised a multi-stakeholder consultation in 2016 to evolve a strategy for revival of millets in Odisha. A key insight that emerged from the consultation was the need to simultaneously act on production, consumption, marketing, and postharvest processing of millets. Various constraints and interlinked aspects of the millet ecosystem were also discussed at this event, and it was decided to put community collectives at the centre of any initiatives on millets.

Based on the learnings from the event, the Department of Agriculture & Farmers' Empowerment, (DA&FE) Government of Odisha, launched the "Special Programme for Promotion of Millets in Tribal Areas," which is popularly known as the Odisha Millets Mission (OMM). It is the first initiative of DA&FE in which consumption was made central to the food system. Other key design elements include: A multi-stakeholder participatory approach; a decentralised operational framework; financial and institutional support from the government backed by convergence of programmes and enabling policies; promotion of rural enterprise in the millet value chain; awareness campaigns for behaviour change; efforts to support household consumption; and, the revival of Indigenous culture.

The Odisha Millet Mission was launched in all the districts with a high tribal population to combat the prevalent poverty, inequality, and malnutrition (OMM, 2021). Millets were seen as key to addressing these challenges, as they comprised the traditional staple of the Indigenous tribes and moreover were easy to grow in the marginal soils of the lands cultivated by these tribes (Gangaiah, n.d.). OMM was initiated in the 30 blocks covering seven districts in 2017; this was extended to 84 blocks covering 15 districts by 2021.[3] The scaling of OMM was proof of success and acceptance by the community of Odisha.

Healthy Diets

Inclusion of millets in government welfare schemes

Improving nutrition in rural communities is the primary goal of OMM. Thus, the government procures millets from farmers at fair prices and distributes them for free or at highly subsidised costs in the following government welfare schemes:

Public Distribution System: In Odisha, finger millet is included in the rations distributed under PDS. 1 kg millet was distributed among 1.6 million households in 2019 and it was distributed to 4.6 million households in 2020.

Integrated Child Development Scheme (ICDS): Finger millet snack was piloted in the ICDS scheme in two districts. Around 0.12 million children continue to be served with this nutritional snack. The Government of Odisha is planning to further scale up the initiative in these districts.

Mid-Day Meal (MDM): The Government of Odisha has advocated for inclusion of millet snacks in the mid-day meal. Accordingly, the Government of India has approved inclusion of millet snacks under the MDM scheme. At present, three tribal districts have been chosen for demonstration of pilot projects.

In addition, under Odisha Millets Mission, there are ongoing discussions to include millets in tribal welfare hostels, government tourist centres, and other public programmes.

Supporting household consumption and Indigenous culture

Food festivals, cooking competitions, awareness campaigns, millet recipe training events, video documentation for building prestige in consumption of millets, and celebration of local millet-based food cultures were undertaken to create awareness on millets. Government department officials from Agriculture, Women and Child Development and schools are closely involved in these awareness campaigns. In the last four years, OMM organised 2,513 training programmes covering 50,260 farmers, 98 block level convergence workshops covering 1,960 participants, 487 food festivals covering 194,800 people, and 732 awareness campaigns covering 585,600 farmers. Overall, to date, it has directly engaged with 0.87 million farmers.

OMM also seeks to map the biocultural diversity of millets in the Indigenous communities. Our contemporary food systems have been de-linked from the spiritual, cultural, and ecological dimensions that once existed in traditional agro-ecosystems. But in Indigenous communities there is a ritual associated with each millet species. It is this cultural dimension of food that has allowed for the rich in-situ conservation of the biodiversity in millet varieties. OMM honours such traditions by supporting Indigenous festivals such as "Burlang Yatra" (an annual Indigenous seed festival in the Kandhamal district) (Niyogi, 2020; Saxena, 2020). Besides the exchange of Indigenous heirloom seeds and knowledge-transfer that happens at such events, the long-term value of such festivals is that they: a) Strengthen the social and the ecological relationships, which are a central behavioural aspect of all food systems; and b) ensure sustainability of in-situ agro-biodiversity through intergenerational participation (Saxena, 2020).

Awareness campaigns for behaviour change to healthy diets

Assured procurement by the government leads to surplus availability of grains in the market. To take care of the surplus and promote healthy diets among the urban population, WASSAN has also sought to popularise millets in urban areas. This effort required concerted multi-pronged efforts to promote a behavioural change for people to adopt sustainable millet-based diets. Through massive awareness campaigns, socio-cultural celebrations like food fairs, and active media engagement, WASSAN has succeeded in changing people's mind-sets from stigmatising millets to being aware about the health benefits of millets. The food fairs are opportunities for demonstrating various millet recipes, training health workers working on women and children nutrition schemes, and making these dishes and showcasing value-added products from millets (Satpathi et al., n.d.). To satisfy urban demand, particularly among the younger generation, new recipes using millets have been introduced for Westernised dishes such as cakes, pizzas, and burgers (Mohanty, 2020). Women Self-Help Groups (WSHGs) help in these endeavours.

A decentralised operational framework

The decentralised operational framework of OMM is due to a committed government-civil society partnership, which as Bharat Rural Livelihoods Foundation (2021) note, is rare in India. In the decentralised approach to project implementation of OMM, in each block (a sub-district), a civil society organisation (CSO) is chosen as the facilitating agency and adopts a cluster-based approach to reach out to farmers and demonstrate agronomic practises that can substantially improve yields. Farmer Producers Organizations (FPOs) and WSHGs also now command power in the value chain, as detailed below. The government's Agricultural Technology Management Agency (ATMA), in each district, serves as the administrative head of the OMM at the district level, helping in monitoring the programme on a monthly basis and overseeing disbursement of funds to facilitating agencies. For research, development, and implementation of OMM, the government has constituted two nodal agencies (NCDS and WASSAN) and allocated distinct responsibilities to each. Overall responsibility lies with a special government committee.

Supportive policies and public investments

Innovations for more efficient utilisation of public funds is not always easy in India, given the fragmentation of responsibilities among various government departments and their autonomous way of working. In OMM, the government acted upon the policy suggestions of NCDS to execute a convergence of programmes between the Department of Agriculture and Farmers' Empowerment, the Tribal Development Cooperative Corporation of Odisha Limited, the Food Supplies & Consumer Welfare Department, among other, which allowed the government to secure the funds needed for OMM (Mishra, 2019). OMM also reached a Memorandum of Understanding with the governmental Department of Women and Child Development in 2020.

In terms of public investments, the Government of Odisha has sanctioned the following funds (OMM, 2021):

- INR 5369.2 million in the FY 2019–2020 for raising the minimum support price (MSP) from INR 6 to INR 29 per kg for finger millet. The revised MSP sets the floor price for purchase of finger millet and makes it worthwhile for the farmers to cultivate this crop.
- INR 327.2 million (about USD 4.5 million) by the governmental Department of Women and Child Development to support millet entrepreneurship and awareness campaigns through Women Self Help Groups (WSHGs) for three years in 14 districts.

The millet supply and value chain

Farmer collectives as service delivery institutions

To strengthen local food systems, millet production and consumption have been decentralised to the block (sub-district) level. OMM envisages a greater role for

Farmer-Producer Organisations (FPOs), beyond aggregation of grain, by making contributions to the input and output supply chain. For delivery of needed services at the block level, such as timely supply of seeds, supply of quality bio-inputs, aggregation and marketing of produce, FPOs will be registered and their capacities built. These FPOs will also anchor the different enterprises by linking with WSHGs and other smaller collectives. Seventy-six FPOs have been formed under OMM till June 2021.

Saving and sharing seeds: One of the key challenges in decentralised agro ecological interventions is to ensure locally suitable quality seeds. In order to achieve this, under OMM, participatory varietal trials were undertaken to understand the performance of local millet landraces as compared to government released varieties. Evaluation by research institutions indicated that there were good local varieties that performed at par with government released varieties and were preferred by the people for their taste or resilience. These seeds were then further purified and multiplied through FPOs to ensure supply of locally suitable seed centres were created in each block for multiplication of seeds of preferred varieties for distribution. Around 185 landraces were evaluated and 72 seed centres were formed.

Setting up decentralised processing facilities: In order to reduce drudgery in postharvest and processing, OMM aimed to set up postharvest and processing machinery at GP and block level. One of the key reasons for the non-realisation of price for millets is poorly-graded grain material. Under the project, through an open process, WSHGs and FPOs are selected to set up postharvest and primary processing units (threshers, destoners, graders, dehullers, and pulverisers). Technical specifications of the machines are finalised through a state level committee, consisting of different experts, on processing. Currently, 415 postharvest processing units have been established under OMM across 15 districts. OMM is the first initiative to develop the standards for the millet processing and value-added machinery, which has simplified the process for adoption by other state governments.

Custom hiring centres: One of the challenges of promoting nature-positive methods is an increase in manual operations. While this helps in greater rural employment, it may hinder large scale adoption. Hence, in order to reduce drudgery of farm operations, implements were supplied to farmers on hiring basis through custom hiring centres. Custom hiring centres were anchored by FPOs at block level and by WSHGs at village or Panchayat level.

Procurement of grain: Procurement of grain is supported by M-PAS, a web-based portal. Government agencies in each block, facilitate the registration of farmers and host millet procurement centres. Regulated Marketing Committees are responsible for creating awareness about procurement prices and Fair Average Quality (FAQ) parameters of the grain eligible for purchase. Procurement of millets was initially a challenge, for all the infrastructure, processes and equipment were designed for the procurement of rice. The rice-procurement system was not suited for millet procurement. Millets are cultivated mostly in hilly rainfed lands and remote villages that are difficult for centralised procuring agencies to access. Communications are an issue due to the diversity of languages spoken by the Indigenous tribes of Odisha. As FPOs in various blocks already supporting millet procurement without remuneration, the government decided to pilot the "empanelment" or formal recognition of Farmer Producer Organisations (FPOs) as Ragi Procurement Agencies at the block level. This is the first intervention by a State Government where millets are being procured through FPOs. Sixteen FPOs have now been selected as procurement agencies. In two years, procurement by FPOs has increased ten-fold.

Currently, there are ongoing training programmes for FPOs to allow them to develop as viable businesses (Balam et al., 2021; Das, 2020). It is expected that, over time, this intervention will result in other fruitful outcomes benefitting both producers and consumers by shortening the supply chain and increasing rural livelihoods. As the government seeks to increase dietary diversity and include other millets or crops for procurement, these FPOs will be well-placed to help in the procurement of the same. (Balam et al., 2021).

Empowering women in the millet value chain

For development of the millet value-chain by women, OMM converged with the Department of "Mission Shakti," the Government of Odisha's nodal department for women's empowerment. OMM entered into a formal agreement with Mission Shakti for leveraging the power of women collectives. The key features of the agreement are: a) Massive awareness campaigns about the nutritional value of millets targeting household consumption of millets; b) operating processing and value-addition units; c) operating millet cafés in the town and cities; and d) preparation of millet recipes and inclusion of millet snacks for the Integrated Child Development Services, a nutritional programme for mothers and children (Balam, 2020).

With rural WSHGs entering the value chain, circular economy loops are created that are benefiting local communities. Food systems in rural India are entwined with rural poverty leading to urban migration. A positive spinoff of OMM has been that it has improved agrarian incomes through diversified livelihood opportunities. Anecdotal data shows that this has reversed the trend of urban migration (Mohanty, 2020).

Adoption of nature positive productivity-enhancement practises

One of the key challenges in the revival of millets in food systems is to address the productivity gap. While average millet productivity ranges from 0.7 to 1 tonnes per hectare, paddy productivity varies from 1.3 to 1.8 tonnes per hectare. At the same time, for sustainable agroecological farming it is important to promote non-chemical intensive practices and diversified cropping systems. OMM adopts nature positive methods of crop productivity, such as system of root intensification, line transplantation, location-specific intercropping, and use of bio-inputs. A reduced incentive structure for a period of three years has also been developed to adopt such agronomic practices. Improved agronomic practices are currently practised by over a million farmers covering 47,399 hectares. To allow for scale, local WSHGs have been trained to prepare bio-inputs for sale to farmers. In some places, to overcome challenges in making timely payments for bio-inputs, farmers and WSHGs have come up with in-kind payment options, namely accepting part of the farmer's harvest as reimbursement. This innovative payment method is likely to lead to local consumption of locally-grown millets.

Conserving biodiversity

PDS requires production and processing of grains at scale. Taking any crop to scale results in agro-ecological trade-offs, namely loss in biodiversity. Comprehensive policy recommendations have been made to include local millets other than ragi (*Eleusine Coracana)* in procurement and public distribution (Jena and Mishra, 2021). To retain biodiversity, the Government of Odisha has taken the following measures:

- It seeks to include minor millets in PDS by decentralising operations to the district and encouraging the development of efficient machines to process minor millets.
- It has established a Centre for Excellence for Agroecology and Agrobiodiversity for active participatory research with local farming communities to take up *in-situ* conservation, maintain the genetic purity of landraces, undertake diverse seed production, and supply of endemic varieties through community managed seed systems. One hundred thirty-three landraces have so far been documented (Balam, 2021).

Conclusion

OMM has received appreciation and accolades from many corners of the world for its potential to transform food systems. NITI Aayog and Ministry of Agriculture and Farmers' Welfare, Government of India have recognised OMM as a model that other states should adopt for promotion of millets. Similarly, the United Nations World Food Programme (WFP) has entered into a partnership with the Government of Odisha for promoting the OMM framework in other millet-growing countries (WFP, 2021). Noting this, the National Food Security Mission (NFSM) now seeks to adapt the OMM operational framework in other states of India (NFSM, 2019). The key innovative features of OMM to transform food systems are:

- A supportive policy framework for disbursal of public funds through convergence of government departments and programmes
- A participatory multi-stakeholder consultative approach and a strong partnership between the government and civil society
- Decentralised governance and research honouring Indigenous knowledge and grassroots experiences allowing for ownership of the programme by farmers
- Ongoing collaboration between the research and programme implementation agencies, ensuring regular monitoring of practises and real-time policy recommendations to make the food system responsive and resilient

- Creating consumer awareness and encouraging household consumption by producers
- Encouraging research and entrepreneurship of the private sector
- Building capacities of local community-based organisations, such as FPOs and WSHGs, to add value to the supply chain and thereby enhance rural livelihoods.

Sustainable diets call for a re-orientation of human behaviour towards the food they consume. A behavioural change to sustainable diets, however, needs to be supported by robust changes in the entire food system, from enabling policy frameworks to decentralised participatory governance, to make such diets accessible to all. OMM is a commendable example of how food systems can be transformed to mainstream sustainable diets. Its innovative features can be adapted, not just in millet-growing countries in South Asia, but agro-ecosystems worldwide.

Notes

1 As estimated by the Population Reference Bureau (2020).
2 The Public Distribution System was instituted as a food welfare system in India in 1942 and received further state support with the establishment of the Food Corporation of India (FCI) in 1965. Under PDS, rice and wheat are supplied to beneficiaries through designated fair price shops (FPS) throughout the country.
3 Unless cited otherwise, all quantitative data about OMM was shared by the erstwhile and current Director of Agriculture & Food Production, M. Muthukumar, Department of Agriculture and Farmers' Empowerment, Government of Odisha who is co-author of this chapter.

References

Balam, D. (2020, September 28). *"Millet Shakti Cafés: Quick service restaurant by WSHGs" to be launched by mission shakti and Odisha millets mission.* Millet Mission Odisha. https://milletmission.wordpress.com/2020/09/28/millet-shakti-cafes-quick-service-restaurant-by-wshgs-to-be-launched-by-mission-shakti-and-odisha-millets-mission/

Balam, D. (2021, January 28). *Community led centre for excellence for agro-ecology and agro-biodiversity launched in Malkangiri district by Government of Odisha.* Millet Mission Odisha. https://milletmission.wordpress.com/2021/01/28/community-led-centre-for-excellence-for-agro-ecology-and-agro-biodiversity-launched-in-malkangiri-district-by-government-of-odisha/

Balam, D., Chaudhury, A., & Sharma, S. (2021, February). FPOs as block level ragi Procurement agencies. *Agriculture World Magazine, 7*(7), 16–20.

Béteille, A. (1991). *Society and politics in india: essays in a comparative perspective.* Oxford University Press.

Bharat Rural Livelihoods Foundation (2021, April 9). *Addressing the nutrition crisis: Reflections from Odisha millets mission.* https://www.ideasforindia.in/topics/agriculture/addressing-the-nutrition-crisis-reflections-from-odisha-millets-mission.html

Burlingame, B., & Dernini, S. (2012). *Sustainable diets and biodiversity directions and solutions for policy, research and action.* FAO.

Das, S. (2020). Supply chain of millets: An FPO perspective (with a special reference to Odisha). *Journal Global Values, 11*(2), 234–257.

DHAN Foundation, & WASSAN (2012). *Supporting millets in India: Policy review and suggestions for Action: Revalorising small millets in rainfed regions of South Asia (RESMISA).*

DHAN Foundation. Retrieved from https://www.dhan.org/smallmillets/docs/report/Millet_Support_Policies.pdf

Food and Agriculture Organization of the United Nations (FAO) (2014). *The state of food and agriculture: Innovation in family farming*. FAO.

Food and Agriculture Organization of the United Nations (FAO) (2018). *Proposal for an international year of millets*. FAO.

Gangaiah, B. (n.d.). *Agronomy: Kharif Crops*. Indian Agricultural Research Institute. Retrieved from http://hrsacademy.in/wp-content/uploads/2016/11/Book-of-Agronomy.pdf

Gussow, J. D., & Clancy, K. L. (1986). Dietary guidelines for sustainability. *Journal of Nutrition Education, 18*(1), 1–5.

Intergovernmental Panel on Climate Change (2019). *IPCC special report on climate change, desertification, land degradation, sustainable land management, food security, and greenhouse gas fluxes in terrestrial ecosystems*. IPCC.

International Seminar on Traditional and New Crops to Meet the Challenges of the XXI Century (2012). *Cordoba declaration on promising crops for the XXI century*. Retrieved from https://www.fao.org/fileadmin/templates/food_composition/documents/Cordoba_NUS_Declaration_2012_FINAL.pdf

Jena, D., & Mishra, S. (2021). Procurement and public distribution of millets in Odisha: Lessons and challenges. *Policy Brief* PB1RRAN0121, RRA Network. Retrieved from https://www.rainfedindia.org/published-page/resources?id=600a6a920d3b4e000a57ec56

Malleshi, N. G., Agarwal, A., Tiwari, A., & Sood, S. (2021). Nutritional quality and health benefits. In Singh, M., and Sood, S. (Eds.), *Millets and pseudo cereals*. Woodhead Publishing.

Mason, P., & Lang, T. (2017). *Sustainable diets: How ecological nutrition can transform consumption and the food system*. Earthscan Routledge.

Meher, R. (2007). Livelihood, poverty and morbidity: A study on health and socio-economic status of the tribal population in Orissa. *Journal of Health Management, 9*(3), 343–367.

Mishra, S. (2019). Ragi procurement in Odisha: Strengthening the farm to plate initiative. *Policy Brief 11*. Bhubaneswar, India, Nabakrushna Choudhury Centre for Development Studies (NCDS). Retrieved from https://ncds.nic.in/sites/default/files/PolicyBriefs/PB11NCDS.pdf

Mohanty, B. (2020, February 17). Odisha millet mission: The successes and the challenges. *Vikalp Sangam*. https://vikalpsangam.org/article/odisha-millet-mission-the-successes-and-the-challenges/

Niyogi, D. G. (2020, July 27). India's millets policy: Is it headed in the right direction? *The Mongabay*. https://india.mongabay.com/2020/07/indias-millets-policy-is-it-headed-in-the-right-direction/

National Food Security Mission Cell (2019). *Minutes of meeting. F. No. 4-1/2019-NFSM (FTS: 72246) 25 July 2019*. Ministry of agriculture and farmers' welfare: Krishi Bhawan.

Nelson, A. R. L. E., Ravichandran, K., & Antony, U. (2019). The impact of the green revolution on indigenous crops of India. *Journal of Ethnic Foods, 6*(1), 1–10.

Odisha Millet Mission (2021). *Odisha millet mission*. Retrieved from http://www.milletsodisha.com/

Population Reference Bureau (2020, July 10). *Population of older adults increasing globally partly because of declining fertility rates*. https://www.prb.org/news/population-of-older-adults-increasing-globally/

Rama Rao, C. A., Raju, B. M. K., Subba Rao, A. V. M., Rao, K. V., Rao, V. U. M., Kausalya, R., Venkateswarlu, B., & Sikka, A. K. (2013). *Atlas on vulnerability of Indian agriculture to climate change*. Central Research Institute for Dryland Agriculture.

Rao, B. D., Bhat, B. V., & Tonapi, A. (2018). *Nutricereals for nutritional security*. ICAR-Indian Institute of Millet Research.

Sage, R. F., & Zhu, X. G. (2011). Exploiting the engine of C4 photosynthesis. *Journal of Experimental Botany, 62*(9), 2989–3000.

Saxena, L. P. (2020). Community self-organisation from a social-ecological perspective: 'Burlang Yatra' and revival of millets in Odisha (India). *Sustainability, 12*(5), 1867.

Satpathi, S., Saha, A., & Basu, S. (n.d.). Millets as a policy response to the food and nutrition crisis–special reference to the Odisha Millets Mission. Retrieved from https://www.brlf .in/wp-content/uploads/2020/01/Odisha-Millet-Mission.pdf

Space Applications Centre. (2016). *Desertification and land degradation atlas of India (based on IRS AWiFS data of 2011–13 and 2003–05).* SRO, Department of Space, Government of India.

UNICEF (n.d.). Tribal nutrition. Retrieved from https://www.unicef.org/india/what-we-do/tribal-nutrition

Vivitha, P., & Vijayalakshmi, D. (2015). Minor millets as model system to study C4 photo-synthesis: A review. *Agricultural Reviews, 36,* 296–304.

World Food Program (WFP) (2021, March 19). WFP and Government of Odisha join forces to improve food and nutrition security. *WFP News Release.* https://www.wfp.org/ news/wfp-and-government-odisha-join-forces-improve-food-and-nutrition-security

30

REFRAMING SUSTAINABLE DIETS AS SUSTAINABLE FOOD CONSUMPTION

Hugh Joseph

Introduction

This chapter traces the evolution of the *sustainable diets* model that centres on what is consumed (i.e., foods and diets) and its intersection with dietary guidance in furthering food system-related sustainability. It explains why diets alone are insufficient to encapsulate sustainability from an end-user perspective and how a more practice-focused approach can address this limitation. Systems-based components of a *sustainable food consumption* (SFC) framework are subsequently outlined.

Sustainable diets and dietary guidelines

What to eat to promote nutrition and personal health and wellbeing has been the central focus of dietary guidance. Dietary Guidelines for Americans (DGA) defines diets as "the combinations and quantities in which foods and nutrients are consumed" (USDA, 2015). The 2020 edition emphasises *dietary patterns* to denote "the combination of foods and beverages that constitutes an individual's complete dietary intake over time" (USDA, 2020). The DGA became the foundation for most other *Food Based Dietary Guidelines* (FBDG), now issued by over 100 countries (FAO, 1996). Such guidelines centre on "what to eat and drink to meet nutrient needs, promote health, and prevent disease" (USDA, 2020).

As noted in Chapter 1, *sustainable diets* is a term introduced in *Dietary guidelines for sustainability* (Gussow & Clancy, 1986), which was a reflection on the DGA, first issued in 1980. Juxtaposing sustainability with public health nutrition and parallel dietary guidance was foundational to the model. "Within the nutrition community there appears to be widespread adoption of the Dietary Guidelines as the basis for nutritional counselling aimed at promoting health; we propose that nutritionists begin to use that same framework to address issues of sustainability" (Gussow & Clancy, 1986, p. 2). Specifically, they made the case for broadening the

DOI: 10.4324/9781003174417-35

nutrition focus of Dietary Guidelines for Americans to incorporate salient food system-related concerns—environmental, macroeconomic, and agricultural—in concert with widespread interest emerging outside the nutrition community.

In so doing, Gussow and Clancy (1986) referred to sustainable agriculture, reflecting a growing focus on alternative farming practices. *Organic farming* was increasingly emphasised, promoted especially by J.I. Rodale since the 1950s, and by his son, Robert Rodale, who also promoted *regenerative agriculture* in the early 1970s. *Sustainable agriculture* per se first appeared at IFOAM in 1977 (Clancy, 1999). Distinctions among these cross-cutting terms were somewhat blurred, but they commonly displayed an ecological emphasis that remains a dominant facet of food system sustainability.

Sustainable development also emerged in this period with publication of *Our Common Future* (Brundtland, 1987). Whereas sustainable agriculture, and likewise sustainable diets, emphasised biophysical environments, Brundtland assigned parallel significance to societal and economic dimensions. It focused somewhat on farming, but not on food systems or diets.[1] Nevertheless, it served as a bridge to a more wide-ranging configuration expressed in FAO's 2009–2010 *Sustainable Diets and Biodiversity*, wherein Tim Lang, a principal author, considered the Brundtland triple focus as not sufficiently precise. In the context of diets, he referenced his "six-headed approach": Quality, environment, social, health, economic and governance—with each further subdivided into more specific elements (Lang, 2014). These were subsequently reflected in FAO's much-cited definition of sustainable diets (Burlingame & Dernini, 2012).

Noteworthy here was a continued broadening of the scope of sustainability from its core environmental footings, particularly in relation to *food systems*—a related construct that evolved during the same period. Circulating food system models have varied considerably in structure and composition. Some focus primarily on the food supply chain per se; others incorporate broader influences, including culture, governance, education, health, and welfare, as did FAO. Scalar distinctions include geographic, temporal, and structural dimensions. Varying worldviews strongly influence the fundamental characterisations and the resulting agendas of sustainability (Hedlund-de Witt, 2014).

But the framing of sustainable diets, including its intersection with dietary guidance, has remained constant. As reflected in his agency-oriented perspective, nutrition expert Jean Mayer was widely quoted for espousing: "Nutrition is not simply a science; it is an agenda" Kennedy et al., 2011). Sustainable diets reflect a similar direction as a change-driven, problem-solving strategy. This strategy originated with Gussow and Clancy (1986) and Tim Lang similarly advocated: "There is already sufficient evidence as to food's impact to warrant the creation of comprehensive sustainable dietary guidelines at national, regional and global policy levels" (Burlingame & Dernini, 2012, p. 25).

Various sustainable dietary guidelines have been issued over the past two decades, some by government-supported bodies (e.g., the Netherlands, Sweden, Germany) and others by NGOs. They typically emulate FBDG, structured as "short, science-based, positive messages" (FAO, 1996) and list diverse recommendations with an environmental emphasis. While issuance of such guidelines has abated more recently, a growing body of sustainable diets-focused research incorporates corresponding recommendations, again prioritising environmental concerns (Jones et al., 2016). The widely circulated EAT-Lancet Commission's *Food in the Anthropocene* report (Willett et al., 2019) employs planetary boundaries as its primary environmental context. Yet as sustainability-related policy guidance has expanded, its underlying diet-centred focus has endured. Specifically, what is eaten (e.g., foods and dietary patterns recommended for personal health and wellbeing) predominates as the backdrop for considering externally contextualised sustainability. Sustainability is thus distinct from public health nutrition. For example, the 2015 US Dietary Guidelines Advisory Committee (DGAC, p. 30) report suggested: "An important reason for addressing sustainable diets, a new area for the DGAC, is to have alignment and consistency in dietary guidance that promotes both health and sustainability." Yet Gussow and Clancy (1986, p. 2) stated, "to 'sustain' is to support in life and health." Similarly, Mason and Lang (2017, p. 4) approached health "as a function of ecological relationships, a web of connections between humans, planet and society." In such contexts, the rationale for inherently separating public health nutrition from this broader public health framework is unclear.

In tandem, optimal nutrition, reflected in recommended dietary patterns, became the standard against which sustainability is considered. This makes sense from a nutrition perspective, but restricting the application of exogenous sustainability concerns to this specific dietary benchmark has limitations:

- Most foods and diets consumed by the public do not meet the nutrition-related specifications, but that should not be a reason to exempt them from sustainability considerations.
- Food attributes such as sensory qualities are also deemed secondary considerations in the context of personal health, but again need not be marginal to sustainability.
- The focus on food intake is inconsistent with addressing sustainability from a systems perspective.

Sustainability is not inherently embedded in foods or diets, any more than is health; rather, it is primarily reflected in the operations and environments within the overall food system from which products and dietary patterns emerge. Gussow and Clancy (1986) applied sustainable agriculture as a corresponding framework for sustainable diets. While food-based agricultural outputs (e.g., crops and animals) end up (directly or indirectly) as products to be consumed, they are not the foundational elements of agricultural sustainability. In *The Pleasures of Eating*, Wendell Berry famously declared, "Eating is an agricultural act" (1990, pp. 145–52). In the parlance of sustainable diets, that declaration might translate as "Diets are agricultural products," which clearly lacks corresponding resonance. Fundamentally, foods (and the diets that incorporate them) simply constitute inputs and outputs transformed throughout the food supply chain, as outlined in Box 30.1.

Box 30.1 Streamlined food supply chain and associated food inputs and outputs

---

Components	Production/ agriculture	Manufacturing/ processing	Distribution/ marketing	Consumption
	↓	↓	↓	↓
Food inputs / outputs	Crops/animals	Formulations/ prepared goods	Provisions/ groceries	Ingredients/dishes Meals/diets

---

Source: Work of the author, Hugh Joseph.

Absent here are the players, activities, and operating environments most germane to sustainability. According to the High-Level Panel of Experts (HLPE) on Food Security and Nutrition report, "A *food system* gathers all the elements (environment, people, inputs, processes, infrastructures, institutions, etc.) and activities that relate to the production, processing, distribution, preparation and consumption of food, and the output of these activities, including socio-economic and environmental outcomes" (HLPE, 2017, p. 23). This definition appropriately characterises food systems as complex adaptive systems, focusing on sustainable processes within an overall structure that incorporates boundaries, scales, composition, and numerous other facets (IOM, 2015).

Thus, reliance on consumer food choices is championed by sustainable diets proponents as a central strategy to influence how and where such items are produced, processed, distributed, and marketed. However, the wide range of operational factors and product transformations throughout the supply chain pose inherent limits to this strategy, and typically result in more generalised advice, such as reducing animal foods (especially beef and dairy) and, alternatively, pursuing a more plant-based diet. Also constrained by this approach is any substantive accounting for the systemic drivers of food utilisation that are reflected in consumption-related activities and associated socio-economic, cultural, and physical environments within which consumers' food decisions are made. These missing elements, in conjunction with diets per se, constitute the overall *consumer* or *consumption* segment of the food chain, as shown in Figure 30.1.

While the term *food consumption* also characterises eating itself, it is constituted here as an often-designated food chain component that, for several reasons, has yet to really come into its own. Some food supply chain depictions terminate with marketing and exclude end users. Similarly, many environmental footprint and life-cycle analyses are weighted to the production end, and this minimises scrutiny of consumption sector impacts. Furthermore, the focus of sustainable diets on food and nutrition per se de-emphasises the potential to incorporate consumption-related practices as associated but independent factors for sustainability.

In turn, *sustainable food consumption* (SFC) expands beyond foods and diets by incorporating these systemic dimensions of consumption. The next step is deter-

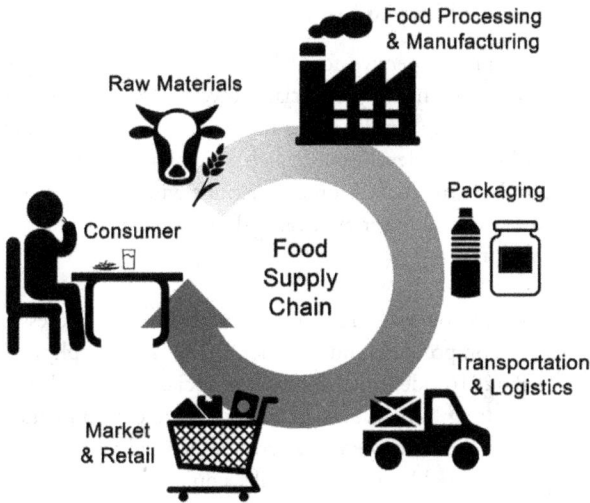

Figure 30.1   Food Supply Chain. Source: Tzounis et al., 2017. Used with permission.

mining which elements therein can best exemplify sustainability from the standpoint of an overall food system framework. A reasonable starting point is to reflect on what has emerged in agriculture: Gussow and Clancy's original reference benchmark.

The agricultural sector

Over the last half-century, sustainable agriculture has evolved into the most developed iteration of a complex adaptive food system. Agriculture itself may be rudimentarily defined as the art, practice, or science of crop and animal production (Hatfield & Karlen, 1994). Considered holistically, it integrates complex, interdependent agronomic, ecological, and social dimensions that scaffold from individual farms to vast global enterprises. These dimensions include soils, farmland and associated environments, resources inputs, product outputs, and all agents involved in the process. Sustainability centres around system dynamics and practices throughout the agricultural spectrum. While there is no single integrated configuration, agroecology has emerged as a widely adopted and fast-evolving sub-discipline. It initially applied to field, farm, and landscape agro-ecosystems, but Miguel Altieri, an original agroecology architect, now contends that it goes beyond this one-dimensional model to embrace an understanding of ecological and social levels of co-evolution, structure, and function (Altieri and Nicholls, 2017). Wezel et al. (2009, p. 505) deem agroecology to be "the integrative study of the ecology of the entire food systems, encompassing ecological, economic and social dimensions, or more simply the ecology of food systems." In short, sustainability does not centre on agricultural products per se nor just on environmental dynamics of production.

Applying sustainability to food consumption

Agroecology provides a structural perspective for delineating other food supply chain components and relevant sustainability criteria. However, its broader systems-level dimensions are still evolving rapidly along multiple, diverse paths. That dynamic, as well as its distinct historical development path, somewhat constrains its applicability for SFC, except where the focus is on associations with food production sustainability. But there are several other applicable domains to help identify and prioritise sustainability factors:

(a) *Sustainable diets and food systems*—This literature primarily addresses upstream food chain components and broader social and biophysical influences. Jones et al. (2016) identified 30 distinct social, ecological, and health determinants among 113 studies, and found that environmental factors, especially GHG, predominated over consumer preferences, food security, and especially cultural heritage and skills, equity, rights, and governance. Nonetheless, applying a SFC approach can bolster diet-related sustainability connections applied throughout the overall food system.

(b) *Sustainable dietary guidelines*—Twenty-six sets of sustainable dietary guidelines issued by government agencies and NGOs between 1996–2016 were compiled (Clarke & Joseph, 2017). Many of these guidelines targeted the public and not just intermediary professionals. The 300+ recommendations centred on food/diet choices linked primarily to environmental and nutritional categories, but they also included local food access, food policy and advocacy, consumer-centred food shopping and preparation, food literacy, and food waste and disposal to nominal extents.

(c) *Ecological frameworks*—This label was proposed by Story et al. (2008) to address the complexities of eating behaviours and settings as they influence dietary choices themselves. Furst et al. (1996) specified major influences, including ideals, personal factors, resources, social contexts, and many other factors. Sobal and Bigsoni (2009) presented a detailed food choice process model that incorporates life course events and experiences, influences, values, and food behaviours. They focused on where routine empowerment is centred ultimately as engaged food citizens—on the potential to control not just what to eat, but the overall decisions of where and what to purchase (or grow); what and how to produce as dishes and meals; and eating as social experiences. The 2015–2020 DGAC report (USDA, 2015) included a lengthy social-ecological model showing how personal, social, organisational, and environmental contexts and systems interact to influence diet and physical activity behaviours and patterns. Sawyer et al. (2021) examined economic, social, physical, and political food environments of low-income groups as a complex adaptive system and developed a "Determinants of Nutrition and Eating" (DONE) framework wherein associations were synthesised into causal loop diagrams to interpret the dynamics underlying food environments and dietary intake.

Within these diverse inventories is a range of options that could also apply to sustainability. However, given the myriad of factors listed, selectivity and prioritisation are called for. Again, working within a SFC construct should facilitate this process.

Underpinnings for a sustainable food consumption framework

Deciding what elements constitute a component representation of a complex adaptive food system is inevitably a subjective and iterative exercise, given evolving and multi-faceted food systems models. Nonetheless, proposing a somewhat holistic, integrated approach is appropriate. According to IOM (2015), this approach includes individual adaptive actors, decentralised behaviour and interactions, and spatial and dynamic complexities. But at this conceptual juncture, an alliterative set of six core properties (a) principles, (b) products/patterns, (c) players, (d) places, (e) practices, and (f) policies—provides a condensed overview.

a) *Principles*: Articulating fundamental SFC principles will reflect underlying values and beliefs that drive sustainability priorities. Models akin to this have been postulated for agroecology (Wezel et al., 2020), for sustainable food systems (IPES-Food, 2015), and elsewhere.

b) *Products and patterns*: Sustainability parameters for foods and associated dietary patterns can expand beyond existing nutrition-related objectives:

   (i) Personal health and wellbeing, reflected in optimal nutrition, can be incorporated into a broader ecological public health framework (e.g., Healthy People, Healthy Planet (OECD, 2017)) and consequently become part of an overall sustainability framework.

   (ii) Sustainability factors now applied to optimal diets can be applied to most foods and dietary patterns that, while less than ideal, reflect corresponding social and environmental concerns. Major food industries, including those manufacturing ultra-processed foods, are actively promoting their products' sustainability (although often seen by others as greenwashing).

   (iii) Sustainability can be applied to other food characteristics beyond nutrition. For example, superior shelf-life and organoleptic attributes of freshly harvested produce, such as flavours, textures, smell, and appearance, are often prioritised by locavores. Taste is a key element of Slow Food's sustainability approach to regain the pleasures of the table (Agrillo et al., 2015; see also the chapter in this collection). The organisation's manifesto urges us to "rediscover the flavours and savours of regional cooking and banish the degrading effects of Fast Food" (Szabo, 2011, p. 548). Credence attributes of food, particularly how products are grown and sourced, provide cues as to provenance, authenticity, and seasonality emphasised by proponents of alternative food systems.

c) *Players*: Everyone engages in food consumption, but certainly not in the same manner.

Gender, social class, culture, religion, and other social and demographic distinctions are typically addressed in food system sustainability assessments; similarly, these strongly influence household food choices and consumption practices, such as who procures and prepares food. Social class, associated income inequalities, and environmental and social justice factors (e.g., racism) certainly influence food security—a major socio-economic dimension for sustainability in general. Sustainable food (and nutrition) security considers the long-term capacity of the food system to provide an adequate amount of nutritious food for everyone (i.e., not just for food insecure populations by traditional measures). This has also been a long-standing priority for community food systems advocates (Feenstra, 1997) and consequently applies to all households within given geographic boundaries in terms of access, food quality, and overall consumption practices.

d) *Places*: Establishing physical boundaries is integral to food system modelling and to defining appropriate food consumption environments, wherein domiciles and their proximate surroundings arguably constitute the central perimeters. Claasen et al. (2015) proposed simplifying socio-organisational scalar dimensions for food systems into micro (individual and household), meso (community), and macro (everything beyond) levels. Food consumption is thus positioned at micro-scale within recursive ecological, geographic, and organisational boundaries across the overall food system.

Households are often peripheral targets when it comes to scaling food system sustainability, but this is the base from which most everyone engages in primary food utilisation—procurement, storage, preparation, and, of course, eating itself. Locality also encompasses wherever food is mainly sourced, such as grocery stores, or eaten away from home, such as at restaurants, schools, and workplaces. Also connected are other physical and social environments that influence consumption practices infrastructures (e.g., media, educational and health institutions, transportation). Relevant sustainability considerations apply here as they do throughout the rest of the food supply chain.

e) *Practices*: These are central to the food consumption approach. Reckwitz (2002) defined practice as a routinised type of behaviour which consists of several elements, interconnected to one another: Forms of bodily activities, forms of mental activities, things and theirs uses, a background knowledge in the form of understanding, know-how, state of emotion and motivational knowledge. Practices here reflect functions of the food consumption sector (per HLPE (2017), but how do they connect specifically to sustainability?

f) *Policies*: Sustainable food consumption is challenged by an increasingly industrialised food system that isn't very amenable to change simply by "voting with your fork." Proponents of sustainability must also consider the need for systemic transformations to support sustainable living for everyone. The regulatory environment and public policies are tools that could be designed to support optimal nutrition, sustainability, while addressing issues of food availability, accessibility, allure, and affordability.

Building capacity for sustainable food consumption

A starting strategy may be to consider the current priorities of sustainable diets research, such as environmental concerns, since these have been relatively downplayed at the consumption level relative to other supply chain components. Azzam (2012) posited that at-home energy consumption constitutes the largest share of total energy use in the US food supply chain, at about 25%. This consumption still underestimates amounts attributable to consumer transportation for food access, and to use of kitchen appliances and household utilities. The use of gas stoves has made media headlines by being a significant source of methane and exuding nitrogen oxides (NOx) as a health damaging household air pollutant (Lebel et al., 2022). Such concerns can be considered within what Barr and Gilg (2006) term "sustainable lifestyles," incorporating energy saving, water conservation, waste recycling, and green consumption as everyday practices. Consumers also account for an estimated 40% or more of overall losses and waste across the food chain (Wunsch, 2021). These findings illustrate the value of incorporating food consumption practices and infrastructures along with diets when examining environmental sustainability, although the relative environmental impacts of varied practices and settings has not been well assessed to date.

Food citizenship considers the public as not just being consumers at the end of the food chain, but also as participants in the overall food system. It emerged from the concept of *food democracy* first introduced by Tim Lang (Renting et al., 2012). It constitutes an empowerment approach that, in broadest terms, supports development of a more democratic, socially, and economically just, and environmentally sustainable food system, including provisioning an adequate, safe, and nutritious food supply (Booth & Coveney, 2015, pp. 13–14).

Social sustainability is a less-considered dimension of food consumption. Foladori (2005) stresses participation as an indicator of democratic governance that can begin at home. In relation to SFC, Booth & Coveney (2015) consider the household as the epicentre of food democracy in action: "It is where people can exercise control in small, but incremental and meaningful ways" (p. 22). Szabo (2011) similarly encourages individuals to "reconnect" or "re-engage" with their food system by also growing food, shopping conscientiously and connecting with food producers. Eating itself can be as prominent as what is on the plate. Macht et al. (2005) explored various physical, social and hedonic pleasures of eating. Mindful eating focuses on how we eat and not just what is consumed (Nhất Hạnh, 2014).

But food consumption certainly does not constitute a uniform set of routines, customs, or overall behaviours, and sustainability-related priorities will vary considerably by population and location. Contrasting representations can exemplify such distinctions. In models of sustainable agriculture, more traditional small-scale farming is often distinguished from large industrial or commodity types of operations when comparing a range of sustainability metrics. A parallel suggested here is active participation and engagement versus passive consumption. For Booth and Coveney (2015) the former occurs in the kitchen in the practice of home cooking or "cooking from scratch," which is from basic ingredients and supports practices embodied in honed meal planning, provisioning, and food preparation skills. This practice can offer greater control over food quality and sustainability priorities, given adequate capacities and time among those involved.

By contrast is a passive "industrial eating" mode, offering much less control over these elements. This can be characterised by degrees of convenience. Convenience (broadly conceived) does benefit everyone, by providing easy access to food produced and supplied by others, and more specifically here when that includes some prior processing of ingredients and having time-saving kitchen equipment for onsite meal preparation. "Ultra-convenience" characterises more extreme situations wherein diets mainly comprise externally-prepared content such as ultra-processed and ready-to-eat foods, along with meals otherwise accessed as take out, home-delivered, or eaten away from home. These types of foods and eating behaviours are generally the least healthy in terms of their nutritional value, are often made from cheap, industrially-farmed and processed ingredients, and are accompanied by wasteful packaging (Jackson & Viehoff, 2016).[2] Such consumers are largely disengaged from understanding the overall food chain, which challenges their potential for eating more sustainably, as outlined above.

Most households of course fall somewhere between these poles. Still, spending on takeout foods and eating away from home now exceeds half of domestic food budgets (Saksena et al., 2018). Consumption of ultra-processed foods comprises the majority of energy intake in the US population and has reached two-thirds of calorie intake among youth (Wang et al., 2021).

Capabilities: There are many challenges to moving the public towards greater SFC given the dominance of an increasingly concentrated and industrialised food system. Capabilities, simply stated, incorporate the freedoms and abilities to execute specified courses of action and to achieve certain outcomes. The concept is infrequently associated with sustainability. However, Foladori (2005) emphasises the importance of social participation and the increase in the capacity of people to construct their future. Perry et al. (2017) applied this approach to consumption practices, including time, resources, skills, knowledge, aptitudes to source, store, prepare, and otherwise manage domestic food responsibilities, to do these in a sustainable manner, and ideally to enjoy eating a nutritious and delicious diet within positive social support systems. But households are far from homogenous in their capacities to engage and perform such activities. Capabilities can be enabled or limited by multiple external influences on the household—employment, education, resources, as well as negative political and socio-cultural forces driving inequality, racism, gender discrimination, and other facets of food and environmental justice that limit the ability of current and future generations to live more sustainably.

Food system literacy is an integral aspect of SFC capabilities. From a practice approach, literacy combines knowledge, skills, and behaviours essential to sourcing, planning, organising, preparing, and eating a diet from a healthier and more sustainable context. This literacy implies an adequate understanding of the broader dimensions of food systems and associated sustainability, including environmental, public health, and social justice concerns. Other chapters on *critical food systems literacy* and teaching and learning explore these themes in greater detail.

Tom Lyson (2004) sought to bridge food justice and literacy through the concept of *civic agriculture*, built on civic community, by calling for active involvement and participation in the food chain. He especially emphasised participation in local and alternative food system engagement (e.g., farmers' markets, food coops, CSAs). Food citizens, in tandem, can become involved with locally based advocacy organisations and similarly promote broader policy changes that impact how food is produced through local, regional, and national supply chains.

Conclusion: Implications for SFC framework development and guidance

This chapter proposes starting points for developing SFC models. As with agroecology and food systems overall, multiple prototypes are indispensable to reflecting diverse values, structural elements, and sustainability criteria, especially when applied to different cultural and geographic settings and to populations with distinct diets, consumption practices, and types of supply chains.

Sustainable food consumption guidance would flow from such models, placing more emphasis on household-level capabilities, including literacy and best practices, as well as on dietary choices. A rare FBDG example from Brazil (Ministry of Health of Brazil, 2015) includes, "Plan your time to make food and eating important in your life" and "Develop, exercise, and share cooking skills." Beyond this, recommendations could cover food attributes besides nutrition as well as addressing food citizenship or equivalent types of political engagement. In many respects, equivalent guidance is needed for all supply chain sectors; without that guidance, consumer food choices, practices, and other actions are limited by the lack of overarching food systems and broader global ecological sustainability. This lack of sustainability certainly cannot be resolved via SFC alone, although it can play a vital role in furthering overall conversion toward a more sustainable world.

Notes

1 The UN Sustainable Development Goals, issued in 2016, support even broader social sustainability goals. See https://sdgs.un.org/goals.
2 Products resulting from "industrial organics" have been critiqued for the modest sustainability benefits they may provide.

References

Agrillo, C., Milano, S., Roveglia, P., & Scaffidi, C. (2015). Slow food's contribution to the debate on the sustainability of the food system. Paper presentation at the 148th seminar of the EAAE, Brussels, Belgium.

Altieri, M., & Nicholls, C. (2017). Agroecology: A brief account of its origins and currents of thought in Latin America. *Agroecology and Sustainable Food Systems, 41*(3–4), 231–237.

Azzam A. (2012, October 31). Energy consumption in the U.S. food system. *Cornhusker Economics, 598*. Retrieved from https://digitalcommons.unl.edu/cgi/viewcontent.cgi?article=1598&context=agecon_cornhusker

Barr, S., & Gilg, A. (2006). Sustainable lifestyles: Framing environmental action in and around the home. *Geoforum, 37*, 906–920.

Berry, W. (1990). *What are people for? Essays by Wendell Berry*. North Point Press.

Booth, S., & Coveney J. (2015). *Food democracy: From consumer to food citizen*. Springer.

Brundtland, G. (1987). *Report of the world commission on environment and development: Our common future*. United Nations General Assembly. Retrieved from https://sustainabledevelopment.un.org/content/documents/5987our-common-future.pdf

Burlingame, B., & Dernini, S. (2012). *Sustainable diets and biodiversity. Directions and solutions for policy, research, and action*. FAO. http://www.fao.org/docrep/016/i3022e/i3022e.pdf

Claasen, N., Namukolo M., Covic, E., Idsardi L., Sandham, A., Gildenhuys, S. (2015). Applying a transdisciplinary mixed methods research design to explore sustainable diets in rural South Africa. *International Journal of Qualitative Methods, 14*(2), 69–91.

Clancy, K. (1999). Reclaiming the social and environmental roots of nutrition education. *Journal of Nutrition Education, 31*(4), 189–246.

Clarke, C., & Joseph, H. (2017). *Compiled table of sustainable dietary guidelines* [Unpublished data]. https://tufts.box.com/s/ipxtrjk1n8ei5eij2f8b9zzhm2e2k3q2

FAO (1996). *Preparation and use of food-based dietary guidelines: Report of a joint FAO/WHO consultation Nicosia, Cyprus*. WHO. Retrieved from http://www.fao.org/docrep/x0243e/x0243e00.HTM

Feenstra, G. (1997). Local food systems and sustainable communities. *American Journal of Alternative Agriculture, 12*(1), 28–36.

Foladori, G. (2005). Advances and limits of social sustainability as an evolving concept. *Canadian Journal of Development Studies, 26*(3), 501–510.

Furst, T., Connors, M., Bisogni, C., Sobal, J., & Falk, L. (1996). Food choice: A conceptual model of the process. *Appetite, 26*(3), 247–65.

Gussow, J. D., & Clancy, K. L. (1986). Dietary guidelines for sustainability. *Journal of Nutrition Education, 18*(1), 1–5.

Hatfield, J. L., & Karlen, D. L., (Eds.). (1994). *Sustainable agriculture systems*. CRC Press.

Hedlund-de Witt, A. (2014). Rethinking sustainable development: Considering how different worldviews envision "development" and "quality of life". *Sustainability, 6*, 8310–8328.

HLPE (2017). *Nutrition and food systems: A report by the High Level Panel of Experts (HLPE) on food security and nutrition*. FAO. Retrieved from http://www.fao.org/3/i7846e/i7846e.pdf

IOM (Institute of Medicine), & NRC (National Research Council). (2015). *A framework for assessing effects of the food system*. The National Academies Press.

IPES-Food (2015, October 15). *Transitioning to a sustainable food system with IPES*. https://www.arc2020.eu/transitioning-to-a-sustainable-food-system-with-the-ipes/

Jackson, P., & Viehoff, V. (2016). Reframing convenience food. *Appetite, 98*, 1–11.

Jones, A. D., Hoey, L., Blesh, J., Miller, L., Green, A., & Shapiro, L. F. (2016). A systematic review of the measurement of sustainable diets. *Advances in Nutrition, 7*(4), 641–664.

Kennedy, E., Webb, P., Walker, P., Saltzman, E., Maxwell, D., Nelson, M., & Booth, S. (2011). The evolving food and nutrition agenda: Policy and research priorities for the coming decade. *Food and Nutrition Bulletin, 32*(1), 60–68.

Lang, T. (2014). Sustainable diets: Hairshirts or a better food future? *Development, 57*(2), 240–256.

Lebel, E., Colin J., Finnegan, C., Ouyang, Z., Jackson, R. (2022). Methane and NOx emissions from natural gas stoves, cooktops, and ovens in residential homes. *Environmental Science & Technology, 56*(4), 2529–2539.

Lyson, T. A. (2004). *Civic agriculture*. Tufts University Press.

Macht, M., Meininger, J., & Roth, J. (2005). The pleasures of eating: A qualitative analysis. *Journal of Happiness Studies, 6*, 137–160.

Mason, P., & Lang, T. (2017). *Sustainable diets. How ecological nutrition can transform consumption and the food system.* Routledge.

Ministry of Health of Brazil (2015). *Dietary guidelines for the Brazilian population.* Ministry of Health of Brazil. Retrieved from https://bvsms.saude.gov.br/bvs/publicacoes/dietary _guidelines_brazilian_population.pdf

OECD (2017). *Healthy people, healthy planet.* https://www.oecd.org/health/healthy-people -healthy-planet.htm

Perry, H., Thomas, H., Samra H. R., Edmonstone, S., Davidson L., Faulkner, A., Peterman, L., Manafò, E., & Kirkpatrick, S. I. (2017). Identifying attributes of food literacy: A scoping review. *Public Health Nutrition, 20*(13), 2406–2415.

Reckwitz A. (2002). Toward a theory of social practices. A development in cultural theorizing. *European Journal of Social Theory, 5*(2), 245–265.

Renting, H., Schermer, M., & Rossi, A. (2012). Building food democracy: Exploring civic food networks and newly emerging forms of food citizenship. *The International Journal of Sociology of Agriculture and Food, 19*(3), 289–307.

Saksena, M. J., Okrent, A. M., Anekwe, T. D., Cho, C., Dicken, C., Effland, A., Guthrie, J., Hamrick, K. S., Hyman, J. and Jo, Y., & Tuttle, C. (2018). *America's eating habits: food away from home.* Washington, DC: USDA. Retrieved from https://www.ers.usda.gov/webdocs/ publications/90228/eib-196.pdf?v=920

Sawyer, A., van Lenthe, F., Kamphuis, C., Terragni, L., Roos, G., Poelman, M., Nicolaou, M., Waterlander, W., Djojosoeparto, S., Scheidmeir, M., Neumann-Podczaska, A., & Stronks, K. on behalf of the PEN Consortium. (2021). Dynamics of the complex food environment underlying dietary intake in low-income groups: systems map of associations extracted from a systematic umbrella literature review. *International Journal of Behavioral Nutrition and Physical Activity, 18*(1), 1–21.

Sobal, J., & Bisogni, C. A. (2009). Constructing food choice decisions. *Annals of Behavioral Medicine, 38*(1 Suppl.), S37–S46.

Story, M., Kaphingst, K. M., Robinson-O'Brien, R., & Glanz, K. (2008). Creating healthy food and eating environments: Policy and environmental approaches. *Annual Review of Public Health, 29*, 253–272.

Szabo, M. (2011). The challenges of "re-engaging with food": Connecting employment, household patterns and gender relations to convenience food consumption in North America. *Food, Culture & Society, 14*(4), 547–566.

Thích Nhất Hạnh. (2014). *How to eat.* Berkeley, CA: Parallex Press.

Tzounis, A., Katsoulas N., Bartzanas T., Kittas, C. (2017). Internet of Things in agriculture, recent advances and future challenges. *Biosystems Engineering, 164*, 31–48.

USDA (2015). *Dietary guidelines for Americans, 2015–2020.* Washington, DC: U.S. Department of Health and Human Services and USDA. Retrieved from https://health.gov/sites/ default/files/2019-09/2015-2020_Dietary_Guidelines.pdf

USDA (2020). *Dietary guidelines for Americans, 2020–2025.* Washington, DC: U.S. Department of Health and Human Services and USDA. Retrieved from https://www.dietaryguide-lines.gov/sites/default/files/2020-12/Dietary_Guidelines_for_Americans_2020-2025 .pdf

Wang, L., Steele, E. M., Du, M., Pomeranz, J. L., O'Connor, L. E., Herrick, K. A., Luo, H., Zhang, X., Mozaffarian, D., & Zhang, F. F. (2021). Trends in consumption of ultra-processed foods among US youths aged 2–19 years, 1999–2018. *Journal of the American Medical Association, 326*(6), 519–530.

Wezel, A., Bellon, S., Doré, T., Francis, C., & Vallod, D. (2009). Agroecology as a science, a movement and a practice. *Agronomy for Sustainable Development, 29*(4), 503–515.

Wezel, A., Gemmill-Herren, B., Kerr, R., Barrios, E., & Luiz, A., Gonçalves, R., & Sinclair, F. (2020). Agroecological principles and elements and their implications for transitioning to sustainable food systems. *Agronomy for Sustainable Development, 40*, 1–13.

Willett, W., Rockström, J., Loken, B., Springmann, M., Lang, T., Vermeulen, S., Garnett, T., Tilman, D., DeClerck, F., Wood, A., Jonell, M., Clark, M., Gordon, L. J., Fanzo, J., Hawkes, C., Zurayk, R., Rivera, J. A., de Vries, W., Sibanda, L. M., … Murray, C. J. (2019). Food in the Anthropocene: the EAT–Lancet Commission on healthy diets from sustainable food systems. *The Lancet, 393*(10170), 447–492.

Wunsch, N. (2021, February 3). *Distribution of food waste in the United States 2018 by segment.* Statista. https://www.statista.com/statistics/1190673/distribution-of-food-waste-in-the-united-states/

PART 6

Transformations and food movements

31

THE INNER DIMENSIONS OF SUSTAINABLE DIETS

Sarah Pittoello and Kathleen Kevany

Introduction

Movements toward sustainable diets have generally focused outward: On the foods consumed, where they come from, how they are grown and distributed, and the waste they create. This chapter aims to flip the spotlight back on ourselves[1] in order to ask: What are the inner landscapes—and the physical, environmental, social and/ or cultural conditions that support them—that create the conditions for nourishment and sustainable choices? What are processes through which this inner landscape may be cultivated? This chapter explores these questions in relation to the inner dimension of sustainable diets; we suggest that while there is an urgency in turning our attention towards eating well, there is joy in it too.

We, the authors, come to this work as white women of European descent, living and working in Mi'kma'ki (Nova Scotia, Canada), which is the traditional and unceded land of the Mi'kmaq people. The tensions and learning between academics, therapeutic practice, activism, and personal practice are reflected in our exploration of inner work and are rooted in, but also limited to, these positionalities.

Systems of disconnection

Global, industrialised food systems are, in large measure, opaque, oppressive, and grossly inequitable (Qualman, 2019; Weis, 2015). And while there are many examples of localised, sustainable, and equitable food systems and initiatives, report after report (IPCC, 2018; Ranganathan et al., 2016; Springmann et al., 2020) reiterates: Industrialised, globalised food systems are known drivers of deforestation, extinction of biodiversity, decimation of coral reefs, and melting of the glaciers, which reverberates down to nutrient deficient soils yielding less nourishing foods.

Extensive scholarship has explored the many dimensions of change necessary to move toward sustainable diets: Food strategies that emphasise decolonisation,

DOI: 10.4324/9781003174417-37

inclusion, local, regional, low carbon, and higher nutrients. However, sustainable diets are often explored in the context of climate change reversal, mitigation, and adaptation, and the study of our climate crisis, widely downgraded to *climate change*, originated in biophysical discourse and solutions have been viewed and treated as such (Wamsler, 2021). This approach places climate change in what Heiftz et al. (2009) described as a *technical problem* that can be solved with increased knowledge, education, expertise and know-how, which he distinguishes from an *adaptive challenge*, in which our mindset—our beliefs, worldview and paradigms—have to adapt and transform, "requir[ing] a new way of viewing both the problems and solutions" (O'Brien, 2018, p. 154). It is this framework of an adaptive challenge that we explore sustainable diets and inner work.

Thus to meet sustainability challenges, attention needs to be directed toward the inner dimensions of change (Cohen et al., 2014; Ives et al., 2020; Wamsler, 2018, 2021; Woiwode, 2016; Woiwode et al., 2021) hand-in-hand with these outer, structural dimensions. While this chapter focuses on the inner dimensions, we see these inner and outer dimensions as necessary complements that can happen alongside each other. Wilber's (2000) AQAL integral model[2] frames change as being supported through four dimensions—subjective "I"; objective "it"; intersubjective "we"; and the interobjective "its"; if we exclude one dimension, we may not see the kind of transformation we desire (Wamsler, 2021). Inner transformation, Wilber's "I" and "we" quadrants, can be connected to Meadow's (1999) systems theory's *deep leverage points.* Meadows (1999) argues that these powerful leverage points are the goals of the system and the mind-set or paradigm out of which the system has evolved its structure, rules, delays, and parameters. Sharma (2007) also highlights the importance of mind-sets in the conditions and process of change:

> Not until we see the global problematic symptoms of a more fundamental deeper-rooted crisis can we begin to mount a more integral and profound response that is likely to move us in a more sustainable way. That crisis is in our individual and shared mind-sets, where psychological and cultural factors reign.
>
> *(para 3)*

Societies shift by groups of people *seeing* the world differently and/or *choosing to act differently*, where acting impacts seeing and vice versa (Wilber, 2000; Woiwode et al., 2021). Geels' (2002 & 2019) often quoted transition theories recommend reducing barriers to social acceptance of new approaches by involving citizens, community organisations, media, government, and industry to unleash ideas, energy, and encourage commitments to sustainability. As we recognise the interconnected nature of our multiple crises—ecological, health, and moral—more individuals, as well as communities and institutions, may be inspired to act: as O'Brien (2018) shares, "Beliefs, values, and worldviews can change within individual lifetimes and over generations, and also through pivotal events" (p. 156): As, for example, the way the complexity of these challenges and shifts were illuminated during the global COVID-19 pandemic.

Cracks revealed through the pandemic

We began this exploration of the inner work of sustainable diets in March 2021, one year after the beginning of the COVID-19 pandemic. The pandemic has been a magnifying glass, exacerbating pre-existing challenges and inequalities, while also highlighting our capacity for community, collaboration, caring, and significant, large-scale action. The value of collective action is particularly evident as we forge a more comprehensive sense of our shared fate and our interconnections (Homer-Dixon, 2020).

Nova Scotia (NS), Canada has seen examples of both this connection and disconnection reflected in relation to food. While there has been a significant increase in eating disorders during the pandemic (Stoodley, 2021), there has also been an increase in growing food, preserving, and baking, evidenced through shortages of seeds, mason jars, flour, yeast, and endless social media posts about sourdough bread. In NS, efforts by the Sipekne'katik band to enact their own self-regulated moderate livelihood lobster fishery were met with violence (Cooke, 2020). Many farmers scrambled when Jamaican and other migrant farm workers were delayed in entering Canada (Tattrie, 2020); one local farmer had to compost thousands of transplants because he didn't have the support to plant, maintain, and harvest them (Thompson, 2020). Feed NS, an organisation addressing food security in Nova Scotia, increased its food box programs, pivoted operations, and offered home delivery with an increase in need of 15% in August alone (Feed NS, 2021). Wolfville Farmers' Market to Go—an online ordering and hub pick up—saw sales increase from 2,704 orders in 2019 to 14,840 in 2020 with total sales just under 1.3 million. This growth was accomplished with the support of 5000 volunteer hours, a 64% increase from the previous year (L. Clowes, personal communication, June 30, 2021). Many other farmers' markets in the province, including the Truro Farmers' Market Cooperative, designed safe in-person and online sales, with 61% offering alternative products and sales arrangements. Revenue was retained for these vendors through community members supporting local (Ells Congdon, personal communication, July 15, 2021). And while these trends are particular to communities in Nova Scotia, they may offer indications of how these patterns may be playing out in others.

Disconnected relationships in our food systems

The pandemic disrupted systems and lives. In various ways, it has invited an examination of those systems. For some, there has been more time to consider what is produced, how, to what ends, and from whom food is purchased. These factors drive food systems and, in turn, dictate lives; that is, *the modes of production and consumption for our material lives are the foundation of our social, physical, political, and intellectual lives.* The pandemic has underscored the role of relationships in these systems and the need for deep change. While we need to reorient the systems and environments in which we operate, we cannot do so without repairing our relationships with and within these systems and environments—this is where inner work is a critical partner in the work of change. These disconnections have resulted in objectifying both food and bodies, which allow food systems to be based on power and oppression, rather than connection and belonging. Hannan (2020) points to the nature of industrialised capitalism as a culprit for the destruction and the disconnection. Yet diets are one of the most basic and ongoing relationships with the more-than-human world. Cohen and colleagues (2014) ask, "how do we change ourselves and our consciousness so that we will not continue the legacy of destructive patterns of being and living?" (p. 23). They answer: "the change for sustainability has to be ontological" (ibid., p. 23). As such, sustainability requires a shift of our ways of being.

Social change through shifting perspectives

At this time, accumulating evidence speaks to the breakdown of social equality, disruption to relations between non-human beings and humans, and a dissolution of the collective (Weber, 2020). In relation to Wilber's (2000) model, each quadrant requires a new "science of understanding" to replace the old "science of manipulation," (Ikerd, 2008) and shifts in cultural drivers and social understanding (Ballard et al., 2010). "The community our social collective belongs to is the collective of life. Our individual existence is granted by partaking in this collective, by taking from and contributing to the mutuality it is built upon" (Weber et al., 2020, p. 14).

Prospects for change become more fertile when individuals sense their abilities and willingness for action, and have access to skills and knowledge needed to usher in more sustainable diets. Thus, personal work is embedded in broader structures and cultural influences that can impede or improve prospects for desirable changes (Ballard et al., 2010; Wilber & Watkins, 2015). For example, we cannot consider building sustainable diets without simultaneously working to shift the way North American culture privileges thin bodies. Diet culture is rampant in North American societies (Harrison, 2019). Emily Nagoski (Nagoski & Nagoski, 2020) describes her experience as she goes on a diet in preparation to give a talk on body acceptance:

> This is so screwed up! On the one hand, I really will be taken more seriously as a professional and an expert if I conform more closely to the aspirational ideal. On the other hand, my efforts to conform to that ideal are in opposition to the very message about which I have been invited to speak, as an expert.
>
> *(p. 127)*

This statement speaks to the complexity of these issues and the genuine struggles of wanting to make suitable choices while adhering to a culture that holds conflicting values.

Inner work is both personal and collective: As part of a larger web, we are both a centre of our own experience and a context for others (Pittoello & Seth, in press). To ignite the scale and type of transformations needed, much more than intellectual or theoretical approaches are needed; commitments to individual inner work need to be supported by substantial cultural and ethical reorientation (Homer-Dixon, 2020; Ives et al., 2020; Woiwode et al., 2021). For new, sustainable approaches to become more widely adopted, social engagement and support systems are essential in order to strengthen individual shifts and collective transformation (Geels, 2002 & 2019; Pretty, 2020).

And while personal and collective inner work are interdependent (Pittoello & Seth, in press), our expertise, and thus the focus of this chapter, is on the individual inner work. Thus, the rest of this chapter is dedicated to exploring individual processes by which we can start to create an inner landscape that can host these kinds of relationships.

Inner dimensions and inner work

Wamsler (2021) describes inner dimensions as "mindsets, including their beliefs, values, world views and associated inner qualities," and inner transformation as "the powerful unleashing of human inner potential to care, commit to and effect change for a better life." Interiority can include intuition, mind-sets, consciousness, values, worldviews, beliefs, emotions, aesthetics, spirituality, and human–nature connectedness (Woiwode et al., 2021). In this chapter, we define inner work as the attention, effort, and care put into building awareness of our internal experience, which may be experienced as body sensations, images, behaviours (both major and subtle), cognition (thoughts, beliefs, dreams, memories), emotions, spiritual impulse/connection, and more. Inner work involves an ongoing, lifetime of work; the material for this work is constantly self-provided and available (Cohen, 2020): We don't have to look outside of our day-to-day experience. This work almost always happens with support in relationships: In our family and partnerships, with pets, with teachers/elders/therapists, with peers and friends, in our communities, in spiritual communities, and in nature. Sufficient support is necessary to make changes that bring us into closer relationships.

Cohen et al. (2014) suggest that "the current environmental and sociopolitical challenges humanity faces as our species' developmental issue" have been "precipitated by the bonding rupture between human and other beings" (p. 23). Various psychotherapeutic approaches and theories—including attachment theory, somatic-centred modalities, polyvagal theory, and parts work, including internal family systems—focus on our need for secure attachment and experiences of safety. These theories explore the ways our experiences are embodied, looking at the impacts overwhelming experiences and/or a lack of consistent support and relationship have on our nervous systems, and the intra-dynamics, or conflict between different parts of ourselves, that emerge from these experiences. These theories can orient us towards acceptance of and curiosity about our patterns and a deep appreciation for the ways we have adapted in order to survive and have our needs met; they also help us identify the way these patterns may not be serving us and offer processes by which we may be able to contemplate, digest, and integrate formative experiences, freeing us to be more available and responsive to present relationships and experiences. These lenses can shift our perspective from critiquing human behaviour to meeting it with compassion and a deep appreciation. This shift may lead to asking questions such as: How might our eating patterns be supporting or hindering our physical, emotional, mental, and spiritual health? What are our contributions to environmental protection, equity, and justice in food systems?

From a therapeutic point of view, this work requires a shift in our internal dynamics; as we shift our relationships to different parts[3] of ourselves, we often experience shifts in our physiology and in our bodily experience. This inner work requires bringing curiosity and acceptance to conflict with (a) part(s) of ourselves or our experiences, and/or exploring a lack of awareness or relationship, in order to cultivate a relationship to these experiences. For example, we could ask: What part of us is driving our behaviour? How is this part motivated and how is it wanting to help? What is it wanting us to know? When we can see "destructive" or "harmful" patterns as protective and recognise the good intention driving these behaviours, rather than only the harmful impact (Schwartz & Sweezy, 2020), there may be an opportunity for change.

Inner relational work is an adaptive rather than a technical solution. The qualities of a relationship that creates the conditions for change includes attention, acceptance, curiosity, and courage, and a process that holds an appropriate pace, sensitivity, multiple of knowing, and support. The emotions, sensations, memories, and drives that create tensions in our inner landscapes, which may include relationships with food, can be in part generational, often due to oppression and trauma[4] (Menakem, 2017; Schwartz and Sweezy, 2020; Treleaven, 2018). The work of exploring, unwinding, and bringing healing to these patterns can only happen at the pace that our bodies and psyches can experience safely. For example, sometimes there is not enough (physical or emotional) safety to make choices that align with our values; in addition to systemic issues of access and availability to "sustainable" food choices, there can also be inner barriers that require attention and care.

In this inner work, we are moving into deeper relationships, in both our inner and outer landscapes (including human and nonhuman animals, earth, community, and culture), and toward having more capacity for choice and, simultaneously, responsibility. Inner work is not prescriptive: It is context-dependent and relational. Sustainable diets may not be defined by the foods we eat, but whether we are nourishing ourselves in ways that respond with care to satisfy the balance between what we need and what is available. As noted in this collection, this complex web of relationships has become strained. This inner work requires navigating the "messy middle" (Nagoski & Nagoski, 2020) of these relational patterns and acknowledging the complexity of conflicting values and goals. As we consider the healing of relationships, and a spectrum of disorder in relationships, it may be instructive to look at holistic approaches to working with disordered eating to recognise patterns of healing.

Disordered eating: A case study

As the pandemic revealed, disordered eating can be a means of emotional regulation and coping; Vuillier and colleagues (2021) in the UK found disordered eating behaviours increased as participants who self-identified as having an eating disorder managed difficult emotions, changes to routine, confinement, and unhelpful social media messages. Based on clinical experience of working with individuals experiencing eating disorders, we can glean insights into possible paths of healing relationships with food. Aspects of this work include:

- Behavioural changes necessary to keep the person safe and sufficient nourishment/energy to have the capacity to do the work.
- Helping to illuminate the social and environmental factors that reinforce their behaviour.
- Working toward the renegotiation of the underlying woundedness in the body and mind, through care, connection, co-regulation, and process.
- Finding a new foundation, connection, and relationship to food and their bodies.
- Cultivating, finding, or building a community that can respect/share these perspectives and practices.

These principles can be applied more broadly in how we heal relationships with food. For example, Costa et al. (2019) found young women working to modify eating patterns to lower carbon and less animal foods also worked at redefining their ways of being in the world through cultivating empathy and emotional awareness; increasing knowledge and understanding of how eating can be healthy and without cruelty; and, consciously choosing animal-free food and fostering relations with others who share similar philosophies around food. Thus, these deeper shifts in relationship to food require inner work and include two important factors: Practices to support inner work and community.

Mindful approaches

How do we create the conditions that might foster a shift? As Cohen (2020) describes, there is "one essential without which you cannot do inner work in any meaningful and productive way: your ongoing process of developed awareness and attention" (para. 5). Mindfulness has been studied in the sustainability literature and found to support the capacity to disrupt routines and promote more congruent consumption behaviours, like nurturing non-materialistic values, enhancing well-being, and fostering pro-social behaviours (Fischer et al., 2017). The effectiveness of mindfulness-based interventions for sustainability have not been sufficiently explored (Geiger et al., 2019), but given the beneficial effects from sustainability-related behaviours and attitudes, the potential benefits of mindfulness training should be more intentionally pursued (Geiger et al., 2019; Sajjad & Shahbaz, 2020; Wamsler, 2018).

From their meta-analysis of literature focused on mindfulness and sustainability, Thiermann & Sheate (2021) identified six areas to build upon for more sustainable behaviour by individuals: a) Increased awareness; b) improved health and subjective well-being; c) greater connection to nature; d) stronger pro-social tendencies like empathy and compassion; e) stronger intrinsic values and ethical decision-making; and, f) greater curiosity and openness to new experience.

Building on the work of Bahl et al. (2016), Thiermann and Sheate (2021), offered helpful framing for inner transformation with mindful approaches to thinking, consuming, living, and being.[5] There is an invitation to *pay attention* to inner sensations, bodily messages, and outer stimuli. This attention is coupled with *non-judgmental acceptance*, an openness, a tenderness, and these may flow into a state of *conscious and compassionate awareness*. This compassion is extended to the former and the current self, as well as to all around us. This awareness can then be infused with the recognition of the transience of situations, of ideas, of worldviews. Strengthening this capacity for *well-formed insight* can aid in moving past temptations, dysfunctions, and suffering. By fostering a more grounded and attuned state of awareness and insight we may then *weaken our attachment* to unconscious connections and unhelpful habits (Bahl et al., 2016; Thiermann & Sheate, 2021). This path of growing our sustainability capacities and sensitivities, we become better prepared to act with care and undertake choices and actions that become more *liberating and sustainable*. Forging healing relationships requires attending to the continually unfolding processes as described above.

Of course, mindfulness is a practice—something that one *does*—and while this framework provides theoretical signposts to watch for, it is always an ongoing and embodied practice. The practice of mindfulness may support our capacity to be present to our experience, which is to be in relationship with ourselves and our environments. Our being present *with*—whether it's worry, the presence of a favourite tree, conflict or care (or both!) with another, or weeding a garden bed—creates the conditions for a sense of self and other as processes of being. This experience supports our sense of belonging, and influences our experience of being present with people, non-human beings, place, food, drink, and nourishment. As such, it may create the appropriate conditions for "healing the wounds of bonding rupture and facilitating the evolution of human consciousness and development of a more mature identity...overcoming materialistic individualisation and moving toward the relational integration of self, community, and world" (Cohen et al., 2014, p. 23).

Sustainability as relationship

This inner work reminds us that sustainability is about the quality of relationships. In addressing the limitations of a mechanistic approach to sustainability, Walsh et al., (2021) suggest, "various authors have argued that a lack of relationality is at the core of many of our current crises, and describe what may be considered an emerging paradigm informed by relational thinking" (p. 74). They explore relational ontology, epistemology, and ethics, "that does not presuppose subject-object and nature-culture binaries" (p. 175), and they support advancing this paradigm in sustainability studies.

The nourishment, or lack of, that we receive is the foundation of, and integral to, our holistic well-being. The inner work of sustainability is about relational healing—relationship within ourselves, with other human and nonhuman beings, and with the environment (Pittoello & Seth, in press). And while we have focused on healing wounds, this inner work is also a commitment to experiences of joy, pleasure, and connection with our food. What might our food choices be if they emerge from joy, pleasure and intentionality? What might be the relationships of our food with our inner worlds, our bodies, our communities, and the land that surrounds us? From a perspective of sustainable diets, foods that are wholesome, organic, local, seasonal, fair trade, may be descriptors of an intentional relationship, though not necessarily. They may also refer to favourite family/friend recipes, made with something you grew/foraged yourself, a gift, or a pleasurable treat: They are opportunities to experience connection and belonging.

Thus, advancing sustainability work necessarily involves forming, maintaining, and protecting the interconnected web of relationships of which we are a part. We may consider these relationships sacred, or in Martin Buber's (1958) language, an "I-Thou" relationship. For some, this deep appreciation might entail appreciating the energy and life force within plant and animal foods that are taken and transformed when they are produced and consumed and finding ways of honouring these interconnections. Indigenous scholars, writers, and activists have long been at the forefront of understanding and describing these deep, interconnected relationships. According to Kimmerer (2013), diets that are sustainable honour mutuality, consent, and care; to honour food, the systems that support food, and ourselves, is to honour the sacred nature of food. Kealiikanakaoleohaililani & Giardina (2016) share that by activating the sacredness of food, multiple dimensions of our human experience engender and engage more care, love, respect, and intimate familiarity with the relationships involved in our producing, accessing, preparing, consuming, valuing, and not wasting food.

Limitations and further areas of research

It is important that scholarship in this area continues to grow and inner work is valued as integral to food sustainability equally and alongside outer dimensions. As we undertake inner work for individual and collective change, we are advised to guard against turning spirituality into an instrument or tool for merely external purposes (Hochachka, 2005, 2020; Woiwode, 2016; Woiwode et al., 2021). O'Brien's model of the three spheres of transformation, based on Monica Sharma's (2007) work, supports "unleashing [individual and shared beliefs, values, worldviews and paradigms] as subjects of change rather than turning them into objects to be changed" (O'Brien, 2021).

Writing this chapter has been an exploration in bringing together some psychotherapy perspectives with sustainability concepts/research, particularly around diets, and this overlap has revealed a number of areas for further research. It encourages us to see inner landscapes as complex, with a range of values, worldviews, and dynamics; the healing of these landscapes will challenge broader patterns of power and oppression. Essential to this line of inquiry is the question: Does inner work, including mindfulness practice, necessarily move us toward more sustainable behaviour? What are the conditions that would allow it to do so? Further research into various counselling theories and modalities may inform the nuances of this question. Also, further explorations into trauma and its impacts on relationships, as well as patterns of power and privilege are also essential to this work. And finally, how does inner work connect into the social spheres and vice versa: How can we take what Sharma (2007) refers to as "secular, sacred, strategic action" (para. 6)?

Conclusion

Multiple drivers—cultural, social, emotional, economic, environmental, neurophysiological, psychological, political, and spiritual—inform our food environments and influence our food choices. Pressing human conditions, including under-, mal-, and over-nutrition, climate crisis, political polarisations, and epidemic levels of infectious and non-infectious diseases, reveal the necessity for greater attention to food policies, philosophies, and practices that are more holistic, grounded, equitable, and accessible. For successful transitions to sustainable diets and, by extension, societies, we must restore our capacities to see and form relations within our human nature, with our natural environments, and to strengthen our abilities for holistic thinking and inner work (Thiermann & Sheate, 2020). Inner work involves engaging in embodied experiences and enabling these experiences to inform our ideas, language, and perspectives based on the care of these relationships, leading us to demonstrate practices of mindful citizenship and consumership.

Joy and compassion are inherent parts of this work, along with cultivating experiences of safety and connection, finding support, and sharing experiences as we work to transform challenging patterns in ourselves. Through a maturing and connected identity (Cohen et al., 2014) we may explore further depths of our inner worlds with curiosity and allow our work to inspire ourselves and others to take steps towards sustainable diets from a new place of relationship.

Notes

1 "We" and "our," while often referring to the authors collectively, also more broadly includes scholars, practitioners, activists, and enthusiasts who are interested in sustainable diets and/or inner work. At times, from a psychotherapeutic point of view, they also refer to inner patterns and dynamics that humans share.
2 One key aspect of Wilber's (2000) Integral theory is the AQAL—all quadrants, all levels—matrix, which describes the physical, mental, and spiritual levels of reality from the perspectives of four quadrants, defined by axes of inner/outer and individual/collective.
3 Many therapeutic approaches have "observed and worked with psychic multiplicity" (Schwartz & Sweezy, 2020, p. 29) or *parts*. Internal family systems characterise parts as follows: "The natural state of the human mind is to contain an indeterminate number of subpersonalities that we call *parts*…we conceptualise them as inner people of different ages, temperaments, talents, and desires who form an internal family or tribe… Multiplicity is the inherent nature of the mind" (Schwartz & Sweezy, 2020, p. 39).
4 While outside of the scope of this chapter, these connections have been more fully explored in Pittoello & Seth (in press).
5 Of course, we recommend not merely reading about these principles but applying these approaches to one's own process.

References

Bahl, S., Milne, G., Ross, S., Mick, D., Grier, S., Chugani, S., Chan, S., Gould, S., Cho, Y.-N., Dorsey, J., Schindler, R., Murdock, M., & Boesen Mariani, S. (2016). Mindfulness: Its transformative potential for consumer, societal, and environmental well-being. *Journal of Public Policy & Marketing, 35*, 198–210.

Ballard, D., Reason, P., & Coleman, G. (2010). Using the AQAL framework to accelerate responses to climate change. *Journal of Integral Theory and Practice, 5*(1), 18.

Buber, M. (1958). *I and Thou* (R. G. Smith, Trans.). New York: Scribner Classics.

Cohen, A. (2020, September 30). *Inner work on your own.* http://www.dravrahamcohen.com/field-notes/inner-work-on-your-own/

Cohen, A., Bai, H., & Rabi, S. (2014). Relationship as teacher of sustainability: Post-individualist education. In F. Deer, T. Falkenberg, B. McMillan, & L. Sims (Eds.), *Sustainable well-being: Concepts, issues, and educational practices* (pp. 23–36). ESWB Press.

Cooke, A. (2020, September 17). *Mi'kmaw fishermen launch self-regulated fishery in Nova Scotia.* CBC. https://www.cbc.ca/news/canada/nova-scotia/mikmaw-fishermen-self-regu-lated-fishery-lower-saulnierville-1.5727920

Costa, I., Gill, P. R., Morda, R., & Ali, L. (2019). "More than a diet": A qualitative investigation of young vegan women's relationship to food. *Appetite, 143*, 104418.

Feed Nova Scotia. (2021). *Our Response to COVID-19.* https://www.feednovascotia.ca/about/our-response-to-covid-19

Fischer, D., Stanszus, L., Geiger, S., Grossman, P., & Schrader, U. (2017). Mindfulness and sustainable consumption: A systematic literature review of research approaches and findings. *Journal of Cleaner Production, 162*, 544–558.

Geels, F. W. (2002). Technological transitions as evolutionary reconfiguration processes: A multi-level perspective and a case-study. *Research Policy, 31*(8–9), 1257–1274.

Geels, F. W. (2019). Socio-technical transitions to sustainability: A review of criticisms and elaborations of the Multi-Level Perspective. *Current Opinion in Environmental Sustainability, 39*, 187–201.

Geiger, S. M., Grossman, P., & Schrader, U. (2019). Mindfulness and sustainability: Correlation or causation? *Current Opinion in Psychology, 28*, 23–27.

Hannan, J. (Ed.). (2020). *Meatsplaining: The animal agriculture industry and the rhetoric of denial.* Sydney University Press.

Harrison, C. (2019). *Anti-Diet: Reclaim your time, money, well-being, and happiness through intuitive eating.* Little, Brown Spark.

Heiftz, R. A., Grashow, A., & Linsky, M. (2009). *The practice of adaptive leadership: Tools and tactics for changing your organization and the world.* Harvard Business Press.

Hochachka, G. (2005). *Developing sustainability, Developing the self—An Integral approach to international and community development.* Victoria, Canada: Trafford. Retrieved from https://integralwithoutborders.org/sites/default/files/resources/DSDS_txt%206x9.pdf

Hochachka, G. (2020). Unearthing insights for climate change response in the midst of the COVID-19 pandemic. *Global Sustainability, 3*, e33.

Homer-Dixon, T. (2020). *Commanding hope: The power we have to renew a world in peril.* Random House of Canada.

Ikerd, J. E. (2008). *Crisis and opportunity: Sustainability in American agriculture.* University of Nebraska Press.

IPCC-Intergovernmental Panel on Climate Change (2018). *Global warming of 1.5° C: An IPCC special report on the impacts of global warming of 1.5° C above pre-industrial levels and related global greenhouse gas emission pathways, in the context of strengthening the global response to the threat of climate change, sustainable development, and efforts to eradicate poverty.* https://www.zotero.org/google-docs/?FMHOe7 IPCC. Retrieved from https://www.ipcc.ch/site/assets/uploads/sites/2/2019/06/SR15_Full_Report_High_Res.pdf

Ives, C. D., Freeth, R., & Fischer, J. (2020). Inside-out sustainability: The neglect of inner worlds. *Ambio, 49*(1), 208–217.

Kealiikanakaoleohaililani, K., & Giardina, C. P. (2016). Embracing the sacred: An Indigenous framework for tomorrow's sustainability science. *Sustainability Science, 11*(1), 57–67.

Kimmerer, R. (2013). *Braiding sweetgrass: Indigenous wisdom, scientific knowledge and the teachings of plants.* Milkweed Editions.

Meadows, D. H. (1999). *Leverage points: Places to intervene in a system.* The Sustainability Institute.

Menakem, R. (2017). *My grandmother's hands: Racialized trauma and the pathway to mending our hearts and bodies.* Central Recovery Press.

Nagoski, E., & Nagoski, A. (2020). *Burnout: The secret to unlocking the stress cycle.* Random House Publishing Group.

O'Brien, K. (2018). Is the 1.5°C target possible? Exploring the three spheres of transformation. *Current Opinion in Environmental Sustainability, 31*, 153–160.

O'Brien, K. (2021, June 8). *The mind, the human-earth connection, and the climate crisis* [Online panel discussion]. 2021 Mind and Life Summer Research Institute.

Pittoello, S., & Seth, C. (in press). *Entangled landscapes: Healing as a path to sustaining food futures.* Routledge.

Pretty, J. (2020). The agroecology of redesign. *Journal of Sustainable and Organic Agricultural Systems, 70*(2020)2, 25–30.

Qualman, D. (2019). *Civilization Critical: Energy, food, nature, and the future.* Fernwood Publishing.

Ranganathan, J., Vennard, D., Waite, R., Lipinski, B., Searchinger, T., & Dumas, P. (2016). *Shifting diets for a sustainable food future.* World Resources Institute. Retrieved from https://files.wri.org/d8/s3fs-public/Shifting_Diets_for_a_Sustainable_Food_Future_1.pdf

Sajjad, A., & Shahbaz, W. (2020). Mindfulness and social sustainability: An integrative review. *Social Indicators Research, 150*(1), 73–94.

Schwartz, R. C., & Sweezy, M. (2020). *Internal family systems therapy* (2nd ed.). The Guilford Press.

Sharma, M. (2007, Winter). Personal to plantary transformation. *Kosmos Journal.* https://www.kosmosjournal.org/article/personal-to-planetary-transformation/

Springmann, M., Spajic, L., Clark, M. A., Poore, J., Herforth, A., Webb, P., Rayner, M., & Scarborough, P. (2020). The healthiness and sustainability of national and global food based dietary guidelines: Modelling study. *British Medical Journal, 370*, Article m2322.

Stoodley, C. (2021, February 7). Nova Scotia sees 400% rise in demand for eating disorder services. *Halifax Today.* https://www.halifaxtoday.ca/local-news/nova-scotia-sees-400-rise-in-demand-for-eating-disorder-services-3361443

Tattrie, J. (2020, May 2). *Jamaican farmers start working in Annapolis Valley after quarantine.* CBC. https://www.cbc.ca/news/canada/nova-scotia/jamaican-farmers-start-working-on-annapolis-valley-farms-after-quarantine-1.5552598

Thiermann, U. B., & Sheate, W. R. (2020). Motivating individuals for social transition: The 2-pathway model and experiential strategies for pro-environmental behaviour. *Ecological Economics, 174*(C), 106668.

Thiermann, U. B., & Sheate, W. R. (2021). The way forward in mindfulness and sustainability: A critical review and research agenda. *Journal of Cognitive Enhancement, 5*(1), 118–139.

Thompson, A. (2020, May 12). *Annapolis Valley farmer to count 26,000 tomato plants among his COVID-19 losses.* Saltwire. https://www.saltwire.com/nova-scotia/business/annapolis-valley-farmer-to-count-26000-tomato-plants-among-his-covid-19-losses-448778/

Treleaven, D. A. (2018). *Trauma-sensitive mindfulness.* W. W. Norton & Company.

Vuillier, L., May, L., Greville-Harris, M., Surman, R., & Moseley, R. L. (2021). The impact of the COVID-19 pandemic on individuals with eating disorders: The role of emotion regulation and exploration of online treatment experiences. *Journal of Eating Disorders, 9*(10), 1–18.

Walsh, Z., Böhme, J., & Wamsler, C. (2021). Towards a relational paradigm in sustainability research, practice, and education. *Ambio, 50*(1), 74–84.

Wamsler, C. (2018). Mind the gap: The role of mindfulness in adapting to increasing risk and climate change. *Sustainability Science, 13*(4), 1121–1135.

Wamsler, C. (2021, June 8). *The mind, the human-earth connection, and the climate crisis* [Online Panel Discussion]. Mind and Life Summer Research Institute.

Weber, A. (2020, September). *Sharing life: The ecopolitics of reciprocity.* India: Heinrich Böll Stiftung. Retrieved from https://in.boell.org/sites/default/files/2020-12/Andreas %20essay%20new.pdf

Weber, A., Chauhan, G. A., & Sahu, C. P. R. (2020). *Sharing life: Animism as ecopolitical practice.* Heinrich Böll Stiftung.

Weis, T. (2015). Meatification and the madness of the doubling narrative. *Canadian Food Studies, 2,* 296–303.

Wilber, K. (2000). *Integral psychology: Consciousness, spirit, psychology, therapy.* Shambhala Publications.

Wilber, K., & Watkins, A. (2015). *Wicked & wise: How to solve the world's toughest problems.* Urbane Publications.

Woiwode, C. (2016). Off the beaten tracks: The neglected significance of interiority for sustainable urban development. *Futures, 84,* 82–97.

Woiwode, C., Schäpke, N., Bina, O., Veciana, S., Kunze, I., Parodi, O., Schweizer-Ries, P., & Wamsler, C. (2021). Inner transformation to sustainability as a deep leverage point: Fostering new avenues for change through dialogue and reflection. *Sustainability Science, 16*(3), 841–858.

32
INSPIRING A PLANT-BASED TRANSFORMATION WITHIN THE FOODSERVICE INDUSTRY

Riana Topan and Stefanie McNerney

Introduction

As one of the major global supply channels of food, the foodservice sector is in a unique position to drive sustainability in both production and consumption. An increasing amount of research has shown that reducing consumption of animal-based foods and shifting towards plant-forward diets are vital to reduce food-related greenhouse gas emissions and meet international climate targets (Bajželj et al., 2014; Hedenus et al., 2014; Springmann et al., 2018). As such, any conversation about sustainability within the foodservice and hospitality industries must focus on the foods being served. Due to both the size and value of the sector overall, foodservice businesses have the potential to catalyse the transformation of our food system by promoting sustainable consumption, driving dietary changes, and shifting eating habits (Wahlen et al., 2011). To achieve these goals, Humane Society International (HSI) strategists have focused on the introduction of plant-based initiatives with institutional partners worldwide.

This kind of change is desperately needed. Industrial animal agriculture is one of the biggest causes of animal suffering; annually, more than 88 billion farmed land animals are raised and slaughtered for food around the world, with many intensively confined in cages and crates so small or crowded that they can barely move (Food and Agriculture Organisation [FAO], 2019). Intensive animal agriculture is one of the leading contributors to climate change, responsible for at least 16.5%[1] of human-induced greenhouse gas emissions (FAO, 2017, Gerber et al., 2013; Twine, 2021), putting meat, egg, and dairy production on par with all forms of transportation combined (IPCC, 2014).[2]

HSI's Global Farm Animal Welfare and Protection program works to improve the living conditions and treatment of animals and to reduce the number of

 DOI: 10.4324/9781003174417-38

animals raised and slaughtered for consumption. HSI works with governments and institutions, both public and private, to reduce reliance on animal products and to increase the availability of appealing, cost-effective, nutritious, and environmentally sustainable plant-based meals.

Given the influence and impact of the foodservice sector, HSI identified it as an important lever for change. It is a vast industry encompassing both a private or commercial sector (including restaurants, bars, fast food outlets, and catering businesses) and public or non-commercial establishments (serving meals at public institutions such as hospitals, schools and universities, residential care facilities, prisons, etc.) (Edwards, 2013). In Canada alone, foodservice companies reported that sales in 2019 exceeded CAD \$93 billion (Restaurants Canada, 2020), and the industry has seen an average growth of 5% per year since 2014 (Food Export, 2020).

The foodservice sector can simultaneously change its sourcing and procurement policies while educating the public about the value of those choices, representing a powerful way to reframe default diets for consumers. One review examining life cycle-based interventions to improve foodservice sustainability measures identified the direct replacement of animal-based meals with plant-based offerings as a hotspot for impact reduction (Takacs & Borrion, 2020). It is one of the most effective strategies to mitigate the environmental impacts of the catering supply chain and could also have widespread influence outside the sector, improving public health (through the benefits associated with increased consumption of fruits, vegetables, whole grains, nuts, and seeds) and reducing the number of animals raised and slaughtered for food (ibid).

Recognising that many foodservice professionals are not equipped with the knowledge and experience to develop plant-based meals that appeal to consumers, HSI and the Humane Society of the United States (HSUS) launched Forward Food. First conceptualised in 2015 in response to a request for culinary assistance from Harvard University, Forward Food has extended efforts in the United States, Canada, and the United Kingdom. The program offers culinary training, recipe and menu development support, guidance on marketing and communications, educational sessions, and greenhouse gas impact assessments, all free of charge to institutions. HSI invites partnering organisations to make a commitment to transitioning at least 20% of their animal-based meals to plant-based offerings. HSI has since introduced this program around the world (although the name varies depending on the region), training hundreds of chefs in countries such as South Africa, Brazil, Vietnam, Singapore, and India, and directly impacting millions of meals.

Making the case for plant-based eating

The farm animal production sector is the single largest anthropogenic user of land: Meat, egg, dairy, and aquaculture production systems use approximately 83% of the world's farmland while providing just 37% of the world's protein and 18% of its calories (Poore & Nemecek, 2018). Animal agriculture is also a major driver of deforestation, biodiversity loss, land degradation, exhaustion of water resources, and pollution (FAO, 2011; Steinfeld, 2006). Research confirms that reducing meat consumption (and thus, demand) can alleviate the environmental impact of our food system while also providing adequate nutrition. The global adoption of a low-meat diet that achieves nutritional recommendations for fruits, vegetables, and caloric requirements is estimated to reduce food-related GHGs by almost 50% and premature mortality by nearly 20% (FAO & World Health Organisation [WHO], 2019) (Figure 32.1).

Increasingly with globalised and industrialised food systems, more countries, particularly in the Global North, but also evident in parts of the Global South, are adopting modern diets, in which individuals increasingly eat large quantities of processed foods, refined sugars, fats, oils, meat, dairy and animal products. These food patterns have been linked in numerous studies to higher risks of non-communicable diseases (NCDs), such as heart disease, cancer, type 2 diabetes, obesity, and more (Melina et al., 2016; Popkin, 2015; Popkin & Du, 2003; Tuso et al., 2013).

Figure 32.1 With mouth-watering recipes such as tiramisù, crème brûlée, and raspberry-filled chocolate mousse cake, the chefs HSI works with are eager to learn how to take eggs and dairy out of desserts while preserving delectable flavours and textures. Trang Dang/HSI. Used with permission from Humane Society International.

In 2000, Health Canada conservatively estimated the economic burden of poor diet and diet-related NCDs in Canada to be CAD $6.3 billion annually, including direct care costs of CAD $1.8 billion (McAmmond, 2000). A 2017 study in the *Canadian Journal of Public Health* revealed that insufficient fruit and vegetable consumption is costing the country CAD $4.39 billion in direct and indirect costs (Krueger et al., 2017). The World Health Organisation's recommendations for a healthy diet to prevent NCDs includes an emphasis on plant-based foods, moderate amounts of dairy or dairy alternatives, modest amounts of fats and oils derived mainly from vegetable sources, and limited (if any) amounts of meat and processed foods (WHO, 2018).

Given the clear benefits to the environment, public health, and animal welfare, it is unsurprising that plant-based eating has grown exponentially in recent years. A 2018 study estimated that over 6.4 million Canadians are reducing or eliminating their meat consumption, and over half of Canadians are willing to reduce their meat intake, with health, followed by environmental sustainability and animal welfare, cited as key drivers (Charlebois et al., 2018). Another survey found that more than 40% of Canadian consumers are actively trying to incorporate more plant-based foods into their diets (National Research Council [NRC] & Agri-food Innovation Council [AIC], 2019). The 2019 Canada's Food Guide reflects both health research and consumer interest in its encouragement for increasing consumption of plant-based foods. It recommends regular intake of vegetables, fruit, whole grains, and plant-based proteins, citing improved health reasons and acknowledging that food choices can impact the environment (Government of Canada, 2019). It is worth noting that although many Canadians are actively trying to consume more plant-based foods, almost 1/3 do not know how to replace meat in their diets (Charlebois et al., 2018).

The foodservice sector has noticed the increasing interest in plant-based eating, with an expanding number of restaurants offering plant-based "meat" options. According to one tally, Toronto currently ranks fifth on the list of the world's most vegan-friendly cities (Spector & Brent, 2019). Major fast-food chains have introduced new plant-based products, whether it be Burger King, Del Taco, and White Castle partnering with Impossible Foods to offer plant-based burgers, or McDonald's, Dunkin', and Carl's Jr. partnering with Beyond Meat to introduce plant-based meals. Latin America and Asia are also experiencing an influx of plant-based products, with companies such as OmniPork in Asia and Fazenda Futuro in Brazil emerging as leaders in the plant-based market. Even traditional animal protein companies are seizing the trend and rebranding themselves as "protein" companies, with major firms such as Tyson (Gartenburg, 2019), Hormel (Reinicke, 2019), Maple Leaf Foods (Schroeder, 2019), and Marfrig (Marfrig Global Foods, 2020a, b) investing in plant-based meats or developing their own line of plant-based proteins.

Recognising the industry's potential, the Canadian government has invested millions of dollars into the booming plant-based protein market (Ho, 2020; Wills, 2020). Food and Consumer Products Canada now has a division dedicated to supporting the regulatory and market interests of plant-based food companies in

Canada (Brown, 2021). Governments elsewhere in the world are similarly taking note of the enormous market potential: In 2018, the German government funded a three-year project to innovate the texture of plant-based meat alternatives (Andrei, 2018). More recently, the European Commission published the Farm to Fork Strategy (FFS), a plan aiming to promote a shift across Europe toward plant-based diets as part of the European Green New Deal. The FFS proposed an estimated €10 billion be used for research in sustainable food sources, including, in part, increasing the availability and source of alternative proteins (European Commission, 2020).

Despite the growing plant-based trend, there is still much work to be done to shift diets. When it comes to foodservice operations *implementing* plant-based initiatives, significant barriers stand out: Lack of education and lack of training on the impacts of food choices and how to create tasty plant-based meals. Culinary education still tends to emphasise animal-based products and techniques, and many chefs have never been trained to prepare plant-based meals that will appeal to consumers. Given that people tend to make food choices based on taste, convenience, and cost, there is an urgent need to help culinary professionals see the benefits of plant-based eating. This illuminates the merit in enabling foodservices to better comprehend their role in building a more sustainable food system and to address the gap in knowledge of plant-based culinary skills (Clark & Bogdan, 2019; Szejda et al., 2020).

Putting research into practice

HSI's plant-based solutions program works directly with foodservice professionals, public and private institutions, and individuals in North America, South America, Europe, Asia, and Africa to increase their knowledge of and access to appealing plant-based options. The program, with regional names, has successfully demonstrated that the foodservice industry appears receptive to change and that this specific model for intervention is sound. Regardless of where it operates, the principles are the same. An experienced local team of program coordinators, chefs, and/or nutrition experts provides resources and services such as a plant-based culinary training program. Managers create 100% plant-based recipes that reflect local cuisine, palate, and ingredients. They provide support with menu development and marketing, educational sessions and workshops, and greenhouse gas impact assessments. All services HSI or HSUS offers are free of charge, with only one ask in return: That all partner organisations, be they a foodservice company, a school, a hotel, or a restaurant, make a meaningful, measurable commitment to transitioning their menus to be more plant-based and less focused on animal-based products.

In the US, where the Forward Food program originated, the HSUS hosted 536 training sessions between 2015–2020 in which instructors trained over 11,000 culinary professionals. Forward Food's first-ever plant-based culinary training occurred at Harvard University in 2015, where, at the time, 40% of the university's meals were vegetarian and 15% were vegan. As a result of this partnership, by 2019, Harvard's offerings were 75% vegetarian and 40% vegan. Harvard serves 5,000,000 meals annually and is a leader in promoting plant-based nutrition among students, faculty, and the community.

In the UK, by the end of 2020, Forward Food had delivered 34 culinary workshops, training more than 230 chefs at prestigious universities such as Oxford and Cambridge, as well as chefs at foodservice companies and chefs in training. In South Africa, HSI/Africa's Green Monday program partnered with the University of Witwatersrand in 2018 to launch 23 plant-based dishes in all six of its residence dining halls. One of these university dining halls is the first in Africa to exclusively offer 100% plant-based dishes on Mondays.

There is also notable progress elsewhere. In 2020, HSI/Brazil's Carnes da Terra (Meats of the Land) program secured commitments from the cities of Americana and Botucatu to implement plant-based initiatives in public school cafeterias. These efforts may result in an estimated two million meals directly transitioned to plant-based per year. This notable initiative engages children in the learning process, acquainting children with plant-based foods, and their benefits. It is anticipated that these eating habits may be continued beyond the school.

A Canadian case study

HSI/Canada launched its Forward Food program in 2017. Since its inaugural event at the University of British Columbia, the organisation has worked with over a dozen of Canada's leading educational institutions, as well as a handful of health-care facilities, restaurants, and major foodservice businesses. Through diligent outreach, a commitment to high-quality services and support, and a professional approach, the program has become popular with its partners and is well-known within the Canadian foodservice industry. Indeed, demand for Forward Food's services has increased steadily each year.

Between May 2017 and February 2020, Forward Food hosted 15 culinary training sessions in institutional kitchens across the country. Partner institutions include Queen's University, the University of Guelph, the University of Ottawa, the University of Toronto (St. George and Mississauga campuses), McGill University, and the Northern Alberta Institute of Technology. After making inroads with higher education foodservice departments, Forward Food began working with restaurants, including fast-casual chain La Prep, and food distributors, including Sysco Canada, the country's largest food distribution company.

Forward Food has expanded from its focus on institutional partners to develop relationships with culinary associations, consumer packaged goods companies, retailers, and, most importantly, contract management companies that operate the foodservice departments of hundreds of institutions and corporations coast to coast.

By early 2021, HSI/Canada had partnered with two contract foodservice companies. Sodexo Canada, has been providing food and facilities management services in Canada for over 40 years and serves one million consumers daily, and Dana Hospitality LP, a Canadian foodservice management firm with over 30 years of culinary expertise. Understanding the environmental, public health, animal

Figure 32.2   Forward Food's culinary training—like this one at Sheridan College in Ontario, Canada—gives chefs the opportunity to learn about interesting plant-based ingredients, explore new culinary techniques, and see how easy, delicious, and fun it can be to create dishes free of animal products. Source: Julia Kuziw/ Compass Group Canada. Used with permission from Compass Group Canada.

welfare, and financial benefits of offering more plant-based options, each company has taken steps to make plants a greater focus of their menus. As the only national plant-based culinary resource program in the country, Forward Food is working closely with both companies to set and fulfil their timelines for plant-based goals, which involve a commitment to transitioning 20% of their menu options (or ingredient purchases) to plant-based at multiple accounts (Figure 32.2).

As with many of HSI's plant-based solutions programs, HSI/Canada's Forward Food resources include culinary trainings (offered virtually during the COVID-19 pandemic), recipe and menu development, marketing guidance, educational sessions and leadership summits, as well as events such as plant-based galas, pop-ups, and tastings. HSI's strategy focuses on demonstrating three fundamental facts to foodservice professionals, including chefs, cooks, managers, and dietitians:

1. Plant-based or plant-forward patterns of eating are a critical component of a sustainable, nourishing food system.

2. Plant-based dishes can be appealing, satisfying, and convenient, especially when prepared and presented well.
3. The demand for plant-based foods is growing quickly and the plant-based food movement is here to stay (Figure 32.3).

Participant feedback in the Forward Food program indicates that the intervention is changing both attitudes and behaviours. Most of these foodservice professionals are interested in expanding their plant-based offerings precisely because their clients are increasingly asking for them. The Executive Chef at one of Forward Food's partner institutions, the University of Waterloo, said of the student response to the campus's first all-vegan restaurant: "They love the ideas, they love the freshness of everything, and in our dining hall the vegan choices are going as well as the meat choices" (De Angelis, 2019). Chefs are impressed with and inspired by the possibilities as well: A participant in Forward Food's training with McGill University said, "I loved this training, it will change the way I will put menus together." Another trainee, from a virtual training, said "I came out of the course with lots of new ideas and a greater appreciation for plant-based menus."

Among Forward Food's many partners, environmental sustainability is the primary reason for making plant-based options available. Foodservice profession-

Figure 32.3 Throughout the COVID-19 pandemic, HSI pivoted its plant-based resources and events to virtual platforms receiving praise from participating chefs and cooks for ease and cost-effectiveness. Above, HSI/Canada Forward Food Chef Amy demonstrates various properties and applications of pulses for chefs from Sodexo Canada. Used with permission from Sodexo Canada.

als and chefs have quickly come to appreciate the opportunities to highlight the reduced carbon footprints associated with their menus. For this reason, Forward Food offers greenhouse gas impact assessments in which the emissions-related environmental benefits of replacing animal products with more plant-based options can be quantified. By analysing a menu or procurement history, Forward Food develops an approximate measure of the reduction in GHGs associated with each institution, helping foodservice professionals understand and communicate the impact of such commitments to their clients.

Measuring success and ensuring future impact

Within Canada and elsewhere, HSI's plant-based programs are committed to achieving meaningful, measurable results. As of the beginning of 2021, HSI had trained more than 2,950 culinary professionals and more than 3,100 culinary students around the world on how to prepare 100% plant-based dishes. The Canadian Forward Food program specifically has trained over 200 culinary professionals and engaged hundreds of others through its Leadership Summits, webinars, presentations, and other initiatives.

While it is difficult to capture the ultimate impact of such engagement, certain metrics do reveal the success of the program. For instance, prominent higher education institutions in Canada typically serve thousands of meals daily. In 2021, for example, McMaster University, which serves 15,000 customers daily, committed to taking the Forward Food Pledge to transition 20% of its food purchases to plant-based within two years. Compared to interventions that address consumer behaviour at the individual level, it is easy to see that shifts within the foodservice industry are much more substantive.

At the start of 2020, HSI began working with partners to track and report the total percent volume of animal products purchased, requesting procurement data at the start of implementation, and following up on an agreed-upon schedule. HSI's goal is to clearly track the reduction in total overall procurement of animal products, support implementation, and address any challenges that arise.

The precise difference in environmental impact will depend on the specific ingredients being used, but research has demonstrated that replacing traditional menu items with plant-based alternatives can reduce GHG emissions by 60% to 85%, reduce habitat loss by 21% to 93% and reduce water use by 46% to 72%, with reductions tending to be higher for lunch and dinner menus compared to those for breakfast (Emery & Molidor, 2019). HSI also estimates that, for every 1,000 meals converted to plant-based, approximately 28 animal lives are saved.[3] When culinary professionals learn of the positive societal impact they can have while still serving food that appeals to and satisfies their clients, they are enthusiastic about joining the movement to build a better food system.

Such enthusiasm is essential for long-term success. As more and more foodservice industry menus at public and private institutions and businesses—including universities, colleges, hospitals, prisons, restaurants, and corporate cafeterias—offer appealing, affordable plant-based options, they help to ensure urgent changes needed reach the scale and impact desired. With 54% of Canadians reporting eating outside of the home at least one time per week (Statistics Canada, 2016) and over 40% of Canadians actively trying to eat more plant-based (NRC & AIC, 2019), it's clear the foodservice sector has tremendous potential as an agent of change in the transition towards a more sustainable food system. While social marketing and behaviour change interventions to nudge consumers to more sustainable food choices at the individual level are undoubtedly critical, they must be accompanied by wider shifts within the food industry itself.

Conclusion

HSI's and HSUS's work with large-scale foodservice operations to advance animal welfare, environmental sustainability, and public health is unique. By shifting the focus to the availability of plant-based foods, significant improvements can be realised in nutrition, food security, and biodiversity preservation, as well as reductions in environmental degradation, animal suffering, and chronic disease. Building relationships with influential stakeholders is a key component of this program. HSI has been successful by making connections with leaders in the foodservice industry, educating foodservice professionals about the benefits of serving more plants and fewer animal products, and providing continuous support to implement plant-based and plant-forward menus. This approach has inspired a change in perception on the part of both culinary professionals and consumers, who realise that plant-based dishes can be enjoyable and easy to prepare. Sustainable eating interventions for the foodservice industry will likely have greater impact if they are based on development of genuine relationships with industry leaders and if they make clear both the "why" and "how" of plant-based eating as well as the tremendous social value it promises to bring.

Of course, this is not to say that this approach is without challenges. While most foodservice operations are eager to collaborate, recognising the need for and value of HSI's plant-based program, some lack the capacity to do so. Meanwhile, staff turnover can result in decreased commitment or learned skills within an operation over time. The COVID-19 pandemic also required a substantial shift in operations, given that in-person events became impossible with lockdowns and limits on public gatherings. Finally, there is difficulty in capturing progress accurately, since many institutions and businesses track internal changes differently (or do not track much at all). Those seeking to implement similar projects elsewhere are encouraged to keep these obstacles in mind.

Transforming the global food system to be more sustainable, nutritious, and equitable is a daunting task, but Forward Food has demonstrated the value of thoughtful and ambitious initiatives that engage constructively with stakeholder institutions on a broad scale. HSI's vision for optimal food production is one in which plants are at the centre of the plate, and a world in which everyone has consistent access to nourishing, sustainable, and satisfying dishes that do not harm our health, animals, or our planet. Forward Food and related programs are grounded in that ideal, but characterised by pragmatic strategies designed to engage, educate, and motivate institutional partners who are ideally positioned to bring about these transformational changes.

Notes

1 The 16.5% figure is calculated by dividing through the ag output calculated by FAO (8.1 GtCO2-eq/yr - FAO, 2017) by total anthropogenic emissions in 2010 (49 ± 4.5 GtCO2-eq/yr, IPCC, 2014).
2 Transport accounts for 7.0 GtCO2-eq/yr (IPCC, 2014).
3 This is a conservative estimate, based on per capita consumption data for Canada from the UN Food and Agriculture Organisation and estimates of edible yield per land animal. It also includes estimates for aquatic animals, using Canadian data from Fishcount. org.uk to approximate wild captured and farmed fish.

References

Andrei, M. (2018, December 5). *Scientists zoom in on more realistic plant-based meat substitutes.* ZME Science. http://www.zmescience.com/science/domestic-science/meat-substitutes-05122018/

Bajželj, B., Richards, K. S., Allwood, J. M., Smith, P., Dennis, J. S., Curmi, E., & Gilligan, C. A. (2014). Importance of food-demand management for climate mitigation. *Nature Climate Change, 4*(10), 924–929.

Brown, D. (2021, March 16). *FCPC launches plant-based foods group.* Canadian Grocer. https://www.canadiangrocer.com/top-stories/fcpc-launches-plant-based-foods-group-82878

Charlebois, S., Somogyi, S., & Music, J. (2018). *Plant-based dieting and meat attachment: Protein wars and the changing Canadian consumer (Preliminary Results)* [PowerPoint slides]. Dalhousie University. https://cdn.dal.ca/content/dam/dalhousie/pdf/management/News/News%20%26%20Events/Charlebois%20Somogyi%20Music%20EN%20Plant-Based%20Study.pdf

Clark, L. F., & Bogdan, A.-M. (2019). The role of plant-based foods in Canadian diets: A survey examining food choices, motivations and dietary identity. *Journal of Food Products Marketing, 25*(4), 355–377.

De Angelis, R. (2019, January 2). *University of Waterloo looking to put more vegan and vegetarian options on the menu.* CBC. https://www.cbc.ca/news/canada/kitchener-waterloo/university-of-waterloo-plant-based-culinary-training-1.4958389

Edwards, J. S. A. (2013). The foodservice industry: Eating out is more than just a meal. *Food Quality and Preference, 27*(2), 223–229.

Emery, I., & Molidor, J. (2019, December). *Catering to the climate: How Earth-friendly menus at events can help save the planet.* Center for Biological Diversity. https://takeextinctionoffyourplate.com/publications

European Commission (2020). *Farm to fork strategy: For a fair, healthy and environmentally-friendly food system.* European Commission. Retrieved from https://eur-lex.europa.eu/resource.html?uri=cellar:ea0f9f73-9ab2-11ea-9d2d-01aa75ed71a1.0001.02/DOC_1&format=PDF

FAO (2011). *The state of the world's land and water resources for food and agriculture (SOLAW): Managing systems at risk.* FAO and Earthscan. Retrieved from https://www.fao.org/3/i1688e/i1688e.pdf

FAO (2017). *Global livestock environmental assessment model version 2.0 (GLEAM).* FAO. www.fao.org/gleam/en/

FAO (2019). *Livestock primary database.* http://www.fao.org/faostat/en/#home

FAO, & WHO (2019). *Sustainable healthy diets: Guiding principles.* FAO. Retrieved from https://www.fao.org/3/ca6640en/ca6640en.pdf

Food Export (2020). *Canada country profile.* Retrieved March 7, 2020, from https://www
.foodexport.org/docs/default-source/country-market-profiles/canada-cmp.pdf

Gartenberg, C. (2019, June 13). *Tyson Foods is getting into the fake meat business.* The Verge.
http://www.theverge.com/2019/6/13/18677396/tyson-foods-alternative-fake-meat-
business-pea-protein-chicken-nugget.

Gerber, P. J., Steinfeld, H., Henderson, B., Mottet, A., Opio, C., Dijkman, J., Falcucci, A. &
Tempio, G. (2013). *Tackling climate change through livestock – A global assessment of emissions
and mitigation opportunities.* Food and Agriculture Organization of the United Nations
(FAO), Rome. https://www.fao.org/3/i3437e/i3437e.pdf

Government of Canada (2019). *Canada's food guide.* https://food-guide.canada.ca/en/

Hedenus, F., Wirsenius, S., & Johansson, D. J. (2014). The importance of reduced meat and dairy
consumption for meeting stringent climate change targets. *Climatic Change, 124*(1–2), 79–91.

Ho, S. (2020, June 23). *Canada: Trudeau announces $100M investment into plant-based protein pro-
duction.* Green Queen. https://www.greenqueen.com.hk/canada-trudeau-announces-
100m-investment-into-plant-based-protein-production/

Intergovernmental Panel on Climate Change (IPCC) (2014). *Climate change 2014:
Mitigation of climate change: Contribution of working group III to the fifth assessment report of the
Intergovernmental Panel on Climate Change.* Cambridge University Press. Retrieved from
https://www.ipcc.ch/site/assets/uploads/2018/02/ipcc_wg3_ar5_full.pdf

Krueger, H., Koot, J., & Andres, E. (2017). The economic benefits of fruit and vegetable
consumption in Canada. *Canadian Journal of Public Health, 108*(2), e152–e161.

Marfrig Global Foods (2019a, August 6). *Marfrig signs deal with ADM to produce plant-based
Burgers in Brazil* [Press release]. Retrieved March 5, 2020, from https://www.refrigerate
dfrozenfood.com/articles/97862-marfrig-signs-deal-with-adm-to-produce-plant-based
-burgers-in-brazil

Marfrig Global Foods (2019b, December 13). *Marfrig introduces Revolution brand of plant-based
burgers* [Press release]. Retrieved March 7, 2020, from https://www.foodbev.com/news/
marfrig-introduces-revolution-brand-of-plant-based-burgers/

McAmmond, D. (2000). *Food and nutrition surveillance in Canada: An environmental scan.* Health
Canada. Retrieved from https://www.canada.ca/content/dam/hc-sc/migration/hc-sc/
fn-an/alt_formats/hpfb-dgpsa/pdf/surveill/environmental_scan-eng.pdf

Melina, V., Craig, W., & Levin, S. (2016). Position of the academy of nutrition and dietetics:
Vegetarian diets. *Journal of the Academy of Nutrition and Dietetics, 116*(12), 1970–1980.

National Research Council (NRC), & Agri-food Innovation Council (AIC). (2019, March).
Plant-based protein market: Global and Canadian market analysis. NRC & AIC. Retrieved
from https://nrc.canada.ca/sites/default/files/2019-10/Plant_protein_industry_market
_analysis_summary.pdf

Poore, J., & Nemecek, T. (2018). Reducing food's environmental impacts through producers
and consumers. *Science, 360*(6392), 987–992.

Popkin, B. M. (2015). Nutrition transition and the global diabetes epidemic. *Current Diabetes
Reports, 15*(9), 64–68.

Popkin, B. M., & Du, S. (2003). Dynamics of the nutrition transition toward the animal
foods sector in China and its implications: A worried perspective. *The Journal of Nutrition,
133*(11, Supplement 2): 3898S–3906S.

Reinicke, C. (2019, September 6). Beyond Meat's blockbuster year has ignited the plant-
based food industry. Here are 6 companies that have launched fake 'meat' products to
compete. *Business Insider.* https://markets.businessinsider.com/news/stocks/top-6-
beyond-meat-competitors-made-by-traditional-food-companies-2019-9-1028506115
#1-tyson-foods1

Restaurants Canada (2020). *Foodservice facts 2020.* Retrieved March 7, 2020, from https://
www.restaurantscanada.org/product/foodservice-facts-2020/

Schroeder, E. (2019, April 8). *Maple leaf to build $310 million plant-based protein plant.* Food
Business News. Retrieved May 14, 2020 from https://www.foodbusinessnews.net/arti-
cles/13590-maple-leaf-to-build-310-million-plant-based-protein-plant

Spector, K., & Brent, E. (2019, November 27). *10 Top vegan-friendly cities.* happycow.net. https://www.happycow.net/vegtopics/travel/top-vegan-friendly-cities

Springmann, M., Clark, M., Mason-D'Croz, D., Wiebe, K., Bodirsky, B. L., Lassaletta, L., de Vries, W., Vermeulen, S. J., Herrero, M., Carlson, K. M., Jonell, M., Troell, M., DeClerck, F., Gordon, L. J., Zurayk, R., Scarborough, P., Rayner, M., Loken, B., Fanzo, J., Godfray, H. C. J.,… & Willett, W. (2018). Options for keeping the food system within environmental limits. *Nature, 562*(7728), 519–525.

Statistics Canada (2016). *Eating out: How often and why?* Statistics Canada. Retrieved from https://www150.statcan.gc.ca/n1/en/pub/11-627-m/11-627-m2019003-eng.pdf

Steinfeld, H. (2006). *Livestock's long shadow: Environmental issues and options.* Retrieved from https://www.fao.org/3/A0701E/a0701e.pdf

Szejda, K., Urbanovich, T., & Wilks, M. (2020). *Accelerating consumer adoption of plant-based meat: An evidence-based guide for effective practice.* Good Food Institute. Retrieved from https://www.gfi.org/images/uploads/2020/02/NO-HYPERLINKED-REFERENCES-FINAL-COMBINED-accelerating-consumer-adoption-of-plant-based-meat.pdf

Takacs, B., & Borrion, A. (2020). The use of life cycle-based approaches in the food service sector to improve sustainability: A systematic review. *Sustainability, 12*(9), 3504.

Tuso, P. J., Ismail, M. H., Ha, B. P., & Bartolotto, C. (2013). Nutritional update for physicians: Plant-based diets. *The Permanente Journal, 17*(2), 61–66.

Twine, R. (2021). Emissions from animal agriculture: 16.5% Is the new minimum figure. *Sustainability, 13*(11), 6276.

Wahlen, S., Heiskanen, E., & Aalto, K. (2011). Endorsing sustainable food consumption: Prospects from public catering. *Journal of Consumer Policy, 35*(1), 7–21.

Wills, L. (2020, December 15). *Canadian government invests $150 million in vegan protein development.* LIVEKINDLY. https://www.livekindly.co/canadian-government-invests-150-million-vegan-protein-industries-supercluster/

World Health Organisation (2018). *A healthy diet sustainably produced: Information sheet.* WHO. Retrieved from https://apps.who.int/iris/bitstream/handle/10665/278948/WHO-NMH-NHD-18.12-eng.pdf?sequence=1&isAllowed=y

33

FOOD MOVEMENTS TO FOSTER ADOPTION OF MORE PLANET-FRIENDLY FOODS AND SUSTAINABLE DIETS

Joseph Tuminello, Stephanie Van, and Kathleen Kevany

Introduction

This chapter provides an appreciative inquiry of four organisations and five campaigns working to increase the adoption of planet-friendly foods[1] with empirical evidence gathered using an online survey tool. These are foods proven to benefit human health, contribute lower greenhouse gas emissions, increase local procurement, be culturally appropriate, and reduce animal suffering and loss of lives. Taken together, these foods could be promoted as a more sustainable diet. Organisations and campaigns designed to educate and transform systems towards more sustainable, planet-friendly food services are currently under-researched. We examine their messages, the means, and methods of conducting their work and assessing their impacts and consider leverage points for systems change. In the discussion of findings, we identify points of contention and divergence, uncovering potential challenges and practical strategies, and areas for further research.

Background

Worldwide, implications of sub-optimal industrial food systems are curtailing longevity and quality of life. There is growing recognition that reducing the prominence of meat and dairy in diets, particularly in Western countries where consumption is highest, is essential for sustainable agro-food systems. This shift could serve to enhance local food production, support connections between farmers and families (Qualman et al., 2018; Qualman, 2019), aid in reversing the climate crisis, and stem the devastating decline of habitats, biodiversity, and ecosystems (Weis, 2019). The

DOI: 10.4324/9781003174417-39

well-being of animals too is appreciatively enhanced through diets that include little to no animal-sourced foods (Cleveland et al., 2020). Such actions also would reduce the burden of non-communicable diseases (NCD) and deliver sufficient and sustainable foods to feed a growing population within finite planetary resources.

Many food movements seek to address industrial agriculture's burdens on humans, animals, and environments. Clark et al. (2021) suggest that food movements are "the networks of people, groups, and organisations that are challenging industrial food systems by experimenting with a variety of alternative ways of producing, harvesting, foraging, processing, distributing, consuming, and, ultimately, governing food" (p. 176). Some examples of food movements include food security, food sovereignty, slow food, sustainable agriculture, urban agriculture, healthy food, veganism, and vegetarianism. Researchers and change agents are experimenting with ways to increase the offering and selection of healthier and friendlier foods in educational and academic settings. These movements share common visions while advocating for distinct goals with diverse audiences.

As an example, with in-school food programs, students demonstrate increased nutritional knowledge, preferences for healthy foods, and a higher intake of nutrient-dense foods, which can have long-term implications for disease prevention, population health, and reduced pressures on healthcare (Acton et al., 2018; Colley et al., 2019; Wethington et al., 2020). Turnwald et al. (2019) led a randomised controlled multi-site intervention for taste-focused labelling; they tested naming menu items to appeal more to pleasure than to health motives. Their study revealed that menu names and food labels focused on taste outperformed other labels that contained descriptive words or lists of ingredients. Other studies have applied "nudge theory" from behavioural economics (Thaler & Sunstein, 2009) to test the capacity to influence healthier, more sustainable food choices. Behavioural researchers argue that what people *do*, more than what they *think*, can be points for interventions like nudging (Thaler & Sunstein, 2009). For example, offering vegetarian meals by default, while ensuring participants can opt for meals containing meat, increases the chances of participants selecting meat-free meals at campus dining halls (Campbell-Arvai et al., 2014) and conferences (Hansen et al., 2021). Strategies for food systems change may include food choice architecture and intentional design of processes and materials (Ensaff, 2021). In settings where vegetables and plant-based foods are competing with less healthy and less sustainable options, innovative approaches to the presentation, description of, and education about the food, are vital.

Methodology

Study design

We chose to frame this work with an appreciative inquiry lens. Appreciative inquiry originated in the field of organisational change as a method to explore the context and factors that contribute to success and practises that may foster desired outcomes (Bleich & Hessler, 2016), rather than focusing on barriers, failures, or dysfunctions (Hammond, 2013). Appreciative inquiry has been applied as a strengths-based approach in a variety of fields such as healthcare research and leadership education (Moore & Charvat, 2007; Priest et al., 2013). As appreciative inquiry matures, it becomes subject to critique and refinement. Investigators applying this method have been encouraged to consider more than generativity and positivity, while also considering the challenges of inequity and injustice (Bushe, 2012).

This appreciative study was designed to potentially expose new perspectives and stories and invite engagement in new conversations, interpretations, and worldviews (Bushe, 2011). Facilitators from an array of food organisations were invited to participate, to appreciate, and to articulate the progress being made by their movements or campaigns. The study was assessed for and deemed in compliance with university research ethics.

Organisations and campaign selection

For this study, purposeful sampling enabled us to 1) identify organisations and campaigns relevant to this chapter and 2) leverage the authors' familiarity with organisations and campaigns working to promote diets that are sustainable or plant-based. Through this purposeful sampling strategy, 18 organisations/campaigns were contacted online. Of those contacted, nine (N=9) responded and participated in the survey, with each respondent representing one of the listed organisations or campaigns.

Survey

We constructed and distributed an online survey using Google Forms. We sought to gather descriptive, qualitative data on the following:

- Organisation or campaign goals
- Strategies to achieve these goals
- Key messages of organisations/campaigns
- Rationale for emphasising key messages
- Platforms used to increase outreach and impact
- Indications of failure and of success
- Methods of evaluating efficacy and impact

Organisation and campaign websites

Survey responses along with information from organisational websites and reports formed the data set. Organisation and campaign websites are listed in the footnotes and included in the references at the end of this chapter.

Findings

Four organisations and five campaigns were analysed for patterns of progress and design. This study included: Green Monday;[2] Farm to Cafeteria Canada;[3]

Plant-Based Canada;[4] Nourish Leadership (formerly Nourish Health).[5] Campaigns, which are distributed movements, often lead by organisations with a broader mandate, in this study included: DefaultVeg;[6] Animal Protection New Mexico's "Promoting Plant-Based Eating Program;"[7] Nourish Nova Scotia's "Nourish Your Roots" Campaign;[8] Forward Food by Humane Society International and the Humane Society of the United States;[9] and Canadian University Campaign.[10] Following are the findings from survey responses as well as information from the nine organisations and campaigns. Henceforth, when not identifying specific organisations or campaigns by name, we use the terms "OCs" to refer to "organisations/campaigns" in general.

Articulated missions

These food campaigns identify precise missions along with being motivated by many driving forces. All identify a commitment to protect the environment by reducing GHG. Eight indicate they intend to improve health, increase food security, and increase equity in resource use. Six of the nine have a desire to increase animal wellbeing, as well as fostering opportunities for broader education on agriculture, food systems, and nutrition. Additional goals held by three OCs include shifting norms and stimulating needed change. Also of significance, but securing only two responses, were increasing community resilience, educating the public and health professionals on benefits of plant-based nutrition, and cultivating cultures of preventative health.

Strategies to achieve goals

For an accessible abbreviation of the goals and approaches see Table 33.1.

Key messages

Each OC lists goals, key messages, and rationale for their efforts, see Table 33.2 for key messages and the frequency of responses. These OCs use a sustainability, ethical, or health framework to inform and inspire their audience to influence change through the power of food. Organisations and campaigns also leverage the ripple effect of schools, universities, healthcare to develop and broadcast the messages for needed change.

Use of platforms for outreach and impact

All nine OCs use various platforms and tools to increase outreach and impact. The most common platforms or tools include websites (eight OCs), email (seven OCs), social media (seven OCs), and published material (five OCs). Other tools and techniques include the use of Facebook, Twitter, blog posts, video or audio with YouTube, podcasts, or webinars.

Table 33.1 Goals and approaches to advance organisational missions and campaigns with response rates

Goals	Responses	Approaches	Responses
Empower and support local food producers and increase local procurement programs	4	Creative use of digital tools like social media, podcasts, newsletters, etc.	9
Expand networks and members through forming partnerships	4	Educational programming	6
Aid in transitioning to sustainable foods	4	Design educational tools	5
Educate health and culinary professionals on the benefits of plant-based nutrition	3	Cultivate leaders and ambassadors to implement strategies	4
Ignite community engagement and problem solving	3	Gather evidence, collect information, and seek policy change	3
Provide funds and tools for organisations and community-wide education	2	Host communities of practice and welcome new cohorts of agents of change	2
Develop new policies, tools, and resources	2	Strategically communicate and share success stories	2
Leverage collective assets	1	Engage across the value chain	1

Source: Work of the authors based on contents from survey responses.

Table 33.2 Key messages from the campaigns and rate of use

Messages	Rate of use #/9
Increase environmental benefits through planet friendly foods	9
Offer healthy foods and support nourishing food environments at schools or universities	6
Food can be the best lever for urgent system change	5
Multiple benefits for humans, animals, and the planet with shifting to sustainability	5
Ignite more compassion for non-human animals	5
By growing more of your food, increase respect for and enjoyment from food	4
Make the connections between justice, equity, the environment, and the types of foods consumed	4
Avoid unnecessary oppression of factory farming	3
Support local farmers	2

Source: Work of the authors based on contents from survey responses.

Methods for evaluating efficacy

The OCs employ an array of metrics to assess their progress. When asked about approaches to assess efficacy of their OC, 5/9 of the OCs invite their stakeholders to contribute to assessing their effectiveness. Respondents to the survey indicate that an important sign of efficacy is the re-orientation of thinking and beliefs based on new evidence. One measure of efficacy noted by five OCs is the number of restaurants, cafeterias, and food services that add plant-based options. Three OCs use the number of downloads of podcasts, e-alerts, and blog posts as indicators of reach and influence. Two consider the number of participants in their programs and services. Two also add in estimates of measures like GHG emissions and water saved from shifts in food. The use of a survey gathering data on their impact is used by two OCs. Two also use focus groups and one-on-one interviews. Other assessments receiving two mentions are developmental evaluation, progress reports, and evaluation frameworks. Additional items measured, as noted by single responses, include assessing the influence on policy and the number of institutional pledges fulfilled.

Indications of failure

The OCs seek to shift culture and practise through delicious food. Respondents were asked to offer any signs of their OCs falling short of their intended goals. It is interesting to note that none of the OCs report that their campaign or organisation failed to advance at least some of their goals. However, signs of falling short of goals include lack of participation and buy-in to workshops, events, or programs. Two OCs note that efforts to modify institutional food are thwarted by problems with logistics and with food purchase or procurement practises. Two are challenged by issues with safely storing food. As well, two OCs note a concern with food being reduced to a tool rather than being seen as a gift or as something sacred. Singular concerns shared are the gap between the level of action and the sense of urgency. In addition, the frequent turnover in staff and participants in some of the programs make capturing results more challenging. In the case of organisations offering instruction on how to enhance the types and tastes of foods being offered, some OCs note that partner institutions worry about possible loss of revenue as a result.

Signs of success

All respondents were able to identify signs of success towards achieving their goals. Growing social media support (e.g., increases in followers on Instagram, Facebook, and Twitter) and media attention also are used as signs of success. Signs of success with the frequency of response are identified in Table 33.3.

Discussion and recommendations

The OCs participating in this research exhibit a variety of overlapping, as well as divergent, strategies for effecting change. These food movements are steering more people towards sustainable diets by offering experiences with growing and

Table 33.3 Signs of success for the movements.

Signs of success	Number of responses
Increasing numbers of participants in programs, attendees for webinars and other events.	6
High level of excitement at changing the narrative to well-being through food	5
Campaigns receive positive responses from institutional and community partners	4
Increased political will to bring about improvements and evidence of culture shifts	3
Government engaging as partners and campaign staff invited to speak at events.	2

Source: Work of the authors based on contents from survey responses.

valuing foods, learning about local sourcing, and engaging in interactive ways to prepare and enjoy foods. To foster conditions for improved health and well-being and lower emissions through food, among other outcomes, these food movements appear to be, what Monterrosa et al. (2020) suggest is, facilitating shared identities and emotions. Some of these campaigns appear to be guided by ethical and emotional drivers and to work to cultivate a socio-emotional link to citizens who share similar ethical notions (Lang, 2010). To develop diets that are sustainable may involve working at the individual level through encouraging the consideration of impacts from food, beliefs about food, and expanding knowledge and skills around food. Individual level change is not sufficient; shifts are needed in policies, in food environments, and institutional reforms. Other researchers have found that substantial changes become supported by political changes that include system-level adjustments, reorientation, and environmental appreciation and protection (Weber et al., 2020).

Exposure to multicultural diversity of planet-friendly foods

People are more motivated to shift to a sustainable diet when, through exposure, they find it tasty, appealing, and satisfying. One example of transitions to planet-healthy is de-emphasising animal-based foods to placing greater attention on plant-based. Unsurprisingly, some factors that can impede the switch from animal-based sources are not liking the taste of meat replacements or not having alternative foods available; such obstacles have been identified as negatively influencing purchasing decisions of plant-based foods (Clark & Bogdan, 2019). Green Monday's Green Common store considers this and makes the most sought after plant-based products (e.g., Beyond Burger and Just Egg) available for online purchasing in Asian markets as well as through nine retail locations in Hong Kong ("Green Common", 2021). The perceived difficulty of preparing plant-based foods has been identified as another barrier to the adoption of a plant-based diet

(Pohjolainen et al., 2015), so campaigns such as DefaultVeg and Forward Food include searchable databases of alluring plant-based recipes. DefaultVeg's recipe tool allows users to search based on various criteria such as the type of course (e.g., main dishes, side dishes, salads, etc.), type of cuisine, and preparation time. Forward Food's database allows users to search for recipes that are most attractive to particular kinds of institutions (e.g., college and university, K-12 schools, healthcare institutions, etc.). Making the learning journey interesting and the food delightful and convenient to prepare are three of the strategies used for food service industry partners to incorporate more plant-based foods in their respective settings ("Recipes", n.d.; "Recipe database", n.d.).[11]

Prominence and placement of plant-based meals

Based on the findings of the efficacy of default meat-free menus at conferences and campus dining halls (Campbell-Arvai et al., 2014; Hansen et al., 2021), increasing the prominence of plant-based meals has become a prominent strategy for multiple OCs working towards institutional shifts. For example, DefaultVeg campaigners ask those making institutional food decisions to "make plant-based food the default and give people the option to opt in for meals with animal products" (DefaultVeg, n.d.). Another strategy is to foster commitments and help build capacity that enables institutional dining programs to shift a proportion of total meals offered to plant-based, such as having 50% of menu offerings in institutions in the United States be plant-based by 2024 (Forward Food, n.d.). These and related organisations, such as Green Monday, also encourage swapping animal products for plant-based products (e.g., offering plant-based instead of conventional mayonnaise as an effective strategy for increasing lower-carbon, friendlier foods in institutional settings).

Reorienting food procurement policies and practises

For systemic change to more sustainable diets, transformations at the level of public policies become essential along with expansion of skills, knowledge, and attitudes towards more sustainability in food choices at the interpersonal and personal levels. Nourish Leadership's 2016–2019 Innovator Cohort validated that furnishing local, sustainable, cultural food for hospital patients aligned with system changes for inclusion and increased personal access to sustainable, culturally appropriate foods. Creative changes are happening with menus, food service models, procurement practises, and provision of Indigenous traditional foods. Members of the Innovator Cohort in Québec have been promoting local and sustainable food sourcing in hospitals and other health centres. They implemented an award-winning room service model, created an interactive sustainable menu website, formed an influential provincial policy team, and as a result, gained media attention. In October 2020, their work paid off at the policy level. Québec's Minister of Agriculture announced a new Institutional Food Policy requiring all schools, hospitals, and seniors' homes in the province to set local healthy food targets for their facilities.

The Nourish Leadership Innovator Cohort was key in creating the pressure to bring about and animate the new policy.

Through building relationships with farmers, community members, and related organisations, Farm to Cafeteria Canada implements innovative approaches to food procurement policies. Their Farm to School initiative assists in primary and secondary schools procuring local food from farms, schoolyard gardens, food distributors, as well as through harvesting wild and traditional foods ("Farm to School", n.d.). Their Farm to Campus initiative integrates similar relationship-building strategies regarding food service at college and university dining facilities, and empowers students to advocate for healthy, affordable, and accessible food options ("Farm to Campus", n.d.).

Nourish Nova Scotia's, "Nourish Your Roots" (NYR) program also focuses on affecting change in food procurement by connecting farms with schools and non-profit childcare centres as a means to ensure boxes of fresh produce are available to support schools' healthy eating programs and schools promote them as a fundraiser. NYR was intentionally designed to help sustain school-wide healthy food initiatives while supporting the local farm and agricultural economy in Nova Scotia. This program adheres to the fundraising guidelines in the Food and Nutrition Policy for NS Public Schools and the Food and Nutrition Standards for Regulated Childcare Centres (Nourish Your Roots, n.d.).

Educating about planet-friendly food systems

The majority of OCs participating in this study use education programs as a means of inspiring more sustainable food practises at the local level, in households, in schools, and in hospitals. Several leverage the power of education in their partnerships to accelerate their capacity within food services to design and deliver more sustainable foods to their clientele. For example, the Canadian University Campaign engages in education about the adverse impacts arising from animal agriculture to spur change in food systems, but also in culture and institutional leadership. Animal Protection New Mexico's "Promoting Plant-Based Eating Program" works to raise awareness of the benefits of plant-based eating by offering plant-based cooking classes, maintaining a New Mexico Vegan Dining Guide, offering a Plant-Based Eating Starter Guide, creating and publishing the podcast, "Teach Me How to Vegan," and maintaining an informative plant-based eating blog (Promoting Plant-Based Eating, n.d.).

As individuals adopt sustainable diets for a variety of reasons like increasing personal health, animal rights/welfare, environmental concerns, weight loss, cost-effectiveness, and religious beliefs (Rosenfeld & Burrow, 2017; Vizcaino, et al., 2021), OCs integrate these drivers into their calls for change. Cultivating respect for food and the life-giving systems that support it, are fundamental strategies. Nourish Leadership underscores the interdependent relationships among climate, health equity, and community well-being. Core to their vision for effecting change, is the vital role of patients, community members, health care institutions, and professionals in fostering desirable transitions. Many of the OCs had significant hopes of facilitating substantial shifts in public discourse and practice. Nourish Leadership looked to spark strategic learning and to track six signals of systems change: Beliefs/culture, relationships, routines, resource flows, policy, and power/authority flows.

Health and wellness can be a motivator for individuals to opt for plant-based diets (Fehér, et al., 2020; Jabs et al., 1998; Mylan, 2018; Rosenfeld & Burrow, 2017). Studies from Fehér et al. (2020) and Mylan (2018) revealed that health and nutrition were the largest motivations for switching to plant-based diets in Hungary and the United Kingdom. Yet only two of OCs surveyed underscored how sub-optimal diets are a leading cause for non-communicable diseases. Plant-Based Canada indicates it works to fill the knowledge and action gaps around healthy menus among health professionals and medical organisations. Plant-Based Canada supports professionals and builds capacity to influence public policy and improve public health by also promoting plant-based living. As noted above, Nourish Leadership also prioritises improved health through culturally appropriate and sustainable food systems. The relatively lower number of OCs focusing on health and nutrition in this study could suggest an opportunity for organisations and campaigns to incorporate more health and nutrition frameworks into their key messaging. However, worth noting is that in the broader population, Turnwald et al. (2017, 2019) found a lower uptake of food options that appeared healthier. Furthermore, research found that ethically-motivated vegetarians and vegans tend to have a stronger drive to continue with their plant-based diet in the long-term compared to health-motivated vegetarians and vegans (Hoffman et al., 2013; Rosenfield & Burrow, 2017). Still, OCs that emphasise health and nutrition in their goals and the importance of reducing global warming and environmental destruction may be better positioned to appeal to and motivate a wider audience, especially given the emerging data regarding consumer perceptions of the compatibility of healthy diets, sustainable diets, and plant-based diets (Van Loo et al., 2017).

In general, in health-related campaigns, social media is used to extend the network, increase motivation, and as a tool to distribute information (Dumas et al., 2018; Sabbagh et al., 2020). Although seven out of nine OCs indicated their use of social media as a tool to increase outreach and impact, only one OC specified expanding their network through specific vlog and blog communities. With the rise of social media and social media influencers, outreach to social media influencers could be a large opportunity for organisations and campaigns to tap into. Campaigns that are looking to gain followers may wish to provide information, support, and edu-tainment to increase their online presence. They may also consider reaching out to social media influencers to increase their audience size.

The use of nudge theory from behavioural economics by Thaler and Sunstein (2009) and choice food architecture with Ensaff (2021) are compelling, yet researchers are to be cautious. Placing responsibilities largely on individuals to change would be insufficient; attention needs to be paid to the shifts required in relationships between people and their food, their food systems, and the broader context in which food is available and consumed. With global health, equity, and environmental crises being highlighted, campaigners are likely to increase interest in public investments in healthier food, economic enticements, and updated regulations for the provision of sustainable food options (Ives et al., 2020; Weber et al., 2020). Substantial shifts to food systems require a blend of strategies that may include food policy changes, taxation, incentivisation, nudging, educating, engaging, and regulating, among other actions. Data to be mounting to justify reduced investments in animal agriculture (Harwatt et al., 2020; Joyce et al., 2014), and increased attention to alternative proteins and plant-based foods (Kevany & Kassam, 2021). Indeed, further research is needed to more fully examine leverage points that ignite needed systems change.

Conclusion

Given the global state of environmental, health, and equity crises, this chapter analyses movements dedicated to leveraging food for optimal human, animal, and planetary well-being. Results from surveying organisations and campaigns provided examples of philosophies, practises, and policies that are advancing shifts to more sustainable food systems. These strategies may aid in accelerating the availability and accessibility of planet-friendly foods and institutions promoting sustainable diets. These case studies revealed efficacious strategies and modes of implementation that acknowledge the urgency and respect the sacredness of food. This work also underscores the importance of framing, messaging, and compelling calls to action. Emerging research shows that restaurant menus that integrated a brief, sustainability message, roughly doubled the percentage of vegetarian dishes that participants ordered with particularly compelling messages being "small changes, big impact" and "joining a movement" (Blondin et al., 2022). Their research also revealed that those exposed to this messaging were inclined to repeat the behaviour in the future in favour of healthier, more sustainable choices. In diverse ways, the organisations and campaigns investigated in this current inquiry, inspire change through a love of food and an appreciation of how to cultivate and care for humans, animals, and the planet, through fortifying food knowledge and skills and activating pro-sustainability attitudes as exemplified by a campaign from Nourish Your Roots that envisions, "champions who demand better—for themselves, their families, the environment, and their communities."

Notes

1 The authors wish to acknowledge and thank all survey respondents for sharing key insights and experiences regarding their respective organisations/campaigns, as well as for their work reviewing and offering feedback on this chapter. Respondents include (but are not limited to) Margo Riebe-Butt, Dr. Zahra Kassam, Eleanor Carrara, Riana Topan, and Ilana Braverman.
2 https://greenmonday.org/en/
3 http://www.farmtocafeteriacanada.ca/
4 https://www.plantbasedcanada.org
5 https://www.nourishleadership.ca/
6 https://defaultveg.org/
7 https://apnm.org/plantbased

8   https://www.nourishns.ca/nourish-your-roots
9   http://friendsofhsi.ca/issues/forward-food/
10 https://onlineacademiccommunity.uvic.ca/defaultveg/canadian-universities/
11 Our discussion of Forward Food focuses mainly on information from Canada and the US, though Forward Food also operates elsewhere.

References

Acton, R. B., Nguyen, N., & Minaker, L. M. (2018). School food policies and student eating behaviors in Canada: Examination of the 2015 Cancer Risk Assessment in Youth Survey. *Journal of School Health, 88*(12), 936–944.

Bleich, M. & Hessler, C. (2016). Appreciative inquiry and implementation science in leadership development. *Journal of Continuing Education in Nursing, 47*(5), 207–209.

Blondin, S., Attwood, S., Vennard, D., & Mayneris, V. (2022). *Environmental messages promote plant-based food choices: An online restaurant menu study.* World Resources Institute. Retrieved from https://doi.org/10.46830/wriwp.20.00137

Bushe, G. (2011). Appreciative inquiry: Theory and critique. In D. Boje, B. Burnes, and J. Hassad, (Eds.). *The Routledge companion to organizational change* (pp. 87–103). Routledge.

Bushe, G. (2012). Stepping cautiously past 'the positive' in appreciative inquiry. *AI Practitioner, 14*(4), 49–53.

Campbell-Arvai, V., Arvai, J., & Kalof, L. (2014). Motivating sustainable food choices: The role of nudges, value orientation, and information provision. *Environment and Behavior, 46*(4), 453–475.

Clark, L. F., & Bogdan, A. M. (2019). The role of plant-based foods in Canadian diets: A survey examining food choices, motivations, and dietary identity. *Journal of Food Products Marketing, 25*(4), 355–377.

Clark, J. K., Lowitt, K., Levkoe, C. Z., & Andree, P. (2021). The power to convene: Making sense of the power of food movement organisations in governance processes in the Global North. *Agriculture and Human Values, 38*(1), 175–191.

Cleveland, D. A., Gee, Q., Horn, A., Weichert, L., & Blancho, M. (2020). How many chickens does it take to make an egg? Animal welfare and environmental benefits of replacing eggs with plant foods at the University of California, and beyond. *Agriculture and Human Values, 38*(1), 157–174.

Colley, P., Myer, B., Seabrook, J., & Gilliland, J. (2019). The impact of Canadian school food programs on children's nutrition and health: A systematic review. *Canadian Journal of Dietetic Practice and Research, 80*(2), 79–86.

DefaultVeg (n.d.). *DefaultVeg.* https://defaultveg.org/

Dumas, A. A., Lapointe, A., & Desroches, S. (2018). Users, uses, and effects of social media in dietetic practice: Scoping review of the quantitative and qualitative evidence. *Journal of Medical Internet Research, 20*(2), e9230.

Ensaff, H. (2021). A nudge in the right direction: The role of food choice architecture in changing populations' diets. *Proceedings of the Nutrition Society, 80*(2), 195–206.

Farm to Campus (n.d.). *Farm to cafeteria Canada.* Retrieved February 17, 2021, from http://www.farmtocafeteriacanada.ca/our-work/farm-to-campus/

Farm to School (n.d.). *Farm to cafeteria Canada.* Retrieved November 26, 2020, from http://www.farmtocafeteriacanada.ca/our-work/farm-to-school-canada/

Fehér, A., Gazdecki, M., Véha, M., Szakály, M., & Szakály, Z. (2020). A comprehensive review of the benefits of and the barriers to the switch to a plant-based diet. *Sustainability, 12*(10), 4136.

Forward Food (n.d.). *Forward Food.* Retrieved August 11, 2022, from http://www.forward-food.org/

Green Common (2021). *Green monday.* https://greenmonday.org/en/greencommon/

Hansen, P. G., Schilling, M., & Malthesen, M. S. (2021). Nudging healthy and sustainable food choices: Three randomised controlled field experiments using a vegetarian lunch-default as a normative signal. *Journal of Public Health, 43*(2), 392–397.

Hammond, S. A. (2013). *The Thin Book of Appreciate Inquiry* (3rd edition). Thin Book Publishing Company.

Harwatt, H., Ripple, W. J., Chaudhary, A., Betts, M. G., & Hayek, M. N. (2020). Scientists call for renewed Paris pledges to transform agriculture. *The Lancet Planetary Health, 4*(1), e9–e10.

Hoffman, S. R., Stallings, S. F., Bessinger, R. C., & Brooks, G. T. (2013). Differences between health and ethical vegetarians. Strength of conviction, nutrition knowledge, dietary restriction, and duration of adherence. *Appetite, 65*, 139–144.

Ives, C. D., Freeth, R., & Fischer, J. (2020). Inside-out sustainability: The neglect of inner worlds. *Ambio, 49*, 208–217.

Jabs, J., Devine, C. M., & Sobal, J. (1998). Model of the process of adopting vegetarian diets: Health vegetarians and ethical vegetarians. *Journal of Nutrition Education, 30*(4), 196–202.

Joyce, A., Hallett, J., Hannelly, T., & Carey, G. (2014). The impact of nutritional choices on global warming and policy implications: Examining the link between dietary choices and greenhouse gas emissions. *Energy and Emission Control Technologies, 2*, 33–43.

Kevany, K., & Kassam, Z. (2021). Rapid, unprecedented change for human, animal, and environmental health. *COVID-19 Pandemic: Case Studies & Opinions, 2*(1), 189–191.

Lang, T. (2010). From 'value-for-money' to 'values-for-money'? Ethical food and policy in Europe. *Environment and Planning A, 42*, 1814–1832.

Monterrosa, E. C., Frongillo, E. A., Drewnowski, A., de Pee, S., & Vandevijvere, S. (2020). Sociocultural influences on food choices and implications for sustainable healthy diets. *Food and Nutrition Bulletin, 41*(2 Suppl.), 59S–73S.

Moore, S. M., & Charvat, J. M. (2007). Promoting health behavior change using appreciative inquiry: Moving from deficit models to affirmation models of care. *Family & Community Health, 30*(1S), S64–S74.

Mylan, J. (2018). Sustainable consumption in everyday life: A qualitative study of UK consumer experiences of meat reduction. *Sustainability, 10*(7), 2307.

Nourish Your Roots (n.d.). *Nourish Nova Scotia.* Retrieved June 29, 2021, from https://www.nourishns.ca/nourish-your-roots

Pohjolainen, P., Vinnari, M., & Jokinen, P. (2015). Consumers' perceived barriers to following a plant-based diet. *British Food Journal, 117*(3), 1150–1167.

Priest, K. L., Kaufman, E. K., Brunton, K., & Seibel, M. (2013). Appreciative inquiry: A tool for organisational, programmatic, and project-focused change. *Journal of Leadership Education, 12*(1), 18–33.

Promoting Plant-Based Eating (n.d.). *Animal protection New Mexico.* Retrieved June 29, 2021, from https://apnm.org/what-we-do/promoting-plant-based-eating/

Qualman, D. (2019). *Civilization critical: Energy, food, nature, and the future.* Fernwood Publishing.

Qualman, D., Akram-Lodhi, A. H., Desmarais, A. A., & Srinivasan, S. (2018). Forever young? The crisis of generational renewal on Canada's farms. *Canadian Food Studies/La Revue canadienne des études sur l'alimentation, 5*(3), 100–127.

Recipe Database (n.d.). *Forward food.* Retrieved June 3, 2021, from http://www.forward-food.org/recipes/

Recipes (n.d.). *DefaultVeg.* https://defaultveg.org/#!/recipes

Rosenfeld, D. L., & Burrow, A. L. (2017). Vegetarian on purpose: Understanding the motivations of plant-based dieters. *Appetite, 116*, 456–463.

Sabbagh, C., Boyland, E., Hankey, C., & Parrett, A. (2020). Analysing credibility of UK social media influencers' weight-management blogs: A pilot study. *International Journal of Environmental Research and Public Health, 17*(23), 9022.

Thaler, R., & Sunstein, C. (2009). *Nudge: Improving decisions about health, wealth, and happiness.* Penguin Books.

Turnwald, B. P., Boles, D. Z., & Crum, A. J. (2017). Association between indulgent descriptions and vegetable consumption: Twisted carrots and dynamite beets. *Journal of the American Medical Association - Internal Medicine, 177*(8), 1216–1218.

Turnwald, B. P., Bertoldo, J. D., Perry, M. A., Policastro, P., Timmons, M., Bosso, C., Connors, P., Valgenti, R. T., Pine, L., Challamel, G., Gardner, C. D., & Crum, A. J. (2019). Increasing vegetable intake by emphasising tasty and enjoyable attributes: A randomised controlled multisite intervention for taste-focused labelling. *Psychological Science, 30*(11), 1–13.

Van Loo, E. J., Hoefkens, C., & Verbeke, W. (2017). Healthy, sustainable and plant-based eating: Perceived (mis)match and involvement-based consumer segments as targets for future policy. *Food Policy, 69,* 46–57.

Vizcaino, M., Ruehlman, L. S., Karoly, P., Shilling, K., Berardy, A., Lines, S., & Wharton, C. M. (2021). A goal-systems perspective on plant-based eating: Keys to successful adherence in university students. *Public Health Nutrition, 24*(1), 75–83.

Weber, H., Poeggel, K., Eakin, H., Fischer, D., Lang, D. J., Von Wehrden, H., & Wiek, A. (2020). What are the ingredients for food systems change towards sustainability? Insights from the literature. *Environmental Research Letters, 15*(11), 113001.

Weis, T. (2019). Agriculture from imperialism to neoliberalism. In Ness, I. and Z. Cope, (Eds.). *The Palgrave Encyclopedia of Imperialism and Anti-Imperialism.* (pp. 69–83), 2nd edition. Palgrave.

Wethington, H. R., Finnie, R. K. C., Buchanan, L. R., Okasako-Schmucker, D. L., Mercer, S. L., Merlo, C., Wang, Y., Pratt, C. A., Ochiai, E., & Glanz, K. (2020). Healthier food and beverage interventions in schools: Four community guide systematic reviews. *American Journal of Preventive Medicine, 59*(1), e15–e26.

34

DO ALTERNATIVE FOOD NETWORKS CHANGE DIETS?

Yuna Chiffoleau and Grégori Akermann

Introduction

Alternative food networks (AFNs) refer to a set of initiatives acknowledged in the 1990s as reactions to industrial agriculture and market globalisation (Goodman et al., 2012). The most commonly cited examples refer to local exchange systems, bringing together producers and consumers in the same town or region (i.e., CSA-Community-Supported Agriculture,[1] farmers' markets, community gardens). Although some of these systems guarantee more sustainable production for the consumer through formal rules (e.g., CSA contracts, which often require a mode of farming respectful of the environment, humans, and animals), the sustainability of the supply circulating in AFNs remains, *a priori*, more assumed than analysed in detail (Chiffoleau & Dourian, 2020). In France, for example, according to the 2010 national agricultural census, short food supply chains[2]—which are not necessarily included in AFNs—incorporated a higher rate of organic farms than long chains (Barry, 2012). Similarly, these networks are often associated with committed consumers who adhere to social and environmental values and practise sustainable diets. However, the literature focuses more on the motivations and characteristics of AFN consumers than on their actual food practices, which will be presented below. Above all, it does not capture the changes in practices generated by the entry of new consumers into these networks. However, this question is becoming a key issue, as local AFNs have developed and diversified considerably in the Global North over the last decade, thus modifying the food environment, which is here considered as the interface where people interact with wider food systems to acquire and consume food (FAO, 2016).

With the support of digital tools and the rise of local food policies, AFNs are gaining popularity and attracting more consumers who are young, have lower levels of education, and/or no experience with sustainable food (Kneafsey, 2015). In this chapter, based on two case studies of local AFNs in the south of France, we

DOI: 10.4324/9781003174417-40

shed light on how and under what conditions AFNs are, or can become, levers for changing food behaviours towards more sustainable diets. By sustainable food or diets, we mean, food or diets mainly composed of fresh, seasonal products from low-input or organic farming. In the first section, we briefly review the literature on AFN consumers and the few studies that address the change of food practices in these networks. In a second section, we present the two AFN cases analysed over time and our theoretical approach. In the third section, we describe three processes supported by social relations that are involved in the change of food practices among consumers with little or no experience of sustainable food. Finally, we discuss the contributions and limitations of our results and emphasise that the role of interpersonal relations should be more considered in the development of food environments favourable to adopting sustainable food practices.

AFN and consumers: A brief literature review

In the literature, the analysis of the relationship between consumers and AFNs has taken three main directions. First, articles emphasise the role of committed consumers in the emergence of AFNs (Goodman et al., 2012), particularly in the case of CSA and equivalent schemes around the world (e.g., Jarosz, 2011). Another series of articles focuses on the motivations for purchasing in an AFN, with the limitation of often confusing AFNs, local food, and organic products making it complicated to capture the specific motivations for conducting purchases in AFNs. The studies show contrasting results, even within the same country, emphasising the importance of personal motivations (e.g., search for fresh, healthy products) for some consumers, and collective or altruistic motivations (e.g., support for producers, desire to preserve the environment) for others (Escobar-Lopez et al., 2019; Zoll et al., 2018). A third set of studies focuses on the socio-demographic characteristics of AFN consumers (Mastronardi et al., 2019), with the same limitation of sometimes confusing AFN, local, and organic. The analyses tend to converge to show the overrepresentation of highly educated, middle-aged women (30–50 years old) (Stanco et al., 2019). Generally, studies rarely provide concrete data on consumers' food behaviours, and even less so on their diets. The richest analyses available are on collective gardens, which have greatly developed in the Global North in the last two decades, thus modifying food environments. Carried out by researchers in nutrition and public health sciences, these analyses showed that gardeners consume more fruits and vegetables than the average, but could not conclude if this is an effect of the garden (Tharrey et al., 2020).

A few studies have addressed the role of AFNs in changing food practices or diets. A few articles suggest that some consumers had little experience of sustainable food when they entered an AFN, and changed their food practices as a result of consumer-producer interactions (Opitz et al., 2017) or peer-to-peer exchanges (Brunori et al., 2012). These studies emphasise learning from experienced insiders (Hassanein, 2008), who act as "cause entrepreneurs" (Dubuisson, 2011). Learning by interacting is the main mechanism put forward in AFN consumer trajectories towards more sustainable consumption, even if these works also highlight or suggest an evolution of identities through these interactions (Brunori et al., 2012). These conclusions call for questioning more largely the possible effects of interpersonal relationships generated by the use of an AFN on food consumption.

Two AFN case studies questioned by economic and network sociology

We addressed this question in two cases of AFNs that we have observed over time and which recently emerged in consumers' food environments. The first case is an open-air market made up of producers, artisans, and traders who mainly offer fresh, seasonal, and low-input food from short food chains. This market was created in 2008 by the town council to socially revitalise a town of 7,000 inhabitants within a metropolis of 500,000 inhabitants in the south of France. The priority given to short chains—not the case for all open-air markets in France—was the result of discussions with our research team. The market takes place weekly, with 600 regular customers in 2021, a number that is continuously growing. According to the local actors and the market stallholders, the majority of these customers were not familiar with sustainable food before they began to attend the market.

The second case is a food cooperative located in the centre of the same southern French metropolis. Based on the model of the Park Slope Food Coop in New York City, this cooperative has more than 3,000 members, including 1,300 regular customers, who buy a share of €100 and spend three hours per month working in the shop. Unlike the open-air market, customers were more experienced with sustainable food before joining the AFN. However, the cooperative offers food from conventional farming and the agro-food industry (e.g., Coca-Cola), alongside fresh, seasonal, artisanal, and organic products. The idea of the Park Slope Food Coop's founders was to extend the membership to less committed consumers and to encourage, through their shop visits, a progressive change in their purchases towards more sustainable food.

Both cases were analysed using the theoretical framework and methodology of economic and network sociology. This branch of sociology considers that any action, such as food practices, is embedded in social relationships (Granovetter, 1985). Following this approach, studies showed how relationships, and social mechanisms supported by them, shape people's practices and encourage them to evolve beyond their socio-demographic characteristics. In both cases, we primarily interviewed consumers who had little or no experience of sustainable food prior to entering the AFN. Interviews focused on food practices (purchasing, cooking, waste), prior to and since entering the AFN, and on relationships in and around the AFN and their possible impact on food practices. We also carried out observations during consumers' shopping periods and while members of the cooperative were fulfilling duty hours. In both these cases, we sought to confirm or attend to the nuance of changes in practices mentioned during the interviews by talking to consumers' families and friends in the case of the market, and by collecting their purchasing receipts over several months in the case of the cooperative, as shown in Table 34.1.

Table 34.1 Methods of data collection in two case studies

Case study Methods	Open-air market	Cooperative
In-depth interviews with less experienced consumers	50 (2016)	25 (2019)
Additional methods	• Short interviews with interviewees' relatives (about 200) • Observation at the market over one year (2016)	• Collection and analysis of purchasing receipts over 13 months or at least five months with a purchase in the AFN at least four times a month including a sample of 525 members (2019) • Participant observation as a working cooperative member over one year (2019)

Source: Work of the authors.

Three relational mechanisms contributing to the food transition

Analysed from the perspective of economic and network sociology, the two cases revealed three relational mechanisms that favoured the evolution of interviewees' practices towards sustainable food: Learning, social control, and identity construction within a system of relations.

Relationships as sources of learning, influenced by activities, and equipment around products

In both cases, we first observed dialogue that circulated knowledge about products, their production, and/or processing methods. According to the less experienced consumers, these verbal interactions allowed them to learn about the "hidden side" of agriculture (i.e., agricultural production models, environmental impacts of these models, subsidies granted according to these models, etc.) as well as of supply chains (i.e., price paid to the producer in short vs. long chains). In the open-air market, we observed these relationships between consumers and producers, as well as between consumers and intermediaries (traders, artisans), who, in the case of this market, operate at least partly in short chains and are thus able to converse about the producers (and the price) from whom they have bought products or raw materials. We also observed dialogue between consumers in the cooperative, where consumers do not have much opportunity to exchange with producers except during in-store deliveries. Also in the cooperative, the duty hours foster exchanges among consumers. The duty rosters are filled out voluntarily, so there is no guarantee that they will mix consumers with different experiences in sustainable food. However, the fact that there are common duty hours for several members, during which they have to receive deliveries, store them on the shelf, and label them, encourages learning interactions about the differences between local, organic, and industrial products, as we observed many times. One interviewee emphasised the contribution of these relations with more experienced people:

> It's always better to have a relationship with Justine, with Anna-Lisa, with people, because they have an activist past: Things to pass on, assurance, confidence, but without making a big deal of it, a feeling of sharing, it opens me up to something I didn't know.
>
> *(Solange, 64 years old, retired, former*
> *librarian, cooperative member)*

In both AFNs, the material and mechanisms involved also favour learning inter-actions. At the market, we observed how the specific colour-coded labelling[3] scheme, indicating seasonal products coming from local and short chains, gener-ated questions from consumers and learning relations with the vendor or other consumers. In the cooperative, the consumers on duty also have to put up signs on the shelves to highlight certain characteristics of the products (organic, local, etc.), which leads to discussions among them. Activities and material equipment around the products thus encourage the development of learning relationships. This observation is in agreement with the results put forward in the actor-net-work theory (Callon, 1984).

According to interviewees themselves, this learning has often had an impact on their food behaviours:

> The supermarket, for fresh produce, is finished [...]. Not only are the products not good and we don't know what's on them, but also, at the price they pay the producers! [...] I didn't know all this before coming to the market. Now I can't see myself doing anything else.
>
> *(Françoise, 47 years old, executive*
> *secretary, market customer)*

In the case of the market, changes have been more radical for consumers with no background in sustainable food. The majority no longer buy non-seasonal fruits and vegetables, and have reduced or even eliminated the purchase of these prod-ucts in conventional supermarkets:

> When I used to go to the supermarket to buy tomatoes in winter, it was the most normal thing in the world. Since going to the market, I realise that it is not very natural. Now I wait until it's in season and I don't buy any more tomatoes at the supermarket!
>
> *(Émilie, 34 years old, specialised*
> *educator, market customer)*

For the cooperative, where consumers were often more experienced before using the AFN, the analysis of purchasing receipts shows that 29% of them have regularly increased their consumption of fruit and vegetables, 32% of bulk products, and 40% of organic products over the period studied (13 or 5 months, see Table 34.1).

Changes in practices brought about through social control generated by the AFN

Engaging in an AFN leads to other relationships besides learning. The two AFNs considered here, particularly when fresh produce is available, encourage a regular consumer presence. Consumers often encounter people they have already met. In these two local AFNs, a large proportion of customers are also neighbours, who are likely to bump into each other outside the AFN and maintain familiar relations. Purchasing in the AFN is not practised as an isolated actor, as in the case of a conventional supermarket where one usually does not know the other consumers. Here, purchases remain individually practised but within a "community" built and maintained around the AFN, and in which many members are also neighbours or friends. Buying non-sustainable products is thus likely to be observed and judged negatively by these "non-anonymous" others. The relationships induced by the AFN, repeated and often combined with neighbourhood relationships, support social control (Coleman, 1990) that is favourable to the evolution of purchases towards more sustainable products:

> Once I went [to the cooperative] to buy dried apricots, there were some very orange ones (not organic) and the others (organic), and I met a friend of a friend who said to me, "What? Do you eat these apricots?" Because I had taken the orange apricots [...] We can afford to make remarks like that at the cooperative.
>
> *(Elisabeth, 57 years old, early-*
> *childhood educator, cooperative member)*

However, larger purchases of sustainable products do not necessarily lead to the radical and complete implementation of a more sustainable diet. A retired customer of the open-air market, for example, makes sure to visit the market weekly and buy products labelled as sourced from short chains "because everyone knows everything in the municipality." She admits, however, that she does not necessarily have the time or motivation to cook them. Sometimes, she even has to throw away some of the products she bought at the market. However, she also acknowledges that her purchases have allowed her to rediscover fresh produce, which she used to hardly consume when she was working and had a busy schedule. Dietary change was not radical, but through the social control induced by the AFN, some non-sustainable consumption routines have been unlocked for this customer and for other not-committed consumers we interviewed.

A new diet encouraged by the construction of rewarding identities

In the tradition of sociologist Pierre Bourdieu, food is classically recognised as a vector of social distinction, differentiating social classes (Bourdieu, 1984). Along these lines, consumption in AFNs has been analysed as a middle class practice, seeking differentiation from the working class culture of consuming cheap products (Paddock, 2015). Economic and network sociology, however, questions the notion of social class and prefers the recognition of specific positions in the networks of interpersonal relations as grounds for identity construction. Without referring to the positioning of one social class against another, our analysis highlights the role of the AFN in activating relationships that allow individuals to construct an identity valued by those around them, as in the case of this interviewee: "When I cook for friends, I think it's great to be able to say to them, 'these are market products, they're all fresh, sustainable, local produce'" (social worker, 29 years old, market customer).

Like this customer, most of the interviewees who had changed their practices since entering the AFN often wanted to share what they had learned and give advice to less-experienced consumers. In this way, even if they do not master all the issues at hand, they are able to build a rewarding identity within the AFN as well as within their social circles: Their family, their municipality, their leisure circles, etc. One market customer, who used to do all her shopping at the supermarket before purchasing in the AFN, says that her family now "congratulates" her and appreciates the "tastier dishes" that she cooks for them with market produce, with the help of "simple recipes" shared by the producers. More broadly, this customer shared with her parents, sister, friends from her sewing class, and her neighbours the benefits of buying from the market, or, if in the conventional supermarket, the importance of paying attention to where the food comes from and whether it is in season. This process of developing social identity through food contributes in turn to reinforcing participation in AFNs, which also become a chosen community affiliation that carries norms and that has become, above all, a carrier of meaning.

Interviewees did not spontaneously recognise this third mechanism of identity construction that is supported by the interpersonal relationships created or enabled by the AFN. However, the accounts collected from consumers and their relatives lead us to emphasise its importance in dietary changes and, even more so, in the consolidation of changes brought about by learning or social control. This result aligns with the analyses developed in the sociologies of the individual, highlighting the importance of identity enhancement in social change within contemporary societies (Corcuff et al., 2005). Moreover, as in these sociologies, our case studies confirm that this enhancement is not achieved at the expense of participation in communities but rather that it is part of them and feeds off them.

Highlighting these relational mechanisms brought about or encouraged by AFNs, which are vectors of changes in food practices, leads us to discuss our results and propose recommendations. First, we propose directions for future research, and second, we recommend better integrating interpersonal relationships in the construction of food environments that are more favourable to changes in food practices.

Discussion and recommendations

Our research aimed to explore the role of interpersonal relationships forged through engaging in an AFN on the evolution of participant food practices. In order to consolidate the analysis, the changes described in the interviews have been compared with declarations of interviewees' relatives in the case of the market and with purchasing receipts for the food cooperative. Our work thus goes beyond studies based only on interviewees' input, that often can be biased by social desirability in surveys about sustainable food (Cerri et al., 2019). The main AFN-induced changes that we observed consisted of buying more fresh and local products, from short chains, but also in cooking more and paying more attention to the origin of products purchased in conventional supermarkets. However, we collected more data on product purchases than on diets, which is a limitation of our work that could be improved with further studies. Data on diets are difficult to produce through interviews, inviting us to think of other approaches, such as citizen science, which allows people to be actors in the production of knowledge and not just subjects (Oakden et al., 2021). Regarding our research question, the two case studies confirmed the importance of learning relationships within the AFN for food transition, as reported in a few previous articles. They also highlighted another social mechanism supported by the relationships, social control, while emphasising the importance of the evolution of identities in AFNs suggested in previous work (Brunori et al., 2012). In this regard, our study could be referred to more classic works in the sociology of food, pointing to the effects of belonging to social groups or strategies of distinction (Cardon et al., 2019). However, whereas the sociology of food emphasises family membership or distinction in relation to one's social class, our study confirms the influence of one's positioning within the system of social relations created by and around the AFN on changes in practices, in line with economic and network sociology. We must still take into account the collective and localised dimension of the two AFNs we studied, in which consumers are often also neighbours, as it could be more favourable to the relational processes highlighted here than other AFNs. For instance, an AFN based on home delivery by an anonymous carrier of a basket of local products ordered online induces *a priori* fewer interactions with the producer or other consumers. This questions the generic scope of our results, which have to be tested in diverse AFNs. Moreover, we do not yet have the means to understand why relational mechanisms generating changes play out for some people and not for others with the same socio-demographic characteristics. This calls for a more in-depth study of consumer trajectories.

Our analysis leads to a discussion in a second direction, related to growing attention to the role of policy in food environments and in shaping healthy diets (HLPE, 2017) But as other chapters have examined food environments, we only mention it here. Turner et al. (2018) distinguish between an exogenous food environment (i.e., the availability of local food supply, the influence of product prices and quality, food-related regulations, etc.) and an endogenous environment (i.e., the individual's accessibility to products, adaptation of available products to his or her tastes, etc.), while pointing out that studies focus mainly on the exogenous environment. However, whether exogenous or endogenous, the considered environment does not include inter-individual relationships around food, which play an important role according to our work. The possibility of developing producer-consumer relationships, as well as relationships with more experienced consumers, could be better taken into account by public programs, in the framework of local food policies in particular, to build environments favourable to more sustainable diets. For instance, while seeking to minimise the environmental footprint of AFNs through travel limitations, it seems important to humanise AFNs and to enhance possibilities for consumers to meet producers, at least periodically. Moreover, concerning the scaling up of AFNs, which is supported or required by policy makers (Kneafsey, 2015), our results argue in favour of multiplying human-sized AFNs that promote face-to-face relationships, rather than merely increasing the size of existing initiatives.

Conclusion

Faced with climate change and the increase in food-related diseases, policy-makers are looking for new levers to encourage more sustainable diets. In this chapter, we show the role that AFNs can play, from two cases that emerged as important spaces for interactions and relationships that contribute to the dissemination and appropriation of new consumption practices towards more sustainable diets. On the one hand, the relationships between producers and consumers allow for learning about products and "what's behind them" within the AFN, but also, as a counterpoint, in conventional supermarkets. On the other hand, relations between consumers also become vectors for learning, advice, social control, and identity enhancement, which motivate or reinforce the change of practices. An AFN is therefore much more than a space of repeated marketing relationships between suppliers and customers, it builds a complex system of relationships in which consumption is embedded and evolves. Analysing the impact of AFNs on food practices has become even more important since the COVID-19 crisis, which has brought these systems to the forefront and increased their audience in the Global North, especially during the initial lockdowns implemented by governments to contain the spread of the virus (Nemes et al., 2021). As the crisis progressed, even though a large proportion of customers who entered an AFN during the lockdown seem to have returned to their habitual purchasing practices in conventional supply chains, some have remained. This motivates further work, in particular longitudinal studies, on the possible role of AFNs in the evolution of these newcomers' practices towards sustainable diets, while considering AFNs' increasing diversity.

Notes

1 System whereby consumers buy shares of a farm's harvest in advance.
2 Since 2009, short food supply chains have been officially defined in France as sales systems involving at most one intermediary between the producer and the consumer, regardless of the geographical distance. They include traditional systems (e.g., open-air markets, on-farm sales, etc.) and more recent ones (e.g., AMAP equivalent to CSA, online platforms, etc.), though not all considered "alternative".
3 See Chiffoleau et al. (2016) for more information on this labelling scheme, which we created with the local authority where this market is located.

References

Barry, C. (2012). Commercialisation des produits agricoles. Un producteur sur cinq vend en circuit court. *Agreste Primeur, 275*, 4.

Bourdieu, P. (1984). *Distinction: A social critique of the judgement of taste*. Harvard University Press.

Brunori, G., Rossi, A., & Guidi, F. (2012). On the new social relations around and beyond food. Analysing consumers' role and action in Gruppi di Acquisto Solidale (Solidarity Purchasing Groups). *Sociologia Ruralis, 52*, 1–30.

Callon, M. (1984). Some elements of a sociology of translation: Domestication of the scallops and the fishermen of St Brieuc Bay. *The Sociological Review, 32*(1), 196–233.

Cardon, P., Depecker, T., & Plessz, M. (2019). *Sociologie de l'alimentation*. Armand Colin.

Cerri, J., Thøgersen, J., & Testa, F. (2019). Social desirability and sustainable food research: A systematic literature review. *Food Quality and Preference, 71*, 136–140.

Chiffoleau Y., & Dourian T. (2020). Sustainable food supply chains: Is shortening the answer? A literature review for a research and innovation agenda. *Sustainability, 12*, 9831.

Chiffoleau, Y., Millet-Amrani, S., & Canard, A. (2016). From short food supply chains to sustainable agriculture in urban food systems: Food democracy as a vector of transition. *Agriculture, 6*(57).

Coleman, J. (1990). *Foundations of social theory*. Belknap Press of Harvard University Press.

Corcuff, P., Ion, J., & de Singly, F. (2005). *Politiques de l'individualisme*. Editions Textuel.

Dubuisson-Quellier, S. (2011). Le consommateur responsable. La construction des capacités d'action des consommateurs par les mouvements militants. *Sciences de la Société, 82*, 105–125.

Escobar-López, S.Y., Espinoza-Ortega, A., Lozano-Cabedo, C., Aguilar-Criado, E., & Amaya-Corchuelo, S., (2019). Motivations to consume ecological foods in alternative food networks (AFNs) in Southern Spain. *British Food Journal, 121*, 2565–2577.

FAO (2016). *Influencing food environments for healthy diets*. FAO.

Goodman, D., DuPuis, E. M., & Goodman, M. K. (2012). *Alternative food networks. Knowledge, practices, and politics*. Routledge.

Granovetter, M. (1985). Economic action and social structure: The problem of embeddedness. *American Journal of Sociology, 91*, 481–510.

Hassanein, N. (2008). Locating food democracy: Theoretical and practical ingredients. *Journal of Hunger & Environmental Nutrition, 3*(2–3), 286–308.

HLPE (2017). *Nutrition and food systems. A report by the high level panel of experts on food security and nutrition of the committee on world food security*. FAO.

Jarosz, L. (2011). Nourishing women: Toward a feminist political ecology of community supported agriculture in the United States. *Gender, Place & Culture, 18*(3), 307–326.

Kneafsey, M. (Ed.). (2015). *EIP-AGRI. Focus group innovative short food supply chain management*. Final report. EIP-AGRI.

Mastronardi, L., Romagnoli, L., Mazzocchi, G., Giaccio, V., & Marino, D. (2019). Understanding consumer's motivations and behaviour in alternative food networks. *British Food Journal, 121*(9), 2102–2115.

Nemes, G., Chiffoleau, Y., Zollet S., Collison, M., Benedek, Z., Colantuono, F., Dulsrud, A., Fiore, M., Holtkamp, C., Kim, T.Y., Korzun, M., Mesa-Manzano, R., Reckinger, R., Ruiz-Martínez, I., Smith, K., Tamura, N., Viteri, M. L., & Orbán, E. (2021). The impact of COVID-19 on alternative and local food systems and the potential for the sustainability transition: Insights from 13 countries. *Sustainable Production and Consumption, 28*, 591–599.

Oakden, L., Bridge, G., Armstrong, B., Reynolds, C., Wang, C., Panzone, L., Schmidt Rivera, X., Kause, A., Ffoulkes, C., Krawczyk, C., Miller, G., & Serjeant, S. (2021). The importance of citizen scientists in the move towards sustainable diets and a sustainable food system. *Frontiers in Sustainable Food Systems, 5*, 596594.

Opitz, I., Specht, K., Piorr, A., Siebert, R., & Zasada, I. (2017). Effects of consumer-producer interactions in alternative food networks on consumers' learning about food and agriculture. *Moravian Geographical Reports, 25*(3), 181–191.

Paddock, J. (2015). Positioning food cultures: 'Alternative' food as distinctive consumer practice. *Sociology, 50*(6), 1039–1055.

Stanco, M., Lerro, M., Marotta, G., & Nazzaro, C. (2019). Consumers' and farmers' characteristics in short food supply chains: An exploratory analysis. *Studies in Agricultural Economics, 121*(2), 292232.

Tharrey M., Sachs A., Perignon M., Simon, C., Méjean, C., Litt, J., & Darmon, N. (2020). Improving lifestyles sustainability through community gardening: Results and lessons learnt from the JArDinS quasi-experimental study. *BioMed Central Public Health, 20*, 1798.

Turner, C., Aggarwal, A., Walls, H., Herforth, A., Drewnowski, A., Coatesf, J., Kalamatianoua, S., & Kadiyala, S. (2018). Concepts and critical perspectives for food environment research: A global framework with implications for action in low- and middle-income countries. *Global Food Security, 18*, 93–101.

Zoll F., Specht K., Opitz I., Siebert, R., Piorr, A., & Zasada, I. (2018). Individual choice or collective action? Exploring consumer motives for participating in alternative food networks. *International Journal of Consumer Studies, 42*, 101–110.

35

SECTORS OF SOCIETY SUPPORTING SUSTAINABLE DIETS

An examination of Slow Food as a pathway towards sustainable diets

Federico Mattei and Eleonora Lano

Introduction

This chapter provides readers with an understanding of how non-governmental organisations (NGOs) and civil society organisations (CSOs) can play key roles in shifting food systems towards greater sustainability through examining Slow Food International as an example. This chapter reviews how Slow Food operates globally, looks at some key methodologies (like the Presidia projects), and outlines how Slow Food works closely with governments, researchers, and communities to increase the sustainability of global and local food systems.

The 2019 study by the International Panel of Experts on Sustainable Food Systems (IPES) indicated that the current global food system causes significant negative outcomes for the social, environmental, and human health sectors (IPES, 2019). In fact, farming systems contribute up to 30% of greenhouse gas (GHG) emissions and use an increasingly large quantity of land and water (Crippa et al., 2021). The effect of food systems on biodiversity has also been extreme with 16.5% of vertebrate pollinators facing the risk of extinction (IPBES, 2016), with the single largest threat factor being the conversion of natural ecosystems to farmland (Tilman et al., 2017).

Food systems also have dramatically affected human health and nutrition, with 1.9 billion adults worldwide being overweight or obese and nearly 800 million undernourished people suffering from hunger (FAO, IFAD, UNICEF, WFP and WHO, 2021). The increasing rate of triple burden of malnutrition (underweight, overweight, and lack of nutrients) also significantly affects human dietary health (Sunuwar et al., 2020). Furthermore, a rising global population will create even

DOI: 10.4324/9781003174417-41

more pressures on ecosystems and on health systems, as highlighted by the recent EAT–Lancet Commission report on food in the Anthropocene, which indicates that anticipated dietary trends, when combined with growth to nearly ten billion by 2050 of the human community, are expected to exacerbate adverse outcomes (Willet et al., 2019).

Food systems with positive and negative externalities

Agriculture provides an essential foundation for the most important livelihood models and one of the most crucial activities that fosters human nourishment and well-being, economic development, and job opportunities, particularly in rural areas. Due to the number of actors who work directly and indirectly in the agricultural sector, increased agricultural output and productivity tend to contribute substantially to the overall economic development of a country. Producers, workers, processors, distributors, and retailers all make up parts of food systems.

The type of food system that is promoted significantly drives investments and impacts in the agricultural sector. Many benefits can arise and ailments averted through enabling community driven food systems that foster meaningful relationships between producers and consumers. Where consumers and producers are connected through short, transparent, direct value chains, producers have an incentive to develop or conserve quality-based production models that are then rewarded by informed consumers. Meanwhile, consumers can access culturally adequate, safe, nutritious food at affordable prices, thus strengthening access to sustainable diets.

Furthermore, sufficiently robust and vibrant food systems, that align with national and international tourism, hold the potential of generating increased local and regional income. Tourism is an important activity and contributes significantly to economic growth, including in remote and rural areas (European Union, 2021). Internal tourism is just as important as international tourism, as it enables and promotes a transfer of money from the cities to rural areas and may help with the transfer of funds from richer to more vulnerable areas nationally. The increased economic benefits of tourism are not limited to the direct actors of the food system, but are shared amongst all the stakeholders in the area. This distribution of benefits also serves as a potent incentive towards retaining rural livelihood models and helps to counteract the gradual displacement of people from rural to urban areas. While the scope of this chapter does not include how food systems and sustainable tourism interact, it is important to mention.

Unfortunately, however, we exist in a global economic paradigm in which a share of the costs and benefits of any economic activity is often externalised. This includes the cost or benefit caused by a producer that is not financially incurred or received by that producer. Unsustainable farming practices often generate high net profits through contributing negative consequences (i.e., erosion of land, water, biodiversity loss, chemical contamination, greenhouse gas emissions) (Olsson et al., 2019; Steinfeld et al., 2006; FAO, 2017). These adverse outcomes often are not reflected in the cost of the food, causing the negative costs to be externalised. In contrast, more sustainable farming practices generate more regenerative natural benefits and substantial advantages for the broader community (Rowntree et al., 2020). Though we can attempt to re-internalise externalities through public intervention in the form of taxation and subsidy schemes, this approach does have limits including the objective difficulty of quantifying externalities and a common reticence amongst policy makers towards taxing food. While the difficulty of the task may be a consideration, it should not cause governments to fail to take appropriate action (Mason & Lang, 2017).

An alternative approach to unsustainable production can be to engage producers and processors in formal certifications or programs that support improved agricultural outcomes. Another approach includes raising awareness around the influence through personal consumption choices, but in this case, it is crucial to ensure that specific production practices, and their positive externalities are adequately measured and transmitted to consumers and negative externalities prevented or incorporated. Formal certification schemes have shown some considerable success with a recent EU study that concluded that the sales value of a product with known certification may on average, secure more than double the sales than similar products without a certification (European Commission, 2020). While increasing sales value may increase economic sustainability for producers it does not guarantee environmental and social sustainability. These need to be intentionally integrated and the focus of producers and consumers.

Evidence suggests that vital food systems that include social justice, ecological regeneration, and are democratically inclusive and contribute to local economies need to engage collaborative food networks (Blay-Palmer et al., 2016). Therefore, it is crucial that CSOs and NGOs support and foster more sustainable food systems. While many approaches and many different organisations are carrying out this work, using Slow Food as an example may serve to extend its reach and inspiration.

Slow Food's approach

Nourishing food systems

Slow Food is a global, grassroots organisation, founded in 1989 to prevent the disappearance of local food cultures and traditions, counteract the rise of globalisation that led to a shift in dietary patterns, and combat people's dwindling interest in the food they eat, where it comes from, and how our (i.e., everyone's) food choices affect the world around us (Slow Food, n.d.b). Since its beginnings, Slow Food has grown into a global movement involving millions of people in over 160 countries, working to ensure everyone has access to good, clean, and fair food, regardless of their income and economic circumstances. Slow Food works on the entire food system in a holistic way with actions and methodologies geared towards specific value chain actors. It carries out significant work with small and medium scale producers. It supports production paradigms to become more resilient and facilitates access to markets of local products to strengthen livelihoods and support rural development. It is a movement committed to the conservation and promotion of agricultural biodiversity while maximising the positive externalities and ecosystem services generated by local producers.

Another principle of Slow Food's strategy relies on promoting sustainable diets and consumption patterns. Slow Food advocates for the provision of clear and concise information to consumers regarding the effect and consequences of their consumption patterns with the desire to leverage consumer power and choices to shape optimal production paradigms. Much of the work in this regard relates to shortening and providing transparency in value chains, raising awareness amongst consumers regarding the importance of agro-biodiversity, the protection of natural capital and, in particular investing in ways to bolster the resilience of more localised food systems.

Slow Food International works from the premise that food is tied to many other aspects of life, including culture, politics, agriculture, and the environment. Through food choices, stakeholders can collectively influence how food is cultivated, produced, and distributed; this collective power holds the potential to lessen some of the global crises and positively influence the world. With each purchase of food, a demand is generated which then influences the actions of the supplier or producer of the food. When consumers choose food and food items with production processes that are not sustainable, they contribute to negative externalities. Conversely, by fostering interest in buying foods that are produced sustainably, this could increase demand for foods whose production process generates more positive externalities.

Increasing trust through transparency

Slow Food aims to improve knowledge about food—from production to consumption to food loss—and to impact people's practices and choices. It does this through designing and implementing educational and training activities and by engaging actors across the food chain to offer consumers more complete and transparent information, Slow Food promotes the narrative label (Slow Food Foundation for Biodiversity, n.d.b), which provides information regarding products (e.g., varieties and breeds, cultivation and processing methods, areas of origin, animal welfare), and promotes food distribution systems that bring small scale producers and consumers into direct contact, such as Earth Markets (Slow Food Foundation for Biodiversity, n.d.a). In many ways, Slow Food's main focus is on building trust-based networks in food systems to promote production and consumption paradigms that generate positive externalities and limit negative externalities, to be an agent for enhanced human health, environmental benefits, and social welfare.

Slow Food uses a variety of methodologies to encourage transparency and build trust, advocate for sustainable food systems and protect and promote biodiversity. The protection of biodiversity is a driving motivation for the organisation. To continue to help people to access sufficient healthy food, maintaining plant biodiversity is critical to support diverse ecosystems and to bolster resilience across the food value chain. When integrated into mixed-crop cultivation systems, fruit trees, which are resilient to climate variability, can provide healthy, nutritionally rich foods all year, thanks to their micronutrients (vitamins and minerals), macronutrients (protein and carbohydrates), and phytochemical substances (antioxidants) (Stadlmayr et al., 2013). For over 20 years, Slow Food has been working to safeguard biodiversity with numerous projects, starting from the Ark of Taste and the Presidia, and over time has constructed a global network of tens of thousands of producers who preserve and share the diversity of food and agriculture in the world.

Ark of Taste

Since starting the Ark of Taste project in 1996, Slow Food has been publicising the risk of extinction of thousands of animal breeds and varieties of fruits, vegetables, and legumes, as well as cheeses, breads, traditional sweets, and all the artisanal knowledge required for their production. For decades, Slow Food has been driven by the importance of environmental, cultural, and economic value being protected and the need to preserve and transform the natural and learned heritage into opportunities for local communities. Over 5,400 products from 150 countries have been welcomed into the Ark of Taste. Behind this cataloguing lies the work of over 100,000 small-scale food producers, who are custodians of biodiversity. The Ark of Taste has brought the attention of the media, public authorities, experts, and many chefs and consumers to heritage that was facing struggles and was in need of saving.

Slow Food Presidia

One of Slow Food's flagship methodologies is the Presidia approach. Presidia, communities of producers focus on a specific breed, crop, or variety, and many have been established all over the world by Slow Food International as a way to protect threatened agricultural species, varieties, or breeds. The specific species or product is intrinsically linked to both the territory where it is produced and the producers and communities whose livelihoods are based on that product. Once Slow Food has identified a product that is at risk of being lost and needs to be protected and supported, it works with producers to develop a protocol defining production methods, values, and associated practices. A narrative is developed that allows the value of the product (and most importantly the positive externalities associated with the production methods) to be transmitted to consumers to generate a price premium on that product. As noted earlier, an example of this approach is Slow Food's narrative label.

In order to measure the impact of the Presidia projects, since 2013 Slow Food has carried out a series of studies on Slow Food Presidia products with the scientific support of Indaco2 (a University of Siena spin-off company that provides environmental consultation and communication services) to measure their carbon footprint and environmental impact. They compare small-scale sustainable food production with products from industrialised systems, by measuring the emissions from the production processes through life-cycle analysis (LCA) and impact (carbon footprint) expressed in the amount of carbon dioxide (CO_2 equivalent). The differences shown by the analyses are significant and, at times, remarkable: All the Slow Food Presidia products studied show emissions savings greater than 30%. This research offers some insights into examples of climate-friendly, healthy diets, and diets that are less sustainable. Distinguishing these are appearing more urgent with the challenges faced by human and animal health and the planet.

Conclusion

As well argued in this handbook, diets that are based on animal protein from factory-farmed meat, sugary drinks, highly processed fatty foods, and refined foods that generate excessive greenhouse gases. In contrast, more climate-friendly diets based mostly on whole, fresh, plant-based foods, cultivated following sustainable or organic practices, are proving to be valuable approaches to production and consumption (Slow Food, n.d.b). In this collection also, many other chapters have stressed the importance of systems change and the necessity for suitable policy to support sustainable diets. While the role of individual consumers and their food choices.is pivotal in this story of social change, the magnitude of changes needed cannot be left up to individuals. All sectors: Government, industry, education, non-governmental organisations play important roles in facilitating needed systems changes.

The challenge the global community faces is the same as millions of farmers across the globe and is of utmost importance to the future of our global society. If unsustainable consumption patterns persist, as noted in this chapter and many others in this collection, the world will continue to see a progressive impoverishment of its ecosystems and cultural landscapes, as well as the associated loss of biodiversity, ecosystem services, and rural livelihoods. The food system is not the only sector that faces this challenge; the global economic system has a role in enabling sufficient and accessible goods and services to an increasing number of consumers without depleting natural capital. The food system may stand alone for its potential to shape our landscapes and ecosystems and for its impact on identity, culture, and traditions.

Civil society organisations and NGOs can, and must, play a crucial role alongside governments in ensuring consumers have access to clear and scientifically robust information on the food products they choose to buy. Producers and producer groups must work together to identify production guidelines and these must, in turn, be communicated to consumers. Consumers will thus be able to decide, through their consumption choices, what production paradigms to support and be fully informed about the positive and negative externalities generated by those production paradigms. Slow Food International seeks to be a global advocate for delicious local food embedded in resilient and durable foodscapes that invite producer and consumer interactions to drive sustainable diets.

References

Blay-Palmer, A., Sonnino, R., & Custot, J. (2016). A food politics of the possible? Growing sustainable food systems through networks of knowledge. *Agriculture and Human Values*, *33*(1), 27–43.

Crippa, M., Solazzo, E., Guizzardi, D., Monforti-Ferrario, F., Tubiello, F. N., & Leip, A. J. N. F. (2021). Food systems are responsible for a third of global anthropogenic GHG emissions. *Nature Food, 2*(3), 198–209.

European Commission (2020, April 20). *Geographical indications: A European treasure worth €75 billion*. European Commission Press Release. https://ec.europa.eu/commission/ presscorner/detail/en/IP_20_683.

European Commission (2021). *Commission staff working document stakeholder consultation: Synopsis report accompanying the document communication from the commission to the European Parliament, the Council, the European Economic and Social Committee and the Committee of the Regions a long-term vision for the EU's rural areas - towards stronger, connected, resilient and prosperous rural areas by 2040*. Retrieved March 22, 2022, from https://eur-lex.europa.eu/ legal-content/EN/TXT/?uri=CELEX:52021SC0167R(01)

FAO (2017). *The future of food and agriculture: Trends and challenges*. FAO.

FAO, IFAD, UNICEF, WFP, & WHO (2021). *The State of food security and nutrition in the world 2021. Transforming food systems for food security, improved nutrition and affordable healthy diets for all*. FAO.

International Panel of Experts on Sustainable Food Systems (iPES Food) (2019). Towards a common food policy for the European Union: The policy reform and realignment that is required to build sustainable food systems in Europe. Retrieved March 22, 2022, from http://ipes-food.org/_img/upload/files/CFP_FullReport.pdf

IPBES (2016). The assessment report of the Intergovernmental Science-Policy Platform on Biodiversity and Ecosystem Services on pollinators, pollination and food production. In S. G. Potts, V. L. Imperatriz-Fonseca, & H. T. Ngo (eds.), Bonn, Secretariat of the Intergovernmental Science-Policy Platform on Biodiversity and Ecosystem Services. Retrieved March 22, 2022, from https://ipbes.net/sites/default/files/downloads/pdf /2017_pollination_full_report_book_v12_pages.pdf

Mason, P., & Lang, T. (2017). *Sustainable diets: How ecological nutrition can transform consumption and the food system*. Routledge.

Olsson, L., H. Barbosa, S. Bhadwal, A. Cowie, K. Delusca, D. Flores-Renteria, K. Hermans, E. Jobbagy, W. Kurz, D. Li, D. J. Sonwa, & L. Stringer. (2019). Land degradation. In P. R. Shukla, J. Skea, E. Calvo Buendia, V. Masson-Delmotte, H.-O. Pörtner, D. C. Roberts, P. Zhai, R. Slade, S. Connors, R. van Diemen, M. Ferrat, E. Haughey, S. Luz, S. Neogi, M. Pathak, J. Petzold, J. Portugal Pereira, P. Vyas, E. Huntley, K. Kissick, M. Belkacemi, & J. Malley (Eds.), *Climate change and land: An IPCC special report on climate change, desertification, land degradation, sustainable land management, food security, and greenhouse gas fluxes in terrestrial ecosystems* (pp. 345–436). Intergovernmental Panel on Climate Change.

Rowntree, J. E., Stanley, P. L., Maciel, I. C., Thorbecke, M., Rosenzweig, S. T., Hancock, D. W., Guzman A., & Raven, M. R. (2020). Ecosystem impacts and productive capacity of a multi-species pastured livestock system. *Frontiers in Sustainable Food Systems, 232*, 1–13

Slow Food Foundation for Biodiversity (n.d.a). Retrieved January 31, 2022 from https:// www.fondazioneslowfood.com/en/what-we-do/earth-markets/

Slow Food Foundation for Biodiversity (n.d.b). Retrieved January 31, 2022 from https:// www.fondazioneslowfood.com/en/what-we-do/what-is-the-narrative-label/

Slow Food International (n.d.a). Retrieved January 31, 2022 from https://www.slowfood .com/

Slow Food International (n.d.b). Retrieved January 31, 2022 from https://www.slowfood .com/a-meatless-healthy-diet-is-a-climate-friendly-diet/

Stadlmayr, B., Charrondiere, U. R., Eisenwagen, S., Jamnadass, R., & Kehlenbeck, K. (2013). Nutrient composition of selected indigenous fruits from sub-Saharan Africa. *Journal of the Science of Food and Agriculture, 93*(11), 2627–2636.

Steinfeld, H., Gerber, P., Wassenaar, T., Castel, V., Rosales, M., & de Haan, C. (2006). *Livestock's long shadow*. FAO.

Sunuwar, D. R., Singh, D. R., & Pradhan, P. M. S. (2020). Prevalence and factors associated with double and triple burden of malnutrition among mothers and children in Nepal: Evidence from 2016 Nepal demographic and health survey. *BioMed Central Public Health, 20*(1), 1–11.

Tilman, D., Clark, M., Williams, D. R., Kimmel, K., Polasky, S., & Packer, C. (2017). Future threats to biodiversity and pathways to their prevention. *Nature, 546*(7656), 73–81.

Willett, W., Rockström, J., Loken, B., Springmann, M., Lang, T., Vermeulen, S., Garnett, T., Tilman, D., DeClerck, F., Wood, A., Jonell, M., Clark, M., Gordon, L. J., Fanzo, J., Hawkes, C., Zurayk, R., Rivera, J. A., Vries, W. D., Sibanda, L. M., Afshin, A., Chaudhary, A., Herrero, M., Agustina, R., Branca, F., Lartey, A., Fan, S., Crona, B., Fox, E., Bignet, V., Troell, M., Lindahl, T., Singh, S., Cornell, S. E., Reddy, K. S., Narain, S., Nishtar, S., & Murray, C. J. L. (2019). Food in the Anthropocene: The EAT–lancet commission on healthy diets from sustainable food systems. *The Lancet, 393*(10170), 447–492.

PART 7

Economics and trade

36

LIVING WAGE AND LIVING INCOME FOR SUSTAINABLE DIETS

Azfar Khan, Richard Anker, and Martha Anker

Background

Following on from the definition of sustainable development,[1] sustainable diets could also be characterised as food regimes that are easily reproducible by individuals and households over successive periods, without adversely impacting the environment.[2] There is, as well, a moral argument about the "right to nutritious food" and elimination of hunger. Thus, it would seem that a goal of sustainable diets to avert hunger and food insecurity, defined as the lack of access to enough nutrition for an active and healthy life. For sustainability in the 21st century, this access means more than merely attaining the basics of a sufficient number of calories that underscore World Bank-inspired poverty lines. For diets to be sustainable, they must be affordable. It can be helpful for nutritionists to provide important public information and specify the amounts and types of foods that people should eat, but this information is insufficient if people cannot afford to purchase what is recommended or do not have access to a wide enough variety of food.[3] To be sustainable diets they must meet the needs of individuals, families, and communities, to access nutritious, palatable, affordable foods in environmentally friendly ways.

In this regard, sustainable diets would mirror the model diets specified in Anker methodology living wage and living income studies, which have been carried out globally under the aegis of the Global Living Wage Coalition (GLWC) and the Anker Research Institute.[4] This model diet is based on the following principles:

1. It should be nutritious beyond caloric requirements (the only criterion for most national and World Bank international poverty lines), have sufficient macronutrients (i.e., proteins, fats, and carbohydrates) and micronutrients (proxied for by sufficient amounts of fruits and vegetables), and limited quantities of sugar and oil.

DOI: 10.4324/9781003174417-43

2. It should include foods from 13 food groups and no soft drinks, snacks, cakes, and confectionaries.

3. It should be relatively low cost for a nutritious diet and generally not include many prepared foods besides prepared cereals, thereby being more affordable and consistent with the cost-conscious shopping habits of typical workers.

4. It should accord with the country's level of development, as people purchase more expensive foods as countries develop and incomes increase.

5. It should be consistent with local food preferences that are intertwined with local culture to be considered palatable.

6. It should be consistent with local food availability to include only food items that are widely available

7. It should allow for additional variety and normal food loss in storing, preparing, and cooking food (Anker & Anker, 2017).

Diets constructed to be nutritious in this way are not only relatively inexpensive and consistent with local food availability but also tend to be earth friendly as they favour locally produced food and include only limited amounts of animal-based products and prepared foods. First, local food preferences and local cooking habits normally rely on locally grown ingredients. Second, imported foods tend to be more expensive than foods grown locally and therefore would not be included in a low-cost model diet if there is a local alternative. In addition, animal products are limited in Anker methodology model diets because they tend to be expensive relative to their nutritional value. Although Anker model diets do contain animal products—milk for children and some dairy, egg, meat, poultry, or fish (depending on relative prices and food customs)—their quantities are limited.

A key feature of any nutritious—and palatable—diet is that it needs to contain a variety of foods. For this reason, Anker living wage methodology adds 10–15% extra to the cost of the model diet for variety. It also adds additional costs for spices, salt, and condiments, and for normal discard and waste. This addition means that a certain amount of cash is needed for families to be able to afford a variety of foods, since it is rare for farmers to produce all that is needed for a nutritious and varied diet.

Diets must be affordable to be sustainable and food is not the only need

There is no absolute scarcity of food in the world: The Food and Agriculture Organisation (FAO) estimated at the World Food Summit in 1996 that world food production was enough to provide every citizen of the world with 2,700 calories per day (Kakwani and Son, 2015).[5] What creates scarcity of food is how it gets distributed and shared: This is the basis of starvation.

There is an interplay of what is a "proper" diet, what food is available, and what can be accessed by various groups in society.[6] These concerns around availability and accessibility emphasise the mutually reinforcing relationship between hunger and poverty that occurs in a market economy because of the inability of some individuals and households to acquire enough food for subsistence, let alone consume a sustainable diet. Consequently, macro indicators of food production and availability are not sufficient in determining what different social and economic groups can actually procure. As Sen convincingly argues, individual hunger can increase at a given level of total food availability if market-based entitlements change due to relative price changes (Sen, 1981).

There is now a general consensus among academics and healthcare professionals that management of hunger, besides considering nutritional adequacy of food supply, must also consider the means of acquiring it. Realising a sustainable healthy diet for all must also take into account the economic status and income generation capacity of various social groups in society. Of course, as Sen emphasised in his *entitlement* approach, there is no technical reason for markets to meet subsistence needs and no moral or legal reason why they should.[7] Under these circumstances, some people go hungry because their economic position does not allow them to access adequate subsistence and a nutritional sustainable diet. As a result of not getting enough to eat, they are less productive and because they are less productive, they remain poor. The cycle reinforces itself (See Dasgupta & Ray, 1990, pp. 191–246; Banerjee & Duflo, 2012, p. 22; Ghosh & Bharadwaj, 1992, p. 146).[8]

It is important to keep in mind that access to food is not the sole concern for households as people also have other needs and desires such as need for healthy housing, adequate health care, children's education, clothing, leisure, transportation, amongst others.[9] These are all critical elements in building up a "decent" existence and they exercise variegated impacts on specific groups of working people and their households. This requires at least a living wage for wage workers and a living income for farmers and the self-employed.[10][11]

Need to consider problems faced by different groups in society in attaining a sustainable diet: A labour status approach

Any examination of diverse societies around the world demonstrates a fundamental inequality between them. Some are better able to satisfy their needs of life and live a "decent" life, while others falter in making ends meet. In market economies, discerning who succeeds and who fails essentially boils down to available opportunities and the income generating capacity of differentiated social groups, where *income is to be understood as a satisfier and an enabler* of the fundamental "need of permanence," or subsistence.

Much of income generation depends upon having remunerative employment in the broader sense of the word. In the developed world, employment means "paid work." The vast majority of those in "paid employment" work for wages and those who are self-employed take in a salary from their businesses. In the Global South, however, the employee "work-for-wages" relationship is not as common or straightforward. A striking feature in these countries is the wide diversity of economic activity and the complex livelihood strategies of families to meet subsistence. "These differences of livelihood," as an eminent researcher in the field suggested, "are related to the nature and organisation of production and exchange and are the basis of social class distinctions" (Crow, 1992, p. 19).

The demarcation of social groups needs to be understood if we, what is meant by "we" here is the concerned readership, academic and researchers, policy makers and the general public, are to find attainable paths to sustainable diets in the Global South, as an injudicious simple analysis would reduce the complex realities of existence and work to popular misconceptions. Determinants of different economic positions are conditioned by multifaceted social and economic relationships that characterise the work process in market economies. For this reason, when we talk of poverty and inability to afford a sustainable diet, it is wrong to look at the poor as one "mass" of helpless victims as implied by typical "poverty line" statistics and poverty rates. Chambers and others encourage us to look beyond this "dehumanising" outlook and connect the (differentiated) *who* with the (differential) *why* of their poverty (Chambers, 1988; Bernstein, 1992).

It is beyond the scope of the chapter to discuss the entire range of work that has been undertaken, but important work and worker categories are worth highlighting. In a typical developing country, a typology would include different worker categories in rural and urban areas. The types of difficulties each of these groups encounters in being able to attain a healthy and sustainable nutritious diet are discussed below.

Rural Areas in the Global South

The work of the majority of rural dwellers in countries of the Global South is characterised by labouring for an income that often yields an income that is insufficient for meeting basic needs, including a sustainable nutritious diet. Furthermore, many remunerative activities in which rural workers engage are characterised by seasonal variations in work, and hence fluctuations in income. This seasonal variation suggests periodicity in the well-being of rural workers and their households, which undermines their ability to attain sustainable nutritious diets over the entire year.

Another factor that undermines the income earnings of rural workers is the environmental and geographical niche they occupy. Living in areas that are essentially dependent on rainfall, and without proper irrigation facilities, adversely impacts what the land makes available. Living in rural areas with infrastructure that is poorly serviced leads to higher costs related to transport for employment, and for purchasing food and other necessities. These concerns further weaken the ability of vulnerable rural groups to generate adequate income.

Seasonality often characterises life in the rural areas of the Global South. Seasonality may necessitate periods of weather differences—like no rain or excessive rain—that rural residents may have to cope with. In peak production periods, if the harvest is robust, obtaining a nutritious diet may become more probable. But harvests fluctuate and a poor harvest may lead to lower incomes and food shortages, which would suggest that poor households may not be able to adequately feed themselves from one harvest to another. In such cases, wage employment, particularly for poor households, becomes important in supplementing income. But finding employment cannot be taken for granted, particularly in times of a downturn in the economy or when the seasonal influences of activities like tourism are more dormant; these are the times when supplemental income is most needed. Thus, if poor rural dwellers cannot find a job to augment their incomes, acquiring a nutritious diet cannot be assured. In other words, fluctuations in income during the year suggest difficulties in procuring sustainable diets.[12]

Land owners/farmers

Land owners/farmers are found in both the Global North and South. A dominant practice in the Global North has involved agricultural operations being consolidated into large farms; these are significantly developed businesses, well beyond the family farm of past eras. In the latter, large landholdings command considerable tracts of cultivable land. Farmers, both in the Global North and South, are habitually linked to global supply chains through multinational corporations. Large farms often attain sufficient income for achieving a sustainable nutritious diet and a decent standard of living.

Landless rural workers

Landless rural workers, mainly found in the Global South, make their living by selling their labour for a wage. Their remuneration can be in cash payments or in-kind. The majority of such workers work in agriculture and are especially needed in times of harvests and for seeding/weeding activities. Given that they cannot generate sufficient income from only their agricultural pursuits, many take on supplementary employment such as selling their labour to contractors for work outside agriculture. Their work normally takes place outside formal contractual relationships and, therefore, they are often locked into a set of coercive obligations. Many migrate to other rural areas and cities, where the probability of finding work and generating cash resources is better. Their incomes are unstable, dependent upon availability of work, and usually quite meagre. Many satisfy their subsistence needs through social networks—social relations of reciprocity—provide some security amidst the flexibility and informality inherent in their existence.[13] They encounter frequent episodes of not being able to fulfil their subsistence and live a "hand-to-mouth" existence, often taking on debt and its obligations.

Sharecroppers/tenant farmers

Sharecroppers/tenant farmers do not own land but rent it from large landowners mainly for farming and livestock rearing. The rental payment is usually in terms of sharing produce; usually half, but arrangements normally favour the landlord. This group does marginally better than landless labourers, but being tied to the land usually puts them at the mercy of the landlords who can exercise control over their labour power through the threat of withdrawal of lease(s). Quite like the landless workers, they also take on supplemental employment, and debt, given the exigent need for raising cash revenues for household emergencies. These conditions mean that affording and consuming sustainable diets throughout the year is difficult.

Subsistence farmers/peasants

Subsistence farmers/peasants are workers or households that own land. A distinction can be made between "poor" and "middle" peasants based on the amount of land owned (Shanin, 1973). Their economic position may be characterised as better than that of the landless workers, sharecroppers, and tenant farmers, as they are "relatively better able to satisfy their basic needs, although it varies over seasons and economic cycles. This variation suggests significant fluctuations in procurement of the necessities of life and an economically insecure existence.

Urban Areas in the Global South

In urban areas in the Global North, working for a salary is the norm, with wage and workplace protection provided under the law. In contrast, in the Global South there are numerous forms of self-employment, as well as work paying wages under diverse conditions of employment, where no regulatory frameworks apply. Thus, wage work, though similar in principle, may involve numerous different practices. The position of a wage worker working under a standard employment relationship with regularity of payment and incumbent social protection under the labour law is drastically different from that of a wage worker employed in the informal economy with no such protection and with erratic remuneration.

Standard wage employment

This type of work is characteristic of the Global North and growing in importance in the Global South with economic progress and industrialisation. Workers in this category in the Global South are typically employed in the public sector and in larger and formal sector private firms and in high-tech services including, high, middle and low-ranking government officers, clerks, accountants, bankers, brokers, university and school teachers, IT specialists, programmers, amongst others. Nevertheless, the situation varies between countries depending upon the development profiles of specific countries. In general, apart from China, this form of work employs less than 15% of the working age population in the Global South.

Workers employed under a standard wage contract enjoy the protection of the labour law. However, given certain protective measures that may apply under formalised arrangements—such as minimum wage and other benefits—those with regulated work contracts are often economically secure and are more than able to fulfil their food and nutritional needs. This ability is not always the case however, especially in countries where the minimum wage is set at a low level. Furthermore, in recent years, with labour flexibilisation and the growing use of external labour (i.e., contract workers, outworkers, homeworkers, agency labour, etc.) that is not supported by legal social protection statutes, the recourse to such labour use compromises the economic security of workers in formal enterprises as they undermine social protection.

Informal economy workers

The informal economy includes a spectrum of work ranging from proto-industrial production processes that can easily be accommodated through subcontracting/outsourcing arrangements to the more ubiquitous activities that, among others, include hawking, peddling, and domestic work. The one significant feature of work in the informal economy is that employment is not covered by regulatory and social protection measures. Informal work, by its very nature, implies insecurity for workers, who are excluded from publicly sanctioned social support and their meagre earnings are far removed from a living wage and living income.[14] Many wage earnings are governed by wage system flexibility, such as time-rate, piece rate, greater use of bonuses, etc.; moving in and out of jobs and shifting employment is a commonplace occurrence. The high degree of labour mobility suggests considerable variations of income over specific periods. Moreover, most informal economy workers are more exposed to economic shocks, particularly inflationary conditions. When money wages do not increase, a rising trend in prices of essential commodities (especially food grains) adversely impacts and exposes the workers and their families to many hardships that exercise a damaging impact on their ability to eke out a sustainable existence. Only a small minority of workers in this category generate earnings close to the living wage over successive time periods (See, for example, Agarwal, 1990; Banerjee & Duflo, 2012; Dreze & Sen, 1990; Ghosh & Bharadwaj, 1992; ILO, 2004; Standing, 2000).

Implications for sustainable diets and decent standard of living

The synoptical analysis of labour activities in the Global South presented above highlighted the wide range of work and labour statuses in diverse workplaces and myriad means by which people manage their livelihoods. This discussion stressed the importance of understanding how people with different labour statuses come by incomes, the volatility and perhaps seasonality of incomes, and whether incomes are sustainable and adequate in enabling a "sustainable diet" and a "decent" standard of living.

When earnings are not sufficient, consumption patterns change and "expenditure switching" between different needs can result in extreme compromises that workers and their households have to make. Income shortfalls may imply that groups with low income cannot afford to: (a) Buy foods which are more expensive per calorie such as fruits and vegetables, dairy, eggs, and meats/fish; (b) acquire a wide enough variety of foods, especially fruits and vegetables; (c) obtain reasonable quality foods that are palatable and socially acceptable; (d) purchase foods in larger economical quantities; (e) effectively store foods and avoid waste of foods going bad, moulding, or being eaten by rodents or infected by insects; (f) access proper education for children that enhances skills development; (g) access proper healthcare and avoid negative impacts on economic capabilities; (h) adequately meet emergencies and avoid debt burden; (i) meet other necessary expenditures, such as transport and clothing; and, (j) secure an acceptable quality of healthy housing. Thus, insecure and fluctuating income not only suggests periods of deprivation and not being able to attain a sustainable diet with long lasting health effects, but also retards human capital development and restricts opportunities for workers and their families to break out of the poverty cycle.

For the majority of workers in the Global South, livelihoods are precarious and unsustainable simply because their earnings do not constitute a living income that would allow them a "decent" existence, including not being able to afford a sustainable nutritious diet year-in and year-out. In challenging times, they have to make difficult decisions and choices that adversely influence not just their food intake but their general standard of living. Apart from food security there is also the need to raise cash revenues for emergencies and securing healthcare, medicines and farm inputs, which are all necessary for sustaining a decent life.

Fostering capabilities of sustainable nutritious diets for all

All human beings have dignity, deserve respect, and are entitled to the goods necessary to satisfy their basic needs, including the right to a sustainable diet. This right to satisfy basic needs takes precedence over all other rights. Commensurately, the objective of a just society should be to ensure that everyone has basic security that ensures access to nutritious food, healthy housing, adequate health services, a decent level of education, and other basic needs of human existence.

Enabling sustainable diets and a decent standard of living are related to concerns around poverty, which in a very basic sense is a situation where people have insufficient income to provide for a minimum standard of living.[15] However, another way to look at poverty is in terms of *capabilities* rather than command over goods. Drèze and Sen advance the notion of "capabilities" as an enabling factor, arguing that to avert poverty "a…reasoned goal would be to make it possible for all to have the capability to avoid undernourishment and escape deprivations."[16] Hence, a basic freedom from poverty may be understood in terms of *what people can do* and *what they can be*. This "freedom," as some have suggested, "means independence from the arbitrary will of another…and this independence always necessitates possession of sufficient material resources to assure one's social existence" (Raventos & Wark, 2018, p. 128).

Living wage for wage workers and living income for farmers and self-employed may then be seen as an enabling element of "capabilities," which could be advocated and implemented given the political will. Of course, it is difficult to implement this for forms of work that fall outside institutional regulations—such as domestic work, informal work, seasonal work, and casual work. In developing countries, these dominant forms of work mean that many individuals in different labour status are subject to periodic reductions in food intake and other entitlements in the face of fluctuations in economic and personal conditions, such as changes in market prices, health related emergencies, loss of a job, low level of earnings in general and particularly in recessionary periods, which leave significant shortfalls in procuring sufficient income.

Thus, there is an argument for provision of food and other basic needs in public policy (Kanbur, 1990). However, though well-intentioned, such social assistance is usually structured on "means-tested" and "selectivity" approaches, which are not very effective in reaching target populations and have very high administrative costs.[17] This is why a "universal" approach, such as Universal Basic Income (UBI) is often advocated, which includes enabling the attainment of a sustainable diet. It could be considered, along with living wage and living income as another enabler of capabilities. The disbursement of UBI to all citizens[18] would work best with policy measures ensuring the implementation of living wages, living incomes, and social investments in education and health.

An enormous amount of money is being spent on poverty-reduction schemes, and yet there are probably more people in poverty and economic insecurity than at any time in human history. Indeed, it may be argued that poor and disenfranchised people can only overcome their impecuniousness and insecure status and attain a sustainable decent standard of living if their basic rights are fulfilled. These rights include access to nutritious sustainable food, proper education, well remunerated income earning opportunities, provision of suitable health services, and equal rights.

Notes

1 The United Nations defines "Sustainable development" as "development that meets the needs of the present without compromising the ability of future generations to meet their own needs."

2 Sustainable diets are defined by the Food and Agriculture Organisation (FAO) as "those **diets** with low environmental impacts that contribute to food and nutrition security and to healthy life for present and future generations."

3 A recent publication (FAO et al, 2021) indicates that more than one in three people (42%) around the world do not have access to an *adequate* diet.

4 The Anker methodology has been used in over 40 countries to estimate living wages of workers based in part on the cost of a low-cost nutritious diet.

5 See FAO (1996). FAO (2020) data suggest that the global average dietary energy supply (DES) is around 2,870 kcal per person per day with intake varying over regions of the world.

6 Drèze and Sen (1993, p. 14) explain this by distinguishing between *nutrition* and *nourishment*, where the former relates to adequacy of "food intake" and the latter to the "state of human being."

7 Sen (1981) reduces food entitlements to four categories: "production-based entitlement" (growing food), "trade-based entitlement" (buying food), "own-labour entitlement" (working for food), and "inheritance and transfer entitlement" (being given food by others).

8 The discussion of "poverty trap" linking hunger with ability to work is a recurring theme in poverty analyses.

9 Article 25 of the Universal Declaration on Human Rights recognises the fundamental right to nutritious food along with health, clothing, housing, medical care and necessary social services as fundamental basic needs.

10 The Global Living Wage Coalition defines living wage as: "the remuneration received for a standard workweek by a worker in a particular place sufficient to afford a decent standard of living. Elements of a decent standard of living include food, water, housing, education, health care, transportation, clothing, and other essential needs including provision for unexpected events." www.globallivingwage.org/.

11 The Living Income Community of Practice has a similar definition of a decent standard of living for smallholder farmers.

12 The poor have various coping strategies. Agarwal (1990) notes five ways of coping: (i) Drawing upon assets; (ii) drawing upon household stores of food, livestock, and other items; (iii) reaching out to social networks; (iv) relying on communal resources; and (v) migrating and taking on employment for diversifying income.

13 A good discourse on "reciprocities" is provided by James Scott (1976).

14 For an interesting discussion on how working conditions and working time are manipulated by employers to the detriment of workers, see Standing (2000).

15 Max-Neef (1986) argues that "one should speak not of poverty, but of poverties...any fundamental need that is not satisfied reveals a poverty: poverty of subsistence is due to insufficient income, food, shelter, etc."

16 See Dréze and Sen (1993), p. 13. They illustrate the point stating that: "If a person does not have the capability of avoiding preventable mortality, unnecessary morbidity, or escapable undernourishment, then it would certainly be agreed that the person is deprived in a significant way."

17 ILO (2004): Chapter 14 "Seeking Income Security" discusses pros and cons of targeting and selectivity approaches versus "universal" provisioning.

18 The issue is extensively discussed in G. Standing (2017).

References

Agarwal, B. (1990). Social security and the family in rural India coping with seasonality and calamity. *Journal of Peasant Studies, 17*(3), 341–412.

Anker, R., & Anker, M. (2017). *Living wages around the world; Manual for measurement*, Edward Elgar Publishing.

Banerjee, A., & Duflo, E. (2012). *Poor economics*. Penguin Books.

Bernstein, H. (1992). Poverty and the poor. In H. Bernstein, B. Crow, & H. Johnson. (Eds.), *Rural livelihoods: Crises and responses* (pp. 13–26). Oxford University Press.

Chambers, R. (1988). *Poverty in India: concepts, research and reality* (Discussion Paper 241). Institute of Development Studies. Retrieved from https://opendocs.ids.ac.uk/opendocs/bitstream/handle/20.500.12413/212/rc269.pdf?sequence=2&isAllowed=y

Crow, B. (1992). Understanding famine and hunger. In T. Allen & A. Thomas (Eds.), *Poverty and development into the 21st century* (pp.15–33). Oxford University Press.

Dasgupta, P., & Ray, D. (1990). Adapting to undernourishment: The biological evidence and its implications. In J. Drèze & A. Sen (Eds.), *The political economy of hunger* (Vol. 1, pp. 191–246). Oxford University Press.

Drèze, J., & Sen, A. (1993). *Hunger and public action*. Oxford University Press.

Drèze, J., & Sen, A. (Eds.) (1990). *The political economy of hunger* (Vol. 1). Oxford University Press.

FAO (1996). *Food for all: World food summit*. FAO. Retrieved from https://www.fao.org/3/x0262e/x0262e00.htm

FAO (2020). *Statistical yearbook 2020*. FAO. Retrieved from https://www.fao.org/3/cb1329en/CB1329EN.pdf

FAO, IFAD, UNICEF, WFP, & WHO (2021). *The state of food security and nutrition in the world 2021: Transforming food systems for food security, improved nutrition and affordable healthy diets for all*. FAO. Retrieved from https://www.fao.org/3/cb4474en/cb4474en.pdf

Ghosh, J., & Bharadwaj, K. (1992). Poverty and employment in India. In H. Bernstein, B. Crow, & H. Johnson, (Eds.), *Rural livelihoods: Crises and responses* (pp. 139–164). Oxford University Press.

ILO (2004). *Economic security for a better world*. ILO. Retrieved from https://www.ilo.org/public/english/protection/ses/info/publ/economic_security.htm

Kanbur, R. (1990). Global food balances and individual hunger: Three themes in an entitlements-based approach. In J. Drèze & A. Sen (Eds.), *The political economy of hunger* (Vol 1., pp. 53–78). Oxford University Press.

Max-Neef, M. (1986). Human-scale economics: The challenges ahead. In P. Ekins (Ed.), *The living economy: A new economics in the making* (pp. 45–54). Routledge and Kegan Paul.

Raventos, D., & Wark, J. (2018). *Against charity*. Counter Punch-AK Press.

Scott, J. (1976). *The moral economy of the peasant*. Yale University Press.

Sen, A. (1981). *Poverty and famines: An essay on entitlements and deprivation*. Clarendon Press.

Shanin, T. (1973). The nature and logic of the peasant economy. *The Journal of Peasant Studies*, *1*(1). (pp. 63–80).

Standing, G. (2017). *Basic income: And how we can make it happen*. Pelican Books.

Standing, G. (2000). *Modes of control: A labour status approach to decent work*. ILO. Retrieved from https://www.ilo.org/public/english/protection/ses/download/docs/modes_of_control.pdf

37

FINANCIALISATION AND SUSTAINABLE DIETS

Phoebe Stephens, Jennifer Clapp, and Ryan Isakson

Introduction

Financial markets, financial actors, and financial motives shape food systems in important ways, and are enmeshed across all stages of food supply chains, though they are not always discernible to the average consumer. The recent development of farmland Real Estate Investment Trusts (REITS), wherein financial enterprises acquire parcels of farmland and pay investors dividends based on the associated rental and mortgage earnings, is one example of how the food system has become increasingly tethered to the world of finance. Another example is ACRE Africa, a micro-insurance enterprise that markets weather-based derivatives to small-scale farmers in the name of "climate proofing" their agricultural operations. At the consumer level, "buy now pay later" apps are expanding into grocery retail— where users pay a fraction of an item's purchase price at the till and spread out the remaining payments over several weeks. Though these examples may appear unrelated, they each demonstrate the growing influence of financial markets and players within food systems. They are part of the broader process of "financialisation," a concept that refers to the growing share of finance in the global economy, including food production, processing, and retail in ways that influence the sustainability of our diets.

The concept of financialisation gained attention at the turn of the 21st century. In arguably the most cited definition, Gerald Epstein (2005), defines financialisation as "the increasing importance of financial markets, financial motives, financial institutions, and financial elites in the operation of the economy and its governing institutions, both at the national and international levels" (p. 3). After the 2007–2008 food crisis, during which dramatic spikes in world food prices coincided with rising financial speculation in agricultural commodities, food scholars began to explore the relationship between financialisation and sustainability outcomes in the food system (Burch & Lawrence, 2013; Clapp, 2009; Clapp & Isakson, 2018; Ghosh, 2010; Isakson, 2014). Scholars identified ways in which financial

DOI: 10.4324/9781003174417-44

investments by large-scale institutional investors—such as pension funds, sovereign wealth funds, and hedge funds—were tied not just to food price volatility but also large-scale investments in farmland. These investments were widely criticised for harming farmer livelihoods and exacerbating food insecurity (Fairbairn, 2020; Sommerville & Magnan, 2015). The growing power and concentration of agrifood corporations is also, in part, fuelled by the growing power of financial actors. Shareholders often pressure corporations to merge or acquire new businesses to increase profits (Clapp, 2019). These strategies concentrate corporate control and increase big food and agri-businesses' leverage over producers and consumers, giving them power to dictate how and what food is produced (Howard, 2016).

As research on financialisation in food economies has deepened, researchers have further teased out the complex relationship between food and finance to identify ways to improve sustainability outcomes (Stephens, 2021a). At its core, the literature on financialisation in the food system highlights the winners and losers so that those seeking to improve outcomes in our food systems can identify leverage points and advance more sustainable solutions. This chapter contributes to this literature by examining how financialisation of food and food systems impact the pace or path of sustainable diets. We demonstrate that pressures to maximise financial returns, coupled with the introduction of novel financial instruments that channel investment into food systems, encourage expanding the scale and deepening the industrialisation of food production while externalising costs. We also discuss how greater financialisation has further marginalised smaller scale sustainable production and distribution systems. Finally, by teasing out the dynamics of financialisation, we emphasise the need for public policies and collective action to ensure more sustainable diets, rather than solely viewing them as an individual responsibility.

In exploring these dynamics, this chapter draws on the list of characteristics and attributes of sustainable diets introduced in chapter one. As we delineate the ways in which financialisation ripples through the food system, it becomes clear that all five characteristics of sustainable diets are affected. By the end of this chapter, readers will have a greater grasp of how the rather abstract concept of financialisation touches on every link in the food supply chain with profound consequences on the sustainability of our diets.

Financialisation and sustainable diets

Often understood as a product of individual choice and responsibility, sustainable diets are being recast as a broader systemic issue—one that considers how diets are shaped by the industrialisation of food systems and the challenges that it poses for social and ecological sustainability (Lang, 2021). Because many of the financial instruments and incentives tied to food system activities have a tendency to encourage industrial practices, financialisation is a relevant focal point for those seeking to understand the forces contributing to unsustainable diets. Likewise, policies that address financialisation in the food system can work to promote more sustainable diets. This conceptual broadening of sustainable diets to include forces that influence a range of activities across food systems allows for linkages to be made between larger structural forces like financialisation and the multi-factor criteria of sustainable diets, as identified in the following:

> The FAO defined sustainable diets as diets with low environmental impacts which contribute to food and nutrition security and to healthy life for present and future generations. Sustainable diets are protective and respectful of biodiversity and ecosystems, culturally acceptable, accessible, economically fair and affordable; nutritionally adequate, safe and healthy; while optimising natural and human resources.
>
> *(FAO, 2012, para 3)*

To begin to tease out some of the ways in which financialisation affects the sustainability of diets, it is instructive to follow the typical categorisation found in the scholarly literature on the ways in which financialisation reshapes economies in general. Specifically, van der Zwan (2014) identifies three general conceptualisations of financialisation, which Clapp and Isakson (2018) have applied in their analysis of financialisation in food systems: (1) The accumulation of capital in the agri-food sector is increasingly tied to financial activities rather than the actual production and trade of food; (2) the everyday activities of food provisioning are increasingly mediated by financial logics and products; and (3) the growing prioritisation of shareholder value by firms in the food system. Conceptualising financialisation in this way can be helpful in terms of grasping the breadth, depth, and complexity of financialisation in the food system. Below we elaborate on these categories and provide concrete examples of the ways that food systems have been shaped by these various aspects of financialisation through the lens of sustainable diets.

Capital accumulation via financial investment in food system activities

The food and agriculture sectors have become a growing target for capital accumulation in the last two decades, as part of a broader trend of rapid growth in financial investment since the 1980s (Greenwood & Scharfstein, 2013; Philippon, 2013). The growth of the financial sector means that corporate profits are increasingly generated through financial channels rather than through the productive activities of the trade and manufacturing sectors. Financial investment in the food sector grew sharply in the context of the 2007–2008 food and financial crises, as the agriculture sector was deemed by many investors as a "safe haven" for investment in an otherwise unpredictable economic climate. Rising and volatile world food prices at that time presented opportunities for speculative investment. Meanwhile, narratives of food scarcity drove up land and food prices further, signalling a potential for significant capital accumulation. These trends have worked to undermine the sustainability of diets.

Agricultural Commodity Markets

There is a long history between finance and agriculture, dating back to the origins of commodity futures markets in the 1600s. However, the nature of that relationship has changed considerably over time, trending in the direction of greater speculative activity. Neoliberal restructuring in the 1980s and 1990s, which included the loosening of regulations on participation in derivatives markets, allowed for the development of new kinds of financial investment vehicles, such as commodity index funds, that allow investors to bet on the direction of an index of commodity prices without having to own physical commodities. Demand for these types of financial investment products increased sharply in the early 2000s, creating upward pressure on commodity prices. The risks of allowing more speculative investors to participate in agricultural commodity markets were realised in 2007–2008 and 2011–2012 when world food prices rose sharply and compromised people's access to food, particularly those who spend a large proportion of their income on food (Clapp et al., 2021).

Though financialisation was one driver of the food crisis, it was not the only one. Other factors like poor wheat harvests at the time and greater demand for biofuels also contributed to rising prices (Jarosz, 2009). The dominant narrative tended towards productionism, a philosophy that takes production as the "sole norm for ethically evaluating agriculture" (Thompson, 2005). Thus, rather than reining in speculative financial investment in agricultural commodity markets in the wake of the crisis, governments sought to remedy the situation by increasing agricultural production and supply in a bid to put downward pressure on food prices (Clapp et al., 2021). They did so by encouraging greater private sector investment in agricultural production, which even further tightened the link between financial actors and food systems. The private sector responded by channelling investments into speculative financial activities like agricultural commodity markets and farmland, contributing to greater financialisation in the food system. Indeed, some argue that "the 2007–09 food crisis can be seen as a key moment when financialisation became more deeply entrenched in the food system" (Clapp et al., 2021, p. 275). This entrenchment has created a more vulnerable global food system, one that is characterised by greater corporate control and an undermining of states' role in safeguarding food security, as became apparent with the disrupted food supply chains during the COVID-19 pandemic (Clapp et al., 2021).

Unstable and rising food prices present obvious challenges for food security. Particularly for poor consumers, an increase in food prices can have devastating impacts on a family's ability to feed itself. In this context, financialisation compromises the sustainable diets requirement of being economically accessible.

Farmland Acquisitions

The 2007–2008 spike in world food prices helped to give rise to the so-called "farmland rush" that began capturing headlines in 2008. Hoping to capitalise on the perceived scarcity of food while also seeking refuge from falling asset values elsewhere in the global economy, financial actors of various stripes developed and invested in a growing array of farmland-based investment vehicles.

Most investments in farmland have been concentrated in the "developed" countries of North America, Australia, and New Zealand. Yet investors have also channelled significant funds into Brazilian farmland and other so-called "transition" countries in South America, Africa, and Eastern Europe (Cotula, 2012). These investments have been facilitated by the creation of a variety of novel investment vehicles in recent years, including private equity funds that specialise in farm-land investment, publicly listed farming ventures, farmland-focused REITs, and "crowdsourced" fractional investment platforms (Fairbairn, 2020).

The emergence of these new investment vehicles and the flow of funds into them have been facilitated by regulatory changes over the past two decades that have loosened restrictions on farmland transactions and ownership. At the same time, financiers' ability to access land is often predicated upon the hardships faced by contemporary farmers. Confronted with rising input prices, increasingly adverse weather, unpredictable commodity prices, the erosion of social benefits, and growing debt burdens, studies suggest that the farmers who transfer their land to investors are generally forced to do so by adverse socio-economic conditions (Clapp & Isakson, 2018; Ouma, 2020; Sippel et al., 2017; Sommerville & Magnan, 2015). Meanwhile, the influx of deep-pocketed investors has driven up farmland prices in many markets, making it prohibitively expensive for existing producers to expand their operations and pricing the aspiring-farmers, particularly those inter-ested in labour-intensive agroecological practices, out of the market. (Desmarais et al., 2017; Magnan & Sunley, 2017; Rotz et al., 2019).

The recent flood of investment in newly developed farmland-based financial products has changed how agricultural producers relate to the land, heightening pressure to maximise returns. In order to squeeze returns sufficient to satisfy fickle investors out of higher-priced farmland, many investment funds have pushed for the intensification of industrial agricultural practices, which are harmful to life enhancing ecosystems. In their rush to generate immediate returns in Eastern Europe and South America, for instance, publicly listed farmland companies deployed cookie-cutter industrial practices that were poorly-suited for local envi-ronmental conditions (Kuns et al., 2016; Sosa Varrotti & Gras, 2021). In Canada, financial landlords prefer to rent their lands to the most capitalised farmers, whom they believe have the most potential to pay high rental fees. Not surprisingly, farmers on rented land have expanded cultivation into sensitive environments in order to maximise returns while opting for short-term fixes like the increased application of fertilisers over longer-term investments in soil health (Rotz et al., 2019). Growing financial control over farmland is also associated with less diversity on farms, especially in a limited number of globally traded "flex crops" like oil-seeds, corn, and sugarcane that can variously be used as animal feed, biofuel feed-stock, or as a handful of the so-called "ingredient crops" that make up the bulk of the highly processed foods on supermarket shelves (Borras et al., 2016; HighQuest, 2010; Rotz et al., 2019).

Financial priorities increasingly shape everyday agricultural and food provisioning activities

Individuals increasingly rely on financial tools and services to achieve the everyday activities of economic provisioning and are progressively motivated by financial incentives in their daily lives through mass marketed tools such as apps for credit and stock-trading. As states and employers scale back social programs and benefits, individuals are increasingly tasked with managing their own security through individual retirement savings and private insurance. This downloading of risk onto individuals is also prevalent in the food and agriculture sector as the neoliberal rollbacks of state support for agricultural producers exposed farmers throughout the world to new risks (Martin & Clapp, 2015). Since the early 2000s, there has been a proliferation of derivative products that purportedly empower farmers to manage the everyday risks of agricultural production (Clapp & Isakson, 2018). Likewise, as workers' wages have remained stagnant in many contexts, food retailers are increasingly providing financial products like credit cards and other banking services that facilitate food purchases. While such products may improve immediate access to food for cash-strapped consumers, the following examples demonstrate how they also contribute to growing levels of household debt and introduce risks that can affect people's ability to access healthy food over time.

Index-based agricultural insurance

Farmers face innumerable risks, some of the more obvious ones being price uncertainty and inclement weather. Recognising the unique value of the food and agriculture sectors to their populations, governments around the world have traditionally intervened in commodity markets to ensure more stable domestic prices and provide safeguards against severe weather events (Martin & Clapp, 2015). However, under neoliberal restructuring governments throughout the world have scaled back their involvement in the sector while market-based approaches have been rolled-out as the preferred approach to agricultural risk management (Clapp & Isakson, 2018). Index-based agricultural insurance (IBAI) is one such market-based tool. Even though it is called "insurance," IBAI is more appropriately understood as a derivative: Indemnity payments are not based on the actual losses that farmers suffer in their fields, but rather an index of environmental measures (e.g., quantity of rainfall) that are *correlated* with agricultural performance. In practice, the correlation between indices and agricultural outcomes can be quite weak, meaning that farmers still bear substantial risks. Nonetheless, IBAI products have been widely marketed to small-scale and poor farmers throughout the Global South for over a decade.

Proponents maintain that IBAI improves poor farmers' access to insurance since it eliminates the need for costly practices like field-level assessments. The lower cost of IBAI, however, comes with reduced security for farmers. An inherent risk of IBAI is that policyholders might suffer losses in their fields, yet weather measurements may not be sufficient to trigger an indemnity payment. Nonetheless, the promise of security that IBAI proponents promote may encourage farmers to take on additional risks that could have the paradoxical effect of exacerbating their vulnerability while simultaneously threatening sustainable diets (Isakson, 2015; Müller et al., 2017). Studies in Kenya (Sibiko & Qaim, 2020), Ethiopia (Vargas Hill & Viceisza, 2012), and Ghana (Karlan et al., 2014), for instance, have linked IBAI with the increased use of chemical fertilisers and other agrichemicals that can compromise soil and water quality. Moreover, a growing number of IBAI products encourage farmers to adopt commercial inputs by bundling insurance contracts with agricultural loans or the purchase of hybrid seed varieties that are less adaptive to environmental change (Mercer et al., 2012). While such changes may contribute to short-term increases in farmers' incomes and agricultural yields, they also increase dependence upon purchased inputs and compromise the functioning of agricultural ecosystems. Thus, the financialisation of the everyday practice of agricultural risk management may very well exacerbate the vulnerability of marginalised agricultural producers to economic and environmental shocks, thereby hindering progress on sustainable diets.

Grocery credit

Big grocery retailers are becoming some of the most powerful actors in food supply chains and have expanded into financial services, which is impacting the way people buy food as well as the types of food that they have access to (Burch & Lawrence, 2005). Large grocery retailers are increasingly offering banking services, mortgages, prepaid debit cards, and credit cards linked to loyalty programs. Retailers view these services as fulfilling a societal function as they benefit populations that are typically underserved by mainstream banks (Clapp & Isakson, 2018). However, the expansion of these programs creates a risk that customers will become dependent on these retailers not only for their food supply but also for access to credit. Grocery retailer credit and loyalty schemes also incentivise consumers to purchase their food exclusively from supermarkets, which tend to stock more processed and packaged foods than traditional markets and therefore influence the types of food consumers eat (Hawkes, 2008).

One of the more recent developments in financialisation of grocery retail is "buy now pay later" (BNPL) schemes. Unlike credit cards, there is no interest on borrowing for BNPL schemes, although they do charge late fees on late payments (Cooke, 2022). BNPL schemes initially emerged to provide credit to users for large, one-off expensive purchases. However, they have recently branched into the food sector. Zilch is one such company that offers BNPL plans, which explicitly encourages users to purchase highly processed foods such as Sainsbury's pizza (Cooke, 2022). This peculiar form of financialisation has the potential to encourage the consumption of unhealthy ultra-processed foods high in salt, sugar, and fat in addition to further fuelling consumer debt, which could threaten people's access to food over time.

The prioritisation of shareholder value

A third category of financialisation relates to shareholder value and pertains to the ways that financial pressures encourage companies that operate across food supply chains to prioritise decisions that will maximise profits for their shareholders. The shareholder revolution arose in the 1980s and tied firm performance to the compensation of executives through stock options. By the late 1990s, the shareholder value model of corporate governance became conventional wisdom and spread around the world, gaining prominence not only in North America but also Europe, Japan, and emerging economies (Blair, 2003, p. 56). As shareholders receive their dividends quarterly, this model tends to result in more short-term value creation rather than long term investments in innovation (Schmidt, 2016). As this section explains, the prioritisation of shareholder value has serious implications for sustainable diets by encouraging firms to undertake strategic decisions that undermine food system sustainability.

Ultra-processed foods

Financial pressure on food manufacturers has informed the quality of food made available to consumers (Clapp & Isakson, 2018). Firms are attempting to increase returns in multiple ways. One way to create "value-added" is through processing and ultra-processing food. Ultra-processed foods are defined as

> industrial formulations made entirely or mostly from substances extracted from foods (e.g. oils, fats, sugar, starch and proteins), derived from food constituents (e.g. hydrogenated fats and modified starch), or synthesised in laboratories from food substrates or other organic sources (e.g. flavour enhancers, colours, and several food additives used to make the product hyper-palatable).
>
> *(Baker & Friel, 2016; Monteiro et al., 2019)*

These foods tend to be high in fat, salt, and glycaemic load and are designed to promote consumption (Baker & Friel, 2016).

In his exposé *Salt, Sugar, Fat: How the Food Giants Hooked Us*, Michael Moss (2014) demonstrated how demanding shareholders have pushed for the production of unhealthy ultra-processed processed foods. He recounted, for instance, how, under pressure from health advocates and government regulators, Campbell had spent several years working to reduce the sodium content of its soups. Yet in 2011, amidst falling share values and complaints from stock analysts, the company hired a new CEO who assured shareholders that she would boost sales by increasing the salt content of their soups. Her announcement earned the praise of financial analysts and immediately boosted the company's share value. The moral of the story is that financial actors want consumers to remain "hooked" on ultra-processed foods that may generate higher returns to shareholders but are also associated with the growing incidence of obesity and diet-related non-communicable diseases around the world (Baker & Friel, 2016).

Supermarketisation

One of the primary strategies big agri-food organisations have taken to maximise shareholder value is via mergers and acquisitions, which have led to a smaller number of powerful corporations dominating the food value chain. As noted above, food retailers in particular have increased their prominence in recent years and are in a uniquely powerful position as they are positioned between consumers and producers (Howard, 2016). The huge size of many food retail firms provides them with considerable influence over the price and quality of food produced. Burch and Lawrence (2005) outline strategies that retailers employ to maximise shareholder value that also undermine the availability of sustainable diets: (1) They streamline sourcing by reducing the number of suppliers; (2) they cut back on labour and increase workloads; and (3) they scale back on their sustainability commitments.

Mergers and acquisitions in grocery retailing have taken place not just in industrialised countries but also in lower and middle income countries, spurring a "supermarket revolution" in the latter, where supermarkets are rapidly replacing more traditional open-air and smaller scale markets (Rischke, 2015). This supermarketisation is influencing diets in these regions in profound ways. Supermarkets tend to supply more highly processed foods, often in larger packaging sizes than what is available in traditional retail formats like wet markets and smaller shops. A growth in supermarketisation can marginalise smaller purveyors who often sell their products in open air markets, impacting livelihoods along the food supply chain.

Supermarketisation has been identified by analysts as a major force in the nutrition transition that is occurring around the world towards more energy-dense and processed foods (Baker & Friel, 2016). The financial pressures driving supermarketisation as a means to satisfy shareholders' desire for investment returns continues to spur mergers and acquisitions in the sector. In 2017, for example, the hedge fund Jana Partners acquired a majority share in the North American grocery chain Whole Foods and, in a successful bid to generate immediate returns on its shares, engineered Whole Foods' acquisition by Amazon. While the change in ownership reportedly netted Jana Partners some US $300 million in profits, analysts have reported a decline in the quality and service of the grocery chain, while it shifted away from local suppliers and focused more on national suppliers (Banker, 2019; Bonazzo, 2018; Thomas, 2017). The grocery retail market is highly concentrated in most countries. In Canada, for example, the five largest grocery chains command over 80% of the food retail market (CBAN, 2020). As the grocery retail sector becomes more concentrated in many countries, these firms have considerable power to shape the kinds of food environments that are available to consumers. By encouraging more mergers and acquisitions in the sector, financialisation is an important driver of detrimental dietary shifts worldwide.

Conclusion

This chapter has mapped out the key ways in which growing financialisation in food systems shapes the availability and accessibility of sustainable diets. The growing roles of financial actors, institutions, and motives in activities all across the food system have resulted in a number of specific dynamics that undermine sustainability. The development of new financial investment tools linked to the food system has increased the participation of speculative investors who are seeking to accumulate capital within the sector. The increase in this type of investment has been associated with higher and more volatile food prices that has placed healthy and sustainable foods out of the reach of many, as well as the global land rush that has displaced farmers and encouraged industrial agricultural production on a global scale. The rise of new kinds of insurance and credit schemes, for both farmers and consumers, by private actors at both the production and retail ends of the food system, have facilitated the entrenchment of financial incentives in daily life while also undermining sustainable food production and access to healthy and sustainable diets. Meanwhile, the prioritisation of shareholder value within large corporations that dominate food systems has encouraged both the growth of ultra-processed foods on offer as well as a global supermarketisation of food retail, both of which undermine the availability of healthy and sustainable food options.

This broader, structural perspective on the role of finance in food systems high-lights that there are limits to focusing on individual responsibility for choosing sustainable diets. A wider view shows that broader forces beyond an individual's control, such as financialisation within food systems, can undermine both the availability and affordability of healthy and sustainable foods. These dynamics are not easily visible for average consumers, which draws attention to the need for wider public policies to address the ways in which such dynamics can weaken public health and sustainability. Such policies might include improved regulations to address the types of financial pressures driving unsustainable production and consumption patterns, such as placing regulatory limits on financial speculation on agricultural commodities and farmland (Clapp & Isakson, 2018). Other measures, such as encouragement of more alternative financial investment initiatives like impact investments that promote more sustainable food systems, can also reduce the pressures of mainstream financialisation that contribute to unsustainable diets (Stephens, 2021a). Impact investments are those that seek financial as well as meas-urable social and environmental returns. Some impact investors are focusing on the food system in an effort to improve sustainability, which could support transitions towards more sustainable diets. However, at present their contributions remain incremental at best as they only represent a small fraction of investments in the food system and do not challenge the structures of the dominant neoliberal regime (Stephens, 2021b). There is a need for more research into how financial invest-ments can be reoriented to contribute to more sustainable diets.

References

Baker, P., & Friel, S. (2016). Food systems transformations, ultra-processed food markets and the nutrition transition in Asia. *Globalization and Health, 12*(1), 80.

Banker, S. (2019). *How amazon changed whole foods.* Forbes. https://www.forbes.com/sites/stevebanker/2019/06/25/how-amazon-changed-whole-foods/?sh=6c8f895878dd

Blair, M. M. (2003). Shareholder value, corporate governance and corporate performance: A post-Enron reassessment of the conventional wisdom. *Social Science Research Network Electronic Journal*, 53–82.

Bonazzo, J. 2018. *Whole foods execs, suppliers bolt over quality concerns after amazon acquisition.* Observer. https://observer.com/2018/03/whole-foods-quality-amazon-acquisition/

Borras Jr., S. M., Franco, J. C., Isakson, S. R., Levidow, L., & Vervest, P. (2016). The rise of flex crops and commodities: Implications for research. *The Journal of Peasant Studies, 43*(1), 93–115.

Burch, D., & Lawrence, G. (2005). Supermarket own brands, supply chains and the transformation of the agri-food system. *International Journal of Sociology of Agriculture and Food*, *13*(1), 1–18.

Burch, D., & Lawrence, G. (2013). Financialization in agri-food supply chains: Private equity and the transformation of the retail sector. *Agriculture and Human Values*, *30*(2), 247–258.

Canadian Biotechnology Action Network (CBAN) (2020). *GMOs in your grocery store: Ranking company transparency*. CBAN & Vigilance OGM. Retrieved from https://cban.ca/wp-content/uploads/GMOs-in-your-grocery-store-report-2020.pdf

Clapp, J. (2009). Food price volatility and vulnerability in the Global South: Considering the global economic context. *Third World Quarterly*, *30*(6), 1183–1196.

Clapp, J. (2019). The rise of financial investment and common ownership in global agrifood firms. *Review of International Political Economy*, *26*(4), 604–629.

Clapp, J., & Isakson, R. S. (2018). *Speculative harvests: Financialization, food and agriculture*. Fernwood Press.

Clapp, J., Collins, A., & Stephens, P. (2021). Legacies of the 2007–2008 food crisis. In M. Koc, J. Sumner, & A. Winson (Eds.), *Critical perspectives in food studies* (3rd ed., pp. 271–286). Oxford University Press.

Cooke, H. (2022, January 20). *Pizza on buy now pay later raises debt concerns*. Financial Times. https://www.ft.com/content/c4da9b2f-5187-4956-931d-0554d4268d4e

Cotula, L. (2012). The international political economy of the global land rush: A critical appraisal of trends, scale, geography, and drivers. *The Journal of Peasant Studies*, *39*(3–4), 649–680.

Desmarais, A., Qualman, D., Magnan, A., and Wiebe, N. (2017). Investor ownership or social investment? Changing farmland ownership in Saskatchewan, Canada. *Agriculture and Human Values*, *34*(1), 149–166.

Epstein, G. A. (2005). Introduction: Financialization and the world economy. In G. A. Epstein (Ed.), *Financialization and the world economy* (pp. 3–16). Edward Elgar.

Fairbairn, M. (2020). *Fields of gold: Financing the global land rush*. Cornell University Press.

FAO (2012). *Shifting to sustainable diets*. United Nations. https://www.un.org/en/academic-impact/shifting-sustainable-diets#:~:text=Sustainable%20diets%20are%20protective%20and,human%20resources%20(FAO%202012).

Ghosh, J. (2010). The unnatural coupling: Food and global finance. *Journal of Agrarian Change*, *10*(1), 72–86.

Greenwood, R., & Scharfstein, D. (2013). The growth of finance. *Journal of Economic Perspectives*, *27*(2), 3–28.

Hawkes, C. (2008). Dietary implications of supermarket development: A global perspective. *Development Policy Review*, *26*(6), 657–692.

HighQuest Partners, US (2010). *Private financial sector investment in farmland and agricultural infrastructure* (Papers no. 33). OECD. Retrieved from https://doi.org/10.1787/5km7n-zpjlr8v-en

Howard, P. (2016). *Concentration and power in the food system*. Bloomsbury Academic.

Isakson, S. R. (2014). Food and finance: The financial transformation of agro-food supply chains. *The Journal of Peasant Studies*, *41*(5), 749–775.

Isakson, S. R. (2015). Derivatives for development? Small-farmer vulnerability and the financialization of climate risk management. *Journal of Agrarian Change*, *15*(4), 569–580.

Jarosz, L. (2009). Energy, climate change, meat, and markets: Mapping the coordinates of the current world food crisis. *Geography Compass*, *3*(6), 2065–2083.

Karlan, D., Osei, R., Osei-Akoto, I., & Udry, C. (2014). Agricultural decisions after relaxing credit and risk constraints. *The Quarterly Journal of Economics*, *129*(2), 597–652.

Kuns, B., Visser, O., & Wästfelt, A. (2016). The stock market and the steppe: The challenges faced by stock-market financed, nordic farming ventures in Russia and Ukraine. *Journal of Rural Studies*, *45*, 199–217.

Lang, T. (2021). The sustainable diet question: Reasserting societal dynamics into the debate about a good diet. *The International Journal of Sociology of Agriculture and Food*, *27*(1), 12–34.

Magnan, A., & Sunley, S. (2017). Farmland investment and financialization in Saskatchewan, 2003–2014: An empirical analysis of farmland transactions. *Journal of Rural Studies, 49*, 92–103.

Martin, S. J., & Clapp, J. (2015). Finance for agriculture or agriculture for finance? *Journal of Agrarian Change, 15*, 549–55.

Mercer, K. L., Perales, H. R., & Wainwright, J. D. (2012). Climate change and the transgenic adaptation strategy: Smallholder livelihoods, climate justice, and maize landraces in Mexico. *Global Environmental Change, 22*(2), 495–504.

Monteiro, C. A., Cannon, G., Levy, R., Moubarac, J. C., Louzada, M., Rauber, F., Khandpur, N., Cediel, G., Neri, D., Martinez-Steele, E., Baraldi, L., & Jaime, P. (2019). Ultra-processed foods; what they are and how to identify them. *Public Health Nutrition, 22*(5), 936–941.

Moss, M. (2014). *Salt, sugar, fat: How the food giants hooked us.* Random House.

Müller, B., Johnson, L., & Kreuer, D. (2017). Maladaptive outcomes of climate insurance in agriculture. *Global Environmental Change, 46*, 23–33.

Ouma, S. (2020). *Farming as financial asset.* Agenda Publishing.

Philippon, T. (2013). An international look at the growth of modern finance. *Journal of Economic Perspectives, 27*(2), 73–96.

Rischke, R., Kimenju, S. C., Klasen, S., Qaim, M. (2015). Supermarkets and Food Consumption patterns: The case of small towns in Kenya. *Food Policy* (52), 9–21.

Rotz, S., Fraser, E. D., & Martin, R. C. (2019). Situating tenure, capital and finance in farmland relations: Implications for stewardship and agroecological health in Ontario, Canada. *The Journal of Peasant Studies, 46*(1), 142–164.

Schmidt, T. P. (2016). *The Political Economy of Food and Finance.* Routledge.

Sibiko, K. W., & Qaim, M. (2020). Weather index insurance, agricultural input use, and crop productivity in Kenya. *Food Security, 12*(1), 151–167.

Sippel, R., Larder, N., & Lawrence, G. (2017). Grounding the financialization of farmland: Perspectives on financial actors as new land owners in rural Australia. *Agriculture and Human Values, 34*, 251–265.

Sommerville, M., & A. Magnan (2015). 'Pinstripes on the Prairies': Examining the financialization of farming systems in the Canadian Prairie Provinces. *Journal of Peasant Studies, 42*(1), 119–144.

Sosa Varrotti, A. P., & Gras, C. (2021). Network companies, land grabbing, and financialization in South America. *Globalizations, 18*(3), 482–497.

Stephens, P. (2021a). Social finance for sustainable food systems: Opportunities, tensions and ambiguities. *Agriculture and Human Values, 38*, 1123–1137.

Stephens, P. (2021b). Social financing for a resilient food future. *Sustainability, 13*(12), 6512.

Thomas, L. (2017, July 19). *Activist investor Jana Cashes out of whole foods in wake of Amazon deal.* CNBC. cnbc.com/2017/07/19/jana-sheds-entire-stake-in-whole-foods.html

Thompson, P. B. (2005). *The spirit of the soil: Agriculture and environmental ethics.* Routledge.

Vargas Hill, R., & Viceisza, A. (2012). A field experiment on the impact of weather shocks and insurance on risky investment. *Experimental Economics, 15*(2), 341–371.

van der Zwan. (2014). Making sense of financialization. *Socio-Economic Review, 12*(1), 99–129.

38

HAZARDS AHEAD? POTENTIAL PITFALLS IN THE SURGE OF PLANT-BASED ALTERNATIVES TO ANIMAL FOODS

Tony Weis and Allison Gray

Introduction: The improving quality and availability of plant-based alternatives to animal foods

The surging growth of plant-based alternatives to animal foods is increasingly apparent across the spaces of contemporary food retailing, from more menu options at fast-food restaurants to increasing space in supermarket freezers. A basic assumption behind this product development is that pulses, nuts, mushrooms, grains, tubers, and vegetable oils can be combined and processed in new ways that respond to the cravings, drive for convenience, and entrenched modes of thinking and cooking associated with livestock-heavy diets. For advocates of these products, close emulation is either: A more plausible pathway to reduce livestock production and consumption than education and moral suasion about the various environmental, ethical, and health problems; or a necessary material accompaniment to it, given how ingrained animal foods are in eating and cooking habits in many parts of the world.

A lot of plant-based product development has focused on some of the central items associated with US fast-food culture, such as hamburgers, hotdogs, and sausages, because of the extent to which they influence dietary aspirations and permeate everyday life (Schlosser, 2012). Fast-food is especially ubiquitous in the US For instance, a survey by the US Centres for Disease Control and Prevention found that 37 percent of American adults consumed a fast-food meal on a typical day between 2013 and 2016 (Fryar et al., 2018). The US is home to most of the world's largest fast-food restaurant chains, including McDonalds, Burger King, KFC, and Subway, which increasingly operate on a world scale and have inspired innumer-

DOI: 10.4324/9781003174417-45

able local imitations. For many, fast-food restaurants and pre-prepared packaged meals represent a partial way to cope with time-stress in hectic lifestyles. They also have contributed to a long-term erosion of food preparation and cooking skills, and for some, a declining willingness to invest time and effort into meals (Howard, 2015; Schlosser, 2012; Winson, 2013). In short, a combination of cultural and structural factors can make it difficult for people to reduce their consumption of animal foods, and the development of plant-based substitutes can help break down these formidable barriers.

Veggie burgers, dogs, sausages, and cold cults are the most prominent examples of plant-based alternatives to animal foods, in both a material (in terms of availability) and symbolic sense (as a broadly recognised cultural referent). Plant-based dairy products also have boomed in recent decades, with milks, cheeses, and yoghourts derived from a range of inputs, including soybeans, rice, almonds, cashews, and oats. Diversification is clearly accelerating, including a growing range of plant-based alternatives to various forms of processed chicken (such as breaded cutlets, nuggets, and wings), fish and shrimp, and whole and liquid eggs. Further, anyone who has been consuming plant-based alternatives to animal foods over a significant period of time is likely to readily acknowledge the significant improvements in taste, texture, and other culinary properties. The first generation of plant-based meats, dairy products, and egg replacers were not only much more dissimilar to their animal-based counterparts but suffered from a variety of culinary and palate problems. For instance, early veggie burgers often crumbled easily, fell through cooking grills, and were generally unconducive to barbequing; soy milk products could be grainy; and veggie dogs, cheeses, and deli slices were frequently rubbery and bland. In contrast, many newer plant-based products not only taste much better but bear a closer sensorial and textural resemblance to the animal foods they are geared to replace, although there is considerable variance in the degree to which plant-based meats emulate the properties of animal flesh (Kazir & Livney, 2021; McClements & Grossmann, 2021).

The rising availability, quality, and diversity of plant-based alternatives to animal foods has considerable potential to contribute to more sustainable and humane diets for two basic reasons: 1) They evade the exploitation of animals and tend to entail much lower resource budgets and pollution loads per unit of nutrition (Fresán et al., 2019; Goldstein et al., 2017; Mejia et al., 2019; Smetana et al., 2015); and 2) they can make the elimination or reduction of animal foods easier through some combination of convenience and appeal to engrained tastes and cooking habits and skills. More details can be found in the chapter on re-meatification: Substituting animal foods with plant-based alternatives. Yet while the potential to reduce interspecies violence and environmental harms can lead some animal and environmental advocates to see the rapid growth of plant-based meats, dairy products, and egg replacers in an unequivocally positive light, it is important to recognise the sizable pitfalls that could divert this momentum. This chapter examines what we identify as the two central pitfalls associated with the trajectory of plant-based product development: First, the growing presence of world's largest livestock-oriented food corporations; and second, the possibility that focusing on close substitution could inadvertently reinforce the centrality of animal foods in diets.

Blunting the edge? Encroachment by livestock-oriented food corporations

The pioneers at the forefront of developing plant-based alternatives to animal foods have tended to be relatively small-scale enterprises (at least initially) and food scientists touting ethical aspirations, to reduce environmental harms and/or animal suffering, which have often infused marketing efforts to varying degrees. The most prominent developers of plant-based meats are US-based Beyond Meat (founded in 2009) and Impossible Foods (founded in 2011), both of which have made the environmental and ethical advantages of deriving protein-dense foods from plants (rather than animals) a central feature of their brand. Both corporations have also similarly invested a great deal of scientific research into understanding the biochemical composition of animal foods, as a basis for developing formulas and processing techniques that allow plant inputs to more closely emulate the taste and feel of major staples of American fast-food culture like hamburgers, hot-dogs, and sausages. Improving emulation is a key factor in the extraordinary speed with which both Beyond Meat and Impossible Foods have moved from niche retailers and into large supermarkets and fast-food chains like Burger King, KFC, A&W, and Carl's Jr.

US-based WhiteWave Foods (established in 1977) was an early leader in the development of plant-based milks, initially based on organic soybeans (and later other plant inputs), and it experienced dramatic growth following the introduction of its Silk product line in the 1990s. Poultry (predominantly chicken) is the fastest growing segment of animal meat production on a world scale, and some enterprises such as US-based Turtle Island Foods and Daring Foods have focused on developing plant-based alternatives to roast birds (the original food associated with Turtle Island's Tofurky brand) and processed chicken products. Enterprises focusing solely on plant-based products have also been at the forefront of developing substitutes for fish and shrimp products, such as US-based Good Catch and New Wave Foods. Compared to other animal foods, there has been relatively less success developing plant-based alternatives to whole eggs in cartons that could be scrambled, fried, or turned into other common egg-related meals, but multiple companies are attempting to develop such a product. However, the biggest way that many people consume eggs is as an ingredient in various processed foods, and here the prospects of substitution are vastly better; US-based Earth Island's "VeganEgg" and Eat Just's "Just Egg" are leaders in reconstituting plant inputs to match the properties (e.g., emulsion, binding) that are valued in processed foods, from mayonnaise and salad dressings to baked goods.

While the US has been at the geographic centre of a lot of plant-based product development, the dynamics of innovation are becoming thoroughly globalised. Some good examples of this globalisation are: Oatly (Sweden-based), a world leader in the boom in oat-based milk; Float Foods (Singapore), pioneering a plant-based whole egg ("OnlyEG"); The Meatless Farm Company (UK); Moving Mountains (UK); Vbites (UK); Quorn (UK); Gold & Green (Finland); Sunfed Meats (New Zealand); Sol Cuisine (Canada); v2food (Australia); Deliciou (Australia), Qishan (China); and Ningbo Sulian Food (China). It is also significant that these products are moving beyond niche retailers, like small health food stores, into a range of larger outlets, from supermarkets to popular restaurant chains, especially in high-income countries.

Although the drive and dynamism of plant-based product development have come from the motivation to replace animal flesh, milk, and eggs solely with plants, this market space has become much more complex. Food corporations are always simultaneously working to shape consumer demand and foster brand loyalty as much as they can, as well as being flexible and responsive to market dynamics they cannot control (Howard, 2015). They are also increasingly dominated by financial capital and the pressures to grow shareholder value (Clapp & Isakson, 2018). The rapid growth of corporations like Beyond Beef, Impossible Foods, and Oatly has made plant-based start-ups increasingly attractive to finance capital. Initial public stock offerings of both Beyond Meat (in 2019) and Oatly (in 2021) drew a lot of media attention, augmented by an array of celebrity investors and endorsers (Evans, 2021; Murphy, 2019). There are also a handful of venture capital funds that are specifically oriented towards companies specialising in plant-based foods (e.g., New Crop Capital, Pivot Food Investment, and Stray Dog Capital), though the great majority of investors focus solely on growth and do not distinguish between livestock and plant-based alternatives in principle. Thus, while corporations whose profitability hinges on animal flesh, milk, and eggs would have derided plant-based alternatives in various ways over time, in the long run it was also inevitable they would seek to subsume as much of this market as possible.

In the 2010s, the world's largest corporations specialising in livestock slaughter and processing—sometimes referred to as "Big Meat" and "Big Dairy"—began to aggressively develop their own plant-based product lines and pursue take-overs of specialised start-ups that maintained or tweaked their original branding (Yaffe-Bellany, 2019). These large corporations include: North America-based giants Tyson (Raised & Rooted), Perdue (Yummy), Cargill (PlantEver), Hormel (Happy Little Plants), Conagra (Gardein), Maple Leaf (Green Leaf Foods, which controls both LightLife and Field Roast brands); Saputo (Bute Island Foods); Brazil-based giants JBS (Vivera; Seara), Marfrig (PlantPlus Foods), BRF (Veg&Tal); China-based WH Group (Pure Farmland); Japan-based NH Foods (NatuMeat); Holland-based Vion (ME-AT); and France-based Danone (WhiteWave). Some of the world's largest processed foods corporations that rely heavily on animal flesh, milk, and eggs have also had a growing interest in developing or acquiring plant-based alternatives to animal foods, including Kellogg's (Morningstar), Kraft-Heinz (Boca Foods), Nestlé (Sweet Earth Foods), and Unilever (The Vegetarian Butcher).[1] As a result, coolers and freezers devoted to plant-based alternatives in large supermarkets now frequently contain a curious mix of products from corporations specialising in plant-based products alongside products from the same corporations that dominate the much larger cooler and freezer space devoted to animal-based foods. Plant-based products from corporations of both sorts can also increasingly be found at the margins of menus at meat-heavy fast-food chains.

The growing presence of "Big Dairy" in plant-based milks clearly illustrates how larger-scale corporations with strong distributional networks and established relationships with retailers can enhance the distribution and visibility of plant-based alternatives. As indicated earlier, WhiteWave Foods was a leader in the development of plant-based milks in the US, led by its Silk brand, and its success eventually made it a target for acquisition by dairy-giant Dean's Food in 2002. WhiteWave was later detached from Dean's Foods, and acquired a number of competitors (Earthbound Farms, Alpro, and So Delicious) before being acquired by Danone, another dairy-giant, in 2016—a course that has dramatically expanded the scale of production, marketing, and supermarket placement for the Silk brand. Some other "Big Dairy" corporations have similarly contributed to the mainstreaming of plant-based milk products, such that in 2020 plant-based milks accounted for 15 percent of all milk sales in the US (Good Food Institute, 2021).

Some might be inclined to see this mainstreaming as a positive outcome irrespective of who is leading it, and hope that it might indicate some level of recognition about environmental or ethical problems among corporations who are at the forefront of livestock production. However, the vastly larger markets and huge amounts of fixed capital invested in animal-based flesh, milk, and eggs make it hard to believe that this growing development or acquisition of plant-based meats, dairy, and eggs could suggest any substantive long-term transitions. It is much more likely that "Big Meat" and "Big Dairy" are racing into the plant-based sector in order to recapture that share of the consumer base which conscientiously objects to their core products, recognising that the more they occupy this market segment the less threat it could poses to their core animal-centric business model in the long-run. Further, there is an emerging approach that aims to capitalise on consumer demand for plant-based products *in tandem with* rather than as *an alternative to* animal foods, with new blended products that are marketed with an emphasis on plant inputs. Two good examples of blended products are the Dairy Farmers of America integrating plant-ingredients into "hybrid" milk and The Better Meat Co. integrating plant ingredients into processed meats.

The ability of large corporations focused on livestock slaughter and processing and animal-heavy processing to acquire and develop successful plant-based product lines has the potential to fortify their overall profitability and attractiveness to investors without destabilising their core business model. Further, the more they encroach into this market, the more the environmental and ethical messaging associated with plant-based meats, dairy, and eggs will be blunted on labels and in marketing campaigns, and the more the oppositional meanings these products carry could fade—a potential pitfall that is also bound up in their taste, look, and feel.

Inadvertent reinforcement? The risks of emulation

Direct emulation and ease of substitution are key reasons why plant-based meats, dairy, and egg products have the potential to contribute to urgent and large-scale reductions in the consumption of animal-based foods, what we refer to as *re-meatification* in this collection. However, laden in this possibility is also the countervailing risk that it could reproduce animal-centred dietary norms. As Sinclair (2016) suggests, plant-based meats—in both their development and marketing—can never be entirely detached from animals, as the palate pleasure they seek to induce hinges on associations with the taste and feel of animal foods. Two clear examples of this are the use of 3D printers to construct muscle-like formations (Kazir & Livney, 2021; McClements & Grossmann, 2021), and the efforts made by Impossible Foods to not only emulate the muscle-like texture of meat but also to try capture the taste and feel of blood in its plant-based burger through a processing technique coupled with a patented genetic modification.

If the explicit referent of what is real meat, milk, cheese, or eggs remains the edible animal, conceptually and materially, it could reinforce or normalise the centrality of animal foods in Western diets (Sexton, 2018) and the elevated status of animal protein, when in fact the meat-centred diets common in high-income countries are a radical and relatively recent transformation in the course of agrarian and culinary history (Weis, 2013). This possibility should be considered in light of evidence suggesting that many people who are willing to replace some of their meat with plant-based meat are not willing to abandon animal flesh altogether (Gray, 2020; Monaco, 2020), and continue to see themselves as omnivores without intending to reduce consumption in a significant way (Twine, 2018). It is notable that a major Beyond Meat distributor estimated that almost 9 out of 10 consumers of these products were self-identified omnivores (Chiorando, 2018).

Another risk associated with the increasing mainstreaming of plant-based alternatives to animal foods is that it could potentially exaggerate both the appearance of change already afoot and faith in the power of consumer choices to affect systemic problems, thereby fostering a sense of complacency about non-livestock related inequalities and environmental instabilities in agro-food systems. Although it is vastly more efficient to use the products of industrial monocultures directly as inputs for plant-based alternatives than it is to funnel much greater volumes of industrial monoculture production through concentrated populations of livestock animals to generate similar amounts of animal-based meat, milk, or eggs, it does not do away with the myriad environmental problems associated with industrial monocultures, including fertiliser and pesticide dependence, chronic soil degradation, GHG emissions, and water pollution (Weis, 2010).

For instance, while soybeans are predominantly used as livestock feed on a world scale, they are also a common input in many plant-based alternatives and processed foods, and there are a range of resource budgets and pollution loads associated with industrial soybean production. As indicated earlier, WhiteWave was an early pioneer in the rise of soy milk in the US, and its original focus on sourcing organic soybeans was abandoned in the course of its growth, acquisitions, and retail mainstreaming (Howard, 2015). Palm oil is also commonly found in plant-based alternatives, and in processed foods more generally, and the rapid expansion of palm oil plantations, especially in Southeast Asia, is notorious for having greatly accelerated deforestation with catastrophic impacts on climate change and biodiversity (Meijaard & Sheil, 2019). Almonds are an important input for some plant-based alternatives, especially dairy substitutes, and large-scale almond production is heavily reliant on the increasing commercialisation of pollination services and the long-distant transhipment of bees, which are bound up in broader declines of pollinator populations (Ellis et al., 2020). It is possible that the rise of plant-based alternatives, from supermarket freezers to menus at fast-food chains, could lead some people to fixate only on the negative impacts of industrial live-stock production while setting aside other inequitable and unsustainable aspects of the corporate-led agro-food system, consciously or unconsciously. To the extent this happens, it could draw some attention away from the need to support more localised and sustainable modes of production and associated farmer livelihoods, and instead serve to reify "technological fixes"—or highly centralised, techno-scientific responses to the biophysical problems that stem from how production is organised under capitalism (Weis, 2010).

A different way that perceptions of change can be exaggerated relates to the skewed nature of plant-based meat product development, with more attention to beef and dairy substitutes and less attention to chicken meat and egg substi-tutes relative to the scale of animal consumption. For decades, chicken production has risen fastest in per capita terms on a world scale, placing it at the forefront of global meatification, with industry champions frequently highlighting both its health advantages, superior feed conversion ratios, and GHG emissions relative to mammalian livestock—points that consistently appear in assessments of livestock production (Anand et al., 2015; Willett et al., 2019). It is possible that the mes-saging that chickens are relatively less harmful could align with their increasing consumption together with the growing plant-based substitution of mammals. If the further orientation of meat production towards chickens does occur it might mitigate some of the health and environmental harms associated with the produc-tion of other species, but it would not do away with the problems of meatifica-tion and would enhance the scale of interspecies violence given the vastly greater populations of individual animal and the nature of their lives in industrial opera-tions (Weis, 2016).

Another reason why the mainstreaming of plant-based alternatives could potentially reinforce animal products relates to the fact that consumers frequently perceive them as being unnatural due to their processed nature and the additives used to enhance flavours and stabilise various plant ingredients to achieve aesthetic and culinary properties (Clark & Bogdan, 2019; Gray, 2020; Román et al., 2017; Slade, 2018; Vainio et al., 2016). Such perceptions will surely only increase the more genetic modification is used to develop plant inputs with particular simula-tion properties (as in the Impossible Burger), the more cultured or "lab-meat" production grows (Mouat & Prince, 2018), and if lines blur between plant-based and cultural meat production (as is evident with Eat Just, a leader in plant-based egg substitutes that is also attempting to develop cultured chicken meat products).[2] Research into consumer perceptions has shown that people tend to strongly asso-ciate foods they see as natural with sustainability (Tobler et al., 2011; Verhoog et al., 2003), which suggests the processed character of plant-based meats, dairy, and eggs could subvert how their environmental benefits are understood and make animal products appear better by association. To the extent that animal meat, milk, and eggs get perceived as simpler and more wholesome products than their plant-based substitutes, the unnatural character of modern livestock production could be obscured. It hinges on such things as systemic antibiotic use, widespread artificial insemination, the specialisation of breeding and growing sites, and the narrowing of animal genetics (involving both the legal ownership of genetic varieties and the breeding stock), as well as the fact that livestock products frequently contain residues of pesticides from concentrated feed, antibiotics, and chemical disinfectants and different sorts of additives.

Ultimately, it is important to be conscious of how the emulation of animal foods in plant-based product lines can lead down some slippery perceptual slopes, at the same time as recognising that cultural conceptions of foods, however strong, are never entirely fixed. It remains possible that the more diversified and accessible plant-based products become, the more they could subvert attitudes about the necessity or superiority of animal protein and make it easier for people to act on their beliefs after they become aware of the problems associated with livestock production.

Conclusions

The increasing accessibility, diversity, and quality of plant-based meat, dairy, and eggs makes it easier for people seeking to avoid or reduce their consumption of animal foods—whether for environmental, ethical, or health considerations—which can ultimately contribute to urgent reductions at a much broader scale. At the same time, it is important not to exaggerate the role that plant-based alternatives can play in de-stabilising industrial livestock production and animal-heavy diets, which are held with a deep sense of entitlement in industrialised countries (Chiles & Fitzgerald, 2018).

This chapter has focused on two significant pitfalls that could potentially reduce the impact that plant-based meat, dairy, and eggs have on dietary change. The first pitfall relates to the growing presence of livestock-oriented corporations in plant-based product markets, which can serve to brace their profitability and attractiveness to investors and diminish the distinctive meanings and significance that people attach to these products, as well as reduce the material and ethical messages conveyed on food labels and in marketing campaigns. The second pitfall relates to the goal of closely emulating key properties of animal flesh and reproductive outputs, which can inadvertently normalise the centrality of animal-based products within diets and reinforce certain positive connotations they carry.

Yet these pitfalls are not destined to undercut all the transformative potential of plant-based alternatives to animal foods. Regardless of who is producing plant-based meats, dairy, and eggs, the animal referent they contain can always provoke, because the question of *why* is inherent in their materiality. That is, the very nature of these foods can challenge people to reckon with possible reasons for consuming processed plants as opposed to the flesh and reproductive outputs of animals. So while the growing presence of plant-based alternatives in mainstream food retail spaces is insufficient to destabilise dietary norms, it does present growing opportunities for provocation, and this cannot be left to the corporations making and marketing these products. On the contrary, much of the work that is needed to substantiate the meanings of plant-based alternatives with respect to environmental impacts and interspecies relations lies beyond the realm of profit-seeking activity.

Notes

1 This is as of January 2022 and is intended to be an illustrative, but not exhaustive, list.
2 Both GM and cultured meats also involve the establishment of intellectual property rights that pose a host of questions beyond the scope of discussion here.

References

Anand, S. A., Hawkes, C., de Souza, R. J., Mente, A., Dehghan, M., Nugent, R., Zulyniak, M. A., Weis, T., Bernstein, A. M., Krauss, R., Kromhout, D., Jenkins, D. J. A., Malik, V., Martinez-Gonzalez, M. A., Mozafarrian, D., Yusuf, S., Willett, W. C., & Popkin, B. M. (2015). Food consumption and its impact on cardiovascular disease: Importance of solutions focused on the globalized food system. *Journal of the American College of Cardiology*, *66*(14), 1590–1614.

Chiles, R. M. & Fitzgerald, A. (2018). Why is meat so important in Western history and culture? A genealogical critique of biophysical and political- economic explanations. *Agriculture and Human Values*, *35*(1), 1–17.

Chiorando, M. (2018, November 2). *86% of consumers of vegan meat products are meat eaters*. Plant Based News. https://www.plantbasednews.org/post/86-consumers-vegan-beyond-meat-meat-eaters?fbclid=IwAR3t873-m0OoqDxnheNRihT-bY1KTqb7wZm0VRpe9q61XAuTipxFH7x5Ge8

Clapp, J., & Isakson, R. (2018). *Speculative harvests: Financialization, food, and agriculture*. Fernwood.

Clark, L. F. & Bogdan, A. (2019). Plant-based foods in Canada: Information, trust and closing the commercialization gap. *British Food Journal*, *121*(10), 2535–2550.

Ellis, R. E., Weis, T, Surayanaman, S., & Beilin, K. (2020). From a free gift of nature to a precarious commodity: Bees, pollination services, and industrial agriculture. *Journal of Agrarian Change 20*(3), 437–59.

Evans, P. (2021, May 23). *Oatly's blockbuster IPO shows healthy appetite for plant-based living is growing*. CBC News. https://www.cbc.ca/news/business/oatly-ipo-vegan-1.6035705

Fresán, U., Maximino-Alfredo, M., Craig, W. J., Jaceldo-Siegl, K., & Sabaté, J. (2019). Meat analogs from different protein sources: A comparison of their sustainability and nutritional content. *Sustainability*, *11*(1), 3231–3241.

Fryar, C. D., Hughes, J. P., Herrick, K. A., & Ahluwalia, N. (2018). *Fast food consumption among adults in the United States, 2013–2016*. National Centre for Health Statistics. Retrieved from: https://www.cdc.gov/nchs/products/databriefs/db322.htm

Goldstein, B., Moses, R., Sammons, N., & Birkved, M. (2017). Potential to curb the environmental burdens of America beef consumption using a novel plant-based beef substitute. *PLos One*, *12*(12), e0189029.

Good Food Institute (2021). *US retail market data for the plant-based industry*. https://gfi.org/marketresearch/

Gray, A. (2020). *Eating in the Anthropocene: Perceptions of dietary-based environmental harm and the role of plant-meat consumption* [unpublished doctoral dissertation, University of Windsor].

Howard, P. (2015). *Concentration and power in the food system*. Bloomsbury.

Kazir, M., & Livney, Y. D. (2021). Plant-based seafood analogs. *Molecules*, *26*(6),1559.

McClements, D. J., & Grossmann, L. (2021). The science of plant-based foods: Constructing next-generation meat, fish, milk, and egg analogs. *Comprehensive Reviews in Food Science and Food Safety*, *20*(4), 4049–4100.

Meijaard, E. & Sheil, D. (2019). The moral minefield of ethical palm oil and sustainable development. *Frontiers in Forests and Global Change*, *2*(22), 1–15.

Mejia, M. A., Fresán, U., Harwatt, H., Oda, K., Uriegas-Mejia, G., & Sabaté, J. (2019). Life cycle assessment of the production of a large variety of meat analogs by three diverse factories. *Journal of Hunger & Environmental Nutrition*, *15*(5), 699–711.

Monaco, E. (2020, Sept 21) *How meat eaters, not vegans, are driving the plant-based foods boom, according to industry experts*. Business Insider. https://www.businessinsider.com/plant-based-meats-flexitarians-vegetarians-vegans-market-revolution-2020-9

Mouat, M. J., & Prince, R. (2018). Cultured meat and cowless milk: On making markets for animal-free food. *Journal of Cultural Economy*, *11*(4), 315–329.

Murphy, M. (2019, May 5). *Beyond Meat soars 163% in biggest-popping U.S. IPO since 2000*. MarketWatch. https://www.marketwatch.com/story/beyond-meat-soars-163-in-biggest-popping-us-ipo-since-2000-2019-05-02

Román, S., Sánchez-Siles, L. M., & Siegrist, M. (2017). The importance of food naturalness for consumers: Results of a systemic review. *Trends in Food Science & Technology*, *67*(9), 44–67.

Schlosser, E. (2012). *Fast food nation: The dark side of the all-American meal* (2nd ed.). Mariner Books.

Sexton, A. E. (2018). Eating for the post-Anthropocene: Alternative proteins and the biopolitics of edibility. *Transactions of the Institute of British Geographers*, *43*(2), 586–600.

Sinclair, R. (2016). The sexual politics of meatless meat: (In)edible others and the myth of flesh without sacrifice. In B. Donaldson, & C. Carter, (Eds.) *The future of meat without animals* (pp. 229–248). Rowman & Littlefield International.

Slade, P. (2018). If you build it, will they eat it? Consumer preferences for plant-based and cultured meat burgers. *Appetite*, *125*, 428–437.

Smetana, S., Mathys, A., Knoch, A., & Heinz, V. (2015). Meat alternatives: Life cycle assessment of most known meat substitutes. *The International Journal of Life Cycle Assessment*, *20*(9), 1254–1267.

Tobler, C., Visschers, V. H. M., & Siegrist, M. (2011). Eating green: Consumer's willingness to adopt ecological food consumption behaviours. *Appetite*, *57*(3), 674–682.

Twine, R. (2018). Materially constituting a sustainable food transition: The case of vegan eating practice. *Sociology*, *52*(1), 166–181.

Vainio, A., Niva, M., Jallinoja, P., & Latvala, T. (2016). From beef to beans: Eating motives and the replacement of animal proteins with plant proteins among Finnish consumers. *Appetite*, *106*, 92–100.

Verhoog, H., Matze, M., van Bueren, E., & Baars, T. (2003). The role of the concept of the natural (naturalness) in organic farming. *Journal of Agricultural and Environmental Ethics* *16*(1), 29–49.

Weis, T. (2010). The accelerating biophysical contradictions of industrial capitalism agriculture. *Journal of Agrarian Change*, *10*(3): 315–341.

Weis, T. (2013). *The ecological hoofprint: The global burden of industrial livestock*. Zed Books.

Weis, T. (2016). Towards 120 billion: Dietary change and animal lives. *Radical Philosophy*, *199*(5), 8–13.

Willett, W., Rockström, J., Loken, B., Springmann, M., Lang, T., Vermeulen, S., Garnett, T., Tilman, D., DeClerck, F., Wood, A., Jonell, M., Clark, M., Gordon, L. J., Fanzo, J., Hawkes, C., Zurayk, R., Rivera, J. A., De Vries, W., Majele Sibanda, L., Afshin, A., … & Murray, C. (2019). Food in the Anthropocene: the EAT-Lancet Commission on healthy diets from sustainable food systems. *The Lancet*, *393*(10170), 447–492.

Winson, A. (2013). *The industrial diet: The degradation of food and the struggle for healthy eating*. UBC Press.

Yaffe-Bellany, D. (2019, October 14) *The new makers of plant-based meat? Big meat companies*. The New York Times. https://www.nytimes.com/2019/10/14/business/the-new-makers-of-plant-based-meat-big-meat-companies.html

39

TOMORROW'S AGRI-FOOD SYSTEM

The connections between trade, food security, and nutrition for a sustainable diet

David Laborde, Valeria Piñeiro, and Johan Swinnen

Introduction

Agricultural development policies used to be concentrated in the agricultural sector alone. Today, the concept of food systems interacts with information that highlights how value chains give new clues to understand diets and design incentives for promoting a sustainable diet.

As noted in the introductory chapter, we can define *sustainable diets* as well as how to measure and promote it. "Sustainable diets are protective and respectful of biodiversity and ecosystems, culturally acceptable, accessible, economically fair and affordable; nutritionally adequate, safe and healthy, while optimising natural and human resources" (Burlingame & Dernini, 2010, p. 8). As suggested by this definition, there is a need for well-formulated and interdisciplinary metrics of the sustainability of diets.

There are two concepts that need to be included in analysing sustainable diets and their promotion. The first is the inclusion of externalities generated by food systems, the *true cost of food*, that includes the pollution of water and air, greenhouse gas emissions, overdrawn aquifers, biodiversity loss, zoonotic diseases,[1] antibiotic resistance, land degradation, and the rise of health illnesses related to food consumption either due to poor quality diets[2] (e.g., diabetes), unsafe food (e.g., foodborne disease), or food production (e.g., exposure to chemical pesticides). The additional social costs, which exceed the private costs, have their origins in market failures, including incomplete information and missing markets, and in particular, negative externalities (Laborde et al, 2020). The second concept is to look at *the quality of the calories*, not only the *amount of calories consumed*. For this, accessibility and affordability are key, with trade playing a key role in both. There are many examples in the literature that show the higher cost of more nutritious foods and

DOI: 10.4324/9781003174417-46

diets compared to basic staples and calorie adequate diets (Headey & Alderman, 2019; FAO, IFAD, UNICEF, WFP & WHO, 2019).

When analysing sustainable diets, it is important to pay attention to the nexus between trade, climate change, and nutrition, as well as highlighting their synergies and limiting the trade-offs within them. Understanding these interactions provides insights on how food systems can be transformed to provide affordable healthy diets to all in a sustainable way. In this context, policy incentives could play an important role in transforming existing systems. The policy environment around a food system has a large influence on this nexus through four drivers: How much we produce, what we produce, how we produce, and where we produce. Policy levers can be used to improve nutrition outcomes through healthy diets using a more climate resilient production system with lower greenhouse gas (GHG) emissions and by preserving natural capital, in particular, biodiversity.

Interconnections between nutrition transition and international trade

The past half-century has seen a significant shift in the quality and quantity of human diets and resulting epidemiology (e.g., underweight, malnutrition, overweight, obesity, and other nutrition-related diseases). Nutrition and associated health and demographic transitions have been heavily influenced by rapid economic and income growth, demographic changes, urbanisation, and globalisation.

At the population level, this malnutrition phenomenon reflects changes that involve a nutrition transition, a demographic transition, and an epidemiological transition. In the last two centuries, these three processes have occurred slowly and in a near-linear fashion in most high-income countries and resulted in intergenerational, incremental, and controlled increases in population height, health, and lifespan, as well as lifestyle. The improved nutrition and higher caloric opportunity led to gradual increases in health. However, as low-income, and particularly, middle-income countries have continued to develop, this process has been accelerated (with these transitions occurring over decades, rather than centuries). The result of this process includes intergenerational changes in diet quality and quantity, leading to a coexistence of undernutrition along with overweight or obesity and other nutrition-related diseases (Demaio & Branca, 2017).

Non-communicable disease (NCD), morbidity, and mortality are a tremendous threat to social and economic development in the current century. Presently, NCDs are responsible for 38 million deaths worldwide annually (Schram et al., 2018). In low-income and middle-income countries, almost five million children continue to die of undernutrition-related causes; yet, simultaneously many of these same populations experience an unprecedented rise in levels of overweight and obesity in childhood (Schram et al., 2018). The result is a triple burden of malnutrition, the coexistence of undernutrition along with overnutrition manifesting as overweight and obesity, and micronutrient deficiencies, also called malnutrition (Gomez et al., 2013).

Increased knowledge of social and structural determinants of health has expanded the breadth of policy areas that are investigated as drivers of health outcomes, including macroeconomic policy areas such as trade and investment agreements. This new policy approach recognises that health behaviours, often framed as individual choices, are in fact conditioned and constrained by policy and food environments (Schram et al., 2018), since they impact food prices, income distribution, and social or cultural preferences. Although for more than a decade the public health community has been actively engaging with trade and investment policy to examine multiple pathways between trade and public health, the literature still lacks a comprehensive review of, and conceptual framework synthesising, the pathways between international trade and health outcomes, especially with regard to the contribution of nutrients that come from the imports of food (Cuevas García-Dorado et al., 2019; Schram et al, 2018). Existing frameworks either have been very broad—such as those examining the larger processes of globalisation on health, at the expense of a detailed exploration of trade and investment provisions—or very specific—providing a sophisticated exploration of the health impacts of trade and investment agreements through one channel, such as food environments (or food systems), at the expense of a more inclusive suite of intervening factors (Schram et al., 2018).

On the other hand, there is a large area of literature examining the patterns of food trade flows and their determinants. Such studies generally focus on explaining or discussing the impact of trade flows in terms of either trade value or volume. In the context of food security—availability, access, utilisation, and stability[3]—however, knowing the value or even volume of imports may not be adequate to understand how countries rely on international trade as a source of food and nutrition (Laborde & Deason, 2015). Where there is evidence on the impact of trade on food availability and prices, there is little written evidence of the direct impact of trade on countries' nutritional offerings. However, there is some evidence from across countries that national availability and prices are linked to national levels of undernourishment (Hawkes et al., 2015; Olper et al., 2018). Food supply diversity in middle and high- (not low-) income countries is also associated with lower rates of several measures of undernutrition and child mortality (Hawkes et al., 2015; Olper et al., 2018). Based on this connection, understanding the nutritional components provided by imported foods to each country is essential to designing a more complete nutritional policy.

Some studies have focused on analysing the impact of trade liberalisation on the food systems of the countries, especially on the quality of the products consumed and in the ability of the large-scale private agro-food industry to conduct business transnationally. The pathways of impact are broadly conceptualised in Hawkes et al. (2015). But even when it has been observed that trade liberalisation has led to a drop in the nutritional quality of the consumed food in low- and middle-income countries, some research suggests that imported food is typically of higher sanitary quality than food in the domestic markets (Hawkes et al., 2015).

The benefits of international trade are embedded in our lives and our meals have been impacted by globalisation. Global improvements in food and nutrition security under an open and inclusive trade regime have contributed to falling levels of undernourishment, better nutrition, greater dietary diversity, and overall

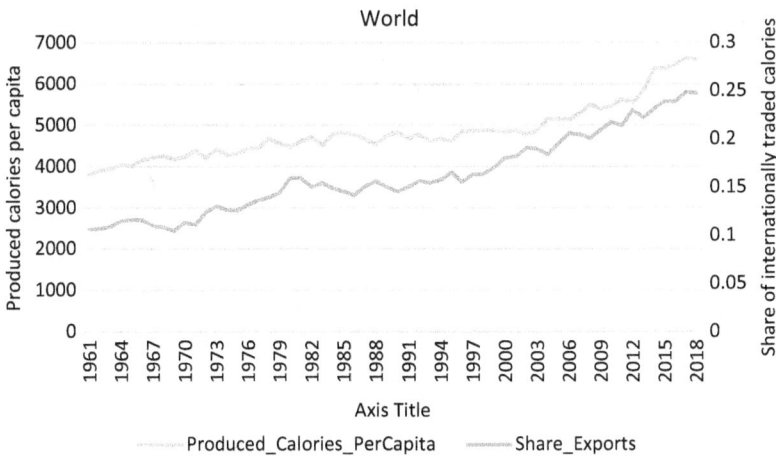

Figure 39.1 Produced calories per capita and share of internationally traded calories. Source: Authors' computations based on FAOSTAT Food Balance Sheets (2022). Intra-EU trade included in the export statistics.

economic development. Trade contributes to the four key requirements of food security (food availability, access, utilisation, and stability of supply). But in today's climate of scepticism about globalisation, the benefits of trade may be forgotten as negative impacts are emphasised by advocates of trade barriers and self-sufficiency (Laborde & Deason, 2015). However, Figure 39.1 shows an increase of 80% in produced calories per capita (for food and non-food use), as well as the share of food, measured in calories, crossing an international border rose from 12.3% to over 25% over the last 40 years. The combined effects lead to 4.2 times more calories crossing a border in 2018 than in 1961.

More needs to be learned about the patterns of international food trade in terms of nutritional content, including caloric, fat, and protein content of traded goods. By looking at patterns of trade in terms of nutritional content, we can identify countries that are heavily reliant on imports as the source of a particular nutritional component. We can also break down trade flows by nutritional composition and see how this evolves over time for any given country or region.

The growing importance of sustainable diets in the international debate

Undoubtedly, the nutritional aspects of food and trade have become an increasingly relevant issue on the international agenda. For example, the Rome Declaration on Nutrition, adopted at the Second International Conference on Nutrition (2014), acknowledged:

> trade is a key element in achieving food security and nutrition and that trade policies are to be conducive to fostering food security and

nutrition for all, through a fair and market-oriented world trade system, and reaffirm the need to refrain from unilateral measures not in accordance with international law, including the Charter of the United Nations, and which endanger food security and nutrition, as stated in the 1996 Rome Declaration.

(Hawkes et al., 2015)

The World Trade Organisation (WTO) also included nutritional aspects at its ninth ministerial meeting in Bali, 2013. The Bali decision on food security represents an opportunity to enhance trade-related food policy space to support both food security and NCDs prevention (Hawkes et al., 2015). Also, in 2016 the United Nations General Assembly adopted a United Nations Decade of Action on Nutrition for the period of 2016–2025. The Decade calls for coordinated action through cross-cutting and coherent policies, programs, and initiatives, including social protection to comprehensively address the triple burden of malnutrition (Demaio & Branca, 2017).

All of these recommendations provide a framework for governments to make commitments to assess the food related nutrition and health impacts of trade policy and to identify trade as an opportunity to improve nutrition. The impacts generated by higher levels of trade will depend on the way these effects are transmitted throughout the food supply chain, how these changes affect food environments, and nutrition and health. Therefore, diagnosing health-related outcomes requires tracing those changes and the incentives that influence them (Hawkes et al., 2015).

International trade: Implications for the food security and nutrition agenda

Agricultural trade is vitally important for achieving the goal of ending hunger by 2030, as enshrined in the second Sustainable Development Goal (SDG). While trade is frequently seen as posing a threat to this important goal, it can play a major role in achieving it in a number of ways. One benefit is the increase in income and economic well-being of the population resulting from globalisation over the last half-century. Low income remains the major driver behind the 3 billion people that cannot afford a healthy diet today (FAO et al., 2019). But agricultural trade in particular allows the world to take advantage of countries' heterogeneity: Differences in endowments (land, water, sun) and technology that translate into different supply potential, food prices, and environmental footprints by generating improvements in the global indicators of sustainability of agri-food systems.

By allowing countries to take advantage of their radically different endowments, with land-abundant countries providing exports and land-poor countries taking advantage of much more efficiently-produced imports (Martin, 2017), which together can be beneficial to the environment, especially with supportive institutional frameworks. Also, by importing products and services from countries with abundant water resources, water-deficient countries can alleviate the pressure on their own water supply (Deng et al., 2021; Pastor et al., 2019).

To analyse the impact of trade on resource use and pollution, a growing literature expresses trade flows in terms of the resource inputs and emission content they carry (virtual resource trade, carbon/land/water footprint). Figure 39.2 shows the water and fertiliser content embedded in international trade for China and the United States, in both exports and imports. Looking at the water content of primary crops, China does not export a considerable amount; however, it does save 200 billion cubic metres of water by importing commodities and 150 billion cubic metres of water in its agricultural imports. The story is very different for the United States, having a higher use of water in its exports compared to their agricultural imports. Fertiliser use follows the same trends as the water consumption for both countries.

Regarding food access and availability, global markets have grown at a higher speed than domestic production (see Figure 39.1) to address the rising gaps between local demand and supply.

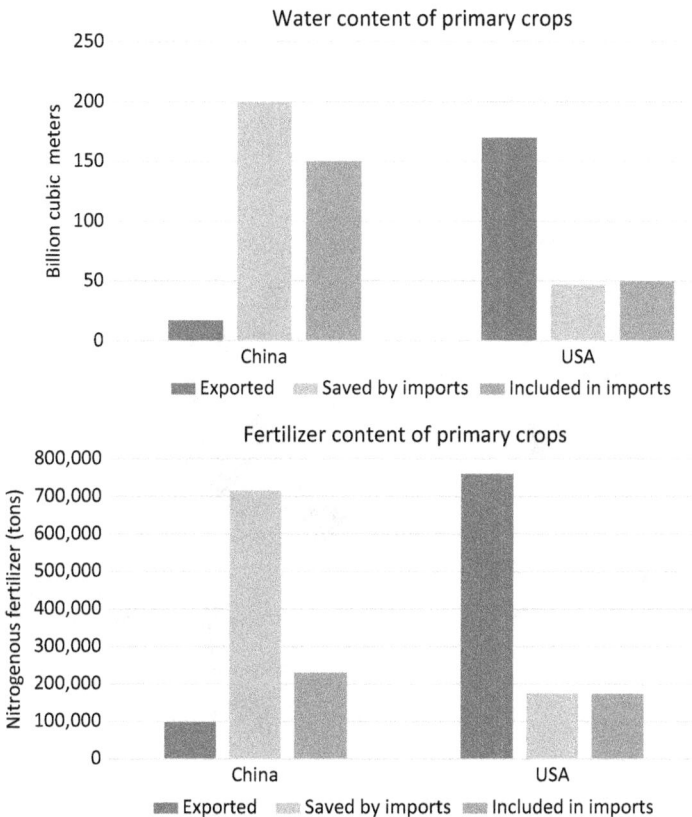

Figure 39.2 Water and fertiliser content embedded in international trade. Source: Original work of the author D. Laborde (2017, unpublished). Note: Water content includes all water used in the production and processing of primary crops.

Beyond calories, Figure 39.3 shows the net imports and food consumption of proteins, highlighting the fact that for some countries trade becomes crucial not only for the quantity of food needed for domestic consumption but also for the access and availability of proteins and nutrients in their diets.

Trade also has considerable potential to improve the diversity and quality of food consumed in a country. Indeed, it is important to look not only at the quantity of calories consumed but also at their quality. For this, accessibility and affordability are key. Different food groups contribute different percentages to the higher cost of more nutritious foods and diets compared to basic staples and calorie adequate diets. Some estimates for the average global cost of healthy diets in 2017 show that protein-rich foods including dairy, fruits, and vegetables, together make up almost 73% of the cost of healthy diets (of which fruits and vegetables account for 40%), while starchy staples and fats account for almost 27% of that cost, although these proportions vary by region (FAO & IFPRI, 2020; Herforth et al, 2020). If people are very poor, they will likely be constrained to consume diets that focus heavily on starchy staples (Masters et al., 2018).

Additionally, Wood and colleagues (2018) found that there is a large disparity between global nutrient production (i.e., nutrients in crops and livestock before accounting for waste, losses, etc.) and nutrient supply (i.e., amount available for consumption), suggesting crop improvements alone would be insufficient to meet nutritional needs without policies to translate it into increased supply. But, at the same time, they found that the current global food supply (i.e., food available for consumption, after waste, losses etc. are accounted for) would meet the caloric and nutritional requirements of the current world population were it to be perfectly

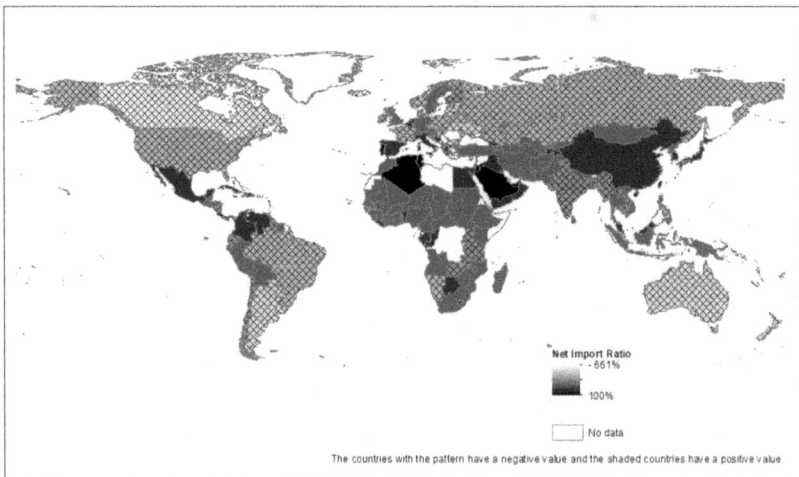

Figure 39.3 Ratio of net Imports and Food Consumption, Proteins. Source: Authors' computations based on FAOSTAT Food Balance Sheets (2022). Note: For the sake of readability, the ratio is capped to -200% and +100% but several economies exceed these bounds.

equally distributed. Additionally, protectionist trade policies that limit the flow of food items would likely have negative consequences for global nutritional security.

Figure 39.4 shows the global food production per capita per day by nine food groups (based on FAOSTAT FBS), compared to global food production per capita per day but adjusted by food loss and waste and the Flexitarian diet from the EAT-Lancet report.[4] The nine food groups are: Animal products excluding dairy, milk, starches, legumes and beans, nuts and seeds, fruits, sugar, vegetable oils, and sugar. This grouping of food products is more aggregated than the one proposed by the EAT-Lancet work to allow for some substitution at the country level (e.g., no need to consume more grains if there is already consumption of starchy roots). By comparing the middle bar that captures the production per capita per day adjusted by food loss and waste and the flexitarian diet (the needed amount of each food group to obtain the flexitarian diet) in Figure 39.4, there is an oversupply of most of the food categories, except for vegetables and fruits at the global level.

However, the story is not the same if we do this comparison at the national level, looking at the number of countries that have the level of current production within their borders to achieve the flexitarian diet. Remans and colleagues (2014) highlight the sharp differences between the nutritional diversity of production and of the food supply in many regions. They highlight the importance of diversity in national food systems for human health and suggest that strategies for addressing nutritional gaps should be tailored to the economic and agricultural conditions in particular countries. For example, low-income countries might target diversification of production, while middle-and high-income countries could focus on using economic capital to purchase nutritional diversity on international markets. The larger variability within regions for production than for supply diversity suggests an important role for regional markets in food trade.

The nutrition dimension of the food security goal is considerably more wide-ranging than the food security dimension (healthy diet vs energy diet (FAO et al., 2019). The link between trade and food quality is a controversial issue. While one

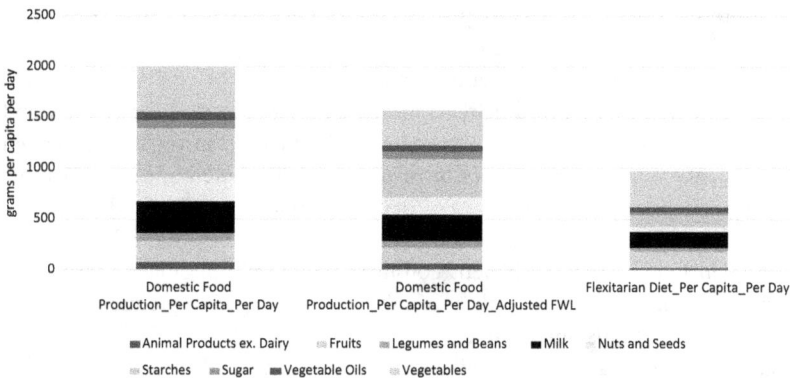

Figure 39.4   Global Agricultural Supply (with and without food loss and waste) compared to per capita requirements from the flexitarian diet. Source: Authors' computations based on FAOSTAT Food Balance Sheets, (2022).

would expect that trade leads to dietary improvements, many have raised concerns about the role of trade in creating nutritional problems, particularly those associated with obesity, by expanding the consumption space for consumers (e.g., easier access to cheaper calories or highly processed food products). See also the chapter on financialisation for further analysis.

Dithmer and Abdulai (2017) looked at 151 countries spanning all income levels and found that trade openness has beneficial effects for dietary energy supply, dietary diversity, and diet quality. Trade plays a role in allowing the availability of a diverse diet all year long, benefiting countries that have different seasons, as well as having a critical role in buffering against variability and volatility, and may contribute to a higher level of food safety when standards and regulations are applied to food trade.

Figure 39.5 shows the number of food groups with insufficient local production to achieve a healthy diet by country adjusted by food loss and waste. Countries in darker colour represent a higher number of food groups with insufficient local production to achieve a healthy diet. For example, Argentina has a level of sufficiency while Madagascar does not. Overall, on an average year, an average country in the world has deficiencies for 5.1 groups, and three-fourths of the countries have deficiencies in four groups or more.

Figure 39.5 underscores the importance of trade in attaining a sustainable diet at a country level. However, as Bouet and Laborde Debucquet (2017) pointed out, political and social configuration make the relationship between trade and food availability more complex due to market failures, inefficient policy interventions, and political economy constraints.

Trade policies and trade reform to achieve sustainable diets

As expected, global food availability increased significantly with trade, but with significant variation between countries and foods. However, the roles of trade policies and trade policy reform are more complex. Trade policies affect medium to long-term trends in the relative prices of foods, the affordability of different categories of food, and the income of everyone involved in production by influencing supply and demand with potentially contrasted effects. In exporting countries, higher integration into world markets means higher food prices for exports, and could deter consumption of such products if the increase in income generated by international trade is not sufficient or unevenly distributed in the population. For importing countries, liberalising agricultural trade will make food, including unhealthy products, cheaper as a consequence of the elimination of tariffs. The decrease in food prices will benefit consumers but could negatively impact the income of smallholders and workers of the import-competing products. In addition, there are no guarantees in a second-best world, where various distortions exist, that removing a subset of trade barriers (what most trade reforms are about) will improve the nutritional outcomes of the populations.

While expanding undistorted trade[5] is still expected to generate a number of positive outcomes, the devil is in the details and careful assessment of each reform

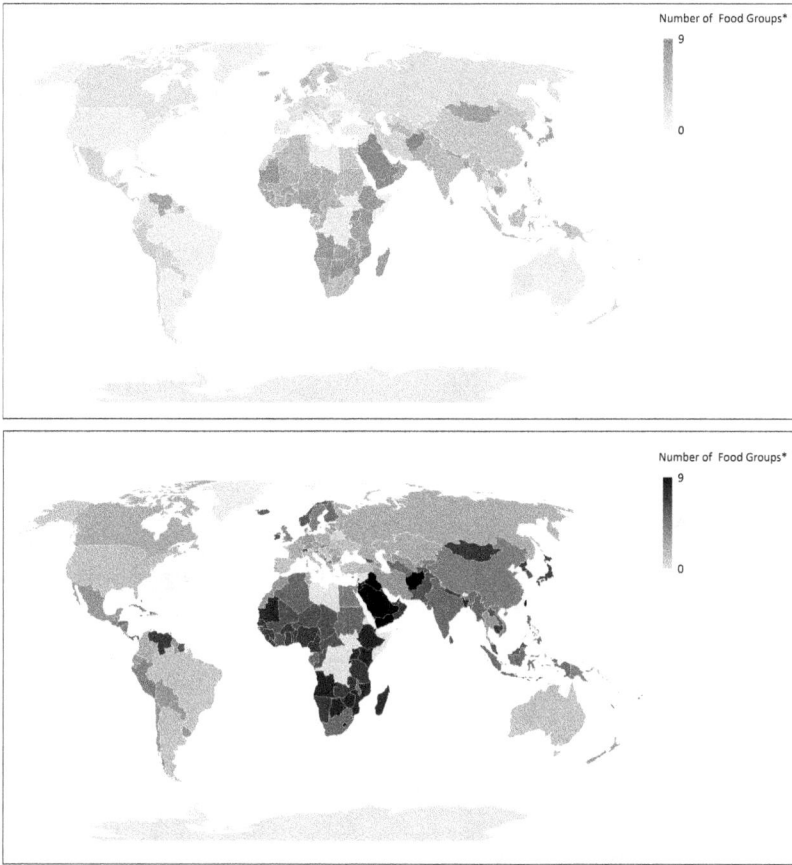

Figure 39.5 Number of Food Groups with insufficient local production adjusted by Food Loss and Waste. Source: Authors' computations based on FAOSTAT Food Balance Sheets (2022) and UNEP (2021) food waste estimates.

should be performed. A comparative study (Hawkes et al., 2015) of the effect of trade reform on food security found significant differences in the effects on food availability between countries. For example, in China per capita supplies of the principal nutrients grew significantly in the post-reform period; in contrast, rates of change were very modest in Malawi, and declined in the United Republic of Tanzania. Swinnen and Beerlandt (2002) looked at key reforms in trade and related economic policies and how they have affected agricultural development, trade, and food security in transition countries, noting that initial conditions played a major role following reform policies. They also noticed that the relative impact of the initial conditions decreased as transition progressed and the heterogeneity of the countries' experiences: China is an example an increase in production, productivity, and farmers' income right after trade and economic reforms started; Russia is an example of decreases in production after the reforms.

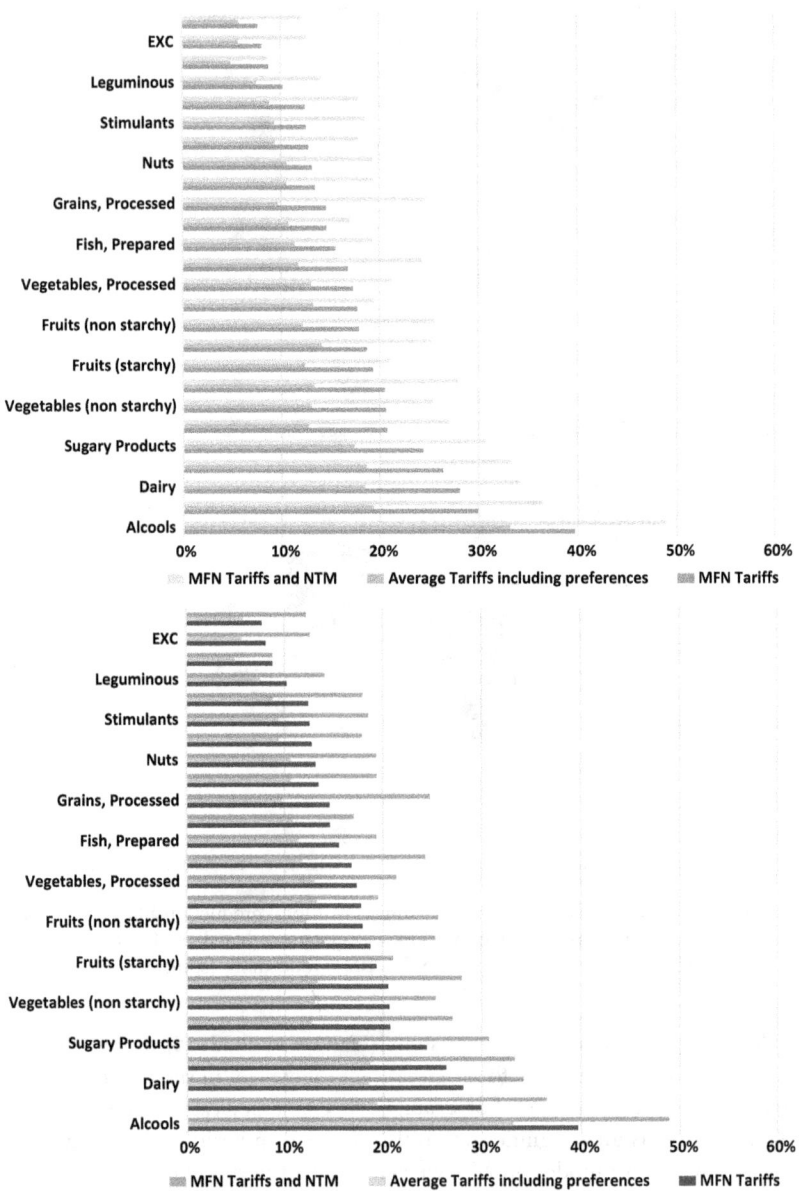

Figure 39.6 MFN, NTM and average tariffs by food group. Source: Authors' computations based on Agricultural Trade Monitor database, MacMAPHS6 and Bao et al. (2020). Note: Tariffs are weighed with global trade.

Observation is needed of the level of distortions—tariffs, with or without preferences, and controlling for non-tariff barriers.[6] Figure 39.6 shows that the international level of trade barriers is far from being homogenous across products. Some "nutritious" food groups—dairy, meat, and vegetables—are heavily taxed, as well as some food groups that are high in caloric content (alcohol, sugar). Similarly, grains and nutritious legumes, are not heavily taxed with a lower percent of Most-Favoured Nation Tariffs (MFN)[7] and Non tariffs Measures (NTM)[8] compared to the other products shown in Figure 39.6.

Of note is that NTM pushes prices to increase but in the same pattern as trade tariffs (e.g., a minor change in the ranking between other fruits and vegetables). Highlighting the high correlation between trade barriers—NTM and MFN tariffs—in this case (e.g., there are high tariffs on tomatoes and meat, as well as high NTMs).

Laborde et al. (2019) found that the use of trade tariffs or barriers to insulate domestic markets from food price shocks may not be effective in protecting the most vulnerable populations groups. During sharp food price spikes, evidence shows that households reduce purchases of nutrient-rich foods, prioritising the consumption of calorie-dense staple foods.[9] Trade policies can be used as part of a portfolio of policies and interventions to enable households, especially poorer ones, to be able to afford and access a sustainable diet.

Conclusion

A sustainable diet for everyone requires sustainable economic growth, from an environmental and climate point of view, compatible with nutritional requirements to eliminate or reduce all forms of malnutrition. Trade in agricultural products is an important factor both in terms of improved global food security but also in its impact on global dietary trends. Trade policies should accompany the country's objective of a sustainable diet—affordability, availability, adequacy, and accessibility—by working within existing trade rules to create better outcomes for nutrition.

In developing the right policies to accomplish sustainable diets, policy makers should look at the primary objectives. If the reduction in the level of consumption of a specific good is sought, a restriction on domestic sales of such food groups through a sales tax could be a more effective way to discourage consumption than only implementing an import tax. A sales tax affects all consumption of that good—imports and domestic production. Policies for sustainable diets are better targeted at producing and consuming the right kinds of foods and with the right production systems. But blocking trade through tariffs and other border measures is misguided and ignores the important role of trade in improving global food sufficiency and security and a better use of global natural resources. Policy makers should identify the domestic nutrition objectives first and then see how trade policies could influence the achievement of those nutritional objectives while analysing the complementary incentives and policies needed to maximise the synergies and minimise the risks of trade policies on sustainable diets.

Acknowledgements

This work was undertaken as part of the CGIAR Research Program on Policies, Institutions, and Markets (PIM) led by the International Food Policy Research Institute (IFPRI). Funding support for this study was provided by PIM. This publication has been peer reviewed and the opinions expressed here are those of the authors and are not necessarily representative of or endorsed by PIM, IFPRI, or CGIAR. We thank Pablo Elverdin for his contributions to this work.

Notes

1 For example, consumption of bushmeat and wet markets are suspected factors in the origins of Ebola outbreaks, HIV/AIDS, and other diseases. Although it is premature to draw conclusions, a wet market may have been a factor in the origin of the current COVID-19 pandemic, contributing to a statement by the Acting Executive Secretary of the Secretariat of the Convention on Biological Diversity, that policy measures may be necessary to mitigate the risk of future pandemics of zoonotic origin (Greenfield, Patrick. "Ban wildlife markets to avert pandemics, says UN biodiversity chief." *The Guardian*, April 6, 2020. www.theguardian.com/world/2020/apr/06/ban-live-animal-markets-pandemics-un-biodiversity-chief-age-of-extinction).

2 As an example, obesity can easily be understood as a socialised cost in countries with public healthcare systems, where taxpayers collectively finance obesity-related health costs. Outside of socialised healthcare systems, the argument around obesity as a socialised cost is more complex.

3 FAO Agricultural and Development Economics Division (June 2006). "Food Security." www.fao.org/fileadmin/templates/faoitaly/documents/pdf/pdf_Food_Security_Concept_Note.pdf

4 A predominantly plant-based diet, which contains small to moderate amounts of animal source foods. It includes at least 500 g/day of fruits and vegetables of different colours and groups (the composition of which is determined by regional preferences), at least 100 g/day of plant-based protein sources (legumes, soybeans, nuts), modest amounts of animal-based proteins, such as poultry, fish, milk, and eggs, and limited amounts of red meat (limit to one portion per week), refined sugar (< 5% of total energy), vegetable oils that are high in saturated fat (in particular palm oil), and starchy foods which have a relatively high glycaemic index. FAO et al., 2019.

5 In the absence of any distortion, such as taxes and subsidies, tariffs and NTBs, externalities, and incomplete information.

6 Trade policies affect medium to long term trends in the relative prices of foods, the affordability of different categories of food, and the income of everyone involved in the production of the products by influencing supply and demand.

7 In current usage, MFN tariffs are what countries promise to impose on imports from other members of the WTO, unless the country is part of a preferential trade agreement (such as a free trade area or customs union). This means that, in practice, MFN rates are the highest (most restrictive) that WTO members charge one another.

8 Non-tariff barriers to trade (NTBs; also called non-tariff measures, NTMs) are trade barriers that restrict imports or exports of goods or services through mechanisms other than the simple imposition of tariffs.

9 Global Panel, 2016.

References

Bao, N., A. Bouët, & Traoré, F. (2020). On the proper computation of ad valorem equivalent of non-tariff measures. *Applied Economics Letters*. Advance online publication. https://doi.org/10.1080/13504851.2020.1864273

Bouët, A., & Laborde Debucquet, D. (Eds.). (2017). *Agriculture, development, and the global trading system: 2000–2015*. IFPRI. Retrieved from https://doi.org/10.2499/978089 6292499

Burlingame, B., & Dernini, S. (2010). Sustainable diets and biodiversity, FAO.

Cuevas García-Dorado, S., Cornselsen, L., Smith, R., & Walls, H. (2019). Economic globalization, nutrition and health: A review of quantitative evidence. *Globalization and health, 15*(1), 15.

Demaio, A., & Branca, F. (2017). Decade of action on nutrition: Our window to act on the double burden of malnutrition. *British Medical Journal - Global Health, 2*, e000492.

Deng, L., Yin, J., Tian, J., Li, Q., & Guo, S. (2021). Comprehensive evaluation of water resources carrying capacity in the Han River Basin. *Water, 13*(3), 249.

Dithmer, J., & Abdulai, A. (2017). Does trade openness contribute to food security? A dynamic panel analysis. *Food Policy, 69*, 218–230.

FAO, IFAD, UNICEF, WFP & WHO (2019). *The state of food security and nutrition in the world 2019*. FAO. Retrieved from https://www.unicef.org/media/55921/file/SOFI-2019-full-report.pdf

Gómez, M., Barrett, C. B., Raney, T., Pinstrup-Andersen, P., Meerman, J., Croppenstedt, A., Carisma, B., & Thompson B. (2013). Post-green revolution food systems and the triple burden of malnutrition. *Food Policy, 42*, 129–138.

Hawkes, C., Grace, D., & Thow, A. (2015). Trade Liberalization, food, nutrition and health. In R. Smith, C. Blouin, Z. Mirza, P. Beyer, & N. Drager (Eds.). *Trade and health: Building a national strategy* (pp. 92–116). WHO. Retrieved from https://apps.who.int/iris/rest/bitstreams/825170/retrieve

Headey, D. D., & Alderman, H. H. (2019). The relative caloric prices of healthy and unhealthy foods differ systematically across income levels and continents. *The Journal of nutrition, 149*(11), 2020–2033.

Laborde, D. (2017, unpublished). International trade and natural resources: A sustainable path towards global food security. Presentation at the US-China Oilseeds and Grains Annual Forum, Beijing, June 17. 2017.

Laborde, D., & Deason, L. (2015). *Trade and nutrition contents*. [Unpublished manuscript].

Laborde, D., Lakatos, C., & Martin, W. (2019). *Poverty impact of food price shocks and policies* (Policy Research Working Paper no.8724). World Bank Group. Retrieved from https://openknowledge.worldbank.org/bitstream/handle/10986/31228/WPS8724.pdf?sequence=1&isAllowed=y

Laborde, D., Parent, M., & Piñeiro, V. (2020). *The true cost of food: Task Force 10: Sustainable energy, water, and food systems*. T20. Retrieved from https://t20saudiarabia.org.sa/en/briefs/Documents/T20_TF10_PB17.pdf

Martin, W. (2017). *Agricultural trade and food security (Asian development bank institute working paper series no. 664)*. ADBI. Retrieved from https://www.adb.org/sites/default/files/publication/228906/adbi-wp664.pdf

Masters, W. A., Bai, Y., Herforth, A., Sarpong, D. B., Mishili F., Kinabo, J., & Coates, J. C. (2018). Measuring the affordability of nutritious diets in Africa: Price indexes for diet diversity and the cost of nutrient adequacy. *American Journal of Agricultural Economics, 100*(5), 1285–1301.

Olper, A., Curzi, D., & Swinnen, J. (2018). Trade liberalization and child mortality: A synthetic control method. *World Development, 110*, 394–410.

Pastor, A. V., Palazzo, A., Havlik, P., Biemans, H., Wada, Y., Obersteiner, M., Kabat, P., & Ludwig, F. (2019). The global nexus of food–trade–water sustaining environmental flows by 2050. *Nature Sustainability, 2*(6), 499–507.

Remans, R., Wood, S., Saha, N., Anderman, T., & DeFries, R. (2014). Measuring nutritional diversity of national food supplies. *Global Food Security, 3*, 107–182.

Schram, A., Ruckert, A., VanDuzer, J. A., Friel, S., Gleeson, D., Thow, A. M., Stuckler, D., & Labonte, R. (2018). A conceptual framework for investigating the impacts of international trade and investment agreements on noncommunicable disease risk factors. *Health Policy and Planning, 33*(1), 123–136.

Swinnen, J., & Beerlandt, H. (2002, July 11–12). Trade and related economic reforms in Transition Economies: What were the impacts of actual policy changes on agricultural development, trade and food security? [Conference presentation] FAO Expert Consultation on Trade and Food Security: Conceptualizing the Linkages, Rome, Italy.

Wood, S., Smith, M., Fanzo, J., Remans, R., & DeFries, R. (2018). Trade and the equitability of global food nutrient distribution. *Nature Sustainability, 1*, 34–37.

40

CIRCULAR BIOECONOMY OF AGRI-FOOD VALUE CHAINS

Innovative, sustainable, and circular business models' contributions to sustainable diets and food systems

Paolo Prosperi

Introduction

According to the High Level Panel of Experts on Food Security and Nutrition (HLPE), the global food system is composed of environmental, societal, institutional, and infrastructural elements, including inputs, processes, and all relevant activities for food production, processing, distribution, preparation and consumption, and the related socio-economic and environmental outcomes (HLPE, 2017). Currently, the globalised food system is characterised by its exposure to multiple drivers of change and by its unsustainability in terms of environmental and socio-economic impacts of food production and consumption. There is a pressing need to orient global strategies towards a more sustainable food system that "delivers food security and nutrition for all in such a way that the economic, social and environmental bases to generate food security and nutrition for future generations are not compromised" (HLPE, 2014, p. 31). Moreover, great amounts of agricultural products and food are regularly wasted; food systems need to be reconsidered to lessen or avoid system inefficiencies.

Food consumption is therefore strongly connected with the sustainability outcomes of food systems, as it can shape the environmental, social, and economic impacts of systems, including the dynamics related to food and waste. A key dimension of sustainable diets endorses practices and techniques from production to consumption that contribute to preserving the ecological environment by maintaining or reducing greenhouse gas emissions, controlling water and land use, nitrogen and phosphorus inputs, and chemical pollution. It also includes efforts to

DOI: 10.4324/9781003174417-47

reduce plastics in food packaging and minimise the loss and waste of agricultural products and food.

Enhanced knowledge and tailored sustainable practices are needed in all phases from production to consumption, including practices that shape consumption behaviours and take into account the efficient management of biophysical materials. These complex behavioural dynamics, involving a plethora of activities and stakeholders, require organisation of practices, activities, and interactions dedicated to sustainability. Businesses and entrepreneurs play crucial roles in building and managing food supply that support sustainable consumption. Thus, sustainable diets require business models that consider the complexity of food systems, the dynamics among different stakeholder roles, and the efficient management of biophysical materials.

How circular bioeconomy contributes to sustainable food systems and diets

The transition to more sustainable food systems and diets strongly builds on the approach of bioeconomy. Bioeconomy is an economic sector whose basic activities of production—transformation and valorisation—and building blocks—materials, chemicals, and energy—originate from living matter and renewable biological resources instead of non-renewable fossil resources (Allain et al., 2022; Diakosavvas & Frezal, 2019; McCormick & Kautto, 2013). In current strategic planning for policies related to economic growth and ecological transition, bioeconomy is increasingly considered by states and intergovernmental organisations as a strategy to support United Nations Sustainable Development Goals (SDGs) and the Paris Climate Agreement. The agricultural and food sectors are central to bioeconomy as they provide the biological resources for production and are a major supplier of biomass (Diakosavvas & Frezal, 2019). In general, the bioeconomy approach represents a critical opportunity to address wicked societal challenges such as food security, climate change, economic development, and limited natural resources. However, because of an increased competition between food supply and non-food biomass production, bioeconomy cannot inherently be defined as sustainable since tensions and economic, environmental, and social trade-offs emerge between the different allocations of food, feed, fuel, and fibre (Allain et al., 2022). Moreover, the bioeconomy concept was originally intended as a component of environmental economics theories and has a weak sustainability stance, since it pledges substitutability between human capital and natural capital and doesn't target a complete change of the dominant economic system (Loiseau et al., 2016).

Bioeconomy is an important opportunity for improving the sustainable development of agri-food systems, by activating interventions around inefficiencies in linear food economies due to loss of productivity, energy, natural resources, and waste production. In particular, these food system inefficiencies generate high levels of pollution, greenhouse gas emissions, and related environmental, social, and economic costs (Jurgilevich et al., 2016). A bioeconomy approach seeks to "contribute(s) to sustainable production and consumption and resource-use efficient agri-food systems in an economically, socially and environmentally sustainable manner" (FAO, 2021, p. 1); consequently, new business models and innovations become essential. In so doing, biomass needs to be valorised across value chains. It is therefore necessary to overcome a linear production approach by the "cascading use" of biomass and the reuse of waste materials (Diakosavvas & Frezal, 2019). These techniques are more specific to a circular economy approach as a set of practices and tools to improve the sustainability of food systems.

According to Haas et al. (2016), circular economy is an economic sector that "improve(s) resource efficiency mainly by closing the resource loop and by stopping the wasteful use of resources" (p. 261). Circular economy follows the steps of industrial ecology and is a concept grounded in theories of ecological economics. It takes a macroeconomic approach that integrates a strong sustainability perspective; in contrast with the bioeconomy, circular economy stands for radical change from the linear economic system (Loiseau et al., 2016) and aims at enhancing the life of materials and resources by increasing efficiency and recycling. With regards to food systems, circular economy aims to reduce agricultural and food waste, lessening the environmental and social impact while preserving economic growth. Circular bioeconomy can significantly reduce the negative impacts of resource extraction and pressure on the environment and can contribute to restoring biodiversity and natural capital. For example, business models of circular bioeconomy keep biomass local as long as possible and replace petrol-based and non-renewable materials with ones that are bio-based. With regards to biodiversity protection and restoration, circular bioeconomy is also characterised by agroecological practices that foster intraspecific crop diversity to manage pests and pathogens, and also build on green manuring, rotations that include legumes, soil cover, integrated plant nutrient management, conservation agriculture, and integrated manure management (Gomez San Juan & Bogdanski, 2021). All flows among the agri-food value chains—including input supply to consumption, waste, and recycling—are to build in closed-loop food systems (Lu & Halog, 2020). It may be more accurate to refer to circular bioeconomy when circular economies overlap with bioeconomy by encompassing production, consumption, and waste valorisation of bio-based resources to minimise environmental impact, to improve efficiency, to harness the full potential of materials, and to create value-added products such as bio-based products, bioenergy, food, and feed (Carus & Dammer, 2018). Circular bioeconomy practices involve sharing, reusing, remanufacturing and recycling, cascading use, utilisation and valorisation of organic waste and side streams, bio-based products, and improving resource efficiency.

In practice, circular bioeconomy is a system that foreground goals of health and environmental sustainability into food systems through orienting public policies and consumer food demand towards the preservation of natural landscapes and by environmentally-friendly and healthy production (Campos & Madureira, 2019). Therefore, food systems based on circular bioeconomy aim to be regenerative, resilient, non-wasteful, and healthy, through techniques that reconnect nutrient loops for restoring degraded soils, minimising the use of fertilisers and pesticides, and thus, cultivating non-toxic, healthier, and less wasteful food supplies (Ellen MacArthur Foundation, 2015). For instance, in circular bioeconomy systems, fertilisers can be obtained from natural production of bacterial biofilms as bio-fertilisers, bio-pesticides can be produced based on organisms such as bacteria, fungi, viruses, yeasts, and others, and bio-plastics for packaging can be obtained through lactic acid fermentation or bacterial polyester fermentation (Gomez San Juan & Bogdanski, 2021). Such practices should be accompanied by tailored business design for less costly production that is directly linked with consumption and regenerative techniques. Digital solutions have an important role in facilitating the coordination of flows of information and materials between stakeholders since they allow data gathering, sharing and analysis to design the most appropriate business models for resource use, efficiency and impact. A transition of food systems towards a circular bioeconomy would, therefore, strengthen their sustainability and resilience, by valorising biomass resources and waste as well as by implementing disruptive social and technological innovations.

Building specific business models for circular bioeconomy within agri-food value chains

While sustainability and circularity are currently considered as coupled principles contributing to the functioning of social-ecological systems, circularity is still difficult to operationalise in the transition of food systems towards sustainability (Wigboldus, 2020). The functioning principles of circular bioeconomy simulate dynamics from natural systems to optimise the efficiency of systems (Ellen MacArthur Foundation, 2015). If the dominant economic model of our food system is linear and follows a "take-produce-consume-discard" logic (Jurgilevich et al., 2016, p. 2), the circular economic model needs to assume that economic growth moves beyond the assumptions of abundant primary resources and unlimited waste disposal, and considers waste as a resource through practices of input reduction, reuse, repair, refurbishing, and recycling of existing materials and products. For circular bioeconomy to substantially influence food systems, circular business models need to be activated to reduce waste, re-use food, use by-products and food waste, and recycle nutrients.

Food systems are an overarching opportunity to apply circular bioeconomy as a new economic model towards SDGs; however, blending theory with practice is not an easy task (Fassio & Tecco, 2019). This challenge must be addressed by combining feasible circular business practices consistent with the circular bioeconomy theory, while monitoring the impacts on efficiency and sustainability of the interventions. For this purpose, coherent and efficient sustainable business models that contribute to sustainable diets are central. A business model describes the logic of a business and it explains how companies and individuals (e.g., farmers, processors, entrepreneurs, industrial organisations, cooperatives, etc.) create, deliver (to customers and consumers), and capture value. Value refers to economic value and, more broadly, involves socio-economic and ecological values (Casadesus-Masanell & Ricart, 2010; Rosenstock et al., 2020; Teece, 2010). More specifically, in relation to business models in circular bioeconomy, according to Nußholz's (2017) definition,

> a circular business model is how a company creates, captures, and delivers value with the value creation logic designed to improve resource efficiency through contributing to extending useful life of products and parts (e.g., through long-life design, repair and remanufacturing) and closing material loops.
>
> *(p. 16)*

These sustainable business models imply significant shifts to generate sustainable agri-food value chains, and require a re-think of how to organise and carry out practices and business while obtaining consistent benefits and return on capital investments (Hilmi, 2018). In that sense, a business, its entrepreneurship, and the connected economic actors and stakeholders of the value chains—and, therefore, their business model organisations—need to function in an efficient way, while reducing impacts on inputs, land, energy, and water resources and providing food sustainability and resilience to stress and shocks (ESCWA, 2014). The constant necessity in circular bioeconomy is to provide and recapture value from business at each stage of the value chain, by lessening the use of ecological assets and alleviating detrimental impacts (or even providing positive outcomes) while also implementing disposal and recycling patterns of generated waste (Hilmi, 2018). Therefore, providing and recapturing value at each stage of the value chain requires a strong coordination for a strategic and efficient value chain management of available material flows, as well as clear governance arrangements for decision and policy making in connection with institutional stakeholders.

Circular bioeconomy needs specific governance for sustainable diets and food systems

Building on these assumptions, it is clear that circular business models are urgently required to shift toward food production and processing that are more sustainable and to consumer dietary patterns that are healthier. While a business model can be conceived at an individual level, it is practically interconnected within an ecosystem of stakeholders of the value chain as well as from institutions. A European study showed that for circular bioeconomy to be effective, the involvement of diverse actors in participative governance was crucial, but still rarely used (Overbeek et al., 2016). For value chain dynamics, governance represents "how various firms across the entire chain are coordinated (or strategically linked) in order to be more competitive and add more value" (FAO, 2014a, p. 9). Moving towards sustainable business models needs to go beyond re-thinking production, consumption, and recycling practices and the related environmental sustainability, economic profitability, and social viability. It requires re-thinking of how organisations, firms, and various private and public stakeholders and institutions throughout the entire chain are coordinated and interact to strategically achieve common sustainability goals. The development of sustainable and circular business models will then need to embed specific governance outputs and dynamics that improve coordination and efficiency among stakeholders interacting in value chains.

According to Kanie et al. (2014), governance is considered the "fourth pillar of sustainable development (complementing the environmental, social, and economic pillars)" (p. 6). Consistently, governance is also included as one of the four dimensions of sustainability for agri-food systems (FAO, 2014b). In food systems, governance includes the interaction of public actors (e.g., local authorities, governments, intergovernmental organisations, etc.), civil society (e.g., non-governmental organisations and social movements), and private sector actors (e.g., businesses, producer organisations, coordinated value chains, etc.) (HLPE, 2020). More specifically, governance within agri-food supply chains "refers to the nature of the linkages both between actors at particular stages in the chain (horizontal linkages) and within the overall chain (vertical linkages)" (FAO, 2014a, p. 10). However, from an operational point of view, the governance approach cannot only refer to business-related elements. Key also for the development and transition to sustainable business models are elements such as information exchange, price determination, standards, payment mechanisms, contracts, market power, lead firms, and wholesale market systems. Governance, with related decision support tools, is one of the main areas of action for implementing food system transformation, together with the economic, political, cultural, and social aspects (Béné et al., 2020). Integrating governance in the design of sustainable business models for food systems implicates the involvement of different actors—especially consumers—in decision making and deliberative dynamics (Wilkins, 2005). For example, stances from consumer initiatives can be integrated in the design of sustainable business models, similar to what occurs in the co-construction of food policy plans, such as in the functional model of The Milan Urban Food Policy Pact. This integration would encourage and strengthen the control of local communities over agricultural and food systems (El Bilali et al., 2021). It is necessary to promote coordination between actions and stakeholders within the cycles of production and consumption and, therefore, coordination between goals for economic growth and environmental and social objectives at different time scales (short and long terms) (Fassio & Tecco, 2019). Basically, what is needed is what Fritsche et al. (2020) define as *sustainability governance* that coordinates the integration of flows of resources, materials, and information in activities of recycling and re-use of residues and waste. This governance would contribute to avoiding risks of lacking availability and access to resources in regions from which biomass would be imported.

Previous research has developed frameworks for circular business models that include holistic considerations of the business environment within which circular bioeconomy activities are carried out (Antikainen & Valkokari, 2016; Joyce & Paquin, 2016). However, specific and systemic governance components—such as fairness, transparency, ethics, accountability, etc.—were not identified and tackled in depth for circular bioeconomy. For these reasons, an innovative design of circular bioeconomy business models is proposed here, within the circular bio-economy dynamics, through adding the governance dimension beside the original Osterwalder and Pigneur's (2010) Business Model Canvas and the Joyce and Paquin's (2016) Triple Layer Business Model Canvas, as shown in Figure 40.1.

The new framework shown in Figure 40.1 illustrates the circular flow within a bioeconomy (production and manufacturing, consumption and use, recycling and reuse) characterised by multidimensional management (economic, environmental, social, governance) of a business model. Knowledge production and the creation of innovation in the field of sustainable development are important for society and emerge from knowledge exchange between the five societal subsystems of the Quintuple Helix model (Carayannis et al., 2012) that are represented by natural capital, economic capital, human capital (i.e., education systems), information and social capital (i.e., media-based and culture-based public), and political and legal capital.

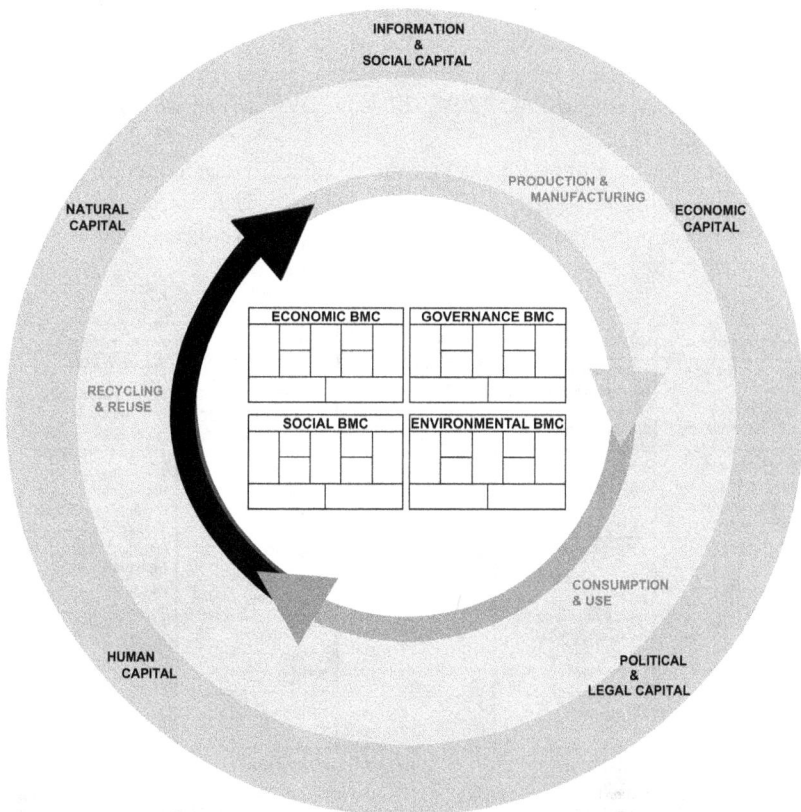

Figure 40.1 Multidimensional circular bioeconomy Business Model Canvas (BMC) framework. Source: Created by the author.

Quadruple Layered Business Model Canvas: Circular bioeconomy business for sustainable diets

The original Business Model Canvas structures the economic functioning of a business in nine organisational components: Value proposition; customer segments; channels; customer relationships; activities; resources; partners; and costs and revenues. The Environmental Business Model Canvas is composed of nine building blocks: Functional value; use phase; distribution; end-of-life; production; materials; supplies and out-sourcing; environmental impacts; and environmental benefits. The Social Business Model Canvas is structured by nine components: Social value, end-user, scale of outreach, societal culture, governance, employees, local communities, social impacts, and social benefits (see Osterwalder & Pigneur (2010), for the original Business Model Canvas, and Joyce & Paquin (2016), for the environmental and social models). While a governance component is included in the latter Social Business Model Canvas, as mentioned above, the specific governance building blocks for a systemic approach to circular bioeconomy business environments are not identified or developed. Therefore, the new Governance Business Model Canvas of circular bioeconomy builds on a novel organisation of a set of governance components, inspired by different literature sources such as FAO (2013, 2014a, b), Devaney et al. (2017), and El Bilali et al. (2021).

The Governance Business Model Canvas of circular bioeconomy is shown in Figure 40.2 and is composed of nine building blocks: 1) Effectiveness and efficiency value; 2) participation and inclusiveness; 3) business ethics, 4) holistic man-

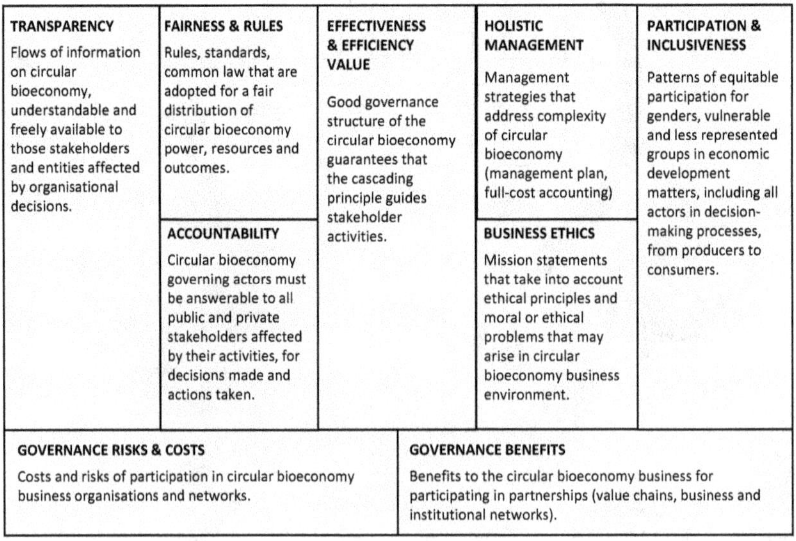

TRANSPARENCY	FAIRNESS & RULES	EFFECTIVENESS & EFFICIENCY VALUE	HOLISTIC MANAGEMENT	PARTICIPATION & INCLUSIVENESS
Flows of information on circular bioeconomy, understandable and freely available to those stakeholders and entities affected by organisational decisions.	Rules, standards, common law that are adopted for a fair distribution of circular bioeconomy power, resources and outcomes.	Good governance structure of the circular bioeconomy guarantees that the cascading principle guides stakeholder activities.	Management strategies that address complexity of circular bioeconomy (management plan, full-cost accounting)	Patterns of equitable participation for genders, vulnerable and less represented groups in economic development matters, including all actors in decision-making processes, from producers to consumers.
ACCOUNTABILITY Circular bioeconomy governing actors must be answerable to all public and private stakeholders affected by their activities, for decisions made and actions taken.			**BUSINESS ETHICS** Mission statements that take into account ethical principles and moral or ethical problems that may arise in circular bioeconomy business environment.	
GOVERNANCE RISKS & COSTS Costs and risks of participation in circular bioeconomy business organisations and networks.			**GOVERNANCE BENEFITS** Benefits to the circular bioeconomy business for participating in partnerships (value chains, business and institutional networks).	

Figure 40.2 Governance Business Model Canvas: Circular Bioeconomy Business. Source: Created by the author.

agement; 5) fairness and rules; 6) accountability; 7) transparency; 8) governance risks and costs; and 9) governance benefits.

The central governance value proposed from a circular bioeconomy business—such as individual or collective business, a business organisation, or a value chain—is identified in the *Effectiveness and Efficiency* of coordination arrangements between economic actors, which allows the best use to be made of resources without harming the environment and ensures that the cascading principle guides stakeholder activity (Devaney et al., 2017). Circular bioeconomy business needs to be based on *Equitable Participation* of genders and vulnerable and emerging groups, including all actors in decision-making processes from producers to consumers (Devaney et al., 2017). *Business Ethics* (or *Corporate Ethics*) in a circular bioeconomy business environment addresses ethical principles and moral or ethical problems (e.g., according to a mission statement) (El Bilali et al., 2021; Gualandris & Kalchschmidt, 2016). *Holistic Management* guarantees the system complexity of circular bioeconomy (e.g., through management plan and full-cost accounting) (El Bilali et al., 2021; Savory & Duncan, 2016). *Rules (and Fairness)* standards (e.g., labelling and criteria for bio-based goods) and common law must be followed for developing circular bioeconomy business and a fair and impartial distribution of bioeconomy power, resources, and outcomes must exist (Devaney et al., 2017). With regards to *Accountability*, public and private actors must be answerable for their decisions and actions to the public, institutional stakeholders, and those affected by their activities (Devaney et al., 2017). With regards to *Transparency*, in a circular bioeconomy business environment information must be understandable and made freely available to those affected by organisational stakeholder decisions (Devaney et al., 2017). Activating and maintaining governance implicates *Risks and Costs* for the business model that invests in network participation and engagement, as well as *Benefits* from partnership integration, such as in value chains and business and institutional networks (FAO, 2014a).

In a nutshell, a circular bioeconomy business model that effectively contributes to sustainable food systems and diets will need to propose not only economic, environmental, and social performance but also effective and efficient performance in terms of systemic governance (i.e., coordination within a business environment composed of many private and public stakeholders). For this to be operationalised, targeted efforts and investments in terms of guaranteeing information transparency, stakeholders' accountability, and the establishment of rules and fairness will support stakeholder participation and inclusiveness, key holistic approaches for management strategies and ethics in business.

Table 40.1 provides the development of a multidimensional business model canvas, with the aim of structuring a rationale that explains how the economic, environmental, social, and governance dimensions of circular bioeconomy contribute to sustainable diets and food systems. The logic of each business model builds on the principles and practices that explain the economic, environmental, and social roles and functioning of circular bioeconomy and the related outcomes of sustainable diets and food systems, as illustrated in the first three sections of this chapter and in Figure 40.1. For the governance dimension, as introduced earlier,

Table 40.1 Quadruple Layered Business Model Canvas: Circular Bioeconomy Business contributing to Sustainable Diets

	Economic Business Model Canvas	Environmental Business Model Canvas	Social Business Model Canvas	Governance Business Model Canvas
Value Proposition	*Value proposition* Supply of accessible and profitable agricultural and food products.*	*Functional value* Supply of agricultural and food products from environmentally friendly practices.*	*Social value* Supply of healthy, nutritious, fair, desirable, culturally respectful agricultural and food products that provide welfare and wellbeing for people, animals, eco-systems.*	*Effectiveness & Efficiency value* Good governance actions that ensure that the cascading principle guides stakeholder activity and coordination.*
Value Delivery	*Customer segments* All people. Consumer diversity according to nutritional, cultural and social needs, and preferences. *Channels* Available channels adapted for each diverse group of consumers (large retail, local/short supply chains, Internet, green public procurement, etc.). *Customer relationships* Affiliation, communication schemes, awareness campaigns, sustainability labels, standards, certifications.	*Use phase* Consumer management of food products (e.g., water and energy consumption). *Distribution* Low environmental impact means of transportations (including packaging), sustainable logistics management. *End-of-life* Consumer behaviour in managing food loss, waste, packaging use, reuse, recycling, disposal (e.g., level of adoption of circular food behaviours).	*End-User* Consumer acceptance and appreciation in terms of taste, physiological needs, cultural preferences. *Scale of Outreach* Diffusion and communication of actions, knowledge and benefits around circular bioeconomy/sustainable diets principles across involved and affected stakeholders. *Social Culture* Consultation and deliberative practices to connect circular bioeconomy/sustainable diets actions more strongly to society (e.g., social responsibility, social dialog platforms).	*Participation & Inclusiveness* Patterns of equitable participation and inclusion of all actors concerned in the agri-food environment. *Business Ethics* Establishment of a business mission statement including ethical problems that might arise in circular bioeconomy/sustainable diets process. *Holistic Management* Implementation of strategies that consider the complexity of circular bioeconomy/sustainable diets dynamics through a system approach.

Value Creation			
Activities	*Production*		*Fairness & Rules*
Agriculture and food related activities (input supply, primary production, processing, packaging, distribution, retail, marketing, logistics).	Production activities based on environmentally friendly practices and on the "cascading use" of biomass and the reuse of waste materials: Sharing; reusing; remanufacturing; recycling; minimisation, utilisation and valorisation of organic waste; improving resource and eco-efficiency; regenerative agriculture; agroecology; packaging with organic or reusable materials).		Normative tools, rules, standards, common schemes that allow equal power within agri-food value chains and connected business and institutional environments.
Resources	*Materials*		*Accountability*
Agricultural land, processing structures, distribution infrastructures, patents, brands, workforce and intellectual resources, financial assets, informatics, etc.	Bio-physical stocks used to render the functional value. Bio-based materials used to provide sustainable food and packaging from environmentally friendly and circular bioeconomy practices.		Traceability and measurement of each stakeholder regarding each role, responsibility and decision made.
Partnerships	*Supplies and outsourcing*	*Governance*	
Network of suppliers and partners that make the business model work (all actors involved in agri-food systems).	Primary resources used such as water, energy.	Forms of organisations of the business involved in circular bioeconomy/sustainable diets. Main principles for establishing fair arrangements with stakeholders (transparency, accountability, etc.).	
		Employees	
		Employees engaged in circular bioeconomy/sustainable diets. Their working and welfare conditions, training, skill development, level of expertise, support programs.	
		Local Communities	
		Relationships—with local communities as resource and materials suppliers—considering the respect of main circular bioeconomy/sustainable diets principles.	
		Transparency	
		Information system (including digital technology) that guarantee traceable, free, and understandable flows of information across agri-food value chain steps.	

(Continued)

Table 40.1 (Continued)

	Economic Business Model Canvas	Environmental Business Model Canvas	Social Business Model Canvas	Governance Business Model Canvas
Value Capture	*Costs* Economic costs from all activities, resources extraction and utilisation, partnerships developed.	*Environmental impacts* Ecological costs from production and distribution activities, materials' utilisation, resource consumption, supplies exploited, emission activities.	*Social impacts* Social costs of resource consumption, natural capital losses, health effects, working conditions, cultural impacts.	*Governance Risks & Costs* Governance negative impacts from lack of coordination or information shared, transaction costs, risks of inefficiency and corruption, maladministration, excessive bureaucracy, unfair competition, power imbalances, exclusion, and top-down approaches.
	Revenues Economic benefits from revenues, sales, trade.	*Environmental benefits* Environmental impact reduction, ecological and regenerative actions from all production, distribution, exploitation, optimisation, consumption actions.	*Social benefits* Social benefits from circular bioeconomy/ sustainable diets actions (e.g., skill development, community partnership, local wellbeing, protection of cultural traditions).	*Governance Benefits* Governance positive impacts from increased accountability, transparency, effectiveness, fairness and participation (including bottom-up and deliberative approaches).

Source: Created by the author.

* From/within circular bioeconomy process / Sustainable and bio-based food products.

the new business model canvas builds on the building blocks shown in Figure 40.2. Here, business is understood in its largest sense, from the micro level of individual business to collective business (e.g., clusters, cooperatives, etc.), as well as large business organisations and entire value chains.

Strategies for effective circular bioeconomy business towards sustainable diets

Science and policy are increasingly acknowledging the key role of innovations in circular bioeconomy for triggering transition and opportunities in agriculture, fishery, forestry, and waste sectors by providing biomass, increasing carbon sinks, and managing land and marine ecosystems sustainably, while improving employment and value creation in rural areas (Fritsche et al., 2020). In this process, strong connections with society can help to shift from an economy based on fossil resources to a circular bioeconomy more oriented towards renewable biological resources (Overbeek et al., 2016). Building on previous works and literature, this chapter has so far argued that for this transformative approach to be operationalised, specific sustainability governance is necessary, also to guarantee and improve inclusiveness through people and stakeholder participation. A participatory governance of circular bioeconomy should consist of proactive involvement of private and public stakeholders through communication, consultation, and deliberative practices. Specifically regarding business, the implementation of circular bioeconomy practices can be supported by policy tools and guidelines that foster participation of all stakeholders, including citizens, as well as the engagement of public and private networks, with particular emphasis on local clusters of enterprises for collective actions. Such networks should be engaged to identify their mission and objectives for the improvement of sustainable diets and food systems through circular bioeconomy, and the instrumental roles, activities and responsibilities of each actor and supportive bodies, such as institutions, consumers, and civil society networks.

The technical aspects of a collective transition to circular bioeconomy requires, in parallel, governance and stakeholder coordination at different scales—through rules and institutions to secure a general equilibrium and viability of social, political, cultural, environmental, and economic factors. In general, a good governance process will allow movement towards improved efficiency in private/public management and administration, modernisation of value chains, transparency, fairness, participation, and avoiding corruption (FAO, 2014a; Devaney et al., 2017). In fact, innovative technologies and food waste reduction alone, within a circular bioeconomy for sustainable diets, cannot reorient market and institutional imperfections against waste and overexploitation of natural resources and environmental and health externalities. Those management imperfections are manifest as social costs emerging from natural capital consumption and loss, as well as from information and communication asymmetries and health impacts. To overcome these pitfalls, it is possible to act on supply and demand by combining market-oriented and technology-driven measures that include: a) Sustainability labels, standards, and certification for new bio-based products; b) green public procurement initiatives;

c) awareness campaigns for consumers about circular bioeconomy food products (claiming the virtuous sustainability character of such products); d) evaluating and improving the sustainability performance of local, healthy food supply chains with less waste; and e) supporting regenerative agricultural practices such as agroecology through large-scale retailing (Diakosavvas & Frezal, 2019; Ellen MacArthur Foundation, 2015). For circular bioeconomy business, strategic efforts need to be oriented towards coordination, integration, participation, trust and knowledge, and information sharing within private and public stakeholder networks. These actions need to be implemented while encouraging supply through policy incentives and regulatory frameworks for different economic uses of biomass such as food, feed, bio-based products, and bio-energy, to enhance the value generated from biomass and creating value chains. In the meantime, the demand for sustainable food from circular bioeconomy needs to be propelled by setting the conditions for informed consumer behaviour—such as comprehensible labels—as well as through communication on the benefits generated across value chain steps and actors from using biobased products (Diakosavvas & Frezal, 2019).

On the demand side, consumer behaviour is crucial in the transition towards circular bioeconomy businesses in food systems as food consumption impacts human health, the environment, economy, and society. It is, therefore, necessary to understand which options of circular food behaviours exist and how consumers perceive them, and what are the related consumer needs, willingness, and degree of acceptance. Food consumption patterns are tightly connected to collective and personal identity, traditions, food culture, and value orientations of citizens (Jurgilevich et al., 2016). In particular, the factors that influence the acceptance of consumers with regards to sustainable diets from circular bioeconomy are multiple, diverse, and can coexist within the same contexts. More specifically, business-to-consumer relationships are considered a pillar for consumer acceptance, with specific regards to food businesses that integrate circular bioeconomy practices. The main drivers of consumer acceptance and participation include the awareness and dependency of food provisioning (strong supplier-consumer relationship), the size of the city where consumption is targeted (in urban contexts recycling activities are more likely to be implemented), convenience of recycling practices (opportunity of cost reduction, or compensation, from recycling), social awareness (social pressure for recycling), and consumers' education and income (Borrello et al., 2020). Awareness campaigns and actions should also consider that there are different levels of adoption even within consumers that already accept and proactively implement practices of circular bioeconomy in their dietary habits (e.g., preventing and reducing food waste, buying foods with upcycled[1] ingredients, purchasing food with packaging made of renewable material, etc.). Do Canto et al. (2021) classified different circular food behaviours that can contribute to sustainable diets as: Linear behaviours (adoption of the available options in the market and reducing food waste to positively impact on the environment), transitioning behaviours (adoption of strongly innovative products, upcycled food products, surplus food, seasonal foods, packaging-free foods, participation in alternative food networks and food waste initiatives), and circular behaviours (directly applying circular practices at home, closing the resource loop through product reuse and recycling). Institutional and research projects, aimed at different scales at involving consumers in circular bioeconomy for sustainable food systems and diets, can be implemented through different participatory tools and techniques, such as circular food design (Sijtsema et al., 2020), building inclusive business models (Rosenstock et al., 2019), persuasive communication on "pro-circular" behaviours and values (Muranko et al., 2018), as well as the implementation of Participatory Guarantee Systems (PGS).[2]

Conclusion

Such diversity and multidimensionality of opportunities, targets, actors, interactions, level of intervention, performance, and outcomes within circular bioeconomy in agri-food sector activities call for designing new policy mixes and new evaluation approaches for sustainable food systems and diets (Galli et al., 2020), supported by a rigorous, transparent and fair flow of knowledge across the Quintuple Helix pillars (i.e., economic system, education system, media and culture, political system and natural environment). The multidimensional circular bioeconomy framework and the Governance Business Model Canvas developed in this chapter aim to contribute to this multifaceted process of knowledge creation, sharing, and innovation for evidence-based evaluation and policy formulation. Therefore, multilevel governance is central for implementing such policy tools—through strong bottom-up approaches and participation of relevant stakeholders—to effectively address challenges in agriculture, food, rural development, environment, forestry, energy, research and innovation, waste, and climate change, and to accelerate the development of circular bioeconomy in agricultural and food systems.

Notes

1 "Upcycled ingredients and food products elevate food that would otherwise be wasted to higher uses and have tangible benefits to the environment and society." (Spratt et al., 2021, p. 7).
2 "Participatory Guarantee Systems (PGS) are locally focused quality assurance systems. They certify producers based on active participation of stakeholders and are built on a foundation of trust, social networks, and knowledge exchange" (IFOAM—Organics International, n.d.).

References

Allain, S., Ruault, J. F., Moraine, M., & Madelrieux, S. (2022). The 'bioeconomics vs bioeconomy' debate: Beyond criticism, advancing research fronts. *Environmental Innovation and Societal Transitions, 42*, 58–73.

Antikainen, M., & Valkokari, K. (2016). A framework for sustainable circular business model innovation. *Technology Innovation Management Review, 6*(7), 5–12.

Béné, C., Fanzo, J., Haddad, L., Hawkes, C., Caron, P., Vermeulen, S., Herrero, M., & Oosterveer, P. (2020). Five priorities to operationalize the EAT–Lancet Commission report. *Nature Food, 1*(8), 457–459.

Borrello, M., Pascucci, S., Caracciolo, F., Lombardi, A., & Cembalo, L. (2020). Consumers are willing to participate in circular business models: A practice theory perspective to food provisioning. *Journal of Cleaner Production, 259*, 121013.

Campos, S., & Madureira, L. (2019). Can healthier food demand be linked to farming systems' sustainability? The case of the Mediterranean diet. *International Journal on Food System Dynamics, 10*(3), 262–277.

Carayannis, E. G., Barth, T. D., & Campbell, D. F. (2012). The Quintuple Helix innovation model: Global warming as a challenge and driver for innovation. *Journal of Innovation and Entrepreneurship, 1*(1), 1–12.

Carus, M., & Dammer, L. (2018). The circular bioeconomy: Concepts, opportunities, and limitations. *Industrial Biotechnology, 14*(2), 83–91.

Casadesus-Masanell, R., & Ricart, J. E. (2010). From strategy to business models and onto tactics. *Long range planning, 43*(2-3), 195–215.

Diakosavvas, D., & Frezal, C. (2019). Bio-economy and the sustainability of the agriculture and food system: Opportunities and policy challenges. *OECD Food, Agriculture and Fisheries Papers, 136.*

Devaney, L., Henchion, M., & Regan, Á. (2017). Good governance in the bioeconomy. *EuroChoices, 16*(2), 41–46.

do Canto, N. R., Grunert, K. G., & De Barcellos, M. D. (2021). Circular food behaviors: A literature review. *Sustainability, 13*(4), 1872.

El Bilali, H., Strassner, C., & Ben Hassen, T. (2021). Sustainable agri-food systems: Environment, economy, society, and policy. *Sustainability, 13*(11), 6260.

Ellen MacArthur Foundation (2015). *Growth within: A circular economy vision for a competitive Europe.* Ellen MacArthur Foundation.

ESCWA (United Nations Economic and Social Commission for Western Asia). (2014). *Green agricultural value chains for improved livelihoods in the Arab region.* ESCWA.

FAO (2013). *Sustainability assessment of food and agricultural system: Indicators.* FAO.

FAO (2014a). *Developing sustainable food value chains – Guiding principles.* FAO.

FAO (2014b). *SAFA Sustainability assessment of food and agriculture systems - Guidelines version 3.0.* FAO.

FAO (2021). *Bioeconomy for a sustainable future.* FAO.

Fassio, F., & Tecco, N. (2019). Circular economy for food: A systemic interpretation of 40 case histories in the food system in their relationships with SDGs. *Systems, 7*(3), 43.

Fritsche, U., Brunori, G., Chiaramonti, D., Galanakis, C., Hellweg, S., Matthews, R., & Panoutsou, C. (2020). *Future transitions for the bioeconomy towards sustainable development and a climate-neutral economy: Knowledge synthesis Final Report.* Publications Office of the European Union.

Galli, F., Prosperi, P., Favilli, E., D'Amico, S., Bartolini, F., & Brunori, G. (2020). How can policy processes remove barriers to sustainable food systems in Europe? Contributing to a policy framework for agri-food transitions. *Food Policy, 96*, 101871.

Gomez San Juan, M. & Bogdanski, A. (2021). *How to mainstream sustainability and circularity into the bioeconomy? A compendium of bioeconomy good practices and policies.* FAO.

Gualandris, J., & Kalchschmidt, M. (2016). Developing environmental and social performance: The role of suppliers' sustainability and buyer–supplier trust. *International Journal of Production Research, 54*(8), 2470–2486.

Haas, W., Krausmann, F., Wiedenhofer, D., & Heinz, M. (2016). How circular is the global economy? A sociometabolic analysis. In H. Haberl, M. Fischer, F. Krausman, & V. Winiwarter, (Eds.), *Social ecology* (pp. 259–275). Springer.

HLPE (2014). *Food losses and waste in the context of sustainable food systems. A report by the High Level Panel of Experts on Food Security and Nutrition of the Committee on World Food Security.* FAO.

HLPE (2017). *Nutrition and food systems. A report by the High Level Panel of Experts on Food Security and Nutrition of the Committee on World Food Security.* FAO.

HLPE (2020). *Food security and nutrition: building a global narrative towards 2030. A report by the High Level Panel of Experts on Food Security and Nutrition of the Committee on World Food Security.* FAO.

Hilmi, M. (2018). A field practitioner's perspective on developing green food value chains. *Food Chain,* 7(1), 1–24.

IFOAM – Organics (n.d.). *Participatory guarantee systems.* Retrieved March 29, 2022, from https://www.ifoam.bio/our-work/how/standards-certification/participatory-guarantee-systems

Joyce, A., & Paquin, R. L. (2016). The triple layered business model canvas: A tool to design more sustainable business models. *Journal of Cleaner Production,* 135, 1474–1486.

Jurgilevich, A., Birge, T., Kentala-Lehtonen, J., Korhonen-Kurki, K., Pietikäinen, J., Saikku, L., & Schösler, H. (2016). Transition towards circular economy in the food system. *Sustainability,* 8(1), 69.

Kanie, N., Zondervan, R., & Stevens, C. (2014). *Ideas on governance 'of' and 'for' sustainable development goals.* Conference Report. United Nations University Institute for the Advanced Study of Sustainability. Retrieved from https://collections.unu.edu/eserv/UNU:6195/conference_report.pdf

Loiseau, E., Saikku, L., Antikainen, R., Droste, N., Hansjürgens, B., Pitkänen, K., ... & Thomsen, M. (2016). Green economy and related concepts: An overview. *Journal of Cleaner Production,* 139, 361–371.

Lu, T., & Halog, A. (2020). Towards better life cycle assessment and circular economy: On recent studies on interrelationships among environmental sustainability, food systems and diet. *International Journal of Sustainable Development & World Ecology,* 27(6), 515–523.

McCormick, K., & Kautto, N. (2013). The bioeconomy in Europe: An overview. *Sustainability,* 5(6), 2589–2608.

Muranko, Z., Andrews, D., Newton, E. J., Chaer, I., & Proudman, P. (2018). The pro-circular change model (P-CCM): Proposing a framework facilitating behavioural change towards a circular economy. *Resources, Conservation and Recycling,* 135, 132–140.

Nußholz, J. L. (2017). Circular business models: Defining a concept and framing an emerging research field. *Sustainability,* 9(10), 1810.

Osterwalder, A. & Pigneur, Y. (2010). *Business model generation: A handbook for visionaries, game changers, and challengers.* John Wiley & Sons.

Overbeek, G.., de Bakker, H. C. M., Beekman, V. (2016). D2.3: Review of bioeconomy strategies at regional and national levels. BioSTEP project. Retrieved from https://bio-step.eu/fileadmin/BioSTEP/Bio_documents/BioSTEP_D2.3_Review_of_strategies.pdf

Rosenstock, T. S., Lubberink, R., Gondwe, S., Manyise, T., & Dentoni, D. (2020). Inclusive and adaptive business models for climate-smart value creation. *Current Opinion in Environmental Sustainability,* 42, 76–81.

Savory, A., & Duncan, T. (2016). Regenerating agriculture to sustain civilization. In I. Chabay, M. Frick, & J. Helgeson, (Eds.), *Land restoration.* Academic Press.

Sijtsema, S. J., Fogliano, V., & Hageman, M. (2020). Tool to support citizen participation and multidisciplinarity in food innovation: Circular food design. *Frontiers in Sustainable Food Systems,* 4, 1–12.

Spratt, O., Suri, R., & Deutsch, J. (2021). Defining upcycled food products. *Journal of Culinary Science & Technology,* 19(6), 485–496.

Teece, D. J. (2010). Business models, business strategy and innovation. *Long range planning,* 43(2-3), 172–194.

Wigboldus, S. (2020). *On food system transitions & transformations; Comprehensive mapping of the landscape of current thinking, research, and action.* Wageningen Centre for Development Innovation, Wageningen University & Research.

Wilkins, J. L. (2005). Eating right here: Moving from consumer to food citizen. *Agriculture and Human Values,* 22(3), 269–273.

41

RETHINKING AND PRACTISING THE 3RS (REDUCE, REUSE, RECYCLE) FOR PREVENTING FOOD LOSS AND WASTE TO INCREASE FOOD SECURITY

Imana Pal

Introduction

Food development and consumer patterns are affected by a fast-changing environment. Attitudes toward food have shifted over time with rising income, demographic transitions, changing food systems, and more diverse lifestyles, and pressing moral and social concerns. Scientific progress and industrial competition have led to changes in the food variety and accessibility in the global food market. Yet questions of food affordability and accessibility remain as paramount today as a decade ago. Hunger causes nine million deaths a year globally and 800 million people to go undernourished (Filho & Kovaleva, 2015).

Over one-third (approximately 1.3 billion tonnes) of all manufactured edible foods are not consumed; this food loss and waste (FLW) in the food supply chain (Gustavsson et al., 2011), is acknowledged as a grave threat to food security, the economy, and the environment (Pagliaccia et al., 2020). *Food loss* refers to food that unintentionally undergoes deterioration in quality or quantity as a result of food spills, spoils, bruising, wilting, degraded by pests, or other such damage as a result of infrastructure limitations at the production, storage, processing, and distribution stages of the food lifecycle (Parfitt et al., 2010). In this chapter, *food waste* means any food and inedible parts of food, removed from the food supply chain, that can be recovered or disposed of. This waste includes food waste that is to be composted, spread to land, treated through anaerobic digestion, combusted for bio-energy production, incinerated, disposed to sewer, sent to landfill, dumped in open dumps, or

DOI: 10.4324/9781003174417-48

discarded to sea. (Jain et al., 2018, p. 9). This substantial amount of FLW is equivalent to 4.4 gigatonnes of carbon dioxide (FAO, 2015a), 250 cubic kilometres of blue water footprint, 28% of the worldwide agricultural area used for agricultural production (FAO, 2013, 2019), and when converted to calories, roughly 24% of all food produced (Gustavsson et al., 2011). Regardless of the social and ecological prices of waste borne by society as a whole, the monetary value of this volume of FLW is calculated to be around USD 936 billion (FAO, 2014). The FLW is sufficient to reduce under-nutrition by one-eighth of the global population and to tackle the global challenge to meet increased demand for food, which by 2050 could reach around 150–170% of current demand (Ishangulyyev et al., 2019).

In spite of food loss and waste being excessively high, few studies quantify the degree of loss and waste, particularly in developing countries. On an annual basis, global food loss and waste is estimated to be roughly 30% of cereals, 40%–50% of root crops, fruits, and vegetables, 20% of oilseeds, meat, and dairy products, and 35% of fish, according to Food and Agriculture Organisation (FAO)-commissioned studies (FAO, 2015b). The FAO claims that over 30% of all edible food produced is lost or wasted somewhere in the food supply chain (from the first phases of household production through the ultimate phases of household utilisation) (Gustavsson et al., 2011).

Food loss and waste is highly contextual and relies on the precise conditions and circumstances of a particular country or culture (FAO, 2015b). In countries with medium and high incomes, rejection of food occurs largely during the consumption phase because of wasteful consumer behaviour due to an excessive amount of food being purchased. In low-income countries, the majority of food waste occurs during the initial and intermediate stages along the supply chain (e.g., throughout the harvest and transportation) but significantly less at the consumer stage. However, in the Global North, far more food is discarded at the household level. On global food loss, the FAO of the United Nations (UN) and the World Resources Institute (WRI) highlight distinct individual variations in FLW (Gustavsson et al., 2011). There are differences in FLW between the Global North (i.e., in North America (NA) and Europe (EU)), and Global South, (i.e., in Sub-Saharan Africa (SSA) and South/Southeast Asia (SEA)), where the regions with the highest rate of malnutrition, with approximately 281 million malnourished people (United Nations, 2018) are located. Food waste is ten times less in the developing regions, like SSA and SEA (yearly 13–24 pound food loss per capita) compared to North America and Europe (yearly between 209 and 253 pounds per person) (Gustavsson et al., 2011).

Methods

This systematic review included an analysis of data spanning the last fifteen years, from 2006–2021. The primary focus was on sustainable diets and development, as well as ways to reduce food waste sustainably. Thus, articles that focused explicitly on food waste valorisation, as well as strategies for directing food waste to incineration and composting were not incorporated. This study included an examination of the literature on the association between food loss and waste and depletion of environmental resources and sustainable diets. With findings from the literature, the focus became one of identifying pathways for the reduction and utilisation of food loss and waste. Science Direct, Scopus, PubMed, Public Health databases, Google Scholar, and other databases were used and the focus of the search was primarily on keywords such as "food loss and waste," "sustainable development," "landfills and greenhouse gas emissions," "depletion of resources," and "food waste in the supply chain." Around 200 papers were analysed. After identifying duplicates and those eliminated due to not meeting the inclusion criteria, 134 articles were included. The analysis of these publications revealed the direct and indirect impacts of food loss and waste, as well as strategies for minimising their reduction throughout the food supply chain. Interesting to note: Countries in the Global South attracted fewer studies and publications than studies about FLW in the Global North.

To comprehend the scope of food loss, and the waste problem as a whole, policies and current initiatives on the issue, subject-specific, appropriate websites, and reports by government agencies (both national and international) also were investigated. Data were collected from the FAO, Environmental Protection Agency (EPA), and United States Department of Agriculture (USDA) government websites. Secondly, papers produced by sustainability research bodies such as the World Resources Institute (WRI), Waste & Resources Action Programme (WRAP), and the Council on Natural Resource Defence were considered.

Food loss scenarios in the Global South

Food loss in lower-income countries are attributed to a variety of managerial and technical constraints associated with harvesting techniques, storage, shipping, handling, preservation facilities, infrastructure, packing, and marketing systems. Africa, Asia, Latin America, and other developing countries face significant difficulties early in the food supply chains (FSC) that account for 70–80% of the food losses due to the shortage of infrastructure. The lower and lower-middle-income economies thus contend with obstacles in food supply before food can be distributed or consumed. Significant obstacles involved in bringing perishable food items to market pose the greatest challenge. These countries frequently lack adequate cold chains to preserve food during transportation and storage. When the nutrients in animal and plant products are exposed to an insufficient temperature, they degrade, resulting in a reduction in food quality or quantity. This degradation has a detrimental effect on food safety as it reduces food supply and nutrient contents (Global Food Security Index, 2014). When exploring the connection between advanced storage technology usage and food protection, invariably, better-quality storage improves food and nutrition security by minimising negative dietary changes by allowing for an extended shelf life of the food's availability (Tesfaye & Titivayi, 2018). Tools and infrastructure help reduce losses during the FSC's early stages.

Another factor is the scale of agronomic production. A non-industrialised agrarian field usually yields small amounts of food at relatively high expense, exacerbating the financial impact of food losses. Small-scale agriculture in the Global South can face a greater risk of food loss during the first stage of the food supply chain due to intensive farming techniques and the small scale of crop production (Davy, 2013). When larger-scale production is aided by availability of resilient seeds, pathways to compost, and technologies to diminish losses early in the food supply chain, these conditions support greater levels of production of food at lower costs (FAO, 2019).

While the benefits of economic progress have been unequally distributed across the populations in developing regions, notably Asia and Sub-Saharan Africa, many in these regions face restricted food supply. This problem is aggravated by substantial loss of food at the early stages of FSC. The inadequate supply of food in the regions leads to food being eaten, even if it is degrading or badly damaged by the time it reaches the consumers (Global Food Security Index, 2014).

Food loss scenarios in the Global North

Countries in the Global North, like in Europe and North America, face distinct challenges with FLW (Global food Security Index, 2014). In general, the distribution and consumer stage are responsible for 40% of food waste in developed countries (i.e., food stores, restaurants, and households at the end of the food supply chain (Gustavsson et al., 2011; Kraemer et al., 2016). Findings in this review of the recent literature show that the EU-28 produce around 100 million tonnes of FLW annually and households are the major contributors (45%). In 2012, households in the United Kingdom lost food equal to almost 7.2 million tonnes (WRAP, 2014). The household food waste of all food purchased in Finland is 30%, Denmark, 23%, Norway, 20%, and Sweden, 10–20% (Gjerris & Gaiani, 2013). In Switzerland, households are the greatest contributors of food waste, as they waste one-third of the food produced (with calorie equivalent) (Beretta et al., 2013). Between 1979 and 2003, the FLW (per capita) in other wealthy nations like the United States climbed by 50% (Hall et al., 2009). In Australia, more than 4.2 million tonnes of FLW are disposed of in landfills each year (Verghese et al., 2013).

For many people with financial security, in much of the Global North, food can be seen as cheap and readily available. Populations are stabilising and people are living longer lives in high-income countries (Global Food Security Index, 2014). The amount of food thrown away in industrialised countries each year (222 million tons) nearly matches the 230 million tons of food produced in Sub-Saharan Africa, highlighting the significant disparities between food waste in the richest parts of the world and food insecurity among the poorest (ibid).

Commitments to reduce food loss and waste

Global food waste has enormous social, financial, and environmental impacts. Within the 17 Sustainable Development Goals (SDGs), the global goals to eradicate poverty and hunger, to preserve our planet, and support well-being were accepted in 2015 (Hanson & Mitchell, 2017). SDG 12.3 is to halve global food waste per capita in households, retail, and public places by 2030. Thus food waste reduction is a challenge for governments, charitable organisations, businesses, as well as individuals to address (Moult et al., 2018). Sustainable development is defined as development that fulfils existing needs without jeopardising future generations' supply. A sustainable food system does not threaten food and nutrition security, but rather ensures it for future generations (economically, socially, and environmentally) (Bilska et al., 2015).

Sustainable development, food security, and malnutrition

The increases in production of food over the last few decades have often come at the cost of damaging ecosystems. As argued in many other chapters in this collection, refinements to agricultural practices are essential to address issues of environmental degradation, ecosystem integrity, food production, and consumption within planetary boundaries and critical concerns with climate change (Flanagan et al., 2019; Ishangulyyev et al., 2019; Shafiee-Jodd & Cai, 2016).

FAO reports indicated that while the world generates sufficient food to feed everyone, an estimated one in every eight people, or approximately 870 million individuals, on the planet go hungry, with another two million (around 12.5% of the population) suffering from malnutrition (FAO, 2013, 2015, 2019). Reduced access to food is a factor that contributes to food insecurity. When food supply chains fail and food becomes physically or economically inaccessible, the most vulnerable are frequently impacted. Food loss suggests systemic difficulties in the agronomic infrastructure required for food security, whether due to inadequate storage for food or a lack of efficient transport from field to market (Global Food Security Index, 2014; Ishangulyyev et al., 2019).

In contrast to the frequent occurrences of hunger and deprivation in many countries of the Global South, food supplies exceed demand in the Global North (KC et al., 2016). The availability of excessive food and less expensive calories are contributing to the development of a crisis with non-communicable disorders, such as for overweight (1.4 billion people), hyperglycaemia, hypertension, and such largely preventable diseases (FAO, 2019; Neff et al., 2015). At the same time, not all have access to good food in the Global North. Approximately 79 million EU people (15% of total citizens) live below the poverty line. Only 16 million (20%) of them benefit from food assistance provided by food redistribution organisations. Sustainable food systems that prevent food loss and waste are crucial to support well-being, inclusiveness, and to address hunger (Bilska et al., 2015).

Threats to the environment

Food waste is frequently classified as unavoidable and avoidable, with the latter referring to food (and eventually beverages) that was edible at some point before being discarded. This avoidable waste involves resource waste, as food requires land, energy, chemicals, and materials to produce and deliver it to the various actors in the food supply chain. Since solid waste is commonly disposed of in landfills due to its simple administration and mass manipulation, the use of landfills is especially crucial in many underdeveloped nations, where waste is managed primarily through biodegradable waste disposal techniques. Researchers in several nations have proven that landfills are a significant source of greenhouse gas (GHG) emissions. Carbon dioxide (CO_2), nitrous oxide (N_2O), and methane (CH_4) are all GHGs that are created through the aerobic and anaerobic breakdown of solid waste. Methane from landfills is the most significant source of GHGs, accounting for 1–2% of overall GHG emissions (Zhang et al., 2019). The Defra waste policy assessment in England projected that food waste at the consumer level was responsible for approximately half of the GHG emissions, with approximately 40% of such waste being disposed of in landfills (Porter et al., 2016). The volume of GHG emissions is expected to increase substantially in the future as a result of waste management in urban areas of developing nations. As global warming is a major environmental concern, numerous studies have been conducted to determine the GHG emissions by waste activities (Friedrich & Trois, 2011). In 2000, the Global South accounted for approximately 29% of these emissions; a figure that is expected to rise to 64% in 2030 and 76% by 2050, with landfills being the primary source of this increase (Flanagan et al., 2019). Due to rising anthropogenic pressures through an array of production and consumption practices, including production of waste, natural resources are being consumed at a rate that could soon exceed Earth's carrying capacity (Kavitha et al., 2020).

Rethinking and practising the 3Rs—Reduce, reuse, recycle

Training in food safety and reducing loss of food

Reducing FLW can result in "multiple victories." Apart from saving money for farmers, businesses, and households, it potentially increases food yields, has the power to enhance food security, and reduces the burden on the environmental resources (Jurgilevich et al., 2016). Food waste represents a significant waste of resources, with adverse environmental impacts that may be avoided or mitigated by the implementation of good waste management practices (Tonini et al., 2018).

Food safety in FLW mitigation is the highest priority, with technical solutions supporting the prevention of food contamination and the maintenance of food quality. The scientific literature underscores agricultural technologies and techniques that serve to diminish losses and improve yields, such as irrigation systems, mechanical harvesting, and a variety of crops resistant to disease and drought (Spang et al., 2019). A wide range of well-developed technologies could be better deployed to avoid food spoilage, such as storage with temperature-control (for example evaporation coolers) and/or energy-efficient refrigerators, together with appropriate storage facilities, metal silos, and hermetic polythene bags (Spang et al., 2019). As well, better storage during transportation, such as using tarps during grain transportation and vented trucks for fresh food and livestock transportation, are available to address the direct pressures of postharvest food losses. Another technological aim is improved infrastructure, particularly highways, as well as market structures and operations (Vilariño et al., 2017). Thus, adopting and investing in technologies that enable efficient handling and storing of fresh fruit and grains in low-income countries may contribute to food loss reduction (Kennard, 2019). To effectively develop and use advanced technological solutions, educational awareness and availability are essential Agricultural extension services, government initiatives, and donor agencies are ideally positioned to support these actions (Lipinski et al., 2013). Training programmes that focus on minimising FLW by using cost effective approaches may be most beneficial (Spang et al., 2019). Also, trainees should be guided in train-the-trainer models to aid in further dissemination of the information (Kennard, 2019). For greater suitability, technologies need to be customised to local settings and designed for adoption by local communities (Kennard, 2019).

Reuse, recovery, and redistribution of food

The issue of food waste by supermarket chains is well known, as they often face pressures to accept from producers only food items meeting specific quality requirements like shape, size, and appearance. With such dynamics of customers wanting blemish-free foods, significant amounts of fresh foodstuffs are rejected (Flanagan et al., 2019). Additionally, several retail chains discontinue products before their expiration dates, stating they are unsellable or create a poor image for the company. Consumer mindset and consuming culture can play significant roles in this dynamic, since consumers reject food that is safe and nutritious but with a less appealing appearance. Due to the volume of transactions through institutions, restaurants, and hotels, these two are influential stakeholders in averting food waste. Creative collaborations with these actors can enable prevention of waste, repurposing safe foods for community use, and avoiding food being sent to land-fill or compost (Ishangulyyev et al., 2019; Mejia et al., 2015).

Numerous organisations are devoted to redistributing quality food generated by supermarkets and customers, through initiatives like food recovery and food redistribution. Recovery of food involves collecting wholesome food from farmers, retail outlets, and food service facilities for the distribution to those in need. The redistribution of food refers to the voluntary donation by consumers and food institutions that would otherwise go to waste and to channel edible foods for use through food banks, community kitchens, and food repurposing programs (Filho & Kovaleva, 2015). Surplus food can be transferred to community organisations to redistribute food while ensuring dignity of participants and serving of healthy food (Parfitt et al., 2010). An additional method applicable in the early production stage involves organising teams of harvesters or "gleaners" to pick remaining crops that would otherwise go unharvested. During the processing phase, when an overproduction may occur, interventions are evolving to salvage the food for community use (Lipinski et al., 2013). These interventions can be done internationally through food assistance/aid or with the help of food banks in the communities (Kennard, 2019). Food redistribution can offer pathways to minimise food waste while helping to reduce food insecurity, ensuring the well-being of all engaged in the food value chain.

Turn waste into wealth: Recycling of food waste

In the last few decades, the world's attention has increased to the necessity to move away from a linear economy centred on the "take–make–consume–dispose" model and toward a circular economy (Pagliaccia et al., 2020). It is believed that the circular economy preserves the value of resources by returning them to the biosphere once they have been used for their original purpose. The goal of a circular economy is to eliminate waste by reusing and recycling products indefinitely rather than throwing them away. This economy necessitates designing systems to be regenerative and to mimic nature. Circularity makes use of a number of natural principles, such as protecting and integrating diversity, recycling, and utilising renewable energy sources through entire systems. The concept of circular economy refers to the reuse, repair, refurbishment, and recycling of materials and goods that may be deemed as waste. Circularity in food systems requires finding more efficiencies in food production, reducing the quantity of food waste generated, creating ways to reuse foods and their by-products and shifting toward more diverse and lower carbon dietary patterns (Jurgilevich et al., 2016).

Foremost in the circular economy is the prevention of waste. Retailers, restaurants, and caterers can apply several ways to improve efficacy, extend food shelf life, and focus on marketable products, while also repurposing materials that otherwise would be discarded, like preparing "take out, ready-made" meals with foods close to their expiration date, thus inspiring immediate consumption by busy consumers (Spang et al., 2019). Appropriate utilisation of food waste and by-products as raw ingredients or food additives can benefit the industry financially, while the integration of these products can aid in the reduction of nutritional deficiencies, improve health, and minimise the ecological consequences of waste mismanagement. Energy, water, and nutrients from organic food waste can be recycled to create high-value products (Xu & Geelen, 2018). Industries worldwide are pursuing zero-waste strategies and approaches to recycling of raw material for novel products and applications. While there are numerous methods and strategies for managing food waste, there remains great opportunities within circular economies for the development of novel procedures for nutrient-reuse (Mahanty et al., 2017).

Education: Waste? Know more!

Food waste can be reduced by adjusting the interaction of consumers with food in the retail environment and by teaching about measures for food safety and wastage prevention (Kennard, 2019). Households can prevent buying too much food with planned purchases, suitable storage techniques, suitable serving sizes, better food preparation methods, and sharing and creative use of leftover foods (Quested et al., 2013). Partnerships among community organisations and education can help engage children and households about the disadvantages of food loss and waste and foster significant behavioural changes through reducing, reusing, and recycling (Lipinski et al., 2013; Neff et al., 2015; WRAP, 2014)

Additional recommendations include application of taxes to reduce consumer wasteful behaviour coupled with broader systemic actions like economic support to upgrade infrastructure for roads and infrastructure, particularly in the Global South (Vilariño et al., 2017). While beyond the scope of this chapter, other authors in this collection offer strategies to reduce barriers to policy development, strategies for coherence in definitions and mechanisms for greater access to reliable and consistent data. These works are underscored by the need for an overall FLW strategy (Koester, 2014; Neff et al., 2015).

Conclusion

Reasons abound for the waste of about one-third of all human food from the production to consumption stages. FLW reduction has enormous potential to enhance food security, conserve natural resources, and contribute to the sustainability of the environment. Efforts should be made across the food chain to minimise food losses and preserve items fit for consumption, including redistribution of food to people in need and raising awareness among consumers, in addition to designing relevant government policies and practices. Raising awareness alone will not result in desired reductions in FLW without the elimination of obstructions and the increase in appropriate technologies, infrastructure, and regulations.

Reducing FLW creates new possibilities for resolving issues in resource-limited locations and malnourished communities as well as igniting needed technical development and innovation.

It is critical that scientific communities, policymakers, food producers, and buyers focus their efforts on reducing FLW in the food supply chain and aid in restructuring the food system locally and globally to ensure greater food availability and affordability, protect nature, promote human and environmental health, and support sustainable development accessible to all.

References

Beretta, C., Stoessel, F., Baier, U., & Hellweg, S. (2013). Quantifying food losses and the potential for reduction in Switzerland. *Waste Manage, 33*(3), 764–773.

Bilska, B., Wrzosek, M., Kołożyn-Krajewska, D., & Krajewski, K. (2015). Food losses and food waste in the context of sustainable development of the food sector. *Chinese Business Review, 14*(9), 452–462.

Davy, T. (2013). *Food wastage: The irony of global gluttony.* FDI. Retrieved from https://apo.org .au/sites/default/files/resource-files/2013-06/apo-nid34673.pdf

FAO (2013). *Food wastage footprint: Impacts on natural resources.* FAO. Retrieved from http:// www.fao.org/3/i3347e/i3347e.pdf

FAO (2014). *Definitional framework of food loss.* FAO. Retrieved from http://www.fao.org/ fileadmin/user_upload/savefood/PDF/FLW_Definition_and_Scope_2014.pdf

FAO (2015a). *Food wastage footprint & climate change.* FAO. Retrieved from http://www.fao .org/3/bb144e/bb144e.pdf

FAO (2015b). *Global initiative of food loss and waste reduction.* FAO. Retrieved from http:// www.fao.org/3/i4068e/i4068e.pdf

FAO (2019). *The State of Food and Agriculture 2019. Moving forward on food loss and waste reduction.* FAO. Retrieved from http://www.fao.org/3/ca6030en/ca6030en.pdf

Filho, W. L., & Kovaleva, M. (Eds.). (2015). *Food waste and sustainable food waste management in the Baltic sea region.* Springer.

Flanagan, K., Robertson, K., & Hanson, C. (2019). *Reducing food loss and waste: Setting a global action agenda.* World Resources Institute. Retrieved from https://files.wri.org/d8/s3fs-public/reducing-food-loss-waste-global-action-agenda_1.pdf

Friedrich, E., & Trois, C. (2011). Quantification of greenhouse gas emissions from waste management processes for municipalities–A comparative review focusing on Africa. *Waste Management, 31*(7), 1585–1596.

Gjerris, M., & Gaiani, S. (2013). Household food waste in Nordic countries: Estimations and ethical implications. *Etikk i praksis-Nordic Journal of Applied Ethics, 1,* 6–23.

Global Food Security Index (2014). *Special report: Food loss and its intersection with food security.* The Economist Intelligence Unit. Retrieved from https://impact.economist.com/sustainability/project/food-security-index/Home/DownloadResource?fileName=EIU%20Global%20Food%20Security%20Index%20-%202014%20Findings%20%26%20Methodology.pdf

Gustavsson, J., Cederberg, C., Sonesson, U., van Otterdijk, R., & Meybeck, A. (2011). *Global food losses and food waste: Extent, causes and prevention.* FAO. Retrieved from http://www.fao.org/3/mb060e/mb060e00.pdf

Hall, K. D., Guo, J., Dore, M., & Chow, C. C. (2009). The progressive increase of food waste in America and its environmental impact. *PLoS One, 4*(11), e7940.

Hanson, C., & Mitchell, P. (2017). *The business case for reducing food loss and waste.* Champions 12.3. Retrieved from https://champions123.org/publication/business-case-reducing-food-loss-and-waste

Ishangulyyev, R., Kim, S., & Lee, S. H. (2019). Understanding food loss and waste: Why are we losing and wasting food? *Foods, 8*(8), 1–23.

Jain, S., Newman, D., Marquez, R. C., & Zeller, K. (2018). *Global food waste management: An implementation guide for cities.* World Biogas Association. https://www.worldbiogasassociation.org/wp-content/uploads/2018/05/Global-Food-Waste-Management-Full-report-pdf.pdf

Jurgilevich, A., Birge, T., Kentala-Lehtonen, J., Korhonen-Kurki, K., Pietikäinen, J., Saikku, L., & Schösler, H. (2016). Transition towards circular economy in the food system. *Sustainability, 8*(1), 69.

Kavitha, S., Kannah, R. Y., Kumar, G., Gunasekaran, M., & Banu, J. R. (2020). Introduction: Sources and characterization of food waste and food industry wastes. In R. Banu, G. Kumar, M. Gunasekaran & S. Kavitha (Eds.). *Food Waste to Valuable Resources.* Academic Press.

KC, K. B., Haque, I., Legwegoh, A. F., & Fraser, E. D. (2016). Strategies to reduce food loss in the Global South. *Sustainability, 8*(7), 595.

Kennard, N. J. (2019). Food waste management. In W. L. Filho, A. M. Azul, L. Brandli, P. G. Özuyar, & T. Wall (Eds.), *Zero hunger encyclopaedia of the UN sustainable development goals.* Springer. https://doi.org/10.1007/978-3-319-95675-6_86

Koester, U. (2014). Food loss and waste as an economic and policy problem. *Intereconomics, 49*(6), 348–354.

Kraemer, K., Cordaro, J. B., Fanzo, J., Gibney, M., Kennedy, E., Labrique, A., Eggersdorfer, M., & Steffen, J. (2016). Food loss and waste: The potential impact of engineering less waste. In M. Eggersdorfer, K. Kraemer, J. B. Cordaro, J. Fanzo, M. Gibney, E. Kennedy, … & J. Steffen (Eds.). *Good nutrition: Perspectives for the 21st Century* (pp. 173–186). Karger Publishers.

Lipinski, B., Hanson, C., Lomax, J., Kitinoja, L., Waite, R., & Searchinger, T. (2013). *Reducing food loss and waste: Instalment 2 of creating a sustainable food future.* World Resources Institute. Retrieved from https://files.wri.org/d8/s3fs-public/reducing_food_loss_and_waste.pdf

Mahanty, T., Bhattacharjee, S., Goswami, M., Bhattacharyya, P., Das, B., Ghosh, A., & Tribedi, P. (2017). Biofertilizers: A potential approach for sustainable agriculture development. *Environmental Science and Pollution Research, 24*(4), 3315–3335.

Mejia, G., Argueta, C. M., Rangel, V., García-Díaz, C., Montoya, C., & Agudelo, I. I. (2015). *Food donation: An initiative to mitigate hunger in the world.* FAO. Retrieved from https://pure.tue.nl/ws/files/11615757/mejiafood2015.pdf

Moult, J. A., Allan, S. R., Hewitt, C. N., & Berners-Lee, M. (2018). Greenhouse gas emissions of food waste disposal options for UK retailers. *Food Policy, 77,* 50–58.

Neff, R. A., Kanter, R., & Vandevijvere, S. (2015). Reducing food loss and waste while improving the public's health. *Health Affairs, 34*(11), 1821–1829.

Pagliaccia, D., Bodaghi, S., Chen, X., Stevenson, D., Deyett, E., De Francesco, A., Borneman, J., Ruegger, P., Peacock, B., Ellstrand, N., Rolshausen, P. E., Popa, R., Ying, S., & Vidalakis, G. (2020). Two food waste by-products selectively stimulate beneficial resident citrus host-associated microbes in a zero-runoff indoor plant production system. *Frontiers in Sustainable Food Systems, 4,* 258.

Parfitt, J., Barthel, M., & Macnaughton, S. (2010). Food waste within food supply chains: quantification and potential for change to 2050. *Philosophical Transactions of the Royal Society B: Biological Sciences, 365*(1554), 3065–3081.

Porter, S. D., Reay, D. S., Higgins, P., & Bomberg, E. (2016). A half-century of production-phase greenhouse gas emissions from food loss & waste in the global food supply chain. *Science of the Total Environment, 571,* 721–729.

Quested, T. E., Marsh, E., Stunell, D., & Parry, A. D. (2013). Spaghetti soup: The complex world of food waste behaviours. *Resources, Conservation and Recycling, 79,* 43–51.

Shafiee-Jood, M., & Cai, X. (2016). Reducing food loss and waste to enhance food security and environmental sustainability. *Environmental science & technology, 50*(16), 8432–8443.

Spang, E. S., Achmon, Y., Donis-Gonzalez, I., Gosliner, W. A., Jablonski-Sheffield, M. P., Momin, M. A., Moreno, L. C., Pace, S. A., Quested, T. E., Winans, K. S., & Tomich, T. P. (2019). Food loss and waste: measurement, drivers, and solutions. *Annual Review of Environment and Resources, 44,* 117–156.

Tesfaye, W., & Titivayi, N. (2018). The impacts of postharvest storage innovations on food security and welfare in Ethiopia. *Food Policy, 75,* 52–67.

Tonini, D., Albizzati, P. F., & Astrup, T. F. (2018). Environmental impacts of food waste: Learnings and challenges from a case study on UK. *Waste Management, 76,* 744–766.

United Nations (2018). *Goal 2: End hunger, achieve food security and improved nutrition and promote sustainable agriculture.* https://unstats.un.org/sdgs/report/2016/goal-02/

Verghese, K., Lewis, H., Lockrey, S., & Williams, H. (2013). *The role of packaging in minimising food waste in the supply chain of the future.* RMIT University. Retrieved from https://www.worldpackaging.org/Uploads/SaveTheFood/RMITRoleofpackagingminimisingwaste.pdf

Vilariño, M. V., Franco, C., & Quarrington, C. (2017). Food loss and waste reduction as an integral part of a circular economy. *Frontiers in environmental science, 5,* 21.

WRAP (2014). *Household food and drink waste: A product focus.* Retrieved from https://wrap.org.uk/sites/default/files/2020-10/WRAP-Product-focused%20report%20v5_3.pdf

Xu, L., & Geelen, D. (2018). Developing biostimulants from agro-food and industrial by-products. *Frontiers in Plant Science, 9,* 1567.

Zhang, C., Xu, T., Feng, H., & Chen, S. (2019). Greenhouse gas emissions from landfills: A review and bibliometric analysis. *Sustainability, 11*(8), 1–15.

PART 8

Design and measurement mechanisms

42

ENABLING AND MEASURING THE ADOPTION OF SUSTAINABLE DIETS

Nicholas M. Holden, Breige McNulty, and Aifric O'Sullivan

Introduction

The global food system threatens the environment while inadequately nourishing people. Despite some progress towards lower rates of undernutrition, certain micronutrient deficiencies persist, and excess weight and diet-related non-communicable diseases (NCDs) are rising (FAO et al., 2017). Examining food consumption from an environmental and public health perspective is critical, but impeded by a lack of metrics and approaches to measure sustainable food consumption (Jones et al., 2016).

A sustainable diet is: nutritious; produced within the ecological ceiling of the planetary boundary; ideally replenishing, rather than depleting, natural capital; and created above a social foundation that ensures life's essentials for those who produce, process, distribute, and eat the food (the "donut economics" concept of Raworth (2017)). This social foundation means foods must be financially viable and nutritious, thus fostering positive health outcomes. The Food and Agriculture Organisation (FAO) says, *sustainable healthy diets* promote all dimensions of an individual's health and wellbeing as well as meeting environmental, accessibility, acceptability, and other criteria (FAO & WHO, 2019). Sustainable nutritious diets must provide enough energy and nutrients to meet, while not exceeding, needs and reduce diet related NCDs, while having low adverse environmental and social impacts. There are trade-offs between social and environmental targets and nutrient requirements, and these differ by subsets of the population (e.g., children, athletes, elderly, culture) and their health outcomes. As decisions must be evidence based, appropriate metrics are essential.

While it is easy to describe components of a sustainable diet in general terms, it is difficult to recognise a sustainable diet, and imagine what the transition to sustainability might look like. Deming (2018) said, "It is wrong to suppose that if you can't measure it, you can't manage it—a costly myth" (p. 6). Therefore, this chapter will focus on evaluating the prospects for developing, delivering, and measuring sustainable nutritious diets with positive health outcomes. It will also consider the technologies needed to provide the necessary data and insights. Rates of technology innovation (Kurzweil, 2006) suggest that these ideas will soon be outdated, but they represent our current understanding of the situation. There is wide scope for misleading environmental (Delmas & Burbano, 2011), nutrition, and health claims (MacFarlane et al., 2020). The transition pathway from our current unsustainable food system (Holden et al., 2018) to a better future, needs us to stop thinking about eco-efficiency, one-off improvements, average nutrition, and general public health outcomes, and start thinking about personalised nutrition, personal health, eco-effectiveness, and system optimisation. Instead of producing large amounts of cheap food, we must produce the right amount of affordable, nutritious food (Drewnowski, 2010). We must recognise that diet is a fundamental tool for maintaining public health, but this will only happen when we can effectively develop, deliver, and measure healthy diets.

Developing nutritious diets while environmental considerations

Nutritious diets must provide sufficient energy and essential nutrients required to maintain good health without compromising the ability of future generations to meet their nutritional needs and without transferring or externalising costs elsewhere in the food system. Whilst dietary patterns with low environmental impacts can be consistent with positive health outcomes, it cannot be assumed that environmentally or socially sustainable diets are by default healthier (Gonzalez Fischer & Garnett, 2016; Macdiarmid, 2012). The relationship between nutrition and sustainability is complex and multidimensional (Johnston et al., 2014).

Modifying diets with a focus on the environment may have unintended effects on nutrition. Meat and dairy have been a focus of these discussions due to conflicting scientific evidence on the nutritional contribution of animal-source foods and the associated health and environmental impacts. For instance, results from a systemic review and meta-analysis suggest processed meats are associated with higher mortality, but authors report no clear relationship with unprocessed meats (Wang et al., 2016). However, links between meat consumption and non-communicable disease risk have been reported by others (Anand et al., 2015; Springmann et al., 2018a; Springmann et al., 2020). Models to find low environmental impact diets that substitute animal-source with plant-based foods have indicated calcium, iodine, zinc, and vitamin B12 intakes are reduced (Scarborough et al., 2012). Hence, maintaining nutritional adequacy must be a focus. Using iso-caloric substitution models with 100% replacement of animal-source with plant-based foods resulted in a 40% reduction in greenhouse gas emissions, with some nutrient intakes below recommendations (Seves et al., 2017). Alternatively, a 30% replacement caused a 14% reduction in greenhouse gas emissions and improved saturated fat, salt, fibre, and vitamin D intakes (Seves et al., 2017). A balanced approach is required, avoiding the complete exclusion of whole food groups. Dietary quality scores based on probability of achieving dietary recommendations or adequate nutrient intakes (Drewnowski, 2005) can be used with environmental parameters for a more balanced multifactor approach that would avoid nutritional deficiencies while improving environmental metrics.

Health and environmental outcomes

Evidence of the health effects of sustainable diets has grown over the past decade. Early observational studies categorised individuals into high or low environmental impact diets or compared existing dietary patterns assumed to be lower-impact (e.g., vegan or vegetarian) with a standard omnivore diet, and then compared disease incidence or disease-related biomarkers between groups (Jarmul et al., 2020). In a step beyond group-based comparisons, modelling studies estimated change in disease risk indicators and greenhouse gas emissions by manipulating the dietary intake data to replace animal-source foods with plant-based foods, while also adding context by accounting for country-specific dietary patterns and disease rates (Springmann et al., 2018a). As evidence thus far is based on observation and modelling, independent validation is required. However, to date, no publications have reported results from a randomised controlled trial on the health effects of a sustainable diet intervention.

The growing research evidence, synthesised by systematic reviews (Aleksandrowicz et al., 2016; Jarmul et al., 2020; Payne et al., 2016), is mixed. There are co-benefits between health and environment but also trade-offs (Springmann et al., 2018a; Jarmul et al., 2020). A more recent analysis of global dietary guidelines suggests reducing beef and dairy will have the greatest impact on the environment, while increasing wholegrains, fruits, vegetables, and legumes will result in positive health outcomes (Springmann et al., 2020). The universal research message, although more empirical data are needed, is that sustainable dietary guidelines must be carefully designed, evidence-based, and adapted to contextual factors. Another important consideration is that studies typically focus on one demographic—adults, children, seniors, or ethnic groups—and lose detail about interactions (e.g., children become adults, families eat the same diet, etc.). Genetics, lifestyle, and the dynamics of nutrient requirements across lifespan greatly complicate our ability to understand the impact of diet on health and the environment. Data collection (sensors, wearables, and digitisation) and curation (networks, communications, and vocabularies) are required to allow computation over the array of datasets that could help answer these questions. The need for empirical validation of current research evidence cannot be over-emphasised.

It is difficult to deliver sustainable healthy eating guidelines that simultaneously lower pressures on the environment and promote health. Region-specific, food based dietary guidelines that encompass health and sustainability are needed to encourage uptake at a population level (Springmann et al., 2020). Guidelines for subsets of the population, or even individuals, are also needed. A recent review found only four countries include sustainability in their guidelines, while some countries publish advice separate to national guidelines (Gonzalez Fischer & Garnett, 2016).

The UK guidelines for health professionals (BDA, 2018) suggest reduced red meat, minimal processed meat, increased plant protein, increased fish (from sustainable sources), moderate dairy consumption, increased fruits and vegetables, and portion control, which is consistent with other countries guidelines (Kromhout et al., 2016; Tetens et al., 2020). Research evidence suggests that reducing animal-source food intakes will have the greatest impact on environmental metrics. However, as described previously, animal-source food can be nutrient rich and aid in minimising nutrient deficiencies. This is particularly true for vulnerable groups like children, who have low requirements of energy but high micronutrient (EFSA, 2010), and for older adults to attenuate age-related muscle loss (Hanach et al., 2019). Furthermore, animal-source foods are often cultural staples, reflecting country or regional culture and farming practises.

The dietary-shift to lower environmental impact diets will require changes in eating behaviours, the types of foods consumed, and in some cases, the introduction of novel foods. To some, such foods may not be so novel as they already consume them regularly, like chickpeas or lentils. For others, innovative foods might be insect protein. Regardless, a complete understanding of social practice is necessary to inform behaviour change interventions that will move consumers to a lower impact diet including novel, less impactful foods (Godin & Sahakian, 2018). As well, delivery must always consider accessibility and affordability of foods, which often are overlooked (Hirvonen et al., 2020).

Personalised nutrition, which involves adjusting diets to the individual, offers potential for effective intervention. New technologies in nutrition research, including genomics, proteomics, and metabolomics, have improved our understanding of the effect of genetics on the interaction between diet, physiology, and metabolism, as well as the role of diet in modifying gene expression. Advances in wearable technologies and data analytics, like recommender systems, are making it possible to collect personal nutrition and health data as well as providing the tools to interpret and respond to that information. A better understanding of the relationships between nutrition and health and the capacity to interpret and make recommendations based on information paves the way for personalised nutrition and health. However, personalised nutrition is in the early stages of development and more evidence is required before widespread deployment (De Toro-Martín et al., 2017). Technologies have expanded capabilities and understanding, but the pipeline from genome or metabolome to personalised dietary advice direct to the consumer is far from being established beyond selected subsets, or elite segments of societies. The other significant roadblock to integrating personal nutrition with environmental metrics is data. To work, the advice must be person specific, but the food must also be source specific, as it has been shown that market is important in determining the impact of a food (Chen et al., 2021). Delivering nutritious diets for positive health outcomes will require commitment from public and private sectors to retain and improve the nutrition value of foods and policies that support dietary change. It will also require a willingness to enable transparent interchange of data that are currently considered commercially sensitive or proprietary and this field will need further study.

Measuring nutritious diets outcomes

The delivery of sustainable public health nutrition strategies could drive a shift towards healthier humans and a healthier planet. Measurement is needed—of food consumption, dietary patterns, mass flows, energy flows, resource consumption, polluting flows, etc.—to understand if beneficial changes are occurring. Assessment of the impact by product, regional dietary patterns, individual choices, and the value chains delivering food to people also need to be digitised to collect and monitor necessary data. A harmonised system and set of core indicators will be needed. A global sustainable diet quality index seems desirable because it would allow comparisons of the healthiness and environmental impact of diets, but regional differences make this difficult to achieve and perhaps even politically undesirable. Furthermore, the effectiveness of tools for eliciting consumers' behaviour through interventions such as information campaigns and product labelling (Latka et al., 2021) needs to be understood.

Measuring sustainable diets requires evidence-based metrics. Many studies use disease risk factors or disease endpoints when modelling the health effects of lower impact diets (Jarmul et al., 2020), but these perhaps over-simplify the complex relationship between food and health. Technologies like genomics, proteomics, and metabolomics, combined with approaches from integrated systems biology might help characterise phenotypes more precisely. Such measurements could provide insight to inter-individual variation and dietary responsiveness, which are central to personalised nutrition. Technologies such as wearables and smart appliances can promote consumer focus by providing feedback on food requirements, consumption, and waste. Personalised nutrition coupled with self-monitoring using smart technologies could help guide consumers on nutrition-health-environment linkages and provide an incentive to change to sustainable diets. There is a clear need for well-defined, sensitive, and precise metrics to measure the nutrition, health, and environmental aspects of diet. These metrics should enable better dietary recommendations and inform food-based dietary guidelines. Collecting, analysing, and sharing these big data sets brings issues of standardised protocols, data annotation, and storage, but will be extremely valuable in targeting resources, developing policies, and tracking accountability (Annan, 2018).

The methods used to calculate the environmental impact of diets are presented elsewhere in this book, including important concepts such as planetary boundaries (Springmann et al., 2018b), life cycle assessment (Roy et al., 2009), natural capital accounting (FAO, 2015), and impact valuation (Lord, 2020). Metrics such as Disability Adjusted Life Years and Minutes of Life Lost offer a means of expressing the impact of diet on humans; there are multiple impact metrics focused on atmosphere, marine and terrestrial ecosystems, hydrosphere, and resource depletion (Wu & Su, 2020). Regardless of the methods used, the key requirements are to collect empirical data that describes: (1) how food is produced, processed, moved, sold, and consumed; (2) the resource consumption and associated emissions; (3) who consumed the food, when, and how much; and (4) health outcomes. The computational challenge is to collect enough of these data, at suitable resolution, and to link the data. To date, it is reasonable to say that impact calculations have been top-down and average values are used to describe products from particular systems of production or country. There has been little progress towards accounting for spatial, temporal, and technical variation in the food system, with most calculations relying on local, regional, or sectoral average data, and the use of background data supplied in databases (e.g., EcoInvent, FeedPrint). Transition to sustainable diets will require a move to high spatial and temporal resolution data of known provenance. Measurement will have to take place throughout the food system (Figure 42.1).

The emergence of digital agriculture is critical at the beginning of the food system, which brings together technologies such as sensing, mechatronics, networks, communications, mapping, and decision support. Digital agriculture is a complete integration of sensors, data, decisions, and actions. To leverage maximum value, the data will have to be interoperable through the whole food chain (Figure 42.1), not restricted to the farm or processor, and will have to start collecting environmental and production related data. Post-harvest and related industries are advancing less rapidly than food processing industries when adopting smart manufacturing and Industry 4.0 technology. Logistics and retail industries have developed smart contact management and secure payments using blockchain: pattern, demand, quality, quantity, and inventory forecasting using artificial intelligence and machine learning; and track and trace using Internet of Things. In the home, the consumer has the potential to generate data while buying, storing, and cooking food using smart phones and smart kitchens. All of these technical developments raise the issue of privacy and security (Mandl & Kohane, 2008). Finally, end-of-food-life also requires monitoring to understand food losses, circular economy opportunities, and food waste, which contribute to diet impact (Chen et al., 2021) and for health monitoring of human waste (Larsen & Wigginton, 2020).

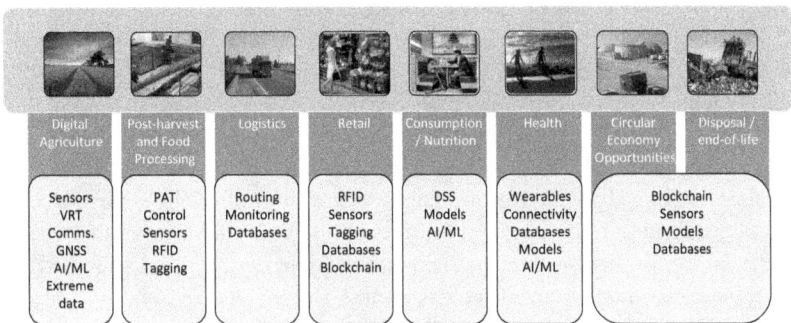

Figure 42.1 Simplified schematic of the food system and the concepts involved in measurement. (VRT = variable rate technology. Comms. = communications. GNSS = Global Navigation Satellite Systems. AI/ML = artificial intelligence/ machine learning. RFID = radio frequency identification. DSS = decision support system). Source: Work of the authors.

Enabling technologies for sustainable, nutritious diets

Digitising and measuring are a small part of the progress needed to enable sustainable, nutritious diets with positive health outcomes (Figure 42.2). A digital revolution has happened across society, in many parts of the world, and the food system is no exception. We are beginning to make digital observations using sensors and tags (*digitise*) to *measure* and, to some extent, *manage* and *use* the data collected. However, technology will have to allow *interconnection* along the food chain (Figure 42.1), which requires sharing of data and systems that can understand each other. Once this becomes widespread it will be possible to use machine learning and artificial intelligence to generate knowledge from big data to *optimise* systems and *automate* recommendations. At this point it will be possible to offer precision agriculture, food traceability, logistics, food safety, personalised nutrition, and complete resource management to minimise waste, reduce environmental impact, and maximise health.

To achieve a technologically enabled sustainable food system some key innovations need to happen:

(1) Sensors need to be shared or use open data standards so systems can be sensor agnostic, agile, and flexible (see standards.theodi.org).
(2) Food items need to be tagged (e.g., RFID (Wang et al., 2006)) or QR coded (Kim & Woo, 2016) to be digitised and recorded on the move from farm to consumer.
(3) Networks are required. Internet of Things (Elijah et al., 2018) to connect many small devices to deliver data and larger networks of interconnected computer systems, such as the GARDIAN ecosystem (gardian.bigdata.cgiar.org). Standards are emerging within parts of the system (e.g., ISO-bus on farm machinery (ISO, 2017), GS1 traceability standards (GS1, 2017)), but more are needed for connectivity.

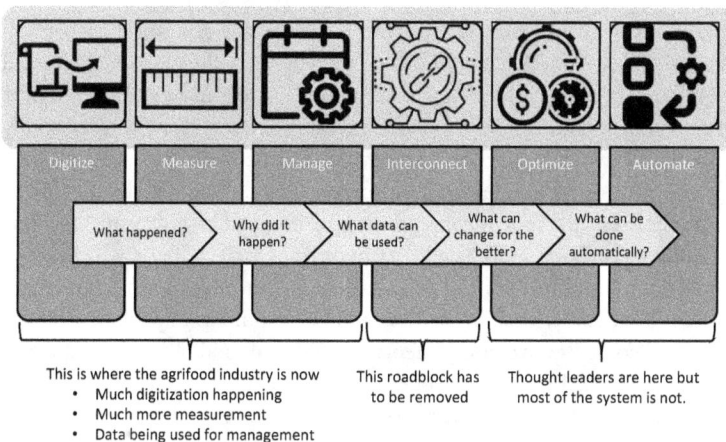

Figure 42.2 Evolution of technical requirements for sustainable diets. Source: Work of the authors.

(4) Interconnection requires APIs to allow software to connect and transfer data. Those data need to be labelled using common vocabularies related to ontologies (Dooley et al., 2018; Subramaniyaswamy et al., 2019) so computer systems can understand data and know how to handle new sources of data without significant manual intervention. Two simple examples illustrate the importance of vocabulary: "harrowing," used in European English to describe seedbed preparation, is known as a "light tickle" in Australia; buttermilk is both the by-product of butter manufacturing and a fermented dairy product. Confusing words makes interoperability impossible and could lead to some very strange dietary recommendations. Innovations like Semantic Web (Berners-Lee et al., 2001) are also required to allow all Internet data to be machine readable, thus providing a wider context for the agri-food and health systems to provide diet recommendations to individuals and communities.

(5) The scale of data storage required will increase exponentially as the food system is digitised. If all data about the supply of food and its impact on the system are stored, the scale of data centres will have to be enormous. This scale raises the question, will running a technology connected food system increase adverse environmental impact because of the data collection, communication, storage, and computation demand it will generate?

(6) To compute over these massive datasets and serve insights to individuals will require a step-change in computing power. There are two routes to achieving this. The current route is to distribute the load to personal devices, such as wearables, phones, and smart appliances. The future will require quantum computing simply to handle the amount of data needed to pull the whole system together (Mohseni et al., 2017).

Conclusions

The enabling of sustainable diets will require technological and social innovation that address acceptance of these systems. Data ownership will have to be regulated (Andreu-Perez et al., 2015). It is unclear whether the current model of free services in exchange for personal data will persist, and whether society will tolerate the current situation. Ownership models must evolve. Privacy must be addressed. The concept is perhaps losing its meaning, but connecting food to health outcomes has many implications for individuals and companies (e.g., liability, equality, bias). Means of ensuring privacy must evolve to be more efficient. Trust will become much more important. Presently, the food system is concerned with authenticity and safety, but in the future trust in the data will be even more important (Siegrist, 2008). Food will comprise two components, the physical food objects and the data describing them. Blockchain may be a key technology to ensure trust in the food system (Zhao et al., 2019). Finally, security, which overlaps with privacy but goes beyond, will be essential.

An alternative view of sustainable, nutritious diets leading to positive health outcomes is to imagine a return to local, seasonal food. This might be possible for a select few in the world, but does not align with current projections of 68% of the world population living in cities by 2050 (UN, 2015), all relying on extended food chains for security of supply. These technological innovations will be necessary to develop, deliver, and measure sustainable, nutritious diets with the goal of leading to positive health outcomes for the vast majority of the world's population, regardless of where they live.

References

Aleksandrowicz, L., Green, R., Joy, E. J., Smith, P., & Haines, A. (2016). The impacts of dietary change on greenhouse gas emissions, land use, water use, and health: A systematic review. *PLoS One*, e0165797.

Anand, S. S., Hawkes, C., de Souza, R. J., Mente, A., Dehghan, M., Nugent, R., Zulyniak, M. A., Weis, T., Bernstein, A. M., Krauss, R. M., Kromhout, D., Jenkins, D., Malik, V., Martinez-Gonzalez, M. A., Mozaffarian, D., Yusuf, S., Willett, W. C., & Popkin, B. M. (2015). Food consumption and its impact on cardiovascular disease: Importance of solutions focused on the globalized food system: A report from the workshop convened by the world heart federation. *Journal of the American College of Cardiology*, 66(14), 1590–1614.

Andreu-Perez, J., Poon, C. C. Y., Merrifield, R. D., Wong, S. T. C., Yang, G. (2015). Big data for health. *IEEE Journal of Biomedical and Health Informatics*, 19, 1193–1208.

Annan, K. (2018). Data can help to end malnutrition across Africa. *Nature*, 555(7694), 7–8.

BDA (2018). *Environmentally sustainable diet toolkit: One blue dot.* https://www.bda.uk.com/resource/one-blue-dot.html.

Berners-Lee, T., Hendler, J., & Lassila, O. (2001). The semantic web. *Scientific American*, 284, 34–43.

Chen, W., Jafarzadeh, S., Thakur, M., Ólafsdóttir, G., Mehta, S., Bogason, S., & Holden, N. M. (2021). Environmental impacts of animal-source food supply chains with market characteristics. *Science of the Total Environment*, 783, 147077.

Delmas, M. A., & Burbano. V. C. (2011). The drivers of greenwashing. *California Management Review*, 54, 64–87.

Deming, W. E. (2018). *The new economics for industry, government, education* (3rd ed.). MIT Press.

De Toro-Martín, J., Arsenault, B. J., Després, J-P, & Vohl, M-C. (2017). Precision nutrition: A review of personalized nutritional approaches for the prevention and management of metabolic syndrome. *Nutrients*, 9, 913.

Dooley, D. M., Griffiths, E. J., Gosal, G. S., Buttigieg, P. L., Hoehndorf, R., Lange, M. C., Schriml, L. M., Brinkman, F. S. L., & Hsiao, W. W. L. (2018). FoodOn: A harmonized food ontology to increase global food traceability, quality control and data integration. *npj Science of Food*, 2, 23.

Drewnowski, A. (2005). Concept of a nutritious food: Toward a nutrient density score. *American Journal of Clinical Nutrition*, 82, 721–732.

Drewnowski, A. (2010). The nutrient rich foods index helps to identify healthy, affordable foods. *American Journal of Clinical Nutrition*, 91, 1095S–1101S.

EFSA (2010). Panel on dietetic products, nutrition, and allergies (NDA). Scientific opinion on principles for deriving and applying dietary reference values. *EFSA Journal*, 8, 1458.

Elijah, O., Rahman, T. A., Orikumhi, I., Leow, C. Y., & Hindia, M. N. (2018). An overview of Internet of Things (IoT) and data analytics in agriculture: Benefits and challenges. *IEEE Internet of things Journal*, 5(5), 3758–3773.

FAO (2015). *Natural capital impacts in agriculture. Supporting better business decision-making.* FAO.

FAO, & WHO (2019). *Sustainable healthy diets. Guiding principles.* Food and Agriculture Organization of the United Nations. FAO.

FAO, IFAD, UNICEF, WFP, & WHO. (2017). *The state of food security and nutrition in the world 2017. Building resilience for peace and food security.* FAO.

Godin, L., & Sahakian, M. (2018). Cutting through conflicting prescriptions: How guidelines inform "healthy and sustainable" diets in Switzerland. *Appetite, 130*, 123–133.

Gonzalez Fischer, C., & Garnett, T. (2016). *Plates, pyramids, planet: Developments in national healthy and sustainable dietary guidelines: A state of play assessment*. FAO & University of Oxford.

GS1 (2017). *GS1 global traceability standard. GS1's framework for the design of interoperable traceability*. Systems for Supply Chains. Retrieved from https://www.gs1.org/sites/default/files/docs/traceability/GS1_Global_Traceability_Standard_i2.pdf.

Hanach N. I., McCullough, F., & Avery, A. (2019). The impact of dairy protein intake on muscle mass, muscle strength, and physical performance in middle-aged to older adults with or without existing sarcopenia: A systematic review and meta-analysis. *Advances in Nutrition, 10*, 59–69.

Hirvonen, K., Bai, Y., Headey, D., & Masters, W. A. (2020). Affordability of the EAT–Lancet reference diet: A global analysis. *The Lancet Global Health, 8*(1), e59–e66.

Holden, N. M., White, E. P., Lange, M., & Oldfield, T. L. (2018). Review of the sustainability of food systems and transition using the Internet of Food. *Npj Science of Food, 2*(1), 1–7.

ISO (2017). *Tractors and machinery for agriculture and forestry — Serial control and communications data network: Part 1: General standard for mobile data communication*. International Standards Organization.

Jarmul, S., Dangour, A. D., Green, R., Liew, Z., Haines, A., & Scheelbeek, P. F. (2020). Climate change mitigation through dietary change: A systematic review of empirical and modelling studies on the environmental footprints and health effects of 'sustainable diets'. *Environmental Research Letters, 15*(12), 123014.

Johnston, J.L., Fanzo, J.C., & Cogill, B. (2014). Understanding sustainable diets: A descriptive analysis of the determinants and processes that influence diets and their impact on health, food security, and environmental sustainability. *Advances in Nutrition, 5*(4), 18–29.

Jones, A. H. D., Blesh, L., Miller, J., Green, L., Fink, A., & Shapiro, L. (2016). A systematic review of the measurement of sustainable diets. *Advances in Nutrition, 7*(4), 641–664.

Kim, Y. G., & Woo, E. (2016). Consumer acceptance of a quick response (QR) code for the food traceability system: Application of an extended technology acceptance model (TAM). *Food Research International, 85*, 266–272.

Kurzweil, R. (2006). *The singularity is near: When humans transcend biology*. Duckworth.

Kromhout, D., Spaaij, C. J. K., de Goede, J., & Weggemans, R. M. (2016). The 2015 Dutch food-based dietary guidelines. *European Journal of Clinical Nutrition, 70*(8), 869–878.

Larsen, D. A., & Wigginton, K. R. (2020). Tracking COVID-19 with wastewater. *Nature Biotechnology, 38*, 1151–1153.

Latka, C., Kuiper, M., Frank, S., Heckelei, T., Havlík, P., Witzke, H.P., Leip, A., Cui, H.D., Kuijsten, A., Geleijnse, J. M., & van Dijk, M. (2021). Paying the price for environmentally sustainable and healthy EU diets. *Global Food Security, 28*, 100437.

Lord, S. (2020). *Valuing the impact of food: Towards practical and comparable monetary valuation of food system impacts*. FoodSIVI.

Macdiarmid, J. I. (2012). Sustainable diets for the future: Can we contribute to reducing greenhouse gas emissions by eating a healthy diet? *American Journal of Clinical Nutrition, 96*, 632–639.

MacFarlane, D., Hurlstone, M. J., & Ecker, U. K. H. (2020). Protecting consumers from fraudulent health claims: A taxonomy of psychological drivers, interventions, barriers, and treatments. *Social Science and Medicine, 259*, 112790.

Mandl, K. D., & Kohane, I. S. (2008). Techtonic shifts in the health information economy. *New England Journal of Medicine, 358*, 1732.

Mohseni, M., Read, P., & Neven, H. (2017). Commercialize quantum technologies in five years. *Nature, 543*, 171–174.

Payne, C. L., Scarborough, P., & Cobiac, L. (2016). Do low-carbon-emission diets lead to higher nutritional quality and positive health outcomes? A systematic review of the literature. *Public Health Nutrition, 19*, 2654–2661.

Raworth, K. (2017). *A safe and just space for humanity*. London: Random House.

Roy, P., Nei, D., Orikasa, T., Xu, Q., Okadome, H., Nakamura, N., & Shiina, T. (2009). A review of life cycle assessment (LCA) on some food products. *Journal of Food Engineering, 90*, 1–10.

Scarborough, P., Allender, S., Clarke, D., Wickramasinghe, K., & Rayner, M. (2012). Modelling the health impact of environmentally sustainable dietary scenarios in the UK. *European Journal of Clinical Nutrition, 66*, 710–5.

Seves, S. M., Verkaik-Kloosterman, J., Biesbroek, S., & Temme, E. H. (2017). Are more environmentally sustainable diets with less meat and dairy nutritionally adequate? *Public Health Nutrition, 20*, 2050–62.

Siegrist, M. (2008). Factors influencing public acceptance of innovative food technologies and products. *Trends in Food Science and Technology, 19*, 603–608.

Springmann, M., Wiebe, K., Mason-D'Croz, D., Sulser, T. B., Rayner, M., & Scarborough, P. (2018a). Health and nutritional aspects of sustainable diet strategies and their association with environmental impacts: A global modelling analysis with country-level detail. *The Lancet Planet Health, 2*(10), e451–e461.

Springmann, M., Clark, M., Mason-D'Croz, D., Wiebe, K., Leon Bodirsky, B., Lassaletta, L., de Vries, W., Vermeulen, S. J., Herrero, M., Carlson, K. M., Jonell, M., Troell, M., DeClerck, F., Gordon, L. J., Zurayk, R., Scarborough, P., Rayner, M., Loken, B., Fanzo, J., Godfray, H. C. J., Tilman, D., Rockstrom, W., & Willett, W. (2018b). Options for keeping the food system within environmental limits. *Nature, 562*, 519–525.

Springmann, M., Spajic, L., Clark, M. A., Poore, J., Herforth, A., Webb, P. Rayner, P., & Scarborough, P. (2020). The healthiness and sustainability of national and global food based dietary guidelines: Modelling study. *British Medical Journal, 370*:m2322.

Subramaniyaswamy, V., Manogaran, G., Logesh, R., Vijayakumar, V., Chilamkurti, N., Malathi, D., & Senthilselvan, N. (2019). An ontology-driven personalized food recommendation in IoT-based healthcare system. *The Journal of Supercomputing, 75*(6), 3184–3216.

Tetens, I., Birt, C. A., Brink, E., Bodenbach, S., Bugel, S., De Henauw, S., ... & Boeing, H. (2020). Food-based dietary guidelines–development of a conceptual framework for future food-based dietary guidelines in Europe: Report of a federation of European nutrition societies task-force workshop in Copenhagen, 12–13 March 2018. *British Journal of Nutrition, 124*(12), 1338–1344.

UN (2015). *World urbanization prospects: The 2014 revision*. United Nations, Department of Economic and Social Affairs, Population Division.

Wang, N., Zhang, N., & Wang, M. (2006). Wireless sensors in agriculture and food industry - Recent development and future perspective. *Computers and Electronics in Agriculture, 50*, 1–14.

Wang, X., Lin, X., Ouyang, Y. Y., Lui, J., Zhao, G., Pan, A., & Hu, F. B. (2016). Red and processed meat consumption and mortality: Dose–response meta-analysis of prospective cohort studies. *Public Health Nutrition, 19*, 893–905.

Wu, Y., & Su, D. (2020). Review of life cycle impact assessment (LCIA) methods and inventory databases. In D. Su (Eds.), *Sustainable Product Development*. Springer.

Zhao, G., Liu, S., Lopez, C., Lu, H., Elgueta, S., Chen, H., & Boshkoska, B. M. (2019). Blockchain technology in agri-food value chain management: A synthesis of applications, challenges and future research directions. *Computers in Industry, 109*, 83–99.

43

MEASURING THE SUSTAINABILITY OF FOOD SYSTEMS

The rationale for Footprint Indicators

Alessandro Galli, Marta Antonelli, and Dario Caro

A brief introduction: Environmental impacts of food systems

"A food system gathers all the elements (e.g., environment, people, inputs, processes, infrastructures, institutions) and activities that relate to the production, processing, distribution, preparation and consumption of food, and the outputs of these activities, including socio-economic and environmental outcomes." (HLPE, 2017, p. 11). Over the past decades we have asked the food system to respond to demographic growth and changing food needs in unprecedented ways, overshooting the planet's capacity and generating a large divide in human society, where at least 690 million people in the world suffer from chronic hunger and two billion people are overweight or obese due to unhealthy diets (FAO et al., 2019, 2020). Worldwide, eight million deaths are attributable to dietary risk factors, while more than three billion people cannot afford a healthy diet, as healthy diets cost 60% more than nutrient adequate diets[1] (FAO, 2020b).

Agriculture is a significant contributor to environmental pressures: From climate change to deforestation and the resultant biodiversity loss; from water resources overuse to air pollution and soil degradation. The agricultural sector accounts for about 23% of total anthropogenic greenhouse gas (GHG) emissions, while accounting for the whole food system (from farm to fork to disposal) increases this share up to 34% (Crippa et al., 2021) or 37% (Arneth et al., 2019) according to recent studies. Between 2000 and 2014, an overall decrease in the emission intensity of food systems (in terms of GHGs emissions per dollar of food production output) was calculated (Mrówczyńska-Kamińska et al., 2021), with significant differences between countries. Moreover, Crippa et al., (2021) revealed a decoupling of food related GHG emissions and population growth worldwide, although regional assessments offer a different picture. Agriculture is also by far

DOI: 10.4324/9781003174417-51

the largest water user globally: Agricultural water abstraction from rivers, lakes, and aquifers account for about 70% of total water withdrawals (FAO, 2020b) and represents over 90% of humanity's water footprint (Hoekstra & Mekonnen, 2012). Nitrogen is understood as essential as a building block throughout the whole food system, moving from farm to fork, as an unbalanced nitrogen cycle contributes to climate change, biodiversity loss, and air and water pollution (Leip et al., 2021).

Different food groups have a different contribution to the overall impacts due to the food sector: It has been suggested that the production of animal-based products generates the largest share (about 75%) of food-related GHG emissions (from manure, enteric fermentation of ruminants, and low efficiency of feed conversion), while it also contributes to 10% of the bluewater used in the food sector, and 20% and 25% of the food-related nitrogen and phosphorus application, respectively (Springmann et al., 2018). Shifting towards more sustainable food systems is thus regarded as an effective measure to contribute to global climate mitigation objectives and not exceed other planetary boundaries (IPCC, 2019; Springmann et al., 2018; Willet et al., 2019), and it will likely require trade-offs between different dimensions (e.g., production and consumption patterns) of food system sustainability (Béné et al., 2019).

In this context, one third of global food production is lost or wasted, with social, economic, and environmental implications. According to FAO (Gustavsson et al., 2011), food losses refer to avoidable edible waste that occur at the agricultural, post-harvest, and processing phases of the food supply chain. Among its causes, there are low infrastructure and investments (Gustavsson et al., 2011). Food waste happens in the last phases of the food supply chain (i.e., at retail and consumption level). According to the United Nations Environment Programme (UNEP) (2021), nearly 570 million tonnes of food waste occur at the household level, with a global average of 74 kg per capita of food wasted each year, remarkably similar across lower-middle and high-income countries. Moreover, in low income countries, 40% of losses are observed at post-harvest and manufacturing levels, while in higher income countries more than 40% of losses happen at retail and consumer levels (UNEP, 2021).

For every US $1 spent on food, US $2 is incurred in economic, societal, and environmental costs related to both production and consumption (Ellen McArthur Foundation, 2019). In Europe, food choices are the most important factors undermining health and well-being: This continent is the most affected by non-communicable diseases (Riccardi and Vitale, 2021), and the cost incurred by the healthcare systems amounts to almost €111 billion for cardiovascular diseases (in 2015) and $181 billion for diabetes (in 2017) (Laurent et al., 2021).

Given the scope of the challenge, there is an increasing urgency to accelerate society-wide response to food system externalities, with a holistic approach that considers social, economic, and ecological dimensions across the food-environment-nutrition nexus. This approach is fundamental to finding and travelling the pathways towards the Sustainable Development Goals (SDGs), as food is directly or indirectly related to all of them, as shown in Figure 43.1. Enabling the transformation of food and agricultural systems has the potential to catalyse

SDG 17 . PARTNERSHIPS FOR THE GOALS
For $1 dollar invested in nutrition, there is a $16 return in economic growth. Collaboration is key to accelerate the transition towards healthier and more sustainable food systems.

SDG 1. NO POVERTY
Poverty restricts access to healthy, good quality food, in adequate quantity to cover biological and social needs.

SDG 16 . PEACE, JUSTICE AND STRONG INSTITUTIONS
The intensification of conflicts causes increased world hunger. Food security is vital for the safety of the countries.

SDG 2. ZERO HUNGER
Almost 690 million people are undernourished. Agriculture and sustainable management of resources are vital to guarantee food for all.

SDG 15. LIFE ON LAND
Forests cover 31% of the earth's surface and host 18% of the global population, but are endangered by agricultural land consumption.

SDG 3. GOOD HEALTH AND WELL-BEING
Adopting healthy diets from childhood, prevents noncommunicable diseases, major causes of death worldwide (over 70%).

SDG14. LIFE BELOW WATER
The survival of millions of people depends on fishing, threatened by ocean pollution and acidification. Overfishing affects 60% of global fish stocks.

SDG 4. QUALITY EDUCATION
Ensuring fair access to all levels of education contributes to food security. Well-fed kids perform better in school.

FOOD and SDGs

SDG 13. CLIMATE ACTION
The food system from farm to fork is responsible for up to 37% of greenhouse gas emissions. The system also suffers the effects of climate change.

SDG 5 .GENDER EQUALITY
Women represent 40% of the agricultural workforce. Where women access to equal resources, agricultural production and food security increase.

SDG 12. RESPONSIBLE CONSUMPTION AND PRODUCTION
One third of the food produced in the world is lost along the supply chain or wasted in stores or by consumers.

SDG 6 . CLEAN WATER AND SANITATION
The access to clean water is essential for preparing/eating food. It is also necessary a more sustainable use of water in agriculture (it consumes 70% of the fresh water drawn).

SDG 11. SUSTAINABLE CITIES AND COMMUNITIES
Il 55% delle persone vive nelle città, nel 2050 si arriverà al 68%. Problemi di accesso a cibo sano e di qualità rendono urgenti politiche alimentari urbane integrate.

SDG 7 . AFFORDABLE AND CLEAN ENERGY
Food systems depend on fossil fuels, responsible for pollution. Renewable energy sources allow to produce food, more sustainably.

SDG 10. REDUCED INEQUALITIES
Poorer and more vulnerable people often suffer from food insecurity, they have less access to land, resources and services needed to produce.

SDG 8. DECENT WORK AND ECONOMIC GROWTH
The agri-food industry offers job opportunities along the supply chain, in the sector development and in the support services.

SDG 9. INDUSTRY, INNOVATION AND INFRASTRUCTURE
Improving infrastructure, access to markets and connections with cities, helps the local economy, improves people's nutrition and slows the exodus from the countryside.

Barilla Center
FOR FOOD
& NUTRITION

Figure 43.1   Food and the Sustainable Development Goals. Source: BCFN 2020, used with permission.

progress in several SDGs, from SDG 2 *Zero Hunger* to SDG 12 on *Sustainable Production and Consumption* and SDG 13 on *Climate Action*.

Against this backdrop, this chapter seeks to provide introductory knowledge on the rationale of Footprint indicators and their possible use in measuring the different stages of food chains, and the multiple environmental impacts associated with food systems, from farm to fork to disposal.

A footprint family for food systems

Footprint indicators measure the pressure placed by humans on the planet because of their use of resources and release of emissions (Vanham et al., 2019). In the last decade, these indicators have become increasingly popular (Vanham et al., 2019): Since the introduction of the Ecological Footprint in the early 1990s (Wackernagel & Rees, 1996), the Footprint rationale has been used by a growing number of researchers and practitioners around the world, applying it to the investigation of different environmental externalities. As a result, multiple Footprint indicators have been conceived, and this proliferation of indicators has triggered attempts to bring them together under a "Footprint Family" of indicators (e.g., Galli et al., 2012; Vanham et al., 2019).

Although the set of indicators to be included in an eventual Family might vary depending on the study or the ultimate scope the suite is intended for, Footprint indicators—usually referred to with the umbrella term *environmental footprints*—adopt life cycle thinking and are all characterised by a consumption-based accounting (CBA) approach. These approaches intend to inform users about the pressures that human final consumption activities place on the Earth's ecosystems or specific ecosystem's compartments (e.g., atmosphere, hydrosphere) and how such pressures are displaced globally (Wiedmann, 2009). Still, Footprint indicators differ from life cycle assessments (LCA) in that they focus on the resources being used and the waste being emitted by human activities (i.e., in a *Pressure-State-Impact* approach, they are *pressure* oriented) whereas LCA assessments are *impact* oriented (Vanham et al., 2019).

Of the multiple existing Footprint indicators (see Vanham et al., 2019), four—Ecological, Water, Carbon and Nitrogen Footprints—are of high relevance in measuring the environmental dimension of food systems sustainability as they deal with four key externalities: Land and water use, as well as carbon and nitrogen release (Springmann et al., 2020). These four Footprints (see Table 43.1) enable the consideration of producer and consumer perspectives on the environmental impacts of human activities and measure either resource appropriation or pollution/waste generation, or both. A quick description of the definition and range of application of each indicator is provided below.

Table 43.1 Research question, related externalities (i.e., types of environmental impact), relevant methodology paper, and key application to the food sector of Footprint indicators

	Externalities addressed	*Key methodology reference*	*Seminal application to food systems*
Ecological Footprint	Bio-productive land demanded for resource generation and carbon sequestration	Borucke et al., 2013	Wada, 1993
Water Footprint	Total volume of freshwater used and polluted	Hoekstra et al., 2011	Gerbens-Leenes et al., 2013
Carbon Footprint	Amount of GHG (CO_2, CH_4, N_2O, HFC, PFC, and SF_6) released in the atmosphere	Wiedmann and Minx, 2008	Clune et al., 2017
Nitrogen Footprint	Reactive nitrogen release into the atmosphere and to water bodies	Leach et al., 2012	Leip et al., 2014

Source: Work of the authors.

Ecological Footprint

Ecological Footprint Accounting (EFA) is comprised of two metrics: *Ecological Footprint* and *biocapacity*. By comparing how fast humans place a demand on the bio-productive terrestrial and marine areas of the Earth (i.e., the *Ecological Footprint*) against how fast these areas provide resources and services (i.e., *biocapacity*), EFA offers a way to measure the resource dimension of the human economy and its various sectors (e.g., food sector) (see Borucke et al. (2013) for details on the calculation equations). Results are expressed in hectare-equivalent units called global hectares and six main types of bio-productive terrestrial and marine areas, which can be appropriated by humans because of their demand for a specific subset of provisioning and regulating services, are considered (Mancini et al., 2018): Cropland to produce food and fibre; grazing land for producing meat-based food; forests for wood and timber products; fishing grounds (both marine and inland) for fish and seafood; build-up land to host residential homes, industries and highways; and, carbon uptake land to sequester excess CO_2 from energy use and fossil fuel combustion (Wackernagel et al., 2002). Galli et al., (2017) have defined the Ecological Footprint of food as the resource provisioning and the regulatory services demanded to provide households with the food they consume.

During the last decade, several studies have investigated the Ecological Footprint of agricultural (Wada, 1993), wine (Niccolucci et al., 2008), and seafood (Kautsky et al., 1997) products—mainly comparing organic and conventional productions (Galli, 2015)—while a few have assessed the sustainability of national diets (Galli et al., 2017). Freely available on-line tools[2] have also been made available for citizens to assess the Ecological Footprint of their individual daily activities (Collins et al., 2020).

Water Footprint

Water Footprint accounts for the appropriation of water resources required for human consumption of goods and services (Hoekstra et al., 2009). The basic methodology has been developed by Hoekstra et al. (2011). Closely linked with the virtual water concept (Allan, 1998), Water Footprint tracks the total volume (direct and indirect) of freshwater that is used to produce a particular product or to support the consumption needs of any group of consumers (e.g., an individual, city, province, state, or nation) or producers (e.g., a public organisation, private enterprise, or economic sector). Three key water components are tracked in its calculation: 1) The consumption of surface and ground water (i.e., blue Water Footprint), 2) the consumption of rainwater stored (as soil moisture) in the soil (i.e., green Water Footprint) and 3) the volume of freshwater required to assimilate the load of pollutants based on existing ambient water quality standards (i.e., grey Water Footprint). Water Footprint has been used in several studies to assess the pressure placed on the hydrosphere by agricultural products (Gerbens-Leenes et al., 2013), the agricultural sector (e.g., Dalin et al., 2017; Fader et al., 2011), food products (Petersson et al., 2021), as well as the overall pressure of national dietary patterns (e.g., Kim et al., 2020; Vanham et al., 2013). Water Footprints have been applied to the Mediterranean and the EU to a considerable extent. A recent study (Vanham et al., 2021) analysed the Water Footprint of 9 Mediterranean countries, pointing out that a dietary pattern consistent with the planetary diet of the EAT Lancet Commission report (Willet et al., 2019) would decrease the Water Footprint within the range of -17% to -48% in all countries considered and accounting for 88% of the region's population. Adherence to the Mediterranean diet instead, would decrease the WF of European Countries, Turkey, Egypt, and Morocco by -4% to -35%. Water Footprint has also been analysed in relation to avoidable food waste in the EU (80% of total), showing that it accounts for 27 litres per capita per day on average (Vanham et al., 2018).

Carbon Footprint

Building on the popularity that the term "Carbon Footprint" was acquiring in the early 2000s within the climate debate, a first attempt at defining the Carbon Footprint concept and its calculation methodology was provided by Wiedmann and Minx (2008). They defined the Carbon Footprint as a measure of the total amount of greenhouse gas (GHG) emissions directly and indirectly caused by an activity or accumulated over the life cycle stages of a product, good, or service. This definition includes activities of individuals, populations, governments, companies, organisations, processes, and industry sectors, amongst others. For all cases, all direct (on-site, internal) and indirect emissions (off-site, external, embodied, upstream, and downstream) are taken into account. Regarding the set of GHG emissions included in a Carbon Footprint assessment, most studies nowadays track the six greenhouse gases identified by the Kyoto Protocol: CO_2, CH_4, N_2O, HFC, PFC, and SF_6. These GHGs are weighted according to their global warming potentials (Wiedmann & Minx, 2008). Carbon Footprint has been used in several studies to assess the pressure placed on the atmosphere by agricultural (Röös et al., 2014) and livestock products (Persson et al., 2015), food products (Clune et al., 2017; Petersson et al., 2021), the wine sector (e.g., Rugani et al., 2013), as well as the overall pressure of national dietary patterns (e.g., Bruno et al., 2019; Gonzalez-Garcia et al., 2018; Scarborough et al., 2014).

Nitrogen Footprint

Nitrogen is an essential nutrient for all living organisms, but its abundant utilisation for human prosperity contributes to several environmental impacts such as climate change, eutrophication, acidification, and biodiversity loss (Erisman et al., 2008; Leip et al., 2015). The first conceptualisation of the Nitrogen Footprint methodology was provided by Leach et al. (2012), as a measure of the emissions of reactive N to the atmosphere and water bodies. However, in several subsequent studies, Nitrogen Footprint also includes emissions of N_2, where the latter does not contribute to any environmental pressure and does not depend on a scarce resource (Peñuelas et al., 2013) but gives a measure for the overall anthropogenic mobilisation of nitrogen (Pelletier & Leip, 2014). Nitrogen Footprint has been used in several studies to assess environmental pressures due to food products (Leip et al., 2014; Noll et al., 2020), farming practices (Majidi et al., 2014), and overall dietary patterns (e.g., Costa Leite et al., 2020; Elrys et al., 2019; Oita et al., 2018).

Applications of the Footprint family to aspects of the food system

Ecological, Water, Carbon, and Nitrogen Footprints have a wide range of applicability: In the food systems debate, their application ranges from single products to the whole sector to the national dietary level. They can also be applied individually, as single indicators, or together, as a suite of indicators, and their results can be addressed to a broad variety of stakeholders, from civil society individuals to industrial stakeholders and decision makers, to policy makers (Vanham et al., 2019).

So far, Footprint indicators have been mainly applied at the product level to assess the environmental impacts caused by specific agricultural or livestock items, as shown in Figure 43.2. Most studies have investigated a single externality, oftentimes limiting their assessment to carbon emissions or water use; more recently, integrated assessments have emerged, again prioritising the joint investigation of water use and carbon release. Ecological and Nitrogen Footprint applications to food products have been limited so far, although several studies have recently focused on the "Land Footprint" concept to track the land use implications of food systems. Such recent increase in the range of food-related externalities investigated has also been coupled with a new focus on the need to understand the impacts of national

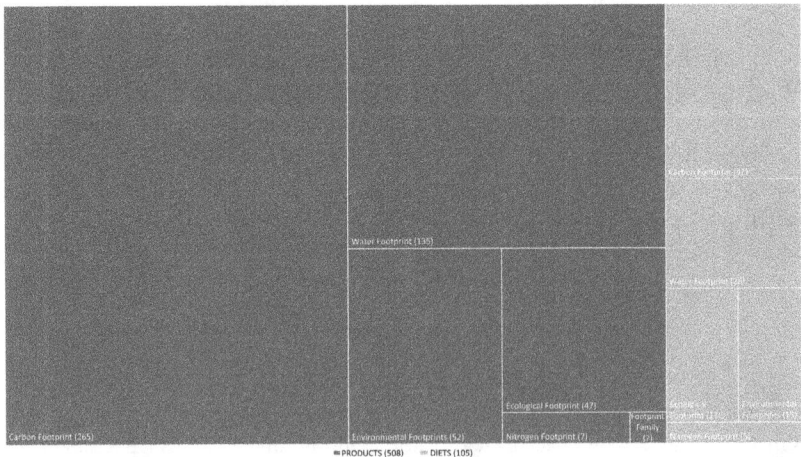

Figure 43.2 Number of articles published on Ecological, Water, Carbon and Nitrogen Footprints applied at product and dietary level from 2000 to 2020 as retrieved from Scopus in May 2021. Studies using multiple Footprints concurrently are distinguished among those using at least two indicators (Environmental Footprints studies) vs. those deploying the whole suite (Footprint Family studies). Source: Work of the authors.

dietary choices, although once again attention has been placed primarily on the water and carbon related externalities (individually or jointly addressed). Still, outcomes of recent studies seem to point towards a clear direction, as described below.

With the increasing understanding and appreciation that local to global solutions for sustainable development need to come from measures at all stages along the supply chain (Foley et al., 2011; Godfray et al., 2010), the use of Footprint indicators seems useful to inform more sustainable and healthier food systems, from farm to fork to disposal. Different studies have compared the environmental impact of different food consumption habits. Carbon Footprints and nutritional quality of diets, for instance, have been analysed by a recent review of 21 studies and 66 dietary choices, corroborating that Mediterranean and Atlantic diets have high nutritional scores and low Carbon Footprints, whereas the diets of the United States and northern and western Europe have the highest Carbon Footprints (González-García et al., 2018). Meanwhile, an application of the Ecological Footprint to the Mediterranean region (Galli et al., 2017) has found food to be the largest economic sector of demand ahead of transportation, in most of the region's countries, although shifting to a calorie-adequate diet or changing dietary patterns towards more plant- and cereal-based products would contribute to an 8–10% reduction in the region's overall Ecological Footprint.

While integrated food-system assessments, scientific analyses, policy formulation, and integrated policy decisions (e.g., taking trade-offs into account) would need multiple Footprint indicators to be applied simultaneously, Figure 43.2 shows that they have until now often been used individually and at the product level, rather than jointly and at the dietary level. Some exceptions include assessments for Argentina (Arrieta et al., 2021), Germany (Helander et al., 2021), and Australia (Ridoutt et al., 2021).

Most measures of food systems sustainability are partial, as they focus on just one environmental compartment and externality, confining land-related externalities to a mere corner in water-food-energy nexus studies, thus not easily allowing trade-offs to be fully investigated and considered in food systems management and governance. Different system boundaries, (cultivation to consumer, as opposed to cultivation to retail or cultivation to farm gate approaches), uncertainties and discrepancies related to data sources, and functional units (a function of the system being calculated so to allow comparisons) are among the main weaknesses identified in existing studies (González-García et al., 2018). Still, the few existing studies—and the joint interpretation of individual Footprint applications—seem to converge in indicating that animal-based products (e.g., seafood and read meat) place a great pressure on the planet's resources as their production (per unit of product) requires more land and water and causes the release of higher CO_2 and nitrogen amounts (i.e., higher Footprint intensities) than that of vegetables, cereals, and dairies, as well as low trophic level fishes and poultry (Clune et al., 2017; Kim et al., 2020; Leip et al., 2014). This information sheds light on possible future pathways for sustainable, nutritionally-viable national dietary patterns, and thus calls for the inclusion of sustainability criteria (and indicators to track progress against them) in national dietary guidelines.

Conclusion: The way ahead for using Footprints to measure the sustainability of food systems

Feeding future generations is among the major challenges of the third millennium. The expected global population growth, coupled with urbanisation trends and changes in food habits, imposes the urgent need to identify solutions to sustainably satisfy future global food demand and provide food for all. As the pressure on the planet due to food systems is increasing, the approaches by which we measure this pressure, and the consequent information generated, become crucial in leading in the right directions in terms of sustainability. While impacts from food systems are investigated via LCA studies, Footprint indicators offer a way to complement this information by tracking the underlying human drivers of pressures on the planet.

Our societies have conceived of the supply chain of food as structured through different steps. However, focusing on single steps, irrespective of how relevant each step can be, limits our understanding of the total burden associated with food. For instance, only looking at the production of food provides important information about the domestic intensity of production, but neglects the potential impacts occurring with trade, consumption, and waste. Therefore, looking at all steps of the supply chain is an informative way to appropriately measure the pressures on the environment associated with food systems, thus preventing the formulation of inadequate recommendations and policies.

Similarly, food systems are responsible for multiple environmental externalities, which are deeply connected with each other. Focusing on a single externality, as most studies have done so far, limits our understanding of the true sustainability of food systems, and ignores the potential trade-offs and synergies among the multiple impacts. We argue that measuring the multiple environmental externalities of our current food systems has now become a priority. In particular, we recommend four externalities to be assessed—carbon, water, land, and nitrogen—and point at Footprint indicators as one of the possible sets of tools for doing that.

In short, assessing the sustainability of food systems requires a holistic approach. Many other chapters in this collection provide substantial rationale for educational programs, and public campaigns, along with the contribution of civil society organisations to support access to sustainable diets. Focusing on single food items, even when they are predominant, such as staple food, or crucial in terms of impact, such as meat, hinders a proper contextualisation with respect to relevant dietary patterns; it narrows the overview needed to properly assess the food systems. A "sustainable diet" is not a new idea; dietary shifts and food item substitution can be compared while accounting for the broad and narrow sustainability aspects. Therefore, when the pressures are assessed, we recommend having a bias towards sustainable diets rather than sustainable foods.

Choosing the right unit of measure for dietary intensities is also an important aspect to take into consideration. For instance, evaluations of pressures and/or impact per kg of food do not provide helpful information and can lead in the wrong direction in terms of sustainability, as dietary patterns are not based on the mass of food. We argue that sustainability assessment of food systems should be developed on the basis of nutritional values of food, which represents the reference unit of a diet. To conclude, we believe that conducting a proper assessment of food systems sustainability is a powerful tool to fill the gap between science and policy. This is crucial to accelerate the pathways towards the SDGs, which are extremely linked with food systems.

Notes

1 According to FAO (2020a), "A nutrient adequate diet meets calorie and nutrient needs (defined by a specific standard for specific populations) but does not necessarily meet dietary guidelines (proportionality between food groups), and does not necessarily satisfy food preferences" (p. 4).
2 See www.footprintcalculator.org.

References

Allan J. A. (1998). Virtual water: A strategic resource, global solutions to regional deficits. *Groundwater, 36*(4), 545–546.

Arneth, A., Denton, F., Agus, F., Elbehri, A., Erb, K., Osman Elasha, B., Rahimi, M., Rounsevell, M., Spence, A., & Valentini, R. (2019). Framing and context. In P. R., Shukla, J. Skea, E. Calvo Buendia, V. Masson-Delmotte, H.-O. Pörtner, D. C. Roberts, P. Zhai, R. Slade, S. Connors, R. van Diemen, M. Ferrat, E. Haughey, S. Luz, S. Neogi, M. Pathak, J. Petzold, J. Portugal Pereira, P. Vyas, E. Huntley, K. Kissick, M. Belkacemi, & J. Malley (Eds.) (pp. 77–129), *Climate change and land: An IPCC special report on climate change, desertification, land degradation, sustainable land management, food security, and greenhouse gas fluxes in terrestrial ecosystems.* IPCC. https://www.ipcc.ch/site/assets/uploads/sites/4/2019/12/04_Chapter-1.pdf

Arrieta, E. M., Geri, M., Coquet, J. B., Scavuzzo, C. M., Zapata, M. E., & González, A. D. (2021). Quality and environmental footprints of diets by socio-economic status in Argentina. *Science of The Total Environment, 801*, 149686.

Béné, C., Oosterveer, P., Lamotte, L., Brouwer, I. D., de Haan, S., Prager, S. D., Talsma E. F., & Khoury, C. K. (2019). When food systems meet sustainability: Current narratives and implications for actions. *World Development, 113*, 116–130.

Borucke, M., Moore, D., Cranston, G., Gracey, K., Iha, K., Larson, J., Lazarus, E., Morales, J. C., Wackernagel, M., & Galli, A. (2013). Accounting for demand and supply of the biosphere's regenerative capacity: The national footprint accounts' underlying methodology and framework. *Ecological Indicators, 24*, 518–533.

Bruno, M., Thomsen, M., Pulselli, F. M., Patrizi, N., Marini, M., & Caro, D. (2019). The carbon footprint of Danish diets. *Climatic Change, 156*(4), 489–507.

Clune, S., Crossin, E., & Verghese, K. (2017). Systematic review of greenhouse gas emissions for different fresh food categories. *Journal of Cleaner Production, 140*, 766–783.

Collins, A., Galli, A., Hipwood, T., & Murthy, A. (2020). Living within a one planet reality: The contribution of personal footprint calculators. *Environmental Research Letters, 15*(2), 025008.

Costa Leite, J., Caldeira, S., Watzl, B., & Wollgast, J. (2020). Healthy low nitrogen footprint diets. *Global Food Security, 24*, 100342.

Crippa, M., Solazzo, E., Guizzardi, D., Monforti-Ferrario, F., Tubiello, F. N., & Leip, A. J. N. F. (2021). Food systems are responsible for a third of global anthropogenic GHG emissions. *Nature Food, 2*(3), 198–209.

Dalin, C., Wada, Y., Kastner, T., & Puma, M. J. (2017). Groundwater depletion embedded in international food trade. *Nature, 543*, 700.

Ellen MacArthur Foundation (2019). *Cities and circular economy for food.* Retrieved from: http://www.ellenmacarthurfoundation.org/publications.

Elrys, A. S., Raza, S., Abdo, A. I., Liu, Z., Chen, Z., & Zhou, J. (2019). Budgeting nitrogen flows and the food nitrogen footprint of Egypt during the past half century: Challenges and opportunities. *Environment International, 130*, 104895.

Erisman, J. W., Sutton, M. A., Galloway, J. N., Klimont, Z., & Winiwarter, W. (2008). How a century of ammonia synthesis changed the world. *Nature Geoscience, 1*, 636–639.

Fader, M., D. Gerten, M. Thammer, Heinke, J., Campen-Lotze, H., Lucht, W., & Cramer, W. (2011). Internal and external green-blue agricultural water footprints of nations, and related water and land savings through trade. *Hydrology and Earth System Sciences, 15*, 1641–1660.

FAO (2020a). *The state of food and agriculture 2020. Overcoming water challenges in agriculture.* FAO.

FAO (2020b). *The state of food security and nutrition in the world 2020. Transforming food systems for affordable healthy diets.* FAO.

FAO, et al. (2019). *The State of food security and nutrition in the world 2019.* FAO.

Foley, J. A., Ramankutty, N., Brauman, K. A., Cassidy, E. S., Gerber, J. S., Johnston, M., Mueller, N. D., O'Connell, C., Ray, D. K., West, P. C., Balzer, C., Bennett, E. M., Carpenter, S. R., Hill, J., Monfreda, C., Polasky, S., Rockstrom, J., Sheehan, J., Siebert, S., Tilman, D., & Zaks, D. P. M. (2011). Solutions for a cultivated planet. *Nature, 478*, 337–342.

Galli, A. (2015). "Footprints". In *Oxford bibliographies in environmental science.* Oxford University Press. https://www.oxfordbibliographies.com/view/document/obo-9780199363445/obo-9780199363445-0046.xml

Galli, A., Wiedmann, T., Ercin, E., Knoblauch, D., Ewing, B., & Giljum, S. (2012). Integrating ecological, carbon and water footprint into a "footprint family" of indicators: Definition and role in tracking human pressure on the planet. *Ecological Indicators, 16*, 100–112.

Galli, A., Iha, K., Halle, M., El Bilali, H., Grunewald, N., Eaton, D., Capone, R., Debs, P., & Bottalico, F. (2017). Mediterranean countries' food consumption and sourcing patterns: An ecological footprint viewpoint. *Science of the Total Environment, 578*, 383–391.

Gerbens-Leenes, P. W., Mekonnen, M. M., & Hoekstra, A. Y. (2013). The water footprint of poultry, pork and beef: A comparative study in different countries and production systems. *Water Resources and Industry, 1–2*, 25–36.

Godfray, H. C. J., Beddington, J. R., Crute, I. R., Haddad, L., Lawrence, D., Muir, J. F, Pretty, J., Robinson, S., Thomas, S. M., & Toulmin, C. (2010). Food security: The challenge of feeding 9 billion people. *Science, 327*(5967), 812–818.

González-García, S., Esteve-Llorens, X., Moreira, M. T., & Feijoo, G. (2018). Carbon footprint and nutritional quality of different human dietary choices. *Science of the Total Environment, 644*, 77–94.

Gustavsson, J., Cederberg, C., Sonesson, U., Van Otterdijk, R., & Meybeck, A. (2011). *Global food losses and food waste.* FAO. https://www.fao.org/3/i2697e/i2697e.pdf

Helander, H., Bruckner, M., Leipold, S., Petit-Boix, A., & Bringezu, S. (2021). Eating healthy or wasting less? Reducing resource footprints of food consumption. *Environmental Research Letters, 16*(5), 054033.

HLPE (2017). *Nutrition and food systems: A report by the high level panel of experts on food security and nutrition of the committee on world food security* (p. 152). Committee on World Food Security.

Hoekstra, A. Y., & Mekonnen, M. M. (2012). The water footprint of humanity. *Proceedings of the national academy of sciences, 109*(9), 3232–3237.

Hoekstra, A. Y., Chapagain, A., Martinez-Aldaya, M., & Mekonnen, M. (2009). *Water footprint manual: State of the art 2009.* Water Footprint Network. https://ris.utwente.nl/ws/portalfiles/portal/5146564/Hoekstra09WaterFootprintManual.pdf://ris.utwente.nl/ws/portalfiles/portal/5146564/Hoekstra09WaterFootprintManual.pdf

Hoekstra, A. Y., Chapagain, A. K., Mekonnen, M. M., & Aldaya, M. M. (2011). *The water footprint assessment manual: Setting the global standard.* Routledge.

IPCC (2019). *Climate change and land: An IPCC special report on climate change, desertification, land degradation, sustainable land management, food security, and greenhouse gas fluxes in terrestrial ecosystems.* IPCC.

Kautsky, N., Berg, H., Folke, C., Larsson, J., & Troell, M. (1997). Ecological footprint for assessment of resource use and development limitations in shrimp and tilapia aquaculture. *Aquaculture research, 28*(10), 753–766.

Kim, B. F., Santo, R. E., Scatterday, A. P., Fry, J. P., Synk, C. M., Cebron, S. R., Mekkonen, M., Hoekstra, A. Y., de Pee, S., Bloem, M. W., Neff, R. A., & Nachman, K. E. (2020). Country-specific dietary shifts to mitigate climate and water crises. *Global environmental change, 62*, 101926.

Laurent, É., Battaglia, F., Marchiori, G. D. L., Galli, A., Janoo, A., Munteanu, R., & Sommer, C. (2021). Toward health-environment policy in a well-being economy. In Laurent, É. (ED.) *The Well-being Transition* (pp. 73–93). Palgrave Macmillan.

Leach, A. M., Galloway, J. N., Bleeker, A., Erisman, J. W., Kohn, R., & Kitzes, J. (2012). A nitrogen footprint model to help consumers understand their role in nitrogen losses to the environment. *Environmental Development, 1*(1), 40–66.

Leip, A., Weiss, F., Lesschen, J. P., & Westhoek, H. (2014). The nitrogen footprint of food products in the European Union. *The Journal of Agricultural Science, 152*(S1), 20–33.

Leip, A., Billen, G., Garnier, J., Grizzetti, B., Lassaletta, L., Reis, S., Simpson, D., Sutton, M., de Vries, W., Weiss, F., & Westhoek, H. (2015). Impacts of European livestock production: Nitrogen, sulphur, phosphorus and greenhouse gas emissions, land-use, water eutrophication and biodiversity. *Environmental Research Letters, 10*(11), 115004.

Leip, A., Bodirsky, B. L., & Kugelberg, S. (2021). The role of nitrogen in achieving sustainable food systems for healthy diets. *Global Food Security, 28*, 100408.

Majidi, A., Shade, J., Leach, A., & Galloway, J. (2014). Farming practice effects on nitrogen footprints. *Building Organic Bridges, 4*, 1131–1134.

Mancini, M. S., Galli, A., Coscieme, L., Niccolucci, V., Lin, D., Pulselli, F. M., Bstianoni, S., & Marchettini, N. (2018). Exploring ecosystem services assessment through ecological footprint accounting. *Ecosystem Services, 30*, 228–235.

Mrówczyńska-Kamińska, A., Bajan, B., Pawłowski, K. P., Genstwa, N., & Zmyślona, J. (2021). Greenhouse gas emissions intensity of food production systems and its determinants. *Plos one, 16*(4), e0250995.

Niccolucci, V., Galli, A., Kitzes, J., Pulselli, R. M., Borsa, S., & Marchettini, N. (2008). Ecological Footprint analysis applied to the production of two Italian wines. *Agriculture, Ecosystems & Environment, 128*, 162–166.

Noll, L. C., Leach, A. M., Seufert, V., Galloway, J. N., Atwell, B., Erisman, J. W., & Shade, J. (2020). The nitrogen footprint of organic food in the United States. *Environmental Research Letters, 15*(4), 045004.

Oita, A., Nagano, I., & Matsuda, H. (2018). Food nitrogen footprint reductions related to a balanced Japanese diet. *Ambio, 47*(3), 318–326.

Pelletier, N., & Leip, A., (2014). Quantifying anthropogenic mobilization, flows (in product systems) and emissions of fixed nitrogen in process-based environmental life cycle assessment: rationale, methods and application to a life cycle inventory. *The International Journal of Life Cycle Assessment, 19*, 166–173.

Penuelas, J., Poulter, B., Sardans, J., Ciais, P., Van Der Velde, M., Bopp, L., Boucher, O., Godderis, Y., Hinsinger, P., Llusia, J., Nardin, E., Vicca, S., Obersteiner, M., & Janssens, I. A. (2013). Human-induced nitrogen–phosphorus imbalances alter natural and managed ecosystems across the globe. *Nature communications, 4*(1), 1–10.

Persson, U. M., Johansson, D. J., Cederberg, C., Hedenus, F., & Bryngelsson, D. (2015). Climate metrics and the carbon footprint of livestock products: Where's the beef? *Environmental Research Letters, 10*(3), 034005.

Petersson, T., Secondi, L., Magnani, A., Antonelli, M., Dembska, K., Valentini, R., Varotto, A., & Castaldi, S. (2021). A multilevel carbon and water footprint dataset of food commodities. *Scientific Data, 8*(1), 1–12.

Riccardi G., & Vitale M. (2021). Dietary habits of European population in relation to the risk of non-communicable diseases. In Barilla Center for Food & Nutrition (Ed.), *Europe and food. Ensuring environmental, health and social benefits for the global transition*. Barilla Center for Food & Nutrition.

Ridoutt, B. G., Baird, D., & Hendrie, G. A. (2021). Diets within planetary boundaries: What is the potential of dietary change alone? *Sustainable Production and Consumption, 28*, 802–810.

Röös, E., Sundberg, C., & Hansson, P. A. (2014). Carbon footprint of food products. In S. S. Muthu (Ed.), *Assessment of carbon footprint in different industrial sectors* (Vol. 1, pp. 85–112). Springer.

Rugani, B., Vázquez-Rowe, I., Benedetto, G., & Benetto, E. (2013). A comprehensive review of carbon footprint analysis as an extended environmental indicator in the wine sector. *Journal of Cleaner Production, 54*, 61–77.

Scarborough, P., Appleby, P. N., Mizdrak, A., Briggs, A. D., Travis, R. C., Bradbury, K. E., & Key, T. J. (2014). Dietary greenhouse gas emissions of meat-eaters, fish-eaters, vegetarians and vegans in the UK. *Climatic Change, 125*(2), 179–192.

Springmann, M., Clark, M., Mason-D'Croz, D., Wiebe, K., Bodirsky, B. L., Lassaletta, L., de Vries W., Vermeulen S. J., Herrero, M., Carlson, K. M., Jonell, M., Troell, M., DeClerck, F., Gordon, L. J., Zurayk, R., Scarborough, P., Rayner, M., Loken, B., Fanzo, J., Godfray, H. C. J., Tilman, D. Rockstrom, J., & Willett, W. (2018). Options for keeping the food system within environmental limits. *Nature, 562*(7728), 519–525.

Springmann, M., Spajic, L., Clark, M. A., Poore, J., Herforth, A., Webb, P., Rayner, N., & Scarborough, P. (2020). The healthiness and sustainability of national and global food based dietary guidelines: Modelling study. *British Medical Journal, 370*. m2322.

UNEP (2021). *Food waste index report 2021*. UNEP.

Vanham, D., Mekonnen, M. M., & Hoekstra, A. Y. (2013). The water footprint of the EU for different diets. *Ecological Indicators, 32*, 1–8.

Vanham, D., Comero, S., Gawlik, B.M., & Bidoglio, G. (2018). The water footprint of different diets within European sub-national geographical entities. *Nature sustainability, 1*, 518–525.

Vanham, D., Leip, A., Galli, A., Kastner, T., Bruckner, M., Uwizeye, A., van Dijk, K., Ercin, E., Dalin, C., Brandao, M., Bastianoni, S., Fang, K., Chapagain, A., Van der Velde, M., Sala, S., Pant, R., Mancini, L., Monforti-Ferrario, F., Carmona-Garcia, G., Marques, A., Weiss, F., & Hoekstra, A. Y. (2019). Environmental footprint family to address local to planetary sustainability and deliver on the SDGs. *Science of the Total Environment, 693*, 133642.

Vanham, D., Guenther, S., Ros-Baró, M., & Bach-Faig, A. (2021). Which diet has the lower water footprint in Mediterranean countries? *Resources, Conservation and Recycling, 171*, 105631.

Wackernagel, M., & Rees, W. (1996). *Our ecological footprint: Reducing human impact on the earth* (Vol. 9). New Society Publishers.

Wackernagel, M., Schulz, N. B., Deumling, D., Linares, A. C., Jenkins, M., Kapos, V., Monfreda, C., Loh, J., Myers, N., Norgaard, R., & Randers, J. (2002). Tracking the ecological overshoot of the human economy. *Proceedings of the National Academy of Sciences, 99*(14), 9266–9271.

Wada, Y. (1993). The appropriated carrying capacity of tomato production: The Ecological Footprint of hydroponic greenhouse versus mechanized open field operations. M.A. Thesis. School of Community and Regional Planning, University of British Columbia: Vancouver, Canada.

Wiedmann, T. (2009). A review of recent multi-region input–output models used for consumption-based emission and resource accounting. *Ecological Economics, 69*(2), 211–222.

Wiedmann, T., & Minx, J. (2008). A definition of 'carbon footprint'. *Ecological Economics Research Trends, 1*, 1–11.

Willett, W., Rockström, J., Loken, B., Springmann, M., Lang, T., Vermeulen, S., Garnett, T., Tilman, D., DeClerck, F., Wood, A., Jonell, M., Clark, M., Gordon Fanzo, J., Hawkes, C., Zurayk, R., Rivera, J. A., de Vries, W., Sibanda, L. M., Afshin, A., Chaudary, A., Herrero, M., Augustina, R., Branca, F., Lartey, A., Fan, S., Crona, B., Fox, E., Bignet, V., Troell, M., Lindal, T., Singh, S., Cornell, S. E., Reddy, K. S., Narain, S., Nishtar, S., & Murray, C. J. (2019). Food in the anthropocene: The EAT–lancet commission on healthy diets from sustainable food systems. *The Lancet*, *393*(10170), 447–492.

44

THE ROLE OF DESIGN IN THE TRANSITION TO SUSTAINABLE DIETS

*Sonia Massari, Margherita Tiriduzzi,
Chhavi Jatwani, and Sara Roversi*

Introduction

More than five years after the global commitment to the 2030 Agenda for Sustainable Development, 690 million people do not have enough food (FAO, IFAD, UNICEF, WFP, and WHO, 2020) and economic forecasts indicate that COVID-19, which exacerbated global food problems and shortages, could add 83 to 132 million undernourished people, hitting the most vulnerable populations hard. Meanwhile, a total of 1.3 billion tons of food is wasted each year (FAO, 2020), using 38% of the total energy consumption of the global food system (FAO, 2019). Unhealthy diets also lead to additional health care costs. Unbalanced diets (i.e., those that provide excessive energy intake, are low in fruits and vegetables, high in saturated fat, grains, sugars, salt, and red and processed meat) represent one of the largest global health burdens (Springmann et al., 2020). These diets contribute to diet-related diseases such as cardiovascular disease, cancer, and type 2 diabetes, which have major impacts on well-being and quality of life and require costly treatment.

In the Global North, healthy diets are more affordable. Affordability is especially compromised in the Global South due to an array of factors. Lower productivity, lower food availability, and diet diversity are at play. Producers work on smaller-scales in fisheries, aquaculture, livestock, and other food production. Food loss is significant at pre and post-harvest stages. Among other impediments to afford-ability are trade policies and employment opportunities and income limitations. To reduce the cost of nutritious foods and make healthy diets more affordable, it is necessary to reorient agricultural priorities toward more nutrition-sensitive food and sustainable production (FAO, 2019).

Recent studies have shown that healthy, sustainable diets can reduce wildlife loss by 46%, premature deaths by at least 20%, and food-related GHG emissions

DOI: 10.4324/9781003174417-52

by at least 30% (WWF, 2020). Sustainable foods can accelerate poverty reduction and social inclusion; increase equity and justice; ensure education and health care for all; and promote biodiversity protection, water security, and climate change mitigation and adaptation—all goals included in the 2030 Agenda for Sustainable Development.

People's food behaviours are inevitably influenced by the design of the food products, food systems, eating and kitchen tools, and services they use. Design methods can enable sustainable behaviours through understanding human drivers, values, and everyday needs and looking at them not as problems but as opportunities for forging suitable designs.

How do designers define sustainable diets?

Other chapters in this volume have revealed that food systems and sustainable diets represent highly interconnected and correlated complexities. Norman (2010) argues that complexity is inherent in the social and natural world. According to Norman, design is an enabling tool that can help address complexities and deliver concrete results to improve the quality of life in our communities. Designers' goals are to understand the values that humanity embeds in our everyday objects and tools (www.ideo.org), including foods that are eaten, and to provide people with appropriate mechanisms and models for navigating systems in life elegantly and effectively. Food Design, as a recognised academic discipline, emerged in the past two decades as a research area at the intersection of food and design studies. However, design studies contain a wealth of documentation that confirms design methods of production have accompanied the production of food, and have especially evolved with the industrialisation of food products and the economic boom (Ferrara & Massari, 2015).

Over time, Food Design moved from interventions and research focused exclusively on cleaning, correcting, and adjusting systems and avoiding any damage, to research models that focus on system thinking, with the aim of innovating within food systems. Interest in Food Design has grown significantly in the past few years, largely due to recognising the consequences of user behaviour on environments as well as the social impacts of technology and design systems. Current efforts focus on the prevention and reduction of behaviours that contribute to health impairments or environmental demise, with a focus to produce positive impacts on the planet from human food choices.

This chapter presents three case studies from food designers who are actively working on food behaviour change for more sustainable diets, such as the Mediterranean Diet. These case studies explain how design methods applied to agri-food system processes have extended from the search for cleaner technologies based on single products or services to the application of more systemic solutions that satisfy healthier and more sustainable scenarios. The chapter ends by identifying the need for a guide for food designers. How can food-related decisions that consumers make on a daily basis be changed? How can the way people think about food be changed by designing new products or tools? What role does design have in the co-construction of conscience in the food community (i.e., producers, distributors, policymakers, consumers, etc.)? How can food design provide new opportunities for helping local and global communities to innovate around more sustainable diet solutions? This chapter's conclusions aim to address these questions by proposing a statement that defines the role of design in the transition to sustainable diets.

Sustainable diets and cognitive choices

Global food systems present a complex and multifaceted set of challenges from farm to table. However, making better food choices at the individual level can be a practical and decisive solution for the well-being of the planet (Tilman & Clark, 2014). Designers cannot overlook the factors that influence decision-making and guide food choices.

Urbanisation and industrialisation of food systems have caused food transitions that have resulted in a reduction in food variety, with individuals' food choices limited due to the prominence and prevalence of ultra-processed, high-calorie-dense foods that are mainly rich in fat and added sugars (Armelagos, 2014). Some elements of design under consideration are: The recognition of nutritional properties and health value (e.g., absence of contaminants) of foods as a factor in determining choices (Kang et al., 2015); information based on nutritional facts, such as sustainability labels and organic identity have been shown to be important factors in defining consumer choices (Barreiro-Hurlé et al., 2010); certification of origin, recyclable packaging, as well as claims of local, traditional, ethical, and environmentally friendly products also affect food choices (Annunziata & Scarpato, 2014); and the design of physical environment, such as local supermarkets or stores are influential in consumers making choices (Glanz et al., 2005). In addition, social interaction is important in making food decisions; for example, eating in the company of other people affects the quality and quantity of food consumed (Pollard et al., 2002).

Reynolds et al. (2019) have identified and summarised food-waste prevention interventions at the consumption/consumer stage of the supply chain via a rapid review of global academic literature from 2006 to 2017. To address healthy diets and waste reduction strategies, they found that school nutrition redesign can reduce vegetable waste by 28%. Over the last decade, designs focused on reducing food waste have worked on: (a) Interventions that changed the size or type of dishes in the hospitality industry; (b) awareness campaigns; and (c) changes in school nutrition guidelines (Reynolds et al., 2019). It is essential for designers to focus on behavioural interventions that change consumer habits, while improving consumer understanding of the consequences of food waste (Massari et al., 2021) and exploring cognitive factors, such as enhancing awareness and empathy (Allievi et al., 2021). Designers also attend to mechanisms to enhance self-concept and emotional intelligence to affect healthier and more sustainable food choices (Asioli et al., 2017).

Sustainable diets and cultural approaches

When considering food systems, health and the environment must be considered simultaneously. To prevent further climate emergencies, demands retooling of current food systems to enhance healthy and sustainable culinary cultures and dynamics with food (Moro & Niola, 2020). Integrated systemic approaches in the food systems can be used to forge more resilient, equitable, and sufficient quantities of healthy, nutritious food for all, while respecting the environment (BCFN, 2021).

The Barilla Centre for Food & Nutrition (BCFN) has recently proposed (BCFN, 2021) a new version of the already-known concept of Double Pyramid (DP), a model based on the Mediterranean Diet that was developed to show the relationship between health and the impact of food on climate. This model aims to explain the complexity of the relationship between individual health and climate, and is a starting point to illustrate the principles of a healthy and sustainable diet (i.e., the impact of choosing fresh, seasonal and local foods; avoiding consuming excessive quantities of food; reducing, reusing and recycling food packaging; supporting social life and community-based initiatives). Geographical and cultural differences also are important for identifying and preserving resources to concretely ground healthier and sustainable diets. These considerations are demonstrated by the foundation's recent research that presents seven experimental Double Cultural Pyramids for a set of regions of the world (i.e., Africa, South Asia, East Asia, the Mediterranean, Nordic countries and Canada, the United States, and Latin America).

Fostering sustainable diets through food design and system mind-sets

Food Design is a concept that recently has become popular and includes innovative solutions for how farmers and industry manage access to food resources (cultivating, harvesting, selecting, transforming, and distributing food), for how consumers access food resources (buying, transporting, storing, cooking, and eating food), for how restaurants create with food (menus, new dishes, solutions to avoid waste), and for the design of multidisciplinary human–food interactions and experiences. Design can enhance the multiple roles that food can play in people's lives, in local communities, and in society in general. Designers are trained to identify connections, to envision opportunities, and to identify how health problems, systemic barriers, and planetary boundaries can guide actionable insights that lead to creative disruption of the status quo (Raworth, 2017).

For most designers who wish to work in the food industry, food design creates the challenge of learning about agriculture, the food industry, culinary processes, and the hospitality industry because many of the current educational design programs do not offer a specialisation in food design (Lee et al., 2020). Similarly, most current designers are unfamiliar with the use of food as a material, even though food products have interesting properties and design challenges (Ayala Garcia, 2015; Bruns Alonso et al., 2013; Perrone and Fuster, 2017; Reissig, 2017).

While designers are trained to understand people's varied and complex needs, and thus to design solutions and services that meet them, most food system actors do not understand these interconnections and relationships and operate within isolated thinking. Food systems, as well as consumer choices, never follow a linear mechanism; they are an eco-system that emerged from human mental patterns and beliefs, which are the basis of both global and local food patterns (Lin et al., 2011). According to a paper on nutrition, education and communication around food are important factors in determining food choices (Future Food Institute, Dole, n.d.). Design methods can be used to create new food literacy and capture inefficiencies in the system and transform them into opportunities (ibid). In 2011, Danish designers claimed in their manifesto that design should answer the needs of "people, profit, and planet" (Valade-Amland, 2011). In this perspective, food products, services, and delivery models can be designed, not only to meet human needs and desires but also to add value and support the regeneration of systems (Acaroglu, 2018).

A mind-set shift is required to realise sustainable actions: Various levels of empathy need to be put in place (Allievi et al., 2021). Individuals benefit from using empathy to recognise oneself in the system while also recognising the point of view of others. When empathy is activated, individuals can see what they can do and ways to cooperate with others to improve the sustainability of things. The co-constructive and collaborative empathic processes as proposed by Allievi et al. (2021) includes the model called EOE—Ego-centric, Other-centric, Eco-centric.

In view of this model of empathy, metabolism of design in the agri-food sector (Massari, 2021) becomes fundamental. Metabolism here means being able to build and design tools, services, products, and systems that lead to healthier and more sustainable food experiences. These can only work in the long term if they are easily metabolised by people and their communities. To do this, novelties will not only have to be assimilated by users, but they will have to become their tools, rituals, and instruments of daily change.

Innovating to make a product or service work better for customers' daily lives is the core of the design-thinking approach, which helps identify what users want, need, prefer, and are willing to pay for. Design thinking is a complex process that helps marketers understand what their customers need and how to create innovative solutions for their problems. It is not just about solving problems, but about finding the right and creative opportunities in actual and real market scenarios. But the Design Thinking approach, putting only human needs at the core of the projects, produces unsustainable side effects that damage the environment and the human consuming them. New design thinking methodologies are currently being developed to prevent these side effects by taking into account the planetary boundaries from the very beginning of the design project. Prosperity thinking (Vignoli et al., 2021) is a hands-on example of a new design-based approach focused on a multi-stakeholder co-design process for change. Prosperity thinking is a methodological approach to designing a world that fulfils all its beings' needs within the planet's ecological means. It aims to enable the design of a better world, starting from a shared, inclusive idea of prosperity that encompasses economic growth and social and environmental well-being.

Case studies: Designing for a sustainable Mediterranean Diet

Recognised by UNESCO (UNESCO, n.d.) as an intangible cultural heritage of humanity, the Mediterranean Diet favours the consumption of vegetables, fruit, cereals, vegetable protein sources, dried fruit, and olive oil, with moderate or limited consumption of eggs, fish, meat, milk, and its derivatives. As mentioned in above, the idea that the Mediterranean Diet is not only good for health but also has social, economic, and environmental benefits is represented and simplified in the graphical and conceptual model of the Double Food Pyramid (Ruini et al., 2013). The Mediterranean Diet is sustainable because it represents a way of life, and thus fits into the definition of sustainable diet proposed by the FAO (FAO, 2012): Focused on the protection and respect of biodiversity and ecosystems; culturally accepted; economically equitable; and accessible, adequate, safe, and nutritionally healthy; and, at the same time, optimises natural and human resources. With the publication of the Met Diet 4.0 in 2016, the International Foundation of Mediterranean Diet identified the four benefits of the Mediterranean Diet, that go far beyond the health aspects proposed by Ancel and Margaret Keys in 1975 (Keys, 1975): Improved health; lower environmental impact and richness of biodiversity; high socio-cultural value; and positive local economic returns.

Three case studies of Food Design projects that were created to support and promote the Mediterranean Diet as a sustainable diet are presented below. The following projects show that supporting a sustainable diet requires producing products that are nutritionally valid for the health of the consumer, creating new narratives that provide the consumer with a more systemic perspective on food, reconnecting the consumer with the entire value chain, and providing it with the tools necessary to preserve and transmit food cultures. These projects have been chosen because of their concrete impact towards creating cultures that are healthier for humans and more sustainable for the planet.

In each case, the design method started from a field observation to understand the existing habits and relationships/networks, both to improve them and to mitigate them. These projects show how designing products, services, and systems means knowing how to enhance the cultural value of the sustainable diet for the specific local community. Supporting sustainable diets means strengthening collaboration between all sectors involved in local food production and distribution, involving people from different sectors, and helping them to work together to find innovative solutions. The following cases show how designers have a significant role in making sustainable diets more concrete, fostering sustainable behaviours and markets, understanding daily consumption needs, and seeing opportunities rather than treating people as a problem.

Case study: Sementino

In 2017, in southern Italy, food designer Antonella Mignacca created "Sementino," (see Figure 44.1), a legume-based snack celebrated for its crunchiness. It was designed to encourage small local communities to develop territorial storytelling and establish new relationships between producers and consumers. Storytelling

Figure 44.1 Sementino. Source: Antonella Mignacca (www.intothefood.eu/team-view/antonella-mignacca/). Used with permission.

represents a new dimension of marketing strategies for small and medium-sized enterprises, which do not always have high budgets and want to increase awareness of the food system. This technique is effective because it leverages narrative skills, involving and working on emotions, which allows the assimilation of information in an effective and enduring manner. It also enables geographies, culture, sustainability, development, and any alternative models of entrepreneurial narratives to be traced.

Sementino storytelling works on different levels:

- The role of local producers: The raw materials, legumes, were picked from local producers, which helped to revitalise some of the local marginalised production and economy.
- The role of designers: Designer Antonella Mignacca involved local artisans, those who transform food, revitalising this sector and putting producers in contact with them.
- The role of local institutions: Sementino Snack was included in schools and made available to children. In this case, the school was a fundamental supporter of and partner on this local product, reinforcing the local identity of the students and the staff of the school. By focusing on a delicious and healthy snack food, the nutritional aspect also was reinforced.

In this case, the food designer was able to connect the whole range of players in the food chain (producers, processors, and consumers), thus creating new virtuous connections for the development of a new and sustainable territorial economy. In fact, Sementino was designed to support the sustainability of consumers' food choices, connect a range of players in the food chain (producers, processors, and consumers), and enhance the value and health of local products. The limitation of this project was in its economic unsustainability, as the designer imagined this type of product for a limited and local distribution, developed by a small food manufacturer or artisan. This severely limited the development of the project. It would be important to inspire food industries in the plant-based market, to design new types of legume-based snacks, committing to maintain three important principles for fostering a sustainable food system and diet: (a) Maintaining the link with the local territory; (b) engagement of local communities; and (c) awareness of situated cultural and historical traditions.

Case study: Frisella Point

Food designer Sara Costantini created "Frisella Point" (see Figure 44.2) in 2018, a store based on the study of the storytelling of Frisella, with a specialised hard bread product common to Puglia, a region of southern Italy. Consumed with fresh tomatoes, olive oil, and oregano, or with garlic or anchovies, it is also an important ingredient in Cialledda or Acqua Sale, a typical Apulian dish. Frisella Point is important for tourism, to cultivate a deeper knowledge in the travellers across the regions, as well as for the local population, to deepen their understanding of their territory, culture, and history.

Over centuries, Frisella has been an important staple because its characteristics allowed it to be preserved, making it an excellent alternative to bread when flour was scarce. The product was designed to be highly suitable for transport where the central hole allowed a thread to be strung through and knotted making it easy to hang and keep them dry. Fishermen had a custom of soaking Frisella in sea water and using it as a base for fish or mussel soups, a practice that was common during fishing trips that lasted several days. Frisella Point was designed as a place to purchase and to eat this local food, and to learn about the history of this product. It was designed to help consumers learn about the origin of the ingredients, their characteristics, nutritional values, history, traditions, and also the rituals and con-

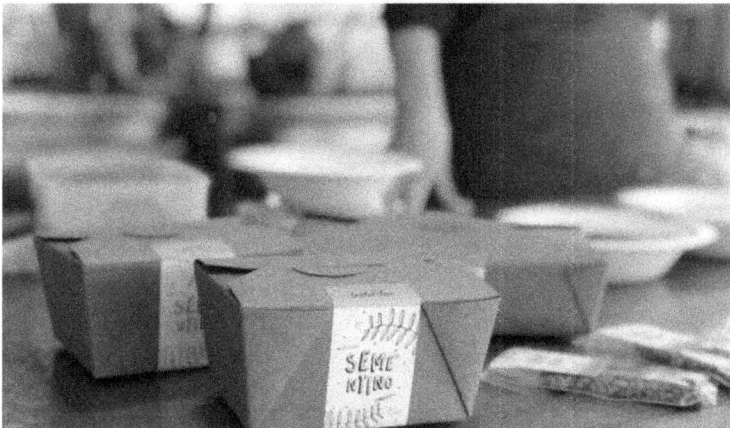

Figure 44.2   Frisella Point. Source. Frisella Point Page (www.facebook.com/fooddesigne xperience/?ref=page_internal). Used with permission.

viviality that arise from its consumption. Gastronomy narratives link environmental and nutritional resources, as well as cultural heritage, with relevant impacts on tourism and hospitality. Frisella Point is not just a simple frisella shop but a place where the frisella is a means to disseminate and preserve the local culture. This illustrates a systemic Food Design project that combines historical research with experimentation and food consumption, with the creation of new ecosystems that include the local municipality's goals, producers, distributors, and environmental resources.

Case study: Legumotti

Legumotti (see Figure 44.3) is a legume-based product, produced by the Barilla company. It is an example of application of design thinking to a product, with the aim of guaranteeing a more nutritious and sustainable food for consumers with plant-based proteins (Fruitbook Magazine, n.d.). This project is the result of a partnership between a brand in the Italian food industry (Barilla) and a large-scale distribution company (Esselunga). Legumotti have been designed as 100%

Figure 44.3 Legumotti. Source: Barilla. Used with permission. (www.barilla.com/it-it/ricerca-prodotti/pasta/gamma/legumotti?sort=alpha&page=1).

legume grains, which provide a high source of fibres and proteins. The product is ready in just nine minutes and is suitable for numerous preparations. It aims to satisfy a growing market that demands healthy, tasty, easy, and quick-to-prepare food products suitable for the whole family. This case is also a great example of a transdisciplinary approach, applied to a co-design project, where food industry professionals from the Esselunga and Barilla company were assisted by a Design Thinking research group (University of Modena and Reggio Emilia) to create an innovative product. For several decades, legumes have been considered "food for the poor," due to their peasant origins. Today they are an attractive product thanks to the recognition of their nutritional properties, their connection to the peasant tradition, and as a tool for protecting biodiversity and the territory. In this project, one of the marketing keys used for sales is that sustainable diets are not only respectful of the environment, but must be healthy and safe, available, accessible to all, and socially, culturally, and traditionally acceptable. This product also has become part of the solutions for weaning babies as it is recommended by paediatricians, and is incorporated into programs of nutrition education.

Reflections and conclusion

Our diets play a crucial role in a movement towards sustainability. Achieving a healthy and sustainable food future is an urgent matter that depends on global collaborative efforts from governments, the private and public sectors, and individuals. Sustainable diets and sustainable food systems are increasingly being explored by diverse scientific disciplines. Meybeck and Gitz (2017) illuminate how sustainable diets are both the results, and the drivers of, food systems. The factors for change by the stakeholders involved are different, and can be triggered by different motivations: Health, environmental, social, and cultural. The authors stressed the need to combine different dimensions and reasons for change to facilitate the transition to sustainable diets, while recognising the regional qualities of the food systems (Meybeck & Gitz, 2017).

Design aims to create experiences, products, tools, and services that have considered the environmental, social, and economic impacts from the initial phase through to the end of life of products. Intentional design applies a systemic approach from the very first phase of research: The designer is able to study, observe, collect, and systemically analyse the cognitive and cultural factors that influence eating behaviour. Designers are trained to understand the macro system, and at the same time are acting on problems at a specific and local level, linking micro-innovation with macro-innovation. Designers take into account the cultural values important to a community, develop a sustainability and transdisciplinary mind-set ready to address the greatest challenges of our era, evaluate problem-solving strategies, and design experience and science-based solutions. Food Design helps to engage consumers during the entire process, introduce radical sustainable product innovation, and foster new food systems development. In this chapter, three recent projects created by food designers were presented. Design methods were applied to the food chain to support and reinforce the idea of the Mediterranean Diet as a sustainable diet.

In conclusion, the food designer has an ethical obligation to design for sustainability. These products also are thought to appeal to *sustainable natives* (Massari, 2015), those who are born and grow up in a world where all products, services, and systems, along with cultural artefacts (physical and cognitive) are designed to support more sustainable lifestyles and sustainable diets. Food Design can help in the process of cultural innovation and bring concreteness, as well as value, to sustainable diets. It is necessary to improve general awareness about the functional role of food designers in the agri-food sector; they can work together with farmers, policy makers, researchers, managers, and educators to prepare the ground for helping a new generation of sustainable natives. Through innovative ideas and creative solutions, the next generation should feel less like part of the problem and more a part of a continually improving system.

References

Acaroglu, L. (2018). *Circular systems design handbook*. The Unschool.

Allievi, F., Massari, S., Recanati, F., & Dentoni D. (2021). Empathy, food systems and design thinking for fostering youth agency in sustainability: A new pedagogical model. In S. Massari (Ed.), *Trans-disciplinary case studies on design for food and sustainability*. Elsevier.

Annunziata, A., & Scarpato, D. (2014). Factors affecting consumer attitudes towards food products with sustainable attributes. *Agricultural Economics, 60*(8), 353–363.

Armelagos, G. J. (2014). Brain evolution, the determinates of food choice, and the omnivore's dilemma. *Critical Reviews in Food Science and Nutrition, 54*(10), 1330–1341.

Asioli, D., Aschemann-Witzel, J., Caputo, V., Vecchio, R., Annunziata, A., Næs, T., & Varela, P. (2017). Making sense of the "clean label" trends: A review of consumer food choice behaviour and discussion of industry implications. *Food Research International, 99*, 58–71.

Ayala Garcia, C. (2015). The basis of processes: Experimenting with food to re-shape the industry language. In Collina, L., Galluzzo, L., & Meroni, A. (Eds.), *The virtuous circle; Design culture and experimentation* (pp. 443–453). McGraw-Hill.

Barreiro-Hurlé, J., Gracia, A., & de-Magistris, T. (2010). Does nutrition information on food products lead to healthier food choices? *Food Policy 35*, 221–229.

BCFN (2021). *One Health: un nuovo approccio al cibo, la Doppia Piramide per connettere cultura alimentare, salute e clima*. Barilla Center for Food and Nutrition. https://www.barillacfn.com/m/publications/a-one-health-approach-to-food.pdf

Bruns Alonso, M., Klooster, S., Stoffelsen, M. A. H., & Potuzakova, D. (2013). Nourishing the design ability throughfood. In Proceedings of the 5th international congress of International Association of Societies of DesignResearch (IASDR), Consilience and Innovation in Design, 26-30 August 2013 (pp. 1–12). Tokyo, Japan.

FAO (2012). *Sustainable diets and biodiversity: Directions and solutions for policy, research and action*. FAO. http://www.fao.org/3/i3004e/i3004e.pdf

FAO (2019). *The state of food and agriculture 2019. Moving forward on food loss and waste reduction*. FAO. http://www.fao.org/3/ca6030en/ca6030en.pdf

FAO (2020). *The state of food and agriculture 2020. Overcoming water challenges in agriculture*. FAO.

FAO, IFAD, UNICEF, WFP, & WHO (2020). *The state of food security and nutrition in the world 2020. Transforming food systems for affordable healthy diets*. FAO.

Ferrara, M. R., & Massari, S. (2015). Evoluzione del concept food design: intersezioni storiche tra cibo, design e cultura alimentare occidentale. *Ais/Design Journal*. https://re.public.polimi.it/retrieve/handle/11311/970601/70952/AIS5_evoluzione%20del%20concept%20food%20design.pdf

Fruitbook Magazine (n.d.). *Co-design fra Barilla ed Esselunga: nascono i Legumotti, naturali e gluten free*. https://www.fruitbookmagazine.it/co-design-fra-barilla-ed-esselunga-nascono-i-legumotti-naturali-e-gluten-free/

Future Food Institute, Dole (n.d.). *Nutrition unpacked*. https://www.nutritionunpacked.com/

Glanz, K., Sallis, J. F., Saelens, B. E., & Frank, L. D. (2005). Healthy nutrition environments: Concepts and measures. *American Journal of Health Promotion, 19*(5), 330–333, ii.

IDEO (n.d.). *About IDEO*. https://www.ideo.com/about

Kang, J., Jun, J., & Arendt, S. W. (2015). Understanding customers' healthy food choices at casual dining restaurants: Using the Value–Attitude–Behaviour model. *International Journal of Hospitality Management, 48*, 12–21.

Keys, A. B. (1975). *How to eat well and stay well the Mediterranean way*. Doubleday.

Lee, Y., Breuer, C., & Schifferstein, H. N. J. (2020). Supporting food design processes: Development of food design cards. *International Journal of Design, 14*(2), 51–64.

Lin, M., Hughes, B., Katica, M., Dining-Zuber, C., & Plsek, P. (2011). Service design and change of systems: Human-centered approaches to implementing and spreading service design. *International Journal of Design, 5*(2), 73–86.

Massari, S. (2015). Sustainable natives, youth manifesto and design approaches: Designing the world for "sustainable natives". In Parasecoli, F., Remein, C., & Rahman, R. (Eds.), Second International Conference on Food Design (pp. 151–162). New York: The New School.

Massari, S. (2021). *Transdisciplinary case studies on design for food and sustainability*. Woodhead Publishing.

Massari, S., Principato, L., Antonelli, M., & Pratesi, C. A. (2021). Learning from and designing after pandemics. CEASE: A design thinking approach to maintaining food consumer behaviour and achieving zero waste. *Socio-Economic Planning Sciences, 82*, 101143.

Meybeck, A., & Gitz, V. (2017). Sustainable diets within sustainable food systems. *Proceedings of the Nutrition Society, 76*(1), 1–11.

Moro, E., & Niola, M. (2020). *I segreti della dieta mediterranea. Mangiare bene e stare bene*. Il Mulino.

Norman, D. A. (2010). *Living with complexity*. MIT Press.

Perrone, R., & Fuster, A. (2017). Food as a system and a material for the creative process in design education. *International Journal of Food Design, 2*(1), 65–81.

Pollard, J., Kirk, S. L., & Cade, J. E. (2002). Factors affecting food choice in relation to fruit and vegetable intake: A review. *Nutrition Research Reviews, 15*(2), 373–387.

Raworth, K. (2017). *Doughnut economics: Seven ways to think like a 21st-century economist*. Random House Business.

Reissig, P. (2017). Food design education. *International Journal of Food Design, 2*(1), 3–13.

Reynolds, C., Goucher, L., Quested, T., Bromley, S., Gillick, S., Wells, V., Evans, D., Koh, L., Carlsson Kanyama, A., Katzeff, C., Svenfelt, A., & Jackson, P. (2019). Review: Consumption-stage food waste reduction interventions: What works and how to design better interventions. *Food Policy, 83*, 7–27.

Ruini, L., Ciati, R., Pratesi, C. A., Principato, L., Marino, M., & Pignatelli, S. (2013). Diete sostenibili: la doppia piramide del Barilla center for food and nutrition. *Barilla Center for Food and Nutrition. Agriregionieuropa, 34*(9). https://agriregionieuropa.univpm.it/it/content/article/31/34/diete-sostenibili-la-doppia-piramide-del-barilla-center-food-and-nutrition

Springmann, M., Spajic, L., Clark, M., Poore, J., Herforth, A., Webb, P., Rayner, M., & Scarborough P. (2020). The healthiness and sustainability of national and global food based dietary guidelines: Modelling study. *British Medical Journal, 370*, m2322.

Tilman, D., & Clark, M. (2014). Global diets link environmental sustainability and human health. *Nature, 515*, 518–522.

UNESCO (n.d.). *Mediterranean diet*. https://ich.unesco.org/en/RL/mediterranean-diet-00884

Valade-Amland, S. (2011). Design for people, profit, and planet. *Design Management Review*, *22*(1), 16–23.

Vignoli, M., Roversi, S., Jatwani, C., & Tiriduzzi, M. (2021). Human and planet centered approach: Prosperity thinking in action. *Proceedings of the Design Society*, *1*, 1797–1806.

WWF (2020). *Bending the curve: The restorative power of planet-based diets.* WWF.

45

TECHNOLOGY, DIGITALISATION, AND AI FOR SUSTAINABILITY

An assessment of digitalisation for food system transitions

Gianluca Brunori, Manlio Bacco, and Silvia Rolandi

Introduction: Sustainable food systems and digitalisation

Transition towards sustainable food systems is widely recognised as one of the most significant challenges facing the planet, and the urgency of such transformations is now irrefutable (Balafoutis et al., 2017). Research is considered a key driver for the transition. In this spirit, the European Commission's Standing Committee on Agricultural Research's (SCAR)—a body that coordinates the member states' research policies in the field of agriculture—launched its 5th foresight exercise with the aim of providing a frame of reference for national research strategies, EU-level partnerships, and joint programming initiatives. The experts were asked to explore the pathways to achieve a "safe and just operating space" for the primary systems in Europe. Three pathways—healthy and sustainable diets for all, circularity in food systems, and diversity—were identified through workshops, expert inputs, and deliberation. These pathways provide different entry points and key driving forces, and require integrated and coherent efforts by a wide range of actors.

The first transition pathway relates to what we eat: Everyone should have access to healthy diets, and at the same time, diets should reflect the need to reduce the pressure on the environment (Van Wassenaer at al., 2021). This pathway implies a change in both food supply and demand. The second pathway relates to circular food economies, which aims at closing the material loops in all the subsystems of the food system, starting from primary production wherein circularity can be managed through ecosystem redesign and the mobilisation of functional biodiversity (Araújo et al., 2021). The third pathway relates to social and biological diversity

DOI: 10.4324/9781003174417-53

to augment ecosystem services and foster more resilient systems. In this context, digitalisation has been highlighted as a key cross-cutting issue.

Given its game-changing role, digitalisation can be a driver of progress and prosperity, and depending on its application, it can cause social inequalities and environmental degradation. Much will depend on the capacity to shape ICT technologies in view of the common good and to design appropriate legal, social, and economic frameworks to control its development. The chapter discusses the role that digitalisation may play in each of these pathways. We provide an overview of the main concepts related to digitalisation, and then identify, for each pathway, current and future ICT applications that may support the process and their potential impact. Discussion and conclusions will follow.

The digital transformation: Main concepts

ICT technologies are the drivers of two distinct, albeit interconnected, processes: *Digitisation* and *digitalisation*. We first explain the former. *Digitisation* is the transformation of analogical signals into digital information. Digitisation increases exponentially the capacity to store, retrieve, exchange, and integrate data, allowing the so-called *datafication*, which is the growing production of data to support decision making (Serazetdinova et al., 2019). The exponential reduction of the costs of computing has made it possible to translate physical information into a digital format, which has generated an unpreceded possibility of access to knowledge and information, raising at the same time issues of privacy and intellectual property.

In more recent years, the development of sensing technologies has allowed a new wave of datafication based on the automatic collection and storage of data with very limited human intervention. Among the most relevant sources of data, we can now count satellites, sensors, and the so-called "Internet of Things" paradigm, mobile phones and apps, censuses and surveys, publications and documents, citizen science, administrative data, and finance data (Jensen & Campbell, 2019).

Digitisation allows the creation of digital representations of the physical world. The Internet allows the circulation of these representations that become part of the cybersphere, a world of informational objects. Through the cybersphere, humans can remotely act on the physical world. To better grasp this interconnectedness, the concept of "cyber-physical system" (CPS) (Cyber Physical Systems Public Working Group, 2016), has been proposed. In a CPS, physical and cyber components are connected and communicate with each other to perform a given function (Bacco et al., 2020). Logically, CPSs can be described as composed by five layers: *Sensing* (data collection through sensors), *data exchange* (through wired or wireless connection), *computing* (data storing and processing), *intelligence* (application with a given purpose), and possibly *actuation* (semi or fully automatic execution of actions). These systems allow "smart processes," as they can adopt flexible decisions and carry out fine-tuned operations based on data. CPSs can radically change the patterns of interaction among people. Communication at a distance allows new forms of sociality, new labour organisation patterns, new lifestyles:

Robots can replace or support human work; such automatic translation opens new possibilities for communication.

At the same time, different social patterns can give CPSs different shapes: The social, the cyber, and the physical domains co-evolve. These changes are captured by the concept of digitalisation. *Digitalisation* departs from the potentialities created by digitisation and regards the transformation of socio-economic-ecological systems in relation to it. The process of digitalisation cannot be captured by the concept of CPSs, where the social component is external to the system. For this reason, we have proposed the concept of *socio-cyber-physical system* (SCPS), which allows technological systems to be described as components of a broader set of relations and interdependencies. The SCPS concept emphasises the two-way relationship between the social and the technological processes, stressing that different social patterns may shape different technological patterns and vice versa. This reciprocity has important implications on innovation policies, as SCPSs approaches stress how to avoid technological determinism and propose an approach that subordinates the technology development to the identification and analysis of social problems, and thus to solutions that adapt technology according to the specific ecological, social, and organisational contexts (Rijswijk et al., 2021).

The impacts of digitalisation on the food system

Digitalisation has already changed the food system, but the technological developments in this field enable us to envision greater and deeper changes. The impact of digitalisation on the food system is briefly analysed in relation to its main actors and activities.

Digitalisation and farming

The impact of digitalisation on the future of farming is of increasing interest, as it promises to bring benefits to farmers. *Productivity gains* can be obtained through better insights into production variables and dynamics, informed decision-making based on insights and scenario analyses, and increased labour productivity. Accurate monitoring and optimisation of processes can bring higher quality and waste reduction. Digitisation can improve *access to markets*, and thus *afford economic gains*, by improving product quality and available information to customers and by alleviating excessive dependency on intermediaries. Digitalisation can also improve the *quality of work*, relieving farmers from heavy or unpleasant tasks, and can improve the communication with administrations. *Benefits to the environment* are also envisioned as digitisation improves the efficiency of the use of external inputs and the monitoring of the impact of agricultural practices on GHG emissions and other environmental indicators (García-Llorente et al., 2018; Garske et al., 2021; Nikolaou et al., 2020). However, biases toward specific agricultural paradigms (specialisation, large-scale) embodied in technologies are problematic, as well as economic barriers to access (high initial investments to adopt the technologies), and technological ecosystems that may encounter inadequate infrastructure or supports for new technologies to work.

Digitalisation and Agricultural Knowledge and Innovation Systems

Agricultural Knowledge and Innovation Systems (AKIS) are about knowledge management: AI, big data, and cloud computing will make the organisation of traditional AKIS obsolete. The mix of face-to-face and remote interaction will be radically reorganised, and the function of face-to-face activities will be much different from the past. Farmers will be able to access information autonomously through specialised sites and increased peer-to-peer communication, and advisors will move from information brokering to data analysis. These services will be growing online, aimed at increasing the value of direct experience and generating trust. As data management will be more relevant for farmers, extension services, and cooperatives, it will change the functions of these food system actors into data managers, analysts, and brokers (Cobby, 2020). Machinery and input suppliers are already "connected" to farmers, and new digital services will be proposed. In terms of risks, lack of ICT skills will create new divides (Akyazi et al., 2020). Another risk that may be related to a technological bias is that extension services may choose to support the technologies that are more digitised. Dismissing traditional services may prompt a further disconnection from the AKIS of farmers lacking the skills to interact with digital services.

Digitalisation and the other actors of the food value chain

Data availability allows the reduction of "information friction" between actors of the value chain. Increased availability of data and of computing capacity will improve the processes of traceability, protecting the identity of products along the chain, reducing the costs of certification, and improving its accuracy. Improved one-to-one or one-to-many communication allows disintermediation from traditional intermediaries, which will reshape the value chain patterns. New business models supported by platforms show that farmers can have direct access to distant markets and to consumers (Sam and Grobbelaar, 2021). Improvement in logistics due to automation and data management will foster new organisational patterns. 3D printing will allow a spatial reorganisation of the food industry, allowing diffused manufacturing, and will also create a divide between 3D printed and "authentic" food (Lipton, 2017). In this new scenario, technology design will be shaped by the actors who have more power in the value chain; actors with greater capacity to collect and analyse data will wield more power. Weak regulation over data property will generate new enclosures and new monopolies and farms left outside the value chains will undergo marginalisation. Unregulated access to sensitive data—such as polluting emissions or quality of soils—may generate inaccurate information and endanger actors' reputations.

Digitalisation and consumers

Digitalisation has given consumers access to an enormous amount of information about products, increasing their power of choice. Information about the product increases responsibility of both producers and users, as information links the choice to its consequences. Labelling and certification, nutritional parameters, and environmental footprints will accompany the product, allowing an assessment of its sustainability and healthiness. Information to consumers will constitute another field of competition between firms. Risks may be related to information overload, trusting wrong sources of information, lack of skills to interpret the data, and the deliberate pollution of information sources. Firms may carefully select the information to be disclosed, thus potentially increasing the level of confusion or the risk of inaccurate claims.

Digitalisation and rurality

Digitalisation reconfigures spatial patterns (settlements, mobility, resource flows, communication), disrupts goods and services provision, and opens new opportunities for natural resources and ecosystem services management. In theory, digitalisation has the potential to reduce distance and isolation in rural areas; in practice, market-led digitalisation may enlarge the gulf between rural and urban areas. To reverse this trend and enable rural areas to harvest such opportunities, a strong effort to rethink the role, resources, and importance of rural areas is needed.

Digitalisation and public administrations

Digitisation offers a multiplicity of tools to policymakers. Improved communication among farmers and administrations will reduce the transaction costs and the power of intermediaries. Better information about production processes will allow public administrations to grant "licences to operate" in a more accurate way. Environmental monitoring may activate performance-based policies (Bikomeye et al., 2021). Monitoring systems for food storage will improve food safety. Data sharing about risks will improve the capacity to respond to crises. Technology will allow the development of "nudging systems" for consumers toward healthy and sustainable diets. Again, inaccurate disclosure of information may create panic, bias, and may undermine the reputation of some actors. Lack of trust in the public administration may hamper data sharing, and excessive nudging could spark a sense of diminishing freedom.

Transitions to sustainable food systems: The role of digitalisation

The role that digitalisation may play in the transition to sustainable food systems will depend on the shape that SCPSs take. Defining transition pathways would include identifying technology pathways.

Pathway 1: Healthy and sustainable diets for all

The first pathway, "healthy and sustainable diets for all" starts from the observation that current diets are both unsustainable and unhealthy. Change in diets will require a change in both demand and supply (Sadhukhan et al., 2020). The

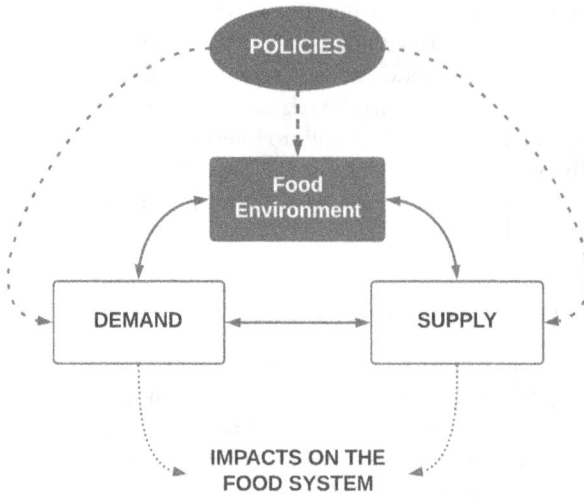

Figure 45.1   Impacts on food systems. Source: Work of the authors.

transition to sustainable diets will also depend on the capacity to shape the food environment, a set of physical and symbolic conditions that affect consumers choice (Bronson, 2020). Change of dietary patterns—on average, reduction of food intake and increase in plant-based food—will stimulate producers to provide food with higher nutritional density and lower pressure on the ecosystem (McClements, 2020). Alternative sources of proteins will be available and diversification will increase the food variety (Figure 45.1).

Reshaping demand and behaviour

Digitisation offers a variety of solutions for sustainable dietary patterns (Smetana et al., 2019). Access to nutrition-related information can be made accessible and apps connected to wearables can provide personalised recommendations. Improved traceability will give consumers information about the impact of what they eat, fostering more sustainable patterns. The growing role of e-commerce will allow retailers to "nudge" consumers, with sustainable solutions being more visible.

Reshaping supply and systems

Digitisation will make it possible—through data provided through several sources—to measure the carrying capacity of agro-ecosystems, and therefore to set levels for sustainable yields per region. Inputs could be distributed in proportion to both environmental and crop needs. Good agricultural practices will be documented, and related information can be communicated to consumers. Environmental certification will accompany food products. Improved access to knowledge and tailored decision support systems can support farmers in their practices (Tsolakis et al., 2021). Food safety can be improved through sensors, smart packaging, blockchain, or other technologies that hold the potential to increase trust and coordination between actors of the chain.

Supporting sustainable trade policies

Trade links together distant social and ecological systems. Sustainable trade occurs when this interaction does not harm the capacity of the systems to provide benefits for the future generations. Increasing evidence shows that a considerable share of international trade is not sustainable, and the increasing availability and accuracy of this information generates a clash between the principle of free trade and sustainability principles. The Lulucf regulation[1] is a key example of policies influencing practice, with telecoupling enabling interaction at a distance.

Digitisation will allow a generalised use of footprinting, which will improve the awareness of the true cost of food and support policy making with substantial evidence (Rincon-Patino et al., 2018). Available information about the impact of distant production processes will inform actors across food systems about problems in, or impacts from, certain patterns of consumption and trade, and will increase the level of responsibility of firms, consumers, and policy makers.

Pathway 2: Full circularity of food systems

The second pathway implies a reorganisation of the food system around principles of circularity. A circular economy is based on: a) Design of the product and production processes to minimise waste and to extend the product life cycle; b) reduce the amount of "funds" by shifting from "goods" to "services"; c) an industrial ecosystem with "functional diversity" that allows industrial symbiosis; d) improving the processes of reuse and recycling e) improving cooperation between firms for reusing and recycling; f) reorganising logistics for optimised flows; and g) connecting the individual stakeholders of the system. Table 45.1 shows the potential impacts of digitisation in this respect. The food system can rely on several naturally circular processes, related to the role of living species in the biogeochemical cycles. The key for a circular economy is the value of biomass to be recycled and reused, and that the cost of biomass procurement and manipulation should be lower than its use value. At the level of primary production, this implies "closing the loops" at the farm or landscape level, with the soil (and living beings therein) having a pivotal role in transforming organic matter; for example, composting (a practice that allows an upgrading of discarded biomass to increase the fertility of soil) is more profitable than burning it. At the level of secondary production, circularity implies reorganising "industrial ecology" to make the reuse and recycle of organic matter profitable (Canali et al., 2020; Gkisakis & Damianakis, 2020).

Redesign products and processes

A circular economy needs an alternative approach to product and process design. Product design should extend the life of products and minimise the disassembling costs. Diversity of input and output multiplies the possibilities that leftovers are

Table 45.1 Current and potential impacts of digitisation with respect to different targets

Target	Impact of digitisation
Design of the product to minimise waste production and to extend its life cycle	Smart packaging and smart labelling (Skinner, 2015), design, simulations
Reduce the amount of "funds" being used in the economy by shifting from "goods" to "services"	B2C platforms to manage durable goods services (Taranic et al., 2016), B2B platforms for sharing durable goods
An industrial ecosystem with "functional diversity" fostering industrial symbiosis	B2B platforms; improved genetic screening of biodiversity. Improved analytics for system biology (Kuzmin et al., 2018)
Redesigning rural areas	e-government; e-services
Improving reuse and recycling processes	Robots, sensors, advanced analytics, artificial intelligence= can be used by all
Improving cooperation between firms for reusing and recycling	Joint databases, big data for characterisation of feedstocks, sharing platforms
Reorganising logistics for optimising flows between firms.	Inter-firm blockchain, robotisation, sharing platforms, IoT, linking transportation planning to agricultural sensors
New consumption patterns	Footprinting, "nudging" software, sensor to monitor waste, collaborative consumption platforms (Sposato et al., 2017), RFID to follow consumers, all information available

Source: Work of the authors.

turned into input for another process. This principle is translated into the construction of industrial ecologies (Marescotti et al., 2020). To maximise the (bio-) value that can be extracted from biomass, its hierarchy should be reflected in its exchange value and proportionally in the costs of procurement. These practices imply a rapid characterisation of feedstocks, proximity between buyers and sellers, improved logistics, and availability of data among potential users. The focus of research is to reduce the costs of manipulation of the feedstocks, of the transformation of biomass, and to reduce the friction between transactions. Robots can help manipulate the feedstock more rapidly and more safely, and sensors allow its rapid characterisation. RFID systems can facilitate the logistics. Sharing such information will reduce the transaction costs between firms. Big data will allow a real-time, locally based, assessment of hierarchies of value between biomass utilisation (Dubè et al, 2018). Open data databases, platforms for B2B collaboration, monitoring tools, and similar solutions can be used to foster circular industrial ecosystems.

Increasing the awareness of firms, consumers, and citizens

Circular economy represents a new socio-techno-economic paradigm, as its main principles are related to the efforts of designing the product, the processes, and the legal/built environment around the optimisation of reducing, reusing, and recycling activities. Adopting this new paradigm is an effort that involves all actors of the economic system, including producers, policymakers, and citizens. This adoption requires strong efforts in the field of education, training, access to information, and support services. Digital education and training courses, open information and experimental data, and dedicated social media for peer-to-peer interaction can strongly support this process. Citizens and consumers fill an important role, choosing products with lower environmental footprint and providing the first step of separation of domestic leftovers. At the consumer level, digitalisation can provide footprint tools, "nudging" software to positively influence the consumers' behaviour, sensors to monitor domestic waste, and collaborative consumption platforms.

Pathway 3: Reversing the erosion of diversity

The third transition discussed in the SCAR foresight exercise focuses on diversity. Diversity is a key to resilience of socio-ecological systems in face of change. In fact, diversity broadens the range of the functions that a system can provide, as well the range of responses to factors of change. This applies to both biological systems and social systems: "In systems with low diversity, there is less chance to create new ideas, components, or connections. Tinkering, mutations, and fortuitous errors are essential to derive new components and links in a system" (Young et al., 2006, p. 311) A substantial increase of biological, social, and economic diversity requires: A change of technical paradigms, consumers and citizens' awareness; access to information; an ecosystem of farms with "functional diversity" surpluses to exchange; and availability of market and technical alternatives. Digitisation can give an important contribution to these efforts, as shown in Table 45.2.

Biodiversity data

One of the first steps of biodiversity conservation is improving information availability. Digital cameras, remote sensing, field data, and citizens' observations can increase such availability. Satellite data can measure biodiversity indicators (Cazzolla Gatti & Notarnicola, 2018), analyse population dynamics, species interaction and community diversity, functional traits and functional diversity, and biodiversity management (Tang et al., 2018). Data coming from biodiversity datasets can be used to train AI in recognising (Cope et al., 2012) and assessing living organisms and their features (Li, 2020). The challenge of developing this field is integrating data from different sources and improving data quality (Zeb et al., 2021). The Global Biodiversity Information Facility provides a single point of access to specimen data from databases of biological surveys and collections. Biologists have access to more than 120 million observations (Yesson et al., 2007). The possibility of creating global databases of biodiversity has generated the measure of global biodiversity, and with it an opening for new avenues for greater sustainability policy and practice (Devictor & Bensaude-Vincent, 2016).

Table 45.2 Impact of digitisation on functional diversity

Target	Impact of Digitisation
A change of technical paradigms	Digitisation can support education and access to information and training. Platforms to share trial data
Raising consumers' and citizens' awareness	Social media to share knowledge; educational tools, support for decision tools
A system of interconnected biodiversity databases	Biodiversity Digitisation (Edwards et al., 2000), Knowledge sharing platforms (Hertz et al., 2016), open big data
Monitoring biodiversity in the systems	Satellites, drones, sensors, cloud, advanced analytics; AI for decision support systems
An ecosystem of farms with "functional diversity" surpluses;	Improved communication between firms: B2B platforms
Availability of market and technical alternatives	Digitisation can substantially accelerate the phenotyping and genotyping (ICT applied to life sciences), improve the capacity to recognise and characterise genetic diversity (open database, AI), monitor biodiversity (sensors, citizens' science) (Duro et al., 2007). Decision support systems based on AI for recognition of biodiversity at farm level (Lin et al., 2019).
A change of business models	Blockchain can help to conserve identity of the genetic variety along the chain (Antonucci et al., 2019). Disintermediation can foster diversification of the B2B relationships to find market channels appropriate to product size and characteristics. Sensors and improved analytics can make social responsibility reports more accurate (Efimova & Rozhnova, 2018)
A change of consumers' practices	Education, improved access to information (footprinting)
Supporting policies	Monitoring impact of policies, mapping biodiversity distribution and risk

Source: Work of the authors.

Building biodiversity-based value chains

Given the characteristics of conventional retailers, products that do not comply with the rules that retailers pose are not accepted. Even if they were accepted, farmers would not be able to communicate to consumers the differences between alternative and conventional varieties. Therefore, the value of biodiversity should be turned into market value through dedicated chains, able to preserve the identity of the product and adequately remunerate all actors. In recent years, the use of e-mail, social networks, or cloud-based tools have made it possible for networks of consumers to organise purchasing groups. E-commerce or B2B platforms have helped farmers to find market channels appropriate to product size and characteristics. Consumers, along with conventional retailers, are increasingly interested in agro-biodiversity. Digitalisation can reduce transaction costs and inform consumers about the identity of both products and producers. Distributed ledgers, for example, can conserve the identity of the genetic variety along the chain (Antonucci et al., 2019).

Supporting policies

Policies for biodiversity need appropriate data collection, analytics, and measurement methodologies for monitoring diversity, mapping biodiversity distribution and risk, assessing the impact of policies, and controlling individual practices (Ehlers et al., 2021).

Knowledge development

The current techno-economic paradigm of conventional agriculture is still influenced by the Fordist principle, according to which efficiency is based on standardisation. For more circularity in systems, deep transformations of production processes and business models are needed. Digitalisation can foster education, access to information, training, and technical support. Moreover, as information about functional diversity is key to its use, sharing of trial data can accelerate dissemination and adoption of innovation. Knowledge about diversity can be enhanced through systematic data collection, open data, and knowledge sharing platforms.

Raising awareness of consumers and citizens

An increasing demand to protect diversity can foster new techno-economic models. Despite the variety of brands that consumers can find in the grocery stores, they may have a limited awareness of the variety of food that nature offers. Access to information can be improved through digital communication channels and through decision support systems on smartphones. By reducing transaction costs between producers and consumers, digitalisation can support biodiversity-based value chains to be clearly differentiated from conventional ones.

Recommendations and conclusion

New generation of ICTs are potential game changers. Their embodiment into existing socio-technical systems will change societal organisation, generating winners, losers, and opponents. The risks related to digitisation and ability to diminish a transition towards sustainable diets can be classified into three types. *Design-related risks* are inherent to the purpose and the quality of the technology design. For example, technologies may "nudge" users to specific patterns of behaviours that may support unsustainable industrial agricultural models (e.g., monoculture, heavy use of external inputs, high degree of mechanisation); they could be vulnerable to cyber-attacks, show bias, give inaccurate responses, lead to missed opportunities, be characterised by fast obsolescence, and cause privacy breaches. Risks could be related to *unequal access* to digital opportunities, which is the distribution of physical, social, and human capital necessary to get access to digital practices. Non-adoption or late adoption may enlarge the gulf between social groups and territories, generating social and economic marginalisation. *Systemic effects* are related to the dynamics activated by the introduction of a given technology into a socio-technical system and generating consequences at a "macro" level. Impact may be related to delayed, cumulative, indirect, or feedback effects. There is also the risk of relying on single technologies and ignoring the full system. One of the issues raised about agricultural ICT is the need of technological ecosystems. Given the interdependency between technologies, the introduction of a new technology would be ineffective without the right contextual conditions. For example, increased productivity at farm level may generate unemployment at industry level; information on pollution may ignite social stigma on those who, even accidentally, pollute; power imbalances in data ownership may bring the creation of monopolies and loss of autonomy by weaker actors.

Considering these potential risks, channelling digitalisation toward societal interests will require deep revisions of the regulatory and governance contexts of food production and consumption. As ICTs can be game changers, anticipation of risks and benefits of digitisation should be strongly encouraged at all levels. When possible, regulation should avoid already known consequences, like breaches to privacy, concentration of power, and cybersecurity (Demestichas et al., 2020). A key area of intervention in the food system is the digital divide. Besides the importance of ensuring regional access to broadband, it is important to prioritise access for disadvantaged groups, to launch broad programs of training to give everyone the opportunity to achieve basic ICT skills, and to develop standards that avoid bias. Research and innovation policies should encourage the development of innovations tailored to small farming and food business. Initiatives such as "smart villages" aim at encouraging rural communities to look for digital solutions for pertinent issues.

Furthermore, a strong emphasis on *open data* would substantially reduce the cost of initiating service-providing start-ups and affordable technological solutions would be possible. Increased information availability, linked to a distributed capacity to produce, use, and communicate knowledge, can increase the level of responsibility of actors.

In conclusion, technological advancement is so far largely separated from the assessment of socio-economic consequences, since the approach to technology design is mainly based on performance remunerated by the market. Ethical considerations should be embodied into innovation processes, and legislation and governance should ensure that potential risk effects are not left only to individual innovators.

Note

1 The LULUCF Regulation implements the agreement between EU leaders in October 2014 that all sectors should contribute to the EU's 2030 emission reduction target, including the land use sector. https://ec.europa.eu/clima/policies/forests/lulucf_en#:~:text=The%20LULUCF%20Regulation%20implements%20the,including%20the%20land%20use%20sector.

References

Akyazi, T., Goti, A., Oyarbide, A., Alberdi, E., & Bayon, F. (2020). A guide for the food industry to meet the future skills requirements emerging with industry 4.0. *Foods, 9*(4), 492.

Antonucci, F., Figorilli, S., Costa, C., Pallottino, F., Raso, L., & Menesatti, P. (2019). A review on blockchain applications in the agri-food sector. *Journal of the Science of Food and Agriculture, 99*(14), 6129–6138.

Araújo, S. O., Peres, R. S., Barata, J., Lidon, F., & Ramalho, J. C. (2021). Characterising the agriculture 4.0 landscape: Emerging trends, challenges and opportunities. *Agronomy, 11*(4), 667.

Bacco, M., Brunori, G., Ferrari, A., Koltsida, P., & Toli, E. (2020). IoT as a digital game changer in rural areas: The DESIRA conceptual approach. *In 2020 Global Internet of Things Summit (GIoTS)* (pp. 1–6). IEEE.

Balafoutis, A., Beck, B., Fountas, S., Vangeyte, J., Van der Wal, T., Soto, I., Gómez-Barbero, M., Barnes, A., & Eory, V. (2017). Precision agriculture technologies positively contributing to GHG emissions mitigation, farm productivity and economics. *Sustainability, 9*(8), 1339.

Bikomeye, J. C., Rublee, C. S., & Beyer, K. M. M. (2021). Positive externalities of climate change mitigation and adaptation for human health: A review and conceptual framework for public health research. *International Journal of Environmental Research and Public Health, 18*(5), 2481.

Bronson, K. (2020). A digital "revolution" in agriculture? In J. Duncan, M. Carolan, J. Wiskerke (Eds.), *Routledge Handbook of Sustainable and Regenerative Food Systems*. Routledge.

Canali, S., Antichi, D., Cristiano, S., Diacono, M., Ferrante, V., Migliorini, P., & Colombo, L. (2020). Levers and obstacles of effective research and innovation for organic food and farming in Italy. *Agronomy, 10*(8), 1181.

Cazzolla Gatti, R., & Notarnicola, C. (2018). A novel multilevel biodiversity index (MBI) for combined field and satellite imagery surveys. *Global Ecology and Conservation, 13*, Article e00361.

Cobby Avaria, R. W. (2020). Searching for sustainability in the digital agriculture debate: An alternative approach for a systemic transition. *Teknokultura, Revista de Cultura Digital y Movimientos Sociales, 17*(2), 225–238.

Cope, J. S., Corney, D., Clark, J. Y., Remagnino, P., & Wilkin P. (2012). Plant species identification using digital morphometrics: A review. *Expert Systems with Applications, 39*(8), 7562–73.

Cyber Physical Systems Public Working Group (2016). *Framework for cyber-physical systems*, Release 1.0. https://s3.amazonaws.com/nist-sgcps/cpspwg/files/pwgglobal/CPS _PWG_Framework_for_Cyber_Physical_Systems_Release_1_0Final.pdf

Demestichas, K., Peppes, N., & Alexakis, T. (2020). Survey on security threats in agricultural IoT and smart farming. *Sensors, 20*(22), 6458.

Devictor, V., & Bensaude-Vincent, B. (2016). From ecological records to big data: The invention of global biodiversity. *History and Philosophy of the Life Sciences, 38*(4), 1–23.

Dubé, L., Du, P., McRae, C., Sharma, N., Jayaraman, S., & Nie, J.-Y. (2018). Convergent innovation in food through big data and artificial intelligence for societal-scale inclusive growth. *Technology Innovation Management Review, 8*(2), 49–65.

Duro, D. C., Coops, N. C., Wulder, M. A., & Han, T. (2007). Development of a large area biodiversity monitoring system driven by remote sensing. *Progress in Physical Geography, 31*(3), 235–260.

Edwards, J. L., Lane, M. A., & Nielsen, E. S. (2000). Interoperability of biodiversity databases: Biodiversity information on every desktop. *Science, 289*(5488), 2312–2314.

Efimova, O., & Rozhnova, O. (2018). The corporate reporting development in the digital economy. In T. Antipova, & A. Rocha (Eds.), The 2018 International Conference on Digital Science (pp. 71–80). Moscow, Russia: Springer.

Ehlers, M. Huber, R., & Finger, R. (2021). Agricultural policy in the era of digitalisation. *Food Policy, 100*, 102019.

Garcia-Llorente, M., Rubio-Olivar, R., & Gutierrez-Briceno, I. (2018). Farming for life quality and sustainability: A literature review of green care research trends in Europe. *International Journal of Environmental Research and Public Health, 15*(6), 1282.

Garske, B., Bau, A., & Ekardt, F. (2021). Digitalization and AI in European agriculture: A strategy for achieving climate and biodiversity targets? *Sustainability, 13*(9), 4652.

Gkisakis, V. D., & Damianakis, K. (2020). Digital innovations for the agroecological transition: A user innovation and commons-based approach. *J Sustainable Organic Agric Syst, 70*(2), 1–4.

Herz, A., Matray, S., Sharifova, H., Wolck, A., Sigsgaard, L., Penvern, S., & Kelderer, M. (2016). EBIO-network: A web-based platform for knowledge sharing on functional agrobiodiversity in organic apple production. In Ecofruit, Germany, 17th International Conference on Organic Fruit-Growing: Proceedings (pp. 277–278). University of Hohenheim: Fördergemeinschaft Ökologischer Obstbau eV (FÖKO).

Jensen, D., & Campbell, J. (2019). *The case for a digital ecosystem for the environment: Bringing together data, algorithms and insights for sustainable development* [Discussion paper]. UN Environment Assembly.

Kuzmin, E., VanderSluis, B., Wang, W., Tan, G., Deshpande, R., Chen, Y., & Koch, E. N. (2018). Systematic analysis of complex genetic interactions. *Science, 360*(6386), 1729.

Li, C. (2020). Biodiversity assessment based on artificial intelligence and neural network algorithms. *Microprocessors and Microsystems, 79*.

Lin, C., Wang, J., & Ji, L. (2019). Notes of Life: A platform for recording species observations driven by artificial intelligence. *Biodiversity Information Science and Standards, 3*, 1–2.

Lipton, J. I. (2017). Printable food: the technology and its application in human health. *Current opinion in biotechnology, 44*, 198-201.

Marescotti, A., Quiñones-Ruiz, X. F., Edelmann, H., Belletti, G., Broscha, K., Altenbuchner, C., & Scaramuzzi, S. (2020). Are protected geographical indications evolving due to environmentally related justifications? An analysis of amendments in the fruit and vegetable sector in the European Union. *Sustainability, 12*(9), 3571.

McClements, D. J. (2020). Future foods: Is it possible to design a healthier and more sustainable food supply? *Nutrition Bulletin, 45*(3), 341–354.

Nikolaou, G., Neocleous, D., Christou, A., Kitta, E., & Katsoulas, N. (2020). Implementing sustainable irrigation in water-scarce regions under the impact of climate change. *Agronomy, 10*(8), 1120.

Rijswijk, K., Klerkx, L., Bacco, M., Bartolini, F. M., Bulten, E., Debruyne, L., Dessein, J, Scotti, I., & Brunori, G. (2021). Digital transformation of agriculture and rural areas: A socio-cyber-physical system framework to support responsibilisation. *Journal of Rural Studies, 85*, 79–90.

Rincon-Patino, J., Lasso, E., & Corrales, J. (2018). Estimating avocado sales using machine learning algorithms and weather data. *Sustainability, 10*(10), 3498.

Sadhukhan, J., Dugmore, T. I. J., Matharu, A., Martinez-Hernandez, E., Aburto, J., Rahman, P. K. S. M., & Lynch, J. (2020). Perspectives on "Game Changer" global challenges for sustainable 21st century: Plant-based diet, unavoidable food waste biorefining, and circular economy. *Sustainability, 12*(5), 1976.

Sam, A. K., & Grobbelaar, S. S. (2021). Research trends, theories and concepts on the utilisation of digital platforms in agriculture: A scoping review. In Conference on e-Business, e-Services and e-Society (pp. 342–355). Matieland, South Africa: Springer.

Serazetdinova, L., Garratt, J., Baylis, A., Stergiadis, S., Collison, M., & Davis, S. (2019). How should we turn data into decisions in AgriFood? *Journal of the Science of Food and Agriculture, 99*(7), 3213–3219.

Skinner, G. A. (2015). Smart labelling of foods and beverages. In Berryman, P. (Ed.), *Advances in food and beverage labelling* (pp. 191–205). Woodhead Publishing.

Smetana, S. M., Bornkessel, S., & Heinz, V. (2019). A path from sustainable nutrition to nutritional sustainability of complex food systems. *Frontiers in Nutrition*, *6*, 39–39.

Sposato, P., Preka, R., Cappellaro, F., & Cutaia, L. (2017). Sharing economy and circular economy. How technology and collaborative consumption innovations boost closing the loop strategies. *Environmental Engineering & Management Journal*, *16*(8), 1797–1806.

Tang, Z., Jiang, M., Zhang, J., & Zhang, X. (2018). Applications of satellite and air-borne remote sensing in biodiversity research and conservation. *Biodiversity Science*, *26*(8), 807–818.

Taranic, I., Behrens, A., & Topi, C. (2016). *Understanding the circular economy in Europe, from resource efficiency to sharing platforms: The CEPS framework* (CEPS Special Reports, no.143). CEPS. https://www.ceps.eu/wp-content/uploads/2016/07/SR%20No143%20Circular%20Economy_0.pdf

Tsolakis, N., Niedenzu, D., Simonetto, M., Dora, M & Kumar, M. (2021). Supply network design to address United Nations Sustainable Development Goals: A case study of blockchain implementation in Thai fish industry. *Journal of Business Research*, *131*, 495–519.

Van Wassenaer, L., van Hilten, M., van Ingen, E., & van Asseldonk, M., (2021). *Applying blockchain for climate action in agriculture: State of play and outlook*. FAO and WUR.

Yesson, C., Brewer, P. W., Sutton, T., Caithness, N., Pahwa, J. S., Burgess, M., Gray, W. A., White, R. J., Jones, A. C., Bisby, F. A., & Culham, A. (2007). How global is the global biodiversity information facility? *PloS One*, *2*(11), 1–10.

Young, O. R., Berkhout, F., Gallopin, G. C., Janssen, M. A., Ostrom, E., & Van der Leeuw, S. (2006). The globalisation of socio-ecological systems: An agenda for scientific research. *Global Environmental Change*, *16*(3), 304–316.

Zeb, A., Soininen, J & Sozer, N. (2021). Data harmonisation as a key to enable digitalisation of the food sector: A review. *Food and Bioproducts Processing*, *127*, 360–370.

PART 9

Food sovereignty and case studies

46

SUSTAINABILITY DIMENSIONS OF INDIGENOUS PEOPLES' FOOD SYSTEMS IN A CHANGING WORLD

Selena Ahmed, Virgil Dupuis, John de la Parra,
Alexandra Adams, and Long Chunlin

Introduction

Acknowledging and honouring Indigenous Peoples' food systems

As place-based food systems, Indigenous Peoples' food systems, or "traditional food systems," were historically primarily based on wild and cultivated food environments (WCFE) and a kinship network of exchange rather than the market economy (Downs et al., 2020; Smith et al., 2019). Food environments are:

> the consumer interface with the food system that encompasses the availability, affordability, convenience, promotion and quality, and sustainability of foods and beverages in wild, cultivated, and built spaces that are influenced by the socio-cultural and political environment and ecosystems within which they are embedded.
>
> *(Downs et al., 2020)*

Collectively, wild and cultivated food environments are natural food environments that function as subsistence areas of food procurement (Downs et al., 2020). Wild food environments are ecosystems including forests, open pastures, and aquatic areas with relatively low-intensity management (Downs et al., 2020); cultivated food environments are agricultural systems including fields, orchards, closed pastures, gardens, and aquaculture from which individuals directly procure food for household consumption rather than for sales (Downs et al., 2020). It is important to note that wild and cultivated food environments are not always distinct from each other but a continuum on the basis of agricultural intensification.

Indigenous Peoples' food systems have evolved over time, including under processes and pressures of cultural adaptation, colonisation, and globalisation. Thus, Indigenous Peoples' food systems are dynamic and diverse with varying levels of access and reliance on wild, cultivated, and built food environments. Recognising this complexity, this chapter explores the characterisations of Indigenous food systems, foods, and associated environmental, spiritual, socio-cultural, and health-giving attributes.

Box 46.1 Key Term: *Indigenous Peoples' Food Systems*

Recognising the great diversity among Indigenous Peoples, we acknowledge the diversity across Indigenous Peoples' food systems. Adapting elements of previous characterisations, we refer to Indigenous Peoples' food systems as: Dynamic and diverse place-based food systems based on procurement of food from wild and cultivated food environments managed by Indigenous people, informed by culturally, historically, and geographically specific knowledge, perceptions, observation, social networks, kinship networks of exchange, practices, innovations, biochemical composition, and desired health outcomes.

Source: Work of the authors.

Indigenous Peoples' food systems have been variously characterised; see Box 46.1 for a description. Kuhnlein and Receveur (1996) characterise them as comprised of foods from the local, natural environment that are culturally acceptable, which includes wild foods that are hunted, fished, and harvested (Kuhnlein & Receveur, 1996; Lemke & Delormier, 2017) and food cultivated through various agricultural practices such as forest and home gardens, agro-forestry, and slash-and-burn farming (Durning, 1992). In addition to the foods themselves, Kuhnlein and Receveur (1996) emphasise that Indigenous Peoples' food systems are comprised of sociocultural meanings, processing techniques, use, composition, knowledge systems, and nutritional consequences associated with food. The Royal Commission on Aboriginal Peoples emphasises traditional food systems as being "part of a cultural heritage" and describes traditional food as that which is "holistically entwined with culture and personal identity, as well as with physical health" (Royal Commission on Aboriginal Peoples, 1996, p. 194). Indigenous Peoples' food systems have further been acknowledged for their innovation through social networks and ecological observation of the cycles of plants and shifts with climate change (Ahmed et al., 2010).

While characterisations vary, Indigenous foods largely are local foods available to a particular Indigenous culture from the local natural environment, including plant and animal species (Gagné et al., 2012; Hanemaayer et al., 2020). Numerous studies highlight the multidimensional valuation of Indigenous foods including taste, freshness, nutritional quality, as a traditional community practice, and as offering a sense of self-sufficiency (Smith et al., 2019a).

Box 46.2 Key Term: *Indigenous Foods*

We acknowledge that what is considered an Indigenous food can differ among people, even within the same cultural group. Here, we refer to Indigenous Peoples' foods as: Foods procured from place-based food systems managed by Indigenous People that reflect local ecology and culture including biodiversity, knowledge, perceptions, and practices that are valued for their cultural meaning, flavour, health attributes, and providing self-sufficiency. It is important to note that many foods that characterise Indigenous Peoples' food systems are consumed by cultures outside of Indigenous cultures.

Source: Work of the authors.

Even within a singular community of an Indigenous socio-linguistic group, what is regarded as an Indigenous food can vary. (See Box 46.2 for a description of Indigenous foods). Hanemaayer et al. (2020) highlight the contextual understanding of Indigenous foods where individuals of a Haudenosaunee community in southern Ontario, Canada had diverse and conflicting understandings of what "traditional foods" encompassed. Foods that have been introduced to Indigenous communities through colonisation, such as flour, salt, sugar, milk, and lard that were introduced through government-issued rations (Gordon et al., 2018), are not generally classified as traditional foods (Hanemaayer et al., 2020). The classification of a food as Indigenous can further differ among various varieties or breeds of the same species. For example, some members of the Haudenosaunee community consider soup made from white corn, but not yellow corn, as a traditional food; others consider yellow corn soup to be a traditional food (Hanemaayer et al., 2020). The preparation of raw ingredients can further influence the classification of foods as to whether it is an Indigenous food (Hanemaayer et al., 2020).

Indigenous Peoples' food systems and Indigenous foods vary greatly based on geography, cultural group, seasonality, management practices, beliefs, knowledge systems, and exogenous influences (Gagné et al., 2012). Over centuries, new foods, production and processing techniques, and dietary perspectives have been introduced into traditional food systems, resulting in dynamic food systems with shifting dietary patterns (Sarkar et al., 2020). Factors of globalisation have further transformed Indigenous Peoples' food systems (Compher, 2006).

Food availability and food choice in Indigenous Peoples' food systems

Availability of adequate amounts and diversity of food is critical for food security, dietary diversity, and food quality for supporting nutritional and health needs.

Indigenous Peoples' food systems provide lessons on the socio-ecological determinants of food availability in wild and cultivated food environments, which is determined by the ecosystem and moderated by a host of factors including climate, technology, and politics (Downs et al., 2020). Ecosystem dynamics along with climate conditions are especially relevant for determining food availability in wild and cultivated food environments including the yields, quality, and diversity (Ahmed & Stepp, 2016). The specific attributes of wild and cultivated food environments ultimately influence the biochemical composition and sensory attributes of Indigenous foods and the diets of Indigenous Peoples.

Food choices in Indigenous Peoples' food systems are dependent on food availability in wild and cultivated food environments coupled with cultural and personal preferences, beliefs, ethnonutrition perspectives, biological needs, education, and media (Byker Shanks et al., 2020; Downs et al., 2020; Kuhnlein & Receveur, 1996). Indigenous Peoples' food systems have generally been based on subsistence models where food affordability extends to the amount of time and energy available to harvest and prepare cultural food items. Procurement of wild foods through hunting, fishing, and foraging can be costly in terms of time, energy, transportation, and equipment.

Positionality

Collectively, the authors of this chapter have carried out research and learned with and from Indigenous communities in multiple countries, including through community-based participatory research projects. One of the authors is a tribal member. Many of the authors have worked in partnership with Indigenous communities to address nutrition and health challenges in ways that integrate community and scientific knowledge. Through our own learnings from long-term research in Indigenous communities as well as from scholarship, we recognise and celebrate the diversity and dynamism that characterises Indigenous Peoples' food systems. We acknowledge that Indigenous Peoples' food systems can mean different things to different people, even within a community. We further recognise the variable impacts of and responses of Indigenous Peoples' food systems to cultural adaptation, colonisation, globalisation, and climate change, including as Indigenous communities are reclaiming their food systems. In the context of the diversity and dynamism that characterises Indigenous Peoples' food systems, we have identified commonalities with regards to environmental, socio-cultural, and human health dimensions of sustainability (Ahmed et al., 2019; Mason & Lang, 2017).

Sustainability attributes of Indigenous Peoples' food systems

Many Indigenous Peoples' food systems are inherently structurally sustainable, promoting environmental, socio-cultural, and human health dimensions of sustainability (Ford, 2009; Ghosh-Jerath et al., 2021; Smith et al., 2019). In recognition of the linkages between human and environmental health, a sustainability approach to food systems and diets is increasingly being adopted within Indigenous communities (Ahmed et al., 2019; Mason & Lang, 2017). In this chapter, we articulate sustainability attributes that are prevalent in many Indigenous Peoples' food systems with the goals to: (1) Enhance our understanding for their preservation and promotion; (2) provide important lessons for sustainable food production and consumption broadly; and (3) advance multiple Sustainable Development Goals (SDGs) of the United Nations Food and Agricultural Organisation (UN FAO). The provision of sustainable diets from sustainable food systems are critically important to multiple SDGs (1, 2, 3, 5, 10, 12, 13, 14, and 15) (Ahmed & Byker Shanks, 2020). Not all sustainability attributes presented here are prevalent in all Indigenous Peoples' food systems, with some elements previously being present and other elements emerging more recently.

Based on environmental sustainability, Indigenous Peoples' food systems are relatively low-intensity systems that support ecosystem services and overall environmental wellbeing. Many Indigenous Peoples' food systems identify a deep connection to the land and include the following ecological attributes: (a) Food procurement from wild and cultivated food environments; (b) utilisation of local seasonal biodiversity; (c) holistic concept of environmental health; (d) stewardship of land; and (e) ecological agricultural practices.

From a socio-cultural perspective, inclusive of spiritual aspects, Indigenous Peoples' food systems support the cultural identity of Indigenous groups and acknowledge practices and socio-cultural meanings of food that have evolved and been transmitted over the generations (Smith et al., 2019). Many Indigenous Peoples' food systems include the following socio-cultural attributes: (a) Sacred and spiritual beliefs; (b) knowledge systems; (c) rituals; (d) display of cultural identity; and (e) food preparation and preservation techniques.

From a human health perspective, traditional food systems support food security and improve health by reducing critical micronutrient deficiencies (Samson & Pretty, 2006; Vinceti et al., 2008) and combatting diet-related chronic diseases (Sarkar et al., 2020). Many Indigenous Peoples' food systems include the following human health attributes: (a) Holistic concepts of health; (b) food as medicine; (c) dietary quality and diversity; (d) natural sources of sugar, fat, and salt; and (e) eating for satiety.

Ecological attributes of Indigenous Peoples' food systems: Connection to the land

Food procurement from wild and cultivated food environments

Indigenous Peoples' food systems have a common theme of having a "connection to the land" (Kuhnlein et al., 2009; Smith et al., 2019) where communities procure food from wild and cultivated food environments (Downs et al., 2020). Numerous studies in the ethnobotanical and ethnobiological literature document interactions of Indigenous Peoples with their surroundings for food.

For example, in Naxi communities of the Sino Himalayan region, home gardens are ecologically and culturally important systems for cultivating food and medicinal plants for wellbeing by healers and farmers (Yang et al., 2014). The

Akha of Baljalpuxeevq in Yunnan Province of Southwestern China have historically procured the majority of their food from a diversified land-use mosaic of forests, agroforests, mixed-crop fields, bamboo and rattan stands, rice paddies, home gardens, grazing pasture, fruit orchards, and bracken land surrounding their village settlement for subsistence farming and foraging, and increasingly for commercialisation in national and global markets (Ahmed et al., 2010b). Cultivation of rice for the Akha was predominantly based on swidden cultivation where forests were cleared for crop production and then allowed to regenerate during a fallow period to replenish soil fertility (Ahmed et al., 2010b).

Indigenous Peoples often have a multidimensional valuation of wild and cultivated food environments, including for contributions to sustainable diets, food security, health, and cultural identity (Smith et al., 2019) along with holistic, spiritual, utilitarian, aesthetic, economic, ecological, social, sovereignty, sensory, and recreational values (Ahmed, Unachukwu, et al., 2010). For example, subsistence farming was valued in an Akha community because of sensory preferences for locally grown food coupled with its contribution to food sovereignty and connection to ancestors and spiritual realms (Ahmed et al., 2010a). Utilisation of WCFE for diets connects communities to their landscape and can facilitate the development of environmental and land ethics, which are fundamental for supporting environmental sustainability.

Utilisation of local seasonal biodiversity

Many Indigenous Peoples' food systems utilise a vast range of local and seasonal edible biodiversity from surrounding wild and cultivated food environments. The ethnobiological literature documents the tremendous edible biodiversity at the genetic and species levels utilised by Indigenous Peoples for food including procurement of numerous species of plants, fish, marine and land mammals, birds, and insects. For example, over 150 species of wild foods and 100 types of cultivated edible plant varieties were documented in Akha home gardens and fields. Recognition of cultivar diversity supports the conservation of genetic diversity and is further associated with diversity of biochemical composition based on variation of the presence and concentrations of specific phytochemicals in different cultivars (Ahmed et al., 2014). Along with biodiversity at the genetic, biochemical, and species levels, Indigenous food systems often maintain seasonal and functional diversity (Ahmed et al., 2014). As many species are only seasonally available, Indigenous Peoples have developed seasonal harvest calendars and stories that shape their seasonal dietary patterns to meet nutrition and health needs across seasons. For example, for many northern Native American tribes, maple sugar moon is the time for harvesting sap and making maple syrup when the days are just warm enough for the sap to run (Kimmerer, 2015).

Conservation of biodiversity in wild and cultivated food environments is crucial for maintaining the benefits from ecosystems along with strengthening resilience (Smith et al., 2019). Biodiversity is broadly defined as the variation among living organisms and their environment (Magurran, 2003). The presence of biodiversity is fundamental to sustaining life on earth by: (a) Providing ecosystem services including nutrient cycling, soil protection, pollination, flood control, and genetic resources (Magurran, 2003); (b) strengthening resilience in response to shocks and change (Gunderson & Holling, 2001), and (c) enhancing farm productivity and farmer livelihoods (Convention on Biological Diversity, 2000). The agricultural biodiversity of cultivated food environments is distinct for its cultural dimensions, being the product of natural and human selection through interactions among the environment, genetic resources, and farmer management practices, participatory processes, and local knowledge (Ahmed & Stepp, 2016). Both the wild and agricultural biodiversity of Indigenous Peoples' food systems are critical for supporting the functioning and resilience of ecosystems.

Holistic concepts of environmental health

Indigenous Peoples' food systems are often based on holistic perceptions of health that include the interdependent wellbeing of humans and the environment (Chalmers & Dell, 2015; Hodge et al., 2009; Mazet et al., 2014). Many Indigenous communities have long acknowledged the benefit of environmental health for human health (Chalmers & Dell, 2015; Mazet et al., 2014). For example, Indigenous Peoples across Canada are unified in their belief that an individual's wellbeing is directly connected to that of the land (Hillier et al., 2021).

Holistic concepts of stewardship of the land for human wellbeing are increasingly incorporated in Western-based public health concepts for supporting sustainability. For example, the notion of One Health was articulated by the World Health Organisation in 2017 for informing policy design and public health interventions. The One Health approach encourages multi-sector stakeholders to collectively consider the combined health outcomes of all intertwined elements (Mazet et al., 2014).

Stewardship of land

Holistic perceptions of health are inextricably linked to the preservation of the land and environment (Chatwood et al., 2017). While human activities are often considered detrimental to biodiversity and the wellbeing of the land, human activities can also have substantial benefits for biodiversity and ecosystem services (Armstrong et al., 2021). For example, the ancient forest gardens once tended by Ts'msyen and Coast Salish peoples along the Pacific coast of Canada contain more biologically and functionally diverse species than surrounding conifer-dominated forests (Armstrong et al., 2021). In particular, the forest gardens contain native fruit and nut trees and shrubs including crab apple, hazelnut, cranberry, wild plum, and wild cherries along with medicinal plants and root foods such as wild ginger and wild rice root in understory layers (Armstrong et al., 2021). In addition, these ancient forest gardens are distinct from surrounding conifer forests in providing habitat for animals and pollinators (Armstrong et al., 2021).

The WHO cites *stewardship* as a critical function of a health system in its One Health concept (World Health Organisation, 2017). Indigenous stewardship of

land, seed, and other ecological resources is recognised as being critical for achieving conservation of biocultural diversity. While stewards of ecological resources
for food vary between communities, a common theme across case studies on
Indigenous Peoples' food systems is women as gatekeepers to family food availability (Kuhnlein et al., 2009).

Ecological agricultural practices

Linked to Indigenous Peoples' preservation of the land is managing cultivated food
environments to mimic natural ecosystems or other low-intensity and regenerative
approaches. Many Indigenous Peoples' food systems are based on no or low input
of agrochemicals and mechanisation. Since time immemorial, Indigenous Peoples
have maintained diverse farming systems such as agroforestry, intercropping, and
polycultures that can be classified as sustainable agriculture and regenerative agriculture. Diversified farming is increasingly recognised as a sustainable agricultural
practice for its contributions to supporting biodiversity and ecosystem services. In
contrast to diversified farming, monocultures degrade the soil, encourage pests and
diseases, and depend on fertiliser and pesticide inputs that further degrade soil and
water resources and harm the beneficial organisms, such as pollinators, upon which
successful agriculture depends (Ahmed et al., 2010a).

For example, planting different species of plants in the same area in a way
that imitates nature, or polycultures, is a practice prevalent in Indigenous Peoples'
food systems. Multiple studies have shown the ecological benefits of polycultures.
The "Three Sisters" polyculture system of maize, beans, and squash in the United
States has been shown to be more efficient at using nutrients, light, and water than
monocultures, irrespective of the species (Pleasant & Burt, 2010; Pleasant, 2016).
Silviculture and agroforestry are other prevalent types of agricultural practices in
Indigenous Peoples' food systems that are recognised for their ecological benefits
(Ahmed et al., 2010a).

North and South American Indigenous Peoples have used biochar as an agricultural soil amendment to increase soil fertility, organic matter, and water holding
capacity (Laird, 2019). Once thought to be a practice in tropical Central American
Indigenous agriculture up to 6,000 years ago, evidence of biochar use is being
documented as far north as Wisconsin 1,000 years ago (Overstreet, 2018). At present, biochar is becoming a leading candidate for sequestering carbon over vast
areas of damaged and depleted agricultural and forest lands (Laird, 2022).

Socio-cultural attributes of Indigenous Peoples' food systems

Sacred and spiritual beliefs

Sacred and spiritual beliefs about food and the environment are prevalent in many
Indigenous Peoples' food systems (Kuhnlein & Receveur, 1996; Kuhnlein et al.,
2009; Stokes & Benedict, 1993; Yang et al., 2011). Many Indigenous Peoples are
animistic and thus may have sacred beliefs about environmental elements. For
example, Dongba, or shamans, in Naxi communities in the uplands of southwest-

ern China traditionally promoted natural resource management practices that respected the biophysical environment through chants and scripts (Yang et al., 2011). Central to the Naxi's Bon belief system is that "nature" and "human beings" are half-brothers. The Dongba developed a set of rules and regulations for respecting the environment and Shu, the Nature God. A series of chants, stories, and pictograms illustrate punishments in Naxi communities for disrespecting the environment and Shu.

Knowledge systems

Indigenous Peoples' food systems are comprised of ecological and cultural knowledge that arose through generations of interacting with local ecosystems and social networks (Kuhnlein et al., 2009), including ecological knowledge regarding the factors that impact food availability and quality such as when and how to harvest edible plants. It further includes knowledge of the attributes of food including their organoleptic properties, sacredness, and healing powers.

Rituals

Many Indigenous Peoples observe food rituals such as honouring the harvest and ancestors with food, feasts and ceremonies, fasting, and communal eating. The offering of fermented beverages from wild plant ingredients serves important roles in interpersonal relationships and cultural events in Shui communities in southwest China. Fresh and fermented tea leaves are eaten during celebrations of Akha and Bulang communities of Yunnan Province, including the annual harvest and nature worship ceremony in agro-forests (Ahmed et al., 2010b). Ceremonies are often held by Native Americans before and after gathering plants such as wild rice, roots, or berries (Geniusz, 2009), using songs and tobacco as offerings and a way of asking for permission to harvest.

Display of cultural identity: Biocultural heritage

Food choices of Indigenous Peoples are often driven by cultural identity and interlinked biological and cultural heritage, or biocultural heritage. Johns and Kokwaro (1991) identified maintenance of cultural identity as one of several factors that affect consumption of wild food plants along with nutritional and medicinal purposes. For example, the harvesting, preparation, and consumption of wild foods was found to be linked to cultural identity based on biocultural heritage on the Flathead Reservation of the Confederated Salish & Kootenai Tribes (Smith et al., 2019).

Food preparation and preservation techniques

In addition to knowledge of food, Indigenous Peoples' food systems are characterised by food processing, storage, and preparation techniques, such as fermentation, that have been passed down and adapted through generations. For example, over 70 ways have been documented for the way the Hopi prepared corn (Hough, 1897).

Shui communities in southwest China have a long history of preparing fermented beverages from wild plant ingredients (Hong et al., 2015). A total of 103 species in 57 botanical families of wild plants were inventoried that are used as fermentation starters (a preparation to initiate the fermentation process) for fermented beverages by the Shui. The majority of the wild plants used as starters by the Shui have multiple uses with medicinal purposes being the most prevalent (Hong et al., 2015). The various ways foods are processed impact their sensory, nutritional, and health attributes (Johns & Kubo, 1988; Kuhnlein & Turner, 1991).

Health attributes of Indigenous Peoples' food systems

Holistic concepts of health

Indigenous Peoples' food systems are often based on a holistic conception of wellbeing that encompasses wellbeing of the individual, the community, and the environment, and entails physical, emotional, mental, and spiritual aspects of health and protection from disease (Hillier et al., 2021; Kuhnlein et al., 2009). The Medicine Wheel of various Indigenous Peoples of North America of the Plains, embodies the Four Directions and cycles of life and presents health as an equal balance between the physical, emotional, spiritual and mental aspects of human beings (Hillier et al., 2021). Alternatively, modern Western conceptualisations of health primarily take a deficit-based approach focused on the absence of illness and disease (Hillier et al., 2021).

Food as medicine

The concept of food as medicine is prevalent in Indigenous Peoples' food systems. Many Indigenous Peoples do not separate species as either food or medicine because the same item can be both at the same time (Geniusz, 2009; Johns, 1994) or fit along a continuum of food and medicine (Pieroni & Price, 2006). The concept of food as medicine views food as not only something to eat but to nourish the body, mind, spirit, soul, and emotions. For example, Tibetan ethnomedicine views food as nourishing the body with vital and physical energies whereas disease is perceived as weakness in the body's vital and physical energies.

Wild plants that are both edible and medicinal are important components of many Indigenous diets and contribute to food security, nutrition, and health. For example, the preparation and consumption of soup made of medicinal plants for promoting health and preventing disease are a key component of diets of the Hakka socio-linguistic group of China's West Fujian Province (Luo et al., 2019). The Hakka use over 40 medicinal plant species for making soup for their flavours and medicinal properties on the basis of their ethnonutrition system. Clearing inner heat, treating inflammation, and counteracting cold in the body are the most prevalent medicinal uses of plants for making soups in Hakka communities (Luo et al., 2019). Informants perceived that the medicinal properties of plants used to make soup are influenced by time of harvest as well as environmental and climate conditions (Luo et al., 2019). Seasonings, herbs, spices, and colour are used in many Indigenous Peoples' food systems to enhance meals with medicinal properties.

Dietary quality and diversity

Many Indigenous Peoples manage biodiversity of wild and cultivated food environments to support household dietary diversity, which they perceive as being crucial for wellbeing. In addition, many Indigenous Peoples select for crop varieties and management and preparation practices that support nutrient-density. For example, nixtamalisation has been recognised to enhance the nutritive quality of corn (Kamau et al., 2020). Multiple studies have documented that Indigenous foods are often nutrient-dense and provide rich sources of macro- and micronutrients while their diversity contributes to complete diets (Sarkar et al., 2020; Vinceti et al., 2008).

Natural sources of sugar, fat, and salt

Indigenous Peoples' food systems that primarily derive from surrounding wild and cultivated food environments generally have limited saturated fat and refined carbohydrates, with non-Indigenous market foods contributing varying amounts to diets. For example, the CINE study in an Inuit community found that when the majority of energy consumed was from traditional food, the average percent of energy as saturated fat was half (6%) versus when no traditional food was consumed (12%) (Kuhnlein et al., 2009). In contexts where Indigenous Peoples' food systems have limited market access, sugars are generally sourced from honey, fruits, vegetables, and whole grains while salt may come from fish, sea greens, and vegetables. Many Indigenous Peoples have extensive use of proteins from non-meat sources including from insects, nuts, legumes, and grains. As a precious limited resource, meat is often eaten in small quantities as a condiment. Such characteristics suggest these patterns align with sustainable diets.

Eating for satiety

With food intake that historically has been limited to surrounding WCFE, Indigenous Peoples have often maintained the principle of eating for satiety rather than for fullness. This practice has resulted in relatively smaller portion sizes than those associated with modern Western diets.

Challenges and opportunities for Indigenous Peoples' food systems

Challenges

Globalisation and other cultural, economic, political, and environmental forces, including nutrition transition and land-use change, are posing threats to Indigenous Peoples' food systems including to foods, habitats, and associated traditional knowledge and systems. The integration of market food environments is increasingly prevalent in many Indigenous communities, with food procured from market food environments being increasingly part of the modern identity of many Indigenous Peoples' food systems (Byker Shanks et al., 2020). In many cases, Indigenous Peoples' food systems have been displaced through the expansion of monocultures of cash crops and the introduction of new foods such as sugar, refined grain flour, and sweetened beverages. A lack of acknowledgment of Indigenous Peoples' food systems along with mass media and advertising of industrialised foods have further disrupted Indigenous resources.

For example, the cultivation of edible and medicinal plants in Naxi communities, along with the associated knowledgebase for maintaining and utilising these resources, is at risk with development, policies, and rapid socio-economic change in China (Yang et al., 2014). Increased commercialisation of natural resources, expansion of monocultures of cash crops, road building, migration, and socio-economic pressures, have transitioned Akha foods systems towards increased reliance on market food environments with greater consumption of processed foods, away-from-home food intake, edible oils, animal products, and sugar-sweetened beverages (Ahmed et al., 2010a).

Shifts in Indigenous Peoples' food systems have resulted in marked disparities in food access and health outcomes, such as the high prevalence of obesity and diabetes mellitus (Byker Shanks et al., 2020). For example, in North America, colonisation of Indigenous Peoples has resulted in a shift away from wild food environments (Compher, 2006; Smith et al., 2019) toward consumption of highly processed foods from market food environments that are high in refined sugars, saturated fats, and salts typical of modern Western diets (Byker Shanks et al., 2020; Popkin, 2001). This nutrition transition has had, and continues to have, a profound impact on the health of Indigenous Peoples (Byker Shanks et al., 2020; Damman et al., 2008).

As place-based food systems dependent on surrounding wild and cultivated food environments, access to land and water, time, and costs for procuring wild foods are major barriers for the consumption of Indigenous foods (Smith et al., 2019). Policies and practices linked to national dominant cultures have reduced opportunities for access and knowledge transmission associated with Indigenous foods, contributing to a decline in their consumption (Booth et al., 1992). National policies in many countries have limited where Indigenous Peoples are allowed to hunt, fish, and forage. Environmental dispossession of Indigenous Peoples from their traditional territories, along with migration of Indigenous Peoples to urban areas, has restricted access to Indigenous foods. Education programming through missionary schools, residential school systems, and public health programs that emphasise the food, language, and culture of the dominant culture have further resulted in the decline of Indigenous Peoples' food systems. A lack of integration of Indigenous Peoples' food systems in contemporary health care systems, including health care education, has eroded confidence in Indigenous Peoples' food systems that promote concepts of food as medicine. (See the food movements chapter in this collection for shared leadership in decolonising health care food systems.)

Land-use change, environmental contamination, and climate change have reduced availability, accessibility, quality, and safety of Indigenous food sources. Climate change has been documented to threaten Indigenous Peoples' food systems globally (Doyle et al., 2013; Ford, 2009). Wild and cultivated food environments are experiencing regional climate variability, including changes in temperature and seasonal patterns (Ahmed et al., 2013; Lynn et al., 2013), which threaten wild food availability, accessibility, and quality (Ahmed et al., 2013; Ford, 2009; IPCC, 2007). For example, thinning and reduced sea ice shortens the hunting season for Inuit communities in the Canadian Arctic (Ford, 2009); members of the Crow Nation in Montana, USA have observed reductions in freshwater fish populations attributable to warming waters (Doyle et al., 2013).

Opportunities

Indigenous Peoples' food systems support food security, health, and cultural identity of communities. These systems can provide strategies to combat malnutrition while supporting Sustainable Development Goals of the United Nations Food and Agricultural Organisation (UN FAO; UN General Assembly, 2015). Over the past 35 years, the FAO has worked to promote traditional foods, including through inter-institutional initiatives with Indigenous Peoples' leadership (FAO, n.d.).

Opportunities to enhance vulnerable food systems of Indigenous Peoples, mitigate the nutrition transition, and improve diet and health outcomes include: (1) Promoting use of wild and cultivated foods in local food environments; (2) selecting foods available in the food environment based on their cultural desirability and food composition; (3) promoting safe and sustainable use of wild foods and underutilised crops; and (3) conserving and promoting sustainable wild and cultivated food environments (Damman et al., 2008; Herforth & Ahmed, 2015). Increased utilisation of wild foods has been recognised to enhance food system resilience by incorporating local biodiversity into diets, thus reducing dependence on market food environments (Ford, 2009). Given the dependence of Indigenous Peoples on biodiversity in their surroundings, and threats from globalisation and climate change, environmental protection of species and habitats that support food security of Indigenous Peoples is needed (Ahmed et al., 2022; Johns, 1994).

The food sovereignty movement has emerged in response to the health outcomes associated with the nutrition transition (Patel, 2009), the negative impacts of climate change, and other challenges that threaten Indigenous Peoples' food systems (Ford, 2009). Food sovereignty movements increasingly are emerging from and manifesting in Indigenous Peoples communities and advocating for the protection and ownership of food environments to enhance food security and human health for all peoples (Patel, 2009).

Documenting Indigenous Peoples' food systems is urgent given the numerous threats they face. Central to understanding, conserving, and promoting Indigenous Peoples' food systems and community nutrition and wellbeing is a systematic and comprehensive documentation of the biochemical composition of Indigenous foods along with associated education programming. The International Network of Food Data Systems (INFOODS) administered by the FAO has led efforts for multiple decades in building regional food composition databases (FAO, 2017). However, there remain gaps for many Indigenous foods due to their limited use, logistics of sampling, and the costs of laboratory analysis (Ahmed et al., 2022). Efforts are emerging, such as The FAO/INFOODS Food Composition Database for Biodiversity and the Periodic Table of Food Initiative, to explicitly include the comprehensive analysis of all foods eaten by people around the world while providing access and benefit sharing to Indigenous communities (Ahmed et al., 2022). Along with biochemical composition data, supporting ethnographic and environmental details are critical to understand the context of Indigenous foods, such as effects of environmental variation on biochemical composition. Knowledge of the comprehensive biochemical composition of Indigenous foods should be made accessible to communities, such as through community nutrition education for youth, families, farmers, and healthcare practitioners.

Conclusion

Indigenous Peoples' food systems should be acknowledged and respected for providing important lessons for sustainable diets, food production, and consumption for society. Ecologically, Indigenous Peoples' food systems have common themes of food procurement from wild and cultivated food environments, utilisation of seasonal edible biodiversity, holistic concepts of health, stewardship of land, and ecological agricultural practices. Socio-culturally, Indigenous Peoples' food systems have common themes of sacred and spiritual beliefs, ecological and food knowledge, food rituals, display of cultural identity, and food preparation and preservation techniques linked to cultural identity. On a health basis, Indigenous Peoples' food systems are grounded in holistic concepts of health with recognition of food as medicine, and valuation of dietary quality and diversity. Factors of global change are posing threats to Indigenous Peoples' food systems including to foods, habitats, and associated traditional knowledge and systems. We call for acknowledgement of Indigenous Peoples' food systems for their inherent values as well as to provide strategies to support multiple Sustainable Development Goals (UN General Assembly, 2015).

References

Ahmed, S., & Byker Shanks, C. (2020). Supporting sustainable development goals through sustainable diets. In Filho, W.L. Wall T., Azul A.M., Brandli L., & Özuyar P.G. (Eds.), *Good health and well-being. Encyclopedia of the UN sustainable development goals*. Springer.

Ahmed, S., & Stepp, J. R. (2016). Beyond yields: Climate change effects on specialty crop quality and agroecological management. *Elementa: Science of the Anthropocene, 4*, 000092.

Ahmed, S., Stepp, J. R., Toleno, R. A., & Peters, C. M. (2010a). Increased market integration, value, and ecological knowledge of tea agroforests in the Akha highlands of Southwest China. *Ecology and Society, 15*(4), 27 http://www.ecologyandsociety.org/vol15/iss4/art27/.

Ahmed, S., Unachukwu, U., Stepp, J. R., Peters, C. M., Long, C., & Kennelly, E. (2010b). Pu-erh tea tasting in Yunnan, China: Correlation of drinkers' perceptions to phytochemistry. *Journal of Ethnopharmacology, 132*(1), 176–185.

Ahmed, S., Peters, C. M., Chunlin, L., Meyer, R., Unachukwu, U., Litt, A., Kennelly, E., & Stepp, J. R. (2013). Biodiversity and phytochemical quality in indigenous and state supported tea management systems of Yunnan, China. *Conservation Letters, 6*(1), 28–36.

Ahmed, S., Stepp, J. R., Orians, C., Griffin, T., Matyas, C., Robbat, A., Cash, S., Xue, D., Long, C., Unachukwu, U., Buckley, S., & Kennelly, E. (2014). Effects of extreme climate events on tea (Camellia sinensis) functional quality validate indigenous farmer knowledge and sensory preferences in tropical China. *PloS One, 9*(10), e109126.

Ahmed, S., Downs, S., & Fanzo, J. (2019). Advancing an integrative framework to evaluate sustainability in national dietary guidelines. *Frontiers in Sustainable Food Systems, 76*, 3.

Ahmed, S., de la Parra, J., Elouafi, I., German, B., Jarvis, A., Lal, V., Lartey, A., Longvah, T., Malpica, C., Vázquez-Manjarrez, N., Prenni, J., Aguilar-Salinas, C. A., Srichamnong, W., Rajasekharan, M., Shafizadeh, T., Siegel, J., Steiner, R., Tohme, J., & Watkins, S. (2022). Foodomics: A data-driven approach to revolutionise nutrition and sustainable diets. *Frontiers in Sustainable Food Systems, 592*(9): 1–8.

Armstrong, C. G., Miller, J. E., McAlvay, A. C., Ritchie, P. M., & Lepofsky, D. (2021). Historical indigenous land-use explains plant functional trait diversity. *Ecology and Society, 26*(2), 6. https://doi.org/10.5751/ES-12322-260206.

Booth, S., Bressani, R., & Johns, T. (1992). Nutrient content of selected indigenous leafy vegetables consumed by the Kekchi people of Alta Verapaz, Guatemala. *Journal of Food Composition and Analysis, 5*(1), 25–34.

Byker Shanks, C., Ahmed, S., Dupuis, V., Houghtaling, B., Running Crane, M. A., Tryon, M., & Pierre, M. (2020). Perceptions of food environments and nutrition among residents of the flathead Indian reservation. *BioMed Central Public Health, 20*(1), 1536.

Chalmers, D., & Dell, C. A. (2015). Applying one health to the study of animal-assisted interventions. *Ecohealth, 12*(4), 560–562.

Chatwood, S., Paulette, F., Baker, G. R., Eriksen, A., Hansen, K. L., Eriksen, H., Hiratsuka, V., Lavoie, J., Lou, W., Mauro, I., Orbinski, J., Pambrun, N., Retallack, H., & Brown, A. (2017). Indigenous values and health systems stewardship in circumpolar countries. *International Journal of Environmental Research and Public Health, 14*(12), 1462.

Compher, C. (2006). The nutrition transition in American Indians. *Journal of Transcultural Nursing, 17*(3), 217–223.

Convention on Biological Diversity (2000). *Report of the Fifth Meeting of the Conference of the Parties to the Convention on Biological Diversity*. Conference of the Parties to the Convention on Biological Diversity.

Damman, S., Eide, W. B., & Kuhnlein, H. V. (2008). Indigenous peoples' nutrition transition in a right to food perspective. *Food Policy, 33*(2), 135–155.

Downs, S. M., Ahmed, S., Fanzo, J., & Herforth, A. (2020). Food environment typology: Advancing an expanded definition, framework, and methodological approach for improved characterization of wild, cultivated, and built food environments toward sustainable diets. *Foods, 9*(4), 532.

Doyle, J. T., Redsteer, M. H., & Eggers, M. J. (2013). Exploring effects of climate change on Northern plains American Indian health. In Koppel Maldonado, J., Colombi, B., & Pandya, R. (Eds.), *Climate change, and indigenous peoples in the United States* (Vol. 120(3), pp. 135–147). Springer.

Durning, A. T. (1992). *Guardians of the land: Indigenous peoples and the health of earth*. Worldwatch Institute. https://www.povertyandconservation.info/en/biblio/b0238

FAO (n.d.). *Indigenous peoples. Policy support and governance gateway.* FAO. Retrieved March 30, 2022, from https://www.fao.org/policy-support/policy-themes/indigenous-peoples/en/

FAO (2017). *FAO/INFOODS food composition database for biodiversity version 4.0 – 22. BioFoodComp4.0.* FAO.

Ford, J. D. (2009). Vulnerability of Inuit food systems to food insecurity as a consequence of climate change: A case study from Igloolik, Nunavut. *Regional Environmental Change, 9*(2), 83–100.

Gagné, D., Blanchet, R., Lauzière, J., Vaissière, É., Vézina, C., Ayotte, P., Déry, S., & Turgeon O'Brien, H. (2012). Traditional food consumption is associated with higher nutrient intakes in Inuit children attending childcare centres in Nunavik. *International Journal of Circumpolar Health, 71*(1), 18401.

Geniusz, W. M. (2009). *Our knowledge is not primitive: Decolonizing botanical Anishinaabe teachings* (Illustrated ed.). Syracuse University Press.

Ghosh-Jerath S., Kapoor R., Barman S., Singh G., Singh A., Downs S., Fanzo J. (2021). Traditional food environment and factors affecting Indigenous food consumption in Munda Tribal Community of Jharkhand, India. *Frontiers in Nutrition, 7*, 1–18.

Gordon, K., Xavier, A. L., & Neufeld, H. T. (2018). Healthy roots: Building capacity through shared stories rooted in Haudenosaunee knowledge to promote Indigenous foodways and well-being. *Canadian Food Studies / La Revue Canadienne Des Études Sur l'alimentation, 5*(2), 180–195.

Gunderson, L. H., & Holling, C. S. (Eds.). (2001). *Panarchy: Understanding transformations in human and natural systems* (Third Printing Used ed.). Island Press.

Hanemaayer, R., Anderson, K., Haines, J., Lickers, K. R. l., Lickers Xavier, A., Gordon, K., & Tait Neufeld, H. (2020). Exploring the perceptions of and experiences with traditional foods among First Nations female youth: A participatory photovoice study. *International Journal of Environmental Research and Public Health, 17*(7), E2214.

Herforth, A., & Ahmed, S. (2015). The food environment, its effects on dietary consumption, and potential for measurement within agriculture-nutrition interventions. *Food Security, 7*(3), 505–520.

Hillier, S. A., Taleb, A., Chaccour, E., & Aenishaenslin, C. (2021). Examining the concept of One Health for indigenous communities: A systematic review. *One Health, 12*, 100248.

Hodge, D. R., Limb, G., & Cross, T. (2009). Moving from colonisation toward balance and harmony: A native American perspective on wellness. *Social Work, 54*, 211–219.

Hong, L., Zhuo, J., Lei, Q., Zhou, J., Ahmed, S., Wang, C., Long, Y., Li, F., & Long, C. (2015). Ethnobotany of wild plants used for starting fermented beverages in Shui communities of southwest China. *Journal of Ethnobiology and Ethnomedicine, 11*(1), 42.

Hough, W. (1897). The Hopi in relation to their plant environment. *American Anthropologist, A10*(2), 33–47.

IPCC (2007). *Climate change 2007: Impacts, adaptation and vulnerability. Contribution of working group II to the fourth assessment report of the intergovernmental panel on climate change.* Cambridge University Press.

Johns, T. (1994). Ambivalence to the palatability factors in wild food plants. In N. L. Etkin (Ed.), *Eating on the wild side* (pp. 46–61). The University of Arizona Press.

Johns, T., & Kokwaro, J. O. (1991). Food plants of the Luo of Siaya district, Kenya. *Economic botany, 45*(1), 103–113.

Johns, T., & Kubo, I. (1988). A survey of traditional methods employed for the detoxification of plant foods. *Journal of Ethnobiology, 849*(1), 81–129.

Kamau, E.H.; Nkhata, S.G.; Ayua, E.O. (2020). Extrusion and nixtamalization conditions influence the magnitude of change in the nutrients and bioactive components of cereals and legumes. *Food Science and Nutrition, 8*, 4.

Kimmerer, R. W. (2015). *Braiding sweetgrass: Indigenous wisdom, scientific knowledge and the teachings of plants* (First Paperback ed.). Milkweed Editions.

Kuhnlein, H., & Turner, N. J. (1991). *Traditional plant foods of Canadian Indigenous Peoples: Nutrition, botany and use* (1st ed.). Routledge.

Kuhnlein, H., Erasmus, B., & Spigelski, D. (2009). *Indigenous peoples' food systems: The many dimensions of culture, diversity and environment for nutrition and health.* Food and Agriculture Organization of the United Nations, Centre for Indigenous Peoples' Nutrition and Environment.

Kuhnlein, H. V., & Receveur, O. (1996). Dietary change and traditional food systems of indigenous peoples. *Annual Review of Nutrition, 16*, 417–442.

Laird, D. (2019). *Sustainable production and distribution of bioenergy for the Central US.* Iowa State University. https://cenusa.iastate.edu/files/cenusa_2019_0060.pdf.

Laird, D. (2022). *Biochar impact on soil carbon sequestration and sustainability of crop residue harvesting for bioenergy.* Department of Energy. Webinar, Bioenergy Role in Soil Carbon Storage.

Lemke, S., & Delormier, T. (2017). Indigenous Peoples' food systems, nutrition, and gender: Conceptual and methodological considerations. *Maternal & Child Nutrition, 13*(S3), e12499.

Luo, B., Li, F., Ahmed, S., & Long, C. (2019). Diversity and use of medicinal plants for soup making in traditional diets of the Hakka in West Fujian, China. *Journal of Ethnobiology and Ethnomedicine, 15*(1), 60.

Lynn, K., Daigle, J., Hoffman, J., Lake, F., Michelle, N., Ranco, D., Viles, C., Voggesser, G., & Williams, P. (2013). The impacts of climate change on tribal traditional foods. *Climatic Change. 120*, 545–556.

Magurran, A. E. (2003). *Measuring biological diversity* (1st ed.). Wiley-Blackwell.

Mason, P., & Lang, T. (2017). *Sustainable diets: How ecological nutrition can transform consumption and the food system.* Routledge.

Mazet, J., Uhart, M., & Keyyu, J. D. (2014). Stakeholders in one health. *Revue Scientifique et Technique, 33*(2), 443–452.

Overstreet, D. (2018). *Unearthing Wisconsin's ancient history.* Sheperd's Press. https://shepherdexpress.com/news/features/unearthing-wisconsins-lost-history/.

Patel, R. (2009). Food sovereignty. *The Journal of Peasant Studies, 36*(3), 663–706.

Pieroni, A., & Price, L. L. (Eds.). (2006). *Eating and healing* (1st ed.). Routledge.

Pleasant, J.M. (2016). Food yields and nutrient analyses of the Three Sisters: A Haudenosaunee cropping system. *Ethnobiology Letters, 7*(1), 87–98.

Pleasant, J. M., & Burt, R. F. (2010). Estimating productivity of traditional Iroquoian cropping systems from field experiments and historical literature. *Journal of Ethnobiology, 30*(1), 52–79.

Popkin, B. M. (2001). The nutrition transition and obesity in the developing world. *The Journal of Nutrition, 131*(3), 871S–873S.

Royal Commission on Aboriginal Peoples (1996). *Report of the royal commission on aboriginal peoples.* Canada Communication Group — Publishing. http://data2.archives.ca/e/e448/e011188230-05.pdf

Samson, C., & Pretty, J. (2006). Environmental and health benefits of hunting lifestyles and diets for the Innu of Labrador. *Food Policy, 31*(6), 528–553.

Sarkar, D., Walker-Swaney, J., & Shetty, K. (2020). Food diversity and Indigenous food systems to combat diet-linked chronic diseases. *Current Developments in Nutrition, 4*(Supplement_1), 3–11.

Smith, E., Ahmed, S., Dupuis, V., Crane, M. R., Eggers, M., Pierre, M., Flagg, K., & Shanks, C. B. (2019a). Contribution of wild foods to diet, food security, and cultural values amidst climate change. *Journal of Agriculture, Food Systems, and Community Development, 9*(B), 191–214.

Stokes, J., & Benedict, D. (1993). *Thanksgiving address: Greetings to the natural world.* Six Nations Indian Museum & The Tracking Project.

UN General Assembly (2015). *Transforming our world: The 2030 Agenda for Sustainable Development* (A/RES/70/1). https://documents-dds-ny.un.org/doc/UNDOC/GEN/N15/291/89/PDF/N1529189.pdf?OpenElement

Vinceti, B., Eyzaguirre, P., & Johns, T. (2008). The nutritional role of forest plant foods for rural communities. In C. J. P. Coler (Ed.), *Human health and forests: A global overview of issues practice and policy.* Earthscan.

World Health Organisation (2017). *World health statistics 2017: Monitoring health for the SDGs, sustainable development goals.* World Health Organisation. https://apps.who.int/iris/handle/10665/255336

Yang, L., Stepp, J., Ahmed, S., Pei, S., & Xue, D. (2011). The role of Montane forests for indigenous dongba papermaking in the naxi highlands of Northwest Yunnan, China. *Mountain Research and Development, 31*, 334–342.

Yang, L., Ahmed, S., Stepp, J. R., Mi, K., Zhao, Y., Ma, J., Liang, C., Pei, S., Huai, H., Xu, G., Hamilton, A. C., Yang, Z., & Xue, D. (2014). Comparative homegarden medical ethnobotany of Naxi healers and farmers in Northwestern Yunnan, China. *Journal of Ethnobiology and Ethnomedicine, 10*(1), 6.

47

FOOD SECURITY, SUFFICIENCY, AND AFFORDABILITY IN SUB-SAHARAN AFRICA

Development of local
sustainable food systems

Galyna Medyna, Mila Sell, Sara Ahlberg, and Antonina Mutoro

Introduction

Food systems encompass the inputs, processes, infrastructures of food as well as the environment and the people that relate to the production, processing, distribution, preparation, and consumption of foods (HLPE, 2017). They are deemed sustainable when they "deliver food security and nutrition for all in such a way that the economic, social and environmental bases to generate food security and nutrition for future generations are not compromised" (FAO, 2018). Current food systems in Sub-Saharan Africa (SSA) are affected by a myriad of factors that include rapid population growth, urbanisation, climate change, as well as a historic burden of colonial times. The reach of the latter is wide, from influencing the crops previously and currently being cultivated and which areas are developed, to how international trade across the continent is organised (Mwanika et al., 2021; Roessler et al, 2020). For example, focus has long been put on monoculture cereal-based systems that have led to local diets being highly dominated by staples and starchy foods such as sugarcane and maize, which provide calories but not the micro-nutrients necessary for a healthy life. This focus is one of the factors leading to high levels of under-nourishment and hidden hunger in many African societies (Ekholuenetale et al., 2020). Moreover, such monoculture systems require large amounts of chemical inputs and result in negative environmental outcomes, including degraded soils, deforestation, loss of ecosystems, and depletion of waterways (Tully et al., 2015), as well as rising greenhouse gas (GHG) emissions (van Loon et al., 2019), globally and in SSA.

DOI: 10.4324/9781003174417-56

At the same time, there is evidence that when local, endemic, agro-biodiversity sources are harvested and consumed, it is possible to fill all micro-nutrient needs of children (see Box 47.1) and generally improve the nutrition at population levels (Nnamani et al., 2015). These sources include underutilised and orphan crops, such as traditional leafy vegetables and fruits that can often be found growing wild. Incidentally, these are the same crops that can be pillars of more resilient and sustainable agricultural and food systems that foster biodiversity, food security, and are able to adapt and rebound from shocks (Borelli, et al., 2020; Tafadzwanashe, et al., 2019).

Adequate nutrition is more complex than just the theoretical availability of sufficient nutritional sources. Indeed, despite the presence of endemic local species that could provide a healthy diet, many consume, and feed their children, staple foods and processed foods of low nutritional content and quality. The reasons for this are manifold and are part of a vicious cycle that starts with prompts for the production or import of these foods, such as from companies hoping to gain new markets. People purchase and consume these products because of targeted advertisement and a sense of status, inadequate awareness of and indifference to nutrition, or perhaps because they cannot afford much else, or they live in a food desert where nothing else is available (Crush et al., 2011). This consumption creates further demand, which keeps the cycle going; it is just one of the many facets of the wicked problems of food systems in SSA.

Food systems today
Food identity, culture, and nutrition

Food is a cornerstone of human identity and is deeply enmeshed in our cultures. The association between humans and food originates from biological (nutritional and physiological benefits) and cultural needs (knowledge, attitude, values, beliefs, rituals, customs, and lifestyles). Moreover, food also fulfils psychological and social needs by creating a sense of belonging and otherness in individuals and groups (Fischler, 1988). Food identity is never set in stone but shifts with trends, technological advancements, and changes in socio-economic status, amongst others.

Currently many African countries are experiencing rapid economic growth and poverty reduction, accompanied by changes in agri-food systems, marketing, and consumption patterns. While, for example, some new markets have opened for healthy organic produce, there has also been a rise in the consumption of animal sourced foods and processed foods high in energy, sugar, saturated fat, and salt (Vorster et al., 2011) because of their increased availability and association with a higher socio-economic image. This shift has been accompanied by an increase in consumption of imported wheat, maize, and rice and a decrease in the consumption of traditional African foods, which are often high fibre staple foods and plant proteins. Besides not being actively promoted as some branded products, these traditional foods are also associated by younger generations with an image of being "poor people's foods" and not fitting the new food identities of many (Voster Ineke et al., 2007). One of the consequences of these factors are increased rates of overweight, obesity, and non-communicable diseases that are attributed to a large and growing burden of death and disability in SSA (Micha, et al., 2020; Steyn & Mchiza, 2014).

Cultural and religious beliefs and practices also influence food choices and inter-household food distribution, often at the expense of women and children. Indeed, in some cultures protein-rich foods, such as eggs and meat products, are primarily reserved for men's consumption, leading to negative effects on the nutrition and health of vulnerable populations, partially explaining the high rates of undernutrition among women and children (Von Grebmer, et al., 2017). While great progress has been made, currently approximately 78% of countries in Africa still have a high prevalence of stunting among children (between 20% and 30%), 55% have a high prevalence of anaemia among women of reproductive age (20–40%), and 58% have a high prevalence of wasting among children under five (> 15%), which is double the global average of 7% (African Leaders for Nutrition (ALN) Initiative, 2020). The negative effects of undernutrition have irreversible repercussions on one's individual health (e.g., a child's mental, physical, and social development), increased maternal, infant, and young child mortality, intellectual ability, and also at a population level (e.g., lower economic productivity and reproductive performance) (De Onis & Branca, 2016). Thus, food identity and culture are strongly linked to the long-term development and well-being of regions.

The role and opportunities for women in food systems

The previous section briefly touched on the impact of gender inequality on nutrition, but its presence has repercussions through all stages of food systems—from farming and processing to consumption, which affect health, well-being, and economic prosperity. Household tasks in SSA, including child rearing and cooking for the family, are still predominantly shouldered by women. Many women directly produce the food eaten by the family (Rapsomanikis, 2015) and they are active buyers at local markets, yet they are rarely present in the middle portion of food value chains, where added value is created and larger profits can be made, and their employment conditions tend to be lower (Maertens & Swinnen, 2012).

This middle portion of the value chain stages includes activities such as transportation, warehousing, processing, production, large retail, and restaurants. Structurally, these sectors are predominantly controlled and dominated by men because of cultural or historical reasons, such as discriminatory access to funding and financing instruments, limited or no access to childcare, and/or perceived lack of adequate skills or physical abilities (World Bank Group, 2019). While some change is occurring (e.g., before 2000, women in Ethiopia were not legally allowed to work outside the home without their husband's permission), it is slow and concerted efforts must be made to ensure equal access to opportunities. In SSA, as in many regions worldwide, men still predominantly meet with peers to discuss activities, build and finance business plans, and are more likely to serve as role models.

One concrete example of difficulties encountered by women trying to enter the middle sections of food value chains can be found in the 100 kg bag that is

Figure 47.1 Photograph of a standard 100kg bag of cassava flour in use in Kenya. By Sara Ahlberg. Used with permission.

the standard for storing and transporting crops and other ingredients in Kenya, as seen in the flour bag photo. This weight can be difficult to lift and handle, yet its lifting and handling are currently expected and often required, for example, in transportation and deliveries. Women may be systemically obstructed through having to contend with this requirement. Moreover, women also have less access to jobs that are perceived as more "dangerous" or requiring a higher skillset, such as operating commercial vehicles and working in warehouses and processing plants (World Bank Group, 2019). The gender gap should especially be considered by large multi-national corporations when they build or expand their value chains in the Global South (United Nations Global Compact and UN Women, 2015) (Figure 47.1).

Production, processing, and food industry

Modern complex value chains largely remain beyond the reach of farmers in SSA, a region where the income of a significant part of the population comes directly or indirectly from farming. Smallholder farms (< 5 ha) are still largely the norm, although there has been an uptick in medium-scale farms in some regions driven by urban investor farmers (Jayna et al., 2016). Crop productivity and the range of crops grown are limited compared to theoretical outputs, leading to large productivity gaps and low and stagnant incomes (Tittonell & Giller, 2013). Indeed, comparatively low yields lead to wasted resources (e.g., land) and lower food security. Diverse crop choices are essential for guaranteeing good incomes: Currently, entire communities and regions cultivate the same crops, leading to an influx of identical crops on the market at the same time, keeping prices down due to competition. The results are lower incomes for farmers, keeping them in poverty.

Farming and production practices are closely linked to the other aspects of the food chains including the set-up for processing in African food industries. In the Global North, food industries have a strong focus on consumers, innovation, and communication (e.g., sustainability labels, novel foods, and processes); in African smallholder farming, the informal market sector, and the production of cash crops for export are at the centre. For example, much of the development aid and projects in SSA look at developing and enhancing local performance of subsistence and smallholder farmers to increase household food security. Increasing local performance is very important to improve livelihoods and nutritional well-being but has limited impacts on linking farmers to formal value chains and lifting households out of poverty and so far has not been linked to a Green Revolution as previously experienced in Asia (Diao et al., 2010). Cash crops are mostly exported before significant value adding processing can happen, thus removing a potential sector of economic development and potential job creation (Keane, 2015; Pinnamang-Tutu & Armah, 2011). Changes to all segments of the African food systems are necessary to ensure well-being, food security, economic prosperity, and long-term environmental sustainability. There is great potential to learn from the pitfalls of food systems in other parts of the world, and described elsewhere in this book, including the reliance on monocropping of very few crops and its promotion through subsidies, extensive consumption of industrialised meats that result in high GHG emissions, and ever-increasing demand for agricultural land for feed, amongst others.

Urbanisation, changing consumption, and practices

Although much of the African population works in agriculture, SSA is the fastest urbanising region in the world, with a projected 1.5 billion urban dwellers in 2050, triple that of 2018 (United Nations—Department of Economic and Social Affairs, 2019). This urbanisation is leading to massive structural changes and challenges, especially since much of the urbanisation is unplanned and follows unregulated growth paths (e.g., no adapted infrastructure can be put in place) (Güneralp et al., 2017). Urban areas play a vital role in economic development, but they are also associated with an increased demand for food, particularly animal products and processed foods, which are linked to the unsustainable practices described above.

The connection between urban centres and agriculture must also be considered. On the one hand, urbanisation can lead to a widespread loss of cropland due to the expansion of cities, which in turn leads to food insecurity because there is limited access to local foods, while food demands increase (Seto & Ramankutty, 2016). On the other hand, the expansion of cities can decrease the distance between some urban and rural areas, which in turn strengthens the connection between urban and rural populations (Christiaensen & Todo, 2014).

In many African regions, the urban population still has strong, close links to their background and relatives in rural areas, which promotes the exchange of information and maintains cultures and traditions. However, with each generation this bond gets weaker (Bah et al., 2003). The rural areas are considered backward and closely connected to the drudgery of farm work, and few young people are interested in such a lifestyle. Those with the option of studying aim at careers within sectors such as IT, healthcare, or business, while those without the means for further education still prefer taking their chances among the diverse livelihoods provided in a growing urban setting (Güneralp et al., 2017). Thus, the divide grows, new consumption practices are adopted, and food identities shift.

Overall, urbanisation is at the core of several challenges facing African food systems. While innovative products can be developed in cities and urban agriculture can provide nourishment, most of the agricultural production must take place elsewhere and remain an attractive sector for younger generations. Moreover, there is a great disparity in income levels in urban areas: From the top 1% to those living in slums, all should have access to attractive, healthy, sustainable, and affordable foods.

Pathways forward

Sustainability is an overarching concern for all current and future actions and innovations within food systems, as most resources are limited, population growth is ongoing, and we are living in a climate emergency. Innovation should help produce more and better while using less (e.g., fewer chemicals, less time spent in the fields, processes suitable for ambient conditions) and, as much as possible, contributing positively to the planet and all its inhabitants. Considering the multifaceted challenges of African food systems, what solutions can be drawn out as potential pathways for SSA to become a leader in sustainable food system transformation?

Several initiatives at different levels aim to create solutions in collaboration with local actors: A key factor for long-term success, as solutions must be location-specific and take local priorities into consideration. One example is the food sovereignty movement: The Alliance for Food Sovereignty in Africa (AFSA)[1] is the largest civil society movement on the continent, and is bringing together different civil society actors to advocate under a single voice for Africa-based solutions based on agroecological and Indigenous approaches. At a consumption level and linking back to food identity, culturally appropriate healthy eating guidelines are also required in many African countries; currently only seven African countries are listed by the FAO as having published food-based dietary guidelines (FAO, 2021). Formal guidance and education on healthy lifestyles are likely to promote the consumption of diverse and nutritious diets. Below we discuss more areas and sectors where innovative solutions are both needed and are actively being developed to provide healthy and sustainable food for all.

Women as full actors of food systems

What would the food industry look like if African women were in decision-making positions, from designing the next consumer food product, establishing school canteen tenders, or writing large regional agricultural policies? Women reaching their full productivity potential will greatly increase their incomes and well-being and that of their families, local communities, and beyond (World Bank & ONE, 2014). Customs, laws, policies, and funding programmes must evolve at all stages of food systems for the world to reap the rewards.

At farm and production levels, many efforts are being put in place to help women have easier access to land ownership, inputs (including farmer credits), agriculture extension services, and social and agricultural networks. These are essential steps to empower women with independent livelihoods, and range from national changes in law to decisions announced by multi-national corporations to empower women farmers in their supply chains (United Nations Global Compact and UN Women, 2015; World Bank Group, 2019). The growth of more complex and modern value chains has also opened employment opportunities outside the home for many women, although equal access to most jobs is not yet achieved and women are under-represented in higher earning positions. Improvements in access to higher education and services such as affordable child care of quality must be a priority, with the latter paving the way to a brighter future both for the working mothers and their children (Hughes et al., 2021).

Future efforts should support existing local women-centred associations and networks that can be found in all regions and provide services such as microfinancing and vocational training (e.g., at CAEPA[2]), as they are able to effectively amplify the voices of women. Moreover, it is essential to understand blind spots and previous oversights that must be rectified, lest past mistakes are repeated. Recent research has highlighted the shortcomings of many data sets used for formulating programmes and policies that aim to tackle gender inequality. For example, women's voices tend not to be reflected in household surveys in households where men are present, and it is especially interesting to note that farmer women and men are affected differently by climate change, and efforts in this crucial area need to be tailored accordingly (Kristjansson et al., 2017) (Figure 47.2).

Figure 47.2 Schoolgirl by Kenyan maize field. By Mila Sell. Used with permission.

Box 47.1 Results from the field

The FoodAfrica (FA) project worked with women in food systems to increase nutrition and well-being. The results show unambiguously that strengthening the role of women is key. Our work is in line with literature suggesting that women plan ahead and use their income to improve the well-being and diets of their families— supporting women to have stronger roles in food systems has significant societal effects. It is also interesting to reflect the changes possible through structure support, rather than targeted and context-specific support. For example, what type of foods would women develop if they had a more substantial role in the food processing and industry sector? More specifically:

- In Benin, it was found that all the necessary micronutrients required by children for healthy development can be provided by integrating underutilised local agrobiodiversity sources and working with women to develop locally acceptable recipes using these ingredients.
- In Uganda, the project found that women were more empowered in households that focused more on food crops, such as leafy green vegetables, rather than cash-crops, such as coffee or sugar cane (usually the domain of the man). This finding suggests that women are empowered when they feel they have the capacity to improve the diets of their children and households. In the same vein, the results also suggest that women who earn an income outside the household

are more empowered than those confined to household and farm work, especially in remote areas.

- In Kenya, maize is commonly grown and consumed by all, including indirectly in animal products (see photo of cornfield in Kenya, in Figure 47.2). The project worked on improving post-harvest practices, including limiting aflatoxins in the maize-dairy chain. The presence of aflatoxins can have negative impacts on health such as acute hepatitis and liver cancer. The project developed local, cost-effective solutions for adequate drying and storage.

Source: Work of the authors.

New approaches to food industry

What should the future African food systems and food industry look like, and how might these efforts improve sustainable diets? New thinking, attitudes, and actors must create the change: The future cannot be built on the repetition of what everyone else is doing, or continuing with current food systems and food industry practices; to get different results, like the sustainable diets and foods; critical and creative actions across all sectors are urgently called for.

The integration of underutilised crops, vegetables, and fruits represents an opportunity like none other for shorter, inclusive, value-adding, and more sustainable value chains that could result in high-tech smart processes for healthy, plant-based food innovations. Local African experts hold the best knowledge of suitable products and processes to deal with African infrastructures and systems, without the need to rely on imported foods or for external innovations. For example, the lack of cold value chains should not be a limiting factor but rather an impulse for innovative processes and products that bypass the cold requirement. For instance, Nigeria-based Vchunks[3] is a locally-produced plant-based meat substitute that requires no refrigeration, an important factor in a country where only 84% of the urban population and 26% of the rural pollution have access to electricity (2019 data[4]).

At the root of a new future is a transformation of the whole system, including a rethinking of values and goals, and a shift in focus. A farmer cannot obtain the skills to develop a new tomato-based superfood if they are only supported in increasing their smallholder output to sell at village-level "sales platforms." Higher education students and entrepreneurs cannot acquire advanced knowledge, update current practices, and create new processes by learning methods and technologies from decades ago (Kirui & Kozicka, 2018). Yet many policies, investments and development projects offer a fragmented landscape and lack realistic long-term focus that would lead to radical change; they prefer instead short-term results and reporting that reinforces the status quo. Perhaps this is preferred by some, as a challenge to current raw material flows would inevitably result in many existing Global North actors in global food chains struggling and having to rethink their activities.

Urban agriculture

Rooftop gardens, vertical farming, and backyard chickens are not just weekend hobbies for many; they provide food security, increase food supply chain resilience, and can generate income. In areas with poor logistics (e.g., lack of cold storage options, bad road infrastructure), urban agriculture can ignite dairy and other animal-based businesses, and provide access to fresh produce in many African cities (Wilson, 2018). The importance of urban agriculture was especially witnessed during the COVID-19 containment measures when food chains were disrupted and urban and peri-urban food prices increased. Broadening engagement in urban agriculture can increase the resilience of food systems in urban areas (Hourssou et al., 2021) and their usefulness is clear to the people benefitting from them, yet they receive relatively little structured investments or are even deemed illegal because of public health concerns over malaria spread, and the ever-increasing land demand for housing and other construction (Poulsen et al., 2015).

Planned and managed urban agriculture initiatives should be included in all urban planning, from neighbourhoods with glistening skyscrapers, to single family home neighbourhoods, and high-density areas with lower income populations. As mentioned previously, urban agriculture can increase city food system resilience and provide healthy, nutritious foods, but they are also urban green spaces that provide ecosystem services such as providing habitats for biodiverse species and carbon sequestration (Lwasa et al., 2015). In addition, community gardens provide opportunities for social innovation, youth employment opportunities, and recreational space. Green areas in urban settings also help lower surrounding temperatures, which is becoming more and more important with our changing climate (Lwasa et al., 2015). To ensure the long-term benefits from urban gardens, they must be recognised through policy and fully integrated into the mesh of cities.

Urban agriculture is also an opportunity for innovation and forward thinking: Several companies and research institutes are exploring commercially viable vertical farming. With decreasing water supplies and a growing population and urban sprawl, vertical farming is one avenue towards more resilient and sustainable urban food systems with shorter supply chains. Leafy vegetables are particularly well adapted for this method of production, and women and youth organisations in urban slums in particular have identified this as a great opportunity for both household nutrition and small-scale businesses. Generally speaking, smaller plants with short growing periods seem to be preferable (Beacham et al., 2019) and several low-technology vertical farming approaches are available for home gardens (e.g., Murungi, 2021). Several vertical farming associations have also sprouted in many regions of Africa, highlighting interest in the technique.

Conclusion

As described above, food systems are complex and all aspects of them are interrelated. Policy and governance are overarching components that play key roles for reaching sustainable food systems in SSA. But meaningful policy cannot be formulated without understanding the cascading impacts to the different actors. Several recommendations have been formulated for policy making based on a "food systems approach" (e.g., Bhunnoo, 2019), yet it is important to consider the specificities of the African context and highlighting the needs of its different actors, with a special focus put on women and youth (Gashu et al., 2019).

SSA has a young population that has great potential for innovation and creativity, also within the food sector. Yet it cannot reach its full potential without providing opportunities for domestic high-quality education for all, supporting small scale infrastructure, and developing vibrant local food industries. A new approach to an emerging food system would require leadership from men *and* women; support for new production, processing, and product design; and strengthening work opportunities outside the home, particularly with fair remunerations for women. Innovation is at the heart of future sustainable food systems in Africa, and thus policy must encourage bold thinkers and support entrepreneurs to flourish so that the future is not merely a repetition of what has been done elsewhere. Tapping into the wealth of traditional knowledge, creativity, and resilience of the African continent to cultivate new pathways may provide innovative and profitable solutions for sustainable diets.

Notes

1 https://afsafrica.org/.
2 https://www.caepacameroon.org/women-and-families.
3 https://veggievictory.com/vegmeat/.
4 https://trackingsdg7.esmap.org/.

References

African Leaders for Nutrition (ALN) Initiative (2020, September 9). *Embedding nutrition within the Covid-19 response and recovery: COVID-19 position paper.* African Union. Retrieved from https://au.int/en/documents/20200909/aln-initiative-embedding-nutrition-within-covid-19-response-and-recovery

Bah, M., Cissé, S., Diyamett, B., Diallo, G., Lerise, F., Okali, D., Okpara, E., Olawoye, J. & Tacoli, C. (2003). Changing rural–urban linkages in Mali, Nigeria and Tanzania. *Environment and Urbanisation, 15*(1), 13–24.

Beacham, A., Vickers, L., & Monaghan, J. (2019). Vertical farming: A summary of approaches to growing skywards. *The Journal of Horticultural Science and Biotechnology, 94*(3), 277–283.

Bhunnoo, R. (2019). The need for a food-systems approach to policy making. *The Lancet, 393*(10176), 1097–1098.

Borelli, T., Hunter, D., Padulosi, S., Amaya, N., Meldrum, G., de Oliveira Beltrame, D. M., Samarasinghe, G., Wasike, V.W., Güner, B., Tan, A. and Koreissi Dembélé, Y., & Tartanac, F. (2020). Local solutions for sustainable food systems: The contribution of orphan crops and wild edible species. *Agronomy, 10*(2), 231.

Christiaensen, L., & Todo, Y. (2014). Poverty reduction during the rural–urban transformation–the role of the missing middle. *World Development, 63*, 43–58.

Crush, J., Frayne, B., & McLachlan, M. (2011). *Rapid urbanisation and the nutrition transition in Southern Africa. Urban Food Security Series, 7.* Queen's University and AFSUN. Retrieved from https://www.alnap.org/system/files/content/resource/files/main/afsun-7.pdf

De Onis, M., & Branca, F. (2016). Childhood stunting: a global perspective. *Maternal & Child Nutrition, 12*(Supplement 1), 12–26.

Diao, X., Hazell, P., & Thurlow, J. (2010). The role of agriculture in African development. *World Development, 38*(10), 1375–1383.

Ekholuenetale, M., Tudeme, G., Onikan, A., & Ekholuenetale, C. (2020). Socioeconomic inequalities in hidden hunger, undernutrition, and overweight among under-five children in 35 sub-Saharan Africa countries. *Journal of the Egyptian Public Health Association, 95*(1), 1–15.

FAO (2018). *Sustainable food systems. Concept and framework.* FAO.

FAO (2021). *Food-based dietary guidelines.* FAO.

Fischler, C. (1988). Food, self and identity. *Social Science Information, 27*(2), 275–292.

Gashu, D., Demment, M., & Stoecker, B. (2019). Challenges and opportunities to the African agriculture and food systems. *African Journal of Food, Agriculture, Nutrition and Development, 19*(1), 14190–14217.

Güneralp, B., Lwasa, S., Masundire, H., Parnell, S., & Seto, K. (2017). Urbanisation in Africa: Challenges and opportunities for conservation. *Environmental Research Letters, 13*(1), 1–8.

HLPE (2017). *Nutrition and food systems. A report by the high level panel of experts on food security and nutrition of the committee on world food security.* FAO.

Hourssou, M., Cassee, A., & Sonneveld, B. (2021). The effects of the COVID-19 Pandemic on food security in rural and urban settlements in Benin: Do allotment gardens soften the blow? *Sustainability, 13*(13), 7313.

Hughes, R., Kitsao-Wekulo, P., Muendo, R., Bhopal, S., Kimani-Murage, E., Hill, Z., & Kirkwood, B. (2021). Who actually cares for children in slums? Why we need to think, and do, more about paid childcare in urbanising sub-Saharan Africa. *Philosophical Transactions of the Royal Society B, 376*(1827), 20200430, 1–5.

Jayna, T., Chamberlin, J., Traub, L., Sitko, N., Muyanga, M., Yeboah, F., & Anseeuw, W. (2016). Africa's changing farm size distribution patterns: the rise of medium-scale farms. *Agricultural Economics, 47*(S1), 197–214.

Keane, J. (2015). *Firms and value chains in Southern Africa.* Overseas Development Institute. Retrieved from https://documents1.worldbank.org/curated/en/840341467999993764/pdf/103071-WP-Box394849B-Keane-Value-Chains-and-Firms-in-SACU-PUBLIC.pdf

Kirui, O., & Kozicka, M. (2018). Vocational education and training for farmers and other actors in the agri-food value chain in Africa. *ZEF Working Paper, 164*, 1–57.

Kristjanson, P., Bryan, E., Bernier, Q., Twyman, J., Meinzen-Dick, R., Kieran, C., Ringler, C., Jost, C. . & Doss, C. (2017). Addressing gender in agricultural research for development in the face of a changing climate: where are we and where should we be going?. *International Journal of Agricultural Sustainability, 15*(5), 482–500.

Lwasa, S., Mugagga, F., Wahab, B., Simon, D., Connors, J., & Griffith, C. (2015). A meta-analysis of urban and peri-urban agriculture and forestry in mediating climate change. *Current Opinion in Environmental Sustainability, 13*, 68–73.

Maertens, M., & Swinnen, J. (2012). Gender and modern supply chains in developing countries. *The Journal of Development Studies, 48*(10), 1412–1430.

Micha, R., Mannar, V., Afshin, A., Allemandi, L., Baker, P., Battersby, J., Bhutta, Z., Chen, K., Corvalan, C., Di Cesare, M. and Dolan, C. & Grummer-Strawn, L. (2020). *2020 global nutrition report: action on equity to end malnutrition.* Development Initiatives.

Murungi, N. (2021, February 6). *Vertical farming: Ugandan company develops solution for urban agriculture.* https://www.howwemadeitinafrica.com/vertical-farming-ugandan-company-develops-solution-for-urban-agriculture/91900/

Mwanika, K., State, A., Atekyereza, P., & Österberg, T. (2021). Colonial legacies and contemporary commercial farming outcomes: sugarcane in Eastern Uganda. *Third World Quarterly, 42*(5), 1014–1032.

Nnamani, C., Oselebe, H., & Igboabuchi, A. (2015). Bio-banking on neglected and under-utilised plant genetic resources of Nigeria: Potential for nutrient and food security. *American Journal of Plant Sciences, 6*(4), 518–523.

Pinnamang-Tutu, A., & Armah, S. (2011). An empirical investigation into the costs and benefits from moving up the supply chain: The case of Ghana cocoa. *Journal of Marketing and Management, 2*(1), 27.

Poulsen, M., McNab, P., Clayton, M., & Neff, R. (2015). A systematic review of urban agriculture and food security impacts in low-income countries. *Food Policy, 55*, 131–146.

Rapsomanikis, G. (2015). *The economic lives of smallholder farmers: An analysis based on household data from nine countries*. FAO.

Roessler, P., Pengl, Y., Marty, R., Titlow, K., & van de Walle, N. (2020). *The cash crop revolution, colonialism and legacies of spatial inequality: Evidence from Africa*. Centre for the Study of African Economies.

Seto, K., & Ramankutty, N. (2016). Hidden linkages between urbanization and food systems. *Science, 352*(6288), 943–945.

Steyn, N., & Mchiza, Z. (2014). Obesity and the nutrition transition in Sub-Saharan Africa. *Annals of the New York academy of sciences, 1311*(1), 88–101.

Tafadzwanashe, M., Chimonyo, V., Hlahla, S., Massawe, F., Mayes, S., Nhamo, L., & Modi, A. (2019). Prospects of orphan crops in climate change. *Planta, 3*, 695–708.

Tittonell, P., & Giller, K. (2013). When yield gaps are poverty traps: The paradigm of ecological intensification in African smallholder agriculture. *Field Crops Research, 143*, 76–90.

Tully, K., Sullivan, C., Weil, R., & Sanchez, P. (2015). The state of soil degradation in Sub-Saharan Africa: Baselines, trajectories, and solutions. *Sustainability, 7*(6), 6523–6552.

United Nations: Department of Economic and Social Affairs (2019). *World urbanisation prospects: The 2018 revision*. United Nations.

United Nations Global Compact and UN Women (2015). *Companies leading the way. Putting the principles into practice*. Retrieved from https://d306pr3pise04h.cloudfront.net/docs/issues_doc%2Fdevelopment%2FCompanies_Leading_the_Way_25_September_2015.pdf

van Loon, M., Hijbeek, R., Ten Berge, H., De Sy, V., Ten Broeke, G., Solomon, D., & van Ittersum, M. (2019). Impacts of intensifying or expanding cereal cropping in sub-Saharan Africa on greenhouse gas emissions and food security. *Global Change Biology, 25*(11), 3720–3730.

Von Grebmer, K., Bernstein, J., Hossain, N., Brown, T., Prasai, N., Yohannes, Y., Patterson, F., Sonntag, A., Zimmerman, S.M., Towey, O. & Foley, C. (2017). *2017 Global Hunger Index: The Inequalities of Hunger*. International Food Policy Research Institute, Welthungerhilfe, Concern Worldwide.

Vorster, H., Kruger, A., & Margetts, B. (2011). The nutrition transition in Africa: can it be steered into a more positive direction? *Nutrients, 3*(4), 429–441.

Voster Ineke, H., van Rensburg Willem, J., Van Zijl, J., & Venter Sonja, L. (2007). The importance of traditional leafy vegetables in South Africa. *African Journal of Food, Agriculture, Nutrition and Development, 7*(4), 1–13.

Wilson, R. T. (2018). Domestic livestock in African cities: Production, problems and prospects. *Open Urban Studies and Demography Journal, 4*(1), 1–14.

World Bank, & ONE (2014). *Levelling the field: Improving opportunities for women farmers in Africa*. World Bank.

World Bank Group. (2019). *Profiting from parity: Unlocking the potential of women's business in Africa*. World Bank.

48

MILLETS AND *KĪRAI*

Two food categories emblematic of the
ability and knowledge of Tamil women
to ensure a healthy and sustainable diet

Brigitte Sébastia and Hélène Guétat-Bernard

Introduction: An endemic condition of malnutrition

Despite its food and health programs, India is failing to reduce the high level of malnutrition affecting its population. The data from the National Family Health Survey, 2019–2020 (NHFS-5)[1] compared to NFHS-4 (2015–2016) reveal a drastic deterioration in almost all malnutrition factors (nutrient and micronutrient deficiencies): Childhood stunting (+3.6%), wasting (+6.8% to severe +16.4%), underweight (-3.9%), and iron anaemia in children (+14.5%) and adults (female: 8.0%; male: +6.5%). These results show that the prevalence of malnutrition remains a serious concern for the country and call into question the implementation of the recent National Food Security Act, 2013. This act was designed to define and regulate food programmes to improve access to a nutritious diet and reduce disparities in food and health.

Many reasons are regularly cited by the media and activists to interpret the poor improvement in malnutrition, including poorly functioning food programs in terms of household inclusion and exclusion (Balani, 2013), poor sanitary conditions, and aversion to breastfeeding. While these factors cannot be overlooked, a major cause of malnutrition is a defective diet that is nutrient imbalanced, low in diversity, and lacking in essential nutritive components. This situation is the result of agricultural and food policies initiated under British rule and developed in the post-colonial period by the launching of the green revolution movement. These actions have reduced the farmer's production or the ability to consume enough of their own product and erected challenges in accessing markets. The increased commercialisation of agriculture, including food crops such as new varieties of rice, had a considerable impact on the diet of the peasantry and urban dwellers. This agricultural movement, with the objectives of stopping famines and making

DOI: 10.4324/9781003174417-57

the country self-sufficient, encouraged crops that provided the population with low-cost calories (cereals, sugar), with insufficient attention to the quality of the food in terms of diversity and nutritional balance. In addition to having increased the dependence of populations on the food market, this new agricultural model has led to the amplification of malnutrition through a drastic increase in metabolic disorders since the 1980s. According to the NFHS-5, 15.1% and 18.3% of women and men respectively have diabetes, and overweight/obesity affects 27.1% and 28.3% of women and men. In comparison, the previous NFHS indicated that 11.9% of men and 8.6% of women were affected by diabetes and 12.6% of men and 9.3% of women were overweight or obese.

Awareness of the intrinsic risks of malnutrition, especially metabolic diseases, emerged in the 1990s among the middle and upper classes. Some ingredients that were traditionally consumed are coming back on the menu. Millets and leafy greens are good illustrations of this movement, as they are experiencing an unprecedented popularity among affluent urban consumers, and consequently, the food industry and marketing. Focusing on these two food categories, the chapter analyses the evolution of their consumption and their perception by the public, which have been subjected to two opposing movements: A drastic depreciation resulting from the lack of support by food and agricultural policies since the colonial period; and a craze for their nutritional value to complement a diet low in animal products. They offer a field of investigation of great relevance because they transcend caste and gender divisions: Women from the lower social strata, considered as the guardians of traditional knowledge, are increasingly sought after by the wealthy, the media, and civil society for information on how to grow and cook them, and how to select the right species for the health needs of the family. The focus on these plant categories is especially relevant when considering food sustainability, as their low chemical input and irrigation requirements keep the soil and environment healthy.

This chapter, which examines millets and leafy vegetables (*kīrai*) in Tamil culture, draws on three sources of information: 1) An ethnographic exploration combining open-ended and semi-structured interviews with individuals and groups (consumers, vendors, farmers, doctors, NGOs, government food program staff), and observations on the marketing of these plant categories and their use in cooking; 2) nutritional archives from the National Institute of Nutrition, Hyderabad; and 3) scientific articles (government food and agricultural policies) and documentation from the media (food program debates, food-related websites). Based on the French Institute of Pondicherry (IFP), the authors conducted interviews within diverse socio-cultural categories of South Indians, paying a special attention to women from lower social strata, as their knowledge and know-how about millets and *kīrai* regarding cultivation, collection, culinary uses, and health properties are cornerstones of healthy and sustainable diets.

The first part of this chapter provides a brief overview of the South Indian dietary pattern. Examined in the second part, is the subject of pioneering nutrition studies during the colonial period to investigate the role of polished rice in the occurrence of beriberi (Arnold, 2010; McCarrison, 1924). In an effort to define the appropriate diet to maintain the body in good health, these studies had urged the colonial government to revamp its food and agricultural policies. The third and fourth parts of this chapter deal with the changing consumption and perception of millets and leafy vegetables by Tamils over the past two decades. The objective of this study is to demonstrate that the recent resurgence of interest in millets and leafy plants can contribute to the reduction of malnutrition and to the improvement of environmental conditions damaged by the excessive use of chemical inputs that endangers the sustainability of agriculture and food. Much attention is paid to the lower caste women, who are able to assist with the popularity of Indian medicines (ayurveda and siddha), especially since the COVID-19 pandemic, as some healthier foods and herbs are prescribed for boosting immunity and fighting viral attacks.

Dietary system in South India: Ill-balanced and undiversified

The food pattern in India is based on cereals: Wheat in the northern and western regions, and rice in the eastern, north-eastern, and southern regions, with some millet/sorghum species added.[2] In the southern States, at noon, steamed rice is eaten with *campar*,[3] a spicy sauce made of vegetables and Bengal gram (*Cajanu cajan*), and a few side dishes, the composition and number of which depend on the socioeconomic status of the families: One or two vegetable side dishes may include some legumes and leafy vegetables, yogurt, or *mōr* (flavoured, watered yoghurt improperly translated as buttermilk), *racam* (spicy broth with some onions, aromatic leaves, and tomatoes), and pickles. At breakfast and dinner, *iṭli* (steamed cakes) or *tōcai* (pancakes), made from a fermented dough of rice and black gram (*Vigna mungo*), are served with *campar* and/or chutney (*caṭni*), a spicy condiment made of fresh coconut, tomato, mint, or peanut. The traditional South Indian *iṭli* and *tōcai* are sometimes replaced by North Indian items such as *chapatti* (whole wheat flatbread) or *upmā* (wheat semolina with vegetables and/or legumes). In poor households, where no grinder is available, steamed rice is eaten at all three meals with pickles, and depending on economic means, with a meagre *campar*.

Despite the common perception that India is a vegetarian country, 70.7% of women and 71.6% of men report consuming the meat/fish/egg category (Census, 2016).[4] However, the amount and frequency of consumption of these ingredients remains low due to economic constraints, and religious observances that impose their abstinence on certain days of the week dedicated to a specific deity, such as Tuesdays and Fridays, and during domestic and calendar celebrations. According to the 68th round of the National Sample Survey (2014a), the average monthly consumption in rural and urban areas for women and men is only 269 g and 382 g of meat (goat/chicken/beef), 266 g and 252 g of fish, and 1.92 and 3.18 eggs, respectively. This consumption contributes only 8% of protein intake, and is not compensated by legumes as these have been declining since 2000 (Roy et al., 2017) and represent 10.5% of protein intake. The consumption of dairy products contributes around 11%. Cereals constitute the main source of protein intake, so their quality and variety are essential to increase the amount and diversity of amino acids.

Nutritional research in colonial period: A need to diversify food and agriculture

In India, malnutrition with episodes of food scarcity and famines have repeatedly hit the population, especially during the colonial times (Amrith, 2008; Siegel, 2014). The pioneer medical scientists, notably McCay in Calcutta and McCarrison in Madras Presidencies,[5] developed nutrition research in the early 20th century to investigate the impact of diet on differences in body shape, physical efficiency, and endurance in Indian communities. McCarrison demonstrated that the diet of Madrassi, based on polished rice and deficient in food categories rich in proteins, minerals, and vitamins, was responsible for their weak physical constitution (McCarrison, 1919, 1944; Sinclair, 1952). His animal experiments revealed that mice fed with hand-husked rice were healthier than those fed with machine-husked rice (such as that imported from Burma) or parboiled rice (such as that processed in India). They also showed greater morbidity in mice fed with polished rice than with millets or wheat (McCarrison, 1924, 1926; Niyogi et al., 1934). In addition to these studies that demonstrated the deleterious effect of polished rice on health, McCay (1912) and McCay et al. (1919) undertook specific investigations into the impact of a rice-based diet on the occurrence of diabetes. They found that middle-class people were at risk of developing diabetes because of their high consumption of polished rice that increased with economic means.

These pioneering studies show that the food patterns observed today in Tamil Nadu are not so different from these previous examples. Today there is high rice consumption at the expense of other cereals and food categories. The recommendations by the researchers to supplement polished rice with wheat and millet and to develop collaborative research between the departments of food, agriculture, and medicine to improve agricultural production and nutrition (McCarrison, 1924, 1926; Niyogi et al., 1934) went unheeded. This work was overshadowed by the green revolution movement, which further skewed food diversification away from local production and sustainable crops.

Millets: Rehabilitation of a cereal for its nutritive values

Practitioners of siddha medicine, the traditional medicine predominantly used in Tamil Nadu, tend to advise their patients with anaemia or metabolic disorders to change their eating habits. In particular, they recommend alternating rice consumption with *kampu* (pearl millet) or *kēḻvaraku* (finger millet), depending on the patient's condition. In siddha medicine, *kampu* is classified as alleviating *pitta* and *kapa tōṣam* (sanskrit: *Pitta* and *kapha doṣam*)[6] (Sambasivam Pillai, 1991). For its ability to diminish *kapa*, practitioners prescribe it to people with diabetes as it is considered a *kapa* disorder. They also recommend *kampu kañji* (porridge) to people suffering from anaemia and weakness. Analysis of the nutritive composition of *kampu* reveals high protein, mineral, and iron content and a low glycaemic index (Gopalan et al., 2007; Parthasarathy et al., 2006), confirming that it is appropriate for treating micronutrient deficiencies and diabetes. Siddha practitioners routinely recommend *kēḻvaraku* to patients with diabetes and those affected by fracture or dislocation. It is ranked as the millet with the lowest glycaemic index and the highest fibre and calcium content (Devi et al., 2014; Gopalan, ibid.; Shobana et al., 2013). Influenced by siddha practitioners, biomedical doctors are increasingly inclined to recommend kēḷvaraku to their patients with diabetes; as a result, it is now sold everywhere from tiny stores to supermarkets. Although kampu and kēḷvaraku are the millets primarily promoted by siddha practitioners because of their availability in the market, some millets are also interesting from a nutritional perspective. Cāmai (small millet) is notable for its high iron content so that, as a porridge, it is a food to fight anaemia (Patil et al., 2015). Tiṉai (foxtail millet) has a high protein content and is prescribed to treat weakness and undernutrition, as well as to bedridden and convalescent people. It is part of the diet of COVID-19 patients in siddha hospitals. A practitioner who recommends the consumption of various millets, explains, "Each millet has its own benefits, so I advise patients to eat all kinds alternately" (Jeyananda, Kanyakumari District, 2011).

From her fieldwork on siddha medicine, Brigitte Sebastia (2015) showed that patients rarely approved of eating millets. Rathna, who accompanied her husband, argued that she was too busy to prepare several dishes, "I cannot prepare kēḷviraku for my husband. My children find the taste too strong and coarse. And I don't have time to cook both rice and kēḷvaraku" (Kanyakumari district, May 2012).[7] Mariyamman complained: "Only rice is available to us in the PDS, sir. How could I buy kampu?" (Kanyakumari District, May 2012).[8] Rathna's reaction echoes the difficulty of reversing dietary change, especially among younger generations who are more eager to experience "foreign food" than the dishes that were eaten by their grandparents and parents. Adults, especially from lower economic backgrounds, often are reluctant to consume millets because they hearken back to times of food scarcity. Although older people abandoned millet consumption when they migrated to the city, they credit them with nutritional values: "When we ate kēḷvaraku, our bodies were strong and we were able to work hard without feeling hungry. The rice fills our stomach, but after two hours, we are hungry." The lack of satiety of rice, especially that sold in the FPSs which is denounced for its substandard quality, is the reason people consume it in large quantities. Mariyamman's complaint, on the other hand, refers to the lack of diversity of ingredients available in the FPSs where the main subsidised cereal is polished rice.

To improve the nutritional status of the population, the National Food Security Act, 2013, stipulated that "coarse grains" (millet and barley) were included among food grains (cereals) in food programs.[9] With the exception of the State of Karnataka, where the population can purchase a limited amount of finger millet or sorghum (Raju et al., 2018), no other southern State offers subsidised millet as levels of production have been unsupported by colonial and post-colonial governments, and now fall far short of the need. From 1950–1951 to 2016–2017, millet production increased from 15.38 to 44.19 million tons, whereas that of rice quintupled from 20.15 to 110.15 million tons (Government of India, 2017). Deficit in millet production generated a vicious circle: The absence of a government policy to purchase millet from farmers discouraged them from growing it, and discouraged its consumption due to its unavailability in the FPSs and its depreciation compared to rice.

Since the 2010s, millets have attracted the enthusiasm of the educated urban middle and upper classes for their nutritional and medicinal properties. These people, who may be affected by diabetes, obesity, and/or cardiovascular diseases, and/or are informed by the media (newspapers, TV, Internet) about diabetes-related diseases, are inclined to change their eating habits. The consumption of millet and a range of vegetables, including leafy greens, popularised for their ability to improve health by preventing and reducing diabetes and micronutrient deficiency disorders, constitutes the most significant change in their diet. Ecological and agricultural benefits of millets are also motivating, as their resistance to drought and pest attacks and their low input requirements make them ideal cereals for ensuring sustainable food.

A growing number of NGOs, through a national millet network, are developing programs to encourage small and marginal farmers to "re-cultivate" and "re-consume" them. In particular, they rely on women farmers who traditionally perform many agricultural tasks (seed selection, sowing, weeding, harvesting, shelling, grain storage, etc.) and are responsible for cooking and caring for the family. Encouraging cultivation for food sovereignty and income, the consumption of traditional dishes, collecting traditional recipes, and educating on healthy and nutritional food, are at the heart of NGOs' concerns and duties. Some of them, such as the Deccan Development Society, and the Women's Collective or the Gandhigram Rural University, organise low-caste women into self-help groups to market the surplus millet harvested from their fields in the form of raw produce, flour, and processed products (biscuits, snacks, salted). Madeline Chera (2017), noting the importance of "snackification" in Indian food habits, opines that "millet advocates can take advantage of the fashion cycle to reintroduce millets into a novel culinary cultural context" (p. 315). Since the 2010s, the food industry has entered the millets market. In addition to raw, unhulled, and flour millet packages, it produces savoury products, biscuits, and ready-to-eat snacks such as *tōcai*, *chapati*, *appalam*, etc., which are distributed in supermarkets and organic stores and through specialised websites, and are popularised at agricultural fairs and food and health conferences. The revitalisation of millets also is exemplified by the development of basic vegetarian restaurants offering millet-based dishes (*tōcai*; *itiyappam*; *poṅkal*; *kitchedi*; *tāli*; etc.) in which protein and vitamin-rich plants such as legumes and leafy vegetables are included. In addition, a new type of street vendor specialising in health food is developing in cities. Some offer *kampu* and *kēḻvaraku* porridges (Patel et al., 2013), with the price depending on the economic status of the consumers in the area. The change in perception of millets has led to better recognition by the government, which now uses the term "nutri-cereals," instead of "coarse grains," in its documents. Indian Prime Minister, Narendra Modi, proposed that the year 2023 be declared the International Year of Millet (IYM). This initiative, supported by 72 countries, was voted by the UN General Assembly in March 2021. It will be interesting to observe if this millet distinction will be accompanied by an agricultural and food incentive policy, as announced in the document of the Ministry of Agriculture and Farmer Welfares.[10]

Kīrai: Vegetables of the poor but sought by well-off for their nutritional values

Kīrai consumption is, to varying degrees, shared by all castes and classes living in Tamil villages, as well as in towns such as Pondicherry where many families have maintained relationships with their native village. Among both rural and urban women over 35, many report consuming *kīrai* two to three times a week (Guétat-Bernard & Sébastia, 2022). While urban women are able to name seven to ten varieties, those in villages who used to gather wild species know between 15 and 25 species for which they are able to indicate their health benefits: Correcting indigestion, purifying the body, regulating *doṣa*, treating ulcers, improving vision, etc. The women explain that they decide the selection, frequency, and rotation of *kīrai* species from discussions at home that take into account the importance of diversifying menus and the health condition of family members. They are familiar with the cooking methods to incorporate them into recipes, and for some, to mitigate the sticky organoleptic characters, such as the bitterness of *vallārai kīrai* (*Hydrocotyle asiatica*), used for its tonic and stimulating properties, or the sourness of *puḷicca kīrai* (*Hibiscus sabdarifa*), a rich source of iron, vitamins, minerals, folic acid, and antioxidants. A marketed *kīrai*, often mentioned for its medicinal values, is *poṉṉāṅkaṇṇi* (*Alternanthera sessilis*). Women add it to *poriyal*, a dish made of varied vegetables, pulses, and spices, to heal stomach pain and liver disorders or improve the condition of the eyes, skin, and hair. This plant, which has antioxidant, anti-cataract, antibacterial properties (Shoda et al., 2017), is used in Indian pharma-copoeia for improving vision and treating vomiting, stomach dysfunction, and malaria. Some pandemic episodes, such as chikungunya, dengue, and COVID-19, have greatly popularised some *kīrai* also used as medicinal plants (*mulikai*) for their high content of immunity-enhancing vitamin C.

In rural areas, a girl is considered well educated if, when she marries, she is able to identify the different *kīrai*, select the cultivated and wild ones according to their health benefits, and prepare them appropriately. Villagers much prefer wild *kīrai* to those grown and sold in markets. They value them for their authentic taste and health benefits provided by the natural environment to which they belong. Eating wild *kīrai* grown on the land where people live ensures good health (Sujatha, 2015), a connection repeatedly emphasised by siddha practitioners to encourage their patients to eat local food, with local being equal to traditional. According to village tradition, women harvest from the cultivated land, fallow land, and pastures where they worked, a bunch of 5 to 7 *kīrai* (*kalavai kīrai*) where the species depend on seasonality and availability. As they consider that *kalavai kīrai* "contain more minerals" than a single *kīrai*, they struggle to diversify the species. Wild *kīrai* are in sharp decline, victims of the modern agricultural model that, with its emphasis on monoculture, chemical inputs, and tractors, compromises the sustainability of food. Some women complain about the difficulty of finding specific *kīrai* and harvesting enough plants for their consumption. This reference to plant scarcity echoes the complaint of some siddha practitioners about their inability to strictly respect the composition of their family remedies as some medicinal plants have disappeared (Sebastia, 2011).

The high value placed on wild *kīrai* is expressed in their intense circulation between women of different castes and classes, and recently, between urban families and their relatives back in the village. Aware of the impact of food on health, city dwellers ask their relatives to collect *kīrai* and prepare dishes that include them. These plants have become the most valuable items to bring back from the village. Their circulation is also lively inside the village, between women of lower castes and of higher status. These transactions, not monetarily driven, allow upper caste women to eat wild *kīrai* that they cannot collect themselves because of their status and limited movement outside the village. Because they are ideologically associated with good health and sustainable eating, the gift of *kīrai* in raw form and/or as a dish offered to guests, metaphorised friendship and mutual trust. As Parvathy, a 35-year-old woman explained in January 2021 during our visit to PS Palayam village (Pondicherry Territory):

> We offer *kalava kīrai* because they contain many minerals and are very good for health. We don't give away purchased *kīrai*, only those we harvest from the land. Exchanging *kīrai* with, or giving to, someone is a sign of kindness and caring that establishes a long relationship between us and the person who receives. I used to send them to my children who live in the city; they consider them as a precious gift because they don't find them in the city. My daughter often calls me to have *kalava kīrai*, raw or cooked, such as pickles, and to get recipes; her husband is very fond of them.

Apart from the wild species that enter the gift economy, *kīrai* are easy to obtain as they are sold in various types of shops or by street vendors and sellers in local markets who are mostly low-caste women. When purchasing, it is not uncommon for some customers to ask the women how to prepare each *kīrai* and what their medicinal and nutritional qualities are. As Sangunthala proudly states when we interviewed her in December 2020 at the Goubert Market in Pondicherry,

> For the past five years, I've had more and more clients, and with COVID-19, their numbers are growing. They come and ask me about *kīrai*. I explain to them the different varieties and recipes (…) Since I have been married for 48 years, I have always been a slave; I had to obey everyone, my husband, my son, my daughter-in-law, my parents-in-law… I was nothing, and nobody listened to me or asked me anything. But today, it's not the same. I sell *kīrai* and talk to rich people. I teach them how to cook *kīrai* so that they stay healthy. I bow to a man because he is rich, but indirectly he also bows to me to have good health.

Sangunthala's words highlight the condition of many women who, even today, in traditional circles, remain highly devalued, even though their role is to take care of the family. In this context, the revival of interest in some traditional foods, such as *kīrai* and millet, for their ability to prevent the metabolic diseases so feared by wealthy city dwellers, has allowed the low-caste women to realise that they are custodians of a precious knowledge.

Conclusion

Low caste women, because of their attachment to certain traditional foods for both agricultural and dietary interests, are significant actors in the sustainability of agriculture and food. However, this new infatuation with traditional foods is not without consequences for sustainability. While low-caste women have always cultivated *kīrai* and millet to supplement their diets, they are increasingly being grown by small farmers for the substantial income they earn from selling them in local or roadside markets and, more recently, by wealthy landowners who produce them on a large scale on irrigated land. *Kīrai* are increasingly available in supermarkets and vegetable chains, where they are sold fresh, but also ready to cook and as condiments such as chutneys and dry powders. Small start-ups promoting organic products and rural women's employment are tapping into the new enthusiasm for *kīrai* and marketing them in packaging that mixes sophistication and simplicity. In recent years, *kīrai* have benefited from the debate that once focused on the importance of consuming millets to improve the nutritional quality of the diet. By becoming commercial products, millets and *kīrai* are becoming less embedded in social relations. They have become a "new trend", moving from home food to outdoor food. With the rapid commoditisation and craze for millets this is leading to a shortage, an increase in price, and deprivation by low-income people who seek to regularly consume them. This story of the rediscovery and enhancement of a food that was perceived as a "famine food" is reminiscent of other foods, such as the famous quinoa, whose popularity due to its health benefits is depriving the populations that traditionally consumed it. What eaters are witnessing is a new narrative of these products, more detached from culinary cultural rules, and reformulated around the attributes of fast, easy, useful, and healthy.

Notes

1 The NFHS-5 data that are now available (June 2021) cover 17 States and five Union Territories (UTs); survey work for the remaining 14 States/UTs having been delayed by COVID-19, is ongoing. Retrieved March 16, 2021 http://rchiips.org/NFHS/NFHS-5_FCTS/NFHS-5%20State%20Factsheet%20Compendium_Phase-I.pdf.
2 In India, millets is the generic term comprising seven species commonly cultivated: Finger millet (*Eleusine coracana*), pearl millet (*Pennisetum glaucum*), foxtail millet (*Setaria italica*), proso millet (*Panicum miliaceum*), kodo millet (*Paspalum scrobiculatum*), barnyard millet (*Echinochloa frumentacae*), and little millet (*Panicum sumatrens*) to which sorghum (*Sorghum bicolor*) is added.

3 The transliteration of vernacular words follows the standardisation defined by the Tamil Lexicon (1982, Madras: University of Madras).

4 India is the country where the rate of vegetarians is the highest in the world. See India census data in the references.

5 Robert McCarrison (1878–1960) was the founder of the Nutrition Research Laboratories at the Pasteur Institute in Coonoor, South India, which was renamed the National Institute of Nutrition in 1969 after its transfer to Hyderabad. His interest in malnutrition arose from the high prevalence of cretinism and goitre he observed while working as an army doctor in India (Aykroyd, 1960; Sinclair, 1952).

6 Siddha and ayurvedic medicines share anatomical and physiological conceptions of the body, among them the theory of *tōṣam/doṣam* (types of energy composed of two primordial elements acting on the body and mind, inaccurately translated as humours), which are of three types: *Pittam/pitta, vātam/vāta*, and *kapam/kapha*. Their correct balance in the body ensures good health, while their imbalance leads to disease.

7 Rathna and her husband are *kūlikkāraṇ*, day labourers. They consulted the practitioner to replace their biomedical treatment for diabetes with siddha remedies. Fear of side effects from long-term medication, greater confidence in siddha remedies, and the hope of saving money are the main motivations for doing so.

8 Mariyamma belongs to the poor families residing in the villages of Kanyakumari district, the southern Tamil district where siddha practitioners are the most numerous and renowned. She accompanied her scrawny daughter who complained of fatigue, dizziness, white discharge, and cessation of menstruation. These symptoms are interpreted by the practitioners as being caused by an anaemic state.

9 Government food programs include the public distribution system, which consists of purchasing cereals and sugar from farmers at a fixed price to supply the FPSs, as well as the midday meals served in public schools, crèches, and canteens for destitute and displaced persons.

10 https://pib.gov.in/PressReleaseIframePage.aspx?PRID=1845652

References

Amrith, S. S. (2008). Food and welfare in India, c. 1900–1950. *Comparative Studies in Society and History, 50*(4), 110–135.

Arnold, D. (2010). British India and the "Beri-beri problem", 1798–1942. *Medical History, 54*, 295–314.

Aykroyd, W. R. (1960). Obituary notice: Major-General Robert McCarrison, Kt, C.I.E., M.A., M.D., D.Sc, L.L.D., F.R.C.P. *British Journal of Nutrition, 14*, 413–18.

Balani, S. (2013). *Functioning of the public distribution system. An analytic report.* PRS Legislative Research. Retrieved from http://www.prsindia.org/administrator/uploads/general/1388728622~TPDS%20Thematic%20Note.pdf

Census India (2016). *Vital statistics.* Retrieved from http://www.censusindia.gov.in/vital_statistics/BASELINE%20TABLES07062016.pdf

Chera, M. (2017). Transforming millets: Strategies and struggles in changing taste in Madurai, *Food, Culture and Society, 20*(2), 303–324.

Devi, P. B., Vijayabharathi, R., Sathyabama, S., Malleshi, G. N., Venkatesan, B. P. (2014). Health benefits of finger millet (Eleusinecoracana L.) polyphenols and dietary fibre: A review. *Journal of Food Science and Technology, 51*(6), 1021–1040.

Gopalan, C., Rama Sastri, B. V., Balasubramanian, S. C. (2007). *Nutritive value of Indian foods.* National Institute of Nutrition.

Government of India (2017). *Pocket book of agricultural statistics.* Government of India, Ministry of Agriculture & Farmers Welfare, Department of Agriculture, Cooperation & Farmers Welfare and Directorate of Economics & Statistics. Retrieved from https://agricoop.nic.in/sites/default/files/pocketbook_0.pdf

Guetat-Bernard, H., & Sebastia, B. (2022). Cuisiner les *kīrai* (légumes feuilles) et soigner les corps. *Anthropology of food, S17*Online since, connection on 20 August 2022. URL : http://journals.openedition.org/aof/13278; DOI : https://doi.org/10.4000/aof.13278.

McCarrison, R. (1919). The pathogenesis of deficiency disease. *Indian Journal of Medical Research, 6*(3), 275–355.

McCarrison, R. (1924). Rice in relation to beri-beri in India. *British Medical Journal, 8,* 414–20.

McCarrison, R. (1926). The nutritive value of wheat, paddy, and certain other food-grains. *Indian Journal of Medical Research, 14*(3), 631–639.

McCarrison, R. (1944). *Nutrition and national health being the cantor lectures delivered before the Royal Society of Arts 1936.* Faber and Faber.

McCay, D. (1912). *The protein element in nutrition.* Edward Arnold.

McCay, D., Banerjee, S. C., Ghosal, L. M., Dutta, M. M., Ray, C. (1919). The treatment of diabetes in India. *Indian Journal of Medical Research, 7*(1), 81–112.

National Simple Survey (NSS 68th round) (2014a). *Household consumption of various goods and services, 2011–2012.* Government of India, Ministry of Statistics and Programme Implementation.

National Simple Survey (NSS 68th round) (2014b). *Nutritional intake in India, 2011–2012.* Government of India, Ministry of Statistics and Programme Implementation.

Niyogi, S. P., Narayana, N., Desai, B. G. (1934). The nutritive value of Indian vegetable food-stuffs. Part V. The nutritive value of ragi (*Eleusine Coracana*). *Indian Journal of Medical Research, 22*(2), 373–382.

Parthasarathy Rao, P., Birthal, P. S., Reddy, B. V. S., Rai, K. N., Ramesh, S. (2006). Diagnostics of sorghum and pearl millet grains-based nutrition in India. *SAT ejournal.icrisat, 2*(1). http://ejournal.icrisat.org/cropimprovement/v2i1/v2i1diagnostics .pdf

Patel, K., Guenther, D., Wiebe, K., Seburn, R.A (2013). Marginalized street food vendors promoting consumption of millets among the urban poor: A case study of millet porridge vendors in Madurai, Tamil Nadu, India. In Food Sovereignty: A Critical Dialogue International Conference September 14–15, 2013. Conference Paper #82 (pp. 1–37). New Haven: Yale University. https://www.tni.org/es/node/1239

Patil, K. B., Chimmad, B. V., Itagi, S. (2015). Glycemic index and quality evaluation of little millet (*Panicum miliare*) flakes with enhanced shelf life. *Journal of Food Sciences and Technology, 52*(9), 6078–82.

Raju, S., Rampal, P., Bhavani, R. V., Rajshekar, S. C. (2018). Introduction of millets into the public distribution system: Lessons from Karnataka. *Review of Agrarian Studies, 8*(2), 120–36.

Roy, D., Joshi, P. K., Chandra, R. (eds.) (2017). Introduction. In D. Roy, P. K. Joshi & R. Chandra (Eds.), *Pulses for nutrition in India. Changing patterns from farm to fork* (pp. 1–20). International Food Policy Research Institute.

Sambasivam Pillai, T. V. (1991). *Tamil-English dictionary of medicine, chemistry, botany and allied sciences.* Directorate of Indian Medicine and Homeopathy.

Sébastia, B. (2011). Competing for medical space. Traditional practitioners in the transmission and promotion of siddha medicine. In V. Sujatha & L. Abraham (Eds.), *Medicine, state and society. Indigenous medicine and medical pluralism in contemporary India* (pp. 165–185). Orient BlackSwan.

Sébastia, B. (2015). Coping with diseases of modernity: The use of siddha medical knowledge and practices for diabetics' care. In K. A. Jacobsen (Ed.), *Routledge handbook of contemporary India* (pp. 474–489). Routledge.

Shobana, S., Krishnaswamy, K., Sudha, V. N., Malleshi, G. (2013). Finger millet (ragi, Eleusine coracana L.): A review of its nutritional properties, processing, and plausible health benefits. *Advances in Food and Nutrition Research, 69,* 1–39.

Shoda, K., Govada, V. R., Anantha, R. K., Verma, M. K. (2017). An investigation into phytochemical constituents, antioxidant, antibacterial and anti-cataract activity of

Alternanthera sessilis, a predominant wild leafy vegetable of South India, *Biocatalysis and Agricultural Biotechnology*, *10*, 197–203.

Siegel, B. R. (2014). *Hungry Nation Food, Famine, and the Making of Modern India*. Cambridge University Press.

Sinclair, H. M. (1952). *The work of Sir Robert McCarrison*. Faber and Faber Ltd.

Sujatha, V. (2015). Is food natural or cultural? Food, body and the mind in Indian medical traditions. In J. Kanjirakkat, G. McQuat & S. Sarrukkai (Eds.), *Sciences and narratives of nature, East and West*, (pp. 506–22). Routledge.

49

CHALLENGES AND OPPORTUNITIES FOR SUSTAINABLE DIETS IN INDIA

A systems strengthening perspective

Manish Anand, Meena Sehgal, and Vidhu Gupta

Introduction

India is home to about 17% of the world's population. In the past 50-plus years Indian agriculture has made progress with aid from the Green Revolution and the widespread adoption of high yielding and improved varieties of wheat and rice, and food and nutrition security have also increased from more resource efficiency through production technologies. These advances have come with significant concerns and critiques as well. Yet a persisting challenge is that 14% of the Indian population continue to be undernourished (FAO, IFAD, UNICEF, WFP and WHO, 2020).

The need for examining the food systems and diets in the Indian context from the perspective of sustainability and public health is becoming more compelling given the concerns about poverty, hunger, malnutrition, and population pressures. There also are calls for a second green revolution to further improve food and nutrition security. Food system critiques are running parallel to the discourse on ecosystem degradation, erosion of forest cover and biodiversity, and loss of Indigenous and traditional knowledge. Additionally, in the context of national development aspirations, there is a sub-narrative relating to loss of employment opportunities and gender-bias in capital-intensive food systems.

It is in this context, the concept of sustainable diets encompasses well-being, health, biodiversity, environment, climate, equity, fair trade, seasonal foods, cultural heritage, skills, food and nutrient needs, food security, and accessibility (Johnston et al., 2014; Lairon, 2012). These analyses provide opportunities to address

DOI: 10.4324/9781003174417-58

burgeoning issues related to sustainability of food systems, food, and nutrition security in a holistic and integrated manner thus also advancing India's commitments to the Sustainable Development Goals (SDG).

Determinants of a sustainable diet

According to Johnston et al. (2014), the significant determinants of sustainable diets include: Agriculture, health, sociocultural, environmental and socioeconomic. Other dimensions also have been discussed in the opening chapter.

Agricultural production

Growth in agricultural output in India has been strong, with about a six times increase in food grain production over the last seven decades: From 50.8 million tonnes in 1950 to an estimated 303.97 million tonnes in 2021–2022. These efforts may be helping ensure food security at the macro-level. However, there appears to be a fine balance between food demand and supply at the macro level in India, except in the case of pulses and oilseeds where there is an acute deficit, as shown in Table 49.1. The introduction of new seed-fertiliser technology during the Green Revolution period in the mid-sixties, and factors like irrigation, high yielding varieties, use of agrochemicals, credit availability, creation of infrastructure facilities such as storage, and remunerative prices, have enabled food production to keep pace with rapid population growth, lowering food prices, raising the health status of the population, and potentially averting widespread famines. However, significant negative consequences have been catalogued in the loss of crop diversity (encouraging monocultures of a few crops like rice and wheat) (Prasad, 2016), in undermining the health and resilience of agroecosystems (Choudhary et al., 2018), and with undermining the diets of rural people in some cases (Nelson et al., 2019). These concerns too are raised in other chapters.

Having the largest area of arable and permanently cropped land in the world, an increase in cropping intensity from 131% in 2000–2001 to 144% in 2016–2017 has imposed an unfavourable high trade-off between quality and quantity dimension of agriculture vis-à-vis natural resource base. A productivity-based approach has resulted in input-intensive and environmentally exploitative agricultural production systems (Chand, 2017). These have manifested in declining land productivity, soil erosion, deforestation, drought, water deficits, and desertification, along with loss of biodiversity. The increased use of agrochemicals has led to over-cropping, which has resulted in depletion of soil nutrients and organic matter at a fast pace. States that contribute more to the food basket of the country consume more synthetic fertilisers, indicating the chemical intensive nature of agriculture (e.g., Punjab—243 kilograms per hectare (kg/ha), Andhra Pradesh—196 kg/ha, Odisha—67 kg/ha, Himachal Pradesh—64 kg/ha) (MoAFW, 2020). Thus, to meet its future food and basic nutritional needs in a sustainable manner, a paradigm shift in existing agricultural practices, services, and policies is required.

Table 49.1 Aggregate food demand and supply estimates (India)—million tonnes

Item	Food demand			Food supply		
	2016–2017	2021–2022	2032–2033	2016–2017	2021–2022	2032–2033
Rice	103.6	109.3	120.8	110.2	122.2	151.7
Wheat	90.2	97.1	113.5	98.4	109.5	138.8
Coarse cereals	40.0	46.9	67.5	44.2	47.9	61.8
Pulses	23.6	26.7	35.2	23.0	24.4	34.0
Food grains	257.4	280.0	337.0	275.7	304.0	386.3
Oilseeds	50.3	61.5	99.6	32.1	39.2	60.0
Milk and products	147.5	181.9	292.2	162.5	202.7	329.7
Eggs, meat, and fish	14.4	18.6	35.5	24.4	34.4	78.6
Vegetables	182.4	224.3	360.8	176.2	221.0	362.9
Fruits	101.5	126.7	203.6	93.7	120.8	202.7
Sugar	36.5	40.3	46.4	31.3	36.7	44.4

Source: Adapted by the authors from NITI Aayog (2018).

Health

Although India is self-sufficient in food grain production, it has about one-quarter of the world's food insecure people. India appears to hold the potential to produce the amount of food necessary to allow all income groups to reach the caloric target. To meet the healthy diet requirements of the total population, demand for food grain for household consumption is projected to rise from 239 million tonnes in 2021 to 277 million tonnes in 2033 (assuming moderate activity as the norm). Based on calorie requirement, there is a food demand gap in almost all food items for household consumption, as shown in Table 49.2.

According to the International Institute for Population Sciences and International Classification of Functioning, Disability and Health macro and micro-nutrient malnutrition is widespread, with 19% of women and 16% of men unable to access enough food to meet basic nutritional needs (IIPS & ICF, 2019–2021). Over 32% of children below five years are underweight. Per capita consumption of calories and protein are falling in rural India, with only one-third of pre-school children meeting protein and calorie adequacy. More than half of India's children and non-pregnant women of reproductive age (67% and 57%, respectively) and 52% of pregnant women are anaemic (ibid).

The consumption patterns have been shifting from strong emphasis on millets, pulses, and vegetables, towards high-value commodities including milk, meat, fish, eggs, fruits and vegetables, particularly in urban areas while consumption of cereals has been declining in spite of increasing output for the period of 2000–2018. Population growth, increase in per capita income, and dietary preferences are leading to faster output growth in higher value-added sectors, resulting in diversification from food crops to non-food crops (such as tea, coffee, etc.). However, the nutritional diversification has not been adequate as food grains account for more than a three-quarter share of the

Table 49.2 Estimated food demand gap for household consumption (India)—actual/requirement (in million tonnes)

Item	2020–2021		2032–2033	
	Normative*	Actual	Normative	Actual
Cereals	200	166	231	178
Pulses	40	16	46	25
Food grains	239	183	277	204
Vegetables	150	152	173	261
Fruits	50	39	58	66
Milk and products	125	102	145	171
Animal foods	30	15	35	25

Source: Adapted by the authors from NITI Aayog (2018).

* Demand as per normative approach based on recommended calorie requirement.

total calorie and protein intake indicating a minimal change in consumption pattern.

This poor quality of diet gets worse among lower income groups. Moreover, it has been observed that per capita, per daily consumption of energy, protein, and fat among the poorest households (i.e., the bottom 30% of monthly per capita consumer expenditure (MPCE) class) of urban India is much lower than the recommended level (around 720–1000 kcal vs. 2090 kcal RDA energy; 20.6–35 g vs. 48 g RDA protein; 11.7–20 g vs. 28 g RDA fat) (MOSPI & WFP, 2019).

Economy

India has 2.3% of the world's land area, 18% of the human population, and 15% of the livestock (MoAFW, 2020). Though agriculture accounts for only about 15% of India's GDP, agricultural lands covering various agro-ecosystems, including irrigated, rainfed, coastal, semi-arid, arid (drylands), and hill and mountain (temperate) regions are becoming increasingly important, as about 46% of India's geographical land is under cultivation (23% under forests and 3.92% under pastures, with wide regional variations) with over 50% of the population dependent on agriculture as their main source of income. The share of agricultural products in exports is also substantial, with 11.9% of export earnings (MoAFW, 2020). While many are employed in agriculture, the income this sizable segment of the population earns, is relatively lower than other sectors and consequently, attention to income in agriculture and the role agriculture plays in development are critical. Gains could be found through improving livelihoods in agriculture, enhancing infrastructure along the agricultural value chain, and through supporting skilled employment in rural areas.

Reducing food loss and waste also are important from the perspective of food and nutrition security as well as natural resource efficiency. The majority of India's losses occur within the post-harvest, processing, and distribution stages of the food chain mostly due to poor management, refrigeration, and preservation practices during storage and transportation. This loss is even more significant for micronutrient-rich commodities such as fruits, vegetables, and animal products. The Central Institute of Post-Harvest Engineering & Technology (CIPHET) estimates that 4.6–6.0% cereals, 6.3–8.4% pulses, 3.0–10.0% oilseeds, 6.7–16.0% fruits, 7.3–12.4% vegetables, 1.1–7.9% plantation crops and spices, and 0.9–10.5% livestock produce are lost during harvest, post-harvest operations, handling, and storage in India amounting to INR 92,651 crore at average annual prices of 2014 (Jha et al., 2015).

Socio-cultural

Indian diets have rich diversity as they continue to be influenced not only by agro-ecological differences but also from religion, economic status, and at other times, caste and gender. These differences are seen in food cultures and dietary habits in terms of consuming, cooking, and conserving different food products. A study by Gupta and Mishra (2014) clearly showed a marked difference in food consumption across all agro-climatic areas. Also, mapping patterns of food consumption clearly showed varied regional food cultures in India. Based on local crop production, the study revealed that rice consumption regions were mostly seen in eastern, southern, and coastal parts of India while wheat consumption regions fall in northern and central areas. Furthermore, Western India preferred eating coarse cereals and wheat. Meat was consumed mainly in the southern, north-eastern, and north-western parts; fish is preferred in Western plains and Ghats, and the Western dry and Lower Gangetic plains. Consumption of vegetables was low across the country compared to other food groups but consumed in good amounts in the country's southern, eastern, and north-eastern sections. This analysis showed that there are food regions in India that may have been made by distinct food cultures and agricultural production systems, a factor that has significant implications for nutrition security but has hardly been factored into India's food policy discourse (Gupta & Mishra, 2014).

A variation in consumption of various products/ food commodities between Indian rural and urban communities has been observed, as shown in Table 49.3. Urban populations have a greater consumption of fruits (banana, mango, apple), milk, and eggs compared to rural populations. A study by Tak et al. (2019) also showed a higher household dietary diversity score (HDDS) in urban households compared to rural households (HDDS of 9.34 food groups in urban households vs. 9.04 food groups in rural households) in 1993–1994. But this score reveals an improving trend when compared to data from 2011–2012 that showed a HDDS of 9.57 in urban households and 9.71 in rural households. This change indicates that rural diets have improved in diversity more than urban diets. The study also reported that cereals dominated the diets in rural areas while vegetables, fruits, oil, sweets, and dairy were consumed in higher quantities in urban areas during 1993–1994 with little change in 2011–2012 (Tak et al., 2019). Such dietary diversity indicates a deficiency in micronutrients but a higher intake of macronutrients such as energy, fat, and protein than the recommendations.

Environment

With economic growth and urbanisation, more land is diverted to non-agricultural uses. The percentage of land area under non-agricultural uses increased from 3.3% (9.36 m ha.) in 1950–1951 to 9% (26.88 m ha.) in 2014–2015. The rise of urbanisation (34% of India's population lives in urban areas; an increase of about 3% since the 2011 Census) is transforming food and land use systems both in terms of physical expansion of urban areas (food supply) and changes in diet (food demand). Some states, such as Tamil Nadu, have over 50% of its population in urban areas, whereas states such as Bihar have 11% of the population in urban areas. Urbanisation will have many consequences for food and land use systems. On the one hand, farmland loss is likely to be substantial as the economy is largely agrarian; on the other, diets of urban and higher-income societies are markedly

Table 49.3 Temporal trends in dietary diversity expressed as per capita consumption of different commodities (quantity in kg/litre/no.) in rural and urban India

Commodities	Quantity consumed in a month in rural India					Quantity consumed in a month in urban India				
	1993–1994	1999–2000	2004–2005	2009–2010	2011–2012	1993–1994	1999–2000	2004–2005	2009–2010	2011–2012
Rice and its products (kg)	6.79	6.59	6.38	3.14	6.13	5.13	5.1	4.71	4.66	4.66
Wheat atta and its products (kg)	4.32	4.45	4.19	4.36	4.43	4.44	4.45	4.36	4.34	4.32
Other cereals (kg)	2.29	1.68	1.55	0.85	0.66	1.03	0.87	0.87	0.38	0.3
All cereals (kg)	13.4	12.72	12.12	11.35	11.22	10.6	10.42	9.94	9.37	9.28
All pulses and pulse products (kg)	0.76	0.84	0.71	0.65	0.78	0.86	1	0.82	0.79	0.9
All edible oil	0.37	0.5	0.48	0.64	0.67	0.56	0.72	0.66	0.82	0.85
Banana (no.)	2.2	2.48	2.37	3.86	4.18	4.48	5	4.14	6.65	6.69
Coconut (no.)	0.32	0.37	0.35	0.46	0.49	0.46	0.51	0.47	0.63	0.61
Mango (no.)	0.06	0.1	0.09	0.11		0.12	0.16	0.11	0.16	
Apples (kg)	0.03		0.03	0.05	0.06	0.11	0.08	0.12	0.16	0.19
Groundnut (kg)	0.03	0.05	0.05	0.05	0.06	0.04	0.06	0.08	0.07	0.09
Vegetables (kg)	2.71	3.3	2.92	4.04	4.33	2.91	3.49	3.17	4.12	4.32
Milk: Liquid (L)	3.94	3.79	3.87	4.12	4.33	4.89	5.1	5.11	5.36	5.42
Egg (no.)	0.64	1.09	1.01	1.73	1.94	1.48	2.06	1.72	2.67	3.18
Fish (kg)	0.18	0.21	0.2	0.27	0.27	0.2	0.22	0.21	0.24	0.25
Goat meat/ mutton (kg)	0.06	0.07	0.05	0.05	0.05	0.11	0.1	0.07	0.09	0.08
Chicken (kg)	0.02	0.04	0.05	0.12	0.18	0.03	0.06	0.09	0.18	0.24

Source: Adapted by the authors based on various rounds from the National Sample Survey (Government of India, 2021, https://mospi.gov.in/web/nss)

different and require additional use of water and energy, not to mention processing and delivery infrastructure.

As noted above, agricultural practices have been criticised as being unsustainable. An estimated 120 million hectares (mha) are being affected by various types of degradation (total geographical area of 328 mha), including water and wind erosion (94 mha), soil acidity (18 mha), and alkalinity/sodicity (3.7 mha) (SAC, 2016). Rainfed farming is a major part of agriculture in India, as 52% of total cropped area is without assured irrigation. There is also significant misalignment in the cropping patterns and available water resources (Sharma et al., 2018). Nearly 86% of Indian agriculture is small-holder agriculture, and a significant part is subsistence agriculture, particularly in dryland areas. A large area under rainfed farming, and the predominance of small-holders with low adaptive capacity, increases the vulnerability of the agricultural sector to climate change. Climate variability, poor marketing, storage, transportation and processing infrastructure, and price volatility for many categories of farm produce (other than wheat, rice, and sugarcane) are making farming unsustainable and non-viable in many parts of the country. Defaults on crop loans, particularly in the wake of droughts or unseasonal rainfall, are becoming a serious issue (GoI, 2018).

To summarise, the determinants of sustainable diets are closely interlinked and there are trade-offs that emerge from such complex interdependence. The different determinants relate to and influence one another and the sustainability of diets. The loss of services from land degradation and its impact on food security and nutrition can result in a decline in agricultural production, accentuated food shortages, food insecurity, threats to rural livelihood, chronic malnutrition, loss of biodiversity, landlessness and social conflicts, rural–urban migration, and increased vulnerability to climate change. Mainstream food security schemes have included public support for the production, marketing, storage, and distribution of food with some emphasis on distribution to poor consumers and efforts to improve the nutritional status of children and women. These approaches have had some success but favour rice and wheat, therefore skew production and consumption choices away from other crops better suited to the agro-climatic conditions in the cultivation area and tend to neglect nutritional quality and food safety.

Progressing towards sustainable and healthy diets: Issues and ways forward

Various government programs have been launched to address the issues of malnutrition and dietary diversity, and thus food and nutrition security as well as sustainability of production systems in the country (as shown in Figure 49.1).

The Green Revolution in India in the 1960s led to an increased production of staple food crops like rice and wheat, which reduced hunger and boosted incomes and some economic growth. However, this progress has been slow to translate from food security (i.e., focused on quantity of food) to nutrition security (i.e., focused on quality of food). India's policymakers have taken a two-pronged approach to food security. On one side focusing on supporting producers by attempting to stabilise the farm prices through assured minimum support prices (MSP), while on the other, providing subsidised food to poor consumers through the public distribution system (PDS), which purchases food from MSP-supported farmers.

While increased production can ensure increased availability of food, it cannot ensure accessibility and affordability of that food, unless growth is widespread and inclusive. Presently the markets for high-value commodities are quite fragmented and small and the existing supply chains are inadequate to handle perishable products. Despite a huge network for delivering key agricultural inputs and services, the outreach is not adequate and the quality of services is poor. Cutting across all the programs is a severe delivery and accountability deficit. Further, the National Mission for Sustainable Agriculture (NMSA), which evolved in the backdrop of increasing realisation of limitations of the Green Revolution and the negative impacts of climate change, did not get enough traction as the institutional and regulatory structures and R&D efforts are still oriented towards irrigated agriculture. There is a need to align these strategies with the SDGs to increase range of adaptation and mitigation. Table 49.4 summarises the existing gaps and obstacles faced in the implementation and execution of the programs, and the likely medium- to long-term solutions in addressing these issues.

These strategies include increasing access to renewable energy to introduce infrastructural enhancements, such as local processing of fresh produce and storage facilities of perishable produce at the community level that support shorter and smarter food supply chains. They also include incentivising growth of renewable energy in food processing and storage through rural micro-food enterprises. These enterprises would ensure reduction in food loss and waste, reduce seasonal fluctuation in food prices, and enable healthy consumption through the year.

Either due to socio-economic characteristics or physiological attributes, some sections of the society are extra vulnerable and need special nutrition policy mechanisms. For such vulnerable communities, government programmes need to include support mechanisms, such as direct cash transfers or reimbursements of expenses for healthy diets. Of paramount importance are infrastructural concerns related to availability of clean water, sanitation, and hygiene to increase the benefits from better nutrition systems.

India has a robust policy environment for addressing nutrition as a key public health challenge and finding solutions to reduce the deleterious effect of undernutrition. A large number of policies and programs have been addressing major areas of public health nutrition, with substantial focus on actions. Some policies (e.g., anaemia control) have been actively revised and integrated with existing programs at large scales such as: ICDS and National Rural Health Mission; with other national programs, such as blindness control; and with nationwide initiatives, such as vitamin A and iron supplementation.

Based on the inability of these policies and programs to provide nutritional security, the following recommendations are designed to create a more balanced food system.

Elevating role of nutrition across the SDGs: In recognition of the critical role of nutrition in achieving several of SDGs, leaders of international financial institutions,

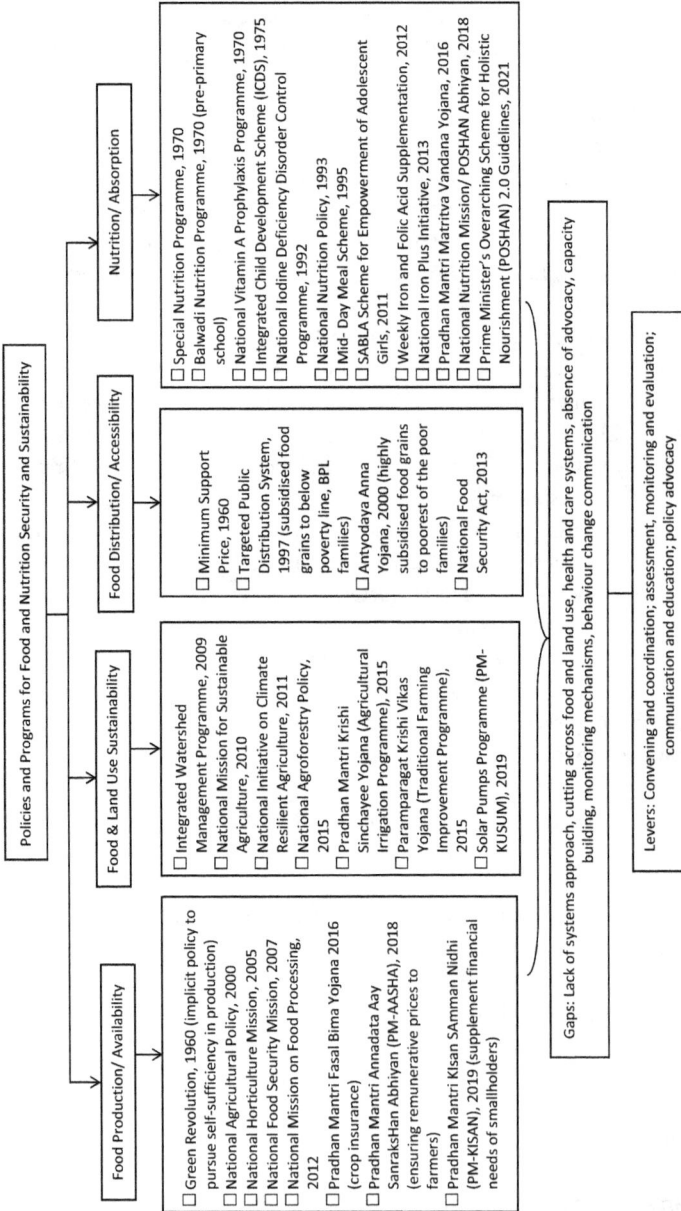

Figure 49.1 Framework of policies and programs for sustainable food and nutrition security in India. Source: Work of the authors.

Table 49.4 Existing gaps, and medium- to long-term solutions

Gaps and obstacles	Medium- to long-term solutions to sustainable food	Actors
Sustainable food production/availability		
Production shortfalls	Better water management through schemes such as Accelerated Irrigation Benefit Program, Integrated Watershed Management Program, Micro Irrigation Mission	Government/public sector, civil society organisations
Productivity gaps	Seed, soil health, pest and disease management, technology and extension education in agriculture	Public and private Sector, civil society organisations
Storage and transportation	Private sector (both domestic and multinational) participation, modern technology for perishable produce to reduce post-harvest losses	Private sector, government
Inadequate food value chain integration	Development of cold chains, strengthening of firm-farm linkages for scaling up processing and retailing operations	Public and private sector, multilateral or bilateral organisations
Financial services	Strengthening of access to formal credit, warehouse receipt systems such as Future groups and Reliance have invested in logistics and building up back-end operations	Private sector, government, civil society organisations
Climate change mitigation	Improve awareness of linkages of climate, with sustainable diets and health, technological innovations	Government
Regional vulnerability to climatic changes	Area-specific strategies and investments	Government
Food distribution/accessibility		
Operational difficulties in PDS	Greater decentralisation, social audits, experimental food coupons, direct cash transfers, including locally grown nutri-dense grains for procurement and Minimum Support Price	Government, private sector, civil society organisations
Identification of beneficiaries and errors in targeting in PDS	Community participation, use of RTI	Government
Issues of choice and preferences	Multiple options for nutrition status and health outcomes, option-based subsidies, targeting the most vulnerable groups	Government

(Continued)

Table 49.4 (Continued)

Gaps and obstacles	Medium- to long-term solutions to sustainable food	Actors
Food absorption		
Clean drinking water and sanitation facilities	Enhanced funding and maintenance of assets at the household level	Government, private sector
Clean cooking fuels	Incentives to use clean energy sources	Government, private sector
Health services	Improved functioning by health care workers and coverage under different schemes or network of virtual consultations	Government, private sector, civil society organisations
Health information system	Tracking trends, identifying vulnerable groups affected by multiple challenges of climate change, and agro-climatic factors	Government, private sector, civil society organisations
Education	Improved functioning and coverage about hygiene and sanitation, dietary issues	Government, private sector, civil society organisations
Gender	Incentivising women's participation in workforce and literacy programs	Government, private sector, civil society organisations

Source: Adapted by the authors from Nandakumar et al. (2010)

the UN, and other national nutrition champions should advocate strongly for the set of SDG nutrition indicators to be included in the indicator set put forward to the UN statistical commission by end of 2015.

1. Improving visibility and addressing malnutrition through a life cycle approach: Malnutrition visibility could be improved through nutrition awareness, advocacy, communication, and capacity building. Malnutrition could be addressed through activities such as strengthening of ICDS by enabling features like flexibility in planning, implementation, and indicator-based defined outcomes, intense monitoring, evaluations, deeper capacity building, training and information education, and communication. Offering home-based counselling for mothers on infant and young child feeding (IYCF) practices especially in high burden districts. Improving health nutrition and hygiene education through promotion of hand washing practices and use of safe drinking water can prevent the cycle of infection and under-nutrition.
2. Multi-sectoral approach in promoting nutrition:
   A unified effort of agriculture, health, and education sectors has a unique and critical role in addressing nutritional problems. For instance, the health sector should include nutrition and sustainable diets in medical education, training of health personnel, primary health care, and surveillance. Similarly, agricultural interventions need to be linked to nutritional

knowledge and behaviour change programmes as well as to women's empowerment.

3. Establishing a nutrition information system:
   This system could use ICDS infrastructure by the National Nutrition Monitoring Bureau (NNMB) team to establish a system of nutrition monitoring, nutrition guidance and standards aligned with sustainable diets, mapping and surveillance enabling area-specific planning, and programming using disaggregated data.

4. Promoting horticultural interventions:
   Promotion of horticultural foods is essential at community and household levels, to supplement school gardens and kitchen gardens. Provision of "cool chambers" at the community/village level are necessary for ensuring the safe storage of perishable foods. Self-help women's groups (SHWG) should be trained in home-scale preservation of fruits and vegetables to preserve produce when it's available in plenty for use during the lean seasons. Propagation of growing green leafy vegetables in empty tins/ as creepers on the roofs of households could also be done.

5. Nutrition resource platform:
   A web portal for easy access to information relating to nutrition will facilitate interactive discussions and provide comprehensive information for all ICDS and related programs. Also helpful for public and professional use would be the creation of a repository of nutrition related resources including research, new publications, government policies, training materials etc.

6. Promoting Indigenous and locally available foods:
   Tribal communities consume wild edible plants, which are nutrient-dense. Similarly, millets are commonly consumed in these communities. The use and consumption of these locally available foods could be promoted among other communities of the region or could be used to prepare low-cost nutritious foods.

7. Imparting basic health and nutrition knowledge for behavioural change:
   Campaigning and awareness-building for healthier consumption would markedly influence dietary food choices. This work needs to be especially targeted at the women of the household, as they are the primary caregivers of children and would be the one to use the knowledge to improve the nutritional status of the family and reduce micronutrient deficiencies.

8. Operationalising convergence through multi-sectoral programmes:
   This operationalisation could be achieved through programmatic convergence, such as the integration of nutrition components in relevant programmes/schemes and allocation of resources, thereby addressing nutritional determinants such as environmental factors.

9. Combating micronutrient deficiencies in a holistic manner:
   A holistic approach needs to be adopted that includes complementary strategies to address micronutrient deficiencies in children, adolescent girls, and women, including child care practices (counselling centres in

health facility and introduction of village report cards), dietary diversification (through health education, information, and communication), horticultural interventions, nutrient supplementation (through Mid-Day Meal (MDM), National Rural Health Mission (NRHM), ICDS), food fortification, and public health measures (i.e., monitoring and surveillance of nutrition programmes).

10. Nutrition education and capacity building:

Nutrition education can be integrated in school education curriculum frameworks at the national level. Nutritional messages could be integrated to ignite behaviour change activities under each national program. Additionally, capacity building could be provided for nutrition education through "change agents." Capacity can be built through training in nutrition counselling, safe water and sanitation issues, addressing familial factors (such as age of marriage, age of first child, access to care in pregnancy), importance of growth monitoring and growth curves. Through change agents, more education opportunities could be planned around the importance of nutritious diets and feeding practices, and the importance of guarding against consumption of junk food and calorie-dense, nutrition-poor foods.

11. Expanding the focus of policy making:

PDS and other social safety net programs can be expanded to include nutrient rich leafy green, millets, and pulses, and promote the consumption of these crops. As well, agricultural policies should shift toward a "crop-neutral" lens that does not incentivise staple crops over other crops. This policy would allow farmers to respond to market signals and demand for non-staple crops like fruits, vegetables, meat, and fish products.

Additional strategies to reduce pressures from climate crises also would be needed.

Conclusion

There are challenges for advancing and operationalising sustainable and healthy diets. The focus of policies and guidelines have been on "provision" and "supplementation," as opposed to advocacy, capacity building, and empowerment, whether for service providers or local self-government. Issues like behaviour change communication are not operationalised or contextualised, so they remain in the domain of policy and planning. Another missing link is "policy coherence"—the lack of coordination of departments has been an issue, and is needed to address complex problems with undernutrition. Monitoring mechanisms are also inadequate. For these reasons, although policies and working group reports refer to convergent actions and outcomes that can be monitored, India's current nutrition scenario is testimony to the fact that such initiatives are few and far between.

All of these issues call for a systems strengthening approach. On the one hand, there is a need to redirect the focus from food production and design to implement agricultural programmes and interventions that foster food literacy and nutrition security. On the other hand, there is a need to shift thinking on diets from a one-dimensional understanding of nutritional adequacy to that of personal and environmental well-being, incorporating aspects of food access and affordability, environmental sustainability, and socio-cultural acceptability (Johnston et al., 2014). An overarching agenda, involving the departments and ministries that address or are well positioned to address sustainable food and nutrition security, is required. It is proposed that a farm-to-fork approach supports more sustainable diets in India given the varied agro-climatic features of the country and the diverse dietary patterns. This chapter provides stepwise holistic and integrated approaches for transformational changes using the levers of convening and coordination; assessment, monitoring, and evaluation; communication and education; and policy advocacy. These actions are critical to advance sustainable and healthy diets in India.

References

Chand, R. (2017). Doubling farmers' income: Strategy and prospects and action plan. *Indian Journal of Agricultural Economics, 72*(1), 1–23.

Choudhary, S., Yamini, N. R., Yadav, S. K., Kamboj, M., & Sharma, A. (2018). A review: Pesticide residue: Cause of many animal health problems. *Journal of Entomology and Zoology Studies, 6*(3), 330–333.

FAO, IFAD, UNICEF, WFP, & WHO (2020). *The state of food security and nutrition in the World 2020: Transforming food systems for affordable healthy diets.* FAO. Retrieved from https://www.fao.org/3/ca9692en/ca9692en.pdf

GoI (Government of India) (2018). *Report of the committee on doubling farmers' income: Volume XIV: Comprehensive policy recommendations.* Ministry of Agriculture and Farmers' Welfare. Government of India. Retrieved from https://farmer.gov.in/imagedefault/DFI/DFI _Volume_14.pdf

Government of India (2021). *National sample survey.* https://mospi.gov.in/web/nss

Gupta, A., & Mishra, D. K. (2014). Food consumption pattern in rural India: A regional perspective. *Journal of Economic and Social Development, 9*(1), 1–16.

IIPS, & ICF (2019–2021). *National family health survey (NFHS-5) 2019–21.* Ministry of Health and Family Welfare. Retrieved from http://rchiips.org/nfhs/factsheet_NFHS-5 .shtml

Jha, S. N., Vishwakarma, R. K., Ahmad, T., Rai, A., & Dixit, A. K. (2015). *Report on assessment of quantitative harvest and post-harvest losses of major crops and commodities in India.* ICAR-CIPHET. Retrieved from https://www.researchgate.net/publication/289637983 _Assessment_of_Quantitative_Harvest_and_Post-Harvest_Losses_of_Major_ CropsCommodities_in_India

Johnston, J. L., Fanzo, J. C., & Cogill, B. (2014). Understanding sustainable diets: A descriptive analysis of the determinants and processes that influence diets and their impact on health, food security, and environmental sustainability. *Advances in Nutrition, 5*(4), 418–429.

Lairon D. (2012). Biodiversity and sustainable nutrition with a food-based approach. In B. Burlingame, & S. Dernini (Eds.), *Sustainable diets and biodiversity: Directions and solutions for policy, research and action* (pp. 30–35). FAO. Retrieved from https://www.bioversityi nternational.org/fileadmin/user_upload/online_library/publications/pdfs/Sustainable _diets_and_biodiversity.pdf

MoAFW (2020). *Pocket book of agricultural statistics 2020.* MoAFW. Retrieved from http://desagri.gov.in/wp-content/uploads/2021/06/Pocket-2020-Final-web-file.pdf

MOSPI, & WFP (2019). *Food and nutrition security analysis, India.* Ministry of Statistics and Programme Implementation & The World Food Programme. Retrieved from http://

www.indiaenvironmentportal.org.in/files/file/Food%20and%20Nutrition%20Security%20Analysis.pdf

Nandakumar, T., Ganguly, K., Sharma, P., & Gulati, A. (2010). *Food and nutrition security status in India: Opportunities for investment partnerships.* Asian Development Bank. https://think-asia.org/bitstream/handle/11540/1390/adb-wp16-food-and-nutrition-security-status-india.pdf?sequence=1

Nelson, A. R. L. E., Ravichandran, K., & Antony, U. (2019). The impact of the green revolution on indigenous crops of India. *Journal of Ethnic Foods, 6*(1), 1–10.

NITI Aayog (2018). *Demand and supply projections towards 2033: Crops, livestocks, fisheries and agricultural inputs.* National Institution for Transforming India (NITI) Aayog. Retrieved from https://www.niti.gov.in/sites/default/files/2019-07/WG-Report-issued-for-printing.pdf

Prasad, S. C. (2016). Innovating at the margins: The system of rice intensification in India and transformative social innovation. *Ecology and Society, 21*(4), 7. http://dx.doi.org/10.5751/ES-08718-210407

SAC (2016). *Desertification and land degradation atlas of India: Based on IRS AWiFS data of 2011–13 and 2003–05.* Retrieved from https://www.sac.gov.in/SACSITE/Desertification_Atlas_2016_SAC_ISRO.pdf

Sharma, B. R., Gulati, A., Mohan, G., Manchanda, S., Ray, I., & Amarasinghe, U. (2018). *Water productivity mapping of major Indian crops.* NABARD and ICRIER. Retrieved from https://www.nabard.org/auth/writereaddata/tender/1806181128Water%20Productivity%20Mapping%20of%20Major%20Indian%20Crops,%20Web%20Version%20(Low%20Resolution%20PDF).pdf

Tak, M., Shankar, B., & Kadiyala, S. (2019). Dietary transition in India: Temporal and regional trends, 1993 to 2012. *Food and Nutrition Bulletin, 40*(2), 254–270.

50

CULTIVATING SUSTAINABLE DIETS IN CHINA

Challenges and opportunities

Ning Dai, Zhenzhong Si, and Steffanie Scott

Introduction

Sustainable diets as a concept has not been widely discussed in Chinese-language academic literature, media reports, or policy documents. Most discussions about dietary transitions in the Chinese language do not use the term "sustainable diets" per se; instead, reports and dietary recommendations use terms such as vegetarian eating (蔬食) and balanced diets (平衡膳食).

Among the existing conceptions of what could be defined as "a good diet" in China, the term "balanced diet" was officially recognised in 2016, when the China Nutrition Society published the fourth version of Dietary Guidelines for Chinese Residents (subsequently referred to as the Guidelines). From a nutrition and health perspective, the Guidelines recommend a balanced diet comprised of sufficient consumption of dairy products, grains, fruits and vegetables, and moderate consumption of meat, eggs, and fish. From a health perspective, this balanced diet aims to reduce the prevalence of non-communicable diseases including cardiovascular disease and diabetes. Although the balanced diet is advocated from a nutrition perspective, it has important environmental implications because of the embedded GHG emissions in meat and dairy products. The balanced diet recommendation suggests moderate meat consumption. In 2016 in China, per capita consumption of meat was 63 kilograms per year, whereas the Guidelines recommended only 14 to 27 kilograms per year (Milman & Leavenworth, 2016). A successful dietary transition towards the Guidelines would require a reduction in meat consumption of over 50% for the average person, which would significantly curtail greenhouse gas emissions, animal farming water use, and soil and water pollution in China. The Guidelines also address the environmental aspect of sustainable diets by discouraging food waste. In addition to this concept of balanced diet, other government

DOI: 10.4324/9781003174417-59

policies and civil society initiatives in China have been promoting sustainable transition of Chinese diets, as we showcase in this chapter.

Table 50.1 summarises some of the key challenges to and opportunities for cultivating sustainable diets in China. From the promotion of balanced diets to the curtailing of food waste, the government of China has taken measures to address the triple burden of malnutrition and the growing ecological hoofprint. At the same time, civil society initiatives are gaining momentum in forging alternative food networks that enable sustainable production, fair trade, and responsible consumption. Therefore, despite the limited recognition of the phrase "sustainable diets," various opportunities are emerging that enhance the sustainability of food production and consumption in China.

This chapter is organised as follows. Following this introduction, section 2 lays out some of the most pressing challenges to dietary sustainability in China, including the shifting dietary patterns, meatification, and long-standing food safety anxieties across the society. Section 3 reviews key government policies and campaigns that promote sustainable and healthy eating, from clean plate campaigns to

Table 50.1 Challenges to and opportunities for cultivating sustainable diets in China. Source: Work of the authors

Challenges	Opportunities
• Food security challenges of malnourishment, over-consumption, micronutrient deficiency • Rising consumption of food away from home, leading to unhealthy diets and excess plastic packaging • Rising consumption of processed food • Westernisation and meatification of Chinese diets • Large scale food waste at restaurants and canteens • High use of agrochemicals leading to unsafe food and water and soil pollution • Loss of traditional food ingredients and agricultural biodiversity which undermines long-term food security	**Government initiatives** • Vegetable Basket Project and Crawling Peg Policy to ensure affordable food access • "Clean plate" campaigns to reduce food waste • Vegetarian canteens to encourage healthy and low-carbon diets • Crackdown on wildlife trade to protect biodiversity and to protect public health • Food safety law & GMO regulations **Civil society initiatives** • Good Food Academy to educate about eating for personal and planetary health • Vegetarian eating campaigns • Clean Plate Campaign to address food waste • Alternative food networks to build trust between producers and consumers • Farmers' Seed Network to preserve heritage seed varieties **Business innovations** • Online platforms to offer marketing training and to enable direct sales for farmers

vegetarian canteens. Section 4 introduces civil society initiatives that educate and contribute to the adoption of sustainable diets. Finally, Section 5 concludes the chapter by highlighting the main contributions of this chapter to knowledge on sustainable diets in China.

Food trends in China that impede sustainable diets

Based on the definition of sustainable diets in the introduction chapter of this book, we delineate sustainable diets through five pillars: Sustainable diets are nutritionally balanced, ecologically sound, culturally appropriate, economically viable, and fair and respectful to actors in the food system. Following the five pillars, below we present the current trends of dietary transitions in China that may pose a challenge to achieving sustainable diets.

Challenges to nutritionally balanced and healthy diets

First, there are a few trends of dietary shifts that hinder the development of nutritionally balanced and healthy diets in China. China faces a triple burden of persistent undernourishment, micronutrient deficiency, and growing over-nourishment (Yuan et al., 2019) among its population. The ongoing dietary shifts in China may compound this burden and warrant investigation on the drivers and the patterns of these dietary changes. Yuan et al. (2019) highlight three important patterns: The increase of food away from home, the increasing consumption of fast food and processed food, and the Westernisation of Chinese diets. Studies show that urban household expenditure on food away from home has significantly risen in recent years (Bai et al., 2010; Ma et al., 2006). The health implications of this trend are not trivial, because food away from home is associated with increased consumption of meat and sugary drinks, and decreased consumption of vegetables and fruits (Yuan et al., 2019). Additionally, a significant portion of food away from home consumption in Chinese cities is fast food consumption (Yuan et al., 2019). Both tendencies lead to the increased consumption of sodium, fat, and added sugar.

Second, consumption of processed food is on the rise among Chinese consumers (Reardon et al., 2012; Zhou et al., 2015). Through a dataset of nearly 15,000 participants, Zhou et al. (2015) found that processed food made up 28.5% of participants' daily energy intake. This ratio is even higher (40.2%) for children and youth from megacities, whose consumption of processed food is linked to being overweight (Zhou et al., 2015).

A third pattern in dietary shifts identified by Yuan et al. (2019) is that higher income is positively associated with the dietary inclusion of Western food in China. In a study on breakfast choices in three large Chinese cities, Bai et al. (2014) found that some Western foods have become increasingly popular for breakfast in urban China. These include fluid (not powdered) milk, bread, cake products, and, to a lesser extent, sausage, cheese, and coffee. This trend is linked to the increasing consumption of meat and dairy products in China.

Challenges to ecologically sound, safe, and respectful diets

Meatification

"Meatification" (Weis, 2013), adds to the ecological footprint of diets in China. Over the past few decades, the average Chinese person has consumed less grain and vegetables, and more animal products. Specifically, grain consumption per capita in rural areas dropped from 257 kg in 1980 to 164.3 kg in 2012, while vegetable consumption per capita decreased from 127 kg to 84 kg (Yuan et al., 2019). Meat consumption per capita in rural areas, however, rose from 7 kg in 1980 to 21.3 kg in 2012. Similar patterns can be identified in the dietary shifts among urban households (Yuan et al., 2019). To meet the increasing meat demand, growth in industrial meat production amplifies the stress on land resources and contributes to greenhouse gas emissions (Weis, 2013).

Online food ordering and delivery

The recent boom of online food ordering and delivery services in Chinese cities has contributed to the total usage of plastic containers, wooden chopsticks, and plastic bags. Alarmingly, much of the plastic waste generated in food delivery services has not been properly recycled but rather buried or burned (Zhong & Zhang, 2019). In 2017, the online food delivery business contributed to 1.5 million metric tonnes of plastic waste in China, a significant escalation from 0.2 million tonnes in 2015 (Song et al., 2018). The magnitude of plastic waste likely increased during the pandemic, when many who faced mobility restrictions turned to online food delivery. The environmental implications of the development of online food ordering warrants further research and policy attention.

Food waste

Food waste is a global challenge to sustainability, since it contributes to waste of land and water in food production, to the cost of waste disposal, and demands for landfill space. Additionally, food waste defies the principle of sustainable diets of respecting all actors in the food system and food itself. There is limited knowledge on the precise data of food loss and food waste in China, due to the insufficiency of official data (Liu, 2014). However, it is estimated that 35 million tonnes of food are lost every year in distribution from farms to consumers (Ren, 2013). Consumers produce food waste through household consumption and eating away from home. The majority of consumer-generated food waste results from China's food service industry (Liu, 2014). It is estimated that restaurants in urban China produce approximately 17–18 million tonnes of food waste per year, enough to feed up 30–50 million people for a year (WWF, 2018). The severity of food waste in China is alarming, and is likely to escalate in light of rising urbanisation, purchasing power, and consumption of food away from home.

Food safety anxiety

The widespread food safety anxiety in China is a longstanding challenge to sustainable and healthful diets. A series of food safety scandals in the 2000s have made food safety one of the utmost concerns for Chinese residents (Si et al., 2018). Through household surveys, Si et al. (2018) found that over three quarters of residents in Nanjing, a relatively well-off city of over eight million people, worry about food safety on a daily basis. Consumers are most worried about fertiliser and pesticide residues in fresh produce, environmental contaminants from soil and water pollution, adulteration and additives in processed foods, and genetic modification (Si et al., 2018). The persisting food safety anxiety has important implications for food policies, dietary choices, and farming practices in China.

Government policies that support sustainable diets

Vegetable basket project and crawling peg policy reinforce food security

Although sustainable diet is a new concept in China, its goals have long been supported in various ways by the Chinese government through policies, regulations, and campaigns. One prominent example is the "vegetable basket project" and the so-called "crawling peg" policy that governs the urban food system (Zhong et al., 2019; Zhong et al., 2021). The vegetable basket project initiated in the late 1980s bestows upon city mayors the responsibility of ensuring fresh produce and meat supply. While the measures taken differ from city to city, the project typically includes strategies to enhance local food production, strengthen the efficiency of food distribution and retailing, and ensure food safety and quality. To support spatial density and appropriate distribution of food retailers in cities, the crawling peg policy has been introduced by the state in addition to the vegetable basket project. It guides and supports the construction of new public food markets (also known as wet markets) in new residential areas to keep up with the growth of urban population. These policies prevent the problem of food deserts and enhance residents' access to fresh and healthy food. They forge a solid foundation for the promotion of sustainable diets in Chinese cities.

"Clean plate" campaigns to discourage food waste

Food waste has long been a severe sustainability concern in China due to a "culture of abundance" that encourages excessive food ordering (Liu, 2014). The scale of food waste is especially shocking in official receptions, business banquets, and school canteens. In 2013, a not-for-profit organisation of 33 members called "IN_33" launched a "clean plate" campaign in Beijing to promote food waste reduction and to call on individuals to finish foods on their plates. The campaign was also launched on the Chinese social media platform *Weibo* to encourage sensible food ordering at restaurants and canteens and to pack leftovers for takeaway.

This group disseminated promotional pamphlets and posters to restaurants and gas stations in Beijing, received widespread support from restaurant owners and diners, and gained attention in the media and social media (Cao & Fang, 2013).

In August 2020, the "clean plate" campaign was re-invented and announced by the central government as a national campaign. In the same vein, the Chinese government enacted a law against food waste in April 2021. This 32-clause law forbids online streamers from generating videos about binge-eating or excessive consumption. It also stipulates a fine for food service providers for wasting significant amounts of food or for encouraging consumers to order excessive amounts of food (Liu, 2021).

Vegetarian canteens

A health-centred campaign named Healthy China was launched nationally in 2019. The promotion of balanced diets is one of the key pillars of Healthy China, which aims to diversify Chinese diets, scale back the dietary intake of sugar, oil, and salt, and increase the consumption of fresh vegetables and fruits (Healthy China Promotion Committee, 2019). This campaign calls on regional governments to design and implement relevant regulations in order to promote balanced diets. In response to this campaign, two healthy vegetarian-only canteens have been established by the Zhejiang provincial government and Qingdao municipal government. Both canteens aim to cut down on sugar, oil, and salt in cooking. The Zhejiang government's vegetarian canteen increases the serving of organic foods, whole grains, and seeds. The Qingdao vegetarian canteen highlights environmental protection as its rationale for serving vegetarian food (Sohu, 2021). According to the reports, both vegetarian canteens have been popular among the government employees.

Crackdown on wildlife consumption

Due to the outbreak of COVID-19 in the city of Wuhan, the Chinese government has cracked down on wildlife trading and closed down all wild animal markets permanently. Before the crackdown, eating game meat was common in many parts of the country, particularly south China, although game meat was not commonly consumed by most Chinese people. In some restaurants, game meat was considered a special treat in fancy banquets due to its novelty and high price. Wild meat is believed to have health benefits and better taste than farmed meat. In some rural areas, people occasionally hunt and eat or sell wild animals. Wildlife farming and trading used to be a large industry in China: In 2020 its value was estimated at 520 billion *yuan* (74 billion USD) while employing over 14 million people (Arranz and Huang, 2020), including farmers and traders who made a living on wildlife transactions, making it an obstacle to wildlife protection and regulation. The ban on eating game meat was initially announced after the SARS outbreak in 2003, but wildlife trading was still a lucrative business until 2020 due to various loopholes in regulations. Driven by the public outcry over game meat consumption after the initial COVID-19 outbreak in Wuhan in early 2020, the Chinese government launched a national educational campaign about the various risks associated with game meat consumption. On February 24, 2020, The Standing Committee of the National People's Congress issued the so-called "comprehensive ban" on illegal wildlife trading and consumption. The ban, though driven by the motive of reducing the risk of zoonotic disease transmission, promotes biodiversity conservation, thus seeking to foster ecological sustainability of Chinese diets.

Regulations of food safety and genetically modified organisms

The Chinese state oversees food production, processing, retailing, and restaurant services in order to mitigate the widespread food safety anxiety. Soon after the outbreak of the melamine milk scandal, the Chinese government enacted the first edition of the Food Safety Law in 2009. Yet, food safety scandals still prevail due to the fragmented food safety regulatory system and the massive scale of food production and retailing in China (Yasuda, 2015). The most recent revision of the food safety law in 2021 remains stringent on permissible food additives and on permissible levels of residues of pesticides, veterinary drugs, and heavy metals in food.

Regulating the production of genetically modified foods has been a critical focus of food safety regulation in China. China is invested in the development of genetic engineering technologies while remaining precautious on granting permission to domestic commercial production of Genetically Modified Organisms (GMOs). Multiple administrative bodies are involved in regulating the safety of GMOs, and they mandate a rigorous five-step process for GMO safety testing before commercialisation: Laboratory research, small-scale pilot testing, medium-scale field testing, large-scale pre-production testing, and biosafety certificate application (Li et al., 2014). As of 2014, only seven genetically modified crops have been approved for commercial production, and only *Bt* cotton and virus-resistant papaya have been produced on a significant scale (Li et al., 2014). As a response to the low public confidence in GM food crops, China adopted mandatory labelling for 17 imported GM products, including soy products, corn, and tomatoes (Lim, 2021).

Civil society initiatives that support sustainable diets

Internationally, civil society actors have been increasingly recognised by researchers and policy makers for their active role in creating innovative solutions to food-related challenges, such as addressing food safety anxiety, reducing the climate impacts of food production and consumption, promoting food education, or preserving traditional breeds and varieties (Andrée et al., 2019; Garnet, Bilali and Strassner, 2018; Koc et al., 2008). Through complex interactions with the state and other actors, civil society actors in China have been able to mobilise consumers, researchers, and entrepreneurs to form a collective force for promoting healthy and sustainable diets (Scott et al., 2018). Actions taken by these civil society initiatives go beyond the sphere of sustainable diets and reproduce rules, norms, and governance of good food to various extents. In this section, we illustrate five recent cases of civil society initiatives that enhance dietary sustainability in China.

Protecting traditional food ingredients and genetic varieties

The loss of genetic varieties of crops and livestock around the globe is a threat to long-term food and nutrition security in a world that is facing greater climate variability (Bélanger & Pilling, 2019; Scherf & Pilling, 2015)). In China, the declining agricultural biodiversity is severe due to the long-term, aggressive promotion of selected commercially profitable seeds and livestock breeds (Song et al., 2012; Si, 2019). For example, according to UNDP/GEF and MFPRC (2005), the varieties of wheat, a major staple grain extensively cultivated in China, have declined from over 10,000 in the 1950s to only around 400 in the 2000s. Recent studies also show that the number of indigenous pig breeds in South China is declining rapidly (Diao et al., 2019). Despite the deterioration of genetic varieties in agriculture, many traditional food varieties are essential to meet local dietary preferences and nutritional needs in diverse cultures. On the flip side, the use of traditional food varieties, especially wild plants and mushrooms (Kang et al., 2013), can sometimes boost local economies through creating job opportunities and entrepreneurial initiatives (Bélanger & Pilling, 2019). For example, the Farmers' Seed Network (see Song et al., 2021), led by a group of agricultural researchers working with farmers, has catalysed the conservation of traditional crop varieties and livestock breeds. These efforts empower farmers through protecting their rights to seeds and improving their livelihoods. In spite of its nascence, this seeking of artisanal flavours is an emerging force to enhance dietary sustainability in China.

Alternative food networks that address agricultural sustainability and food safety anxiety

Another key dimension of dietary sustainability is the sustainability of farming approaches such as intercropping, composing, reducing chemical agrochemical inputs, and integrated pest management (see Luo, 2016; Scott et al., 2018). Since the 1990s, overuse of pesticides has caused massive food safety problems in China's food system. As food safety anxiety has increased since the 2000s, demand for ecologically produced food (i.e., food produced without using agrochemicals, growth hormones, or antibiotics) has risen dramatically (Scott et al., 2014; Shi et al., 2011). Alternative food networks (AFNs), such as community supported agriculture (CSA), ecological farmers' markets, urban agriculture, and consumer buying clubs, emerged within this socioeconomic context as trustworthy food sourcing schemes (Krul & Ho, 2017; Si et al., 2015).

Through rebuilding producer-consumer connections and trust, AFNs may aid in addressing the concerns of food safety and promote the development of ecological agriculture. As of 2018, China has more than 1000 CSA farms, dozens of ecological farmers' markets (CSA Network, 2018), and a growing number of ecological farmers' cooperatives. Agroecological practices are also becoming increasingly common at farms and gardens in and around Chinese cities (Luehr et al., 2019). These civil society initiatives have joined forces with consumers, entrepreneurs, and academics to reshape the landscape of urban food systems in major cities. They constitute "good food movements" with Chinese characteristics (Zhang, 2018), in terms of using food safety as a window through which to gain people's interest and then educate them about the environmental and social values of participating in AFNs (Schumilas & Scott, 2016; Si, 2017; Si et al., 2015). AFNs have also become a new force in promoting sustainable diets in China through consumer education programs, on-farm activities, media engagement, social media activities, and training workshops.

Food education about healthy and sustainable eating

Food education about healthy and sustainable eating has been a key task among many AFNs in China. One prominent case is the Beijing Farmers' Market, the most influential ecological food market in China. At its store and office in Beijing, the market organises open seminars that feature speakers with diverse backgrounds in food sustainability. Regular gatherings greatly enhance the emotional bond between keen consumers and the market while disseminating knowledge about sustainable and healthy eating. In 2017, managers of the Beijing Farmers' Markets founded a social media platform called "Foodthink" that features hundreds of articles about sustainable eating and many other related themes. These articles focus on the social and environmental implications of people's food choices. Foodthink also launched the Lianhe Program to fund projects that aim to promote sustainable diets and farming in local contexts.

The Good Food Academy is another example, established in 2017 by the China Biodiversity Conservation and Green Development Foundation, this initiative focuses on mitigating the impacts of industrial diets on personal health, planetary health, and animal welfare through food education. The Good Food Academy invites interested institutions and individuals to commit to the initiative, and in so doing, they would abide by at least five of the initiative's eight core principles: Prioritising plant-based proteins, animal welfare, choosing healthy foods, reducing food waste, sourcing local and seasonal food, reducing single-use plastics, protecting biodiversity, and enrol employees for food education (Good Food Foundation, 2019).

Vegetarian eating campaign

In 2015, the international environmental organisation Wildaid founded a vegetarian eating campaign in China. This campaign is motivated by the health and environmental benefits of plant-rich diets. The name of the campaign is Shushi Zhuyi (蔬食主义), a wordplay on vegetarianism (素食主义). The purpose of this wordplay is to differentiate the campaign from vegetarianism (Nita, 2019), which has not been well-received by all Chinese eaters. Instead of promoting meat-excluding vegetarianism, the Shushi campaign encourages residents in China to increase vegetable consumption. To reach a wide audience, this campaign has featured celebrities in educational videos and collaborated with restaurants to promote low-carbon cuisines.

Fair trade

Fair trade is a relatively new concept in China. Despite nascent production in fair trade tea and coffee in China (see Qiao et al., 2016), access to and consumer awareness about registered fair trade food products remain low (Gomersall & Wang, 2012). However, since 2020, some food retail businesses in China have provided support to farmers in order to relieve farmers' economic stress caused by supply chain interruptions during the pandemic. For example, online retail businesses such as Pinduoduo have been connecting farmers to consumers through direct sales and live streaming, while offering marketing training for farmers and offering access to fertilisers (Liang & Cheah, 2020). In lieu of the fair-trade certification, online platforms promote direct sales, which may provide a fairer distribution of sales revenues to farmers than the conventional supply chain. In addition, alternative food networks such as the Beijing Farmers' market have demonstrated the principle of fair trade in their business model by enabling small-scale ecological farmers to sell their products at fair prices.

Conclusion

Overall, this chapter highlights four challenges to fully implementing sustainable diets in China: (a) The increasing food waste; (b) rising food-away-from-home and online delivery; (c) persistent food safety anxiety and ecological threats posed by the over application of agrochemicals and food additives; and (d) the loss of traditional food varieties and agricultural biodiversity. In response, the government has been encouraging healthy and balanced eating in dietary guidelines and has enacted the food safety law to address food safety concerns. At the same time, civil society initiatives are emerging across the country to address food safety and environmental concerns through sustainable food production, food education, and building producer-consumer networks.

While the term "sustainable diets" is not widely used in China, the practices of purchasing and eating healthy and ecologically sourced food have been promoted by recent government policies and civil society initiatives. Due to the sheer scale of China's food consumption and its significance in world food trade, these ongoing dietary shifts in China have important global implications. For example, China's soybean imports from Brazil as livestock feed are linked to the deforestation of the Amazon rainforest (Schneider, 2011). This chapter's analysis of the challenges and opportunities of sustainable diets in China allows us to reimagine the global agenda of advocating for sustainable food systems. The Chinese experience can provide some inspiration for government interventions that educate about and cultivate sustainable diets. This chapter encourages policy makers to consider the multiple dimensions of sustainable diets in the conception of food policies, laws, and guidelines. This will allow food challenges, such as food safety risks, food waste, and agricultural pollution, to be holistically addressed rather than treated as separate issues. In addition, the civil society food initiatives in China may provide inspiration for international organisations working on the promotion of sustainable diets. It will be helpful for civil society food initiatives to adopt and to promote the language of sustainable diets, in China and beyond. Thus, creating space for collaboration and solidarity between food organisations with different foci to proactively extend the reach and impact.

References

Andrée, P., Clark, J. K., Levkoe, C. Z., & Lowitt, K. (Eds.) (2019). *Civil society and social movements in food system governance*. Routledge.

Arranz, A., & Huang, H. (2020, March 4). *China's wildlife trade*. South China Morning Post. https://multimedia.scmp.com/infographics/news/china/article/3064927/wildlife-ban/index.html

Bai, J., Wahl, T., Lohmar, B., & Huang, J. (2010). Food away from home in Beijing: effects of wealth, time and 'free' meals. *China Economic Review, 21*(3), 432–441.

Bai, J., McCluskey, J., Wang, H., & Min, S. (2014). Dietary globalization in Chinese breakfasts. *Canadian Journal of Agricultural Economics, 61*(3), 325–341.

Bélanger, J., & Pilling, D. (Eds.) (2019). *The state of the world's biodiversity for food and agriculture*. FAO Commission on Genetic Resources for Food and Agriculture Assessments. Retrieved from https://www.fao.org/3/ca3129en/CA3129EN.pdf

Cao, P., & Fang, M. (2013, February 1). *gōngyì 2013–cānyù huìjù néngliàng:"guāngpánxíngdòng" bèi zhuǎnfā 5000 wàncì* 公益2013–参与汇聚能量:"光盘行动"被转发5000万次 *[Philanthropy 2013: Accumulating the positive energy: "Clean Plate" campaigns were reblogged 50 million times]*. People.com.cn. http://dangjian.people.com.cn/n/2013/0125/c117092-20321311.html

CSA Network (2018, December 15). *zhōngguó shèhuì shēngtài nóngyè CSA dàhuì shízhōunián* 中国社会生态农业CSA大会十周年 *[The 10th anniversary of Chinese socio-ecological agriculture and community supported agriculture convention]*. http://csanetwork.cn/index.php/cms/show-67.html

Diao, S., Huang, S., Xu, Z., Ye, S., Yuan, X., Chen, Z., Zhang, H., Zhang, Z., & Li, J. (2019). Genetic diversity of indigenous pigs from South China area revealed by SNP array. *Animals, 9*(6), 361.

Gernert, M., Bilali, H. E., & Strassner, C. (2018). Grassroots initiatives as sustainability transition pioneers: Implications and lessons for urban food systems. *Urban Science, 2*(1), 23.

Gomersall, K., & Wang, M. Y. (2012). Expansion of fairtrade products in Chinese market. *Journal of Sustainable Development, 5*(1), 23.

Good Food Fund (2019). *liángshí yìtí rùmén jīngxuǎn* 良食议题入门精选 *[Selected subjects to learn about good food]*. The Good Food Fund of China Biodiversity Conservation and Green Development Foundation. Retrieved from https://brightergreen.org/wp-content/uploads/2020/04/Good-Food-Summit-Primer_%E8%89%AF%E9%A3%9F%E5%9F%BA%E9%87%91%E9%A3%9F%E7%89%A9%E8%AE%AE%E9%A2%98%E5%85%A5%E9%97%A8.pdf

Healthy China Promotion Committee (2019, July 15). *jiànkāng zhōngguó xíngdòng(2019–2030 nián)* 健康中国行动（2019–2030年）*[Healthy China campaign (2019-2030)]*. http://www.gov.cn/xinwen/2019-07/15/content_5409694.htm

Kang, Y., Łuczaj, Ł., Kang, J., & Zhang, S. (2013). Wild food plants and wild edible fungi in two valleys of the Qinling Mountains (Shaanxi, central China). *Journal of Ethnobiology and Ethnomedicine, 9*(26), 1–19.

Koc, M., MacRae, R., Desjardins, E., & Roberts, W. (2008). Getting civil about food: The interactions between civil society and the state to advance sustainable food systems in Canada. *Journal of Hunger & Environmental Nutrition, 3*(2–3), 122–144.

Krul, K., & Ho, P. (2017). Alternative approaches to food: Community supported agriculture in urban China. *Sustainability, 9*(5), 844.

Li, Y., Peng, Y., Hallerman, E. M., & Wu, K. (2014). Biosafety management and commercial use of genetically modified crops in China. *Plant Cell Reports, 33*(4), 565–573.

Lim, G.Y. (2021, April 22). *GM labelling: Chinese consumers willing to pay for traceability codes and allergen presence for soybean oil: Survey*. Food Navigator-Asia. https://www.foodnavigator-asia.com/Article/2021/04/22/GM-labelling-Chinese-consumers-willing-to-pay-for-traceability-codes-and-allergen-presence-for-soybean-oil-Survey

Liang, H., & Cheah, S. M. (2020, November 25). *Pinduoduo: Empowering farmers with an e-commerce platform*. Pingduoduo.com. https://stories.pinduoduo-global.com/articles/pinduoduo-empowering-farmers-with-an-e-commerce-platform

Liu, G. (2014, April 3). *Food losses and food waste in China: A first estimate*. OECD Publishing. Retrieved from https://www.oecd-ilibrary.org/docserver/5jz5sq5173lq-en.pdf?expires=1648719680&id=id&accname=guest&checksum=2FAE4BA58B4E8A53BCBFD5DBD61785A4

Liu, C. (2021, April 29). *China adopts law against food waste; binge eating, excessive leftovers to face fines*. Global Times. https://www.globaltimes.cn/page/202104/1222490.shtml

Luehr, G., Glaros, A., Si, Z., & Scott, S. (2019). Urban agriculture in Chinese cities: Practices, motivations and challenges. In A. Thornton (Ed.), *Urban food democracy and governance in North and South* (pp. 291–309). Palgrave.

Luo, S. (2016). Agroecology development in China: An overview. In S. Luo & S. R. Gliessman (Eds.), *Agroecology in China: Science, practice, and sustainable management* (pp. 3–35). CRC Press.

Ma, H., Huang, J., Fuller, F., & Rozelle, S. (2006). Getting rich and eating out: consumption of food away from home in Urban China. *Canadian Journal of Agricultural Economics, 54*(1), 101–119.

Milman, O., & Leavenworth, S. (2016, June 20). *China's plan to cut meat consumption by 50% cheered by climate campaigners*. The Guardian. https://www.theguardian.com/world/2016/jun/20/chinas-meat-consumption-climate-change

Nita (2019, March 21). *shūshízhǔyì láile, zhège gōngyìzǔzhī xiǎng ràngnǐ yòng měiyìkǒu gǎibiàn shìjiè* 蔬食主义来了，这个公益组织想让你用每一口改变世界 *[Vegetarianism is coming, this non-profit organization wants you to change the world one bite at a time]*. Socialbeta. https://socialbeta.com/t/case-wildaid-shushi-program-campaign-2019-03

Qiao, Y., Halberg, N., Vaheesan, S., & Scott, S. (2016). Assessing the social and economic benefits of organic and fair trade tea production for small-scale farmers in Asia: A comparative case study of China and Sri Lanka. *Renewable Agriculture and Food Systems, 31*(3), 246–257.

Reardon, T., Timmer, C. P., & Minten, B. (2012). Supermarket revolution in Asia and emerging development strategies to include small farmers. *Proceedings of the National Academy of Sciences of the United States of America, 109*(1), 12332–12337.

Ren, Z. (2013, February 17). *guójiā liángshí júzhǎng: cānzhuō làngfèi měinián gāodá 2000 yìyuán* 国家粮食局长：餐桌浪费每年高达2000亿元 *[Head of the state administration of grain: Food waste at the table amounts to 200 billion yuan per year]*. Xinhua News Agency. https://china.caixin.com/m/2013-02-18/100491651.html.

Scherf, B. D., & Pilling, D. (Eds.) (2015). *The second report on the state of the world's animal genetic resources for food and agriculture*. FAO Commission on Genetic Resources for Food and Agriculture Assessments. Retrieved from www.fao.org/3/a-i4787e.pdf.

Schneider, M. (2011). *Feeding China's pigs: Implications for the environment, China's smallholder farmers and food security*. Institute for Agriculture and Trade Policy. Retrieved from https://repub.eur.nl/pub/51021/Metis_198351.pdf

Schumilas, T., & Scott, S. (2016). Beyond 'voting with your chopsticks': Community organising for safe food in China. *Asia Pacific Viewpoint, 57*(3), 301–312.

Scott, S., Si, Z., Schumilas, T., & Chen, A. (2014). Contradictions in state-and civil soci-ety-driven developments in China's ecological agriculture sector. *Food Policy, 45*, 158–166.

Scott, S., Si, Z., Schumilas, T., & Chen, A. (2018). *Organic food and farming in China: Top-down and bottom-up ecological initiatives.* Routledge.

Shi, Y., Cheng, C., Lei, P., Wen, T., & Merrifield, C. (2011). Safe food, green food, good food: Chinese community supported agriculture and the rising middle class. *International Journal of Agricultural Sustainability, 9*(4), 551–558.

Si, Z. (2017). Rebuilding consumers' trust in food: Community-supported agriculture in China. In J. Duncan & M. Bailey (Eds.), *Sustainable food futures: Multidisciplinary solutions* (pp. 34–45). Routledge.

Si, Z. (2019). *Shifting from industrial agriculture to diversified agroecological systems in China.* The Centre for Agroecology, Water and Resilience. https://www.coventry.ac.uk/globalassets /media/documents/research-documents/coventry-china-agriculture-aw-new-style.pdf

Si, Z., Regnier-Davies, J., & Scott, S. (2018). Food safety in urban China: Perceptions and coping strategies of residents in Nanjing. *China Information, 32*(3), 377–399.

Si, Z., Schumilas, T., & Scott, S. (2015). Characterizing alternative food networks in China. *Agriculture and Human Values, 32*(2), 299–313.

Sohu (2021, January 7). *zhòngbàng! zhèjiāngshěngzhèngfǔ kāishè sùshí jiànkāng cāntīng, sùshí fēngcháo zài zhèngfǔjīguān xīngqǐ!* 重磅！浙江省政府开设素食健康餐厅，素食风潮在政府机关兴起！ *[Breaking news! The provincial government of Zhejiang opens a vegetarian healthy canteen, vegetarianism is spreading across government institutes!].* https://www.sohu .com/a/443138872_689357

Song, G., Zhang, H., Duan, H., & Xu, M. (2018). Packaging waste from food delivery in China's mega cities. *Resources, Conservation and Recycling, 130*, 226–227.

Song, X., Li, G., Vernooy, R., & Song, Y. (2021). Community seed banks in China: achieve-ments, challenges and prospects. *Frontiers in Sustainable Food Systems, 5*, 108.

Song, Y., Li, J., & Vernooy, R. (2012). China: Designing policies and laws to ensure fair access and benefit sharing of genetic resources and participatory plant breeding products. In M. Ruiz, & R. Vernooy (Eds.), *The custodians of biodiversity* (pp. 94–120). Earthscan.

UNDP/GEF and MFPRC (Ministry of Finance and the People's Republic of China) (2005). *China biodiversity conservation national capacity self-assessment report (third draft).* GEF. Retrieved from https://www.thegef.org/sites/default/files/ncsa-documents/377_0.pdf

Weis, T. (2013). The meat of the global food crisis. *The Journal of Peasant Studies, 40*(1), 65–85.

World Wide Fund for Nature (WWF). (2018). *Report on Food Waste in the Food Catering Industry in Chinese Cities.* Accessed from http://www.zjsef.cn/upload/editor/1573/ 1537495761555.pdf (in Chinese).

Yasuda, J. K. (2015). Why food safety fails in China: The politics of scale. *The China Quarterly, 223*, 745–769.

Yuan, M., Seale Jr, J. L., Wahl, T., & Bai, J. (2019). The changing dietary patterns and health issues in China. *China Agricultural Economic Review, 11*(1), 143–159.

Zhang, J. Y. (2018). Cosmopolitan risk community in a bowl: a case study of China's good food movement. *Journal of Risk Research, 21*(1), 68–82.

Zhong, R., & Zhang, C. (2019, May 28). *Food delivery apps are drowning china in plastic.* The New York Times. https://www.nytimes.com/2019/05/28/technology/china-food-delivery-trash.html#:~:text=Scientists%20estimate%20that%20the%20online,jump% 20from%20two%20years%20before

Zhong, T., Si, Z., Crush, J., Scott, S., & Huang, X. (2019). Achieving urban food security through a hybrid public-private food provisioning system: the case of Nanjing, China. *Food Security, 11*, 1071–1086.

Zhong, T., Si, Z., Scott, S., Crush, J., Yang, K., & Huang, X. (2021). Comprehensive food system planning for urban food security in Nanjing, China. *Land, 10*(10), 1090.

Zhou, Y., Du, S., Su, C., Zhang, B., Wang, H., & Popkin, B. (2015). The food retail revolution in China and its association with diet and health. *Food Policy, 55*, 92–100.

51

BIO-CULTURAL DIVERSITY IN SOUTH AMERICA

Overcoming agro-extractivism linked to unhealthy diets

Theresa Tribaldos, Johanna Jacobi,
Aymara Llanque, and Maria Teresa Nogales

Agro-extractivism and its implications for local food systems

Over the last decades, as researchers in these regions, we have observed an accelerating process of large-scale land conversion from forests and small-scale agricultural production to large monocultures producing commodities for export. These commodities are used for biofuel production, feed production, and human consumption. Soy requires special attention in this regard, as it is used for all three purposes. Yet the majority of soy production goes into feed and biofuel production to satisfy a growing global demand for meat and dairy products and alternatives for fossil fuels.

Brazil, Bolivia, and Argentina are particularly affected by this emphasis on production for feed and fuel due to state support and favourable conditions for agribusiness and land conversions (Ye et al., 2020). The associated effects of this process range from destructive environmental implications to human rights violations, the displacement of rural small land holders, and the destruction of whole communities. In combination, these create severe and devastating impacts on local food security and sustainable food systems. This large-scale land conversion diminishes biodiversity and agro-biodiversity from rich ecosystems and habitats to deserts that are occupied by very few plants and organisms. Accordingly, clearing out whole landscapes also limits livelihoods in such places because it leaves no room for diverse food systems that support a variety of activities along the food chain. These developments make food imports necessary as they lead to a collapse of local food crops and breeds and traditional fisheries and hunting. Well-known outcomes include simplified, unhealthy diets, an increased dependency on supermarket offers, and higher prices for healthy food.

DOI: 10.4324/9781003174417-60

The exploitation of nature, labour, and society has reached such dimensions that scholars have used the concept of *agro-extractivism* (Delgado Wise, 2020; McKay, 2017; Ye et al., 2020). This term refers to the creation of wealth and power along agricultural commodity chains through the process of extraction (traditionally used in mining and logging) from nature in combination with technological innovations. McKay (2017) defines the concept based on the case of soy production in Bolivia, along with the following criteria: The extraction of large volumes for export; a concentration of power and wealth; and degrading environmental and societal impacts. Ye et al. (2020) stress that extractive practices mean the natural resources and wealth are taken from less prosperous economic regions and these goods and services are transferred to wealthier regions largely for export without compensating those bearing the burden of this model and without practices of regeneration for the ecology. The concentration of wealth in the hands of few beneficiaries, and especially the consumptive use of resources without replenishing, means jeopardising resources for future generations. Furthermore, such highly concentrated agricultural commodity chains are also opaque in terms of their networks of interest that steer power dynamics and decision-making behind the scenes, which have the means and capacity to undermine democratic decision-making in food systems.

In essence, agro-extractivism as an agricultural production model destroys the various dimensions of diversity—biodiversity, agrobiodiversity, and cultural diversity—through the conversion of ecosystems and small-scale diverse farming systems to monocultures. Agro-extractivism also curtails lively culture in communities through migration and displacement. This diversity is not just central for healthy and resilient ecosystems and communities, but it also provides the basis for diverse healthy diets and human wellbeing. And relatedly, agro-extractivism also has indirect effects on dietary injustices, shifts in food availability and contributes to rising health costs.

In the remainder of this chapter, we explore possible responses to agro-extractivism and dive into opportunities that diversity offers along the whole food chain for food systems and health. We then introduce two transformative initiatives from Bolivia and Brazil that demonstrate such responses, before explaining the meaning of enabling environments and by-passing agro-extractivism through short value chains and more direct producer consumer networks. We conclude the chapter by stressing the need for context-specific solutions to food system problems at different scales and in different places.

Promising opportunities through diversity

Diversity in food systems can be seen as the backbone that connects healthy and resilient ecosystems to human health through healthy diets. In this section, we expand on this connection through highlighting the role diversity plays in the different parts of food systems. This argument helps to better understand why agro-extractivism as a production model for agricultural commodities is eroding the foundation of sustainable food systems and depleting social-ecological resources.

The impacts of modern industrialised diets on human health in the form of obesity, non-communicable diseases (Poti et al., 2017), and mental health problems (O'Neil et al., 2014) have been investigated for some time. It is also well known that the quality of drinking water and air are crucial in this regard (Fattore et al., 2011; Levallois & Villanueva, 2019) and more recent research is increasingly uncovering the dangerous impacts of pesticides on human health (Kim et al., 2017). Now, how does this relate to diversity in food systems?

While guidelines for healthy diets differ in detail from country to country and among different institutions, they all include a balance of fresh fruit and vegetables, legumes, whole grains, few animal proteins, unsaturated fats, and only occasionally sugary and processed food items (Gonzalez Fischer & Garnett, 2016). A healthy person's gut profits from more diversity in the diet as this increases its microbiome. Accordingly, a variety of fresh and affordable food should be available to the whole population if the right to food as a fundamental human right is to be fulfilled. Attempts to link healthy diets to a healthy planet are also growing. Among the most prominent of these is the planetary health diet proposed by the EAT Lancet Commission (Willett et al., 2019). While the goal of this proposal—to achieve healthy nutrition stemming from a healthy environment—is rather uncontested, the two targets—sustainable food production and healthy diets—raise questions as to their implementation. The main question is how to translate global targets into individual contexts. For example, what does it mean to have an acceptable extinction rate of biodiversity loss and which species can be done without in a given context? Who decides upon this? And how can such a blueprint take into account the cultural and traditional richness of historically diverse foodscapes? Other critiques include the affordability of the proposed diet (Hirvonen et al., 2020) and the feasibility of increasing the amount of produced nuts required for it (Braich & Martinez, 2019).

Human constructed food systems should be there to serve the people, and healthy nutrition for people within planetary boundaries should be put at the centre of the discussion. The five strategies for a great food transformation stipulated by the EAT-Lancet commission include some highly controversial points. The most problematic ones include that a sustainable intensification of agriculture seems to rely in large part on fine tuning conventional agricultural practices and does not mention agroecological or highly diverse agricultural production systems. The proposal also propagates the half-earth approach, which demands that 50% of the earth's surface is set aside for the protection of intact natural ecosystems. This approach is highly problematic with regards to ethical questions around what to protect, and for whom, as many of these areas are also populated by people. It is also unclear what "intact" means and who decides about it. The proposed planetary health diet addresses primarily agricultural production and consumption while neglecting other parts of food systems and making no mention of unequal power relations.

The EAT-Lancet commission's demand for a Great Food Transformation (Willet et al., 2019) seems at a first glance like a reference to Polanyi's "The Great Transformation" (Polanyi, 1944), but the "Transformation" stipulated in the

EAT-Lancet report falls short on exactly the questions behind its own goals. It was the political-economic "Great Transformation" famously described by Polanyi (1944), with its commercialisation of land and labour, social disintegration, racist and sexist subordination, and loss of values, that brought the food system to where it is now. By the increased commodification of all aspects related to the food system—be it seeds, land, or labour—rural populations abandon food production and make way for agro-extractivism that promotes flex crops, increasing food insecurity linked to inequality, and the depletion of soils, water, forests, and biodiversity. It seems naive to apply the same concept to the much-needed food systems transformation today. The trajectory of the same way of thinking that has created the problems in the first place can also be observed in an increasing "nutritionism," or treating symptoms of homogenous diets rather than the root causes such as the lack of diversity that creates the lack of nutrients. On the contrary, alternatives that exist (and many have existed for a long time) need to become more prominent and more common.

Hence, while the idea of introducing diets that are good for humans and nature deserves support, the actual proposal on how to implement it needs further reflection. Healthy diets are diverse diets, which must be based on sustainable food and agroecosystems, of which biodiversity and agrobiodiversity are essential components. But the importance of diversity also continues in other parts of the food chain. Diverse local and regional processing and distribution facilities support more diverse options for livelihoods, create jobs in different occupations, and support the shaping of local specialities. The interplay between such diverse food and agroecosystems and local culture traditionally emerged in many places and can be conceptualised as bio-cultural diversity: The knowledge, innovation, and practises applied to safeguarding and sustaining healthy natural environments in symbiosis with local livelihoods (Bridgewater & Rotherham, 2019). Such symbiosis entails profiting from such environments through healthy diets and natural resources while sustaining their high quality and health.

If we consider bio-cultural diversity as an essential link between human and nature health for sustainable food systems, it becomes clear that healthy diets must build on the diversity of local contexts and hence, cannot be approached from a universal perspective on diets for the whole planet. In addition, the consideration of bio-cultural diversity in food systems as a context- and place-specific phenomenon raises questions about decision-making in general. Who should have the right to decide about different food system activities and what should such processes look like? The next section will introduce two cases of transformative initiatives in Bolivia and Brazil, which shed light on questions of diversity within food systems and how to establish processes that support greater diversity.

The cases of Bolivia and Brazil

Community garden Lak'a Uta at 4000 m.a.s.l. in La Paz, Bolivia

The Huerto Orgánico Lak'a Uta is dedicated to urban, diversified, and organic food production. Founded in 2014 and set in the mountainside of the city of La

Paz, the garden promotes urban agriculture, fosters community values and outdoor recreation, and encourages neighbours to adopt healthy diets (Figure 51.1).

The garden is located in what had been a densely populated neighbourhood that was buried in a landslide in the 1990s. Abandoned for close to 20 years, the park was frequented by persons consuming alcohol and known for violent incidents. Today, the park holds 40 plots in which families grow more than 30 different varieties of fruits, vegetables, and herbs. Fundación Alternativas, the organisation behind this initiative, provides technical support in ecological urban farming methods, lends tools, keeps a seed bank, and provides materials, which local businesses donate to support the garden. The participating families have become active change agents in their communities, where they practice and promote urban agriculture as a means to improve food security. In 2018, after having visited the garden on several occasions, the local government of La Paz gave way to the adoption of the Municipal Autonomous Law 321 for the Promotion of Urban Food Gardens in the Municipality of La Paz. The first of its kind, this law provides municipal land to the public so that people can initiate food gardens in their communities.

The Huerto Orgánico Lak'a Uta offers visitors and community members a place for healthy recreation as well as interaction and reconnection with Mother Earth (*Pachamama*)—as expressed by the Indigenous peoples of the Andes, this Earth is considered as a sacred living being. The garden has adopted low-cost production models based on recycled material and focuses not only on ensuring each family can produce fruits and vegetables to complement their regular diet but also in recovering community values and customs that are increasingly lost in an urban world. To this end, citizens have, for example, re-adopted the ancient tradition of swapping (*trueque*). In tandem, and in an effort to bolster community development,

Figure 51.1 The Huerto Orgánico Lak'a Uta at 4000 m.a.s.l. in La Paz, Bolivia. Source: Fundación Alternativas. Source of image: Johanna Jacobi. Used with permission.

Fundación Alternativas organises diverse activities based on ancestral Andean traditions of reciprocity such as community potlucks (*apthapis saludables*), collaborative work days (*ainis*), and provides an array of workshops on food security and agroecological food production. These activities enable people to come together to work collaboratively, share meals as a community, and continuously engage in learning experiences. Most of the plots are run by women from nearby neighbourhoods. They produce healthy, pesticide-free food for their families. They also build community by exchanging knowledge and produce with others.

Yearly evaluations conducted by Fundación Alternativas track members' perceptions regarding the importance of the garden. Interestingly, interviews conducted in 2018 with 40 involved families revealed that 100% of participants indicated their primary motivation to participate in the garden is having the opportunity to grow their own food; 64% enjoy the peace and quiet that the garden gives them; and 54% enjoy the opportunity to spend time in nature. Likewise, 51% stated that they enjoy the opportunity to engage socially with other members of the garden. Future research should consider the volume of food produced and the contribution the gardens can make to household food security and sufficiency. Over the years, the garden has also become an educational platform where students of all ages come to learn about possibilities to grow food in the city and at this altitude. Adults, children, and youth visit the garden to learn different growing techniques as well as how to make organic fertilisers, grow food in small spaces, and use organic pest control. Likewise, since 2017, Fundación Alternativas has been conducting workshops specifically designed for public school teachers and community educators in an effort to ensure they learn about the multidimensional impacts of food insecurity and how to bridge these topics in the classroom. With the active participation of teachers, the organisation has developed classroom material to ensure students from kindergarten through twelfth grade receive appropriate instruction on such important topics. To date, more than 10,000 people have visited the garden and participated in activities designed to encourage citizens to become more aware of the importance of caring for the environment, improving their diet, and reconnecting with Mother Earth.

Today and over the course of seven years, the Huerto Orgánico Lak'a Uta has become a well-known local initiative that has given way to citizen's appreciation for urban agriculture, its ability to serve the environment, and ensure people can easily access fresh and healthy food.

Ecovida Network's guarantee certification system for food sustainability in southern Brazil

The Ecovida agroecology network is an initiative born in the south of Brazil (Santa Catarina and Rio Grande do Sul) and extends to the states of Paraná, São Paulo, and Bahia, with a proposal for sustainable production, transformation, distribution, and consumption. Rede Ecovida was created in 1998 as a result of a progressive aggregation process of different local NGOs, whose goal was to find an alternative to the Green Revolution (Cabral et al., 2022; da Costa et al., 2017;

da Silveira, 2013). The network promotes the exchange of agroecological products through a participatory guarantee system—a methodology for verifying the conformity of products without market intermediation in the management of the seals in order to face certification by auditing private companies.

The Ecovida network brings together small family farmers, traditional communities, local and state NGOs, consumer organisations, global initiatives such as the Via Campesina, the landless movement (a social movement in Brazil advocating for land reform to provide poor workers with access to land for food production), and Slow Food, as well as universities and some local governments. For the last 30 years, they have been strengthening agroecological production initiatives combined with technical training spaces, as well as massive meetings between farmers, where they exchange experiences, seeds, production, and innovative solutions for their food systems. The support of this initiative by NGOs and social organisations is relevant because their subsidies enable farmers to receive seeds, production training, marketing training, and training to develop strategies to reach markets in other districts. NGOs were fundamental to advance the recovery of agricultural land and to strengthen agroecological practices in the production processes. For example, NGOs such as CEPAGRO (Center for the Study and Promotion of Group Agriculture), the Federal University of Santa Catarina, and the FIA Inter-American Foundation, have created financial support for small family farming linked to the Ecovida Network, which has been working for more than 15 years.[1] Gradually, the financing of this initiative has been complemented with funds generated by the network itself.

The farming families involved in the network form nuclei by municipality and region. Their aim is to produce "real food," first for their families, then for their neighbours and ultimately for the market. The network operates in a decentralised manner based on the organisation of the families and their own local/territorial dynamics. Each nucleus has a person responsible for regional coordination, who makes decisions at the municipal level among Ecovida Network farmer groups. There are also supra-municipal spaces, between states, where the local nuclei participate through a representative. The logic of organisational representation makes it possible to carry the voices of the nuclei forward to other decision-making levels and is also a way of delegating Ecovida Network management to a coordinating group.[2] A relevant factor in the organisation of the Ecovida network is social cohesion.

Transparency and accountability are important values for the Ecovida Network; its organisational capacity depends on building trust between nuclei. For example, the reputation of the members in the network has high value, and the fear of losing reputation prevents farmers from breaking the rules (da Silvieira, 2013). Each nucleus has specific forums for deliberation and decision-making, namely Group Meetings, Core Plenary, State Plenary, General Plenary, and Extended Meeting (a larger meeting space for network members that takes place every 2 years).

The relationship between producers and consumers is relatively close with a short supply chain, as an important part of the food is consumed locally, and the surpluses are distributed in the region. Furthermore, most of the bio-inputs are technologies generated by the producers themselves. One of the principles of the Ecovida Network is to promote the social dynamics that arise from the integration of multiple actors involved in the production, processing, and consumption of food. Credibility is generated from the seriousness of the farming family and is socially legitimised as each farming family involved assumes a commitment to the Ecovida Network to produce agro-ecological food without the use of pesticides, and to use the Participatory Guarantee Seal (SPG). In order to use the seal, it is necessary to go through an internal peer review process based on the requirements set by a group of farmers who are also interested in using the Ecovida seal.

The Ecovida Network is a platform that supports diversity in the territories where agroecological production takes place. It gives priority to integrated production with biological products for agroecological pest management. In addition, the network includes permanent points of sale for traditional local foods as a way of recovering culturally valuable diets.

An illuminating example of the social cooperation among the network members is the Enlarged Meeting in 2017, a three-day event for the exchange of experiences and participatory decision-making, where the nuclei of the network participated as groups of associates. In the Enlarged Meeting, there was space for political discussions, on the one hand about the structural reasons for the dominance of the agro-industrial food system in the southern region of the country, and on the other about ongoing alternatives made in the nuclei of the network. One of the most relevant activities in the meeting was the Sabores e Saberes (flavour and knowledge) fair—a showcase of products offered by all participants, members of the Network, and guests. The Sabores e Saberes fair also included the offer of food made with agroecological ingredients and cooked by Slow Food, a political organisation that promotes the consumption of agroecological food. For some illustration, see the images in Figures 51.2 and 51.3.

Another characteristic of the network is the feminist approach of organised women with the slogan: "without feminism there is no agroecology." Within the network there are women's initiatives called peasant feminism, which promote a view on the role of women in building agroecological processes that consider the context and challenges they face within the territories.[3]

The main challenges of the network are related to state and national regulations, since a good part of their products are not considered as compliant with national regulations on food safety, which usually requires specific industrial procedures (Oliveira, 2020). Furthermore, the distribution system organised by the Ecovida network includes trucks managed by farmers' organisations, which are not compliant with transport regulations. The complex regulations for heavy transport, especially associated with safety and volume standards, make the transport service more expensive and thus, unrealistic, for farmers' organisations.

Reconnecting diverse social-ecological systems and healthy diets

States set the regulatory framework for the food system activities of different food system actors, which should include responsibilities in guaranteeing healthy diets

Figure 51.2 The mystique of the encounter: The mystique consists in the preparation of a ritual table, at the beginning of each meeting of the Ecovida network. This display shows the diversity of seeds, foods, and traditions of peasant families in the southern region of Brazil. Source: Aymara Llanque-Zonta. Used with permission.

Figure 51.3 Red Ecovida 2017 with a fair within the extended meeting. Participants are exposed to local foods, have opportunities to exchange knowledge, and to sell their products. Source: Aymara Llanque-Zonta. Used with permission.

for the whole population and the preservation of natural resources. Such responsibilities and obligations are based on several foundations, including the Right to Food and the UN Declaration on the Rights of Peasants and Other People Working in Rural Areas, which are especially important in Latin America due to the high inequalities tied to food insecurity in many rural areas. The Right to Food as a fundamental human right was recognised in 1999 underlining the urgency to eliminate malnutrition and hunger (UN, 1999). A broad interpretation of this right entails not only the access, availability, and sufficiency of food suitable for different genders, ages, and cultures but extends to stability over time and thus, includes social, environmental, and economic sustainability in food systems (De Schutter, 2014). The UN Declaration on the Rights of Peasants and Other People Working in Rural Areas was adopted in 2018 and is meant to protect peasants and rural people engaged in small-scale agricultural production, their access to land, and the right to use natural resources for their livelihoods. Special attention is paid to the non-discrimination of women. Additional analysis of international agreements and commitments to the right to food also are embedded in other chapters in this collection.

Although these rights are recognised by the UN to remind states of their obligations to create enabling environments for sustainable food systems, not all states show the political will or are making sufficient progress in fulfilling these rights. This non-compliance is often closely linked to networks of interest between political elites and corporate power with substantial financial influence. In view of the shortcomings of some states with regards to the aforementioned rights and the lack of enforcement instruments at the international scale, communities in many states are increasingly starting their own initiatives to regain control over their food systems, as shown in the examples from Bolivia and Brazil. The aim of such initiatives includes providing fresh and healthy food for their members, creating a sense of community and places for social activities, and/or collectively implementing new small-scale business ideas. Van der Ploeg (2018) describes these initiatives as the emergence of new nested markets, which exist alongside conventional food markets and "bypass" conventional power structures by building their own networks of producers, processors, and consumers. Such nested markets show a certain distinctiveness in terms of price, quality, or production process, and thus, stand out from conventional food markets. They have in common that, apart from creating local specialty products, they also contribute to sustaining social capital and ecological resource bases, and they can preserve cultural links between food production and consumption.

A special case of nested markets can be seen in *prosumption*—different forms of involving consumers in the food production process and a reorganisation of labour in food systems with the aim of rebuilding trust in producer-consumer relations (Podda et al., 2021). With clear institutions and rules guiding prosumption structures, they are able to create added local value and sometimes even social innovations with new economic spaces (Podda et al., 2021).

Conclusion

Overcoming complex problems of current food systems and agro-extractivism requires a diversity of solutions. Bio-cultural diversity can help to look for local, context-specific solutions, which are manifold. Examples show us that change is possible but needs both policy support and civil society applying and demanding change. It is clear that the aim cannot be to promote a globally uniform, sustainable and healthy diet—void of culture, tradition, and connection—for all, but instead create spaces for place-specific food systems, based on a respectful valorisation and application of local knowledge and bio-cultural diversity.

The community garden in La Paz shows how the links between community, recreational space, and healthy diets can be made. The community garden has set an important precedent regarding how to transform under-utilised spaces into green and productive areas that foster the resurgence of a sense of community. The Ecovida network exemplifies how conventional food structures can be overcome by building trust among participants while creating the necessary institutions and rules to set up fair food systems that are able to provide sustainable, healthy, and affordable diets for their members. It also shows that connections between regions and beyond local markets are possible if the rules allow them to pursue common goals such as the participatory guarantee system established among the Ecovida network. Remarkable in this network is the emphasis on democratic decision-making and deliberative practices to reach mutual understanding.

Such alternatives to conventional food systems with all their shortcomings have the potential to add value and diverse economic opportunities to communities, support thriving agricultural landscapes, and create well-being for humans and nature instead of exhausting social and natural capital through the model of agro-extractivism.

Notes

1 https://cepagroagroecologia.wordpress.com/.
2 http://ecovida.org.br.
3 https://agroecologia.org.br/2018/09/05/sem-feminismo-nao-ha-agroecologia-2/.

References

Braich, G., & Martinez, J. D. (2019, February 11). *The EAT-Lancet diet: Is it all nuts?* The Nature of Food. Retrieved from https://medium.com/the-nature-of-food/the-eat-lancet-diet-is-it-all-nuts-2db9b0d7579f

Bridgewater, P., & Rotherham, I. D. (2019). A critical perspective on the concept of bio-cultural diversity and its emerging role in nature and heritage conservation. *People and Nature, 1*(3), 291–304.

Cabral, L., Pandey, P., & Xu, X. (2022). Epic narratives of the green revolution in Brazil, China, and India. *Agriculture and Human Values, 39,* 249–267.

da Costa, M. B. B., Souza, M., Júnior, V. M., Comin, J. J., & Lovato, P. E. (2017). Agroecology development in Brazil between 1970 and 2015. *Agroecology and Sustainable Food Systems, 41*(3–4), 276–295.

da Silveira, S. M. P. (2013). Rede Ecovida de agroecologia: uma inovação estratégica para o desenvolvimento territorial sustentável na zona costeira catarinense? *Revista Internacional Interdisciplinar INTERthesis, 10*(2), 181–213.

De Schutter, O. (2014). *Final report: The transformative potential of the right to food.* UN General Assembly, Human Rights Council. Retrieved from https://www.ohchr.org/EN/HRBodies/HRC/RegularSessions/Session25/Documents/A_HRC_25_57_ENG.DOC

Delgado Wise, R. (2020). Capitalism on the frontier of agroextractivism. In H. Veltmeyer & E. Lau (Eds.), *Buen Vivir and the challenges to capitalism in Latin America* (pp. 50–70). Routledge.

Fattore, E., Paiano, V., Borgini, A., Tittarelli, A., Bertoldi, M., Crosignani, P., & Fanelli, R. (2011). Human health risk in relation to air quality in two municipalities in an industrialised area of Northern Italy. *Environmental Research, 111*(8), 1321–1327.

Gonzalez Fischer, C., & Garnett, T. (2016). *Plates, pyramids, planet: Developments in national healthy and sustainable dietary guidelines: A state of play assessment.* FAO & FCRN. Retrieved from https://www.fao.org/3/i5640e/i5640e.pdf

Hirvonen, K., Bai, Y., Headey, D., & Masters, W. A. (2020). Affordability of the EAT-Lancet reference diet: A global analysis. *The Lancet Global Health, 8*(1), e59–e66.

Kim, K. H., Kabir, E., & Jahan, S. A. (2017). Exposure to pesticides and the associated human health effects. *Science of The Total Environment, 575*, 525–535.

Levallois, P., & Villanueva, C. M. (2019). Drinking water quality and human health: An editorial. *International Journal of Environmental Research and Public Health, 16*(4), 631.

McKay, B. M. (2017). Agrarian extractivism in Bolivia. *World Development, 97*, 199–211.

O'Neil, A., Quirk, S. E., Housden, S., Brennan, S. L., Williams, L. J., Pasco, J. A., Berk, M., & Jacka, F. N. (2014). Relationship between diet and mental health in children and adolescents: A systematic review. *American Journal of Public Health, 104*(10), e31–e42.

Oliveira, D. (2020). Inovações e novidades na construção de mercados para a agricultura familiar: os casos da Rede Ecovida de Agroecologia e da RedeCoop. *Redes, 25*(1), 135–163.

Podda, A., Loconto, A. M., Arcidiacono, D., & Maestripieri, L. (2021). Exploring prosumption: Reconfiguring labour through rural-urban food networks? *Journal of Rural Studies, 82*, 442–446.

Polanyi, K. (1944). *The great transformation: Politische und ökonomische Ursprünge von Gesellschaften und Wirtschaftssystemen* (H. Jelinek, Trans.). Suhrkamp Verlag.

Poti, J. M., Braga, B., & Qin, B. (2017). Ultra-processed food intake and obesity: What really matters for health: Processing or nutrient content? *Current Obesity Reports, 6*(4), 420–431.

United Nations (1999). *General comment no. 12: The right to adequate food (art. 11 of the covenant).* CESCR. Retrieved from https://www.refworld.org/docid/4538838c11.html

van der Ploeg, J. D. (2018). *The new peasantries: Rural development in times of globalisation* (2nd ed.). Routledge.

Willett, W., Rockström, J., Loken, B., Springmann, M., Lang, T., Vermeulen, S., Garnett, T., Tilman, D., DeClerck, F., Wood, A., Jonell, M., Clark, M., Gordon, L. J., Fanzo, J., Hawkes, C., Zurayk, R., Rivera, J. A., Vries, W. D., Sibanda, L. M., & Murray, C. J. L. (2019). Food in the anthropocene: The EAT-lancet commission on healthy diets from sustainable food systems. *The Lancet, 393*(10170), 447–492.

Ye, J., van der Ploeg, J. D., Schneider, S., & Shanin, T. (2020). The incursions of Extractivism: Moving from dispersed places to global capitalism. *The Journal of Peasant Studies, 47*(1), 155–183.

52

CLIMATE TRANSFORMATIONS

Evolving food security, migration, and alternative livelihood strategies in Panama

Alyson Dagang

Introduction

It is widely regarded that change can be a source of innovation. However, when multiple transformations occur in parallel and at accelerated rates, the impacts have the potential to become more prejudicial than beneficial (Paudel Khatiwada et al., 2018; Hatfield et al., 2020). Broad-based evidence suggests that recent changes in climate patterns, historically relied upon by mountain and coastal communities, have triggered fundamental changes in livelihood strategies among many groups across the Neotropics (Peri & Sasahara, 2019). Recent research has also established unequivocal links between climate change and food security. On account of new, unpredictable precipitation regimes, traditional and long-established agricultural practices are becoming less productive (Li & Ford, 2019). Decreased localised food production jeopardises food availability, reduces access to healthful food, and can precipitate a cascade of adversity for communities (Giller et al., 2021; Tidd et al., 2022). Changes in food supply can also heighten human out-migration patterns, creating a self-perpetuating cycle of declining food production, reduced agricultural labour availability, increased food insecurity, and intensified rural-to-urban migration (Espindola et al., 2005).

Trends in food security

The Food and Agriculture Organisation (FAO) (2003) established that food security exists, "when all people, at all times, have physical and economic access to sufficient, safe, and nutritious food that meets their dietary needs and food preferences for an active and healthy life" (FAO, 2003, p. 27). In 2020, approximately

DOI: 10.4324/9781003174417-61

768,000,000 people experienced hunger worldwide (FAO, 2020). In Latin America and the Caribbean (LAC), food security has experienced important fluctuations in the last two decades. Between 2000 and 2018, 20 million people in the region improved their situations and reached food security. Indeed, by 2018, the region was 3.5% below the world average of undernourished people, as shown in Table 52.1 and child undernourishment and wasting was also less than the world average. However, between 2014 and 2018, the prevalence of moderate food insecurity rose in LAC from 26.2% to 31.1%—close to 5% in just five years (see Table 52.1). The renewed increases in moderate food insecurity, despite the advances made in the last two decades, are a clear cause for concern. Growing food insecurity in LAC casts a confounding juxtaposition, as it contrasts growing hunger with the region's robust agricultural sector. In fact, recent agricultural production data indicate that the LAC region produces more than 3000 calories per capita per day, which is 50% more than the minimum human daily requirement, clearly demonstrating Latin America's capacity for self-sufficiency in its food supply (Morris et al., 2020). With skyrocketing food sales in 2019 and its position among the top exporting regions of the world (OECD-FAO, 2019), LAC's status as a *net* agricultural exporter calls into question the circumstances of its 187 million citizens who currently struggle with food insecurity (Boliko, 2019). The discrepancy between more-than-sufficient food production and prevalent food insecurity highlights the challenges to overcoming hunger in the region. If food production is ample and hunger is growing, food availability and food distribution should be key foci for research and action.

Given the forecast for forthcoming change to climate patterns that will directly affect agricultural production and human livelihoods (Rosegrant et al., 2021), conditions call for special attention. Decreasing food security will likely be exacerbated by looming transformations, and clearly require action to mitigate potentially perilous circumstances. In Panama, changing climatic, commercial, and social conditions are catalysing transformation across geography and human livelihoods. Climate forecast research in Panama predicts significant changes in heat and rain patterns during the 21st century (Fabrega et al., 2013). It is well known that alterations in climate render changes that cut across multiple sectors of society. Increased precipitation and temperature bring about flooding, heavy pest loads, and seasonal drought, which create new conditions for people to manage and navigate (Fand et al., 2018). This situation is particularly acute in Panama where a significant sector of the work force and GDP depend on the rain-dependent Panama Canal. More than 50% of Panama's energy matrix is produced through hydroelectric power, and the majority of agricultural production is rainfed (Jordan, 2021; Chacon et al., 2019). The degree to which Panamanian well-being (employment, GDP, crop production) depends on predictable climate patterns, and adequate rain in general, is arguably unparalleled in the region. From such widespread dependence emerges the question of whether Panama will be able to muster the required capacity and robust ability to adapt to the pending and overarching climate changes and their potential consequences.

Table 52.1 Moderate and serious food insecurity: World, Latin America, Mesoamerica, South America, prevalence (%) and millions of people, 2014–2016 and 2016–2018. *Note.* Data adapted are from FAOSTAT (2020). Used with permission for open access

Location	Serious food insecurity				Moderate and serious food insecurity			
	Prevalence (%)		Millions		Prevalence (%)		Millions	
	2014–2016	2016–2018	2014–2016	2016–2018	2014–2016	2016–2018	2014–2016	2016–2018
World	7.9	8.7	584.6	654.1	23.5	25.4	1736.8	1915.1
Latin America	7.3	8.9	43.2	53.7	26.2	31.1	154.6	187.0
Mesoamerica	10.5	10.6	18.2	18.8	32.2	31.7	55.5	56.2
South America	6.0	8.2	25.0	34.9	23.8	30.8	99.1	130.8

Climate adaptation and food security in Panama

The inclination toward or away from adaptation to climate change and its impacts will be determined by Panama's 4.3 million inhabitants and the wherewithal of government institutions (Mann et al., 2021). Social and institutional well-being will be key to Panama's ability to adapt. As a country, Panama is often characterised as *middle-income*; however, the reality is different (Schappert, 2007). Panama possesses a severely skewed distribution of wealth, evidenced by a Gini Index of 49.8 (consistently among the top three most-skewed economies in the LAC region, shared with Brazil and Colombia). This erred and distorted view of the Panamanian reality and perceived well-being of most Panamanians has deleterious impacts on Panama's ability to access aid and programs for structural reform. With an average poverty rate of 23%, similar to that of Bangladesh and Ghana, Panama is often not a recipient of aid, as countries in similar situations are, due to the perception that it is a middle-income country (FAO, 2019; Meyer & Martin, 2021; World Bank, 2021). Coupled with unequal income distribution, persistent poverty is weighed down by a deficient educational system. In 2019, Panama took 71st place out of 79 countries that participated in the OECD's Program for International Student Assessment (MEDUCA, 2020). Moreover, among school-aged children and adolescents, there are approximately 84,000 who do not attend school, totaling 36% of the adolescent population who do not attend secondary school (UNICEF, 2018). Children bear a sizable brunt of the situation; 19.1% of school-aged children in Panama suffered from undernourishment in 2018 (FAO, 2018), while malnutrition is the third leading cause of death among children between one and four years of age (MINSA, 2014). Child undernourishment is a symptom of a larger, deleterious situation regarding food insecurity and deficient societal and institutional well-being in Panama.

Undernourishment, and food insecurity in particular, are of special concern in Panama. The Panamanian Constitution explicitly recognises the right to food as a fundamental right of its citizenry, in which it specifies the State's "obligation to ensure *optimal nutritional status* for the entire population, with the promotion of the availability, consumption, and optimal utilisation of adequate food" (Gaceta Oficial No. 25,176. Article 110. November 2004, p. 20). According to national statistics, in 2005, 27% of the population, 856,000 people, were undernourished. Since then, undernourishment has decreased by more than half (SENAPAN, 2017), mainly through the establishment of direct transfer social protection programs. According to the most recent available statistics, there are an estimated 400,000 undernourished people in Panama, approximately 10% of the population, 40% of whom live in Indigenous territories. Undernourishment can be part of a suite of indicators of food security challenges (Ayala & Meier, 2017). However, in Panama, food security specifically is not tracked as part of routinely monitored socioeconomic indicators, rather rates of food insecurity are derived from data gathered on extreme poverty. Not measuring food security directly makes it unlikely to adequately capture the entirety of the food insecurity panorama in Panama. In addition, information, research, and attention to urban food insecurity are severely lacking and urgently needed. As noted above, 23% of the Panamanian population lives in poverty (MEF, 2017), which points to the possibility of a wider food security problem than revealed by general poverty statistics. More definitive research directly focused on food security is vital for the creation of adequate and appropriate policies and actions that address the challenges to, and pursuit of, well-being in Panama.

Like LAC, incongruence shapes the food picture in Panama. Despite it not being an agricultural powerhouse, Panama's 248,650 growers produce an average of 2733 kcals per capita per day, which exceeds the per capita daily requirement defined by the national government (2,336 kcals) (SENAPAN, 2017). Growers' success is commendable in this regard, given that 90% of them do not receive governmental technical or extension services, and 30% of national production experiences post-harvest loss (Chacon et al., 2019). While data reveal that sufficient food is already produced, and given that 10% or more of the population is undernourished and urban food insecurity levels are unknown, food distribution and post-harvest management are important bottlenecks for achieving widespread food security. Nevertheless, the majority of public policies do not focus on improved food distribution or postharvest loss reduction, rather they foment increases in agricultural production, which does not seem to be lacking (Collado et al., 2018). Government initiatives support technological intensification and expansion of agricultural production without focusing on distribution—side-lining an important aspect of the food availability landscape as well as alienating small-scale producers who may not be apt to adopt technological packages.

In Panama, the population is particularly vulnerable to the effects of the *distancing* (Clapp, 2014) of the food supply chain. Without abundant fertile soils, its location as a trading hub, and weak agricultural protection policies and supports, Panama's markets experience frequent agricultural *dumping*. In general, Panama depends heavily on imports, potentially an artefact of political-economic design rather than need. In 2015, it was estimated that 60% of Panama's food supply originated outside of the country (Mann et al., 2021). Moreover, between 2005 and 2019, national rice production, the primary staple of the Panamanian diet, fell by 37% (INEC, 2019). According to the National Rice Growers Association, cheap rice imported by private intermediaries has hurt local producers and provoked a reduction in rice planting nationwide ("Productores de Chepo se unen," 2018). Furthermore, in 2016, total agricultural imports for human consumption (which excludes animal feed and other products) totalled 8%, yet four years later, in 2020, imports for human consumption leaped to 24% of total imported agricultural goods (Mann et al., 2021).

While rice planting has decreased due to private sector imports, governmental price controls were placed on commonly consumed food items, including specific cuts of beef, chicken, tomatoes, tubers, and other basic items, in recognition of the barriers to access to food in 2014 (Redacción, 2014). While freezes on food prices undoubtedly have had positive impacts on food access for Panamanians, for the 400,000 people living in extreme poverty, it is questionable whether these policies substantially impacted their food security considering the prevalent obstacles to food availability and distribution that exist in Panama. While some public policies have been beneficial, undernourishment persists among the population, including 17.7% of children nationwide under five, 57% of first graders in Indigenous communities who are underweight, as well as 23% of pregnant women nationwide, and 65% of pregnant women in Guna Yala specifically, who suffer from anaemia (De Leon et al., 2018; MINSA, 2015). Specific, directed programs are needed to pinpoint, raise awareness, and support these vulnerable, at-risk populations.

Transformation in Guna Yala

Climate unpredictability has changed the abundance of food production, and consequently, lifestyles in some Indigenous rural communities in Panama. In the Guna Yala Comarca, an autonomous Indigenous territory on Panama's northeast coast, changes in marine ecosystem health have had detrimental effects on the Guna diet. Fish stocks, the primary protein source for the Guna (Rivera et al., 2012), have experienced a significant decrease in abundance due to rising ocean temperatures and, in some cases, high fishing pressures (Brown, 2018). It is widely accepted that ongoing increases in ocean water temperatures, resulting from climate change, trigger changes in ocean habitats and reductions in fish populations (Nurse, 2011). Consequently, in Guna Yala, local fish catches have become smaller and less frequent (Hoehn & Thapa, 2009). To catch sufficient fish to satisfy their dietary needs, Guna fishers have to travel farther distances to access larger fish populations. However, travelling long distances in open ocean is usually prohibitive for the vast majority of fisher families who rely on dugout canoes and home-crafted sails for transportation to fishing grounds (A. Velez, personal communication, March 8, 2018). As a result, fisher families' access to quality protein has been reduced. Insufficient protein sources are reflected in the alarming 61% of Guna children who are chronically undernourished (De Leon et al., 2018). While reduced local fish stocks have negatively impacted the population's diet, climate fluctuations have also affected Guna well-being via increases in malaria outbreaks. Hurtado and colleagues (2018) found a significant increase in malaria cases resulting from ENSO climate events in Guna Yala. While it is regarded that Guna Yala is among the Indigenous territories with better access to government health services, overcoming malaria is complicated by obstacles to access medical services as well as intercultural barriers (Hurtado et al., 2018).

Given the changing conditions, the people of Guna Yala have been pushed toward seeking alternative livelihood strategies. Climate transformation, a compromised capacity to satisfy their dietary needs compounded by greater threats of illness, as well as a territory-wide 91% poverty rate are some of the factors that are heightening the rate at which change is occurring (MEF, 2017). While in 2015 close to 80% of the population participated in subsistence agriculture and fishing (MEF, 2017), during the pre-pandemic period, subsistence agricultural production began to wane. Decreased crop productivity was evidenced by increases in the presence of plantain trucks in Guna Yala ports and increases in imports of rice and other foodstuffs into the territory (A. Castillo, personal communication, July 12, 2021). Descending pre-pandemic crop production in Guna Yala follows a national trend. In fact, between 2005 and 2015, nationwide crop production decreased 15% while livestock production increased 18% during the same period (SENAPAN, 2017). Changes in land use from crops to livestock illustrates farmers' strategy toward adaptation to confront climate impacts (Lawrence et al., 2019). Crop production involves higher risk and stakes in the face of climate change, while contending with livestock raising brings its own challenges as well discussed in chapters in this collection. Producers in Panama are often inclined to change their agricultural activities in order to minimise potential risk (Dagang, 2017). In pre-pandemic Guna Yala, reduced agricultural productivity dissuaded Guna youth (who make up 50% of the Guna population (MEF, 2017)) from pursuing agricultural endeavours.

Migration as adaptation

In response to the confounding factors and pressures of reduced food availability, alternative livelihood strategies have emerged among many Panamanian communities (Doering, 2016). Accelerated labour migration, including migration of youth at younger ages and larger numbers of people migrating for income earning opportunities, have become prevalent in *campesino* and Indigenous rural communities. Rural-rural labour migration has a century-long history in Panama (Rudolf, 1999), including the movement of field workers to cut sugar cane and pick coffee during harvest seasons, as well as rural-urban migration to work in cities in domestic roles. However, reduced agricultural production from unpredictable climate events, and the consistent decline in farming operations nationwide, including the diminishing number of small-scale growers and the related expansion of large farms, have expedited the migratory outflow and transformation of rural inhabitants to urban labourers (INEC, 2019). Through the migratory process, rural dwellers create alternative livelihood strategies and, as described by Black et al. (2011), through migration seek the means to build resilience in the face of threats brought about by environmental change, as well as diversify their income-producing strategies, and reduce risk.

Through the migration process, remittances from migrants to rural households inevitably transform household budgets and have been known to impact rural diets (Moniruzzaman, 2022), while increased available income from remittances also can improve diet quality (Musah-Surugu et al., 2017). In the case of Panama, remittances have been key to increasing household food availability, reducing seasonal scarcity, and expanding healthy diet options (Rudolf, 2021). However, the recent phenomenon of increased household food budgets coupled with nationwide reductions in smallholder food production call into question the nature of the changes in rural diets and merit further inquiry. In preliminary research in a north-western Indigenous territory on Panama's Caribbean coast, Dagang (2017) found that increases in urban-rural remittances precipitated changes in land use and resource decision making—from crop production to livestock husbandry (cattle raising)—which decreased local availability of staple crops. Preference for cattle raising, and in some cases plantain (*Musa* spp.) cash crops, signalled a change in land use and cultivation from crop production for local consumption to crop production for cash generation and savings, resulting in less available land for crop cultivation for local consumption. Reduced availability of local crops and the use of local land for non-traditional production activities can have important impacts on the maintenance of traditional diets in terms of the distancing of households from traditional, culturally relevant dietary practices. It can also bring about the substitution of local crops for non-traditional purchased foods of non-local origin as Keding (2016) and Khomje and Qaim (2019) have reported in different parts of the world. In fact, households that face previously unknown prosperity in addition to nutritional transitions are known to experience the "dual (and triple) nutritional burden" as a result of changes in income and food availability (Doak et al., 2005). Further research is necessary to identify whether this phenomenon has begun to unfold in Panama (McDonald et al., 2015).

Indigenous tourism as an alternative livelihood strategy

In the face of adversity, certain groups have turned toward tourism as an alternative livelihood strategy. Tourism activities can imply risks to cultural well-being and imposed transformation. In the case of the Guna, tourism has brought challenges and achievements. Armed with strict internal controls and steadfast protection of their cultural heritage, the Guna have embarked upon tourism on their own terms, free of participation or intervention from outside actors, generating alternative sources of income for their communities. Through a well-established, albeit potentially fragile balance, Guna tourism has emerged as a model for autonomous tourism endeavours in which activities are governed and managed by communities (De Leon, 2016; Pereiro, 2015). While it is broadly recognised that the Guna have proudly achieved wide success in maintaining and driving their autonomy as purveyors of beach and island tourism, there is growing interest in understanding the potential long-term impacts of tourism on Guna society. As more Guna become involved in tourism, specifically youth, understanding the impacts on Guna traditions, particularly food production and diet, has become a topic of interest among Guna leaders. With increased food imports, fewer youth farmers, and greater pressure placed on Guna fish stocks to satisfy tourists' appetites, there is concern regarding the medium- and long-term impacts of tourism on Guna society, particularly regarding the traditional Guna diet and consequent health impacts (Representative of the Instituto de Desarrollo e Investigación Guna, personal communication, July 12, 2021). By way of comparison, in Yucatan, Mexico, Leatherman et al. (2020) found preponderant dietary delocalisation resulting from international tourism activities in which fundamental nutritional transitions occurred among Maya children and adults in the region. Undesirable health effects have been a consequence of successful tourism initiatives among the Maya in which traditional cultivation practices and communal food preparation have been abandoned in some cases.

In the case of the Guna, decreases in agricultural activity and production resulting from participation in alternative income generating activities associated with tourism have triggered changes in food acquisition and dietary selection. Increases in income, paired with reduced agricultural activity, have created concomitant increases in food imports to the territory. Consequent transformation of the food system and diets, from traditional and local to imported and purchased, could impact local food availability and food sovereignty. As Leatherman et al. (2020) and Clapp (2014) have noted, commoditisation of food systems is among the most influential phenomena in precipitating traditional diet transformation. In this context, control of the food supply is ceded to *ex situ* actors who supplant local stakeholders at the decision-making table (Isakson, 2014). As communities move away from their own agricultural practices, whether geographically due to migration or professionally due to changes in livelihood strategies (e.g., tourism), as is the case of the Guna, so do they risk distancing themselves from their dietary traditions and control over their food supply.

Conclusion

Understanding human adaptation and decision making in the face of climate change impacts are imperative. Ecological transformation resulting from climate change is present, and demands the concomitant transformation of human systems to respond and adapt. While it is evident that alternative livelihood systems optimise for income generation, and can successfully do so, traditional diets and food sovereignty, among other factors, can be overlooked and lost in the process. Research and action are needed to understand the ways in which communities can be respectfully supported so that livelihood adaptations to climate change optimise for well-being and communities aren't forced to sacrifice sustainability, diet, or food security in the process.

References

Ayala, A., & Meier, B. M. (2017). A human rights approach to the health implications of food and nutrition insecurity. *Public Health Reviews, 38*(1), 1–22.

Black, R., Bennett, S. R., Thomas, S. M., & Beddington, J. R. (2011). Migration as adaptation. *Nature, 478*(7370), 447–449.

Boliko, M. C. (2019). FAO and the situation of food security and nutrition in the world. *Journal of Nutritional Science and Vitaminology, 65*(Suppl.), S4–S8.

Brown, A. (Director). (2018). *A Fishing Story* [Video file]. Available from https://filmfreeway .com/AFishingStory

Castillo, A. (2021, July 12). Personal Communication [Personal Interview].

Chacón, Á., Dutra, T., Egas, J. J., Shik, O., & De Salvo, C. P. (2019). *Análisis de políticas agropecuarias en Panamá*. Inter-American Development Bank. Retrieved from https://publications.iadb.org/publications/spanish/document/An%C3%A1lisis_de_pol%C3%ADticas _agropecuarias_en_Panam%C3%A1__es_es.pdf

Clapp, J. (2014). Financialization, distance and global food politics. *The Journal of Peasant Studies, 41*(5), 797–814.

Collado, E., Fossatti, A., & Saez, Y. (2018). Smart farming: A potential solution towards a modern and sustainable agriculture in Panama. *AIMS Agriculture and Food, 4*(2), 266–284.

Dagang, A. (2017). *Adoption of cattle ranching among two indigenous groups in Northwest Panama: A preliminary study*. Latin American Studies Association.

De Leon, J., Gonzalez, E., Barba, A., Sinisterra, O., & A. Atencio (2018). *Segundo monitoreo nutricional en las instalaciones de salud del MINSA*. MINSA, INCAP-OPS/OMS.

Retrieved from https://nutricionistaspanama.com/wp-content/uploads/publicaciones/INFORME_MNINUT.pdf.

De León Smith Inawinapi, C. (2016). Resignificación política del manejo de los recursos naturales en una comunidad indígena de Panamá: Los Gunas y el turismo. *Ecología Política*, *52*, 45–48.

Doak, C. M., Adair, L. S., Bentley, M., Monteiro, C., & Popkin, B. M. (2005). The dual burden household and the nutrition transition paradox. *International Journal of Obesity*, *29*(1), 129–136.

Doering, L. (2016). Necessity is the mother of isomorphism: Poverty and market creativity in Panama. *Sociology of Development*, *2*(3), 235–264.

Espíndola, E., León, A., Martínez, R., & Schejtman, A. (2005). *Poverty, hunger and food security in Central America and Panama*. ECLAC. Retrieved from https://repositorio.cepal.org/handle/11362/6099

Fábrega, J., Nakaegawa, T., Pinzón, R., Nakayama, K., & Arakawa, O. (2013). Hydroclimate projections for Panama in the late 21st Century. *Hydrological Research Letters*, *7*(2), 23–29.

Fand, B. B., Tonnang, H. E., Bal, S. K., & Dhawan, A. K. (2018). Shift in the manifestations of insect pests under predicted climatic change scenarios: key challenges and adaptation strategies. In S. K. Bal, J. Mukherjee, B. U. Choudhury, & S. K. Dhawan (Eds.), *Advances in crop environment interaction* (pp. 389–404). Springer.

FAO (2003). *Trade reforms and food security: Conceptualising the linkages*. FAO. Retrieved from https://www.fao.org/documents/card/en/c/bcab5bfe-a5d9-5c78-9c74-7f5fbf7be2be/

FAO, IFAD, UNICEF, WFP, & WHO (2020). *The State of Food Security and Nutrition in the World 2020. Transforming food systems for affordable healthy diets*. FAO. Retrieved from https://www.fao.org/documents/card/en/c/ca9692en/

FAO, OPS, WFP, & UNICEF (2018). *Panorama de la seguridad alimentaria y nutricional en América Latina y el Caribe*. FAO. Retrieved from https://iris.paho.org/handle/10665.2/49616

FAO, OPS, WFP, & UNICEF (2019). *Panorama de la seguridad alimentaria y nutricional en América Latina y el Caribe*. FAO. Retrieved from https://www.fao.org/documents/card/en/c/CA6979ES/

FAOSTAT (2020a). *Food and agriculture data*. https://www.fao.org/faostat/en/#home

FAOSTAT (2020b). *Suite of food security indicators* [Dataset]. FAO.

Giller, K. E., Delaune, T., Silva, J. V., van Wijk, M., Hammond, J., Descheemaeker, K., van de Ven, G., Schut, A., Taulya, G., Chikowo, R., & Andersson, J. A. (2021). Small farms and development in sub-Saharan Africa: Farming for food, for income or for lack of better options? *Food Security*, *13*(6), 1431–1454.

Hatfield, J. L., Antle, J., Garrett, K. A., Izaurralde, R. C., Mader, T., Marshall, E., Nearing, M., Robertson, G. P., & Ziska, L. (2020). Indicators of climate change in agricultural systems. *Climatic Change*, *163*(4), 1719–1732.

Hoehn, S., & Thapa, B. (2009). Attitudes and perceptions of indigenous fishermen towards marine resource management in Kuna Yala, Panama. *International Journal of Sustainable Development & World Ecology*, *16*(6), 427–437.

Hurtado, L. A., Calzada, J. E., Rigg, C. A., Castillo, M., & Chaves, L. F. (2018). Climatic fluctuations and malaria transmission dynamics, prior to elimination, in Guna Yala, Republic of Panama. *Malaria Journal*, *17*(1), 1–12.

INEC (Instituto Nacional de Estadísticas y Censo) (2019a). *Encuesta agrícola de arroz, maíz y frijol de bejuco*. INEC. Retrieved from https://www.inec.gob.pa/archivos/P0705547520 2002070840251.pdf

INEC (Instituto Nacional de Estadísticas y Censo) (2019b). *Estadística panameña: Situación Demográfica, Boletín N°8: Estimaciones y proyecciones de la población total en la Republic of Panama, por provincia y comarca indígena, según sexo y edad, periodo 1990–2030*. INEC.

Isakson, S. R. (2014). Food and finance: The financial transformation of agro-food supply chains. *The Journal of Peasant Studies*, *41*(5), 749–775.

Jordan, W. (2021, January 3). *Hidros aportan mas del 70% de la energía.* La Prensa. https://www.prensa.com/impresa/economia/hidros-aportaron-mas-del-70-de-la-energia/

Keding, G. (2016). Nutrition transition in rural Tanzania and Kenya. *World Review of Nutrition and Dietetics, 115,* 68–81.

Khonje, M. G., & Qaim, M. (2019). Modernization of African food retailing and (un) healthy food consumption. *Sustainability, 11*(16), 4306.

Lawrence, T. J., Stedman, R. C., Morreale, S. J., & Taylor, S. R. (2019). Rethinking landscape conservation: Linking globalised agriculture to changes to indigenous community-managed landscapes. *Tropical Conservation Science, 12,* 1–19.

Leatherman, T., Goodman, A. H., & Stillman, J. T. (2020). A critical biocultural perspective on tourism and the nutrition transition in the Yucatan. In H. Azcorra, & F. Dickinson (Eds.), *Culture, environment and health in the Yucatan Peninsula* (pp. 97–120). Springer.

Li, A., & Ford, J. (2019). Understanding socio-ecological vulnerability to climatic change through a trajectories of change approach: a case study from an Indigenous community in Panama. *Weather, Climate, and Society, 11*(3), 577–593.

Mann, C. G., Ramírez, C., & Marcus, J. (2021). ¿Conculcación de derechos? *Iustitia et Pulchritudo, 2*(1), 31–61.

McDonald, A., Bradshaw, R. A., Fontes, F., Mendoza, E. A., Motta, J. A., Cumbrera, A., & Cruz, C. (2015). Prevalence of obesity in Panama: some risk factors and associated diseases. *BioMed Central: Public Health, 15*(1), 1–12.

MEDUCA (Ministerio de Educación), & Gobierno de Panamá (2020). *Pisa Panamá Programa Para La Evaluación Internacional de Estudiantes.* MEDUCA. Retrieved from https://www.oecd.org/pisa/pisa-for-development/Panama_PISA_D_National_Report.pdf

MEF (Ministerio de Economía y Finanzas, & Gobierno de Panamá (2017). *Índice de pobreza multidimensional de Panamá.* MEF. Retrieved from https://www.mides.gob.pa/wp-content/uploads/2017/06/Informe-del-%C3%8Dndice-de-Pobreza-Multidimensional-de-Panam%C3%A1-2017.pdf

Meyer, P. J., & Martin, R. (2021). *U.S. foreign assistance to Latin America and the Caribbean: FY2021 appropriations.* Congressional Research Service. Retrieved from https://fas.org/sgp/crs/row/R46514.pdf

MINSA (Ministerio de Salud). (2014). *Indicadores de Salud Básicos. Indicadores Básicos de País.* Gobierno Nacional.

MINSA (Ministerio de Salud). (2015). *Monitoreo Nutricional En Las Instalaciones De Salud Del MINSA (MONINUT).* Gobierno Nacional.

Moniruzzaman, M. (2022). The Impact of remittances on household food security: Evidence from a survey in Bangladesh. *Migration and Development, 11*(3)352–371.

Morris, M., Sebastian, A. R., & Perego, V. M. E. (2020). *Future foodscapes: Reimagining agriculture in Latin America and the Caribbean.* World Bank Group. Retrieved from https://documents1.worldbank.org/curated/en/942381591906970569/pdf/Future-Foodscapes-Re-imagining-Agriculture-in-Latin-America-and-the-Caribbean.pdf

Musah-Surugu, I. J., Ahenkan, A., Bawole, J. N., & Darkwah, S. A. (2017). Migrants' remittances: A complementary source of financing adaptation to climate change at the local level in Ghana. *International Journal of Climate Change Strategies and Management, 10*(1), 178–196.

Nurse, L. A. (2011). The implications of global climate change for fisheries management in the Caribbean. *Climate and Development, 3*(3), 228–241.

OECD/FAO (2019). *OCDE-FAO Perspectivas Agrícolas 2019–2028.* FAO & OECD Publishing.

Paudel Khatiwada, S., Deng, W., Paudel, B., Khatiwada, J. R., Zhang, J., & Wan, J. (2018). A gender analysis of changing livelihood activities in the rural areas of central Nepal. *Sustainability, 10*(11), 4034.

Pereiro Pérez, X. (2015). Reflexión antropológica sobre el turismo indígena. *Desacatos, 47,* 18–35.

Peri, G., & Sasahara, A. (2019). *The impact of global warming on rural-urban migrations: Evidence from global big data*. National Bureau of Economic Research. Retrieved from https://www.nber.org/system/files/working_papers/w25728/w25728.pdf

Redacción (2014, July 1). *Varela establece control de precios de emergencia a 22 productos*. Capital Financiero. https://elcapitalfinanciero.com/varela-establece-control-de-precios-de-emergencia-22-productos/

Representative of the Instituto de Desarrollo e investigación de Guna Yala (2021, July 12). Personal Communication [Personal Interview].

Rivera, V. S., Borrás, M. F., Gallardo, D. B., RL, C., Ochoa, M., Castañeda, E., & Castillo, G. (2012). *Regional study on social dimensions of MPA practice in Central America: Cases studies from Honduras, Nicaragua, Costa Rica and Panamá*. International Collective in Support of Fishworkers. Retrieved from https://www.iccaconsortium.org/wp-content/uploads/2015/08/example-regional-study-mpa-central-america-rivera-2012-en.pdf

Rosegrant, M. W., Wiebe, K. D., Sulser, T. B., & Willenbockel, D. (2021). Climate change and agricultural development. In K. Otsuka & F. Shenggen (Eds.), *Agricultural development: New perspectives in changing world* (pp. 629–660). IFPRI. Retrieved from https://ebrary.ifpri.org/utils/getfile/collection/p15738coll2/id/134119/filename/134324.pdf

Rudolf, G. (1999). *Panama's poor: Victims, agents, and historymakers*. University Press of Florida.

Rudolf, G. (2021). *Esperanza speaks: Confronting a century of rural change in Panama*. University of Toronto Press.

Schappert, J. (2007). Poverty reduction in Panama's Indigenous communities. *Perspectives on Business and Economics, 25*(4), 25–32.

SENAPAN (Secretaría Nacional para el Plan Nacional de Seguridad Alimentaria y Nutricional), & Gobierno de Panamá (2017). *Plan Nacional de Seguridad Alimentaria Panamá 2017–2021*. MIDES. Retrieved from https://www.mides.gob.pa/wp-content/uploads/2017/03/Plan-SAN-Panam%C3%A1-2017.pdf

Tidd, A. N., Rousseau, Y., Ojea, E., Watson, R. A., & Blanchard, J. L. (2022). Food security challenged by declining efficiencies of artisanal fishing fleets: A global country-level analysis. *Global Food Security, 32*, 100598.

UNICEF (2018). *UNICEF strategic plan 2018–2021 country and regional education profiles [Data set]*. UNICEF. https://data.unicef.org/resources/unicef-strategic-plan-education-country-profiles/

Velez, A. (2018, March 8). Personal Communication [Personal Interview].

World Bank (2021, April 6). *Panama overview*. World Bank Group. https://www.worldbank.org/en/country/panama/overview

53

A REFLECTION ON GLOBALISATION INFLUENCING RESILIENCE OF THE CONTEMPORARY KOSRAEAN FOOD SYSTEMS

David Fazzino and Ashley Meredith

Introduction

In charting potential pathways towards sustainable diets, researchers' perspectives matter. Sachs (1999) outlined three perspectives of sustainable development: Contest, astronaut, and home. Contest perspective proponents emphasise integrating environmental concerns within the market, such that those in the "developed" world would provide training and capital investment to the "underdeveloped" world to enhance their capacity and achieve sustainability (Sachs, 1999). This business-as-usual approach incorporates sustainability language and frameworks to create rubrics that are allegedly achievable and will allow for necessary shifts without overcorrecting and unnecessarily, if at all, impeding growth. The metrics for sustainability look different if we take this second astronaut perspective and imagine Earth as a spaceship and the astronaut's perspective through which Earth is a variety of biogeochemical processes. These processes can be effectively measured and managed through the lead of developed countries deploying robust coordination of global surveillance of Earth's processes and tracking anthropogenic-induced change. Since the 1970s, proponents of this approach have often called for limits to continued growth to fend off impending catastrophic and neo-Malthusian consequences, while others have argued for more thorough incorporation of environmental outcomes into the market (through cap and trade and labelling schemes directed towards the individual consumer). The third, home perspective is the most divergent of the three with a focus on the continuation of local livelihoods (Sachs, 1999). The focus on local livelihoods has often been dismissed as anachronistic given the very situations, structures, and institutions that proponents

DOI: 10.4324/9781003174417-62

of market-based approaches have helped to foster, suggesting that there is no alternative. The goods and foods produced in local settings are viewed as infinitesimally small in an era of efficiency, intense population pressures, and globalisation. There are a number of ethnocentric assumptions inherent in dismissing Sach's (1999) home perspective as untenable, including, but not limited to: (1) False notions of efficiency that mistakes efficiency with the substitution of externally-generated technologies, petroleum, and its products for human labour and locally-generated tools and technologies to adapt and problem-solve (Stone, 2007); and (2) power masquerading as beneficent and divinely-ordained calls to feed the world (Fazzino, 2003; 2012). The home perspective offers humans the most time-tested approach to confront issues of sustainability. This home perspective can also work in unison with contest or astronaut perspectives, as many tools are needed to answer the call to producing and procuring sustainable foods and diets. This complementarity is illustrated by Hornorg's (2007) call for creating a coupon system to foster exchange of local goods and services.

Narratives and frameworks of sustainability, measured along a variety of metrics, are part of globalisation. The viewpoints are expressed in the context of a global management of resources to influence and shape policy directives towards a limitation on the adverse impacts of food systems along a variety of parameters both socially and ecologically (Clark et al., 2020; Fears et al., 2019; Kevany & Kassam, 2021). These are perceived as valuable in arresting the most deleterious impacts at the country and global level. They are also perceived as a necessity in the face of imminent collapse of the existing and supposedly non-sustainable global systems of consumption. Frameworks such EAT-Lancet discuss food in the Anthropocene, placing an emphasis on achieving targets of global adoption of the healthy reference diet, with an emphasis on individual consumers and arguing for the necessity of using, "a full range of policy levers, from soft to hard" (Willett et. al, 2019, p. 448). The globalisation of sustainability as a response to global crises produces contexts within which the logics of colonialism are likely to play out. As responsibility for climate change is foisted upon all of us, flattening uneven levels of consumption and, invariably through the same logics of market-based exchanges, seeking out markets and opportunities to mitigate and reduce impacts. As noted by Young (2020) in the context of climate change, "By responsibilizing all of humanity for climate change, instead of the colonial processes that have actually driven it, Anthropocene discourses risk further masking and naturalizing this colonialism" (p. 231).

At the onset of the COVID-19 pandemic, author David Fazzino (DF) received a call from a close friend who had long worked in the field of sustainability asking, "Is this the end?" DF replied that although there were some disruptions in the supply chain, coupled with panic buying for some, but not all, commodities, that the utilities, including internet connectivity, were for the most part, still operating normally. In hindsight, at the time of this writing, the familiar trope of (imminent) system collapse of the global industrial production and distribution systems did not come to pass but for a time resonated with greater intensity as images of hoarding and resultant empty grocery store shelves and New York City hospitals as war zones flashed across our screens. The increasingly likely scenario of system collapse demonstrated the inherent failure and hence unsustainability of the contest and astronaut perspectives, leaving only the home perspective to potentially provision for the sustainable future of humanity. It is within these contexts that we began to consider the potential implications that disruptions to come might have on the future of sustainable diets in Kosrae, FSM. Nonetheless, the home perspective, as a lens by which to consider and advocate for sustainable diets, is not beyond reproach as illustrated below.

Kosrae in global context

Kosrae is the easternmost of four States in The Federated States of Micronesia (FSM). Kosrae is located in the Caroline Islands of the Pacific region, situated approximately halfway between Sydney and Honolulu.

Kosraean encounters with Europeans occurred intermittently between 1824 and 1910 with observers noting the importance of respect relationships present in Kosrae's political system (Cordy, 1993; Sarfert, 1919). Kosraean political organisation included a clear division of labour. Commoners, governed by respect behaviour, supplied their high chief or ruler "with almost daily food tribute, with labour on houses and canoes, with shell valuables, with labour on fishing expeditions, and so on" (Cordy, 1993, p. 100). Commoners competed with one another in contributing to feasts to demonstrate their prowess (Cordy, 1993). Traditional political organisation on Kosrae was radically altered in the late 1800s with drastic population declines from 3,000–6,000 to 300–400 and conversion to Christianity by American and Hawaiian missionaries (Cordy, 1993, pp. 96–97, 101; Ritter, 1981).

The Japanese held FSM through World War II and continued the work of the Europeans, further developing Kosrae for economically-valuable exports, including agricultural production, while also building military installations (Poyer et al., 2001). The Japanese time in FSM, particularly during WWII, was one marked by continued militarisation, disciplining of everyday life, and food rationing, including limiting access Micronesians had to locally-produced foods (Poyer et al., 2001).

After World War II, FSM was held as part of the Trust Territory of the Pacific Islands (TTPI) by the United Nations, administered by the United States. Kosrae's economy remained stagnant during the so-called "Navy Period" from 1945–1947 with little movement of goods or people through the islands (Falgout, 1995, p. 103). This "hands-off" approach began to shift with the arrival of researchers, particularly anthropologists (Falgout, 1995, p. 104). Micronesia quickly became the most anthropologically studied group from 1947–1960 (Falgout, 1995, p. 99). Anthropologists were agents of change in the development of democracy, while also working to preserve traditional customs (Falgout, 1995, p. 100).

This preservation took the form of creation and codification of customary practices manifest in the "nine volumes on land tenure, agricultural practices, economic institutions, and naming, authored by district anthropologists and their Micronesian assistants" edited by staff anthropologist John DeYoung (Falgout, 1995, p. 108). The utilisation of this information by administrators was problematic as it: (a) Flattened the cultural complexities present; and (b). resulted in minor changes or tweaks to American-based democratic institutions producing results in the emerging legal system that were not in line with customary practices (Falgout, 1995, p. 108). The impacts of these efforts persisted, as Falgout (1995) noted, "Seeking to enter the global world but also valuing their unique pasts, Micronesians today continue trying to blend the two in ways begun by American anthropologists and American administrations 50 years earlier" (p. 109). This pattern remains the case in contemporary times, based on our research in FSM over the past five years.

As a part of the concerted effort to "Americanize" Micronesians, which began in earnest by the United States in 1960, the FSM food system was altered by promoting imported foods for consumption. This took several forms, particularly pertinent was the United States Department of Agriculture's (USDA) supplementary feeding programs starting in the 1960s, accelerating in the 1970s, and continuing through the 1990s (Englberger et al., 2003). The USDA surplus foods included rice and tinned foods and were distributed through school lunch programs at a cost of $5 million in 1985 alone (Englberger et al., 2003).

FSM secured its independence in 1986, and entered into a bilateral agreement with the United States called the Compact of Free Association (COFA), which included large sums of money (relative to FSM's GDP) for programming. While there was discussion of promoting local food production, most development efforts ultimately focused on production of export crops (Englberger et al., 2003). This focus contributed to continued nutrition-related issues in FSM, shifting diets towards imported foods and away from local foods rich in vitamin A and other key nutrients (Englberger et al., 2003).

Peoples' (1985) observation of an overall dependency on foreign aid and wage/government employment in FSM remains largely true today, with nearly half of the workforce employed in the public sector. Not only does the money earned goes into buying imported foods (Connell 2015) that are desired but not needed to maintain health, but it also supports a mixed subsistence food production system, wherein money is spent on imported technologies that assist in contemporary subsistence. These technologies include outboard motors, fibreglass boats, fishing line and lures, gasoline, and "green machines" or weed eaters. These purchased goods replace or modify historic and traditional agriculture, hunting, and fishing technologies and associated knowledge, while in many instances, increase the time efficiency in procuring foods.

The increased interdependence of subsistence and globalisation was well illustrated on a walk we took through the path connecting residences throughout the coastal community of Walung in the Tafunsak Municipality in summer 2019. There, in the shade of coastal palms, three generations of men talked over their next steps in repairing an outboard motor. This technology has been commonplace in Kosrae for decades. The use of outboard motors connects food systems to the reliance on global distribution while also supporting mixed subsistence economies. This is one of the many interconnections between global and local knowledge and practices that dominate the contemporary Kosraean food system. We consider the contemporary Kosraean food system broadly to include historically-derived practices and products through the lens of the fusion of cultures and colonial encounters, resulting in foods that are locally collected and produced. This system includes the incorporation of techniques and technologies that have arrived on the Kosrae over time and through a variety of means.

While wage labour can certainly take time away from subsistence production, the public sector hours of the Kosraean work week allows for ample time to pursue subsistence pursuits, particularly when assisted by time-saving technologies (i.e., automobiles to access upland farms and outboard motors to access fish). In addition to working in a mixed-subsistence context, there remains a desire for learning and utilising traditional knowledge and practices. Kosraeans balance the use of both outside resources, while they contend with extensive pressures to consume imported foods, and those found in their local communities. Many appear successful at achieving some degree of lifestyle balance through blending global sourcing with some local production and consumption of foods. This balance is something we have noted in the context of our continued consulting and ethnographic fieldwork projects in Kosrae (2017–2021).

The Kosraean food system in lived experience

Kosrae illustrates the complexity of considerations necessary to holistically consider food systems in the context of historical change and continuity, as well Kosraean hopes and dreams for the future of food and life. Kosraean explanations of their food and lifestyles blend reflections on the importance of maintaining local and traditional foods while recognising the continuity of change, as Kosrae has been at the crossroads of contact from a series of outsiders with varying degrees of interest in the island's offerings (Matsuda, 2012; Peattie, 1992; Rainbird, 2004). Globalisation processes defy any singular definition and are rather a culmination of continued interventions viewed and informed by shifting perspectives and priorities. This section delves into the complexities revealed from representative semi-structured interviews conducted in summer 2019 on Kosraean perceptions of dietary change and healthy diets.[1] Although experiences are not universally shared by Kosraeans, as intracultural variation is present in reported behaviours and desires regarding diet, these interview excerpts highlight the shift towards imported foods in historical memory for older adults, narratives of their return to locally-produced foods, and the desire for imported foods amongst young adults. As was the case for Tohono O'odham (Fazzino, 2008), the interviews of Kosraeans revealed shifting preferences and perspectives as well as a portrayal of the struggle to confront the settler society and new struggles focused on traditional foods and ways of being to address health challenges.

Residential schooling provided a training ground for young Kosraeans to come to know and experience diversified ways of living and eating. The requirement to attend these schools took Utwe and Malem youth away from their families for the entire school week. The commute to the school was also difficult and time consuming, as children would, in some instances, walk three to six miles to arrive by

Sunday evening and remain until Friday evening. As middle-aged adults and elders reflected on what, for many, were their first sustained encounters with imported foods, the school was the primary site where they came to know and love the new and exciting, if nutritionally inferior and expensive, imported foods. As recalled by an Utwe elder as a 10-year-old,

> We didn't really go for the local. We liked the new ones that were intro-duced, those were new…They were new and they tasted better than what we had…I liked imported foods because we had peaches, pears, peanut butter, and pancakes.

The children requested imported foods at home but were often told that they were too expensive. The dichotomy of diet—local foods at home and imported foods at the school—did not always hold true even in the same meal. As Gilbert, an older male from Lelu described, "My mom used to fix bananas with flour. They call it tempura…it's flour mixed with banana, [like a] banana pancake." This example illustrates a mixing of local and imported ingredients and Japanese styles of cooking.

While residential school youth relayed that they were well-fed most of the time, particularly with their often new-found access to imported foods, the timing of the meals left much to be desired. Food became a way to discipline youth to time-based regimes of workdays; the cook's workday ended at 4:30 PM, meaning dinner was served at 4 PM. According to Gilbert, children would be hungry by 8 PM and would have to forage for food off campus, violating the rules and facing physical punishment if caught.

The movement away from the residential school program did not mean a return to local foods, as the school continued to be a site where youth could access imported foods that were provided by the school's lunch. Local foods would be consumed as a less desired substitute in the instances when rice was not available. According to Andrea, a middle-aged woman from Lelu, "The school provided lunch. Until I graduated high school, we still had lunch at school…sometimes…when there was no rice…the school gave us taro and breadfruit."

Similarly, a middle-aged woman from Utwe recalls her fondness for the imported foods during her formative years and her transition to desiring more locally-sourced foods,

> I loved [imported foods]. I remember when I was still in high school I would go home after school and ask my mom, "What's the food?" And if she told me that, "We have breadfruit," and I will go to my neighbours' family and look for rice. And they were always teasing me when I'm growing old that they still remember me eating rice. But nowadays when I have local, I feel full, healthy. If I'm eating rice it's like … for me rice is like ramen, just for a few hours then gone. I need to eat. I will starve again. But when I'm eating local it could be for many hours.

The taste and perception of what food is considered a good meal does generally vary between young adults and their elders. Alberta, a young woman from Tafunsak, shares that she generally eats canned meat paired with rice for lunch, as, "Rice always has to be included." When breadfruit is substituted for rice she reported not getting full.

The lunch program, a major source of imported foods in the past, is no longer provided to most school-aged children. However, this change has not resulted in a return to locally-based diets. Wendy, a young woman from Lelu, explains that students and teachers get ready-made meals and snack foods from stores located near the school rather than bringing in local foods. Wendy described her ideal meal as, "Super Green Ramen with tropical Kool-Aid…soda if [I had] enough money" Despite the changing nature of foods, the cultural values of sharing and commensality remain highly valued. Wendy explains, "we sometimes donate so we can buy for all of us. We'll save up until Thursday or Friday, then we buy everything we want… We'll share our food [brought from home] and then save the money." Wendy and many of the youth shared their minimal participation in manual subsistence work, which sharply contrasts with the lived experiences of many elders who were accustomed to subsistence.

Gilbert, an elder, recalls that as a ten-year-old during Trust Territory times he and his siblings made regular weekly trips to provision the household with subsistence foods, particularly banana and breadfruit. These trips would take place after school. As he got older, in the seventh grade, he began regularly helping his father collect copra, which was a major source of income for his household, to be exported as ships arrived at three month intervals. He shared, "The process of it was very hard. I know it's not for me. That's why I wanted to go to school." The requirements of manual labour were diminished for children while at school but included gathering firewood and performing maintenance and cleaning tasks on the school grounds and buildings. The weekend began on Friday with some relatively short tasks at home like feeding pigs. Saturday was typically a big work day for the entire family as they prepared for their religious and social obligations on Sunday. Today outboard motors, generally, though not always, on fibreglass boats, aid greatly in the process of subsistence fishing and farming. Previously, the commute to and from the upland farms proved challenging, and for some relied on the tides, which extended the workday. Today, the road network and automobiles shorten the commute to farmlands. Labour-saving electric stoves and take-out foods have, in some instances, replaced the traditional *um* or earthen oven cooking that typically took place on Saturdays.

Gilbert also shared that alongside these technological and social shifts, his own dietary preferences shifted. When he desired imported foods in his youth, his parents and grandparents reminded him of the importance of consuming local foods to keep him strong, healthy, and food secure by avoiding "dying in the future… [when there] will be no more rice [or] canned meat." Gilbert's return to the old traditions, "to the old way of eating local…was difficult," but satisfying. The lessons that Gilbert learned apply to his life today as he works his taro patch in, equating the sweat and hard work to grow the foods with a healthy life. Just as his elders convinced him of the importance of eating the local foods, so too does he work to impress their importance upon his children, lamenting that they are not always receptive. Just as he did, Gilbert's children prefer imported foods, shunning local foods; despite these preferences, he works in unison with his wife to change their minds. The interviewees suggest that over the life course change in diet towards more traditional and locally-based foods is coming, just not at the pace that parents and grandparents would like to see it occur. In general, there is a sense amongst those in older generations, that the youth of today are too easily swayed by outside ideas that lead to changes in proper forms of respect and a work ethic. Many of these same elders reported being enamoured with Western foods when they were younger.

COVID-19 and the deceleration of globalisation

Kosraeans are interested in building the local capacity of the food system based not only on past practices of fishing, agroforestry, animal husbandry, and to a lesser extent hunting but also by adopting new technologies, such as greenhouse and annual crop farming initiatives, to promote local food. Agencies and individuals on Kosrae are using the Kosraean Moon Calendar in subsistence management. These efforts are coupled with continued interest in maintaining Kosraean language, community outreach programs designed to foster traditional ecological knowledge amongst youth, and historic preservation initiatives. These efforts demonstrate the sense of pride many Kosraeans have in their shared identity and sense of place. These assertions of sovereignty and cultural revitalisation are resources to grapple with Kosrae's colonial past. Subsistence practices and knowledge remains central to most household economic strategies, which includes wage employment and utilisation of outmigration and subsequent remittances in order to purchase imported goods, including food and materials used for subsistence.

COVID-19 disrupted and decelerated global movement of people and goods into the FSM; it also raised questions about reliance on imported goods. As Love and Wu (2020) note, "90 percent of all material products and goods, including survival essentials, are still transported by seafarers who are 'quarantined' at sea because of concerns about the coronavirus and the need for labour to ship supplies" (p. 55). This disruption resulted in shortages of items originating from the U.S., namely toilet paper and bottled water. COVID-19 related delays in the arrival of container ships portended a collapse of the system. Of course, this tale of global collapse, and the resultant waves of disturbance and an increased reliance on local production were overstated, as local desires for both local and imported goods and foods remained relatively unchanged from 2019 to 2021 (Meredith & Fazzino, 2021). Issues with supply chain logistics is something that Micronesians have dealt with before. They pivoted, making do without or finding alternatives. Hence, our initial lens of disruption, fuelled by media coverage of resultant impacts including plummeting global markets, was one informed by our own outlook. In other words, our own beliefs in the eventual collapse of business-as-usual production, distribution, and consumption patterns in food and other goods towards more sustainable, locally-based products and processes, shaped our readings of the fall-out of COVID-19. Our commitments to the home perspective underscores our emphasis on the promotion of food sovereignty.

Conclusion

The Kosraean case study reveals tensions and contradictions in Sach's (1999) home, contest, and astronaut perspectives. Researchers adopt these perspectives, either implicitly or explicitly, in order to frame and then attempt to address global issues such as the sustainability of diets. The Kosraean food system highlights the potential pitfalls of adopting theoretical approaches to confront the complexities and contradictions present in historical transformations, globalisation, and the lived experiences found in diverse locales. The COVID-19 pandemic illustrated several vulnerabilities in globalisation, making a shift to more socially and ecologically sustainable locally-based diets more appealing. The choice and potential for local foods to make up all of the Kosraean diet still remains with subsistence production widely practised, particularly amongst older generations. Kosraeans ultimately maintained their contemporary food system, blending both imported and locally-produced foods in a variety of ways, albeit with some generational differentiation. Nevertheless, the transformative potential of locally-based sustainable food systems and hence diets, cannot be overstated, as it offers an alternative path away from vulnerability and dependency (Hunn, 1999). Kosraeans, particularly those of the older generation, remain aware of this transformative potential by encouraging youth to make the choice to eat local.

Note

1 Research ethics approval—BU IRB# 2019-14-COLA.

References

Clark, M. A., Domingo, N. G., Colgan, K., Thakrar, S. K., Tilman, D., Lynch, J., Azevedo, I. L., & Hill, J. D. (2020). Global food system emissions could preclude achieving the 1.5° and 2°C climate change targets. *Science*, *370*(6517), 705–708.

Connell, J. (2015). Food security in the island Pacific: Is Micronesia as far away as ever? *Regional Environmental Change*, *15*(7), 1299–1311.

Cordy, R. 1993. Respect behaviour on Kosrae: An ethnohistoric study. *The Journal of Pacific History*, *28*(1), 96–108.

Englberger, L., Marks, G. C., & Fitzgerald, M. H. (2003). Insights on food and nutrition in the Federated States of Micronesia: A review of the literature. *Public Health Nutrition*, *6*(1), 5–17.

Falgout, S. (1995). Americans in paradise: Anthropologists, custom, and democracy in post-war Micronesia. *Ethnology*, *34*(2), Special Issue on Politics of Culture in the Pacific Islands: Part I 99–11.

Fazzino, D.V. (2003). The meaning and relevance of food security in the context of current globalization trends. *Journal of Land Use & Environmental Law*, *19*(2), 435–449.

Fazzino, D.V. (2008). Continuity and change in Tohono O'odham food systems: Implications for dietary interventions. *Culture & Agriculture*, *30*(1–2), 38–46.

Fazzino, D. V. (2012). The will to end hunger in the age of security. In K. Coulter & W. Schumann (Eds.), *Governing cultures: Anthropological perspectives on political labor, power and government* (pp. 183–208). Palgrave Macmillan.

Fears, R., Canales, C., Ter Meulen, V., & von Braun, J. (2019). Transforming food systems to deliver healthy, sustainable diets: The view from the world's science academies. *The Lancet Planetary Health*, *3*(4), e163–e165.

Hornborg, A. (2007). Learning from the Tiv: Why a sustainable economy would have to be multicentric. *Culture & Agriculture*, *29*(2), 63–69.

Hunn, E. S. (1999). The value of subsistence for the future of the world. In V. D. Nazarea (Ed.), *Ethnoecology: Situated knowledge/located lives* (pp. 23–36). University of Arizona Press.

Kevany K, & Kassam Z. (2021). Rapid, unprecedented change for human, animal, and environmental health. *COVID-19 Pandemic: Case Studies & Opinions*, *2*(1), 189–191.

Love, S., & Wu, L. (2020). Are we in the same boat? Ethnographic lessons of sheltering in place from international seafarers and Algerian Harraga in the age of global pandemic. *Anthropology Now*, *12*(1), 55–65.

Matsuda, M. K. (2012). *Pacific worlds: A history of seas, peoples, and cultures*. Cambridge University Press.

Meredith, A., & Fazzino, D. (2021). Learning from the past? Sovereign space and recreating self-reliance in Kosrae, federated states of micronesia. In Y. Campbell & J. Connell (Eds.), *COVID in the Islands: A comparative perspective on the Caribbean and the Pacific* (pp. 177–192). Palgrave Macmillan.

Peattie, M. R. (1992). *Nanyo: The rise and fall of the Japanese in Micronesia, 1885–1945. Pacific Islands Monograph Series* (Vol. 4). University of Hawaii Press.

Peoples, J. (1985). *Island in trust: Cultural change and dependence in a Micronesian community*. Westview Press.

Poyer, L., Falgout, S., & Carucci, L. M. (2001). *The typhoon of war: Micronesian experiences of the Pacific War*. University of Hawaii Press.

Rainbird, P. (2004). *The archaeology of Micronesia*. Cambridge University Press.

Ritter, P. L. (1981). Population of Kosrae at contact. *Micronesica*, *17*(1–2), 11–28 (December).

Sachs, W. (1999). Sustainable development and the crisis of nature: On the political anatomy of an oxymoron. *Living with Nature*, *23*(1), 23–42.

Sarfert, E. (1919). *Kosrae. Vol. 1, Ethnography: General information and material culture. Results of the South seas expedition 1908–1910*. L. Friederichsen, de Gruyter & Co., Hamburg [translated by Carmen Petrosian-Husa, 2008].

Stone, G. D. (2007). Agricultural deskilling and the spread of genetically modified cotton in Warangal. *Current Anthropology*, *48*(1), 67–103.

Willett, W., Rockström, J., Loken, B., Springmann, M., Lang, T., Vermeulen, S., Garnett, T., Tilman, D., DeClerck, F., Wood, A., Jonell, M., Clark, M., Gordon, L. J., Fanzo, J., Hawkes, C., Zurayk, R., Rivera, J. A., De Vries, W., Majele Sibanda, L., Afshin, A., Chaudhary, A., Herrero, M., Agustina, R., Branca, F., Lartey, A., Fan, S., Crona, B., Fox, E., Bignet, V., Troell, M., Lindahl, T., Singh, S., Cornell, S. E., Srinath Reddy, K., Narain, S., Nishtar, S., & Murray, C. J. L. (2019). Food in the anthropocene: The EAT-lancet commission on healthy diets from sustainable food systems. *Lancet*, *393*(10170), 447–492.

Young, J. C. (2020). Environmental colonialism, digital indigeneity, and the politicization of resilience. *Environment and Planning E: Nature and Space*, *4*(2), 230–251.

PART 10

Calls to action

54

MONITORING FOOD ENVIRONMENTS AND SYSTEMS FOR SUSTAINABLE DIETS IN AFRICA

Lessons from Ghana, Kenya, Senegal, and South Africa

Amos Laar, Hibbah Araba Osei-Kwasi, Matilda Essandoh Laar, William K. Bosu, Silver Nanema, Gershim Asiki, Mark Spires, Adama Diouf, Julien Sodiba Manga, Jean Claude Moubarac, Elom K. Aglago, Ali Jafri, Phyllis Ohene-Agyei, Michelle Holdsworth, and Stefanie Vandevijvere

Introduction

Over the past half century, globalised food systems have increasingly transformed diets from basic food commodities or minimally processed to highly processed foods that are high in energy, saturated fats, sugars, and salt. This transformation, and other enablers, have hastened the pace of the nutrition transition globally. Paradoxically, these systems, which are feeding a greater proportion of the world population excessive amounts of calories, are unable to nourish billions with the nutrients required for optimal health. Globally, over 690 million people are estimated to suffer from hunger (FAO, IFAD, UNICEF, WFP, & WHO, 2020); about two billion are overweight or obese, and some two billion suffer from micronutrient malnutrition (Fanzo et al., 2019). The global burden of disease is now predominantly diet-related, with obesity, hypertension, diabetes, and other nutrition-related non-communicable diseases (NCDs) among the top ten causes of disability and death (Afshin et al., 2019). It is estimated that diets high in salt and low in fresh fruits and whole grains account for approximately half of deaths and two-thirds of diet-related disability-adjusted life years (DALYs) (Afshin et al., 2019). By

DOI: 10.4324/9781003174417-64

2030, NCDs are predicted to become the leading cause of death in Africa (GBD Obesity Collaborators, 2017). The rate at which NDCs are increasing in sub-Saharan Africa (SSA) is alarming. An analysis spanning 1975–2016 showed that six of 60 nations in the world with the fastest-rising rates of adult obesity are in Africa (NCD Risk Factor Collaboration, 2016).

Currently, Africa faces a new challenge of the coexistence of a triple burden of malnutrition—a surge in obesity and other diet-related NCDs while undernutrition and micronutrient deficiencies persist. For several decades, health strategies in Africa have focused on addressing undernutrition, communicable diseases, and maternal and child health challenges. Efforts by African governments to address NCDs, particularly using food environments approaches, are nascent and limited in scope and depth (Asiki et al., 2020; Booth et al., 2021; Laar et al., 2020). For example, over the years, high level continental agriculture, nutrition, health and development policies/strategies such as the 2003 Maputo commitments (Africa Union, 2003), the 2014 Malabo Declaration (Africa Union, 2014), the Africa Region Nutrition Strategy 2015–2025 (Africa Union, 2015a), and the African Union's Agenda 2063 (Africa Union 2015b) have all focused more on hunger and food insecurity, than on NCDs. Africa needs policies that regional, national, and local actors can use to promote healthy food environments and assure healthy and sustainable diets.

Food environments and food environment policies

It has long been recognised that the lived physical and social environments, as well as the commercial and legal environments that shape where we live, work, and eat, are critical determinants of our health. Hence, there have been recent calls to improve our food environments to promote healthier diets and public health nutrition outcomes. Food environments have been conceptualised to include physical, economic, policy, and socio-cultural surroundings, opportunities, and conditions that influence peoples' food consumption patterns, and are acknowledged as determinants of health (Swinburn et al., 2013a). Others divide food environments into personal environments (including food accessibility, food affordability, convenience, and desirability); and external environments (encompassing food availability, prices, vendor, and product properties, marketing, and regulation) (Turner et al., 2018). Unhealthy food environments, particularly the greater availability of, and access to heavily marketed ultra-processed food products, especially sugar-sweetened beverages (Green et al., 2020), play a significant role in creating unhealthy diets (Reardon et al., 2021). Food environments are part of complex food systems; therefore, policy efforts to improve the healthiness of food environments need to be initiated at multiple levels and engage multiple actors across diverse sectors that account for the coexistence of multiple forms of malnutrition, i.e., with "double duty" actions) (World Health Organisation, 2017).

While evidence of the implementation of these policies is sparse (especially in Africa), articulations exist that reveal the actual and potential impacts of food environment policies that may serve to make unhealthy foods unattractive while increasing the availability of healthy foods and beverages in Africa (Laar, 2021; World Cancer Research Fund International, 2020). Presenting evidence from a review study, Booth et al. (2021) offer a logic model to elucidate the connection between implementation of food environment policies and availability of healthy diets. They argue that if Governments implement comprehensive policy measures that serve to limit the availability of unhealthy food products, while intervening to avail healthy ones to consumers, the production, processing, and promotion of unhealthy diets will be reduced, reducing the availability, attractiveness, and consumption of unhealthy foods.

Other researchers have developed a variety of conceptual frameworks to facilitate understanding and organisation of interventions to influence consumer behaviours within their food environments (Bailey & Harper, 2015). Bailey and Harper (2015) organise these policies and interventions into three groups: Those that *inform and empower,* those that *guide and influence,* and those that *incentivise, discourage or restrict.* Inform and empower policies are predicated on the assumption that better informed individuals will behave responsibly within their food environments and make better-informed choices. It is suggested that providing people with information on the behaviours of others (negative or positive) can encourage them to make certain choices (a phenomenon often termed *normative feedback*). These policies motivate the implementation of food environment policy measures such as dietary or nutrition guidelines, labelling, and certifications. *Guide and influence* strategies aim to change contextual cues, alter the prominence of different options, or change default options. For instance, if a governmental policy objective is to improve diet, efforts may be made to provide a cue to purchase fruit and vegetables in supermarkets by: Including a designated space in shopping trolleys; increasing the prominence of healthy foods and beverages in canteens or on supermarket shelves; changing the default options in restaurants and cafeterias from animal to plant-based foods, and from chips to salad or green vegetables; or, reducing plate sizes in restaurants to encourage lower consumption. Policies that aim to *disincentivise, discourage, or restrict* consumption of unhealthy and/or unsustainable foods can help by adding to the cost as relative to alternatives, for example through added taxes. Conversely, they can *incentivise* consumption of healthy and/ or sustainable foods by lowering the relative price (e.g., subsidies). Government can also impose restrictions and limits on the sale or use of undesirable foodstuffs or ban them outright, as in the case of Ghana where standards for meat products outlawed importation of low-quality fatty meat (Annan et al., 2018). However, a particular gap in Africa is the lack of attention to a major component of the food environments—the informal retail sectors.

The World Health Organisation (WHO) policies for the prevention and control of NCDs focus on the four key NCD risk factors (tobacco, harmful use of alcohol, unhealthy diet and physical inactivity); four disease areas (cardiovascular disease, diabetes, cancer and chronic respiratory disease) are another set of policies (some of which aim to improve food environments). Referred to as "Best

'Best buys': Effective interventions with cost effectiveness analysis (CEA) ≤$100 per DALY averted in LMICs	• Reduce salt intake through the reformulation of food products to contain less salt and the setting of target levels for the amount of salt in foods and meals[1] • Reduce salt intake through the establishment of a supportive environment in public institutions such as hospitals, schools, workplaces and nursing homes, to enable lower sodium options to be provided[1] • Reduce salt intake through a behaviour change communication and mass media campaign • Reduce salt intake through the implementation of front-of-pack labelling[2]
Effective interventions with CEA >|$100 per DALY averted in LMICs	• Eliminate industrial trans-fats through the development of legislation to ban their use in the food chain[2] • Reduce sugar consumption through effective taxation on sugar-sweetened beverages
Other recommended interventions from WHO guidance (CEA not available)	• Promote and support exclusive breastfeeding for the first 6 months of life, including promotion of breastfeeding • Implement subsidies to increase the intake of fruits and vegetables • Replace trans-fats and saturated fats with unsaturated fats through reformulation, labelling, fiscal policies or agricultural policies • Limit portion and package size to reduce energy intake and the risk of overweight/obesity • Implement nutrition education and counselling in different settings (for example, in preschools, schools, workplaces and hospitals) to increase the intake of fruits and vegetables • Implement nutrition labelling to reduce total energy intake (kcal), sugars, sodium and fats • Implement mass media campaign on healthy diets, including social marketing to reduce the intake of total fat, saturated fats, sugars and salt, and promote the intake of fruits and vegetables

Figure 54.1 WHO Best Buys and other recommended interventions-focus on unhealthy diets. Source: WHO, 2017, accessible through Open Access. Notes: 1. Requires multi-sectoral actions with relevant ministries and support by civil society; 2. Regulatory capacities along with multi-sectoral action is needed.

Buys" and other recommended interventions, these policies comprise a total of 88 interventions. Out of the 88 interventions, 16 qualify as "Best Buys." Considered to be the most cost-effective and feasible for implementation were those with an average cost-effectiveness ratio of ≤ $100/DALY 2 averted in low and lower middle-income countries (World Health Organisation, 2017). Others interventions considered effective or recommended are those with cost-effectiveness ratio of >$100/DALY averted, and are recommended despite lacking the evidence of cost-effectiveness, as shown in Figure 54.1.

Sustainable diets

Diets, the foods that people are recommended to consume, can be characterised as safe, nutritious, healthy, and sustainable. Sustainability encompasses all of the above and more; it requires meeting present needs without diminishing the ability of others to meet their present needs, or future generations to meet their needs (Fanzo, 2019). While the introductory chapter offers some valuable insights, currently no agreed upon definition has been finalised on sustainable healthy diet. There is, however, a broad consensus that diets that are lower in meat and dairy, ultra-processed foods, and beverages, and higher in whole grains, fruits, vegetables, and legumes can provide good nutrition at lower environmental costs (FAO and WHO, 2019). The FAO defines sustainable diets as, "dietary patterns that promote all dimensions of individuals' health and wellbeing; have low environmental pressure and impact; are accessible, affordable, safe and equitable; and are culturally acceptable" (FAO and WHO, 2019). So defined, sustainable healthy diets aim to: Achieve optimal growth and development of all individuals; support functioning and physical, mental, and social wellbeing at all life stages for present and future generations; contribute to preventing all forms of malnutrition (i.e., undernutrition, micronutrient deficiency, overweight, and obesity); reduce the risk of diet-related NCDs; promote planetary health; and support the preservation of biodiversity.

Monitoring food environments and sustainable diets

As outlined above, food environments are critical for ensuring equitable access to foods that are healthy, safe, sustainable, and affordable. The physical, economic, and policy components of the food environment can all be acted on to promote sustainable healthy diets (Drewnowski et al., 2020). Physical spaces can be modified to improve relative availability of food outlets that carry nutritious foods in low-income communities. To address economic access, certain actions may improve affordability, such as fortification, preventing food loss through supply chain improvements, and commodity specific vouchers for fruits, vegetables, and legumes (Drewnowski et al., 2020). Other potential policy actions that address accessibility to sustainable healthy foods are comprehensive marketing restrictions and easy-to-understand front-of-pack nutrition labels.

The International Network for Food and Obesity/NCDs Research, Monitoring and Action Support (INFORMAS) has a framework of *process modules* that make it possible to monitor the policies and actions of the public and private sectors; *outcome modules* that monitor dietary quality, risk factors, and NCD morbidity and mortality; and *impact modules* that monitor the key characteristics of food environments (Swinburn et al., 2013b; Vandevijvere and Swinburn, 2014). Some of the above indicators are currently featured in the INFORMAS Healthy Food Environment Policy Index (Food-EPI) module (Swinburn et al., 2013a). This module was originally developed to assess the extent of implementation of recommended food environment policies by national governments compared with international best practices, and to derive concrete priority actions to fill identified implementation gaps (Swinburn et al., 2013a). It comprises two components—policies and infrastructure support—and currently over 40 good practice indicators (see indicators in Table 54.1). By helping governments to recognise the impact from and shortcomings of food environments, and actions to implement in response, the Food-EPI process can monitor food environments and, to some extent, sustainable healthy diets. Appropriate sustainability indicators are required and still under development.

A recent review for assessing sustainable diets identified a set of indicators from the health/nutrition, environmental, and socio-economic viewpoints (Aldaya et al., 2021). Nutrition and health indicators include indicators such as: The prevalence of diet-related morbidity/mortality; food's contribution to energy, nutrient,

Amos Laar et al.

Table 54.1 Implementation of food environment policies in selected African countries

Country	Ghana	Kenya	Senegal	South Africa
Food Environment Policy Domains				
FOOD COMPOSITION				
1 Food composition standards established *for processed foods*	+	+	+	++
2 Food composition standards established *for out-of-home meals in food service outlets*	x	x	-	-
FOOD LABELLING				
1 *Ingredient lists and nutrient declarations in line with Codex recommendations*	++	+	+	+++
2 *Robust, evidence-based regulatory systems are in place for approving/reviewing claims on foods*	+	+	-	+
3 *A single, consistent, interpretive, evidence-informed front-of-pack labelling is applied to packaged foods*	x	+	+	-
4 *A consistent, single, simple, clearly-visible system of labelling the menu boards of quick service restaurants (i.e., fast food chains) is applied by the government*	x	x	-	-
FOOD PROMOTION				
1 *Effective policies are implemented by the government to restrict exposure and power of promotion of unhealthy foods to or for children through broadcast media (TV, radio)*	+	+	-	-
2 *Effective policies are implemented by the government to restrict exposure and power of promotion of unhealthy foods to or for children through non-broadcast media*	+	+	-	-
3 *Effective policies are implemented by the government to ensure that unhealthy foods are not commercially promoted to or for children in settings where children gather (e.g., preschools, schools, sport and cultural events)*	+	+	-	-
4 *Effective policies are implemented by the government to restrict the marketing of breastmilk substitutes*	+++	++	x	x
FOOD PRICES				
1 *Taxes or levies on healthy foods are minimised to encourage healthy food choices where possible (e.g., low or no sales tax, excise, value-added or import duties on fruit and vegetables)*	+	-	+	++
2 *Taxes or levies on unhealthy foods (e.g., sugar-sweetened beverages, foods high in nutrients of concern) are in place*	+	-	+	+
3 *The intent of existing subsidies on foods, to favour healthy rather than unhealthy foods*	+	+	+	+
4 *The government ensures that food-related income support programmes are for healthy foods*	+	+	+	-

FOOD PROVISION

1	The government ensures that there are clear, consistent policies (including nutrition standards) implemented in schools and early childhood education services for food service activities (canteens, food at events, fundraising, promotions, vending machines etc.) to provide and promote healthy food choices.	+	+	+	+
2	The Government ensures that there are clear, consistent policies in other public sector settings for food service activities to provide and promote healthy food choices.	X	+	+	-
3	The Government ensures that there are good support and training systems to help schools and other public sector organisations and their caterers meet the healthy food service policies and guidelines	+	+	+	+
4	Private company nutrition policy	x	x	x	-

FOOD RETAIL

1	Zoning laws and policies are robust enough and are being used, where needed, by local governments to place limits on the density or placement of quick serve restaurants or other outlets selling mainly unhealthy foods in communities, and to encourage the availability of outlets selling healthy options such as fresh fruit and vegetables	+	x	-	-
2	The Government ensures existing support systems are in place to encourage food stores and food service outlets to promote the availability of healthy foods and to limit the promotion and availability of unhealthy foods	x	+	-	-
3	Food hygiene policies are robust enough and are being enforced, where needed, by national and local governments to protect human health and consumers' interests in relation to food.	X	x	x	x

FOOD TRADE AND INVESTMENT

1	The Government undertakes risk impact assessments before and during the negotiation of trade and investment agreements, to identify, evaluate and minimise the direct and indirect negative impacts of such agreements on population nutrition and health	+	x	+	+
2	The government adopts measures to manage investment and protect their regulatory capacity with respect to public health nutrition	++	+	+	+

Source: Work of the authors

Key: Level of implementation in country very little (-) || low (+) || medium (++) || high (+++) x = not assessed

and biocompound requirements; food rations adjusted to nutrient/energy requirements (e.g., serving size by age/physical activity); dietary diversity, including by
composition and level of processing); and energy intake from sustainable sources.
Environmentally, a sustainable diet may be assessed using two main approaches—
life cycle analysis (LCA) (Rose et al., 2019) and environmental footprints (i.e., carbon footprint, water footprint, land footprint, nitrogen, phosphorus, and particulate
matter footprints, chemicals/pesticides footprint, biodiversity footprint) (Graham
et al., 2019; Springmann et al., 2020). Socio-economic viewpoints of sustainability
are based on indicators that affect supply (e.g., production costs, availability, scalability, societal factors) and demand (e.g., accessibility, affordability, acceptability),
and the necessary value chains that connect them (Aldaya et al., 2021).

Monitoring policies for food environments and sustainable diets: Lessons from Ghana, Kenya, Senegal, and South Africa

This section draws on available country level Food-EPI scorecards (see Table 54.1)
and prioritised policy actions for the case study countries (Ghana, Kenya, Senegal,
and South Africa) to gauge the level of implementation and priorities for food
environment policy. The Food-EPI assessments were done between 2016 and 2020;
that is, in 2016 (South Africa), 2017 (Ghana and Kenya), and in 2020 (Senegal).
In addition to the Food-EPI scorecards, we consulted country progress reports
included in the WHO NCD Monitor for 2020 (World Health Organisation,
2020). Finally, we gauged the potential of the outlined food environment actions
(from the Food-EPI assessments and the WHO Best Buys) in monitoring sustainable diets.

Case 1: Ghana

The Ghana Food-EPI exercise identified several gaps in the implementation of
food environment policies (Laar et al., 2020). Of note, only food policy indicators
(n = 16) are featured in this chapter; the infrastructure-support related indicators
(n = 27) are not included here. Ghana was rated "low"/"very little" in implementation on most (three-quarters) of all good practice indicators (Laar et al., 2020;
also see Table 54.1). Restricting the marketing of breast milk substitutes was the
only indicator rated "very high." Current government actions were low on food
prices (e.g., taxes and subsidies), food retail, food provision, and unhealthy food
promotion to children; policy action to establish ingredient lists/nutrient declarations was assessed as "low," as were the government's efforts to protect regulatory
capacities regarding nutrition, in particular through setting standards for maximum
fat contents in beef, pork, mutton, and poultry (Laar et al., 2020). Government has
expressed intentions for a policy limiting the level of trans fats and salt in indus-
trially processed food, as well as a future intention to create nutritional standards
for out-of-home meals. Hints included in the policies of the possibility of taxing unhealthy food resulted in a surge in advocacy/campaigning by public health
professionals and civil society nudging the government to introduce sugar sweet-

ened beverage tax. Despite these efforts, several policy areas remain sub-optimally developed. These include legislative action for a labelling system to better inform consumer food choices, and actions to provide and promote healthier food in public places. In addition to these "policy desert" areas, Ghana lacks the requisite resources for effective enforcement of already enacted policies.

Where common indicators exist (e.g., food composition and marketing policies), the Food-EPI assessment compares favourably with the country's efforts captured in the 2020 WHO NCDs Monitor (WHO, 2020). For instance, both the Food-EPI and the NCDs Monitor report that salt/sodium policies targets, saturated fatty acids and trans-fats policies, and marketing to children restrictions are sub-optimally implemented. Marketing of breast-milk substitutes restrictions, however, is fully achieved (Laar et al., 2020; WHO, 2020).

On the basis of these findings, the Ghanaian Food-EPI expert panel proposed policy actions for creating healthier food environments in Ghana. Details of the prioritisation process are reported elsewhere (Laar et al., 2020). These priority recommendations require the government to enact and implement food promotion, labelling, and provisioning policies, as shown in Table 54.2.

Case 2: Kenya

Using the Food-EPI process, the extent of implementation of food environment policies in Kenya were recently assessed (Asiki et al., 2020). Nearly all the best practice indicators of the policy domains were rated as "very little" or "low" implementation (as shown in Table 54.1), such as, on *food labelling* (i.e., ingredient lists and nutrient declaration in line with Codex Alimentarius on packaged foods, regulatory systems for approving claims, and evidence informed front of pack nutrition information systems). Similarly, three sub indicators on *food promotion* were rated mostly at a low level of implementation (i.e., restrictions of unhealthy foods for children through broadcast media, restrictions of unhealthy foods to children through non broadcast media, and restrictions of promotion of unhealthy foods to children in settings they gather). The only sub indicator that scored as "medium" level of implementation in Kenya was policies to restrict the marketing of breastmilk substitutes. On food prices, the indicators were rated either as "very low" (taxes or levies on healthy foods and unhealthy foods) or "low" level of implementation (intent of existing subsidies for healthy foods, food related income supporting programmes for healthy foods). Food retail, food trade, and investment were rated as "low" in implementation. There was no evidence of any government action documented for five policy areas of good practice including: Food composition standards/targets for out-of-home meals in food service outlets; front-of-pack or menu board labelling systems; risk assessments for trade agreements; or, zoning laws on the density/location of healthy/unhealthy food service outlets.

As is the case with Ghana, the Kenya Food-EPI assessment was compared with the country's status report regarding level of effort at implementing: Salt/sodium policies (not initiated); marketing to children restrictions (not initiated); saturated fatty acids and trans-fats policies (suboptimal implementation); and marketing of

Table 54.2 Priority actions for implementation to improve food environment in study countries

Food environment policy domain	Ghana	Kenya	Senegal	South Africa
Food composition	–	Ensure that food standards for processed foods include the information on the energy density for different target groups	Define a model of good practices and nutritional standards focusing on the content of nutrients of concern in foods offered in catering establishments (fast-food restaurants and mass restaurants)	–
Food labelling	Support nutrition advocacy (e.g., with financial support, knowledge and research development, capacity planning)	Ensure that food policy includes international best practices to eliminate trans fats and where some percentage is included to label and issue traffic lights Ensure that food labelling is standardised and explicit to the nutrition profile of the processed food	Introduce obligatory legislation for nutrients (energy, saturated fat, total fat, protein, carbohydrates, sugars and salt) on packaging	Enact regulations towards the implementation of a standardised format for easy-to-read front of pack labelling and logos format

Food promotion	Pass legislation to regulate the promotion, sponsorship, advertisement and sale of food and drink with added sugar, and other nutrients of concern (saturated fatty acids/trans-fat, salt) in the school environment and other child-laden setting, enforceable with fine, and in print and electronic media, enforceable with fines Enforce legislation to regulate the promotion, sponsorship, advertisement and sale of food and drink with added sugar, and other nutrients of concern (saturated fatty acids/trans fats, salt) in print and electronic media, enforceable with fines	Ensure regulations and standards on advertising and marketing are enforced and a policy framework on how to engage with commercial processed food producers is in place	Promote healthy eating during peak viewing hours through the media (public and private)	Devise regulations that would support municipal infrastructure for healthier retail, especially to informal vendors – these may include zoning relaxations for informal vendors selling healthy food vendors; and subsidies for these vendors
	–	–	–	–
Food prices	–	Establish tax policies that favour production and consumption of healthy foods and discourage unhealthy foods Increase taxes on unhealthy foods and drinks and on restaurants that sell fast foods to increase their prices	–	Introduce subsidies for vendors who sell healthy foods –

(Continued)

Table 54.2 (Continued)

Food environment policy domain	Ghana	Kenya	Senegal	South Africa
Food provision	Implement a requirement for caterers involved in school feeding programmes to pass a training course on healthy meal planning	Introduce policy to provide health foods in government funded food programmes	Develop a regional school menu based on local products. Establish nutritional standards for school meals	Provide workplace support for breastfeeding, e.g., raising awareness of the legal rights of breastfeeding women to have 2 x ½ hour breaks for breastfeeding
Food in retail	–	–	–	Display and shelving regulations – having cut off limits for an upper limit on shelf space allocated to unhealthy foods vs. A minimum allocation for healthy foods. Enact regulations that would support municipal infrastructure for healthier retail, especially to informal vendors e.g., zoning relaxations for informal vendors selling healthy food vendors
Food trade and investment	–	Ensure there is a proper trade policy to ensure the importation of healthy foods and regulate unhealthy foods related to NCDs	–	–

Source: Work of the authors.

breast-milk substitutes restrictions (suboptimal implementation) (WHO, 2020). Informed by these gaps, Kenyan stakeholders recommended policy implementation focusing on food promotion, labelling, food composition, trade and investment, food prices, and food provision (see Table 54.2).

Case 3: Senegal

In Senegal, the Food-EPI exercise revealed that efforts aimed at creating healthy food environments were suboptimal. Overall, 45 priority actions were identified, including ten priority actions recommended to the Government of Senegal with high importance, high capacity, and high potential effect to improve food environments and reduce the triple burden of malnutrition (Manga et al., 2022). Both the Food-EPI assessments and the WHO NCD Monitor report for Senegal show national inaction on salt/sodium policies targets, saturated fatty acids and trans-fats policies, and marketing to children restrictions (World Health Organisation, 2020). Regarding restrictions of breast milk substitutes marketing, the 1994 Inter-ministerial Decree has been revised and a draft legislation on the commercialisation of complementary foods incorporating specific guidelines on labelling practices and the regulated promotion within and outside of health facilities has been developed and submitted to the government (Manga et al., 2022). Furthermore, assessment of children exposure to unhealthy food marketing in public spaces was initiated to provide evidence to support policies aimed at restricting food marketing of unhealthy foods. To respond to the identified gaps, Senegalese stakeholders identified, prioritised, and recommended food environment policy actions (food composition, food provision, food labelling, and promotion) for implementation (see Table 54.2).

There is consensus in Senegal that these actions would be realised if a deliberate effort is made to integrate existing food environment's good practice indicators into the monitoring and evaluation system of the Multisectoral Strategic Plan for Nutrition that implements the National Nutrition Development Policy 2015–2025. To facilitate this integration, Senegal has initiated a process of developing a programme called "Resilient food systems towards healthy diets for people vulnerable to malnutrition in Senegal." This programme involves the National Nutrition Development Council, FAO, Solidarity Union Cooperation, and other technical and financial partners. One of the strategic components of this programme aims to "integrate good practices in healthy food environments and the recommendations of national Food Based Dietary Guidelines into institutional and sectoral nutrition and food security strategies and policies" (Senegal Food-EPI as cited by Manga et al., 2022).

Case 4: South Africa

In South Africa, the extent to which food environment policies have been developed and implemented was assessed using the Food-EPI process (Spires et al., 2017). Overall, experts rated the levels of policy implementation across most indicators as "very little/if any" to "low" (see Table 54.1). The only indicator that was rated "high" was the implementation of a regulatory system to ensure that labels on packaged foods have ingredient lists and nutrient declarations that were in line with Codex recommendations. In 2010, the South African Food Labelling Regulations (R146/2010) was introduced and outlines the details on how ingredients should be listed on all packaged food labels, including lettering size, order, naming of ingredients, the inclusion and itemised listing of additives, quantitative ingredient declaration, the listing of class names of included fats and oils, and guidelines for nutrient declarations.

Although significant efforts have been made to develop legislation on food composition standards for processed foods, the guidelines for implementation have not been equally established at the time the Food-EPI was conducted. Currently, in South Africa, 19 specified basic foodstuffs (including brown bread, dried beans, maize meal, milk eggs, fish, and some fruits and vegetables) are considered zero rated supplies (taxable supplies on which value-added tax is levied at a rate of 0%) and the government has enacted a levy on sugar-sweetened beverages. In addition, a planned strategic action for the prevention and control of obesity in South Africa for 2015 to 2020 was to improve demand for healthy food by exploring the expansion of rebates on healthy food purchases.

Although appreciable effort has been made in the implementation of the above policies, experts identified significant gaps in implementation of other policies. For example, there are currently no policies regulating menu boards at fast food outlets and there is a notable gap in government policy implementation to ensure that unhealthy foods are not commercially promoted to children in settings where children gather. No evidence of zoning laws and policies that support the availability of healthy foods exists.

Generally, the South African Food-EPI assessments are in line with the WHO NCD Monitor report for South Africa (WHO, 2020). The Monitor reports that salt/sodium policies targets, saturated fatty acids and trans-fats policies are partially achieved, and marketing to children restrictions are not achieved. The implementation of marketing of breast-milk substitutes restrictions is shown as fully achieved by the WHO NCD Monitor; this indicator was not assessed as part of the Food-EPI efforts in South Africa (Spires et al., 2017). Actions prioritised for implementation in South Africa focused on food retail, food pricing, and food provision (see Table 54.2).

Potential for food environment indicators to monitor sustainable diets

The indicators for assessing sustainable diets embody many dimensions—including at least health/nutrition, environment, and socio-economic equity. The current measures of healthiness of food environments (see Table 54.1) can be applied to monitor nutritional/health and socio-economic sustainability dimensions. The shared indicators include prevalence of diet-related morbidity/mortality, food's contribution to energy, food rations adjusted to nutrient/energy requirements, and dietary diversity (including composition and level of processing). Similarly, socio-economic indicators of sustainability, such as measures of food availability, resilience/stability, acceptability, production costs, and impact on farmers' livelihoods (Aldaya et al., 2021), are food environment indicators. Indicators of food environments include: Guidelines on consuming processed foods; food labelling with ingredient lists; taxes or levies on unhealthy foods; zoning laws to limit density and placement of food outlets selling processed foods; and supports for outlets selling healthy foods; restrictions for exposure to and promotion of unhealthy foods (particularly to children) and levels of nutrition.

However, environment-related indicators for sustainable healthy diets are indirectly and inadequately addressed by the current food environment monitoring protocols (Swinburn et al., 2013a; Swinburn et al., 2013b). For example, life cycle analysis of individual foods/beverages and environmental footprint measures (i.e., encompassing water, land, nitrogen/phosphorus, and biodiversity) (Aldaya et al., 2021; Springmann et al., 2020) are not covered in the food environment protocols. In recent years, scientists (Global Syndemic Lancet Commission, EAT Lancet Commission) as well as international organisations (WHO, FAO, Nordic Council of Ministers, Committee on World Food Security, UNICEF) and other global organisations (GAIN) have published increasing recommendations for policies to create sustainable food systems, with a major focus on government actors. Based on these recommendations and available evidence regarding impact on nutrition-related outcomes, nutrition-related inequalities, and environmental sustainability outcomes (i.e., reducing greenhouse gas emissions, land use, water use, nitrogen and phosphorus use, biodiversity, etc.), the existing Food-EPI could be transformed into an index measuring the actions of governments to create healthy diets from sustainable food systems. Indicators with double and triple duty potential could be prioritised for inclusion in such an index.

Challenges

Several multifaceted challenges hinder the implementation and monitoring of food environments and sustainable diets in specific African contexts. First, is a lack of comprehensive models/framework and data: Whereas models such as those of the INFORMAS provide a useful set of indicators for monitoring nutrition and health indicators, models to monitor the environmental effects of food production and consumption or to monitor socioeconomic sustainability are generally lacking. Many studies that provide the basis of these monitoring indicators tend to be conducted in middle and high-income countries. Models for monitoring often exclude informal vendors, even though they serve a significant proportion of the community in Africa. Second, is a lack of awareness and capacity in the food systems: there are a number of systemic, structural, programmatic, and operational challenges to transitioning to sustainable diets. These challenges include weak capacity of regulatory agencies, weak enforcement of laws, prioritisation of undernutrition and food security to the disadvantage of NCDs-interventions, low financial investments aligned to the burden of NCDs, national strategic plans lacking defined targets aligned to global targets, aggressive advertising of sweetened food and beverages, and low coverage of social protection programmes (e.g., school feeding programmes, cash and food transfers). Finally, regarding food composition, there is a need to improve and fully implement proper strategies, as processed foods and out-of-home food consumption are steadily increasing in Africa (Reardon et al., 2021). To maximise health benefits, the Food-EPI components on food prices should be implemented on both healthy and unhealthy foods (Blake et al., 2021), although potential challenges to implementation exist (Laar, 2021).

Conclusion: Opportunities and way forward

This chapter has highlighted policies associated with food environments and sustainable diets. This study considered lessons from monitoring food environments in select African countries and the extent of monitoring of sustainable diets along with the challenges, opportunities, and recommendations for improved food environments and systems for sustainable diets. It is evident that more research in Africa is needed to define models for monitoring sustainable diets. Studies will be most useful if they engage policy makers, programme, and community actors from the start and on a continuing basis, learning from the cost of hunger studies, INFORMAS and MEALS4NCDs (Laar, 2021). As articulated by Swinburn et al. (2015), accountability systems should be strengthened to support government leadership and stewardship, hold private sector actors accountable, equip civil society to create demand for healthy food environments, and to monitor progress in meeting the global nutrition and NCD targets.

A range of recommendations and initiatives include the Codex Alimentarius, United Nations Decade of Action on Nutrition (2016–2025), Global Nutrition for Growth; the Nutrition for Growth (N4G) Summits, the United Nations Food Systems Summit Dialogues, the SUN Movement, and the Initiative for Food and Nutrition Security in Africa (IFNA). The recent set of *nutrition and health, environment, and socio-economic indicators* proposed by Aldaya et al. (2021) provide a useful framework for the development of a core set of indicators on sustainable healthy diets for Africa. Countries can build on existing multi sector initiatives, such as the national and regional agricultural investment plans that cover food and nutrition security, *one health* approach, and the sanitary and phytosanitary measures. African countries can and should take advantage of the momentum from current global and regional initiatives to sustainably tackle malnutrition in all its forms.

References

Afshin, A., Sur, P. J., Fay, K. A., Cornaby, L., Ferrara, G., Salama, J. S., Mullany, E. C., Abate, K. H., Abbafati, C., Abebe, Z., Afarideh, M., Aggarwal, A., Agrawal, S., Akinyemiju, T., Alahdab, F., Bacha, U., Bachman, V. F., Badali, H., Badawi, A., … & Murray, C. J. L. (2019). Health effects of dietary risks in 195 countries, 1990–2017: A systematic analysis for the Global Burden of Disease Study 2017. *The Lancet*, *393*(10184), 1958–1972.

African Union Commission. (2003) African Union declaration on agriculture and food security in Africa. Addis Ababa (Ethiopia): African Union Commission. Contract No.: Assembly/AU/Decl.4-11 (II).

African Union Commission. (2014) Malabo declaration on accelerated agricultural growth and transformation for shared prosperity and improved livelihoods. Addis Ababa (Ethiopia): African Union Commission.

African Union Commission. (2015a) African Regional Nutrition Strategy 2015-2015. Addis Ababa (Ethiopia): African Union Commission.

African Union Commission. (2015b) Agenda 2063: The AfricaWeWant. Addis Ababa (Ethiopia): African Union Commission.

Aldaya, M. M., Ibañez, F. C., Domínguez-Lacueva, P., Murillo-Arbizu, M. T., Rubio-Varas, M., Soret, B., & Beriain, M. J. (2021). Indicators and recommendations for assessing sustainable healthy diets. *Foods, 10*, 999.

Annan, R. A., Apprey, C., Oppong, N. K., Petty-Agamatey, V., Mensah, L., & Thow, A. M. (2018). Public awareness and perception of Ghana's restrictive policy on fatty meat, as well as preference and consumption of meat products among Ghanaian adults living in the Kumasi Metropolis. *BioMed Central Nutrition, 4*(1), 1–8.

Asiki, G., Wanjohi, M. N., Barnes, A., Bash, K., Muthuri, S., Amugsi, D., Doughman, D., Kimani, E., Vandevijvere, S., & Holdsworth, M. (2020). Benchmarking food environment policies for the prevention of diet-related non-communicable diseases in Kenya: National expert panel's assessment and priority recommendations. *Plos One, 15*(8), e0236699.

Bailey, R., & Harper, D. R. (2015). *Reviewing interventions for healthy and sustainable diets.* Research paper. The Royal Institute of International Affairs.

Blake, C. E., Frongillo, E. A., Warren, A. M., Constantinides, S. V., Rampalli, K. K., & Bhandari, S. (2021). Elaborating the science of food choice for rapidly changing food systems in low-and middle-income countries. *Global Food Security, 28*, 100503.

Booth, A., Barnes, A., Laar, A., Akparibo, R., Graham, F., Bash, K., Asiki, G., & Holdsworth, M. (2021). Policy action within urban African food systems to promote healthy food consumption: A realist synthesis in Ghana and Kenya. *International Journal of Health Policy and Management, 10*, 828–844.

Drewnowski, A., Monterrosa, E. C., de Pee, S., Frongillo, E. A., & Vandevijvere, S. (2020). Shaping physical, economic, and policy components of the food environment to create sustainable healthy diets. *Food and Nutrition Bulletin, 41*(2_suppl), 74S–86S.

Fanzo, J. (2019). Healthy and sustainable diets and food systems: The key to achieving sustainable development goal 2? *Food Ethics, 4*(2), 159–174.

Fanzo, J., Hawkes, C., Udomkesmalee, E., Afshin, A., Allemandi, L., Assery, O., Baker, P., Battersby, J., Bhutta, Z., Chen, K., Corvalan, C., Di Cesare, M., Dolan, C., Fonseca, J., Grummer-Strawn, L., Hayashi, C., McArthur, J., Rao, A., Rosenzweig, C. and Schofield, D. (2019). *2018 Global Nutrition Report.* Global Nutrition Report.

Food and Agriculture Organization and World Health Organization. (2019). *Sustainable healthy diets: Guiding principles.* Rome, Italy.

FAO, IFAD, UNICEF, WFP, & WHO. (2020). *The state of food security and nutrition in the world 2020: Transforming food systems for affordable healthy diets.* FAO.

GBD Obesity Collaborators (2017). Health effects of overweight and obesity in 195 countries over 25 years. *New England Journal of Medicine, 377*(1), 13–27.

Graham, F., Russell, J., Holdsworth, M., Menon, M., & Barker, M. (2019). Exploring the relationship between environmental impact and nutrient content of sandwiches and beverages available in cafés in a UK university. *Sustainability, 11*, 3190.

Green, M. A., Pradeilles, R., Laar, A., Osei-Kwasi, H., Bricas, N., Coleman, N., Klomegah, S., Wanjohi, M. N., Tandoh, A., Akparibo, R., Aryeetey, R. N. O., Griffiths, P., Kimani-Murage, E. W., Mensah, K., Muthuri, S., Zotor, F., & Holdsworth, M. (2020). Investigating foods and beverages sold and advertised in deprived urban neighbourhoods in Ghana and Kenya: A cross-sectional study. *British Medical Journal Open, 10*(6), p.e035680.

Laar, A. (2021). The role of food environment policies in making unhealthy foods unattractive and healthy foods available in Africa. *EClinicalMedicine, 36.* https://doi.org/10.1016/j.eclinm.2021.100908

Laar, A., Barnes, A., Aryeetey, R., Tandoh, A., Bash, K., Mensah, K., Zotor F., Vandevijvere S., & Holdsworth, M. (2020). Implementation of healthy food environment policies to prevent nutrition-related non-communicable diseases in Ghana: National experts' assessment of government action. *Food Policy*, *93*, 101907.

Laar, A., Kelly, B., Holdsworth, M., Quarpong, W., Aryeetey, R., Amevinya, G., Tandoh A., Agyemang C., Zotor F., Laar M. E., Mensah K., Laryea D., Asiki G., Pradeilles R., Sellen, D., L'Abbe, M. R., & Vandevijvere, S. (2021Providing measurement, evaluation, accountability, and leadership support (MEALS) for non-communicable diseases prevention in Ghana: project implementation protocol. *Frontiers in nutrition*, 504.

Manga, J. S., Diouf, A., Vandevijvere, S., Diagne, M., Kwadjod, K., Dossou, N. I., Thiam, E. H. M., Ndiaye, N., & Moubarac J. C. (2022). Evaluation and prioritisation of actions on food environments to address the double burden of malnutrition in Senegal: Perspectives from a national expert panel. *Public Health Nutrition*, *25*(8), 2043–2055–.

NCD Risk Factor Collaboration (2016). Trends in adult body-mass index in 200 countries from 1975 to 2014: A pooled analysis of 1698 population-based measurement studies with 19· 2 million participants. *The Lancet*, *387*(10026), 1377–1396.

Reardon, T., Tschirley, D., Liverpool-Tasie, L. S. O., Awokuse, T., Fanzo, J., Minten, B., Vos, R., Dolislager, M., Sauer, C., Dhar, R., Vargas, C., Lartey, A., Raz, A., & Popkin, B. M. (2021). The processed food revolution in African food systems and the double burden of malnutrition. *Global Food Security*, *28*, 100466.

Rose, D., Heller, M. C., & Roberto, C. A. (2019). Position of the society for nutrition education and behaviour: The importance of including environmental sustainability in dietary guidance. *Journal of Nutrition Education and Behaviour*, *51*(1), 3–15.

Spires, M., Sanders, D., & Swart, R. (2017). Creating healthy food environments through the benchmarking of nutrition-related government policies in South Africa: Local expert recommendations for improved practice. Poster presented at *IUNS 21st ICN*, Buenos Aires, 15-20 October 2017.

Springmann, M., Spajic, L., Clark, M. A., Poore, J., Herforth, A., Webb, P., Rayner M., & Scarborough, P. (2020). The healthiness and sustainability of national and global food based dietary guidelines: Modelling study. *British Medical Journal*, *370*, m2322.

Swinburn B., Sacks G., Vandevijvere S., Kumanyika S., Lobstein T., Neal B., Barquera S., Friel S., Hawkes C., Kelly B., L'Abbé M., Lee A., Ma J., Macmullan J., Mohan S., Monteiro C., Rayner M., Sanders D., Snowdon W., & Walker C. (2013a). INFORMAS (International network for food and obesity/non-communicable diseases research, monitoring and action support): Overview and key principles. *Obesity Reviews*, *14*(Supplement 1), 1–12.

Swinburn, B., Vandevijvere, S., Kraak, V., Sacks, G., Snowdon, W., Hawkes, C., Barquera, S., Friel, S., Kelly, B., Kumanyika, S., L'Abbé, M., Lee, A., Lobstein, T., Mal, J., Macmullan, J., Mohan, S., Monteiro, C., Neal, B., Rayner, M., ..., & Walker, C. for INFORMAS (2013b). Monitoring and benchmarking government policies and actions to improve the healthiness of food environments: A proposed government healthy food environment policy index. *Obesity Reviews*, *14*, 24–37.

Swinburn, B., Kraak, V., Rutter, H., Vandevijvere, S., Lobstein, T., Sacks, G., Gomes F., Marsh T., & Magnusson, R. (2015). Strengthening of accountability systems to create healthy food environments and reduce global obesity. *The Lancet*, *385*, 2534–2545.

Turner, C., Aggarwal, A., Walls, H., Herforth, A., Drewnowski, A., Coates, J., Kalamatiaou S., & Kadiyala, S. (2018). Concepts and critical perspectives for food environment research: A global framework with implications for action in low-and middle-income countries. *Global Food Security*, *18*, 93–101.

Vandevijvere, S., & Swinburn, B. (2014). International network for food and obesity/non-communicable diseases (NCDs) research, monitoring and action support (INFORMAS): Towards global benchmarking of food environments and policies to reduce obesity and diet-related non-communicable diseases: Design and methods for nation-wide surveys. *British Medical Journal Open*, *4*(5), e005339.

World Cancer Research Fund International (2020). *Building momentum: Lessons on imple-menting evidence-informed nutrition policy.* https://www.wcrf.org/int/policy/our-publica-tions/building-momentum-lessons-implementing-evidence-informed-nutrition

World Health Organisation (2017b). *Double-duty actions for nutrition: Policy brief* (No. WHO/NMH/NHD/17.2). WHO. https://apps.who.int/iris/bitstream/handle/10665/255414/WHO?sequence=1

World Health Organisation (2017a). *Tackling NCDs: 'Best buys' and other recommended inter-ventions for the prevention and control of noncommunicable diseases.* WHO. https://apps.who.int/iris/handle/10665/259232

World Health Organisation (2020). *Noncommunicable diseases progress monitor 2020.* WHO. https://movendi.ngo/wp-content/uploads/2020/02/Noncommunicable-Diseases-Progress-Monitor-2020.pdf.s

55

CALLS TO ACTION FOR SUSTAINABLE DIETS

Kathleen Kevany and Paolo Prosperi

Introduction

Nothing less than arousing calls to action can be the focus of this final chapter. We are inspired by the conviction of food systems actors worldwide working towards sustainable diets; we also are compelled by the profound arguments put forth in this handbook, yet we are dismayed by the insufficient will to protect the common good. This concluding chapter provides readers, all who are ready to take action, a comprehensive map of strategies that if deployed could dramatically increase sustainability and put us on a course for reversing global warming, diet-related non-communicable diseases, and oppression and racism and cultivate the nourishing food systems so urgently needed.

Framing and vision

Framed by definitions for sustainable diets, the first thematic section of the handbook deals with the central issues of dignity, justice, right to food, and governance systems across food systems. In Chapter 2, Francis Adams demonstrates the necessity of involving local communities in the design, implementation, and monitoring of projects and food initiatives; this active engagement guarantees improved outcomes and ensures that those who produce, distribute, and consume food are the key decision makers. The importance of local actors is made evident in smallholder agriculture, artisanal-fishing, pastoralist-led grazing, and food production systems that are socially and environmentally sustainable. To sustain these high functioning systems, policies and practices must ensure that the lands, waters, seeds, livestock, and biodiversity are in the hands of those who produce food in local communities. Authors Lucy Hinton and Caitlin Scott agree, and in Chapter 4 they explain how food systems are characterised by jurisdictional, political, and corporate-level policies, and that existing power and financial imbalances ought to be displaced and

DOI: 10.4324/9781003174417-65

mitigated by collaborative and collective policies and governance. According to Gabriel R.Valle, in Chapter 3, these efforts, in collaboration with civil society and communities, should be oriented towards guaranteeing food democracy—forms of engagement that challenge existing power relations and serve to equalise participation. Such purposeful efforts could avert recurrent dynamics of power imbalances and create the necessary conditions for healing from oppressive practices that have shaped food systems (e.g., impacting marginalised peoples, cultures, and ways of living). Priority should be placed on consuming foods that support sustainable farming practices, farmers, a living wage for farmworkers, cultural values, traditions, and the sacredness of food.

Environmental strategies

Tremendous signs of resilience are noted in the face of many challenges affecting food systems such as climate crises, land degradation, water scarcity, extreme weather events, pest and epidemic outbreaks, and animal and vegetal biodiversity loss. Authors Francis Adams (Chapter 5), Roland Ebel and co-authors (Chapter 6), Amanda Shankland (Chapter 7), and Paul Manning and Jennifer Marshman (Chapter 8) illustrate several creative strategies based on the principles and practices of agroecology, agroforestry, and agrobiodiversity. Adams illustrates the efficacy of agroecology and agroforestry to contribute to carbon sequestration and how these are key approaches to increase food security and preserve natural resources. Through international coordination, more conservation efforts are needed. Ebel and co-authors stress the importance of education and evidence to drive policy action for restoring agrobiodiversity, and guarding food systems from monotonisation, in all countries, but particularly in lower-income countries where small farmers are struggling with the political and financial power of the agri-food industry. In this vein, Shankland, in Chapter 7, further highlights the key role of agroecology for Indigenous and peasant farmers in food sovereignty and affording greater sustainability in food systems. She clearly illustrates how following the natural cycles of agroecology includes respecting the connections between biodiversity, cultural suitability, development practices, landscape management, optimising and recycling, and greater localised production and consumption. Manning and Marshman, highlight the beneficial functions of insects in natural ecosystems and agroecosystems, with particular importance placed on agroecosystem adaptations to provide insects appropriate habitats and reproductive spaces.

In Chapter 9, readers also learn about insects' importance, as author Bruno Borsari illustrates the opportunities and contributions of eating insects to sustainable diets. He explores potential consumer barriers to insect consumption, as well as strategies to overcome them, such as using alluring communications, offering delightful dishes, as well as recipes with opportunities for tasting at public events. Other alternative foods also are explored by Anna K. Farmery and Jessica R. Bogard who, in Chapter 10, show the significant opportunities from diverse vegetal and animal aquatic foods for sustainable diets, reminding readers of the importance of taking into account the accessibility and affordability of those foods in regions around the world. On aquatic food acceptability, L. Sasha Gora, in Chapter 11, provides an elegant account and cultural analysis of cultivating, marketing, and eating of oysters. Whether that eating is considered elegant, that is left to the eater, or the reader, to decide.

Additional environmental strategies for sustainable diets are then illustrated by Jason Hannan (Chapter 12) and Tony Weis and Allison Gray (Chapter 13), in which authors catalogue the litany of ecological impacts when animals are produced for food compared to plant-based food systems. In Chapter 12, Hannan not only highlights how plant-based food systems can reduce environmental impact on greenhouse gas emissions and water pollution but also stresses the importance of avoiding pain for animals raised for food and improving the hard working conditions for humans in slaughterhouses. In Chapter 13, Weis and Gray explain how a large array of plant-based food choices can aid in accelerating the adoption of plant-based diets, as these foods often mimic more familiar, animal-based foods, while allowing consumers to adopt more sustainable and kinder food behaviours.

Health and well-being

The health and well-being section of this collection starts with the contribution of Gabriella Luongo and co-authors who, in Chapter 14, identify that the most effective strategies to shape healthy food environments, inspire consumer behaviours, and reduce the burden of disease are through policy interventions rather than information sharing strategies alone. The interconnectedness between human well-being, animal health, the environment, and livelihood are then presented by Anna Okello and co-authors in Chapter 15. These authors highlight the impact of the global burden of foodborne and zoonotic disease on society, economy, and the environment and propose the development of more robust methodologies to assess risk, as well as understanding the contribution of safe food and animal health to global food and nutrition security and socio-economic and environmental well-being. In Chapter 16, Sarah Elton and Donald Cole further strengthen the multidimensional integration of policy shaping healthy food systems. The authors develop a conceptual framework—the Ecological Determinants of Health—for sustainable diets that considers local ecosystems, cultural appropriateness, and supports food security by intertwining health and environmental outcomes of food systems.

The interlinks among food, health, and the environment are is further explored in Chapter 17 by Joseph W. Dorsey and Marian E. Davidove as they reveal how breastfeeding can contribute to global food security, reducing greenhouse gas emissions, and biodiversity loss, as a critical component of sustainable diets. The strategies they illustrate take into account cultural diversity, economic factors, the active involvement of nursing mothers, along with the value of familial and social supports in developing sensitive and contextualised approaches to breastfeeding. With regards to nutrition and food related behaviours of young populations, in Chapter 18 Fiorella Picchioni and co-authors, through a case study in India, demonstrate the advantages of a holistic approach to study sustainable lifestyles (comprising diets and physical activity) to address health challenges, and explore the nutrition transitions rural adolescents must contend with in rapidly transforming societies.

Education and public engagement

The section on education and public engagement begins with Chapter 19 from Liesel Carlsson who stresses the important role of registered dieticians in supporting communities to know sustainable diets through developing a common language and nurturing meaningful collaboration between actors of multiple disciplines and sectors, to more expeditiously devise local and regional strategies for the uptake of sustainable diets. In Chapter 20, Alicia Martin and co-authors highlight how sharing information, knowledge, and a common language on sustainable diets must be supported by government driven initiatives, such as the identification of national dietary guidelines, designed to promote food literacy, health, and sustainability. Martin and colleagues recommend applying interdisciplinary approaches to determine indicators that could describe and cultivate food systems literacy. In Chapter 21, Liz Nix and Chris Fink advocate for increased literacy about food systems and knowledge of food production and consumption to usher in an era of sustainable diets. They argue that food systems thinking should be promoted for all food system stakeholders. This collective approach to spreading knowledge and information is also identified in Chapter 22 by Roland Ebel and co-authors who explore collective action as a pedagogical strategy, together with systems thinking, as a framework for educating and inspiring students—the future key players of food systems—on sustainable diets. A valuable perspective on educational approaches for sustainable diets is articulated in campus food growing spaces. In Chapter 23, Michael Classens and Nicole Burton identify the need to go beyond classrooms to learn with the land, partner with the community, and address the "unbearable whiteness of alternative food." They call for standing to displace colonialism, racism, and all forms of discrimination while demonstrating solidarity with producers, workers, and consumers of food, by integrating pedagogical activities with students leading learning from diverse cultural contexts. Still in a pedagogical perspective, in Chapter 24 Daniela Soleri and co-authors find that food gardens across the community increase availability and consumption of healthy foods, and enhance physical and mental well-being through social interactions, organising community events, sharing resources, and promulgating ecological stability.

The section is rounded off by profound reflections from Terry Gibbs and Tracey Harris in Chapter 25 in which they argue that honest examinations and re-considerations of nonhuman animal issues in the food system would improve human sensibilities and foster compassion to address social injustices as well as enhance capacities for better relationships with humans and nonhumans.

Social policies and food environments

In Chapter 27, Selena Ahmed and co-authors clearly expose the key challenges of food environments towards sustainability. Food environment frameworks must embed sustainability principles and practices to enable retailers and consumers to appreciate where and how foods and beverages are produced and to foster cultural shifts in the way foods are valued. With the complexity and multidimensionality of sustainability, adequate indicators need to be incorporated into food environments to enable consumers to recognise and seek out sustainability attributes beyond marketing and brand labels. To extend the purposeful thinking and design around food environments, Rachel Mazac and co-authors (Chapter 28) show that, in general, national governments need to implement food-based dietary guidelines that simultaneously include considerations of the environmental implications from agriculture, food security, economics and trade, along with sociocultural and political components.

When social policies and food environments are informed by critical food pedagogy and public engagement, the lack of consumer demand for information on sustainable foods could be replaced with an approach for a wider array of foods with sustainability attributes (Chapter 27). To overcome these pitfalls, Hugh Joseph, in Chapter 30, structures a food consumption framework that considers the sustainable functioning of food environments and the interactional dynamics between principles, products, players, places, practices, and policies. Joseph also echoes calls for enhanced food literacy training to bolster consciousness of food choices and sustainability challenges. The goal is not passive consumption, rather active food citizenship, to build co-ownership of the construction of sustainable food systems. To make this possible and operational, Francesco Cirone and co-authors, in Chapter 26, provide an analysis of the fragility of food sovereignty in cities that has been further exacerbated by COVID-19 pandemic. They highlight the power to rebuild, strengthen, and disseminate knowledge on sustainable food systems through a harmonisation of local governments and city-region policies that encourage and activate small-scale local production through partnerships across sectors and citizen leadership, by fully valuing local resources and targeting local food consumption.

The case of the Odisha Millet initiative in India, illustrated by Saurabh Garg and co-authors in Chapter 29, shows how a millet production initiative activated a participatory multi-stakeholder consultative approach and a strong partnership between the government and civil society to create consumer awareness, build capacities of local community-based organisations, add value to the supply chain, and enhance rural livelihoods. In this case, multifaceted efforts of decentralised community engagement, involving production and consumption stakeholders, contributed to developing affordable and accessible healthy diets at local levels.

Transformations and food movements

Eating mindfully is a point of advice offered to citizens through Canada's Food Guide. In Chapter 31 Sarah Pittoello and Kathleen Kevany encourage readers to recognise the complexity of drivers—cultural, social, emotional, economic, environmental, neurophysiological, psychological, political, and spiritual—behind food choices. To foster more sustainable food systems and societies, greater attention on relationships would activate honouring the human condition, respecting natural environments, and enhancing the abilities of citizens to think holistically and to undertake the inner work needed to transition to sustainability. High functioning, collaborative relations among agricultural and food producers are key to encouraging commitments to sustainability and communicating the importance of these practices to consumers are well noted by Federico Mattei and Eleonora Lano in Chapter 35. These authors explain how to equip consumers with knowledge and tools for choosing what production paradigms to support and to be informed about the positive and negative externalities. Such social relationships established between and within producers and consumers can strongly contribute to the adoption of more sustainable food consumption behaviours.

Yuna Chiffoleau and Grégori Akermann in Chapter 34 use the case of the first COVID-19 lockdown in France to demonstrate how several new consumers participated in alternative food networks (AFN). While a large number returned to their habitual purchasing habits after the crises, many remained as new members in the AFN thanks to the social relationship dynamics. As established throughout this handbook, the forming of collaborative and genuine relationships is key to moving food systems to more sustainability. Through a survey of civil society organisations and movements seeking to transform food systems for inclusiveness and sustainability, Joseph Tuminello and co-authors in Chapter 33 demonstrate the localised and broader impacts being realised through these focused efforts. Examples are provided to encourage replication of successes like relationships fostered between Indigenous leaders and food services in health care, which brought about substantial improvements in serving culturally appropriate and healthy foods for in-patient care. Other findings reveal that compelling communications of sustainability messages on food products can ignite a shift of consumer behaviours towards pro-sustainability dishes. In Chapter 32, Riana Topan and Stefanie McNerney highlight the importance of building genuine relationships with food industry leaders to spark the active participation in training by chefs and food procurement managers on new menu options. Through Forward Food, an international programme of the Humane Society, they have increased the amount of plant-based and sustainable eating interventions in the foodservice sector. Other food movements surveyed actively champion the cause for school food programs, changing course on the climate crises, or on animal welfare to substantially transform food systems.

Economics and trade

Economic and financial challenges can strongly impact the sustainability of diets and food systems. Policy measures for a living wage are needed to ensure people have access to sufficient and healthy food, as Azfar Khan and co-authors highlight in Chapter 36. The majority of workers in rural areas and the urban informal economy of the Global South struggle with low incomes, which impedes the procurement of sustainable diets and generates tremendous costs to society with preventable diseases, lower productivity, and decreased opportunities in learning and innovating. In addition to such policies, according to Phoebe Stephens and co-authors in Chapter 37, regulation around financial speculation on food, agricultural commodities, and farmland are imperative. In fact, financial systems have an inordinate amount of influence and largely dictate food provisioning; consequently, consumers are largely confined when facing choices about food when so much is controlled by industrialised food agents. Therefore, alternative financial investments should be conceived and reoriented to ensure the development and protection of sustainable diets. Similar efforts of finding alternative investments consistent with the aim of building more sustainable food systems are also claimed in Chapter 38 by Tony Weis and Allison Gray. These authors argue the need to go beyond the production of plant-based foods to educate consumers and governments about leveraging and selecting plant-based alternatives, as they hold the potential of more positive health outcomes, environmental impacts, and enhanced interspecies relations.

According to David Laborde and co-authors in Chapter 39, trade regulations should also consider national nutrition objectives and identify trade measures that trigger positive nutritional outcomes and lessen the risk for consumers to be exposed to nutritionally poor and unhealthy food products. While trade is deemed by these authors as a key tool to regulate the quality of the food supply, ill-conceived policies can undermine the important role of trade in improving global food sufficiency. Thus policy makers are advised to incorporate trade policies that meet identified nutrition objectives, while including complementary incentives to invite the maximum synergies with minimal risks in cultivating foods for sustainable diets.

Questions about the right to food and the necessity for food aid also are addressed in this collection. Both international and local plans need to be updated around food assistance and food aid, particularly should overproduction occur so measures are in place to avoid food waste. By paying attention to the prevention of food loss and waste, Imana Pal, in Chapter 41, calls upon the circular economy approach for sustainable food systems. In particular, she explains how circular economy practices can be better operationalised in food systems by investing in training programmes that show how cost effectiveness approaches are most beneficial on minimising food loss and waste, and argues the need to adapt circular economy technologies to local communities. Furthermore, for circular economy to be effective in food systems, Paolo Prosperi offers a framework in Chapter 40 that integrates governance dimensions in the construction of circular business models that foster improved partnership, participation, coordination, trust, efficiency, ethics, and equity among stakeholders and consumers within agri-food value chains.

Design and measurement mechanisms

In Chapter 44, Sonia Massari and co-authors argue the importance of applying food design to shape a cultural innovation process that brings value and contributes to the sustainability of diets. In particular, they explain that understanding human drivers, values, and needs can provide opportunities to design sustainable food systems with more transparency, choice architecture, and select product placement to enable more sustainable behaviours. In three examples of food design applications to food chains they illustrate ways to support and reinforce the Mediterranean Diet as a sustainable diet. In addition to designing sustainable food systems, it is important to be able to assess the sustainability of food systems holistically and to appreciate and map internal and external interactional dynamics. In Chapter 43, Alessandro Galli and co-authors illustrate the application of Footprint indicators to monitor progress towards the Sustainable Development Goals and planetary boundaries, especially in complex (food) systems. In particular, the authors demonstrate that sustainability assessment of food systems should be developed on the basis of nutritional values of food, representing the reference unit of a sustainable diet.

Therefore, data availability and quality are crucial challenges and opportunities for the future development of sustainable food systems and diets. In Chapter 42, Nicholas Holden and co-authors argue that food will involve two components—the physical food objects and the data describing them. For instance, the use of blockchain technology can help to ensure trust across value chain actors, with regards to food product information, characteristics, quality, and information about transactions. These authors illustrate how such technological innovations, when harnessed, enable the development, delivery, and measurement of sustainable, nutritious diets. In this respect, in Chapter 45, Gianluca Brunori and co-authors introduce and develop the application of the socio-cyber-physical systems' paradigm. Through this approach, the authors demonstrate the need to consider ethical aspects within technological innovation processes, as well as the role of legislation and governance to avoid risks on society due to insufficient and unregulated digital innovation. With specific regards to the agri-food system dynamics and rural contexts, the authors highlight the importance of ensuring easy access to broadband internet to avoid inequity, and to provide all stakeholders training opportunities on ICT skills.

Food sovereignty and case studies

This section offers case studies on sustainable diets, largely focused on the Global South. In Chapter 46, Selena Ahmed and co-authors show how inspiring Indigenous Peoples' food systems can be, with food procurement from wild and cultivated food environments, utilisation of seasonal edible biodiversity, holistic concepts of health, stewardship of land, and ecological agricultural practices. Additionally, these systems recognise food as sacred, retain ecological and food knowledge, and share rituals that can be valuable in protecting and celebrating cultural identity. The authors call for acknowledgement of Indigenous Peoples' food systems for their inherent values and ways to support multiple Sustainable Development Goals (SDGs). In Chapter 47, Galyna Medyna and colleagues argue that exploring in-depth traditional knowledge, creativity, and resilience in Sub-Saharan Africa can provide innovative and profitable solutions for sustainable diets.

The case of the rediscovery of millet in India is presented in Chapter 48, by Brigitte Sébastia and Hélène Guétat-Bernard. Recently perceived as a "famine food," millet is now popular food thanks to its health and sustainability benefits, even though a localised presence of globalised food systems has been found to displace such valuable traditional foods for a large segment of the population. Chapter 49 from Manish Anand and co-authors explore building sustainable food systems in India, highlighting the importance of thinking about diets well beyond nutritional adequacy, to integrate the personal, social, and environmental dimensions of well-being, including socio-cultural acceptability of food. The case study from China described by Ning Dai and co-authors in Chapter 50 explains efforts to increase understanding of sustainable diets and the important role filled by civil society food initiatives in advancing health and sustainability goals. In Chapter 51, Theresa Tribaldos and co-authors present the cases of Bolivia and Brazil, in which the effectiveness of food system activities includes the involvement of local communities and strengthening of intra-communal links. From Central America, Alyson Dagang presents in Chapter 52 a case from Panama, illustrating the contradictions that emerge from increasing the tourism economy, which, in turn, provokes changes in food and dietary habits because of shifting from agricultural activity to tourism, thus changing lifestyle and income.

In Chapter 53, David Fazzino and Ashley Meredith share findings of recent research on nutrition transition in Kosraean Food System. Their findings suggest significant vulnerability and dependency within the contemporary food system, so they recommend not overlooking the transformative potential of locally-based sustainable food systems through blending both imported and locally-produced foods. These findings also illustrate how facing vulnerabilities can increase the appeal for more locally based diets that are socially and ecologically sustainable.

Lessons driving calls to actions

1) Framing and vision

   Dignity and justice re-established by anti-racist and de-colonial work must be prominent in facilitating the wide adoption of sustainable diets. Priority must be given to community-led initiatives, around the right to food, and multisectoral approaches that foster policy coherence and enabling environments that activate interests in more local, smaller scale production for increased local consumption.

2) Environmental strategies

   Committing to reversing GHG emissions, particularly throughout the Global North, is an imperative, not only to stabilise sustainable diets but also to encourage more peaceful societies. Governments, businesses, and civil society organisations, along with citizens, should use every tool in the toolkit to reduce emissions: Agroecology; agroforestry; agrobiodiversity; reducing and revalorising food waste; emphasising plant rich and supporting lower carbon food systems; and removing subsidies from significant emitters, particularly in the fields of animal agriculture.

3) Health and well-being

The health of a population ought to be the priority of governments, and strategies are needed that acknowledge the intertwining dynamics among the health of humans, animals, and ecological systems. Focused research and policy interventions that support One Health strategies, National Dietary Guidelines, and food environment standards can influence nutritious food availability, affordability, and accessibility while also supporting lower emission foods.

4) Education and public engagement

Focused actions would help people recognise sustainable diets and strengthen food system literacy, including more critical understanding of biodiversity loss, decline of water availability, weather destabilization, and dysfunctions arising from the animal-industrial complex. Education and engagement can cultivate desires to ensure that food production and consumption nourishes life, in the fullest senses of the word, while fostering resiliency and sovereignty.

5) Social policies and food environments

Citizens can only eat what is available; they cannot eat what is lacking. Food environment policies and practices are essential to encourage making the healthy choice the attractive and easy choice. Local and collaborative food chains are ones that value and respect food, more effectively reveal food sources, and promote multiple ways to prevent and reduce waste.

6) Transformation and food movements

It is essential to leverage the positions and power of institutions and other food system actors to bring about transformational change. Farmers, producers, and processors can be urged to reduce externalities, practise greater transparency, and encourage citizens to support food options that are inclusive, local and slow, when feasible, and delicious and nutritious.

7) Economics and trade

Trade mechanisms and regulatory schemes must support the economic and financial profitability of local and national food producers, and regulate multinational bodies' actions for the benefit of local workers, consumers, and sustainable business. As well, the viability, accessibility, and affordability of food production for consumers is integral to sustainable diets, commencing with living wages. Economic goals to be harnessed are impact investments, international trade, the circular bioeconomy, and efforts to reduce and reuse resources in ways that improve global food sufficiency and security.

8) Design and measurement mechanisms

Fostering trust in the data is key. Using tools like blockchain and artificial intelligence to advance the SDGs, and holistic and fulsome measures of progress and gaps in sustainable diets, are required. Design and data have to be built on integrity, scrutiny, and accountability.

9) Food sovereignty

The importance of centring voices of Indigenous Peoples, women, and youth is paramount to accelerate the adoption of sustainable diets by attending to cultural nuances and local varieties, and supporting food security initiatives.

10) Calls to action

Countries in Africa, by way of a well presented critical study by Amos Laar and co-authors (Chapter 54), have illustrated strategies for food guidance and regulations that have aided in food safety, security, and sustainability as well as areas where insufficient actions by governments have stymied progress. Sustainability must not be overlooked; human and planetary well-being are in peril.

New, recurrent, and unexpected world crises, such as the most recent COVID-19 pandemic or the aggression in Ukraine, among several political and violent wars and crises, should not excuse any governments to step back from sustainability goals; all actions are needed now in their different forms and scales, from local endeavours to transnational initiatives. Delays cannot be justified; collaborative action plans are essential. Policy makers are well advised to invest in improving the sustainability of agriculture and food, and to utilise the set of tools and techniques as have been well demonstrated in this handbook, for improving the health of humans, animals, and the planet as urgent pathways for increasing equity, justice, well-being, and peace.

INDEX

Page numbers in bold denote tables, those in italic denote figures.

For Product Safety Concerns and Information please contact our EU
representative GPSR@taylorandfrancis.com
Taylor & Francis Verlag GmbH, Kaufingerstraße 24, 80331 München, Germany

www.ingramcontent.com/pod-product-compliance
Lightning Source LLC
Chambersburg PA
CBHW060447240326
41598CB00088B/3672